Geodynamics of the Western Pacific-Indonesian Region

Geodynamics Series

Inter-Union Commission on Geodynamics
Editorial Board
  A. L. Hales, Chairman
  R. D. Russell, Secretary
  O. L. Anderson
  F. M. Delany
  C. L. Drake
  J. Sutton

American Geophysical Union/Geological Society of America
Editorial Board
  Kees Dejong
  C. L. Drake
  D. H. Eckhardt
  E. Irving
  W. I. Rose
  Rob Van der Voo

The Final Reports of the International Geodynamics Program sponsored by the Inter-Union Commission on Geodynamics.

Geodynamics of the Western Pacific-Indonesian Region

Edited by
Thomas W. C. Hilde and Seiya Uyeda

Geodynamics Series
Volume 11

American Geophysical Union
Washington, D.C.

Geological Society of America
Boulder, Colorado

1983

Final Report of Working Group 1, Geodynamics of the Western Pacific-Indonesian Region, coordinated by Thomas W.C. Hilde and Seiya Uyeda on behalf of the Bureau of Inter-Union Commission on Geodynamics

American Geophysical Union, 2000 Florida Avenue, N.W.
  Washington, D.C. 20009

Geological Society of America, 3300 Penrose Place, P.O. Box 9140,
  Boulder, Colorado 80301

Library of Congress Cataloging in Publication Data

Main entry under title:

Geodynamics of the Western Pacific-Indonesian region.

   (Geodynamics series; v. 11)
   Includes bibliographies.
   1. Geodynamics--Addresses, essays, lectures.
2. Geology--Pacific area--Addresses, essays, lectures.
3. Geology--Indonesia--Addresses, essays, lectures.
I. Hilde, Thomas W. C. II. Uyeda, Seiya, 1929-   .
III. Inter-Union Commission on Geodynamics. Working
Group 1--Geodynamics of the Western Pacific-Indonesian
Region. IV. Series.
QE501.G426 1983   551   83-14396
ISBN-0-87590-500-5

Copyright 1983 American Geophysical Union. Figures, tables and short excerpts may be reprinted in scientific books and journals if the source is properly cited; all other rights reserved.

Printed in the United States of America

CONTENTS

FOREWORD 1
C.L. Drake and A.L. Hales

PREFACE 3
T. Hilde and S. Uyeda

PART 1 GENERAL STUDIES

I. FUNDAMENTAL ASPECTS OF WESTERN PACIFIC DYNAMICS

SEISMOTECTONICS OF THE WESTERN PACIFIC REGION 5
T. Seno and T. Eguchi

LATE QUATERNARY VERTICAL CRUSTAL MOVEMENTS IN AND AROUND THE PACIFIC AS DEDUCED 41
FROM FORMER SHORELINE DATA
N. Yonekura

BASIN FORMATION; THE MASS ANOMALY AT THE WEST PACIFIC MARGIN 51
R.C. Bostrom, K.K. Saar, and D.A. Terry

DEPTH ANOMALIES OVER MESOZOIC CRUST IN THE WESTERN PACIFIC 63
J. Mammerickx

TRENCH DEPTH: VARIATION AND SIGNIFICANCE 75
T. Hilde and S. Uyeda

ENTRAPMENT ORIGIN OF MARGINAL SEAS 91
Z. Ben-Avraham and S. Uyeda

II. SEISMIC STRUCTURE OF THE CRUST AND LITHOSPHERE

EXPLOSION SEISMOLOGICAL EXPERIMENTS ON LONG-RANGE PROFILES IN THE NORTHWESTERN 105
PACIFIC AND THE MARIANAS SEA
T. Asada, H. Shimamura, S. Asano, K. Kobayashi, and Y. Tomoda

VELOCITY ANISOTROPY EXTENDING OVER THE ENTIRE DEPTH OF THE OCEANIC LITHOSPHERE 121
H. Shimamura and T. Asada

SEISMIC CRUSTAL STRUCTURE AND THE ELASTIC PROPERTIES OF ROCKS RECOVERED BY 127
DRILLING IN THE PHILIPPINE SEA
R.L. Carlson and R.H. Wilkens

III. PETROLOGY, CHEMISTRY, AND AGE OF THE IGNEOUS ROCKS

SUMMARY OF GEOCHRONOLOGICAL STUDIES OF SUBMARINE ROCKS FROM THE WESTERN PACIFIC 137
OCEAN
M. Ozima, I. Kaneoka, K. Saito, M. Honda, M. Yanagisawa, and Y. Takigami

PHENOCRYST ASSEMBLAGES AND H_2O CONTENT IN CIRCUM-PACIFIC ARC MAGMAS 143
M. Sakuyama

PETROLOGY AND GEOCHEMISTRY OF OPHIOLITIC AND ASSOCIATED VOLCANIC ROCKS ON THE 159
TALAUD ISLANDS, MOLUCCA SEA COLLISION ZONE, NORTHEAST INDONESIA
C.A. Evans, J.W. Hawkins, and G.F. Moore

MAGMATIC EVOLUTION OF ISLAND ARCS IN THE PHILIPPINE SEA 173
R.B. Scott

PART 2 REGIONAL STUDIES

I. OKHOTSK-KURIL-KAMCHATKA COMPLEX

GEOLOGY AND PLATE TECTONICS OF THE SEA OF OKHOTSK — 189
L. Savostin, L. Zonenshain, and B. Baranov

SEISMOFOCAL ZONES AND GEODYNAMICS OF THE KURIL-JAPAN REGION — 223
R.Z. Tarakanov and C.U. Kim

HEAT FLOW AND GEODYNAMICS PROBLEMS OF THE TRANSITION ZONE FROM ASIA TO THE NORTH PACIFIC — 237
P.M. Sychev, V.V. Soinov, O.V. Veselov, and N.A. Volkova

THE TECTONICS OF THE KURIL-KAMCHATKA DEEP-SEA TRENCH — 249
H. Gnibidenko, T.G. Bykova, O.V. Veselov, V.M. Vorobiev, and A.S. Svarichevsky

II. JAPAN

CYCLES OF SUBDUCTION AND CENOZOIC ARC ACTIVITY IN THE NORTHWESTERN PACIFIC MARGIN — 287
K. Kobayashi

THE ROLE OF OBLIQUE SUBDUCTION AND STRIKE-SLIP TECTONICS IN THE EVOLUTION OF JAPAN — 303
A. Taira, Y. Saito, and M. Hashimoto

VERTICAL CRUSTAL MOVEMENTS OF NORTHEAST JAPAN SINCE MIDDLE MIOCENE — 317
N. Sugi, K. Chinzei, and S. Uyeda

HIGH MAGNESIAN ANDESITES IN THE SETOUCHI VOLCANIC BELT, SOUTHWEST JAPAN AND THEIR POSSIBLE RELATION TO THE EVOLUTIONARY HISTORY OF THE SHIKOKU INTER-ARC BASIN — 331
Y. Tatsumi

CROSS SECTIONS OF GEOPHYSICAL DATA AROUND THE JAPANESE ISLANDS — 343
T. Yoshii

COMPRESSIONAL WAVE VELOCITY ANALYSES FOR SUBOCEANIC BASEMENT REFLECTORS IN THE JAPAN TRENCH AND NANKAI TROUGH BASED ON MULTICHANNEL SEISMIC REFLECTION PROFILES — 355
Y. Aoki, T. Ikawa, Y. Ohta, and T. Tamano

DEEP SEISMIC SOUNDING AND EARTHQUAKE PREDICTION AROUND JAPAN — 371
M. Hayakawa and S. Iizuka

III. TAIWAN

GEOTECTONICS OF TAIWAN--AN OVERVIEW — 379
V.C. Juan, H.J. Lo, and C.H. Chen

IV. INDONESIA

COMPLICATIONS OF CENOZOIC TECTONIC DEVELOPMENT IN EASTERN INDONESIA — 387
J.A. Katili and H.M.S. Hartono

GEOLOGICAL-GEOPHYSICAL PARADOXES OF THE EASTERN INDONESIA COLLISION ZONE — 401
J. Milsom, M.G. Audley-Charles, A.J. Barber, and D.J. Carter

EARTHQUAKE STRESS DIRECTIONS IN THE INDONESIAN ARCHIPELAGO — 413
H.D. Tjia

V. TONGA TRENCH AND NEW ZEALAND

NEW ZEALAND HORIZONTAL KINEMATICS — 423
H.W. Wellman

FOREWORD

After a decade of intense and productive scientific cooperation between geologists, geophysicists and geochemists the International Geodynamics Program formally ended on July 31, 1980. The scientific accomplishments of the program are represented in more than seventy scientific reports and in this series of Final Report volumes.

The concept of the Geodynamics Program, as a natural successor to the Upper Mantle Project, developed during 1970 and 1971. The International Union of Geological Sciences (IUGS) and the International Union of Geodesy and Geophysics (IUGG) then sought support for the new program from the International Council of Scientific Unions (ICSU). As a result the Inter-Union Commission on Geodynamics was established by ICSU to manage the International Geodynamics Program.

The governing body of the Inter-Union Commission on Geodynamics was a Bureau of seven members, three appointed by IUGG, three by IUGS and one jointly by the two Unions. The President was appointed by ICSU and a Secretary-General by the Bureau from among its members. The scientific work of the Program was coordinated by the Commission, composed of the Chairmen of the Working Groups and the representatives of the national committees for the International Geodynamics Program. Both the Bureau and the Commission met annually, often in association with the Assembly of one of the Unions, or one of the constituent Associations of the Unions.

Initially the Secretariat of the Commission was in Paris with support from France through BRGM, and later in Vancouver with support from Canada through DEMR and NRC.

The scientific work of the program was coordinated by ten Working Groups.
   WG 1 Geodynamics of the Western Pacific-Indonesian Region
   WG 2 Geodynamics of the Eastern Pacific Region, Caribbean and Scotia Arcs
   WG 3 Geodynamics of the Alpine-Himalayan Region, West
   WG 4 Geodynamics of Continental and Oceanic Rifts
   WG 5 Properties and Processes of the Earth's Interior
   WG 6 Geodynamics of the Alpine-Himalayan Margins
   WG 9 History and Interaction of Tectonic, Metamorphic and Magmatic Processes
   WG 10 Global Synthesis and Paleoreconstruction
   These Working Groups held discussion meetings and sponsored symposia. The papers given at the symposia were published in a series of Scientific Reports. The scientific studies were all organized and financed at the national level by the national committees even when multi-national programs were involved. It is to the national committees, and to those who participated in the studies organized by those committees, that the success of the Program must be attributed.

Financial support for the symposia and the meetings of the Commission was provided by subventions from IUGG, IUGS, UNESCO and ICSU.

Information on the activities of the Commission and its Working Groups is available in a series of 17 publications: Geodynamics Reports, 1-8, edited by F. Delany, published by BRGM; Geodynamics Highlights, 1-4, edited by F. Delany, published by BRGM; and Geodynamics International, 13-17, edited by R. D. Russell. Geodynamics International was published by World Data Center A for Solid Earth Geophysics, Boulder, Colorado 80308, USA. Copies of these publications, which contain lists of the Scientific Reports, may be obtained from WDC A. In some cases only microfiche copies are now available.

This volume is one of a series of Final Reports summarizing the work of the Commission. The Final Report volumes, organized by the Working Groups, represent in part a statement of what has been accomplished during the Program and in part an analysis of problems still to be solved. This volume from Working Group 1 was edited by Thomas W. C. Hilde and Seiya Uyeda.

At the end of the Geodynamics Program it is clear that the kinematics of the major plate movements during the past 200 million years is well understood, but there is much less understanding of the dynamics of the processes which cause these movements.

Perhaps the best measure of the success of the Program is the enthusiasm with which the Unions and national committees have joined in the establishment of a successor program to be known as:

Region, East
  WG 7 Geodynamics of Plate Interiors
  WG 8 Geodynamics of Seismically Inactive
Dynamics and evolution of the lithosphere: The framework for earth resources and the reduction of the hazards.

To all of those who have contributed their time so generously to the Geodynamics Program we tender our thanks.

C. L. Drake, President, ICG, 1971-1975
A. L. Hales, President, ICG, 1975-1980

Members of Working Group 1 and Working Group 1 Study Groups:

M. Audley-Charles	R.L. Larson
A. Balce	A. Macfarlane
J.A. Brooks	D.I.C. Mallick
D.A. Christoffel	R.T. Murphy
P. Cook	G. Packham
H. Davies	Y.S. Pan
D. Denham	R. Richmond
J. Dubois	T. Rikitake
I.B. Everingham	J.G. Sclater
F.F. Evison	S.J. Su
A. Ewart	G.H. Sutton
S.A. Fedotov	P.M. Sychev
F.C. Gervasio	R.Z. Tarakanov
G. Grindley	R.B. Thompson
N.S. Haile	H.D. Tjia
W. Hamilton	Y.B. Tsai
L. Hawkins	G.B. Udintsev
P. Hedervari	S. Uyeda
T. Hilde	Y. Wang
V.C. Juan	H.W. Wellman
J.A. Katili	M. Yasui
K. Kobayashi	M.T. Zen
L.C. Kroenke	

PREFACE

The contributions to this volume have been organized into two categories: General Studies, which deal with the dynamic processes, and the systematic geological and geophysical relationships of importance to the region as a whole; and Regional Studies, which focus on specific areas and features of the Western Pacific. The composition of the volume reflects the research efforts undertaken in the Western Pacific within the framework of the Geodynamics Project, by the countries of the region and by other countries concerned with the geodynamics of the Western Pacific. The new understanding of the dynamics and development of the trench-arc-backarc systems, and the dynamic relationship of the Western Pacific basin lithosphere and ancient continental margins with trench-arc-backarc development, brought about through the Geodynamics Project, is a major advancement in the geosciences. Much of these new findings is contained in this volume. The editors wish to thank all of the authors for their outstanding contributions, and for their patience with the editing process and assembly of the publication.

In addition to the contributors to this volume, the entire Working Group 1 membership and the study group members of Working Group 1 have contributed greatly to the success of the Western Pacific portion of the Geodynamics Project. Their participation in the identification of outstanding problems and design of research activities at the many Working Group meetings, their efforts to implement the Working Group 1 recommendations, and their individual research constitute the true success of the Western Pacific Geodynamics Project. All members of Working Group 1 and its study groups are listed in alphabetical order on the preceding page. T. Rikitake served as Chairman of Working Group 1 at the beginning of the Geodynamics Project and was succeeded by S. Uyeda.

T. Hilde and S. Uyeda

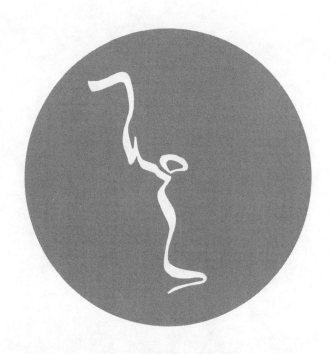

SEISMOTECTONICS OF THE WESTERN PACIFIC REGION

Tetsuzo Seno[1]

Department of Geophysics, Stanford University, Stanford, California 94305

Takao Eguchi[2]

Earthquake Research Institute, University of Tokyo, Bunkyo-ku, Tokyo, Japan

Abstract. The western Pacific shows a great variety of modes of plate consumption. In this paper, we summarize the overall pattern of seismicity and modes of plate consumption associated with earthquakes in the western Pacific region.

Recent shallow seismicity delineates the major plates involved in the tectonics of the region. Seismicity in the northern island arcs has a sufficient resolution to discriminate between the activity beneath the trench axis, along the thrust zone of plate subduction, and in the back-arc region. Possible decoupling of southeast Asia from Eurasia is discussed based on the shallow seismicity; the speed of southeastward migration of southeast Asia relative to Eurasia was estimated to be 1-1.5 cm/yr on the basis of the scheme proposed by Molnar and Tapponnier (1977).

The mode of deformation accompanying great shallow earthquakes in the western Pacific region is mostly underthrusting beneath island arcs. Normal faults at trench axes are the second feature. Intermediate great earthquakes, whose mode of deformation is less documented than the shallow ones, represent huge ruptures within descending slabs such as hinge faulting or decoupling from the oceanic plate. The pattern of occurrence of great shallow earthquakes shows a drastic variation between in the northern and southern halves of the region. In the northern half, subduction zones have been filled with rupture zones of great earthquakes without significant overlap. In contrast, in the southern half, the occurrence of great shallow events is only infrequent. Seismic slip rates along island arc segments are calculated based on the seismic data during the past 84 years. From Kamchatka through Kurile and northern Japan to Izu-Mariana, the fraction of seismic slip in the plate convergence decreases gradually from about 90 per cent to 0 per cent. In the subduction zones of the southern half of the region, seismic slip takes up only a small fraction of plate convergence except for the Philippines and western New Guinea.

Almost all reliable focal mechanism solutions of recent shallow earthquakes are listed and their displacement types are shown and discussed in connection with kinematics of major and minor plates in the western Pacific region. These solutions can mostly be incorporated into the present kinematics of the plates; however, more data are needed to permit thorough descriptions of the current tectonics of the Taiwan, Luzon Straits, eastern Sunda arc, and New Guinean regions.

The geometry of the Wadati-Benioff zone and associated focal mechanisms are described in each region. All deep events show down-dip compression type mechanisms. Along the Kurile-northern Honshu arc, the dual zone of down-dip compression and down-dip tension shown in the intermediate activity and focal mechanisms is recognized unambiguously. At Hokkaido, central Japan, and the eastern Banda Sea, the stress axes of intermediate and deep events show localized effect of contortion, lateral bending, or disruption of slab at these junctions of arc segments.

Seismotectonic studies of the type reviewed in the present paper contribute much to the understanding of the tectonic process currently occurring along the subduction and collision zones in the western Pacific region, and will provide a basic set of data for forecasting the tectonic movements which will occur in the future.

Introduction

The western Pacific is the place where most plates are currently being consumed. In this area we can see a variety of phenomena associated with plate consumption, such as subduction of oceanic lithosphere beneath island arcs, collision of an island arc with another island arc or with a continental mass, and opening of marginal basins. The tangled pattern of island arcs, marginal basins, continental blocks, and seamounts in the western Pacific presents us with a great variety of modes of plate consumption which are taking place at present.

Seismicity and the mode of deformation accom-

---

[1] On leave from International Institute of Seismology and Earthquake Engineering, Building Research Institute, Ministry of Construction, Ohho-machi, Tsukuba, Ibaraki Pref., Japan 305

[2] Present address: National Research Center for Disaster Prevention, Science and Technology Agency, Ten-nodai, Niihari-gun, Ibaraki Pref., Japan 305

panying earthquakes are likely to be two of the most useful guides to understand the tectonic processes which now occur in the western Pacific and which will occur in the future, because most of the mechanical energy now spent at the surface of the earth is released within a few narrow orogenic belts that are affected by important deformation accompanying a strong seismic activity (e.g., Le Pichon et al., 1973), and because the mode and amount of deformation caused by earthquakes can be estimated quantitatively by using long-period body and surface wave data. However, modern seismicity data only cover the past several decades at most and this puts sometimes a severe limitation on the efforts to understand the tectonic processes.

In this paper, we overview the seismic activity in the western Pacific region in order to understand the tectonic processes occurring there. In the process, we come to understand the pattern and regularity of seismic manifestations from the viewpoint of tectonics. We will focus our efforts on describing the overall pattern of seismic activity in the western Pacific region and interpreting it in terms of plate tectonics. Thus, we present a summary of plate geometries and kinematics consistent with the mode of deformation accompanying earthquakes. Long term earthquake prediction is an important byproduct of seismotectonic studies; however, such predictions are of course extremely imprecise.

The region treated in this study is bounded by 90°E and 165°E and 10°S and 60°N. Thus the Sunda arc will be included, but we exclude the area of southwestern Pacific, i.e., Papua New Guinea, New Britain-Solomon Islands, and New Hebrides, from the region.

We first present the overall pattern of seismicity including small-magnitude ($m_b \geq 4.5$) earthquakes in this region. Next, we discuss occurrence of great earthquakes and their mode of deformation, because a major portion of the relative motion along most of the convergent boundaries is taken up by those great earthquakes. However, along some portions of the boundaries aseismic slip is the predominant mode of slip. Finally, mode of deformation accompanying earthquakes including smaller-magnitude ($m_b \geq 5.6$) earthquakes are presented and discussed together with plate tectonics in each area of the region.

Most of the geographical names cited in each section in this paper are shown in Figs. 7, 8 and 10. The reader should refer to these figures when needed. The region treated is vast; thus, for the details of regional tectonics, the reader should refer to other papers in this volume or previous seismotectonic studies.

Modern Seismicity 1964-1975

Figs. 1 and 2 show the epicentral map of shallow (focal depth = 0-60 km), and intermediate and deep (focal depth = 61-700 km) earthquakes; data are from the ISC bulletins from 1964 through 1975. Those earthquakes of $m_b = 4.5$ and above, for which twenty or more P-arrival times are reported, are plotted. We define shallow focus events as having a focal depth of 60 km or less, instead of 0-69 km as in Gutenberg and Richter (1954). This distinction does not make a large difference in the seismicity map. We feel the lower limit of 60 km better defines the contact zone between plates along consuming plate boundaries because thrust type earthquakes occur only to the depth of 60 km along the northern Honshu, Kurile, and central Aleutian arcs (Yoshii, 1975, 1979a; Seno and Pongsawat, 1981; Kawakatsu and Seno, 1982; Veith, 1974; Topper and Engdahl, 1978). The depth range of 40-60 km has been used to define the aseismic front beneath the northern Japan arc, which is considered to coincide with the seaward edge of the low V and low Q mantle beneath the Japanese Islands (see Fig. 3 and Yoshii, 1975; 1979b). Thus we believe that the proposed depth range of 0-60 km for shallow focus events has more geophysical significance.

Shallow focus earthquakes form narrow zones of activity along island arcs, which is a well known fact leading to the plate tectonic hypothesis. On the basis of the fundamental premise of plate tectonics, these zones are the place where interaction between plates occurs, and we can easily identify major plates involved in the tectonics of the western Pacific region on the basis of Fig. 1: they are the Pacific, Eurasian, Indian-Australian, and Philippine Sea plates. The diffuse activity seen within China and the Mongol-Baikal area presents a problem; is it appropriate to treat this activity in terms of the differential plate motion along plate boundaries or not (Molnar et al., 1973; Das and Filson, 1974; Molnar and Tapponnier, 1975, 1977; Chapman and Solomon, 1976)? This point will be discussed in the next section. Another diffuse zone is the area from north of Luzon through the Molucca Sea to the Banda Sea. This diffuse activity is likely to be caused by the multiplicity of plate boundaries and the complex mode of plate convergence along them, which will be discussed in later sections.

Seismic activity within the ocean floors and the marginal basins is minor compared with the activity along the island arcs. However, a few features are noted. First, activity in the Andaman Sea shows a clear lineament along the ridge and troughs in the basin; this differs from the activity which is seen in other marginal basins or active back-arc basins such as the Mariana Trough but is similar to the activity along the mid-oceanic ridges. This suggests the existence of currently active spreading centers in the Andaman Sea and another plate northwest of the spreading centers (see Eguchi et al., 1979 and Curray et al., 1979). Second, there is a low level of scattered activity in the area of Caroline Islands; Weissel and Anderson (1978) proposed the existence of a Caroline plate based on marine geophysical and seismological evidence. Third, there appears to be a very low level of activity from north of Hokkaido, Japan, through Sakhalin Island to eastern Siberia; Chapman and Solomon (1976) proposed that the North American-Eurasian plate boundary is located along this zone of low activity.

We plotted only earthquakes having well-constrained hypocenters by restricting the events used to those of $m_b \geq 4.5$ which have twenty or more P arrival times reported (Figs. 1 and 2). This, along with the definition of 60 km as the lower limit for shallow focus events, permits us to resolve a more detailed feature in the spatial distribution of shallow earthquakes along the Kamchatka-Kurile-northern Honshu arcs than previously shown in seismicity maps of the western Pacific region.

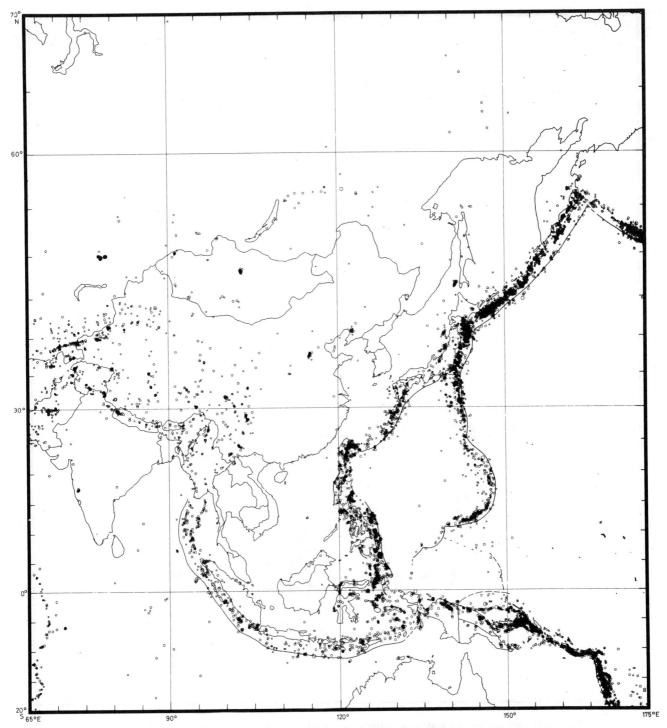

Fig.1 Shallow seismicity (focal depth ≦60 km) map of the western Pacific and Asia during the period from 1964 through 1975, based on the ISC data. Events with magnitude mb ≧4.5, for which twenty or more P-arrivals are reported, are plotted. Larger circles indicate larger-magnitude events.

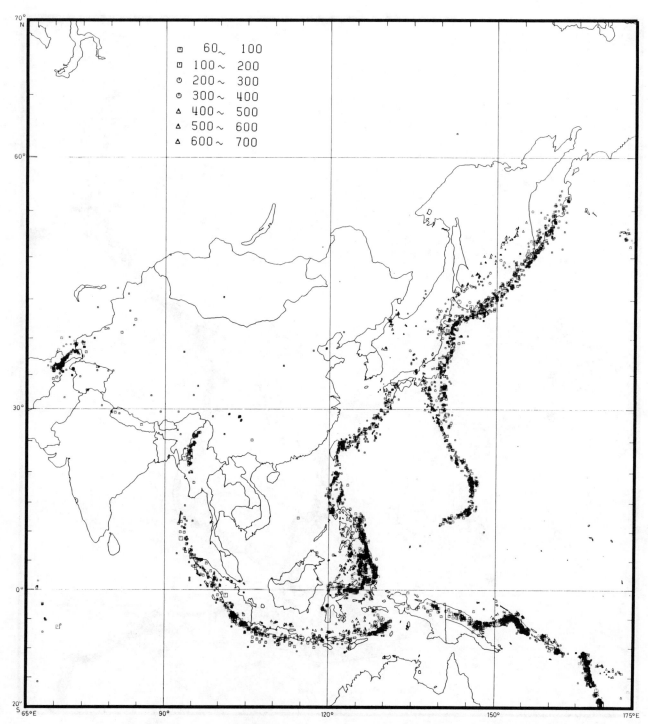

Fig.2 Intermediate and deep seismicity (focal depth>60 km) map of the western Pacific and Asia during the period from 1964 through 1975, based on the ISC data. Selection of the events is same as in Fig.1.

First, the activity beneath the trench axis is clearly segregated from the activity at the thrust zone farther inland (see Fig.1). Second, the landward edge of the activity at the thrust zone is rather sharp, clearly defining the aseismic front along the Kamchatka-Kurile-northern Honshu arc. Yamashina et al. (1978) proposed an "aseismic belt" which is defined as a seismically inactive area in the crust along the frontal arc between the aseismic front and the volcanic front. We, however, cannot recognize the "aseismic belt" clearly in Fig.1 except for the Izu-Bonin arc. Fig.3 shows a schematic illustration which represents the seismicity and type of earthquake mechanisms in the cross-section perpendicular to the northern Honshu arc after Yoshii (1979a). The activity shallower than 60 km includes three groups: the intraoceanic activity beneath the trench axis, the intense activity along the thrust zone, and the intracontinental activity landward of the volcanic front. These three groups of activity are successfully segregated in Fig.1 along the northern island arcs of the region.

Along the southern island arcs of the western Pacific region, the above distinction is not as obvious. This is very likely because the accuracy of hypocentral determinations in these areas is not as precise as along the northern island arcs and also possibly because the activity at the thrust zone between the trench and the forearc is not as great for these areas. The pattern of occurrence of great earthquakes, as will be discussed in a later section, also indicates less seismic activity at the thrust zones for these areas.

Fig.2 shows the epicentral distribution of intermediate and deep earthquakes; epicenters are indicated by different symbols according to the following depth ranges: 61-200 km, 201-400 km, and 401-700 km. The intermediate and deep seismicity is probably confined within the subducting slabs beneath the island arcs because no thrust type earthquakes occur in this depth range. Thus Fig.2 provides us with information on the geometry and stress state of the slabs.

Activity deeper than 400 km is only seen along the Kamchatka-Kurile-Japan-Izu-Bonin-Mariana arcs, the eastern Sunda arc, and beneath the Celebes Sea-Mindanao. Along the Kamchatka-Kurile-Japan-Izu-Bonin arcs, activity in the depth range of 200-400 km is less intense than the shallower or deeper activity, thus this represents an apparent gap in activity in the Wadati-Benioff zone along these arcs. The appearance and disappearance of the activity deeper than 400 km in the Celebes basin and along the Sunda arc is rather abrupt.

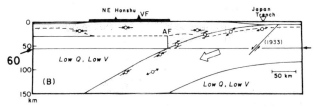

Fig.3 A schematic figure which represents seismicity and focal mechanism type in a cross-section perpendicular to the northern Honshu arc (from Yoshii, 1979a). Stippled areas indicate seismically active regions. Arrows show mode of deformation caused by earthquakes. AF and VF denote the aseismic and volcanic fronts, respectively.

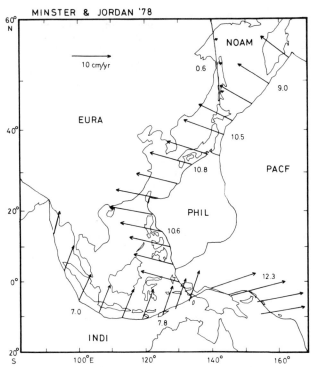

Fig.4 Plate convergence vectors along the plate boundaries of the western Pacific region, calculated from the RM-2 model of Minster and Jordan (1978). In this figure, the Philippine Sea plate is not incorporated into the plate interaction in this region.

Source mechanisms of earthquakes within the slabs beneath island arcs provide basic information on the stress state within the slabs (e.g., Isacks and Molnar, 1971; Fujita and Kanamori, 1981). Isacks and Molnar (1971) proposed that the slab acts as a stress guide which transmits various forces acting on the slab such as negative buoyancy and resistive forces at the bottom or sides of the slab in the mantle. The most spectacular finding in recent years on the Wadati-Benioff zone is the double-planed structure of intermediate seismicity and associated focal mechanisms, which have been found along the Kurile-northern Honshu arcs (Tsumura, 1973; Veith, 1974; Umino and Hasegawa, 1975; Hasegawa and Umino, 1978; Stauder and Mualchin, 1976; Barazangi and Isacks, 1979; Yoshii, 1979a; Seno and Pongsawat, 1981; Suzuki et al., 1981). Fujita and Kanamori (1981) showed that the arcs in which double seismic zones are associated are very rare and a subtle force balance within the slab may be necessary for the stress to produce the double seismic zones. The details of the Wadati-Benioff zones and associated earthquake mechanisms in each region will be presented and discussed in the regional tectonics section.

Kinematics of Major Plates

Fig.4 shows the major plates that are involved in the current tectonics of the western Pacific region. Plate convergence vectors between all but the

Pacific-Philippine Sea plates, calculated from the model RM-2 (Minster and Jordan, 1978), are indicated by the arrows. Plate boundaries in this figure show only a gross configuration of the major plates; the Eurasian-North American boundary follows that proposed by Chapman and Solomon (1976). The Philippine Sea plate is not involved in the kinematics presented in this figure; instead the relative motion of the Pacific to Eurasia is presented along the western margin of the Philippine Sea. Plate convergence vectors based on Chase's (1978) model do not differ significantly from those presented in this figure.

We note that the plate convergence vectors shown in Fig.4 were obtained without using data for relative plate motions for the western Pacific, such as strikes of transform faults, earthquake slip vectors, or ridge spreading rates, except for the data on the slip vectors along the Aleutian-Kurile arc (see Minster and Jordan, 1978, Fig.1). Thus the plate convergence vectors between the major plates shown in Fig.5 put no constraints on the relative motions of minor plates to adjacent major plates or the plate motions between minor plates. They do, however, provide boundary conditions for the relative motions between all the minor and major plates in the western Pacific region. For example, the difference between the observed slip vectors of shallow thrust type earthquakes and the Pacific-Eurasian convergence vectors along the Nankai Trough-Ryukyu Trench boundary implies that the Philippine Sea plate really has a relative motion to the Eurasian plate different from that of the Pacific plate and this difference provides data for constraining the Philippine Sea plate motion relative to the adjacent major plates. This kind of procedure is necessary for seismotectonic studies of the western Pacific region in order to find the motions of minor plates (e.g., Fitch, 1972; Seno, 1977; Cardwell et al., 1981), which will be described in more details in later sections.

When we investigate the present motions of so-called minor plates in the western Pacific region using the above method, attention should also be payed to deformations within the plates. Major intraplate deformations are the opening of back-arc basins, strike-slip faulting along conspicuous transcurrent faults subparallel to the arcs, and deformation within Asia. Some of these intraplate deformations might be treated more effectively by introducing even more minor plates. For example, the opening in the Andaman Sea resembles sea-floor spreading at mid-oceanic ridges and, along with the strike-slip movement at the Semangko fault in Sumatra, can be incorporated into a rigid plate interaction by introducing a Burma plate (Curray et al., 1979; see also discussions in Regional Tectonics). However, spreading in some back-arc regions of the western Pacific may not be of a rigid nature (e.g., the Mariana Trough, see Karig et al., 1978). When the rate of back-arc spreading is large, it may seriously bias the observed slip vectors from those predicted by rigid interaction. Transcurrent faults within the overriding plates near oblique convergent plate boundaries are also likely candidates for resolving the oblique convergence between major plates into strike-slip motion along the faults and normal convergence along the trench (Fitch, 1972).

Intracontinental deformation within Asia is important in the tectonics of the western Pacific region because it may decouple southeast Asia from the Eurasian mainland. This decoupling should be taken into account in the interaction between plates along the southeastern boundaries of southeast Asia (Rodolfo, 1969; Cardwell and Isacks, 1978; Katili, 1970; Hamilton, 1979; Cardwell et al., 1981). There is a wide area of scattered seismic activity in western China, Mongolia, and Baikal-Primorye (see Fig.1). The southern boundary of this activity is sharply defined along the Himalayan-Burma arc which is a suture zone between the Indian subcontinent and Eurasia (e.g., Molnar and Tapponnier, 1975). Some authors have attempted to interpret the scattered activity in Asia by introducing smaller plates (Morgan, 1972; Das and Filson, 1974). However, the diffuse seismic activity, the complexity of related deformation, the lack of knowledge about geological ruptures in China, and the small rate of the Baikal rifting make the usefulness of introducing small plates questionable (Molnar et al., 1973; Molnar and Tapponnier, 1975, 1977; Chapman and Solomon, 1976). Tapponnier and Molnar (1978) and Molnar and Tapponnier (1977) demonstrated that a slip-line theory of a rigid-plastic medium (Asia) indented by a rigid die (India) can be effectively

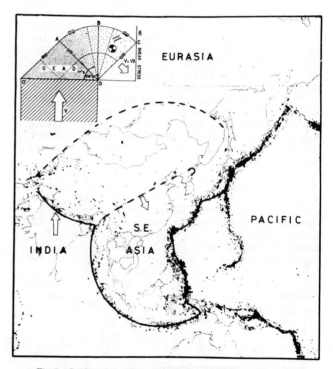

Fig.5 Outline of the boundary of the southeast Asian plate and the interaction with the rest of Asia and India based on the scheme of Molnar and Tapponnier (upper inset). The Indian subcontinent is colliding with Asia in a north-south direction along the Himalayan-Burma arc. Shallow seismicity is from Fig.1. The region between the broken lines is the area of internal deformation. The diagram in the upper inset (Molnar and Tapponnier, 1977) is based on the slip line theory on the rigid-plastic material. Diagonal-ruled region is an indenter (India). Density of dots decreases from areas of high compressive stress and crustal thickening to areas of tensile stress and crustal extension. Triangle at the right (southeast Asia) moves southeastward relative to Eurasia.

applied to the deformation of western and northeastern Asia. Their proposed scheme is shown in the inset of Fig.5. In this scheme, the southeastern boundary of southeast Asia, i.e., the Sunda arc and the Ryukyu-Philippines, is idealized as a "stress free" boundary. According to these investigators, a part of the relative motion between the Indian and Eurasian plates is absorbed in the strike-slip motion along major transcurrent faults in western China and in the normal faulting in northeastern China and Mongol-Baikal area. They suggested that the aseismic southeast Asian block is squeezing out from the rest of Asia towards the "stress-free" consuming plate boundaries on the southeast (Fig.5).

It seems important to estimate quantitatively the motion of southeast Asia relative to Eurasia when we discuss the tectonics of the Philippine and Indonesian region. Thus we present here a quantitative estimate for the speed of migration of southeast Asia from Eurasia, according to the scheme proposed by Molnar and Tapponnier (1977). From the mass conservation (see the inset of Fig.5), the following relation can be derived:

$$_{indi}V_{eura} \times L_{hb} = {}_{se.a}V_{eura} \times L_{se.a} \quad (1)$$

, where $_{indi}V_{eura}$ and $_{se.a}V_{eura}$ are the velocities of migration of the contact zones between India and Eurasia and between southeast Asia and Eurasia, respectively. L_{hb} and $L_{se.a}$ are the lengths of the contact zones along the Himalaya-Burma and along the northwestern boundary of southeast Asia. Note that $_{indi}V_{eura}$ is not equal to the relative motion between the Indian and Eurasian plates, but the part of the relative motion which is not consumed as underthrusting motion of India beneath Eurasia. Molnar and Tapponnier (1975) estimated the value of $_{indi}V_{eura}$ as one or two thirds of the relative motion between India and Eurasia mainly based on the geological evidence. In contrast, Fitch (1970b) estimated seismic slip rate along Himalaya as 6.8 cm/yr using historical earthquake data, which implies that almost all of the relative motion is taken up by the underthrusting motion of India beneath Eurasia. However, re-calculation of the seismic slip rate along this section (Seno, unpublished data, see the section of Seismic Slip Rates about details of the calculation) gives a rate of 2-3 cm/yr. The relative motion between India and Eurasia is about 5 cm/yr at the middle of the Himalaya arc (Minster and Jordan, 1978; Chase, 1978). Thus the value of $_{indi}V_{eura}$ as 2-3 cm/yr is in good agreement with the estimation by Molnar and Tapponnier (1975). The northwestern boundary of the southeast Asia cannot be clearly defined; one possible boundary is drawn in Fig.5 on the basis of seismicity. Using the above value of $_{indi}V_{eura}$ and assuming that the ratio of $L_{se.a}$ to L_{hb} is about 2, a value of 1-1.5 cm/yr is derived for $_{se.a}V_{eura}$. Although this value does not seem to seriously affect the plate convergence vectors along the southeastern boundary of southeast Asia, this motion of southeast Asia with respect to Eurasia cannot be neglected and can really be detected in the deviation of observed slip vectors from those expected for the Indian-Eurasian interaction along the Sunda arc (Cardwell et al., 1981), as will be shown in the regional tectonics section.

Great Earthquakes in the Western Pacific Region

Great earthquakes represent the major portion of deformation along consuming boundaries in the western Pacific region. They can provide extremely valuable information on present-day and recent tectonic processes for the following reasons: 1) detectability of such great earthquakes during the last about one hundred years is uniform (Gutenberg and Richter, 1954); 2) the analysis of source mechanisms of such great events is easy because large data sets are available; 3) a large portion of the slip along major converging boundaries is taken up by seismic slip associated with those great earthquakes (e.g., Kanamori, 1977a; Davies and Brune, 1971). Note also that the time span of recurrence for great earthquakes along the northern converging boundaries in the western Pacific region is on the order of one hundred years. Thus it is possible to discuss characteristic patterns of occurrence of great earthquakes on the basis of the available data.

In the following subsections, we shall describe the pattern of activity and deformation related to the great earthquakes. Great earthquakes are defined here as events for which at least one of either the surface wave magnitude M_s, body wave magnitude m_B, which is determined on long-period instrument records at the period of 7-12 sec, or M_w, which is a magnitude scale for very long periods (100-500 sec), based on the seismic moment (Kanamori, 1977b), is 7.8 or greater. Table 1 lists great earthquakes in the western Pacific region during the past eighty-four years from 1897 through 1980. Fig.6 shows the associated rupture zones or, if rupture zones are not known, epicenters of these earthquakes.

In Table 1, earthquakes are numbered in order of their location from north to south, and they are grouped by region. Hypocentral parameters and origin times are from Gutenberg (1956) for the period 1897-1903, Gutenberg and Richter (1954) for 1904-1952, ISS (International Seismological Summary) for 1953-1958, and PDE (Preliminary Determination of Epicenters) or EDR (Earthquake Data Report) for 1959-1980. When re-calculated hypocenters and origin times were derived in studies of the source mechanisms for individual earthquakes, the re-calculated parameters are cited. References to studies of individual events are listed in the last column of Table 1. Values of M_s and m_B in Table 1 are mostly from Geller and Kanamori (1977), Geller et al. (1978), Kanamori and Abe (1979), and Abe and Kanamori (1979, 1980); otherwise they are from the studies of individual events. Values of M_w are mostly from Kanamori (1977b) and Abe and Kanamori (1980). Otherwise they are calculated based on the seismic moment, M_o, from the studies of individual events. Source dimensions S ($= W \times L$) are mainly from the studies of individual events; for the Kamchatka earthquakes of 1923, 1959, and 1969, they are estimated on the basis of the rupture zones as described by Kelleher et al. (1973) and Kurita and Ando (1974). The values of M_w or S estimated in this study are indicated by the superscript "t" at the right of their values in Table 1.

Rupture zones shown in Fig.6 are from Kelleher et al. (1973) for the Kamchatka events of 1923, 1959, and 1971; Fedotov (1965) for the 1904 Kamchatka and 1918 Kurile Islands events; Kurita and Ando (1974) for

TABLE 1. List of Great Earthquakes in the Western Pacific Region, 1897-1980.

No.	Region	Date	time	Location N°	E°	Depth	M_s	$m_B(T)$	M_w	M_o (Aki)	W×L (km)	Ref.
1	Kamchatka	1969 11 22	23:09:39	57.7	163.6	38	7.3	7.0(7)	7.8	7	60×110	KA74
2		1917 01 30	02:45:36	56 ½	163	s	7.8					
3		1971 12 15	08:29:55	56.0	163.3	33	7.8		7.9^t	10	60×120^t	KA74
4		1923 02 03	16:01:41	54	161	s	8.3	7.7(7)	8.3		90×220^t	
5		1959 05 04	07:15:42	53.2	159.8	74	7.7	7.8(9)	8.2		110×160^t	
6		1897 11 23	09:49	53	159	s	7.9					
7		1952 11 04	16:58:26	52.7	159.5	s	8.2	7.9(8)	9.0	350	200×650	Kn76
8		1904 06 25	14:45.6	52	159	s	7.9	7.8(8)				
9		1904 06 25	21:00.5	52	159	s	8.0	7.7(6)				
10		1904 06 27	00:09.0	52	159	s	7.9	7.5(7)				
11	Kurile Is.	1915 05 01	05:00.	47	155	s	8.0	7.7(11)				
12		1918 09 07	17:16:13	45 ½	151 ½	s	8.2	7.6(8)				
13		1963 11 13	05:17:51	44.8	149.5	60	8.1	7.7(14)	8.5	75	150×250	Kn70
14		1901 04 05	23:30:45	45	148	s	7.9					
15		1958 11 06	22:58:08	44.4	148.6	32	8.1	8.0(7)	8.4	44	80×150	FF79
16		1969 08 11	21:27:40	43.2	147.5	33	7.8	7.9(10)	8.2	22	85×180	Ab73
17		1978 12 06	14:02:01	44.6	146.6	90			7.8^t	7	30×100^t	Sd79
18	Hokkaido	1973 06 17	03:55:03	43.1	145.8	49	7.7	7.3(7)	7.8	7	100×60	Sm74
19		1952 03 04	01:22:43	42.5	143	s	8.3	8.0(9)	8.2	23	100×130	Ks75
20	N. Honshu	1968 05 16	00:48:57	40.8	143.2	7	8.1	7.6(12)	8.2	28	100×150	Kn71a
21		1933 03 02	17:30:56	39.2	144.5	10	8.3	8.2(11)	8.4	43	100×185	Kn71d
22		1897 08 05	0.2	38	143	s	>8.1		8.0^t	12	50×100	Sn79a
23		1938 11 05	10:50:17	37.2	141.8	45	7.8		7.8	5	60×100	Ab77
24	C. Honshu	1953 11 25	17:48:56	34.1	141.6	50	8.0	7.7(8)	7.9	9	55×75	AS82
25		1923 09 01	02:58:32	35.4	139.2	0-10	8.2	7.7(9)	7.9	8	70×130	Kn71c
									8.0	11	55×85	An74
26	S. Honshu	1944 12 07	14:35:39	33.7	136.5	30	8.2	7.8(15)	8.1	15	80×120	Kn72a
									8.2	28	75×190	Is81
27		1946 12 20	19:35:39	33.1	135.8	30	8.2	7.6(10)	8.1	15	80×120	Kn72a
									8.5	63	85×300	An75
28	Ryukyu Is.	1911 06 15	14:26:15	29	129	160	8.1					
29	Taiwan	1920 06 05	04:21:28	23 ½	122	s	8.0					
30	Mariana Is.	1914 11 24	11:53:30	22	141	110		7.9(7)				
31		1902 09 22	01:46.5	18	146	s	7.9					
32	Yap Is.	1911 08 16	22:41:18	7	137	s	7.8					
33	Phil. Is.	1897 10 18	23.8	12	126	s	8.0					
34		1924 04 14	16:20:23	6 ½	126 ½	s	8.3					
35		1943 05 25	23:07:36	7 ½	128	s	7.7	7.8(7)				
36		1948 01 24	17:46:40	10 ½	122	s	8.2					
37	W. Mindanao	1897 09 20	19.1	6	122	s	>8.1					
38		1897 09 21	5.2	6	122	s	>8.1					
39		1976 08 16	16:11:07	6.3	124.0	33	7.8		8.1	19	80×160	SC78
40		1918 08 15	12:18.2	5.7	123.5	s	8.0	7.6(7)				
41	Talaud Is.	1913 03 14	08:45.0	4 ½	126 ½	s	7.9					
42	Molucca S.	1932 05 14	13:11:00	0 ½	126	s	8.0					
43	N. Celebes	1905 01 22	02:43:54	1	123	90		7.8(5)				
44		1939 12 21	21:00:40	0	123	150		7.8(9)				
45	Seram	1965 01 24	00:11:14	-2.4	126.0	6	7.5	7.8(12)				
46	New Guinea	1914 05 26	14:22:42	-2	137	s	8.0					
47		1971 01 10	07:17:05	-3.1	139.7	33	7.9	7.5(10)	7.8			
48		1935 09 20	01:46:33	-3.5	141.8	s	7.9					
49	Banda Sea	1938 02 01	19:04:18	-5.3	130.5	s	8.2	8.0(6)	8.5	70	80×235	Bm77
50		1963 11 04	01:17:11	-6.9	129.6	100		7.8(12)	8.3	30	70×90	OA81
51	Java	1977 08 19	06:08:55	-11.1	118.5	33	8.1		8.3	40	100×160^t	GK80
52		1903 02 27	00:43.3	-8	106	s	7.9					
53	Andaman Is.	1941 06 26	11:52:03	12 ½	92 ½	s	7.7	8.0(8)				

T in $m_B(T)$ is the period at which the body wave magnitude was determined. W and L are fault width and length, respectively. Superscript t at M_w and W×L indicates that those values are determined in this study. Ref. indicates the authors and the year of publication of the paper from which the various parameters are cited; Ab, Abe; An, Ando; AS, Ando and Seno; Bm, Ben-Menahem; FF, Fukao and Furumoto; GK, Given and Kanamori; Is, Ishibashi; KA, Kurita and Ando; Ks, Kasahara; Kn, Kanamori; OA, Osada and Abe; SC, Stewart and Cohn; Sd, Sudo; Sm, Shimazaki; Sn, Seno.

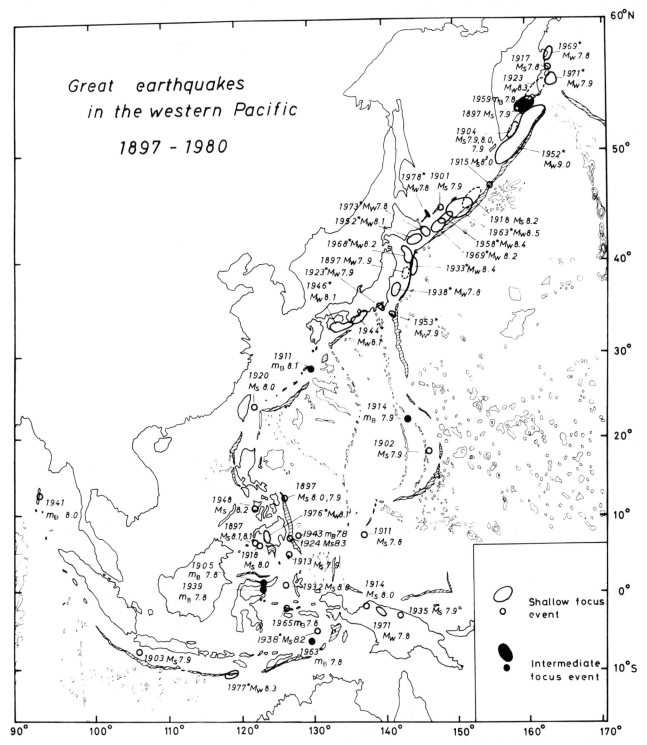

Fig.6 Rupture zones or epicenters of great earthquakes in the western Pacific region during the period 1897-1980. The various parameters of the events are listed in Table 1. For the intermediate depth events, rupture zones or epicenters are filled.

the 1969 Kamchatka event; and Seno (unpublished data) for the 1971 New Guinean and 1977 Indonesian events. The remainder are from the studies of individual events.

Fifty-three earthquakes are listed as great earthquakes in the western Pacific region during the period 1897-1980. The earthquakes of August 9, 1901 (twin M_s=7.8 and 7.9), July 6, 1905 (M_s=7.8), Nov. 18, 1941 (M_s=7.8), all of which are located near the coast of the Japanese Islands, are not included in Table 1 or Fig.6 because they are not considered to be truly great according to the JMA magnitude scale (Utsu, 1979b) and the size of tsunamis caused by those events (Hatori, 1971, 1975; Utsu, 1979b). Out of fifty-three earthquakes, only seven events have intermediate focal depths and no earthquake has a deep focal depth. Intermediate depth events are indicated by the filled rupture zones or filled circles in Fig.6.

Source Mechanisms of Great Earthquakes

Out of forty-six great shallow events, source parameters for twenty-three events have been analysed using long-period body and surface waves, or other information such as tsunamis or crustal deformations, and their tectonic significance has been interpreted. Most of the events whose source mechanisms were analysed occurred since the year 1963, when long-period records at the WWSSN (World Wide Standardized Seismograph Network) stations became available. However, in the vicinity of the Japanese Islands, a number of events which occurred before 1963 have also been analysed using old seismograms or other data sources such as crustal deformation and tsunami data.

Most of the shallow earthquake mechanisms which have been analysed are thrust type and are interpreted as indicating underthrusting of the oceanic lithosphere beneath island arcs. The slip vectors of these earthquakes are consistent with the convergence vectors of major plates along the northern island arcs shown in Fig.4 except where minor plates are involved as in southwest Japan. Another type of mechanism for shallow events is normal faulting beneath or seaward of the trench axis. The 1933 Sanriku earthquake off the Pacific coast of northern Honshu and the 1977 Sumbawa Island earthquake in Indonesia are the only two examples of this type. Kanamori (1971a), Given and Kanamori (1980) and Stewart (1978) contended that the rupture caused by these earthquakes probably cut the entire thickness of the lithosphere, most likely driven by the gravitational pull of the descending slab. If this is true, these events should be discriminated from the earthquakes beneath and near the trench axis caused by bending stress within the oceanic lithosphere (Stauder, 1968; Chapple and Forsyth, 1979). However, Fitch et al. (1981) posed a question on the rupture extended over the entire thickness of the plate because the larger aftershocks of the 1977 Indonesian event distributed only down to 24 km depth. Chapple and Forsyth (1979) had interpreted these events as also due to bending stress.

The following earthquakes are ones whose mode of deformation does not belong to either of the above two types. The 1953 earthquake off central Honshu is an unique event which was located at the TTT (a) type triple junction connecting the Japan and Izu-Bonin Trenches with the Sagami Trough. This event shows predominantly normal faulting on a fault plane striking northwest and dipping steeply to the southwest; this event was interpreted as hinge faulting within the Pacific plate with the motion of the southern part of the slab downward relative to the northern part (Ando and Seno, 1982). One hinge faulting event beneath central Honshu (Isacks and Molnar, 1971) is similar to the mechanism of the 1953 event; both are in accord with the supposed disruption of the slab at the junction between the northern Honshu and Izu-Bonin arcs (Aoki, 1974).

The 1965 Seram earthquake was located at the westernmost margin of the Seram Trough, between the Sula and Buru islands in northeastern Indonesia. Only the P-wave first motion diagram has been obtained for this event (Fitch, 1970a); it shows reverse faulting with one nodal plane dipping to the southwest-south and another to the northeast. These nodal planes are poorly constrained (Fitch, 1970a, Fig.19-A). The 1965 earthquake occurred in the complex region where the plate boundaries between southeast Asian, Indian-Australian, Pacific (or Caroline), Philippine Sea plates, and other minor plates meet forming a mega-junction. Crustal blocks in this region are likely to be fragmented by intervening transcurrent faults (see Hamilton, 1979; Cardwell et al., 1980; Fitch, 1970a). The 1965 event probably represents a crustal shortening in this complex region, although a definite interpretation seems impossible at the present time. The small number of aftershocks associated with this event (five reported during the two days after the mainshock) and the small M_s value of 7.5 suggest that the event was not a great earthquake. The large m_B value of 7.8, being larger than the value of M_s, may reflect a high stress drop.

The 1938 Banda Sea earthquake was located at the west of the Weber basin, eastern Banda Sea. Ben-Menahem (1977) showed that surface wave data are consistent with right-lateral strike-slip motion on a northwesterly striking fault (Ritsema and Veldkamp, 1960) and obtained an amount of slip of 8.8 m with a fault dimension of 80×235 km. These values were obtained using data recorded only at one station (PAS); thus they do not seem very reliable.

The 1971 New Guinean earthquake was located to the north of the medial mountain range of West Irian (Indonesian New Guinea). A preliminary study of P-wave first motion data for this event (Kaplan and Seno, in preparation, 1982) shows reverse faulting with one nodal plane steeply dipping to the northeast; a nearly vertical aftershock distribution striking northwest indicates that the above nodal plane is the fault plane. A south-facing island arc collided with southern New Guinea in Miocene time forming a suture zone in the medial mountains (e.g., Hamilton, 1979). The high angle reverse faulting cited above presumably represents the differential motion between southern New Guinea and the Caroline plate; the rupture zone is located at the paleo-island arc north of the medial mountains in New Guinea.

Out of seven great intermediate earthquakes in the western Pacific region, only three earthquakes have been analysed for source mechanisms. The 1959 Kamchatka m_B=7.8 earthquake has a mechanism of down-dip tension type, with one vertical nodal plane striking northeast (Seno, unpublished data). The focal depth of this event was determined as 74 km by ISS. This event is likely to indicate a decoupling of the subducted slab from the oceanic plate due to gravitational pull, which may be the intermediate version of the great normal faults at the trench axis (Y. Fukao, personal communication).

The 1978 M_w=7.8 southern Kurile Islands earthquake represents reverse faulting along a nearly vertical fault plane that strikes northwest normal to the arc trend and dips along the slab with motion of the northeastern block downward with respect to the southwestern block (Sudo et al., 1979). Mechanisms of a similar type are also found in smaller magnitude events at the same location and are interpreted as indicating a contortion of slab at the junction between the Kurile and northern Honshu arcs (Sasatani, 1976b).

The 1963 Banda Sea m_B=7.8 earthquake was located beneath the southwestern end of the Weber Basin at the depth of 100 km, where the contour of the Wadati-Benioff zone is strongly curved to the northeast from the east-west strike along the Java Trench (see Cardwell and Isacks, 1978). This event represents reverse faulting on a steep fault plane that strikes east-northeast with motion of the northern block downward relative to the southern block (Osada and Abe, 1981). Osada and Abe (1981) interpreted this event as a large fracture within the slab and suggested that repeated deformation of this type might cut off the slab entirely in the future. The mechanism of this event is similar to those of smaller magnitude intermediate events in this area (Cardwell and Isacks, 1978; Fitch, 1970a). Their P-axes parallel to the local strike of the arc seem to reflect the lateral bending of the slab at the eastern Banda arc (Cardwell and Isacks, 1978).

Those earthquakes for which source mechanisms have been analysed are indicated by the asterisk in Fig.6. About half of the great earthquakes listed in Table 1 have been analysed for their source mechanisms. Modes of deformation associated with shallow great earthquakes have been well documented as stated above; however, the faulting nature and cause of intermediate earthquakes are poorly known at the present time because of the scarcity of data. For some of the intermediate earthquakes for which source mechanisms are not known, their mechanisms may be similar to those of smaller earthquakes which occurred in the same location; for example, because all the intermediate earthquakes south of the northeastern arm of Celebes (Sulawesi) are of down-dip tension type (Cardwell et al., 1980), we can guess that two great intermediate events which occurred there (Fig.6) may possibly have the same mechanism type.

Pattern of Occurrence of Great Earthquakes

Fig.6 shows that there is a large variation in the pattern of occurrence of great earthquakes in the western Pacific region during the past 84 year period. In the northern half of the region, i.e., from Kamchatka to northern Honshu, most of the subduction boundaries have developed great earthquake rupture zones during this period, without any significant overlap. By contrast, in the southern half of the region, the distribution of great earthquakes is very sparse. Most sections of the Izu-Bonin-Mariana-Yap-Palau arc, Ryukyu and Sunda arcs lack records of great earthquakes. The reader might think that the seismic record for the past 84 years is of insufficient length, and the anomalous quiescence along the southern arcs represents only a temporary feature during a period of strain accumulation. However, we feel that the quiescence of those subduction zones is more likely a long-term seismic pattern, because there is a strong contrast in the mode of occurrence of great or large earthquakes between the northern and southern island arcs. Accepting a kind of ergodic theorem that sampling over time at one specific rupture zone can be replaced by sampling over space during a short time period along an arc section much longer than a dimension of one rupture zone, the southern island arcs are sufficiently sampled for occurrence of earthquakes over space. This suggests that the aseismic character in the southern island arcs is a longer-term seismic pattern. The above ergodicity does not hold if the occurrence of great earthquakes is not random but has a strong coherency over an arc section or if the characteristic pattern of occurrence of earthquakes varies rapidly within one arc section. Some kinds of coherency have been noticed in the pattern of occurrence of great earthquakes in the western Pacific region (Mogi, 1974, 1977). However, the return period of the coherent activity for large earthquakes along the seismic belts is less than or equal to the time period treated in this study (84 years); thus this would not strongly affect the statistics for great earthquakes in the western Pacific region. Moreover, the correlations of the spatial distribution of bathymetric highs on the subducting ocean floor with the large earthquake activity (Kelleher and McCann, 1976; McCann et al., 1979) and of the age of the subducting oceanic lithosphere with the large earthquake activity (Ruff and Kanamori, 1980; Peterson and Seno, 1982) strongly suggest that the quiescence in the southern half has tectonic origins and not a result of a relatively brief seismic record as was discussed by Kelleher and McCann (1976). This point will be discussed in the next section.

Most segments along the northern island arcs have been ruptured by the great earthquakes during the period 1897-1980 without significant overlap. This phenomenon has lead previous investigators to the concept of a "gap" in activity for great earthquakes, which has been used to predict locations of future great earthquakes (e.g., Fedotov, 1965, 1970; Mogi, 1968, 1969; Kelleher et al., 1973; Utsu, 1972). Along a segment where at least two great earthquakes are known to have occurred, the concept of recurrence can be substantiated. The regions for which these recurrence intervals are known are the southern Kurile-Hokkaido-northern Honshu arc and the southwestern Honshu arc; the former has recurrence times of 80-100 years (e.g., Utsu, 1968, 1972) and the latter has those of 100-200 years (e.g., Ando, 1975b).

In the northern region, it is easy to point out gaps in activity for great earthquakes on seismicity maps such as Fig.6. The northern Kurile arc, between 46°N and 49°N, presently shows the most prominent gap. The 1915 (M_s=8.0) earthquake was located within the gap; however, its location close to the trench axis (see

Fig.6) suggests that the event possibly involved normal faulting; thus no known historic earthquake can unequivocally be related to rupture of the thrust zone in this segment. Besides the above gap, two segments in Kamchatka have been listed as likely locations for future events; the segment which includes the location of the 1917 event and the segment trenchward of the 1923 earthquake (Kelleher et al., 1973). In the northern Honshu arc, the segment which ruptured in 1897 has been nominated as a likely location for a future great shock (Seno, 1979; Utsu, 1979a). Off central Honshu, there are three conspicuous gaps; the zone south of the rupture zone of the 1938 earthquake and landward of the southernmost Japan Trench, the zone between the 1923 and 1953 earthquakes along the Sagami Trough, and the zone between the 1923 and 1944 earthquakes along the Nankai Trough. In the first segment, no great historic earthquake except for the large tsunami-genic earthquake in 1677 is known to have ruptured the segment. It is very likely that the plate motion at this zone involves a large portion of aseismic slip so the "gap" represents the aseismic character.

In the zone between the rupture zones of the 1923 and 1953 earthquakes, the 1703 Genroku earthquake is the only event which unequivocally ruptured a part of the gap (Matsuda et al., 1978). However, note that the 1953 earthquake is not an interplate thrust type event but a normal fault within the slab at the triple junction. Thus, the concept of gap cannot be applied in a straightforward manner to this section. The characteristic pattern of occurrence of great earthquakes along the entire section of the Sagami Trough has not yet been understood. The gap between the rupture zones of the 1923 and 1944 earthquakes is the most prominent gap nominated as a likely location for a future great shock in the vicinity of the Japanese Islands (e.g., Mogi, 1969; Utsu, 1974; Ando, 1975a; Ishibashi, 1981). The 1854 Ansei I and 1707 Hoei great events are likely to have ruptured the zone (Ishibashi, 1981).

In the southern half of the western Pacific region, the distribution of great earthquakes in space or time is sparse and the concepts of "gap" or "recurrence" for great earthquakes cannot be established. During the time period treated, the activity of great earthquakes is higher in the region from the central Philippines through the Molucca Sea to northern New Guinea than in other sections of plate boundaries. However, the tectonics of the region is very complicated. There are multiple subduction boundaries and collision zones of island arcs with other island arcs or with continental blocks, as will be shown in the regional tectonics section. Where the tectonics is less complicated, i.e., along the Ryukyu, Izu-Bonin-Mariana, and Sunda arcs, great earthquakes occur only infrequently. Large M_s=7-7.5 earthquakes do occur rather frequently along sections of the northernmost Ryukyu arc near Kyushu, Taiwan, Sumatra, the Philippines, the Molucca Sea and western New Guinea (Rowlett and Kelleher, 1976; McCann et al., 1979). Although these are not great earthquakes, they very likely play an important role in the tectonics in these areas.

Seismic Slip Rates in the Western Pacific Region

Because there is a large variation in frequency of great earthquakes in the western Pacific region, that part of the relative plate motion taken up by seismic slip at the time of great earthquakes may be one of the most relevant quantities with which to characterize variations in the mode of plate consumption (e.g., Kanamori, 1977a). Davies and Brune (1971) estimated the rate of seismic slip at the circum-Pacific and trans-Atlantic-Mediterranean seismic belts by summing up seismic moments over an island arc segment and dividing it by the dimension of the section and the time interval 1897-1968. Their estimate does not seem accurate enough for discussing the characteristics of seismic slip rates in the western Pacific, primarily because they used the magnitude values listed by Duda (1965), which are systematically larger than the surface wave magnitudes given by Gutenberg and Richter (1954) (Geller and Kanamori, 1977). Also, the magnitude versus seismic moment relation which they used was not appropriate for the data. Recently, values of M_s for great shallow earthquakes have become more complete (Geller and Kanamori, 1977; Geller et al., 1978; Abe and Kanamori, 1980; Kanamori and Abe, 1979). Thus, we re-estimate the seismic slip caused by great shallow earthquakes in the western Pacific, using the more relevant relation between M_s and seismic moment. For many earthquakes, we can use the values of seismic moment as listed in Table 1. For the earthquakes for which seismic moments are not known, we used the relation between the magnitude M_s and the seismic moment M_0;

$$Log M_0 = 1.5 M_s + 16.1. \qquad (2)$$

This relation is obtained by substituting M_s for M_w in the relation between M_w and M_0 (Kanamori, 1977b). This is reasonable because M_w agrees very well with M_s for many earthquakes with a rupture length less than about 100 km (Kanamori, 1977b).

Only the earthquakes which are considered to represent the differential motion between the plates are used. Thus the 1933 Sanriku, 1953 Boso-Oki, 1977 Indonesian, and 1938 Banda Sea earthquakes are excluded from the calculation. The 1905 Kurile Islands and 1943 Philippine earthquakes are also excluded because these events are located close to or seaward of the trench axis, thus they have possibly involved normal faulting. We assumed that the thrust zone at the subducting-overriding plate interface is 0-60 km in depth (see Fig.3 and previous discussions), and estimated the width of the thrust zone using the average dip of the inclined seismic zones in this depth range in each arc segment. The summed seismic moment is divided by the rigidity (assumed as 5.0×10^{11} dyn/cm^2), the length of the arc, the width of the thrust zone, and the 84 year time interval (1897-1980) to give seismic slip rates. Table 2 lists the calculated seismic slip rate in each arc segment of the western Pacific region, along with the summed seismic moment, length of the arc segment, and the average dip of the thrust zone. The convergence rates between the plates calculated from the models of Minster and Jordan (1978) and, for the boundaries of the Philippine Sea plate, Seno (1977), and the ratio of the seismic slip rate to the convergence rate are presented in the last two columns in this Table.

The fraction of seismic slip in the plate convergence shows a remarkable variation from one arc section to another and quantitatively describe the variation in the pattern of occurrence of great earthquakes shown in Fig.6. The seismic slip rates

TABLE 2. Seismic Slip Rates in Island Arcs of the Western Pacific

Region	M_o (10^{27} dyn-cm)	Length (km)	Dip (deg)	V_{seis} (cm/yr)	V_{plate} (cm/yr)	V_{seis}/V_{plate} (%)
Kamchatka	448	1100	30	8.1	9.0	90
Kurile	196	1100	30	3.6	9.3	38
N.E. Honshu	56	550	25	1.7	10.5	16
S.W. Honshu	94	550	25	3.5	3.5	100
Ryukyus	0	1500	25	0.0	5.5	0
Izu-Marianas	9	3000	20	0.0	5.0	0
Philippines	149	1900	30	1.6	8.0	19
New Guinea	28	1000	75	1.6	12.0	10
Andaman-Sunda	5	6000	20	0.0	7.0	0

Length is the arc length; Dip is the average dip of upper 60 km of subduction zone. V_{seis} is the seismic slip rate. Relative plate motions V_{plate} are calculated from RM-2 model Minster and Jordan (1978) averaged over the each island arc. For the island arcs around the Philippine Sea, the plate motions are from Seno (1977).

estimated here are greatly different from those estimated by Davies and Brune (1971), except for New Guinea. This difference is partly due to the crudeness for their estimation and partly due to the fact that we do not include earthquakes of M_s (or M_w)≤7.7. If we assume the b-value to be 1.0, the contribution to the seismic slip from the earthquakes of M_s (or M_w)≤7.7 is 32 per cent when maximum M_w is 9.0, and 46 per cent when maximum M_w is 8.8.

The results in Table 2 show that seismic slip accounts for almost all of the relative plate motion along the Nankai Trough off southwest Japan. Nankai Trough is the segment of the subduction boundary for which the most complete data set for historical events exists. Historical earthquakes have recurred in a regular pattern with an interval of 100-200 years (Ando, 1975b). The seismic slip rate along this section has also been estimated by dividing the amount of slip at the time of the historic great earthquakes by recurrence intervals (Kanamori, 1972a, 1977a; Seno, 1977). These studies show that the seismic slip rate is 3-3.5 cm/yr, which is in good agreement with the value of 3.5 cm/yr in Table 2. Furthermore, the drag rate at the plate interface during the interseismic period has been calculated fitting theoretical deformation patterns to the observed crustal deformation data at two localities on the Philippine Sea coast of southwest Japan, giving values of 3-4 cm/yr (Seno and Ishibashi, 1982). This rate is also in good agreement with the above seismic slip rate. The regular pattern of occurrence of great earthquakes and the fact that the plate convergence rate is almost all taken up by seismic slip along the southwest Japan arc have been discussed in connection with the idea that subduction of the Philippine Sea plate beneath southwest Japan has started only very recently, resulting in a strong coupling of the plates at this subduction boundary (Kanamori, 1972a, 1977a; Seno, 1977). Kanamori (1972a) estimated the time of onset of the subduction as 2 myBP, and Seno (1977) and Seno and Matsubara (1980) as 5-6 myBP.

In the Kamchatka arc, most of the plate motion is taken up by the seismic slip. However, it should be noted that the seismic moment of the 1952 earthquake is extremely large, accounting for 78 per cent of the total seismic moment along this arc. Thus, the sampling time interval may be insufficient for this arc. Kanamori (1977a) and Lay and Kanamori (1982) estimated that the amount of seismic slip ranges from one to two thirds of the total plate convergence, dividing the seismic slip at the time of the 1952 earthquake by the recurrence interval of 100-200 years.

From Kurile through northern Honshu, the seismic slip rate is 4-2 cm/yr, which takes up about 40-20 per cent of the total convergence rate. If we include the seismic slip caused by smaller-magnitude earthquakes, total seismic slip would amount up to 60-30 per cent of the total plate convergence. Kanamori (1977a) estimated the seismic slip rate at the southern Kurile arc as 2.5 cm/yr dividing the amounts of slip at the time of the 1963 and 1969 earthquakes by the recurrence interval of about 100 years. Shimazaki (1974a) and Seno (1979) estimated the drag rate at the plate interface off and beneath eastern Hokkaido and northern Honshu, respectively, as about 3-4 cm/yr, fitting the finite element calculations to the observed geodetic data. These values are in good agreement with the seismic slip rate estimated in Table 2.

The seismic slip rate caused by great earthquakes is only a small fraction of the total convergence rate in the southern half of the western Pacific region, contrasting sharply with the northern half (see Table 2). The Philippine and New Guinea convergence zones are the only segments where a considerable fraction of the relative plate motion is taken up by seismic slip. In the Philippines, subduction of the adjoining marginal basins occurs from both sides of the Philippine Islands (e.g., Hamilton, 1979; Seno and Kurita, 1978; Stewart and Cohn, 1979; Cardwell et al., 1980). The calculation of the seismic slip rate was made assuming a single boundary here; thus the seismic slip at each boundary of the Philippines is less by a factor of about two than that shown in table 2.

Along the Kamchatka-Kurile-northern Japan-Izu-Bonin-Mariana arc, the ratio of seismic slip to plate motion systematically decreases from north to south, although the change between the northern Japan and Izu-Bonin arcs is rather abrupt. This result is roughly in accordance with the previous investigation of seismic slip around the circum-Pacific belt by Kanamori (1977a). Kanamori (1971b, 1977a) explained the systematic variation of seismic slip from the Chile through Aleutian to Mariana arcs with his tectonic evolution model for subduction zones, which explains the variation of seismicity in terms of variation of coupling-decoupling at the plate interface, assuming that the strength of coupling decreases as the subduction zone evolves since the onset of subduction. However the situation is more complex in the southern half of the western Pacific, because marginal basins, most of which have been formed in the Tertiary, are subducting there and the mode of subduction shows rapid change in a short distance between arc segments. For example, along the western margin of the Philippine Sea, the Shikoku Basin and West Philippine Basin are currently subducting along the Nankai Trough, Ryukyu Trench, and Philippine Trench. Note the difference in seismic slip rate between the Ryukyu arc and the Philippines, beneath both of which the West Philippine Basin is subducting. The onset time of current subduction beneath these arcs is likely to be almost the same (Seno and Kurita, 1978).

Along the southern arcs, bathymetric highs on subducting lithosphere, temperature, thus age, of the slab, or history of subduction in the early stages may play a more important role in controlling the mode of subduction. The modification of seismicity by subduction of aseismic ridges has been discussed by Vogt et al. (1976), Kelleher and McCann (1976), Rowlett and Kelleher (1976), and McCann et al. (1979). They suggested that buoyancy of aseismic ridges has possibly modified the subduction to be aseismic. Fig.6 represents aseismic ridges on the ocean and marginal basin floors in the western Pacific region. We can see a good correlation between the lack of rupture zones of great events and the existence of aseismic ridges on the oceanic floor adjacent to the subduction zones as was discussed by Kelleher and McCann (1976) and others. Although it seems difficult to quantitatively correlate the seismicity with the aseismic ridge subduction, it would be a possible factor which controls variation of the seismic slip along the island arcs in the western Pacific region.

Ruff and Kanamori (1980) showed that the age of the subducting lithosphere and the rate of convergence can be correlated with the maximum value of M_s of the earthquakes in each arc section. The slower the convergence rate, and the older the lithospheric age, the smaller the characteristic M_s magnitude. They assigned M_s of 8.0 to the Ryukyus. However, this value seems to be overestimated because no great earthquakes have occurred along the Ryukyus during the last 200 years. Recently, Peterson and Seno (1982) showed that, within subduction zones which belong to the same oceanic plate, there is a relation between the rate of seismic moment release per unit arc length and the age of the subducting lithosphere, with the moment release decreasing as the age increases. For example, the variation of the seismic moment release from southwest Japan through the Ryukyus to the Philippines has a correlation with the age of 25, 60, and 40 myBP of the Philippine Sea along these subduction zones. The correlation also holds for the Kamchatka-Kurile-northern Honshu-Izu-Mariana arc and for the Sumatra-Java arc, where the Pacific and Indian-Australian plates are subducting, respectively, with age increasing in the order above. They, however, did not find a relation between the convergence rate of the plates and the seismic moment release rate, in contrast to the results by Ruff and Kanamori (1980).

Although subduction seismicity along the circum-Pacific region seems roughly controlled by the above the buoyant features on the sea floor, the details of subduction seismicity in the western Pacific region could also be affected by the features pertinent to regional tectonics.

In this section, we have treated only seismic slip caused by great earthquakes. We have estimated the amount of the seismic slip contributed from smaller-magnitude events as at most about 50 per cent; however, in order to discuss characteristic patterns of seismicity in the region in more detail, we have to take into account the earthquakes of $M_s \leq 7.7$. In the southern half of the region, these $M_s = 7$ class earthquakes possibly play an important role in tectonics because great earthquakes occur only infrequently there. Although Kelleher and McCann (1976), Rowlett and Kelleher (1976), and McCann et al. (1979) have discussed the characteristic pattern of occurrence of earthquakes including $M_s = 7$ class events in the western Pacific region, their studies are mostly from the viewpoint of forecasting the locations of future large shocks. Additional quantitative studies may be necessary in order to fully understand the relationship between magnitude, frequency of occurrence, and the nature of related deformation.

Focal Mechanisms and Regional Tectonics

The focal mechanism is one of the most useful pieces of information derived from earthquake analyses with which to elucidate the mode of deformation occurring along a seismic belt. We shall compare the spatial distribution of focal mechanisms of earthquakes including smaller magnitude ($m_b \geq 5.6$) events in features pertinent to the various western Pacific convergence zones with the regional and global tectonics.

Types of focal mechanisms of shallow (focal depth ≤ 60 km) earthquakes are schematically represented for three regions of the western Pacific in Figs.7,8, and 10. These focal mechanisms are from previous studies. We selected only reliable solutions which were derived using long-period body or surface wave data. Only solutions for which we can assess the reliability directly in the original works are listed. Where a number of spatially close mechanisms of similar type were available, we excluded some of them to clarify the illustration. Where a thrust type earthquake is interpreted as indicating the differential movement between the two plates, we showed its slip vector by an arrow diverging from its epicenter. P-axes of reverse fault type earthquakes, which are not directly linked with the relative plate motion, and T-axes of normal fault type earthquakes are projected horizontally; they are represented by two opposing arrows converging on or diverging from their epicenters, respectively. For strike-slip solutions, the strike of the two nodal planes and sense of motion on these planes are indicated. If it was possible to choose a preferred orientation for the fault plane on the basis of aftershock distribution, geological structure, or some other evidence, only the preferred strike is shown. The date, location, focal depth, fault type, trend of slip vectors, P or T axes, or fault strikes, and

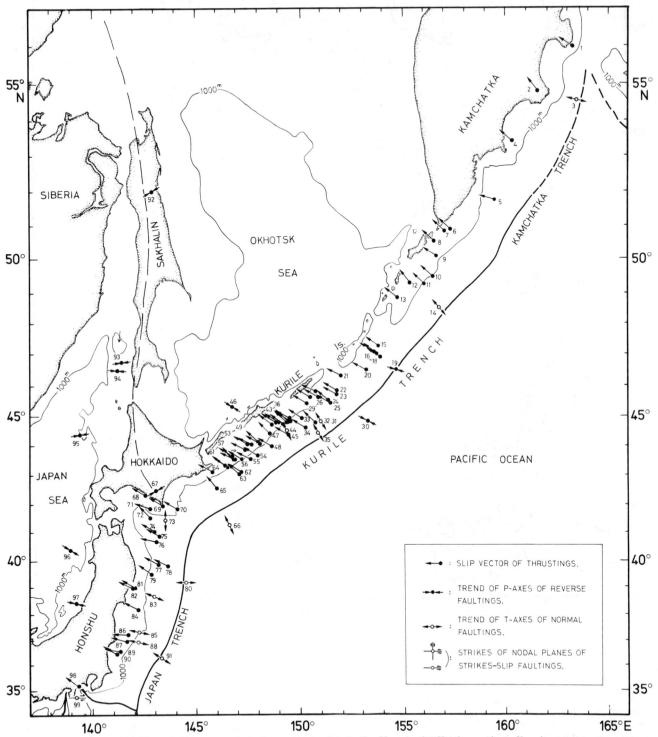

Fig.7 Focal mechanism type of shallow earthquakes in the Kamchatka-Kurile-northern Honshu arc. Slip vector of thrust type events, projection of P or T axes of reverse or normal fault events, and strike of nodal planes of strike-slip fault type events are shown. Numbers associated with the events refer to Table 3.

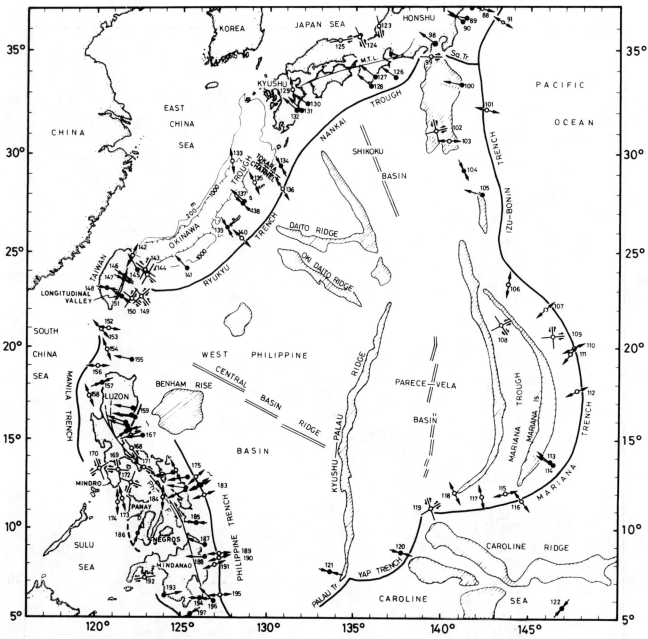

Fig.8 Focal mechanism type of shallow earthquakes in the region around the Philippine Sea. PH.F. denotes the Philippine fault.

papers from which the mechanism data were obtained are listed in Table 3.

In the following subsections, we will describe seismic activity and modes of deformation associated with shallow, intermediate and deep earthquakes in each region of the western Pacific from the viewpoint of regional and global tectonics. The present kinematics of the Philippine Sea plate is also discussed.

Kamchatka-Kurile Arc

The Kamchatka-Kurile Trench constitutes part of the northwestern boundary of the Pacific plate. The Pacific plate is being subducted beneath this trench as indicated by the northwestward dipping zone of intermediate and deep seismicity (Fig.2) and by the slip vectors of shallow thrust type earthquakes (Fig.7). Though there are some ambiguities about the configuration of the plate boundary around the Sea of Okhotsk, the preferable model is that the Kamchatka-Kurile Trench corresponds to the border between the North American and Pacific plates (Chapman and Solomon, 1976). Shallow seismicity along this trench is high (Fig.1). Historical great shallow earthquakes have occurred frequently and some of them were accompanied by tsunamis. Focal mechanism studies for shallow earthquakes based on body wave data (e.g., Stauder, 1962; Udidas and Stauder, 1964; Stauder and Bollinger, 1966; Shimazaki, 1972; Veith, 1974; Stauder and Mualchin, 1976) and analyses of source parameters of large or great shallow earthquakes based on long-period surface wave data (Kanamori, 1970, 1976; Abe, 1973; Shimazaki, 1974b; Kurita and Ando, 1974; Fukao and Furumoto, 1979) make it possible to understand the basic nature of deformation along this boundary. It consists of underthrusting at or near the plate interface and varied deformation within the oceanic and continental plates.

As can be seen in Fig.1, the epicentral distribution of shallow earthquakes along the Kamchatka-Kurile arc seems to be grouped into two different linear zones parallel to the trench axis, although the seaward zone at or near the trench axis is much less active. Shallow events in the zone beneath or seaward of the trench axis are mostly of normal and reverse fault types with subhorizontal T and P-axes normal to the trench axis (Hanks, 1971; Stauder and Mualchin, 1976; Chapple and Forsyth, 1979). Some of these earthquakes are likely to be caused by the bending stress within the oceanic plate beneath and seaward of the trench; thus they are sometimes called "bending earthquakes". Chapple and Forsyth (1979) cited seven earthquakes (five tension type and two compression type) as bending earthquakes for the Kurile-Kamchatka Trench (Fig.7, Nos.3,14,19,30,31,35 and 68). However, some of them (Nos.30,31 and 35) have P or T axes dipping landward at an angle of about 45 degrees and No.31 has a relatively deep focus (60 km, PDE). Thus more detailed studies on their locations and depths are needed to identify them as bending earthquakes.

Bending of the oceanic plate near the trench axis has been discussed in connection with the topography and gravity anomaly (the "outer trench high") seaward of the trench axis. Hanks (1971) first suggested that a horizontal compressive stress of several kilobars is acting on the elastic part of the oceanic lithosphere normal to the Kurile Trench to sustain the deflection of the Pacific plate at the the bulge seaward of the trench. However, Parsons and Molnar (1976) and Caldwell et al. (1976) showed that the deflection of the oceanic lithosphere can also be explained with small or no horizontal compressive stress if a vertical load is applied on the subducting lithosphere landward of the trench. Since the work of Hanks (1971), the bending of oceanic plates at trenches has been modeled with various rheologies and stresses such as elastic or elastic-plastic models with or without horizontal compressional stress, and viscous models (Caldwell et al., 1976; Turcotte et al., 1978; McAdoo et al., 1978; Chapple and Forsyth, 1979; De Bremaecker, 1977; Melosh, 1978). However, a conclusive model has not yet been selected, since the detailed process of subduction and the rheological properties of the oceanic lithosphere and ambient asthenosphere are not clearly known.

Focal mechanisms of shallow earthquakes in the zone of intense seismic activity between the frontal arc and the trench axis are mostly thrust type. Directions of slip vectors of these earthquakes at the Kamchatka-Kurile Trench are uniform over the entire length of the arc; their averaged values have azimuths of about N304±7°E, as seen in Fig.7. The direction of the relative motion between the Pacific and North American plates calculated from the rotation vector of Minster and Jordan (1978) is about N308°E at the Kamchatka and about N293°E at the northern Japan arc, showing a good agreement with the seismic slip vectors.

Geometry of the Wadati-Benioff zone and associated earthquake mechanisms along the Kamchatka-Kurile arc have been studied by Sykes (1966), Isacks and Molnar (1971), Veith (1974), Stauder and Mualchin (1976), Sasatani (1976a,b) and Isacks and Barazangi (1977). The Kurile arc is one of the arc segments where the so-called double seismic zones were found first. Sykes (1966) noted that hypocenters near 44°N, 149°E in southern Kurile seem to delineate at least two very narrow zones of activity within the broader zone of earthquakes that dips under the arc in the depth range of 100-180 km. Veith (1974) confirmed, by relocation of hypocenters using source-region/station corrections, that at many sections along the Kurile arc intermediate depth earthquakes constitute a double zone of down-dip compression in the upper part of the slab and down-dip tension about 30 km below the upper zone. Stauder and Mualchin (1976) also showed that intermediate depth earthquakes of down-dip compression type occur very close to those of down-dip tension type in the central and northern Kurile arc. Recently, the data recorded at the microearthquake seismometer network of Hokkaido University revealed a double seismic zone beneath central Hokkaido (Suzuki et al., 1981). Although along the Kamchatka arc, the cross-sections of earthquakes by Veith (1974) and Fedotov (1968) do not show an unambiguous double-seismic zone, it is plausible that there is also a double zone beneath Kamchatka (e.g., Fujita and Kanamori, 1981).

Veith (1974) proposed that the volume change associated with phase transition could result in down-dip tensional stress reaching a critical value within the interior of the slab while down-dip compressional stress could also reach critical values on the other side of the phase transition boundary, near the outer surface of the plate. Fujita and Kanamori (1981) discounted the olivine-spinel phase change as a mechanism for the double zone because a plausible upward shift of the phase change boundary within the slab is at most 100 km. Isacks and Barazangi (1977) proposed the "unbending" of the plate as a mechanism for the double zone beneath the Kurile arc. The finding of the dual zone of stress orientations within the slab changed the view of Isacks and Molnar (1971) that the slab simply transmits axial stresses induced by gravitational sinking or by resistive force at the mesosphere.

Large deep earthquakes along the Kamchatka-Kurile arc are all of the down-dip compression type (Isacks and Molnar, 1971; Veith, 1974; Stauder and Mualchin, 1976; Sasatani, 1976a). However, beneath the northern part of Hokkaido and the Sea of Okhotsk north of Hokkaido, mechanisms do not show any down-dip orientation of stress axis; T axes dip north-northeast and P axes dip west-northwest with a low angle (Stauder and Mualchin, 1976; Sasatani, 1976b). Sasatani (1976b) and Stauder and Mualchin (1976) attributed these orientations of stress axes to contor-

TABLE 3. List of Shallow Earthquakes of which Focal Mechanism Types
are presented in Figs. 7,8 and 9.

Event	Date	Location °N	°E	Depth (km)	Mechanism Type	Trend	Ref.
1	1971 12 15	56.0	163.3	33	Th	300	SM76-2
2	1973 03 04	54.8	161.6	32	Th	320	SM76-2
3	1970 02 06	54.6	163.5	43	Nm	281	Cm75
4	1970 06 28	53.4	160.4	23	Th	310	SM76-5
5	1952 11 04	52.7	159.5	s	Th	284	Kn76
6	1973 04 12	50.9	157.4	52	Th	310	SM76-7
7	1973 03 12	50.8	157.1	54	Th	310	SM76-8
8	1972 02 28	50.5	156.6	27	Th	310	SM76-9
9	1973 03 12	50.1	156.7	49	Th	300	SM76-10
10	1965 10 03	49.5	156.5	33	Th	310	SM76-11
11	1972 08 04	49.2	156.1	54	Th	310	SM76-14
12	1971 08 19	49.3	155.4	33	Th	320	SM76-13
13	1968 05 20	48.8	154.8	40	Th	305	SM76-15
14	1962 09 15	48.5	156.8	33	Nm	138	UB69
15	1964 07 24	47.2	153.8	33	Th	305	SM76-19
16	1964 07 24	47.1	153.6	33	Th	301	SM76-20
17	1964 07 24	47.0	153.7	33	Th	308	SM76-21
18	1964 07 24	46.9	153.9	33	Th	303	SM76-22
19	1963 03 16	46.5	154.7	26	Rv	106	SB66
20	1963 06 28	46.5	153.2	33	Th	298	SB66
21	1967 04 01	46.3	152.0	40	Th	297	SM76-24
22	1967 04 01	45.8	151.8	40	Th	300	SM76-25
23	1967 04 01	45.7	151.8	40	Th	298	SM76-27
24	1967 03 25	45.5	151.4	41	Th	310	SM76-32
25	1967 08 30	45.4	151.5	33	Th	300	SM76-33
26	1964 04 08	45.8	150.8	40	Th	294	SM76-26
27	1969 08 01	45.6	150.9	38	Th	300	SM76-30
28	1963 10 13	45.6	150.5	33	Th	300	SM76-31
29	1967 08 10	45.4	150.3	37	Th	300	SM76-34
30	1971 12 02	44.8	153.3	24	Rv	296	SM76-72
31	1963 10 14	44.8	151.0	60	Nm	329	SM76-73
32	1963 10 20	44.7	150.7	25	Th	311	SB66
33	1968 05 21	44.9	150.1	33	Th	302	SM76-35
34	1968 05 20	44.6	150.3	49	Th	300	SM76-45
35	1971 09 09	44.4	150.9	7	Nm	329	SM76-76
36	1970 06 10	44.9	149.5	57	Th	298	SM76-36
37	1964 07 05	44.8	149.6	54	Th	302	SM76-37
38	1963 10 13	44.8	149.5	60	Th	313	SB66
39	1972 12 10	44.8	149.4	13	Th	310	SM76-38
40	1972 12 17	44.7	149.2	50	Th	305	SM76-43
41	1963 10 12	44.8	149.0	40	Th	302	SB66
42	1970 03 10	44.8	148.9	40	Th	300	SM76-41
43	1965 06 11	44.7	148.7	47	Th	305	SM76-44
44	1967 09 22	44.5	149.4	60	Nm	334	SM76-75
45	1964 10 16	44.3	149.5	33	Th	300	SM76-46
46	1962 05 07	45.3	146.7	25	Rv	302	SM76-69
47	1958 11 06	44.4	148.6	32	Th	315	FF79
48	1969 08 12	43.9	148.7	26	Th	300	SM76-51

Mechanism types Th, Nm, Rv, and Ss denote thrust, normal fault, high angle reverse fault, and strike-slip fault, respectively. Trend indicates the azimuth of slip vector, T-axis, P-axis, and nodal plane strike for Th, Nm, Rv, and Ss, respectively. Ref. indicates the authors and year of publication of the paper to which the paramters in the preeceding columns are referred, and the event number in each paper; AA, Acharya and Aggarwal; Ab, Abe; CF, Chapple and Forsyth; CI, Cardwell and Isacks; CIK, Cardwell, Isacks, and Karig; Cm, Cormier; CS, Chapman and Solomon; Eg, Eguchi; EUM, Eguchi, Uyeda, and Maki; FF, Fukao and Furumoto; Ft, Fitch; JM, Johnson and Molnar; Kn, Kanamori; KS, Katsumata and Sykes; SB, Stauder and Bollinger; SC, Stewart and Cohn; Sd, Sudo; SK, Seno and Kurita; Sm, Shimazaki; SM, Stauder and Mualchin; Sn, Seno et al.; SP, Seno and Pongsawat; Ss, Sasatani; SS, Shimazaki and Somerville; St, Stewart; UB, Udidas and Bollinger; WA, Weissel and Anderson; WGS, Wang, Geller and Stein; Wu, Wu; YA, Yoshioka and Abe; YM, Yamashina and Murai; Ys, Yoshii.

TABLE 3 (continued)

Event	Date	Location °N	°E	Depth (km)	Mechanism Type	Trend	Ref.
49	1963 03 30	44.1	148.0	33	Th	312	SM76-47
50	1969 08 13	44.0	148.1	33	Th	300	SM76-48
51	1969 08 13	44.0	147.7	33	Th	295	SM76-49
52	1964 10 23	44.0	147.5	45	Th	300	SM76-50
53	1970 10 08	43.8	147.4	15	Th	295	SM76-52
54	1969 08 12	43.6	148.0	33	Th	300	SM76-53
55	1975 06 15	43.50	147.65	55	Th	295	Ys79b
56	1968 01 29	43.5	147.2	36	Th	300	SM76-56
57	1969 01 29	43.6	146.7	40	Th	303	SM76-55
58	1970 11 20	43.5	146.9	36	Th	295	SM76-57
59	1964 05 31	43.5	146.8	48	Th	319	SM76-58
60	1973 06 24	43.3	146.4	50	Th	305	SM76-61
61	1973 06 26	43.2	146.6	50	Th	305	SM76-60
62	1968 01 30	43.1	147.2	28	Th	300	SM76-63
63	1973 06 26	43.0	147.1	39	Th	310	SM76-64
64	1973 06 17	43.1	145.8	49	Th	297	Sm74
65	1973 06 18	42.5	146.0	29	Th	320	SM76-65
66	1965 07 25	41.2	146.6	11	Nm	334	Sm72
67	1970 01 20	42.5	143.0	46	Rv	240	SM76-115
68	1968 09 21	42.2	142.6	33	Th	300	SM76-113
69	1965 06 13	41.9	143.4	32	Th	301	Sm72
70	1966 11 12	41.8	144.1	33	Th	298	Sm72
71	1964 01 10	41.85	142.78	40	Th	288	Ys79b
72	1968 05 16	41.5	142.7	33	Th	315	SM76-109
73	1971 08 02	41.4	143.5	51	Nm	359	SM76-116
74	1968 06 17	41.0	143.0	48	Th	300	SM76-112
75	1968 05 16	40.8	143.2	7	Th	300	Kn71a
76	1965 03 29	40.7	143.2	40	Th	283	Ss73
77	1968 05 16	39.8	143.1	37	Th	297	Kn71a
78	1968 05 16	39.7	143.6	29	Th	295	SM76-110
79	1968 06 12	39.47	142.89	s	Th	305	YA76
80	1933 03 02	39.2	144.5	10	Nm	270	Kn71d
81	1965 02 16	38.97	142.01	55	Th	300	SP81
82	1973 11 19	38.99	141.93	49	Th	295	Ys79b
83	1968 04 21	38.68	142.99	41	Nm	291	SP81
84	1978 06 12	38.15	142.22	28	Th	295	Sn80
85	1938 11 06	37.33	142.18	17	Nm	280	Ab77-4
86	1938 11 05	37.24	141.75	45	Th	272	Ab77-3
87	1938 11 05	36.97	141.71	30	Th	285	Ab77-2
88	1938 11 06	36.91	142.19	33	Nm	280	Ab77-5
89	1938 05 23	36.58	141.34	40	Th	280	Ab77-1
90	1974 07 08	36.44	141.17	41	Th	300	Ys79b
91	1975 06 14	36.31	143.30	18	Nm	301	Ys79b
92	1964 10 02	51.95	142.92	s	Rv	246	CS76
93	1971 09 06	46.76	141.39	15	Rv	263	Ys79b
94	1971 09 27	46.41	141.16	20	Rv	94	Ys79b
95	1940 08 01	44.35	139.46	33	Rv	262	FF75
96	1964 05 07	40.39	139.05	22	Rv	121	FF75
97	1964 06 16	38.40	139.26	16	Rv	282	Ab75
98	1923 09 01	35.4	139.2	0-10	Th	305	Kn71c
99	1978 01 14	34.76	139.20	4	Ss	270	SS78
100	1972 02 29	33.38	140.97	33	Th	280	Ys79b
101	1974 08 25	32.0	142.3	45	Nm	282	CF79-54
102	1976 12 02	31.0	139.5	48	Ss	166 256	Eg80
103	1965 11 12	30.59	140.36	21	Nm	270	KS69-13
104	1964 01 15	29.19	141.15	50	Rv	154	KS69-6
105	1965 12 28	27.95	142.21	36	Th	286	KS69-14
106	1976 12 22	23.42	143.71	47	Nm	189	Eg81
107	1966 10 27	22.15	145.94	21	Nm	49	KS69-23
108	1973 06 29	21.02	143.27	20	Ss	243 325	Eg81
109	1966 02 10	20.77	146.38	38	Ss	264 356	KS69-16

TABLE 3 (continued)

Event	Date	Location °N	°E	Depth (km)	Mechanism Type	Trend	Ref.
110	1967 04 05	20.00	147.35	34	Nm	245	KS69-24
111	1974 05 11	19.73	147.34	47	Nm	220	CF79-52
112	1970 09 01	17.70	147.65	50	Nm	66	CF79-45
113	1966 05 20	13.76	146.19	58	Th	308	KS69-18
114	1966 05 23	13.67	146.47	35	Th	302	KS69-19
115	1970 03 04	12.1	143.7	30	Nm	256	Ft72-33
116	1964 07 04	11.72	144.63	10	Nm	324	KS69-7
117	1971 10 24	11.86	142.38	25	Nm	348	Eg81
118	1967 08 26	12.23	140.82	53	Nm	334	KS69-25
119	1966 06 07	11.32	139.57	53	Ss	161 238	KS69-20
120	1969 01 27	8.7	137.7	5	Rv	105	Ft72-52
121	1976 12 19	7.7	133.7	33	Rv	284	Eg81
122	1968 08 20	5.6	146.9	33	Rv	220	WA78
123	1948 06 28	36.1	136.2	13	Ss	345	Kn73
124	1927 03 07	35.6	135.1	13	Ss	335	Kn73
125	1943 09 10	35.5	134.2	13	Ss	80	Kn73
126	1966 01 11	33.65	137.33	32	Th	303	KS69-15
127	1944 12 07	33.7	136.2	30	Th	301	Kn72a
128	1946 12 20	33.1	135.8	30	Th	310	Kn72a
129	1975 04 20	33.14	131.29	11	Nm	350	YM75
130	1968 04 01	32.5	132.2	30	Th	306	Ft72-27
131	1969 04 21	32.1	131.8	30	Th	312	Ft72-30
132	1970 07 25	32.1	131.6	30	Th	311	Ft72-34
133	1977 12 22	29.58	127.89	39	Nm	352	Eg81
134	1972 09 02	29.41	130.64	41	Rv	331	Ys79b
135	1969 12 31	28.5	129.1	30	Nm	155	Ft72-27
136	1976 12 14	28.27	130.67	39	Nm	148	Eg81
137	1972 10 26	27.48	128.57	48	Th	312	Eg81
138	1968 11 12	27.5	128.4	30	Rv	311	Ft72-29
139	1966 04 07	26.25	127.57	44	Rv	148	Eg81
140	1968 08 03	25.6	128.4	19	Nm	133	Ft72-28
141	1966 07 10	24.09	125.21	32	Th	315	KS69-22
142	1964 11 26	24.92	122.03	17	Nm	207	Wu78-3
143	1966 03 12	24.09	122.65	42	Ss	125 209	Sd72b
144	1966 03 23	23.75	122.91	51	Ss	130 211	Sd72b
145	1972 04 17	24.10	122.44	48	Th	327	SK78-7
146	1972 11 09	23.87	121.61	22	Rv	124	SK78-12
147	1972 04 24	23.60	121.55	29	Rv	116	SK78-8
148	1964 01 18	23.09	120.58	18	Rv	99	Wu78-2
149	1975 05 23	22.70	122.57	6	Ss	140 230	Wu78-21
150	1972 01 04	22.50	122.07	6	Ss	126	SK78-4
151	1968 02 26	22.76	121.47	8	Rv	302	Wu78-11
152	1965 04 26	21.0	120.68	29	Nm	98	Wu78-4
153	1972 01 08	20.95	120.26	36	Nm	136	Wu78-15
154	1963 06 06	19.97	120.62	0	Nm	158	KS69-3
155	1972 02 08	19.36	122.06	54	Th	280	CIK80-1
156	1974 02 03	18.93	120.13	30	Nm	90	CIK80-2
157	1970 08 26	18.02	120.48	50	Rv	68	CIK80-3
158	1972 05 22	16.60	122.19	41	Nm	156	SK78-23
159	1968 08 01	16.30	122.11	30	Th	280	Ft72-44
160	1972 05 22	16.60	122.19	41	Th	275	CIK80-5
161	1968 11 22	16.17	122.17	60	Th	280	Ft72-47
162	1970 04 07	15.78	121.71	50	Th	288	Ft72-51
163	1971 07 04	15.60	121.85	46	Th	288	CIK80-9
164	1968 08 28	15.55	122.02	25	Rv	69	Ft72-53
165	1970 04 08	15.43	121.75	7	Rv	236	CIK80-11
166	1970 04 12	15.08	122.01	25	Ss	251	SK78-3
167	1970 04 15	15.11	122.71	31	Th	264	CIK80-12
168	1966 12 20	14.57	122.17	32	Ss	323	Ft72-41

TABLE 3 (continued)

Event	Date	Location °N	°E	Depth (km)	Mechanism Type	Trend	Ref.
169	1966 08 15	13.28	121.36	23	Ss	170 272	Ft72-43
170	1972 04 25	13.38	120.34	37	Ss	158 240	CIK80-17
171	1973 03 17	13.41	122.87	44	Ss	118	CIK80-16
172	1970 02 05	12.58	122.09	8	Ss	202 288	Ft72-54
173	1975 10 21	11.65	121.58	20	Nm	164	SK78-21
174	1973 08 18	11.45	121.38	22	Nm	178	CIK80-37
175	1973 07 05	13.18	124.65	20	Nm	154	CIK80-19
176	1964 12 24	13.12	124.58	25	Th	258	Ft72-40
177	1964 12 27	12.92	125.37	1	Th	261	Ft72-39
178	1975 11 15	12.91	125.88	35	Nm	229	CIK80-23
179	1975 10 31	12.47	126.01	49	Nm	59	CIK80-24
180	1975 11 06	12.50	126.06	30	Nm	61	CIK80-25
181	1973 07 03	12.27	125.32	37	Th	276	CIK80-27
182	1973 07 03	12.21	125.29	36	Th	247	CIK80-28
183	1965 01 09	11.88	126.26	27	Nm	250	Ft72-38
184	1970 11 13	11.78	123.97	18	Nm	0	CIK80-35
185	1967 08 19	10.36	125.87	55	Rv	93	Ft72-37
186	1975 12 14	9.87	122.56	33	Rv	19	AA80
187	1970 11 08	9.16	126.41	41	Rv	288	CIK80-32
188	1976 11 07	8.48	126.38	60	Rv	270	CIK80-41
189	1969 03 20	8.69	127.35	43	Nm	264	Ft72-55
190	1971 03 16	8.43	127.23	51	Nm	253	CIK80-42
191	1971 05 10	7.99	126.95	37	Nm	70	CIK80-46
192	1976 08 17	7.25	122.94	22	Ss	260	SC79-2
193	1976 08 16	6.26	124.02	33	Th	80	SC79-1
194	1971 06 05	6.02	126.11	52	Rv	272	CIK80-61
195	1973 03 09	6.32	127.38	41	Nm	266	CIK80-59
196	1968 10 24	6.06	126.97	43	Th	309	Ft72-46
197	1965 05 16	5.26	125.57	36	Rv	240	Ft72-33
198	1968 11 25	4.9	126.8	30	Rv	68	Ft72-66
199	1969 02 03	4.9	127.3	30	Rv	260	Ft72-69
200	1969 01 31	4.1	128.0	30	Th	282	Ft72-68
201	1969 02 02	3.89	128.34	38	Th	282	CIK80-76
202	1969 02 17	3.8	128.4	14	Th	242	Ft72-70
203	1964 10 12	3.0	126.7	59	Th	270	Ft70a-31
204	1968 11 09	2.4	126.8	30	Rv	258	Ft72-76
205	1973 03 16	2.18	126.65	11	Nm	285	CIK80-99
206	1974 05 11	1.87	126.48	37	Rv	294	CIK80-101
207	1968 08 11	1.5	126.1	30	Rv	100	Ft72-73
208	1968 08 17	1.3	126.3	30	Rv	281	Ft72-74
209	1975 05 13	1.03	126.02	25	Rv	112	CIK80-107
210	1969 01 25	0.8	126.0	24	Rv	297	Ft72-77
211	1971 04 07	2.45	129.13	41	Nm	237	CIK80-85
212	1971 10 13	2.9	132.8	47	Ss	29 117	WA78
213	1970 06 12	-2.9	139.1	30	Th	26	JM72-5
214	1967 04 27	-1.8	138.7	33	Rv	22	JM72-4
215	1963 11 06	-2.6	138.3	0	Rv	31	Ft70a-23
216	1969 03 09	-4.1	135.5	14	Ss	136 226	JM72-1
217	1968 06 29	-0.2	133.4	11	Rv	196	Ft72-59
218	1963 04 16	-0.9	128.1	0	Ss	96 186	Ft70a-21
219	1963 06 04	-1.2	127.3	20	Ss	63 167	Ft70a-20
220	1965 04 26	-1.7	126.6	21	Nm	246	Ft70a-18
221	1966 08 18	-0.2	125.1	51	Rv	114	Ft70a-25
222	1967 10 26	-0.2	125.2	42	Rv	130	Ft70a-26
223	1964 10 11	-0.6	121.8	45	Nm	206	Ft70a-29
224	1966 04 23	-0.9	122.4	45	Nm	28	Ft70a-28

TABLE 3 (continued)

Event	Date	Location °N	°E	Depth (km)	Mechanism Type	Trend	Ref.
225	1964 06 30	-0.6	122.6	0	Nm	240	Ft70a-27
226	1974 11 12	2.27	121.06	60	Th	180	CIK80-133
227	1970 03 27	0.3	119.3	8	Ss	136 228	Ft72-57
228	1968 08 14	0.1	119.7	23	Nm	193	Ft72-56
229	1968 06 07	-1.7	120.1	19	Ss	69 164	Ft72-60
230	1969 02 23	-3.1	118.8	13	Rv	253	Ft72-61
231	1965 10 07	12.5	114.5	5	Rv	320	WGS79
232	1965 01 24	-2.4	126.0	6	Rv	213	Ft70a-19
233	1972 09 11	-3.29	130.75	37	Th	238	CI78-4
234	1972 03 08	-3.74	131.39	29	Th	267	CI78-5
235	1964 04 23	-5.4	133.9	0	Nm	266	Ft70a-24
236	1972 09 24	-6.22	131.15	33	Ss	164 255	CI78-16
237	1969 02 24	-6.1	131.0	38	Ss	149 240	Ft72-62
238	1974 03 06	-6.6	129.98	26	Ss	136 226	CI78-15
239	1967 04 22	-5.56	126.81	33	Rv	146	Ft70a-16
240	1970 06 28	-8.7	124.2	30	Th	298	Ft72-63
241	1972 11 05	-9.82	122.17	45	Ss	183 268	CI78-11
242	1968 01 26	-8.8	120.4	29	Ss	164 265	Ft72-64
243	1963 03 24	-9.7	120.6	0	Rv	90	Ft70a-12
244	1973 04 10	-9.81	119.29	55	Rv	80	CIK78-8
245	1970 01 09	-9.27	117.25	58	Rv	45	CI78-7
246	1977 08 19	-11.16	118.41	78	Nm	0	St78
247	1972 05 28	-11.05	116.97	45	Nm	164	CI78-2
248	1967 03 30	-11.0	115.5	32	Nm	0	Ft70a-13
249	1963 04 07	-4.9	103.2	46	Th	24	Ft70a-8
250	1969 11 21	2.0	94.6	20	Ss	112 203	Ft72-65
251	1967 08 21	3.6	95.8	33	Th	23	Ft70a-7
252	1967 04 12	5.3	96.6	55	Rv	12	Ft70a-5
253	1964 04 02	5.8	95.5	16	Ss	150 241	Ft70a-3
254	1964 11 30	6.8	94.8	32	Ss	146 236	Ft70a-4
255	1967 06 02	8.7	93.8	30	Ss	171 261	Ft72-80
256	1964 09 15	8.9	93.1	37	Th	51	Ft70a-2
257	1973 06 09	10.7	92.6	56	Ss	135 227	EUM79-5
258	1971 03 29	11.2	95.1	17	Ss	188 280	EUM79-4
259	1971 08 12	12.5	95.1	20	Ss	168 257	EUM79-1
260	1964 07 28	14.3	96.2	33	Nm	160	Ft70b-9
261	1967 09 06	14.8	93.6	33	Nm	106	Ft70b-12

tion of the slab at the junction between the Kurile and northern Honshu arcs. For the intermediate depth earthquakes beneath western Hokkaido and the southernmost Kurile Islands, they also found many earthquakes the stress axes of which have no simple relation to either the dip or the strike of the slab. These are also likely indicators of a local effect of slab contortion near the junction.

Northeast Japan Arc

The Pacific plate meets with the Eurasian plate along the Japan Trench. Only one-fifth of the relative plate motion seems to be accounted for by seismic slip caused by great earthquakes along this segment, as was shown in the former section. The slip vectors of the shallow thrust type earthquakes roughly coincide

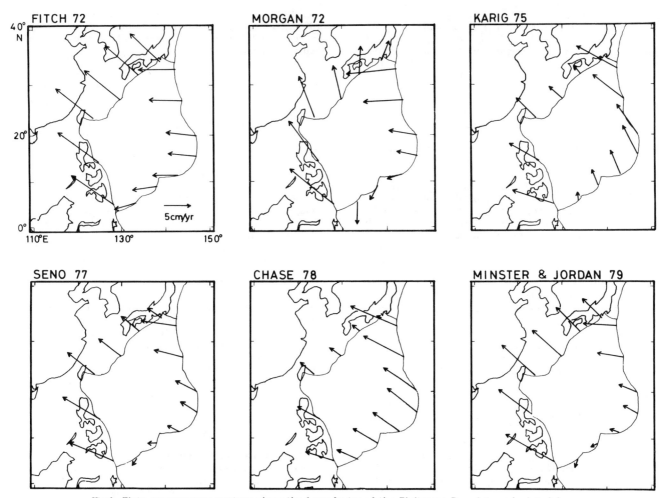

Fig.9 Plate convergence vectors along the boundaries of the Philippine Sea plate, calculated from various models by Fitch (1972), Morgan (1972), Karig (1975), Seno (1977), Chase (1978), and Minster and Jordan (1979).

with the N290°E direction of relative plate motion calculated from the rotation vector of Minster and Jordan (1978). It is possible that the seismic slip vectors represent the Pacific-southeast Asian plate convergence because the Japanese Islands may not be a part of the Eurasian plate but a part of southeast Asia which is decoupled from the Eurasian plate, as was discussed in the former section (see Fig.5). However, the difference in motion between the Eurasian and southeast Asian plates along this segment of boundary would be very small, even if it exists, because there is no systematic difference between the observed seismic slip vectors and the Pacific-Eurasian convergence vector (Eguchi, 1980).

The Wadati-Benioff zone, extending beneath northeastern Japan to Primorye, has a length of about 1200 km and dips to the west-northwest at an angle of about 30 degrees. The deep events along the northern Honshu arc are all of the down-dip compression type (Isacks and Molnar, 1971; Mikumo, 1971; Koyama, 1975, 1978; Stauder and Mualchin, 1976). Isacks and Molnar (1971) reported earthquakes beneath central Japan having a distinctly different orientation of P or T axis and suggested that they reflect tearing or contortion of the slab at the junction between the northern Honshu and Izu-Bonin arcs.

The microearthquake seismometer network of Tohoku University located in northern Honshu has clearly revealed a double seismic zone beneath this segment of island arc (Umino and Hasegawa, 1975; Hasegawa et al., 1978). The depth range of the double seismic zone beneath northern Honshu is 60-180 km and the distance between the upper and lower zones is about 30 km. These zones can also be seen by plotting earthquakes from the bulletin listings for which the hypocentral parameters are of good quality (Yoshii, 1979a; Barazangi and Isacks, 1979; Seno and Pongsawat, 1981; Kawakatsu and Seno, 1982). Focal mechanisms of earthquakes which occurred in the upper plane are characterized by down-dip compression, and in contrast, those in the lower plane by down-dip tension (Umino and Hasegawa, 1975;

Hasegawa and Umino, 1978; Yoshii, 1979a,b; Seno and Pongsawat, 1981; Kawakatsu and Seno, 1982). The seaward end of the double seismic zone corresponds approximately to the interface of the aseismic front and the thrust zone (see Fig.3); thrust type events occur down to the depth of 60 km right beneath the aseismic front (Yoshii, 1975, 1979a; Seno and Pongsawat, 1981; Kawakatsu and Seno, 1982). In some areas along the northern Honshu arc, the double seismic zone extends trenchward of the aseismic front, forming a triple-planed structure along with the thrust type events; these areas are near the rupture zones of recent or expected large $M_s \geq 7.5$ earthquakes at the deeper thrust zone of the depth range of 40-60 km (Seno and Pongsawat, 1981; Seno, 1981; Kawakatsu and Seno, 1982). The coupling of the plates at the deeper (40-60 km) thrust zone prior to large earthquakes may result in stress buildup within the slab near those rupture zones and be the cause of the anomalous position of the double seismic zone seaward of the aseismic front (Seno and Pongsawat, 1981; Kawakatsu and Seno, 1982). Goto and Hamaguchi (1978, 1980) and House and Jacob (1982) proposed a thermal stress within the slab due to heating by the ambient mantle as the cause for the double seismic zone beneath northern Honshu.

Tsumura (1973) was first to note the double seismic zone beneath eastern central Honshu, using the data recorded at the microearthquake network of University of Tokyo, although it was not so conspicuous as that which was later found beneath northern Honshu.

Shallow earthquakes that occur westward of the volcanic front in the continental plate are mostly of reverse fault type with subhorizontal P-axes in the east-west direction, indicating the predominance of a compressive stress regime throughout the back-arc area of northern Honshu (see Fig.7, Nos.96 and 97). Actually, many active fault systems of the dip-slip reverse fault type, striking nearly in the north-south direction, are developed there. At the eastern margin of the Japan Sea, shallow seismic activity is relatively high (see Fig.1); large earthquakes have occurred there, accompanied by destructive tsunamis, about ten times since the year 1500 (Hatori and Katayama, 1977). In contrast, at the southwestern margin of the Japan Sea, few large shallow earthquakes are known to have occurred in the offshore area; large earthquakes have mainly occurred on the Japan Sea coast of southwest Japan. This contrast in activity of large shallow earthquakes at the above two margins of the Japan Sea suggests a different stress state or a different cause for large intraplate earthquakes in each region in the Japanese Islands. Assuming that the horizontal compressive stress, which is most likely to originate from the underthrusting of the Pacific plate along the Kurile-Japan Trench, is large enough to fracture the boundary between the Japan Sea and northern Honshu, a new trench system may eventually be initiated at this border. This type of trench initiation is likely to be seen only at the margins of marginal basins where strong compressive tectonic stresses exist. At the western margin of the Philippines, marginal basins such as the South China Sea, Sulu and Celebes Basins are currently being subducted from the west (Ludwig et al., 1967; Hayes and Ludwig, 1967; Stewart and Cohn, 1979; Seno and Kurita, 1978; Hamilton, 1979; Cardwell et al., 1980; Acharya and Aggarwal, 1980). This example of marginal basin subduction may represent a future stage of the activity at the eastern margin of the Japan Sea.

Izu-Bonin-Mariana-Yap-Palau Arc and Caroline Islands

The Pacific plate meets with the Philippine Sea plate along the Izu-Bonin-Mariana-Yap-Palau trench system. As stated previously, large or great shallow earthquakes have occurred very infrequently at this plate boundary. Beneath the central part of the Mariana arc, the oceanic lithosphere is vertically subducted and reaches a depth of greater than 600 km. One mechanism solution of the deepest events beneath the Mariana arc indicates down-dip compression like other events in the deepest subducted oceanic slabs (Katsumata and Sykes, 1969). Deep earthquakes along the Izu-Bonin arc are all of down-dip compression type (Isacks and Molnar, 1971; Katsumata and Sykes, 1969; Mikumo, 1971). Focal mechanisms for intermediate events are sparse. One intermediate event in the central part of the Izu-Bonin arc shows lateral compression parallel to the arc (Katsumata and Sykes, 1969). Wilson and Toldi (1978) reported down dip tensional stress, Samowitz and Forsyth (1981) reported one down-dip compression and one down-dip tension event, and Katsumata and Sykes (1969) reported one lateral tension type event for the intermediate activity at the Mariana arc.

Several mechanism solutions for shallow events have been reported by Katsumata and Sykes (1969), Fitch (1972), Chapple and Forsyth (1979), and Eguchi (1981). Shallow earthquakes near the trench axis are of the normal fault type, with subhorizontal T-axes normal to the trench axis. Most of these events may be caused by the tensional stress in the upper layer of the oceanic lithosphere where it bends downward near the trench axis (Jones et al., 1978; Hilde and Sharman, 1978). However, since the hypocenters of the tensional events near the southern Mariana Trench (Fig.8, Nos.115-118) are not constrained well, there is some ambiguity as to whether these events occur within the subducting oceanic lithosphere or within the Philippine Sea plate at its margin near the trench. If the latter case is correct, these earthquakes may be related to a tensional stress regime in the southern part of the Mariana Trough. Only four shallow thrust type earthquakes have been reported along the Izu-Bonin-Mariana-Yap-Palau arc. This fact, together with their small magnitude, suggests that the coupling at this plate boundary is weak as maintained by Kanamori (1971b, 1977a) and Uyeda and Kanamori (1978). Beneath the back-arc of the Izu-Bonin arc at about 30-31°N, one normal fault type event with an east-west striking T-axis and one strike-slip fault type event occurred (Fig.8, Nos.103 and 102). The normal fault type event suggests that the stress state behind this arc is not compressive. Behind the Mariana Trench, active spreading has been substantiated by marine geophysical surveys such as the measurements of magnetic anomalies, heat flow and gravity, seismic reflection profiles, and deep sea drillings (e.g., Karig, 1971; Anderson, 1975; Hussong and Uyeda et al., 1978; Karig et al., 1978). Eguchi (1981) obtained a

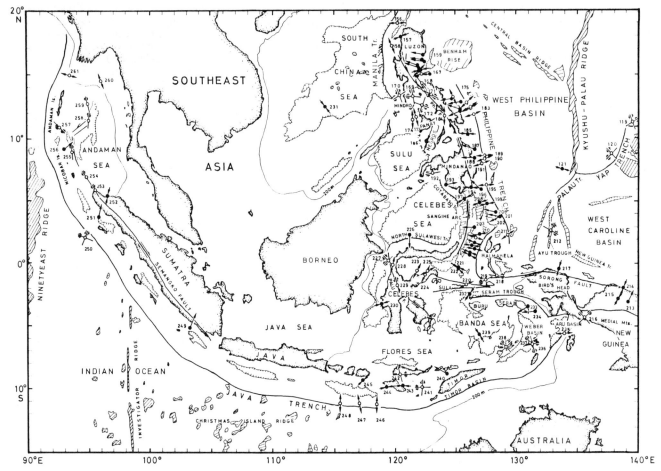

Fig.10 Focal mechanism type of shallow earthquakes in the Philippines, Indonesia, and New Guinea.

strike-slip solution for one shallow earthquake which occurred in the northernmost part of the Mariana Trough (Fig.8, No.108). One nodal plane is parallel to and the other normal to the local trend of the arc. Karig et al. (1978) proposed that opening has occurred at the axial zone of the trough in a direction normal to the trend of the arc. Eguchi (1981) interpreted this event as left-lateral strike-slip faulting on a transform fault which offsets the spreading center. Bathymetric features in the central part of the trough indicate that the spreading centers are offset by many transform faults (Karig et al., 1978).

At the Yap Trench, a large buoyant mass, the Caroline Ridge, is colliding with the trench. Two strike-slip type events have occurred at the central part and at the intersection with the Mariana Trench (Fig.10, Nos.120 and 119). The former event (No.120) contains considerable portion of dip-slip component; thus the P-axis of this event is indicated in Fig.8 instead of strikes of nodal planes. The two strike-slip events could be interpreted as indicating the Pacific-Philippine Sea relative plate motion by selecting one of the nodal planes trending subparallel to the arc as a fault plane (see Seno, 1977, Fig.5). On the other hand, Karig's (1975) Pacific-Philippine Sea relative motion is not in accord with any of the nodal planes of these solutions.

On the basis of marine geophysical and seismological data, Weissel and Anderson (1978) proposed the existence of a Caroline plate, decoupled from the Pacific plate to the east of the Yap-Palau Trench. Although several focal mechanism solutions are available for the margin of this proposed plate, there is generally a low level of seismicity at the boundary and the kinematics of the Caroline plate with respect to the surrounding plates are not well understood at the present time. The pole of the Caroline-Philippine Sea relative plate motion was located at the east of the Palau Trench by Weissel and Anderson (1978); this also does not explain the two strike-slip solutions at the Yap Trench (Nos.119 and 120 in Fig.10) as the differential motion between these two plates.

Southwest Japan

As stated in the former section, great earthquakes have recurred quite regularly along the Nankai Trough off southwest Japan. Along the thrust zone, where the great earthquakes recur, smaller magnitude activity ($M_s < 7.5$) is very low (see Fig.1 and also

Fig.8). No bending earthquakes have been reported beneath the axis of the Nankai Trough. This pattern of seismicity along the Nankai Trough is quite different from the other subduction zones in the western Pacific region. The recent commencement of Shikoku Basin subduction beneath southwest Japan (2-6 myBP) (Kanamori, 1972a; Seno, 1977; Matsubara and Seno, 1980) and the young age of the Shikoku Basin (17-30 myBP, Kobayashi and Nakata, 1978) may explain this anomalously strong coupling between the plates. In the overriding plate, large earthquakes ($M_s \leq 7.5$) have occurred infrequently on the Japan Sea coast. Here, the seismic events are mostly of strike-slip fault type with P-axis trending northwest or west-northwest (Fig.8).

Ryukyu Islands

Ryukyu Trench appears to be connected with the Nankai Trough off Kyushu, Japan. However, the pattern of seismicity along the Ryukyu Trench is quite different from that along the Nankai Trough. Therefore, it is inferred that the tectonic history for these two sections of the northwestern Philippine Sea margin may be quite different. Great or large earthquakes occur very infrequently along Ryukyus (Kelleher and McCann, 1976; Rowlett and Kelleher, 1976). At the junction of the Ryukyu Trench and the Nankai Trough, large earthquakes ($M_s = 7$-7.5) have occurred with a 20-30 year frequency (e.g., Utsu, 1974; Hatori, 1971). Recent earthquakes for which focal mechanism solutions are available are all of thrust type (see Fig.8 and Shiono et al., 1980). In the Ryukyu arc, only a few thrust type solutions have been obtained. Most events indicate shallow normal or reverse faulting near the Ryukyu Trench and further inland in the central part of the arc. One of the normal fault type events occurred in the back-arc with a T-axis in a north-south direction (Fig.8, No.133). Eguchi (1981) demonstrated that this event may indicate back-arc opening in a north-south direction, subparallel to the trend of the arc at this point. At the southwestern part of the Okinawa Trough, north-south opening of the basin has been substantiated by marine geophysical surveys (Herman et al., 1979; Lee et al., 1980).

Geometry of the Wadati-Benioff zone and associated focal mechanisms have been studied by Katsumata and Sykes (1969), Mikumo (1971), and Shiono et al. (1980). The depth of the Wadati-Benioff zone is shallower than 300 km along this arc. Shiono et al. (1980) showed a distinct contrast in the dip of the zone and focal mechanisms between north and south of the Tokara Channel at about 29°N, 130°E. North of the channel, the Wadati-Benioff zone dips at an angle of 70 degrees in the deeper part and all mechanism solutions are down-dip tension type. South of the channel, the Wadati-Benioff zone dips at an angle of 40 degrees and all solutions are down-dip compression type. Shiono et al. (1980) explained the differences in seismic activity and focal mechanisms by the difference in the convergence velocity of the slab or in the viscosity of ambient mantle between the two segments.

Taiwan

Seismic activity in the vicinity of Taiwan during this century is very high compared with the two adjacent sections of the plate boundary, i.e., the Ryukyus and Luzon Straits. Although there have occurred many $M_s = 7$-7.5 earthquakes during this century along and off the eastern coast of Taiwan (Rowlett and Kelleher, 1976, Fig.5), only one great $M_s = 8.0$ earthquake in 1920 is known in the vicinity of Taiwan (see Fig.6). Focal mechanisms of major earthquakes which have occurred since the installation of the WWSSN network, including four $M_s = 7$-7.5 earthquakes, have been studied by Wu (1970, 1978), Sudo (1972b), and Seno and Kurita (1978). These studies indicate that the major mode of rupture is: strike-slip faulting with P-axis trending nearly east-west off the eastern coast of Taiwan, and high angle reverse faulting along the eastern coast.

The strike-slip events off the northeastern coast include the large $M_s = 7.5$ earthquake of March 12, 1966 (Fig.8, No.143). A number of events there represent strike-slip faulting with nodal planes and P-axes similar to that shown in Fig.8 (Sudo, 1972b; Wu, 1970, 1978). The epicenters of these events are distributed in an elongate fashion in a roughly northwest direction. Wu (1970, 1978) argued that the faulting of the 1966 event was a right-lateral strike-slip on the northeast trending nodal plane on the basis of the aftershock distribution, bathymetric features off the coast and the asymmetry in the radiation pattern of G waves. With left-lateral shear motion along the Longitudinal Valley, which is a major active fault along the east coast (Allen, 1962; Kaneko, 1970), he concluded that a sliver of oceanic plate is moving northward and downthrusting at the Ryukyu-Taiwan junction. In contrast, Sudo (1972a,b) preferred a left-lateral strike-slip motion on the nodal plane with a northwest strike for this event based on the multiple rupture mode and proposed that this event indicates a differential movement between the Philippine Sea plate and the adjacent Asian continental block northeast of Taiwan. The aftershock distribution does not clearly indicate the preferred fault strike (Sudo, 1972b).

The strike-slip events off the southeastern coast, including the $M_s = 7.5$ event of Jan. 25, 1972, show similar mechanisms to those off the northeastern coast (Nos.149 and 150 in Fig.8, Seno and Kurita, 1978; Wu, 1978). Seno and Kurita (1978) considered these events to be left-lateral ruptures striking in a northwest direction caused by supposed left-lateral shear deformation off the coast due to the gradual transition of the consuming boundary from the eastern coast of Taiwan to the eastern coast of Luzon. The aftershock area of the 1972 event is elongated in a northwest direction (Wu, 1978), suggesting a left-lateral strike-slip motion for this event. Karig (1973) inferred that the area is dominated by left-lateral shear deformation in a north-northeast direction on the basis of bathymetric features observed there; however the sense of the proposed shear is inconsistent with the focal mechanisms (Fig.8, Nos.149 and 150). More detailed marine geophysical studies are necessary to resolve the ambiguity in the interpretation of the current tectonics of the area.

The northern extension of the west-facing Luzon arc has been colliding with Taiwan since late Pliocene time (Chai, 1972; Karig, 1973). The contact zone between the plates in Taiwan is considered to be located along the thrust zone in western Taiwan (e.g., Bowin et al., 1978); however, activity of large earthquakes in western Taiwan is far less than that along the Longitudinal Valley along the east coast. Although a left-lateral shear motion has been reported along the Longitudinal Valley based on geological evidence (Allen, 1962; Kaneko, 1970) and based on composite focal mechanism solutions for smaller earthquakes (Wu, 1978), no strike-slip type solution has been obtained along this fault system since the WWSSN became operative. High angle reverse fault type solutions as shown in Fig.8 suggest that a fairly large fraction of the relative plate motion between the Philippine Sea and southeast Asian plates is being consumed in the dip-slip motion along this fault. A normal fault type event at northeastern coast of Taiwan and a thrust type event at the junction of the Ryukyus and Taiwan are probably related to the extensional stress regime at the westward edge of the Okinawa Trough and the subduction of the Philippine Sea plate at the junction, respectively.

Luzon Straits

The region between northern Luzon and Taiwan shows diffuse seismic activity (Fig.1). All solutions obtained for the events east of the Manila Trench are of normal fault type except for two events located further eastward. These normal fault type solutions are problematic because the solutions near Taiwan strongly suggest compression as was shown previously. Seno and Kurita (1978) interpreted these normal faults as representing incipient opening of a zone between the trench and forearc which is likely to be caused by a change of stress regime from compression to tension due to the Taiwan-Luzon arc collision just north of the area. However, this interpretation does not seem straightforward. More detailed studies including marine geophysical surveys are necessary to elucidate the current tectonics of this area.

Philippine Islands

The Philippine convergence boundary is one of the most complex boundaries in the western Pacific region. Along at least three sections of the western margin of the islands, marginal basins are currently being subducted under the Philippines. These sections are: the Manila Trench - North Luzon Trough system, the Negross Trench at the eastern margin of the Sulu Basin, and the Cotabato Trench at the northeastern margin of the Celebes Basin (Ludwig et al., 1967; Hayes and Ludwig, 1967; Ludwig, 1970; Murphy, 1973; Karig, 1973; Seno and Kurita, 1978; Stewart and Cohn, 1979; Hamilton, 1979; Cardwell et al., 1980; Acharya and Aggarwal, 1980). This western subduction boundary seems to partly decouple the Philippines from southeast Asia. The subduction zone is interrupted by the continental fragments which had collided with the Philippines at Mindoro-Panay and Zamboanga, the western peninsula of Mindanao (see Fig.10, Hamilton, 1979). Since, only one thrust type earthquake has been reported along the western subduction zone (Fig.10, No.193), it is impossible to describe the motion of each part of the Philippine Islands with respect to southeast Asia by using seismic slip vectors along the western margin of the islands. The slip vectors of thrust type events along the subduction boundary east of the islands, i.e., along the Philippine Trench, deviate from the Philippine Sea-Eurasian convergence vector (e.g., Seno, 1977). The reader might think that the decoupling of southeast Asia from Eurasia is the cause of this deflection of slip vectors; however, the counter-clockwise rotation of the observed seismic slip vectors with respect to the convergence vectors cannot be explained by this decoupling because southeast Asia is moving approximately southeastward with respect to Eurasia (Fig.5). Therefore, the decoupling at the western margin of the islands from southeast Asia or strike-slip motion along the Philippine fault is most likely the cause for the deflection of slip vectors. Fitch (1972) interpreted the observed deflection of slip vectors as caused by decoupling of the oblique convergence into part thrusting and part right-lateral strike-slip motion along the Philippine fault, which is a major active fault system subparallel to the arc. Rowlett and Kelleher (1976), Cardwell et al. (1980), Acharya and Aggarwal (1980) and Acharya (1980) also concluded that the Philippine fault absorbs a considerable portion of the relative motion between the Philippine Sea and southeast Asia. However, Seno (1977) and Seno and Kurita (1978) discounted the role of the Philippine fault because of its low slip rate estimated from geological data and low level of seismicity associated with the fault in recent years. They attributed the observed deflection of seismic slip vectors to the decoupling of all the islands from Eurasia along the western subduction boundary and suggested that the Philippine Islands are moving to the north or northwest relative to Eurasia.

Both of the views represented by Fitch (1972) and Seno (1977) seem too simplified to explain the observed deformation along the Philippine Islands. Seismicity within the islands is diffuse (see Fig.1) and the mode of deformation accompanying earthquakes along the western margin of and within the islands (Fig.8) cannot be simply interpreted as a blockwise movement of the islands or as simple decoupling along the Philippine fault (e.g., Cardwell et al., 1980). Rutland (1967) and Hamilton (1979) disputed the northward and southward projection of the Philippine fault into northern Luzon and into Mindanao, respectively. Acharya (1980) estimated the seismic slip rate along the Philippine fault as 6.9 cm/yr by summing up the seismic moments of large earthquakes which are likely to be associated with the fault. However, recalculation using the method described previously and the same earthquakes as he used gives 1.2 cm/yr, which is comparable to the slip rate estimated from geological data. It is thus most likely that the relative motion of the Philippine Sea and southeast Asia is taken up by both subduction along the western and eastern boundaries and internal deformation of the islands including the seismic slip along part of the Philippine fault. Acharya and Aggarwal (1980) showed that the seismicity along the Philippine fault is higher along the part adjacent to and east of Mindoro and Panay where the eastern subduction boundary is interrupted by a continental mass than in the other part of the fault.

Subduction seismicity associated with the Philippine Trench and Manila-Negross-Cotabato Trenches is mostly shallower than 200 km. The deep seismicity beneath Mindanao and north of it (Fig.2) should not be related to the subduction of the Philippine Sea plate, but is associated with the subducted slab of the Molucca plate (Seno and Kurita, 1978; Hamilton, 1979; Cardwell et al., 1980; see the subsection of the Molucca Sea).

Present Kinematics of the Philippine Sea Plate

There are difficulties in obtaining rotation vectors for the Philippine Sea plate relative to adjacent major plates partly because the only data available are the slip vectors of shallow thrust type events around the Philippine Sea plate margin and they are limited in number (see Fig.8), and partly because of the tectonic complexity in the region from Taiwan through the Philippines as stated above. Nevertheless, several instantaneous rotation vectors for the Philippine Sea plate motion relative to adjacent major plates have been proposed. These are reviewed below and convergence vectors based on these models are shown in Figs.4 and 9.

Kanamori (1971b) proposed a model for the evolution of subduction zones along the circum-Pacific belt. In his model, the Izu-Bonin-Mariana arc corresponds to the latest stage of evolution in which the sinking lithospheric slab becomes detached from the remaining seaward part of the oceanic lithosphere at the trench axis. Kanamori (1972a) suggested that there is currently no convergence between the Pacific and Philippine Sea plates at the Izu-Bonin-Mariana arc because of the supposed detachment of the slab. He argued that evidence for this is the parallelism of the seismic slip vectors along the Nankai Trough to the convergence vector of the Pacific plate relative to the Eurasian-North American plate based on McKenzie and Parker's (1967) pole. Kelleher and McCann (1976) and McCann et al. (1979) also suggested a termination of the plate convergence at the Bonin-Mariana arc due to the interaction of bathymetric highs on the oceanic plate with the overriding plate at the subduction zone. If the above case is correct, the Philippine Sea plate motion is described by the Pacific plate motion as shown in Fig.4.

Fitch (1972) obtained the rotation vectors using the seismic slip vectors along the Nankai Trough - Ryukyu Trench, the strike along part of the Philippine fault, and the convergence rate of 8 cm/yr at Nankai Trough as estimated by Fitch and Scholz (1971). The procedure for obtaining the direction of the Philippine Sea plate motion at one section of the Philippine fault was not clear in his paper. His model produces rapid convergence of about 8 cm/yr along Nankai and Sagami Troughs (see Fig.9).

Morgan (1972) obtained the rotation vectors when he obtained absolute motions of major plates based on the hot spot hypothesis. Because the details on the data he used for obtaining the Philippine Sea plate motion are not yet available, it is not possible to evaluate his results. However, the convergence direction based on his model (see Fig.9) does not seem to be compatible with the observed seismic slip vectors.

Karig (1975) estimated the rotation vectors from the seismic slip vectors along the Nankai Trough - Ryukyu Trench and assuming convergence rates at two localities: 2 cm/yr at DSDP Site 298 off Kyushu, and 5 cm/yr at the south end of the Ryukyu Trench or 10 cm/yr at the central Philippine Trench. The convergence rate at the DSDP Site was that inferred from the relation between the convergence rate and sedimentation rate (Karig, 1975). Since the Philippine Sea - Eurasian pole was located near central Honshu, this model gives a very small convergence rate at Sagami and Nankai Troughs (Fig.9).

Seno (1977) determined the rotation vectors using all the available seismic slip vectors along the boundaries of the Philippine Sea plate and incorporating the Pacific-Eurasian relative motion of Minster et al. (1974) into the calculation. This model gives the convergence rates of 3 cm/yr and 3-4.5 cm/yr at Sagami and Nankai Troughs, respectively (Fig.9).

Recently, Chase (1978) and Minster and Jordan (1979) obtained the rotation vectors by inverting seismic slip vector data around the margin of the Philippine Sea incorporating their solutions for other major plates. This is essentially the same method as Seno (1977). In spite of the methods being nearly same, Chase (1978) obtained quite different solutions from the other two models; it resembles Karig's (1975) solution (Fig.9). On the other hand, Minster and Jordan's (1979) solution resembles Seno's (1977). It is likely that the selection of slip vectors from the two nodal planes of mechanism solution for the events along the Yap Trench is different in these models. Minster and Jordan's (1979) model gives a slip rate of 5 cm/yr at Sagami Trough and Chase's model gives a very small slip rate there.

The above models should be tested against various seismological and geological data along the boundaries of the Philippine Sea plate. Although crucial data resolving the various models for the Philippine Sea plate motion are limited, we shall discuss here some aspects of the above models on the basis of the available data.

The most primitive data to test the models are the seismic slip vectors along the Nankai Trough (Fig.8). The direction of the Pacific-Eurasian convergence vector is about N75°W at Nankai Trough, which is significantly different from the observed seismic slip vectors of N50-55°W. Thus the models by Kanamori (1972a) and Kelleher and McCann (1976) are rejected. Possible decoupling of southwest Japan from Eurasia may be a cause of this deflection; however, the difference in motion between the Eurasian and southeast Asian plates in the vicinity of the Japanese islands is very small as stated in the former section, and thus it is not likely to produce such a large difference in slip vector along the Nankai Trough. Morgan's (1972) model produces convergence vectors largely different from the seismic ones; thus this model is also not appropriate. Other models are automatically adjusted to the seismic slip vectors along the Nankai Trough because they use these vectors as the data to obtain the rotation vectors.

Seismic slip rate, length of the slab and the onset time of subduction are the second data set to test the convergence rate of the various models. First, the convergence rate must be larger than or equal to seismic slip rate. The seismic slip rate at Nankai Trough is at least 3-4 cm/yr as previously shown; thus the models by Chase (1978) and Karig (1975), which give very small convergence rate at this section, seem inappropriate for the present Philippine Sea plate motion. On the other hand, the length of slab will give a constraint on the maximum convergence rate when considered with the onset time of subduction. Southwest Japan lacks intermediate and deep seismic activity (see Figs.1 and 2). Kanamori (1972a) estimated that the slab does not reach asthenosphere beneath the central part of southwest Japan based on the subcrustal seismicity and P-wave travel time anomalies. Shiono (1974) also estimated the length of the slab beneath southwest Japan as 150-200 km based on the same kind of data as Kanamori (1972a). However, it should be noted that the absence of a Wadati-Benioff zone does not necessarily means the absence of a slab as has been suggested in the Juan de Fuca subduction zone (e.g., McKenzie and Julian, 1971), and the travel time anomaly data in these studies are limited to the region along the frontal arc of southwest Japan. Recently, Nakanishi (1980) obtained evidence for a sharp velocity contrast at the depth of 50-70 km beneath two seismographic stations located about 240-270 km inland from the axis of Nankai Trough using ScSp phases observed at these stations. He inferred that the velocity contrast indicates the interface between the top of the slab subducted beneath southwest Japan and the surrounding mantle, although no seismic activity is seen in the mantle. If this part of the slab is continuous to the slab which is currently being subducted beneath the frontal arc of southwest Japan, the total length of the slab would be at least 280 km (Nakanishi, 1980, Fig.11). The distribution of Quaternary volcanoes on the Japan Sea coast of southwest Japan (Aramaki and Ui, 1978) seems to reconcile this longer slab. Since ScSp phase data have been obtained only at the two stations in central southwest Japan, more detailed studies are necessary to substantiate the length of slab beneath southwest Japan. At the present time, the slab seems no longer than 300 km, at most. On the other hand, the onset time of the present subduction at Nankai Trough is not more recent than 3 myBP because the trough became deep enough to trap the turbidite deposits by that time (Ingle, Karig et al., 1975). Thus, based on the above two data sets, the convergence rate at Nankai Trough should be less than 10 cm/yr, which gives no constraints for the various models for the Philippine Sea-Eurasian relative motion. Although more seismological and geological studies on the length of slab and the onset time of subduction along Nankai Trough-Ryukyu Trench are needed, the convergence rate of 3-4 cm/yr seems most appropriate at the present time considering the seismic moment release and tectonic deformation at the collision zone between the Nankai and Sagami Troughs (Ishibashi, 1980; Somerville, 1978; Nakamura and Shimazaki, 1981). Thus, in the comparison with the seismic slip rate (Table 2), we used the convergence rate by Seno (1977) which gives 3-4.5 cm/yr rate along Sagami and Nankai Troughs.

For the models which do not use the seismic slip vectors along the Izu-Mariana-Yap arc as the data, those slip vectors, although only a few in number, can be used in testing the convergence vector along this arc section. However, the back-arc spreading in the Mariana Trough and the possible existence of the Caroline plate make a direct comparison questionable. Both of the convergence vectors of Fitch's (1972) and Karig's (1975) models deviate from the observed slip vectors along the Bonin-Mariana arc; the sense of deviation of Karig's model is consistent with the sense of opening of the Mariana Trough (Karig, 1975) and Fitch's is not. However, these deviations seem to be within the error of the determination of the focal mechanism solutions.

Molucca Sea

The Molucca Sea is a region where two opposing island arcs, the Sangihe arc to the west and the Halmahera arc to the east, are currently colliding. The Molucca plate, which is mostly hidden under the melange zone in the Molucca Sea, is being subducted beneath both of these arcs; an intermediate and deep seismic zone is dipping to the west beneath the Celebes Basin and western Mindanao and an intermediate one is dipping to the east beneath Halmahera (Cardwell et al., 1980; Silver and Moore, 1978; Hamilton, 1979). The axial zone between these two arcs is one of the most active areas for shallow seismicity in the western Pacific; the focal mechanism solutions are mostly of the reverse fault type (Fig.10), suggesting strong compression and crustal shortening in the melange zone in this arc-arc collision zone. The deep earthquakes beneath the Celebes Basin and Mindanao are all of the down-dip compression type (Cardwell et al., 1980). Now two subduction zones are being newly created along the southern extensions of the Philippine Trench and the Cotabato Trench, according to the arc-arc collision in the Molucca Sea (Hamilton, 1979; Cardwell et al., 1980).

Intermediate activity is very high southeast of the north arm of Celebes (see Fig.2); there have occurred two great intermediate events in 1905 and 1935 (see Fig.6). The intermediate earthquakes in recent years show down-dip tension type mechanisms in a nearly vertical seismic zone (Cardwell et al., 1980); the two great earthquakes may have similar mechanisms. There is currently southward subduction along the North Sulawesi Trench beneath the north arm of Celebes. However, the intermediate activity southeast of the arm is likely to be representing the Molucca plate subducted from the southeast (Cardwell et al., 1980).

Indonesian New Guinea

In this paper, we discuss only the western half of New Guinea (Indonesian New Guinea, or West Irian) west of 141°E. There is a zone of shallow activity about 100 km inland from the northern coast of New Guinea (Fig.1). Great earthquakes have occurred along this zone (Fig.6). Most of the seismic activity along the zone is the aftershocks of the 1971 M_s=7.8 earthquake (Fig.6). In the westernmost part around Bird's Head, the large northwestward projection of New Guinea west of 135°E, shallow activity is diffuse; however, two trends in activity can be seen; one along the northern coast of Bird's Head and the other elongated southwestward from 4°S, 135°E to the Aru Basin west of the Banda Sea (Fig.1). Between Bird's Head and the northern coast of the New Guinea mainland, the Solong fault system is active as evidenced by the recent large (M_s=7.7) earthquake associated with this fault (Kaplan and Seno, in preparation, 1982).

Only a small number of mechanism solutions have been obtained in Indonesian New Guinea (Fig.10; Johnson and Molnar, 1972; Fitch, 1970a). Three events in the seismic zone of the northern mainland (Nos.213-215 in Fig.10) have thrust type solutions with one steeply dipping nodal plane of a west-northwesterly strike and another poorly constrained nodal plane with a shallow southward dip. The aftershocks of the 1971 great earthquake distribute nearly vertically trending northwest (Kaplan and Seno, in preparation, 1982); this suggests that the steeply dipping nodal plane with northwesterly strike is likely to be the fault plane for those earthquakes. The zone of shallow activity in northern New Guinea is located in the lowland between the medial mountain range in central New Guinea and the coastal mountain range along the northern coast; Quaternary through Miocene sediments have filled the lowland basin, which was originally a back-arc of a south-facing paleo-island arc during the late Cretaceous through middle Miocene (Hamilton, 1979).

The medial mountain range is the zone where the south-facing island arc had collided with the southern New Guinean continental shelf in Miocene time (Hamilton, 1979). Off the northern coast, the New Guinea Trench is likely to have initiated subduction as identified from topography and seismic reflection profiles (Hamilton, 1979). However, it is noted that the zone of shallow activity in northern New Guinea is not associated with either of the above two features but is located between them. There is no doubt that a considerable portion of the Caroline - Indian-Australian relative motion is accommodated by deformation in the shallow seismic zone in northern New Guinea because rupture zones of great earthquakes are located along this zone (Fig.6). However, it is difficult to obtain the convergence vector between these plates because the fault movement is likely to be occurring on the nearly vertical plane while the other shallow dipping nodal plane, which represents the slip vector, can hardly be determined. More detailed studies of earthquake mechanisms and seismicity are needed in Indonesian New Guinea to permit more thorough discussions of the current tectonics in this region.

Andaman Sea

Several lines of geological and geophysical evidence from the Andaman Sea suggest that it is a young back-arc basin (Rodolfo, 1969; Fitch, 1970a, 1972; Curray et al., 1979; Eguchi et al., 1979). A zone of seismicity can be recognized extending from the western border of the Andaman Sea to Sumatra (see Fig.1). This activity may be related to the highly oblique convergence of the Indian plate beneath the Andaman Sea area. Distinctly separate from this zone is a linearly distributed zone of shallow epicenters located in the northern part of the Andaman Sea. This shallow activity possibly marks the locus of crustal opening in the Andaman Sea.

Although only a small number of large earthquakes have occurred in the Andaman Sea region, several focal mechanism solutions have been obtained (Fig.10, Fitch, 1970a, 1972; Eguchi et al., 1979). Along the Semangko fault in Sumatra, right-lateral shear displacement is documented geologically (Katili, 1970). This is one of the major transcurrent faults subparallel to island arcs in the western Pacific region and has been interpreted as representing a decoupled zone associated with oblique convergence (Fitch, 1972). The submarine continuation of this fault may extend to the Nicobar island arc (Fitch, 1972). The Nicobar arc bounds the southwestern margin of the Andaman Sea. The Central Andaman Sea Rift Valley, centered at about 11°N and 95°E and trending in a northeast direction, has been an active spreading center since the Middle Miocene (Curray et al., 1979). The north-south trending seismic zone and the right-lateral strike-slip mechanisms along 95.2°E at the east of the valley (Fig.10, Nos.258 and 259) suggest the existence of a north-south striking transform fault system there. This fault system may connect the Central Andaman Sea Rift Valley with another possible spreading center further to the north in the vicinity of 14-15°N and 94-96°E. The latter spreading center, although not apparent in the submarine topography, is inferred from the normal fault type earthquakes with north-northwest trending T-axes (Fig.10, No.260). The recent tectonic regime of the Andaman Sea thus can be represented by segments of active spreading centers and right-lateral transform fault system. Curray et al. (1979) incorporated this and the right-lateral shear motion along the Semangko fault in Sumatra into a rigid plate scheme by introducing a Burma plate. The Burma plate is a sliver of plate whose boundaries are marked by the trench system from the Andaman Islands to Sumatra on the west and by the Semangko fault in Sumatra and the ridge-transform fault system in the Andaman Sea on the east. The linear seismic zone in the Andaman Sea (Fig.1) which closely resembles mid-oceanic ridge seismic zones supports this idea.

Eguchi et al. (1979) speculated on the tectonic history of the Andaman Sea on the basis of the evolution of the eastern Indian Ocean, Greater India's northward movement away from Australia and Antarctica and the recent seismotectonics described above. They concluded that the Andaman Sea has developed in the area between Burma and western Sumatra as a combined result of subduction of actively spreading mid-oceanic ridges beneath the western part of the old Sunda arc and collision of Greater India with Asia. They indicated that these events took place at about 20-10 myBP and 30-10 myBP, respectively.

Sumatra-Java-Banda Sea

Focal mechanisms of shallow earthquakes in this region have been studied by Fitch (1970a, 1972) and Cardwell and Isacks (1978). No thrust type earthquakes have been reported except for the two events along Seram Trough (Fig.10, Nos.233 and 234) and the two events along Sumatra (Fig.10, Nos.251 and 249). The north-northeasterly slip vectors along Sumatra roughly coincide with the Indian - Eurasian convergence vector (see Fig.4). However, it should be noted that there are the Burma and southeast Asian plates between the above two plates, as described previously. Thus these slip vectors indicate the relative motion between the Burma and Indian plates. Recently, Cardwell et al. (1981) obtained several thrust type mechanism solutions south of Java, which indicate the northward Indian plate motion relative to southeast Asia. Thus the systematic difference in slip vectors between Sumatra and Java can be attributed to the right-lateral strike-slip motion along the Semangko fault in Sumatra associated with the northwestward motion of the Burma plate relative to the southeast Asian plate (Cardwell et al., 1981). Furthermore, the counterclockwise deflection of the observed seismic slip vectors south of Java from the north-northeasterly Indian - Eurasian convergence vector (Fig.4) is consistent with the southeastward motion of southeast Asia relative to Eurasia with a speed of 1-1.5 cm/yr previously obtained.

In the eastern half of the arc, i.e., along the Banda arc, a number of reverse and strike-slip fault type solutions have been obtained in the frontal and back arc area (Fig.10). The interpretation of these solutions is not very straightforward, unlike those of shallow earthquakes in the back-arc region of the Japanese Islands. There seems to be no systematic pattern in the distribution of P-axes or nodal plane strikes for these shallow events in either the frontal or back arc areas. For example, two strike-slip fault type solutions at the eastern margin of the Weber Basin (Fig.10, Nos. 236 and 237) have completely reversed senses of motion with almost the same nodal plane strikes. However, the P-axes are generally more perpendicular to the arc trend than the T-axes. Three normal fault type events are located at the eastern end of the Java Trench (Fig.10, Nos.246-248); the largest event of these, the 1977 Indonesian Ms=8.1 earthquake (No.246), has been interpreted as indicating a large fracture which took place over almost the entire thickness of the descending slab (Given and Kanamori, 1980), or as indicating a fracture limited in the upper 24 km of the elastic part of the plate with a high stress drop of 500 bars (Fitch et al., 1981). The very low level of activity for thrust type events along the entire Banda arc, along with the occurrence of a great normal fault at the eastern end of the Java Trench, may indicate the modification of subduction seismicity along this arc segment due to subduction of the continental shelf beneath the eastern half of the arc (Cardwell and Isacks, 1978; Cardwell et al., 1981).

Geometry of the Wadati-Benioff zone and associated focal mechanisms have been studied by Fitch and Molnar (1970), Cardwell and Isacks (1978), Hamilton (1974, 1979), and Isacks and Molnar (1971). Beneath Sumatra, the maximum depth of the Wadati-Benioff zone is 200 km. In contrast, along the Java Trench-Timor Trough, it reaches 650 km. The oblique subduction along Sumatra explains only part of the difference in length of the zone (e.g., Fitch, 1970a; Hamilton, 1979). Cardwell and Isacks (1978) suggested that a north-northwesterly subduction along the Sunda arc can account for more of the difference; however, this direction of convergence, which was inferred from the geometry of the slab and the fixed outline of the trench along the Banda arc, is not supported by the observed seismic slip vectors along Sumatra and Java. The age of the subducting lithosphere is younger (~60myBP) along the trench south of Sumatra than that south of Java (~130myBP). Furthermore, the ocean floor is more smooth south of Sumatra than south of Java. Thus other factors which could control the deep seismic activity, such as temperature of the slab or interaction between a bathymetric high on the subducting plate and the overriding plate at subduction zone (Vlaar and Wortel, 1976;

Molnar et al., 1979; Geller et al., 1981), should be taken into account to explain the abrupt change of the depth of the Wadati-Benioff zone between the two arc segments.

The arc trend bends approximately 180 degrees around the Banda Sea from the Timor Basin to the Seram Trough (Fig.10). Cardwell and Isacks (1978) postulated a transform fault north of the Aru Basin and suggested that the slab beneath the south Banda Sea subducting along the Timor Basin and Aru Basin and the slab beneath the Seram and Buru Islands are disconnected on the basis of the well located hypocenters beneath the Banda Sea. Hamilton (1974, 1979) opposed this and drew a contour line of the Wadati-Benioff zone as spoon shaped indicating continuous subduction along the Timor-Seram Troughs. Cardwell and Isacks (1978) assumed a fixed outline of the trench along the Banda arc, rejecting a spoon shape as impossible from geometrical considerations, and, in contrast, Hamilton (1979) permitted trench migration and deformation of the subducted sheet of slab. This dispute is not yet resolved.

Two intermediate earthquakes beneath Sumatra show down-dip tension like other intermediate events of island arcs which have no deep seismicity (Fitch and Molnar, 1970; Isacks and Molnar, 1971). Along the eastern Banda arc, where the trend of the arc and the Wadati-Benioff zone shows a northerly bend, intermediate events are all of down-dip tension type and P-axes of many of them are subhorizontal and parallel to the local strike of the arc. Four deep events in the bent edge of the zone beneath the Banda Sea also have subhorizontal P-axes in a northeast direction. These events seem to reflect the effect of bending at the western edge of the slab subducted along the Banda arc (Fitch and Molnar, 1970; Isacks and Molnar, 1971; Cardwell and Isacks, 1978). Deep events beneath the Banda Sea north of Java and northwest of Timor are all of the down-dip compression type (Fitch and Molnar, 1970; Cardwell and Isacks, 1978).

Conclusions

Mode of deformation caused by shallow earthquakes in the western Pacific region can be mostly interpreted in terms of plate interaction along subduction or collision zones. Major features are underthrusting at the thrust or collision zones, decoupling or bending of the oceanic plate at trenches, and internal deformation in the back arc area in the overriding plate.

Seismic slip vectors representing underthrusting motion can be used effectively to find minor plates, which interact with major plates or among them, and to obtain their motions. Such minor plates involved in the current tectonics of the western Pacific region are the southeast Asian, Burma, Philippine Sea, and Caroline plates. The Philippine islands are also partly decoupled from southeast Asia. Since the seismic slip vectors are almost the only available data used to obtain their plate motions, they have not been uniquely determined yet. Further collection of data is needed to precisely determine their motions with respect to other plates.

Decoupling of the oceanic plate at the trench axis has been suggested by great normal fault type earthquakes which occurred there. The number of such events which are documented is quite few; the 1933 Sanriku, northern Honshu, and 1977 Indonesian earthquakes are the only two events in the western Pacific region. However, because the precise depth extent of their ruptures is still unresolved, it seems difficult to attribute their origin to the decoupling and not to the bending stress. There occurred at least two intermediate great events which are likely to have caused a huge rupture in the deeper part (70-150km) of the slab in Kamchatka and Banda Sea. Beneath or seaward of the trench, a number of normal fault type events with smaller magnitudes are reported. They are likely associated with the bending stress within the oceanic plate before subduction. A pair of normal and reverse faulting found in close location at appropriate depth can substantiate this cause; however, such cases are not abundant in the western Pacific region (see Chapple and Forsyth, 1979).

Internal deformation within the overriding plate is due to reverse faulting, normal faulting or strike-slip motion depending on the stress regime which dominates the back-arc area. All of these types are found in the back arc areas of the western Pacific; each of the areas has a dominant mode of deformation.

Intermediate earthquakes along the Kamchatka-Kurile-northern Honshu arc show a dual zone of activity and focal mechanism type. Down-dip compression and down-dip tension are associated with the upper and lower zones of activity, respectively. In the other island arcs in the western Pacific, down-dip tension is the dominant mechanism type for intermediate events except for the southern and central Ryukyus, for which events show down-dip compression type. Hypocentral locations have not been determined as precisely in the southern half of the western Pacific region as in the northern half. We cannot preclude the possibility that further installation of seismic networks in the southern region will make a double seismic zone visible in some arc segments in this region.

Deep seismic activity is seen along the Kurile-northern Honshu-Izu-Mariana arc, the eastern Sunda and Sangihe arcs. Although almost all mechanism solutions associated with deep events are down-dip compression type, in some areas, anomalous distribution of stress axes for deep and intermediate events are found and are very likely to represent the effect of slab contortion or disruption at the junction between two arc segments.

Much work on the seismotectonics in the western Pacific region has been done as reviewed so far in this study. However, much more work is needed to elucidate current tectonics in each region and permit more detailed discussion of the interaction of the plates, rheology, stress state, and seismic - aseismic behavior of the descending or pre-descending plates, and the stress state, modes of deformation, and their causes in the back-arc areas along the island arcs in the western Pacific region.

Acknowledgements. We thank Seiya Uyeda and Tom Hilde for their critical review of the manuscript and continuous encouragement throughout this study. We also thank Kunihiko Shimazaki and Rick Schult for their critical review of the manuscript. Discussions with Yoshio Fukao and Rich Cardwell were helpful. Satoru Honda, Makoto Yamano, Junko Endo, Noriko Sugi, Kazuto Kodama, and Hitoshi Kawakatsu helped us prepare the manuscript in the early stage of this work. This work was supported by the National Science Foundation under Grant EAR 81-08718.

References

Abe, K., Tsunami and mechanism of great earthquakes, *Phys. Earth Planet. Inter.*, 7, 143-153.

Abe, K., Re-examination of the fault model for the Niigata earthquake of 1964, *J. Phys. Earth*, 23, 349-366, 1975.

Abe, K., Tectonic implications of the large Shioya-Oki earthquake of 1938, *Tectonophys.*, 41, 269-289, 1977.

Abe, K., and H. Kanamori, Temporal variation of the activity of intermediate and deep focus earthquakes, *J. Geophys. Res.*, 84, 3589-3595, 1979.

Abe, K., and H. Kanamori, Magnitudes of great earthquakes from 1953 to 1977, *Tectonophys.*, 62, 191-203, 1980.

Acharya, H. K., Seismic slip on the Philippine fault and its tectonic implications, *Geology*, 8, 40-42, 1980.

Acharya, H. K., and Y. P. Aggarwal, Seismicity and tectonics of the Philippine islands, *J. Geophys. Res.*, 85, 3239-3250, 1980.

Allen, C. R., Circum-Pacific faulting in the Philippine-Taiwan region, *J. Geophys. Res.*, 67, 4795-4812, 1962.

Anderson, R. N., Heat flow in the Mariana marginal basin, *J. Geophys. Res.*, 80, 4043-4048, 1975.

Ando, M., Seismo-tectonics of the 1923 Kanto earthquake, *J. Phys. Earth*, 22, 263-277, 1974.

Ando, M., Possibility of a major earthquake in the Nankai district, Japan and its pre-estimated seismotectonic effects. *Tectonophys.*, 25, 69-85, 1975a.

Ando, M., Source mechanisms and tectonic significance of historical earthquakes along the Nankai trough, Japan, *Tectonophys.*, 27, 119-140, 1975b.

Ando, M., and T. Seno, The Boso-Oki earthquake of Nov. 26, 1953: Lithospheric hinge faulting at a triple junction, *in preparation*, 1982.

Aoki, H., Plate tectonics of arc-junctions at central Japan, *J. Phys. Earth*, 22, 141-161, 1974.

Aramaki, S., and T. Ui, Major element frequency distribution of the Japanese Quaternary volcanic rocks, *Bull. Volcanol.*, 41, 390-407, 1978.

Barazangi, M., and B.L. Isacks, A comparison of the spatial distribution of mantle earthquakes determined from data produced by local and by teleseismic networks for the Japan and Aleutian arc, *Bull. Seismol. Soc. Am.*, 69, 1763-1770, 1979.

Ben-Menahem, A., Renormalization of the magnitude scale, *Phys. Earth Planet. Inter.*, 15, 315-340, 1977.

Bowin, C., R. S. Lu, C. S. Lee, and H. Schouten, Plate convergence and accretion in the Taiwan-Luzon region, *Bull. Am. Assoc. Petr. Geol.*, 62, 1645-1672, 1978.

Caldwell, J. G., W. F. Haxby, D. E. Karig, and D. L. Turcotte, On the applicability of an universal elastic trench profile, *Earth Planet. Sci. Lett.*, 31, 239-246, 1976.

Cardwell, R. K., and B. L. Isacks, Geometry and the subducted lithosphere beneath the Banda sea in eastern Indonesia from seismicity and fault plane solutions, *J. Geophys. Res.*, 83, 2825-2838, 1978.

Cardwell, R. K., B. L. Isacks, and D. E. Karig, The spatial distribution of earthquakes, focal mechanism solutions, and subducted lithosphere in the Philippine and northeastern Indonesian islands, in *The Tectonic and Geologic Evolution of Southeast Asian Seas and Islands*, Geophys. Monogr., 23, edited by D. E. Hayes, AGU, Washington, D. C., 1-35, 1980.

Cardwell, R. K., E. S. Kappel, M. B. Lawrence, B. L. Isacks, Plate convergence along the Indonesian arc, *EOS*, 62, 404, 1981.

Chai, B. H. T., Structure and tectonic evolution of Taiwan, *Am. J. Sci.*, 272, 389-411, 1972.

Chapman, M. E., and S. C. Solomon, North-American-

Eurasian plate boundary in northeast Asia, *J. Geophys. Res.*, *81*, 921-930, 1976.

Chapple, W. M., and D. W. Forsyth, Earthquakes and bending of plates at trenches, *J. Geophys. Res.*, *84*, 6729-6749, 1979.

Chase, C. G., Plate kinematics: the Americas, east Africa and the rest of the world, *Earth Planet. Sci. Lett.*, *37*, 355-368, 1978.

Cormier, V. F., Tectonics near the junction of the Aleutian and Kuril- Kamchatka arcs and a mechanism for middle tertiary magmatism in the Kamchatka Basin, *Geol. Soc. Am. Bull.*, *86*, 443-453, 1975.

Curray, J. R., D. G. Moore, L. A. Lawver, F. J. Emmel, R. W. Raitt, M. Henry, and R. Kcieckhefer, Tectonics of the Andaman Sea and Burma, *Am. Assoc. Petrol. Geol. Mem.*, *29*, 189-198, 1979.

Das, S., and J. Filson, A plate model of Asian tectonics, *EOS*, *55*, 301, 1974.

Davies, G. F., and J. N. Brune, Regional and global fault slip rates from seismicity, *Nature*, *229*, 101-107, 1971.

De-Bremaecker, J. C., Is the oceanic lithosphere elastic or viscous?, *J. Geophys. Res.*, *82*, 2001-2004, 1977.

Duda, S. J., Secular seismic energy release in the circum-Pacific belt, *Tectonophys.*, *2*, 409-452, 1965.

Eguchi, T., Relative plate motion between the Eurasian plate and the Pacific plate at the Japan trench, *J. Seismol. Soc. Japan*, *33*, 95-97, 1980 (in Japanese).

Eguchi, T., Seismotectonics of back-arc basins and some problems at plate boundaries, *Ph.D. Thesis*, Univ. of Tokyo, 1981.

Eguchi, T., S. Uyeda, and T. Maki, Seismotectonics and tectonic history of the Andaman Sea, *Tectonophys.*, *57*, 35-51, 1979.

Engdahl, E. R., and C. H. Scholz, A double Benioff zone beneath the Central Aleutians and unbending of the lithosphere, *Geophys. Res. Lett.*, *4*, 473-476, 1977.

Fedotov, S. A., Regularities of the distribution of strong earthquakes in Kamchatka, the Kuril islands, and northeast Japan, *Trudy Inst. Fiz Zemli., Acad. Nauk. SSSR*, *36*, 66-93, 1965 (in Russian).

Fedotov, S. A., On deep structure, properties of the upper mantle, and volcanism of the Kuril-Kamchatka island arc according to seismic data, *Geophys. Monogr.*, *12*, AGU, Washington D. C., 131-139, 1968.

Fedotov, S. A., A. M. Bagdasarovc, I. P. Kuzin, and R. Z. Tarakanov, *Earthquakes and the deep structure of the south Kuril island arc, Israel Program for Scientific Translations*, *6*, Jersalem, 249 pp., 1971.

Fedotov, S. A., N. A. Dolbikina, V. N. Morozov, V. I. Myachkin, V. B. Preobrazensky, and G. A. Sobolev, Investigation on earthquake prediction in Kamchatka, *Tectonophys.*, *9*, 249-258, 1970.

Fitch, T. J., Earthquake mechanisms and island arc tectonics in the Indonesian-Philippine region, *Bull. Seismol. Soc. Am.*, *60*, 565-591, 1970a.

Fitch, T. J., Earthquake mechanisms in the Himalayan, Burmese, and Andaman regions and continental tectonics in central Asia, *J. Geophys. Res.*, *75*, 2699-2709, 1970b.

Fitch, T. J., Plate convergence, transcurrent faults, and internal deformation adjacent to southeast Asia and the western Pacific, *J. Geophys. Res.*, *77*, 4432-4460, 1972.

Fitch, T. J., and P. Molnar, Focal mechanisms along inclined earthquake zones in the Indonesia-Philippine region, *J. Geophys. Res.*, *75*, 1431-1444, 1970.

Fitch, T. J., R. G. North, and M. W. Shields, Focal depths and moment tensor representations of shallow earthquakes associated with the Great Sumba earthquake, *J. Geophys. Res.*, *86*, 9357-9374, 1981.

Fitch, T. J., and C. H. Scholz, Mechanism of underthrusting in southwest Japan: A model of convergent plate interactions, *J. Geophys. Res.*, *76*, 7260-7292, 1971.

Fujita, K., and H. Kanamori, Double seismic zones and stresses of intermediate depth earthquakes, *Geophys. J. R. Astr. Soc.*, *66*, 131-156, 1981.

Fukao, Y., Tsunami earthquakes and subduction process near deep-sea trenches, *J. Geophys. Res.*, *84*, 2303-2314, 1979.

Fukao, Y., and M. Furumoto, Mechanism of large earthquakes along the eastern margin of the Japan Sea, *Tectonophys.*, *26*, 247-266, 1976.

Fukao, Y., and M. Furumoto, Stress drops, wave spectra and recurrence intervals of great earthquakes - implications of the Etorofv earthquake of 1958, November 6, *Geophys. J. R. Astr. Soc.*, *57*, 23-40, 1979.

Geller, R. J., Scaling relations for earthquake source parameters and magnitudes, *Bull. Seismol. Soc. Am.*, *66*, 1501-1523, 1976.

Geller, R. J., and H. Kanamori, Magnitudes of great shallow earthquakes from 1904 to 1952, *Bull. Seismol. Soc. Am.*, *67*, 587-598, 1977.

Geller, R. J., H. Kanamori, and K. Abe, Addenda and corrections to "magnitudes of great shallow earthquakes from 1904 to 1952", *Bull. Seismol. Soc. Am.*, *68*, 1763-1764, 1978.

Geller, R. J., F. R. Schult, S. C. Wang, and J. L. Morton, Why deep earthquakes are'nt occurring where they do'nt, *EOS*, *62*, 403, 1981.

Given, J. W., and H. Kanamori, The depth extent of the 1977 Sumbawa Indonesia earthquake, *EOS*, *61*, 1044, 1980.

Goto, K., and H. Hamaguchi, A double-planed structure of the intermediate seismic zone - thermal stress within the descending lithospheric slab, *Abstr. Ann. Meet. Seismol. Soc. Japan*, *2*, 36, 1978 (in Japanese).

Goto, K., and H. Hamaguchi, Distribution of thermal stress within the descending lithospheric slab, *Abstr. Ann. Meet. Seismol. Soc. Japan*, *2*, 48, 1980 (in Japanese).

Gutenberg, B., Great earthquakes 1896-1903, *Trans. Am. Geophys. Union*, *37*, 608-614, 1956.

Gutenberg, B., and C. F. Richter, *Seismicity of the earth and associated phenomena*, 2nd ed., Princeton University Press, pp. 130, 1954.

Hamilton, W., Earthquake map of the Indonesian region, *U. S. Geol. Survey Misc. Inv. Ser. Map.*, 1-875C [1975], 1974.

Hamilton, W., Tectonics of the Indonesian region, *U. S. Geol. Survey Professional Paper 1078*, pp. 335, 1979.

Hanks, T. C., The Kuril trench - Hokkaido rise system: large shallow earthquakes and simple models of deformation, *Geophys. J. R. Astr. Soc.*, *23*, 173-189, 1971.

Hasegawa, A., and N. Umino, Spatial distribution of hypocenters and earthquake mechanisms under northeastern Japan, *Abstr. Ann. Meet. Seismol. Soc.Japan*, *1*, 34, 1978 (in Japanese).

Hasegawa, A., N. Umino, and A. Takagi, Double-planed structure of the deep seismic zone in the northeastern Japan arc, *Tectonophys.*, *47*, 43-58, 1978.

Hatori, T., The magnitude of tsunamis generated in Hyuganada during the years 1926-1970, *J. Seismol. Soc. Japan*, *24*, 95-106, 1971 (in Japanese).

Hatori, T., Tsunami magnitude and wave source regions of historical Sanriku tsunamis in northeast Japan, *Bull. Earthquake Res. Inst., Univ. of Tokyo*, *50*, 397-414, 1975 (in Japanese).

Hatori, T., and M. Katayama, Tsunami behavior and source areas of historical tsunamis in the Japan Sea, *Bull. Earthquake Res. Inst., Univ. of Tokyo*, *52*, 49-70, 1977 (in Japanese).

Hayes, D. E., and W. J. Ludwig, The Manila Trench and west Luzon Trough-II. Gravity and magnetic measurements, *Deep-Sea Res.*, *14*, 545-560, 1967.

Herman, B. M., R. N. Anderson, and M. Truchan, Extensional tectonics in Okinawa Trough, *Am. Assoc. Petrol. Geol. Mem.*, *29*, 119-208, 1979.

Hilde, T. W. C., and G. F. Sharman, Fault structure of the descending plate and its influence on the subduction process, *EOS*, *59*, 1182, 1978.

House, L. S., and K. H. Jacob, Thermal stresses in subducting lithosphere can explain double seismic zones, *Nature, 295*, 587-589, 1982.

Hussong, D. S., S. Uyeda et al., Leg 60 ends in Guam, *Geotimes, 23*, 19-22, 1978.

Ingle, J. C., D. E. Karig et al., Site 297, *Init. Rep. Deep Sea Drilling Project, 31*, U. S. Government Printing Office, Washington, D. C., 275-316, 1975.

Isacks, B. L., and M. Barazangi, Geometry of Benioff zones. Lateral segmentation and downwards bending of the subducted lithosphere, in *Island Arcs Deep Sea Trenchs and Back-arc Basins, Maurice Ewing Series, 1*, edited by M. Talwani and W. C. Pitman III, AGU, Washington, D. C., pp. 99-114, 1977.

Isacks, B. L., and P. Molnar, Distribution of stresses in the descending lithosphere from a global survey of focal-mechanism solutions of mantle earthquakes, *Rev. Geophys. Space Phys., 9*, 103-174, 1971.

Ishibashi, K., Modern tectonics around the Izu Peninsula, *The Earth Monthly, 2*, 110-119, 1980 (in Japanese).

Ishibashi, K., Specification of a soon-to-occur seismic faulting in the Tokai district, central Japan, based upon seismotectonics, in *Earthquake Prediction, Maurice Ewing Series, 4*, edited by D. W. Simpson and P. G. Richards, AGU, Washington, D. C., pp. 297-332, 1981.

Jones, G. M., T. W. C. Hilde, G. F. Sharman, and D. C. Agnew, Fault patterns in outer trench wall and their tectonic significance, *J. Phys. Earth, 26 Suppl.*, 85-102, 1978.

Johnson, T., and P. Molnar, Focal mechanisms and plate tectonics of the southwest Pacific, *J. Geophys. Res., 77*, 5000-5032, 1972.

Kanamori, H., Synthesis of long-period surface waves and its application to earthquake source studies - Kurile islands earthquake of October 13, 1963, *J. Geophys. Res., 75*, 5011-5040, 1970.

Kanamori, H., Focal mechanism of the Tokachi-Oki earthquake of May 16, 1968: Contortion of the lithosphere at a junction of two trenches, *Tectonophys., 12*, 1-13, 1971a.

Kanamori, H., Great earthquakes at island arcs and the lithosphere, *Tectonophys., 12*, 187-198, 1971b.

Kanamori, H., Faulting of the great earthquake of 1923 as revealed by seismological data, Bull. Earthquake Res. Inst., Univ. of Tokyo, 49, 13-18, 1971c.

Kanamori, H., Seismological evidence for a lithospheric normal faulting the Sanriku earthquake of 1933, *Phys. Earth Planet. Inter., 4*, 289-300, 1971d.

Kanamori, H., Tectonic implications of the 1944 Tonankai and the 1946 Nankaido earthquakes, *Phys. Earth Planet. Inter., 5*, 129-139, 1972a.

Kanamori, H., Mechanism of tsunami earthquakes, *Phys. Earth Planet. Inter., 6*, 346-359, 1972b.

Kanamori, H., Mode of strain release associated with major earthquakes in Japan, *Ann. Rev. Earth Planet. Sci., 1*, 213-239, 1973.

Kanamori, H., Re-examination of the earth's free oscillations excited by the Kamchatka earthquake of November 4, 1952, *Phys. Earth Planet. Inter., 11*, 216-212, 1976.

Kanamori, H., Seismic and aseismic slip along subduction zones and their tectonic implications, in *Island Arcs, Deep Sea Trenches and Back-arc Basins, Maurice Ewing Series, 1*, edited by M. Talwani and W.C. Pitman III, AGU, Washington, D. C., pp. 163-174, 1977a.

Kanamori, H., The energy release in great earthquakes, *J. Geophys. Res., 82*, 2981-2987, 1977b.

Kanamori, H., and K. Abe, Reevaluation of the turn-of-century seismicity peak, *J. Geophys. Res., 84*, 6131-6139, 1979.

Kaneko, S., Transcurrent buckling and some notes on neotectonics in Taiwan, *J. Geol. Soc. Japan, 76*, 215-222, 1970.

Karig, D. E., Structural history of the Mariana island arc system, *Geol. Soc. Am. Bull., 82*, 323-344, 1971.

Karig, D. E., Plate convergence between the Philippines and the Ryukyu islands, *Marine Geol., 14*, 153-168, 1973.

Karig, D. E., Basin genesis in the Philippine Sea, *Init. Rep. Deep Sea Drilling Project, 31*, U. S. Government Printing Office, Washington, D. C., 857-879, 1975.

Karig, D. E., R. N. Anderson, and L. B. Bibee, Characteristics of back arc spreading in the Mariana trough, *J. Geophys. Res., 83*, 1213-1226, 1978.

Kasahara, M., Fault model of the 1952 Tokachi-Oki earthquake, *Abstr. Ann. Meet. Seismol. Soc. Japan, 2*, 90, 1975 (in Japanese).

Katili, J. A., Large transcurrent faults in southeast Asia with special reference to Indonesia, *Geol. Rundschau, 59*, 581-600, 1970.

Katsumata, M., and L. R. Sykes, Seismicity and tectonics of western Pacific: Izu-Mariana-Caroline and Ryukyu-Taiwan regions, *J. Geophys. Res., 74*, 5923-5948, 1969.

Kawakatsu, H., and T. Seno, Triple seismic zone and the regional variation of seismicity along the northern Honshu arc, *J. Geophys. Res., submitted*, 1982.

Kelleher, J., and W. McCann, Buoyant zones, great earthquakes and unstable boundaries of subduction, *J. Geophys. Res., 81*, 4885-4896, 1976.

Kelleher, J., L. R. Sykes, and J. Oliver, Possible criteria for predicting earthquake locations and their application to major plate boundaries of the Pacific and the Carribean, *J. Geophys. Res., 78*, 2547-2585, 1973.

Kobayashi, K., and M. Nakata, Magnetic anomalies and tectonic evolution of the Shikoku inter-arc basin, *J. Phys. Earth, 26, Suppl.*, 391-402, 1978.

Koyama, J., Source process of Vladiostok deep focus earthquake of Sep. 10, 1973, *Sci. Rep. Tohoku Univ., Ser.5, Geophys., 23*, 83-101, 1975.

Koyama, J., Seismic moment of the Vladiostok deep-focus earthquake of Sept.29, 1973, deduced from P waves and mantle Rayleigh waves, *Phys. Earth. Planet. Inter., 16*, 307-317, 1978.

Kurita, K., and M. Ando, Large earthquakes in Kamchatka and the motion of the Pacific plate, *Abstr. Ann. Meet. Seismol. Soc. Japan, 1*, 14, 1974 (in Japanese).

Lay, T., H. Kanamori, and L. Ruff, The asperity model and the nature of large subduction zone earthquake occurrence, *Earthq. Predict. Res., submitted*, 1982.

Lee, C. S., G. L. Shor, Jr., L. D. Bibee, R. S. Lu, and T. W. C. Hilde, Okinawa Trough: Origin of a back-arc basin, *Marine Geol., 35*, 219-241, 1980.

Le Pichon, X., J. Francheteau, and J. Bonnin, *Plate Tectonics*, Am. Elsvier, New York, pp. 302, 1973.

Ludwig, W. J., The Manila trench and West Luzon trough - II. Seismic-refraction measurements, *Deep-Sea Res., 17*, 553-571, 1970.

Ludwig, W. J., D. E. Hayes, and J. I. Ewing, The Manila trench and West Luzon trough - I. Bathymetry and sediment distribution, *Deep-Sea Res., 14*, 533-544, 1967.

Matsubara, Y., and T. Seno, Paleogeographic reconstruction of the Philippine Sea at 5 m.y.B.P., *Earth Planet. Sci. Lett., 51*, 406-414, 1980.

Matsuda, T., Y. Ohta, M. Ando, and N. Yonekura, Fault mechanisms and recurrence time of major earthquakes in southern Kanto district Japan as deduced from coastal terrace data, *Geol. Soc. Am. Bull., 89*, 1610-1618, 1978.

McAdoo, D. C., J. G. Caldwell, and D. L. Turcotte, On the elastic-perfectly plastic bending of the lithosphere under generalized loading with application to the Kuril Trench, *Geophys. J. R. Astr. Soc., 54*, 11-26, 1978.

McCann, W. R., S. P. Nishenko, L. R. Sykes, and J. K. Krause, Seismic gaps and plate tectonics: Seismic potential for major boundaries, *Pure Appl. Geophys., 117*, 1082-1147, 1979.

McKenzie, D., and B. Julian, Puget Sound, Washington, earthquake and the mantle structure beneath the northwestern United States, *Geol. Soc. Am. Bull., 82*, 351-3524, 1971.

McKenzie, D., and R. L. Parker, The north Pacific: an example of tectonics on a sphere, *Nature, 216*, 1276-1280, 1967.

Melosh, H. J., Dynamic support of the outer rise, *Geophys. Res. Let., 5*, 321-324, 1978.

Mikumo, T., Source process of deep and intermediate earthquakes as inferred from long-period P and S waveforms 2. Deep-focus and intermediate-depth earthquakes around Japan, *J. Phys. Earth, 19*, 303-320, 1971.

Minster, J. B., and T. H. Jordan, Present-day plate motions, *J. Geophys. Res., 83*, 5331-5354, 1978.

Minster, J. B., and T. H. Jordan, Rotation vectors for the Philippine and Rivera plates, *EOS, 60*, 958, 1979.

Minster, J. B., T. H. Jordan, P. Molnar, and E. Haines, Numerical modeling of instantaneous plate tectonics, *Geophys. J. R. Astr. Soc., 36*, 541-576, 1974.

Mogi, K., Sequential occurrences of recent great earthquakes, *J. Phys. Earth, 16*, 30-36, 1968a.

Mogi, K., Some features of recent seismic activity in and near Japan (1), *Bull. Earthquake Res. Inst., Univ. of Tokyo, 46*, 1225-1236, 1968b.

Mogi, K., Some features of recent seismic activity in and near Japan (2) Activity before and after earthquakes, *Bull. Earthquake Res. Inst., Univ. of Tokyo, 47*, 395-417, 1969.

Mogi, K., Active periods in the world's chief seismic belts, *Tectonophys., 22*, 265-282, 1974.

Mogi, K., Seismic activity and earthquake prediction, *Proc. Earthq. Predict. Res. Symp. (1976)*, 206-214, 1977.

Molnar, P., T. J. Fitch, and F. T. Wu, Fault plane solutions of shallow earthquakes and contemporary tectonics in Asia, *Earth Planet Sci. Lett., 19*, 101-112, 1973.

Molnar, P., D. Freedman, J. S. F. Smith, Lengths of intermediate and deep seismic zones and temperatures in downgoing slabs of lithosphere, *Geophys. J. R. Astr. Soc., 56*, 41-54, 1979.

Molnar, P., and P. Tapponnier, Cenozoic tectonics of Asia: Effects of a continental collision, *Science, 189*, 419-426, 1975.

Molnar, P., and P. Tapponnier, Relation of the tectonics of eastern China to the India-Eurasia collision. Application of slip-line field theory to large-scale continental tectonics, *Geology, 5*, 212-216, 1977.

Morgan, W. J., Plate motions and deep mantle convection, *Geol. Soc. Am. Memoir, 132*, 7-21, 1972.

Murphy, R. W., The Manila Trench - west Taiwan fold belt: A flipped subduction zone, *Geol. Soc. Malaysia Bull., 6*, 27-42, 1973.

Nakamura, K., and K. Shimazaki, The Sagami and Suruga Troughs and plate subduction, *Kagaku, 51*, 490-498, 1981 (in Japanese).

Nakanishi, I., Precursors to ScS phases and dipping interface in the upper mantle beneath southwest Japan, *Tectonophys., 69*, 1-35, 1980.

Osada, M., and K. Abe, The Banda Sea earthquake of November 4, 1963 - A large fracture of lithospheric plate at the depth of 100 km, *Phys. Earth Planet. Inter., 25*, 129-139, 1981.

Parson, B., and P. Molnar, The origin of outer topographic rises associated with trenches, *Geophys. J. R. Astr. Soc., 45*, 707-712, 1976.

Peterson, E. T., and T. Seno, Seismic moment release, seismic slip rate, and coupling in subduction zones, *preprint*, 1982.

Ritsema, A. R., and J. Veldkamp, Fault-plane mechanisms of southeast Asian earthquakes, *Ned. R. Meteorol. Inst., Publ. 76.*, 1960.

Rodolfo, K. S., Bathymetry and marine geology of the Andaman basin, and tectonic implications for southeast Asia, *Geol. Soc. Am. Bull. 80*, 1203-1230, 1969.

Rowlett H., and J. Kelleher, Evolving seismic and tectonic patterns along the western margin of the Philippine Sea plate, *J. Geophys. Res., 81*, 3518-3524, 1976.

Ruff, L., and H. Kanamori, Seismicity and the subduction process, *Phys. Earth Planet. Inter., 23*, 240-252, 1980.

Rutland, R. W. R., A tectonic study of part of the Philippine fault zone, *Quat. J. Geol. Soc. Lond., 123*, 293-325, 1968.

Samowitz, I. R., and D. W. Forsyth, Double seismic zone beneath the Mariana island arc, *J. Geophys. Res., 86*, 7013-7021, 1981.

Sasatani, T., Source process of a large deep-focus earthquake of 1970 in the Sea of Okhotsk, *J. Phys Earth, 24*, 27-42, 1976a.

Sasatani, T., Mechanism of mantle earthquakes near the junction of the Kurile and the northern Honshu arc, *J. Phys. Earth, 24*, 341-354, 1976b.

Seno, T., The instantaneous rotation vector of the Philippine Sea plate relative to the Eurasian plate, *Tectonophys., 42*, 209-226, 1977.

Seno, T., Intraplate seismicity in Tohoku and Hokkaido and large interplate earthquakes: A possibility of a large interplate earthquake off the southern Sanriku coast, northern Japan, *J. Phys. Earth, 27*, 21-51, 1979.

Seno, T., Seismicity and focal mechanisms prior to the 1978 Miyagi-Oki, Japan, earthquake, *preprint*, 1981.

Seno, T., and K. Ishibashi, Plate convergence along the Nankai Trough as deduced from the geodetic data, *in preparation*, 1982.

Seno, T., and K. Kurita, Focal mechanisms and tectonics in the Taiwan - Philippine region, *J. Phys. Earth, 26 Suppl.*, 249-263, 1978.

Seno, T., and B. Pongsawat, A triple-planed structure of seismicity and earthquake mechanisms off Miyagi prefecture, northeastern Honshu, Japan, *Earth Planet. Sci. Lett., 55*, 25-36, 1981.

Seno, T., K. Shimazaki, P. Somerville, K. Sudo, and T. Eguchi, Rupture process of the Miyagi-Oki, Japan, earthquake of June 12, 1978, *Phys. Earth Planet. Inter., 23*, 39-61, 1980.

Shimazaki, K., Focal mechanism of a shock at the northwestern boundary of the Pacific plate: Extensional feature of the oceanic lithosphere and compressional feature of the continental lithosphere, *Phys. Earth Planet. Inter., 6*, 397-404, 1972.

Shimazaki, K., Preseismic crustal deformation caused by an underthrusting oceanic plate in eastern Hokkaido, Japan, *Phys. Earth Planet. Inter., 8*, 148-157, 1974a.

Shimazaki, K., Nemuro-Oki earthquake of June 17, 1973: A lithospheric rebound at the upper half of the interface, *Phys. Earth Planet. Inter., 9*, 314-327, 1974b.

Shiono, K., Travel time analysis of relatively deep earthquakes in southwest Japan with special reference to the underthrusting of the Philippine Sea plate, *J. Geosci. Osaka City Univ., 18*, 37-59, 1974.

Shiono, K., T. Mikumo, and Y. Ishikawa, Tectonics of the Kyushu-Ryukyu arc as evidenced from seismicity and focal mechanism of shallow to intermediate-depth earthquakes, *J. Phys. Earth, 28*, 17-43, 1980.

Silver, E. A., and J. C. Moore, The Molucca collision zone, Indonesia, *J. Geophys. Res., 83*, 1681-1691, 1978.

Somerville, P., The accommodation of plate collision by deformation in the Izu block, Japan, *Bull. Earthquake Res. Inst., Univ. of Tokyo, 53*, 629-648, 1978.

Stauder, W., S-wave studies of earthquakes of the north Pacific, Part I: Kamchatka, *Bull. Seismol. Soc. Am., 52*, 527-550, 1962.

Stauder, W., Tensional character of earthquake foci beneath the Aleutian trench with relation to sea-floor spreading, *J. Geophys. Res., 73*, 7693-7710, 1968.

Stauder, W., and G. Bollinger, The S-wave project for focal mechanism studies, earthquakes of 1963, *Bull. Seismol. Soc. Am, 56*, 1363-1371, 1966.

Stauder, W., and L. Mualchin, Fault motion in the larger earthquakes of the Kurile-Kamchatka arc and of the Kurile-Hokkaido corner, *J. Geophys. Res., 81*, 297-308, 1976.

Stewart, G., Implications for plate tectonics of the Aug. 19, 1977 Indonesian decoupling normal fault earthquake, *EOS, 59*, 326, 1978.

Stewart, G., and S. N. Cohn, The 1976 August 16, Mindanao, Philippine earthquake (Ms=7.8) - evidence for a subduction zone south of Mindanao, *Geophys. J. R. Astr. Soc.*, *57*, 51-65, 1979.

Sudo, K., Two distinct phases in the initial P-wave group of the Niigata earthquake of 1964 and of the Taiwan-Oki earthquake of 1966, *J. Phys. Earth*, *20*, 111-125, 1972a.

Sudo, K., The focal process of the Taiwan-Oki earthquake of March 12, 1966, *J. Phys. Earth*, *20*, 147-164, 1972b.

Sudo, K., T. Sasatani, and M. Kasahara, Source process of the intermediate large earthquake of Dec.6, 1978 beneath Kunashiri Channel (II), *Abstr. Ann. Meet. Seismol. Soc. Japan*, *2*, 93, 1979 (in Japanese).

Suzuki, S., Y. Hontani, and T. Sasatani, Double seismic zone beneath Hokkaido, *Abstr. Ann. Meet. Seismol. Soc. Japan*, *1*, 12, 1981 (in Japanese).

Sykes, L. R., The seismicity and deep structure of island arcs, *J. Geophys. Res.*, *71*, 2981-3006, 1966.

Tapponnier, P., and P. Molnar, Slip-line field theory and large-scale continental tectonics, *Nature*, *264*, 319-324, 1978.

Topper, R. E., and E. R. Engdahl, Fine structure of the double Benioff zone beneath the central Aleutian arc, *EOS*, *59*, 1195, 1978.

Turcotte, D. L., Flexure, *Advances in Geophysics*, *21*, 51-86, 1979.

Turcotte, D. L., D. C. McAdoo, and J. C. Caldwell, An elastic-perfectly plastic analysis of the bending lithosphere, *Tectonophys.*, *47*, 193-205, 1978.

Tsumura, K., Microearthquake activity in the Kanto district, *Spec. Publ. for the 50th Anniv. of the great Kanto earthquake*, Earthquake Res. Inst., Univ. of Tokyo, 67-87, 1973.

Udidas, A., and W. Stauder, Application of numerical method for S-wave focal mechanism determinations to earthquakes of Kamchatka-Kurile Islands region, *Bull. Seismol. Soc. Am.*, *54*, 2049-2065, 1964.

Umino, N., and A. Hasegawa, On the two-layered structure of deep seismic plane in northeastern Japan arc, *J. Seismol. Soc. Japan*, *28*, 125-139, 1975 (in Japanese).

Usami, T., *Descriptive Catalogue of Disaster Earthquakes in Japan*, Univ. of Tokyo Press, Tokyo, pp. 327, 1975 (in Japanese).

Utsu, T., Seismic activity in Hokkaido and its vicinity, *Geophys. Bull. Hokkaido Univ.*, *20*, 51-75, 1968.

Utsu, T., Large earthquakes near Hokkaido and the expectancy of the occurrence of a large earthquake off Nemuro, *Rep. Coord. Comm. Earthq. Predict.*, *7*, 7-13, 1972 (in Japanese).

Utsu, T., Space-time pattern of large earthquakes occurring off the Pacific coast of the Japanese islands, *J. Phys. Earth*, *22*, 325-342, 1974.

Utsu, T., Some remarks on a seismic gap off Miyagi prefecture, *Rep. Coord. Comm. Earthq. Predict.*, *21*, 44-46, 1979a (in Japanese).

Utsu, T., Seismicity of Japan from 1885 through 1925. - A new catalogue of earthquakes of M>6 felt in Japan and smaller earthquakes which caused damage in Japan -, *Bull. Earthquake Res. Inst., Univ. of Tokyo*, *54*, 253-308, 1979b (in Japanese).

Uyeda, S., and H. Kanamori, Back-arc opening and the mode of subduction, *J. Geophys. Res.*, *84*, 1049-1061, 1979.

Veith, K.F., The relationship of island arc seismicity to plate tectonics, *Ph.D. Thesis*, Southern Methodist Univ., pp. 67, 1974.

Vlaar, N. J., and M. J. R. Wortel, Lithospheric aging instability and subduction, *Tectonophys.*, *32*, 331-351, 1976.

Vogt, P., A. Lowrie, D. Brace, and R. Hey, Subduction of aseismic oceanic ridges: Effects on shape, seismicity, and characteristics of consuming plate boundaries, *Geol. Soc. Am. Spec. Pap.*, *172*, pp. 59, 1976.

Wang, S. C., R. J. Geller, and S. Stein, An intraplate thrust earthquake in the South China Sea, *J. Geophys. Res.*, *84*, 5627-5631, 1979.

Weissel, J. K., and R. N. Anderson, Is there a Caroline plate?, *Earth Planet. Sci. Lett.*, *41*, 143-158, 1978.

Wu, F. T., Focal mechanisms and tectonics in the vicinity of Taiwan, *Bull. Seismol. Soc. Am.*, *60*, 2045-2056, 1970.

Wu, F. T., Recent tectonics of Taiwan, *J. Phys. Earth*, *26 Suppl.*, 265-299, 1978.

Yamashina, K., and I. Murai, On the focal mechanism of the earthquakes in the central part of Oita prefecture and in the northern part of Aso of 1975, especially, the relations to the active fault system, *Bull. Earthquake Res. Inst., Univ. of Tokyo*, *50*, 295-302, 1975 (in Japanese).

Yamashina, K., K. Shimazaki, and T. Kato, Aseismic belt along plate subduction in Japan, *J. Phys. Earth*, *26 Suppl.*, 447-458, 1978.

Yoshii, T., Proposal of the "aseismic front", *J. Seismol. Soc. Japan*, *28*, 365-367, 1975 (in Japanese).

Yoshii, T., A detailed cross-section of the deep seismic zone beneath Japan, *Tectonophys.*, *55*, 349-360, 1979a.

Yoshii, T., Compilation of geophysical data around the Japanese islands, *Bull. Earthquake Res. Inst., Univ. of Tokyo*, *54*, 75-117, 1979b (in Japanese).

Yoshioka, N., and K. Abe, Focal mechanism of the Iwate-Oki earthquake of June 12, 1968, *J. Phys. Earth*, *24*, 251-262, 1976.

LATE QUATERNARY VERTICAL CRUSTAL MOVEMENTS IN AND AROUND THE PACIFIC
AS DEDUCED FROM FORMER SHORELINE DATA

Nobuyuki Yonekura

Department of Geography, University of Tokyo, Tokyo, Japan 113

Abstract. Present heights of former shorelines provide critical data for clarifying the nature and rate of vertical crustal movements of coastal regions in the late Quaternary. Former shorelines formed in and around the Pacific at the last interglacial maximum (ca. 125 $\pm$ 10 kyr ago) and at the middle Holocene (5 $\pm$ 1 kyr ago) are chosen as the basic data. The result indicates that the rate of coastal uplift is much larger in continental arc and collision zones than in island arcs: the average rate of maximum uplift in the former ranges 1-10 mm/yr, while it is below 0.5 mm/yr in the latter. Oceanic islands (atolls and volcanic islands) show least vertical movements as expected from their stable tectonic environment. The remarkable difference in vertical movement among collision zones, continental arcs and island arcs is probably related to the degree of interaction between converging plates.

Introduction

The trench-arc system is one of the most active mobil belts in the world. The fundamental process occurring in trench-arc systems is the subduction of oceanic plates. Great shallow thrust-type earthquakes along the trench-arc systems give direct evidence of the interaction between the subducting and overthrusting plates. The Chilean earthquake of 1960 and the Alaskan earthquake of 1964 are good examples of these great thrust-type earthquakes at subduction zones (Plafker, 1972). None of these really great shallow earthquakes have occurred at island arcs such as the Mariana arc. This indicates that subduction is going on virtually aseismically at these arcs (Kanamori, 1978).

Uyeda and Kanamori (1979) classified trench-arc systems according to the nature of their back-arc regions, "continental arcs" and "island arcs". By definition, the former have no back-arc basins and the latter do. They proposed that there are two fundamentally different modes of subduction. Two convergent plates press hard against each other along continental arcs, while the subducting plate sinks without exerting strong compression on the upper plate along island arcs. Using the names of the most typical end members of the two modes, the former are called Chilean-type and the latter Mariana-type modes of subduction. According to their classfication, uplift of the outer or frontal arc region is expected to be more pronounced at Chilean-type subduction boundaries than the Mariana-type.

The purpose of this paper is to review late Quaternary coastal uplift in and around the Pacific on the basis of former shoreline data and to interprete tha rate of vertical crustal movements, mainly uplift, in terms of the tectonic setting of the region. It is expected that this review will be a useful check of the idea proposed by Uyeda and Kanamori (1979).

Studies of marine terraces and raised coral reefs in Japan and other regions of the world have shown that the present heights of former shorelines or paleo-sealevels provide important data for clarifying the nature and rate of vertical crustal movements of coastal regions in the late Quaternary. Most recent studies are based on radiometric age data, which enable one to make correlations among remote regins and to compare the regional differences in the mode of tectonic deformation. The present heights of the former shorelines formed at the last interglacial maximum (ca. 125 $\pm$ 10 kyr ago) and at the middle Holocene (ca. 5 $\pm$ 1 kyr ago) are chosen as the basic data. The reason for choosing these ages will be discussed in the next section.

This paper does not present the complete data set of the former shorelines. For such information, the reader is refered to a map compilation of the world shoreliens which was started in 1972 by the Commission on Quaternary Shorelines of the International Union for Quaternary Research. The first map for the Pacific and Indian Oceans sector was published in 1980 (Commission on Quaternary Shorelines, 1980). The Sea-Level Project of the Internatinal Geological Correlation Programme (project 61) has also been acive since 1974 and an atlas of sea-level curves during the last 15,000 years has been published (Bloom , 1977).

The main part of this paper was presented at the symposium on "Sea-level changes and tectonics" organized by A. Sugimura and Y. Ida and held on

May 24-25, 1979 at the Ocean Research Institute, University of Tokyo (Yonekura, 1979).

Sea-level changes in the late Quaternary

A number of curves of sea-level changes have been reconstructed for the period from the last glacial maximum to the present. The post glacial rise in sea-level has been well documented in the world and it is considered to correspond to the melting of continental ice sheets, mostly in the northern hemisphere (Bloom, 1971).

The sea-level project of IGCP (project 61) initially aimed to establish a graph of the trend of mean sea-level during the last deglacial hemicycle (about 15,000 years). As the project proceeded, it became clear that differences in regional sea-level histories were so large that one single global curve was difficult to obtain. It is now generally accepted that, in addition to the global eustasy, other factors such as isostacy (Walcott, 1972) and local tectonics (Pirazzoli, 1977; Bloom, 1980) play important roles in the world-wide distribution of former sea-level features. Bloom (1980) conveniently categorized Quaternary tectonic movements as slow subsidence (0.1 mm/yr or less), slow uplift (about 0.1 mm/yr) and rapid uplift (equal or larger than 1.0 mm/yr) and discussed a variety of coasts in the coral reef zones of the Southwest Pacific Ocean in terms of tectonic diversity. The records of late Quaternary sea-level will give us fundamental data for detecting the nature of vertical movements in various regions and for estimating such parameters as the strength and elasticity of the earth's outer layers (Clark et al., 1978). The post glacial rise in sea-level was very rapid during the period from 18,000 to 7,000 years ago. Since about 6,000 years ago, changes in the volume of sea water became smaller and sea-level has gradually attained the present height. Sea-level at 5 $\pm$ 1 kyr ago is chosen as the datum for the present purpose, because uplifted shorelines in non-glaciated regions began to emerge at about that time.

Cyclic changes in sea-level during the late Quaternary are recorded in various forms such as former shorelines along coasts, marine sediments in basins, and even isotopic changes in deep-sea sediments. Multiple episodes of relatively high sea-level stands are recorded at the same altitude on the land, if the water melted from continental ice sheets had the same volume for respective interglacial periods. Therefore, it is difficult to distinguish multiple episodes of relatively high sea-level stands from the records on the land, especially in tectonically stable regions. On the other hand, studies on elevated coral reefs or marine terraces in tectonically uplifted regions have demonstrated multiple episodes of relatively high sea-level stands. Thus, uplifted regions are suitable for studies of sea-level changes in the late Quaternary.

Changes in sea-level during the late Quaternary have been brought about mainly by glacial eustasy. High stands of sea-level correspond to interglcial or interstadial periods and low stands of sea-level correspond to glacial periods. Results of oxygen isotope analyses of deepsea sediments have given much information about the past sea surface temperature and volume of continental ice. Shackleton and Opdyke (1973) demonstrated that cyclic changes in the oxygen isotopic composition of ocean sediments were brought about by changes in the volume of terrestrial ice, that is, glacial changes in sea-level. This enables us to use the isotopic variations in deep-sea sediments for constructing a standard sea-level curve.

Shackleton and Opdyke (1973) indicated that there were eight stages of sea-level maxima for the last 700 kyr, and delineated a glacio-eustatic sea-level curve for the past 130 kyr. This result shows that prominent sea-level maxima occurred at about 120 kyr ago and during the Holocene, and that the relatively high sea-level stands at about 100 kyr and 80 kyr ago were significantly lower.

Radiometric dating of shallow water biogenic calcium carbonates and oolites such as corals and Tridacna shells provided valuable data for dating former sea-levels in coral reef regions. Fission tranck dating of volcanic materials gave an absolute time scale for tephrochronology in volcanic regions. Studies on elevated coral reefs on Barbados, West Indies (Breocker et al., 1968), Huon Peninsula, Papua New Guinea (Veeh and Chappell, 1970; Chappell, 1974; Bloom et al., 1974), and Kikai-jima, Ryukyus (Konishi et al., 1974) show that relatively high sea-level stands occurred at about 125 kyr, 100 kyr, and 80 kyr ago during the last interglacial period. Similar resutls were obtained from South Kanto, Japan, on the basis of tephrochronological studies of marine terraces (Machida, 1975). The 120 kyr high stand of sea-level has also been demonstrated from the coral reefs of oceanic islands (Veeh, 1966; Ku et al., 1974).

As the results of dating grew in number, the age of sea-level maximum during the last interglacial period began to reveal a considerable divergence ranging from ca. 135 kyr to ca. 118 kyr ago. This may indicate the existence of two peaks in sea-level during the last interglacial maximum as indicated on Barbados (Mesolella et al., 1969) and on Papua New Guinea (Chappell, 1974; Bloom et al., 1974; Aharon et al., 1980). However, it is difficult to distinguish the two peaks in other reliable data, an age of 125 $\pm$ 10 kyr ago is, therefore, assumed, in this paper, for the age of sea-level maximum during the last interglacial period.

Present heights of the former shorelines in and around the Pacific

To evaluate the rate of vertical crustal movements in the late Quaternary, data on the present height of the former shoreliens or the sea-levels of 125 $\pm$ 10 kyr and 5 $\pm$ 1 kyr ago are reviewed for various regions in and around the Pacific. The regions are classified into three types in terms of tectonic setting.

(1) Oceanic islands in the Pacific. They are located on the oceanic plate and, therefore, considered to be relatively stable in the tecotnic sense. They include both volcanic islands (high islands) and atolls (low islands).

(2) Arcs in and around the Pacific. Arcs are characterized by subduction of an oceniac plate. They are divided into two classes; island arcs (having back-arc basins) and continental arcs (without back-arc basins).

(3) Accretion or collision zones around the Pacific. These are the areas of convergence where buoyant features, such as island arcs and microcontinents, have arrived at the subduction zones and collided with the landward plate.

Oceanic islands

Oceanic islands located in the Pacific provide useful data for studying the eustatic changes in sea-level, because they are considered to have remained tectonically relatively stable during the time under consideration. The heights of the former sea-levels in oceanic islands will, thus, give us a frame of reference for evaluating the vertical tectonic movements of the more mobile regions.

Extensive carbon-14 dates from core holes through fringing coral reefs within Hanauma Bay, Oahu, Hawaii, indicate that sea-level has risen smoothly from -7.9 m to the present level during the last 6,000 years. Sea-level at Oahu was about -4 m at about 5,000 years ago and it never stood higher than at present during the last 3,500 years (Easton and Olson, 1976). This result corresponds closely to the curve of sea-level change obtained from the eastern Caroline Islands (Curry et al., 1970; Bloom 1970). However, studies on reef features at Eniwetok atoll indicate that sea-level reached its present height no later than 4,000 years ago and stayed more than 1 m above the present height from about 3,500 to about 2,000 years ago (Buddenmeier et al., 1975). In the Gilbert Islands, sea-level followed a curve similar to that for Eniwetok atoll and reached a maximum of about +2.4 m at 2,760 C-14 yBP (Schofield, 1977). Labeyrie et al. (1969) reported that sea-level at Mururoa atoll was -6 m at 5,600 yBP and then reached a maximum level of +3 m at 3,000 yBP.

Thus, the maximum height of Holocene sea-level is variable in both time and in space. The height of sea-level around 5,000 years ago ranges from -4 m to +2 m in different regions. These regional differences may be explained in terms of hydro-isostatic movements (Clark et al., 1978).

Extensive dating of the fossil corals associated with the Waimanalo shoreline on the island of Oahu, Hawaii, has shown that at about 120 kyr ago the ocean was approximately 7.6 $\pm$ 2 m above its present level (Ku et al., 1974). Similar ages of emerged coral reefs at a like elevation (+2 to +9 m) were reported from the Tuamotu Islands and Cook Islands (Veeh, 1966). Although the choice of a specific value between 0 and 10 m is not critical, the value of +6 m at 125 kyr ago has been used as a datum to estimate the absolute heights of sea-levels at periods younger than 125 kyr ago (Broecker et al., 1968; Mesolella et al., 1969; Veeh and Chappell, 1970; Bloom et al., 1974; Chappell, 1974).

Radiometric ages of coral formatins of about 6 kyr and 120 kyr ago were obtained from drilling samples at Eniwetok atoll (Thurber et al., 1965). Samples from less than 12 m below the surface gave an age less than 6 kyr and samples from between 12 and 21 m depths were dated close to 120 kyr ago. The hiatus in the develoment of coral between 6 kyr and 120 kyr ago on Eniwetok atoll suggests that the ocean level was considerably lower during this period than at present. The present depth (-12 m) of the depositional surface formed at the last interglacial period indicates that Eniwetok atoll has been affected by a slow subsidence and that this subsidence of Eniwetok atoll has amounted to about 20 m relative to volcanic islands such as Oahu, Hawaii, for the
period of the last 125 kyr; the average rate of subsidence being about 0.16 mm/yr. Similar results come from Muraroa atoll, in which coral formation at the last interglacial period is presently between 6 and 12 m in depth (Lalou et al., 1966).

Islans arcs

The Tonga and Mariana arcs are typical island arcs with active back-arc basins; the Lau Basin and the Mariana Trough, respectively. On Tongatapu and Eua, the two main islands of the Tongatapu block of the Tonga frontal arc, the last interglacial reefs are 5.5 m above high-tide level and gave ages of 135 $\pm$ 15 kyr and 133 $\pm$ 12 kyr, respectively. Considering the 1.2 m mean daily tidal range, the maximum emergence since the last interglacial period is concluded to be 6.7 m (Taylor et al., 1977). Both islands also have mid-Holocene reefs and notches that indicate 2.2 m of emergence. The heights of reefs formed durng both the Holocene and last interglacial periods in the Tonga arc are so close to those in oceanic islands that it is reasonably clear that vertical tectonic movements have been negligible during these time intervals.

Extensive terraces and emerged reefs are observed in the Mariana Islands, such as on Guam (Tracey et al., 1964), but radiometric ages are not available except for a few cases. A coral head from an emerged reef at +1.4 to +1.8 m was dated at 2,880 $\pm$ 110 yBP (Curray et al., 1970). A beach deposit at +2 m above low tide in Adorius Island, Palau Group, was dated at 6,500 $\pm$ 200 yBP and 5,133 $\pm$ 80 yBP (Easton and Ku, 1980). These results suggest that emergence since mid-Holocene time has been about 2 m. Although further investigations are required, the small amount of existing data suggest that, like the Tonga arc, there has been little vertical tectonic movement since late Quaternary time.

Although the New Hebrides arc is also an island arc with an active back-arc basin, the Fiji plateau, the polarity of subduction is to the east and seaward, which is reversed compared to other arcs with spreading back-arc basins. In this case, the Australian plate is subducting beneath the "Pacific" plate. Uplifted reef terraces are well developed on Santo and north Malekula islands in the central New Hebrides arc and several have been dated (Taylor et al., 1980; Jouannic et al., 1980). The Holocene uplift rate of Santo ranges from 1 to 5.5 mm/yr, generally increasing from east to west. A high rate of 5.2 mm/yr was obtained from Araki Island, south Santo, in which elevated coral reefs at +27 m were dated at 5,430 ± 200 and 5,470 ± 160 yBP. The main surface of the eastern limestone plateau of Santo is assumed to correspond to the 125 kyr paleo-sea level (Jouannic et al., 1980). If so, the highest altitude of the surface attains +300 m in the Queiros Peninsula.

According to Bloom et al. (1978), Efate Island, central New Hebrides, has been uplifted at the average rate of 1 mm/yr for the last 200 kyr. Radiometric ages for coral from uplifted terraces at Port Havannah, Efate, indicate that either the terrace at +85-90 m (114-130 kyr) or +110-130 m (124-141 kyr) is correlative with the time of the last interglacial high sea-level. The altitudes for Holocene dated samples on Efate range from 2 to 6 m above high-tide level and gave ages of 2,865 to 6,800 yBP (Bloom et al., 1978). Thus, islands in the central New Hebrides arc have been significantly uplifted in the late Quaternary and the average rate of uplift is especially large in Santo and north Malekula. Santo and north Malekula islands are located where the New Hebrides Trench is apparently divided into two sections by a topographic high. It appears likely that the subduction of the D'Entrecastaux Fracture Zone on the Australian plate has affected the tectonic uplift of this part of the New Hebrides arc.

The Kurile and Northeast Japan arcs are usually considered to be island arcs with marginal basins. The back-arc basins behind these arcs were at some stage actively spreading, but are believed to have been inactive for the last several million years (Uyeda and Kanamori, 1978). In the case of the Aleutian arc, a major part of its back-arc basin, the eastern Bering Sea, is thought to have formed by entrapment of a part of the Kula plate (Cooper et al., 1976). Therefore, these arcs are considered to be at an intermediate state between typical island arcs and typical continental arcs. The author has not been able to locate useful data on the former shorelines of the Kurile arc. Most presumed Pleistocene marine terraces in the Aleutian arc are not directly dated except for Amchitka Island in the Aleutians. An average age of 127 ± 8 kyr for fossil shells and bones was obtained from a deposit of the South Bight II marine transgression at Amchitka Island. The altitude of the marine terraces of the South Bight II varies from +37 to +49 m (Szabo and Gard, 1975). The +2-5 m Holocene marine level is found on most islands of the Aleutian arc such as Umnak, Adak, Amchitka and Attu islands. During the mid-Holocene, sea-level was +2-5 m on many Aleutian Islands, and returned gradually to its present level after about 4,000-3,000 years ago (Black, 1980).

The Japanese arcs, including the Southwest Honshu and Ryukyu arcs, and New Zealand will be treated in the next section.

Continental and Intermediate arcs

The Andean and Alaskan arcs are typical continental arcs without back-arc basins. In these arcs, we find rapidly uplifting coasts in the focal regions of the Chilean earthquake of 1960 and the Alaskan earthquake of 1964.

Mocha Island lying near the edge of the continental shelf southwest of the Arauco Peninsula, central Chile, has been rapidly uplifted in the late Quaternary. The island was suddenly uplifted by 1.0 to 1.8 m during the Chilean earthquake of 1960 (Plafker and Savage, 1970) and the highest Holocene shoreline reaches +33 m. Carbon-14 dates for beach deposits on this terrace suggest that the average rate of uplift has been about 5.5 mm/yr during the last 4,000 years, and the age of the highest shoreline is estimated to be about 6,000 years, if the average rate of uplift is extrapolated (Kaizuka et al., 1973).

The Pleistocene marine terraces are well developed on both Mocha Island and the Arauco Peninsula. Although radiometric ages are not available for these terraces, the Cañete Surfaces, the youngest and widest of the Pleistocene surfaces, is estimated to be of the last interglacial age on the basis of geomorphic features. The Cañete Surface reaches +225 m on the western part of the Arauco Peninsula (Kaizuka et al., 1973). We can not find marine terraces along the Peruvian coast except in the Nazca region where the Nazca Ridge intersects with the Peru-Chile Trench. The Nazca coast seems to have been uplifted by the subduction/collision of the Nazca Ridge.

Successive uplifts of Middleton Islands, 80 km off the mainland coast of Alaska, are recorded by a flight of marine terraces. The youngest terrace was formed by the sudden uplift of about 3.3 m during the 1964 great earthquake. Based on radiocaron dates from older terraces on Middleton Islands, the average rate of uplift since the island emerged from the sea about 4,500 years ago is estimated as about 10 mm/yr (Plafker, 1972). The marine terrace on the island reaches +46 m and it is the highest one of the Holocene marine terraces around the Pacific.

Along the east coast of Kamchatka Peninsula, former shorelines at +200 to +500 m are suggested to be of the last interglacial age (Commission of Quaternary Shorelines, 1980). Most of them do not seem to be controlled by radiometric dating, although the author has not been able to check all of the original data.

The marine terrace in South Kanto gives a stratigraphic standard of the late Quaternary in Japan. The middle to late Pleistocene changes of sea-level are analysed by establishing the stratigraphic relations between the marine formations and the tephras. Based on the dated tephras from the marine terraces in South Kanto, sea-level maxima are estimated at 120-130 kyr, 100 kyr, 80 kyr, and 60 kyr ago in the late Quaternary (Machida, 1975). Data on the heights of the former shorelines of the last interglacial maximum and the Holocene were compiled for the Japanese coastal areas by Ota and Naruse (1977) and Ota and Yoshikawa (1978).

The height of the shorelines of the last interglacial maximum ranges form +10 to +200 m at various coastal regions in Japan. It reaches +160 m in Oiso Hills, South Kanto (Machida, 1973) and +200 m in Muroto Peninsula, Shikoku (Yoshikawa et al., 1964). The Holocene marine terraces in Japan are normally less than +10 m, but attain +13 m in South Shikoku (Kanaya, 1978) and +26 m in South Kanto (Yonekura, 1975). Numerous radiocarbon dates indicate that these uplifted marine terraces have emerged from the sea bout 6,000 years ago. Thus, the average rate of maximum uplift is 1.5 mm/yr for the last 125 kyr and 4mm/yr for the last 6 kyr along the Japanese coasts. Deformation of the former shorelines in Japan shows large regional differences (Ota, 1975; Ota and Yoshikawa, 1978). The Pacific coast of Southwest Japan, facing the Sagami and Nankai Troughs, is characterized by a landward tilting with rapid uplift. This type of coastal uplift is considered to originate from tectonic deformation associated with shallow thrust-type earthquakes along the troughs (Yonekura, 1975; Matsuda et al., 1978).

Similar studies of coral terraces have been conducted on many islands in the Ryukyu arc (Konishi et al., 1974; Nakata et a., 1978; Ota and Hori, 1980). Kikai-jima, located close to the Ryukyu Trench, is capped with raised coral reefs in which the highest reef surface attains +224 m. Corals at +170 m on the highest terrace were dated at about 125 kyr (Konishi et al., 1974). The Holocene fringing reef of Kikai-jima is also elevated up to +13 m for the last 6,000 years (Nakata et al., 1978; Ota et al., 1978). These results indicate that Kikai-jima has been uplifted at an average rate of 2 mm/yr for the last 125 kyr. A westward tilting from Kikai-jima toward the eastern part of Amami-Ohshima is considered to represent a similar mode of deformation to that along the Pacific coast of Southwest Japan (Yonekura, 1975; Ota and Hori, 1980). The rapid uplift of Kikai-jima may be related to the possible collision of topogaphic features on the Philippine Sea plate. The west Philippine Sea off this part of the Ryukyu arc is abundant in apparently buoyant features, such as Amami Plateau, Daito and Oki-Daito Ridges.

Remarkable uplift is also observed in the North Island of New Zealand. The Otamaroa marine terrace along the Bay of Plenty is estimated to have formed at about 125 kyr ago on the basis of fission track ages from tephra formations (Yoshikawa et al., 1980). The altitude of this terrace increases eastward from +10 m to +300 m between the Bay of Plenty and East Cape. The Holocene marine terrace reaches +20 m on the northeast coast of of the region. These features indicate a 2-3 mm/yr average rate of maximum uplift for the last 125 kyr. A series of raised beach ridges indicate recent uplifts due to active folding and faulting in the southern tip of the North Island of New Zealand. The highest raised beach ridge attains +25 m on Turakirae Head, which indicates an uplift rate of 4 mm/yr in the Holocene (Wellman, 1967). The amount of coastal uplift in the South Island is not so large (Ota, 1977).

Thus, coastal uplift of frontal arcs in the late Quaternary is significantly large in continental and intermediate type arcs. The maximum height of the Holocene marine terraces range from +13 to +33 m except in Alaska, where a + 46 m terrace is found. The average rate of maximum uplift in respective arcs is about 2 to 5.5 mm/yr with an exception of 10 mm/yr (Alaska) for the last 6,000 years. The highest former shoreliens on the marine terrace formed at ca. 125 kyr ago are over +200 m and reach +300 in some places. The average rate of maximum uplift in respective arcs ranges from 1.5 to 2.5 mm/yr for the last 125 kyr.

Accretion or Collision zones

Taiwan Island is considered to represent a young arc-continent collision zone in the Ryukyu-Taiwan-Philippine system. Collision is believed to have started in the late Pliocene and still be actively going on (Chai, 1972; Karig, 1973; Wu, 1978). Radiometric ages of raised coral reefs in Taiwan are reported by many authors such as Konishi et al. (1968), Lin (1969), Hashimoto (1972), Peng et al. (1977) and Pirazzoli (1978). The altitudes of dated corals are mostly less than +25 m. The highest point of the Holocene surface, however, is about +40 m at the Tainan Tableland and the Milun Tableland (lin, 1969) and +46 m at the Milun Tableland (Pirazzoli, 1978). A coral at +35 m from Taitung area, east Taiwan, gives an age of $6,132 \pm 184$ yBP (NTU-151) (in Pirazzoli, 1978), which suggests that the average rate of uplift has been about 6mm/yr.

Konishi et al. (1968) estimated that the average uplift rate of raised coral reefs is high (6-9.7 mm/yr) for the Hualien area (the northern tip of the Coastal Range in east Taiwan) as compared to 1.8-4.8 mm/yr for other parts of Taiwan. Peng et al. (1977), however, estimated that the average uplift rates of the Hengchun Peninsula, south Taiwan (5.3 ± 0.2 mm/yr), the Tainan area (4.3 ± 0.4 mm/yr) and the Coastal Range (5.0 ± 0.4 mm/yr) do not differ greatly from one another for the last 9,000 years, based on the altitude of the raised coral reef samples and the eustatic sea-

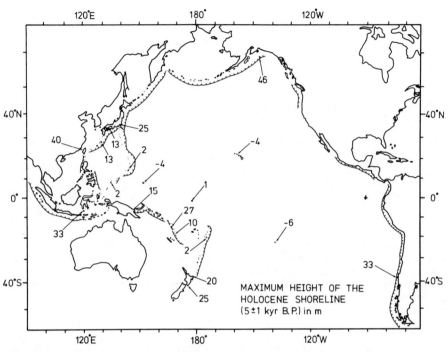

Fig. 1. Maximum height of the Holocene shoreline (5 ± 1 kry) in meters in and around the Pacific.

level rise. Although some discrepancies remain among different studies, these results clearly indicate that Taiwan has been uplifted at a high rate of about 5 mm/yr or more duirng the Holocene. The amount of uplift during the Pleistocene is not available from marine terrace data. The high uplift rate of Taiwan is consistent with the high seismicity and its tectonic setting as an active collision zone.

Northern New Guinea is considered to be another collision zone, between the Australian plate and an island arc on the Pacific plate, with collision having started in Miocene time (Dewey and Bird, 1970). A flight of coral terraces rising to over +700 m exists on the northeast coast of Huon Peninsula, Papua New Guinea. Results of extensive radiometric dating established a history of sea-level change in this area for the last 220 kyr (Veeh and Chappell, 1970; Chappell, 1974; Bloom et al., 1974). The particular coral terrace correlative to a high sea-level at 125 kyr ago varies its elevation laterly from +33 m to +400 m. The average rate of maximum uplift is about 3mm/yr (Veeh and Chappell, 1970). A Holocene reef has emerged as much as +12-15 m at Huon Peninsula (Bloom et al. 1974) and the crest of the Holocene reef has formed around 6,500 years ago (Chappell and Polach, 1976.)

Timor Island is located in a collision zone between the Sunda arc and the Australian plate. Raised coral terraces on the north coast of East Timor and at Atauro Island, north of Timor, were studied by Chappell and Veeh (1978). They concluded that the uplift rate along the north coast of Timor is close to 0.5 mm/yr on the basis of the maximum height (+65 m) of the raised coral reefs formed at 125 kyr. This slow rate of uplift greatly differs from the high rates of uplift in Taiwan and New Guinea. However, the position of the plate boundary between the Sunda arc and the Australian plate is not definitive in this area, and Timor Island is defined either as an outer arc to the north of the subduction zone or as the northern margin of continental Australia. Studies on the south coast of Timor will be useful to solve these problems.

Discussion and conclusion

Coastal uplifts duirng the Holocene and late Quaternary time, since the last interglcial maximum, in various regions in and around the Pacific, outlined in the previous chapters, can be summarized as shown in Figs. 1 and ‥. It is evident in these figures that the amount of vertical movements is significantly larger in continental arcs and collision zones than in oceanic islands and some island arcs.

One of the most interesting points is that uplift is negligible for the last 125 kyr in island arcs such as the Tonga and Mariana arcs. The heights of both the Holocene and the last interglacial maximum shorelines in the Tonga arc are almost the same as those on ocenaic islands of volcanic origin. Although we need further studies in the Mariana arc, it seems likely that the Mari-

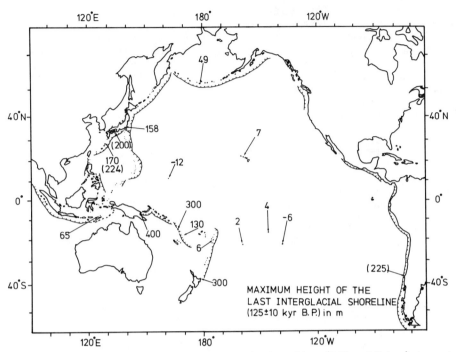

Fig. 2. Maximum height of the last interglacial shoreline (125 ± 10 kyr) in meters in and around the Pacific. Numerals in parentheses indicate the heights which are not controlled by radiometric dates.

ana arc has a history similar to the Tonga arc. If the tectonic uplift of coasts at subduction zones is assumed to be due to upthrusting of the landward plate in response to the subduction of the seaward plate, this result implies that an island arc with an active back-arc basin is weakly coupled to the subducting plate.

If a net uplift or subsidence of 0.1 mm/yr continued for the last 130 kyr, it resutls in either a 13 m elevation or submergence of former coastlines. This rate is apparently enough for preserving islands above the sea or forming atolls in the ocean. For instance, the average rate of subsidence of Eniwetok and Mururoa atolls is about 0.1 or 0.2 mm/yr for the last 125 kyr. The rate of movement for the longer term is smaller. Subsidence of Eniwetok atoll is estimated at only 0.03 mm/yr for the last 45 million years based on the thickness of the limestone. Similarly, an average uplift of only 0.03 mm/yr is required for the coral limestone on Eua Island, Tonga, to emerge 170 m in the 5.2 million years since the end of the Miocene (Taylor and Bloom, 1977). These facts indicate that a rate of uplift as small as 0.1 mm/yr can maintain islands above the sea, at least those included in this study.

Island arc between a subducting oceanic plate and an actively spreading back-arc basin show a similar magnitude of vertical movements as oceanic islands. In this respect, they may be especially referred to as "oceanic arcs". On the other hand, Santo and Malekula islands of the New Hebrides arc indicate that a subducting plate with a buoyant body such as a fracture zone or an aseismic ridge can produce rapid uplift of the frontal arc. The rate of such uplift ranges form 2 to 5.5 mm/yr, which is nearly the same order of magnitude as the average rate of maximum uplift in continental arcs. The Ryukyu arc has an active and almost incipient spreading back-arc; the Okinawa Trough. It may be to considered to be in transition from a continental arc to an oceanic arc. The relatively high uplift rate of Kikai-jima, however, may be related to the collision process.

The average rate of uplift for the Holocene is large in typical continental arcs such as Chile (5.5 mm/yr) and Alaska (10 mm/yr) and also in collision zones such as Taiwan (at least 5 mm/yr). These rates of uplift are larger by a factor of 2 or 3 than those in Southwest Japan (2-4 mm/yr), the Ryukyus (2 mm/yr) and North Island, New Zealand (3-4 mm/yr). The subduction of oceanic plates is at a shallow dip beneath the Chilean and Alaskan arcs and this may produce a wider interactive face of coupling between the two plates. This implies that rapid uplift originates from strong compression between the converging plates in this type of subduction zones.

Even though our data are incomplete for a through evaluation of Pleistocene uplift, the average rate of maximum uplift for the last 125 kyr is about 1.5 to 2.5 mm/yr in continental arcs. Papua New Guinea has a higher rate of 3.2 mm/yr. The uplift rate in the Holocene seems to be equal to or higher than that in the late Pleistocene. These values of average uplift rates, ranging 1 to 10 mm/yr, are at least an order of magnitude larger than those in such oceanic arcs as the Tonga arc and are of the same order as the average slip rate of the class A (1-10 mm/yr) active faults in Japan (Matsuda, 1977). It seems that island (oceanic) and continental arcs can be roughly divided by the average uplift rate of 0.5 mm/yr.

Thus, oceanic islands (atolls and volcanic island), island arcs (including oceanic arcs and arcs in transition between island arc and continental arc), continental arcs, and collision zones can be arranged in this order in terms of increasing rates of late Quaternary vertical crustal movememnts. This order seems to be related with the degree of interaction of converging plates. The degree of interaction at plate boundaries may depend on various factors, such as the angle and the rate of convergence, the dip of subduciton and the absolute velocity of overthrustnig plates (Uyeda and Kanamori, 1979; Ruff and Kanamori, 1980), but the nature of converging plates should also play an important role (Kelleher and McCann, 1976; Molnar and Atwater, 1978). The interaction at converging plate boundaries would be weak between oceanic plates, strong between oceanic and continental (or buoyant) plates, and stronger between two continental (or buoyant) plates.

Another interesting observation from uplift studies is that tectonic uplift is not uniform within the same tectonic terrain. The most prominent feature in a frontal arc is a landward tilting or a reverse tilting with rapid uplit. Offshore islands or peninsulas facing oceanic trenches are rapidly uplifted, while islands or coasts located closer to the axis of the arc are more slowly uplifted. These features are evident in Chile (Kaizuka et al., 1973), Southwest Japan (Yonekura, 1975), the Ryukyus (Konishi et al., 1974; Ota and Hori, 1980), and North Island of New Zealand (Yoshikawa et al., 1980). Such landward tilting is common in continental arcs and it has been considered to originate from overthrusting of a frontal arc on a subducting plate.

Yonekura and Shimazaki (1980) pointed out that the rapid uplift and landward tilting of the frontal arc in subduction zones is mainly due to imbricated thrust faults branching upward from the megathrust between the oceanic and continental plates. A surface fault which appeared on Montague Island at the time of the Alaskan earthquake of 1964 is considered to be an example of these faults (Plafker, 1972). Surface tectonic features in collision zones such as in Taiwan and the Himalayas suggest that imbricated thrust faults are a characteristic mode of surface deformation along the foothills.

It is concluded that the difference in the stress state at interplate boundaries is manifested in the nature and rate of late Quaternary vertical crustal movements. The proposed classification of subduction boundaries by Uyeda and Kanamori (1979) is apparently valid and useful for understanding the vertical tectonic movements associated with subduction. In addition, an expanded classification of converging boundaries, including subduction zones and accretion or collision zones, is possible in terms of Quaternary vertical tectonic movements.

Acknowledgments. I would like to express my thanks to Profs. A. Sugimura and Y. Ida for giving the opportunity of presenting this paper at the symposium of "sea-level changes and tectonics" held on May 24-25, 1979 at the Ocean Research Institute, University of Tokyo. I am grateful to Prof. S. Uyeda for carefully reviewing this paper and Prof. K. Shimazaki for useful discussions and suggestions.

References

Aharon, P., J. Chappell, and W. Compston, Stable isotope and sea-level data from New Guinea supports Antarctic ice-surge theory of ice ages, Nature, 283, 649-651, 1980.

Black, R.F., Isostatic, tectonic, and eustatic movements of sea level in the Aleutian Islands, Alaska, Earth rheology, isostasy and eustasy (Morner, N.A. ed.), 231-248, John Wiley & Sons, Chichester, 1980.

Bloom, A., Paludal stratigraphy of Truk, Ponape, and Kusaie, Eastern Caroline Islands, Geol. Soc. Amer. Bull., 81, 1895-1904, 1970.

Bloom, A., Glacial-eustatic and isostatic controls of sea level since the last glaciation, The late Cenozoic Glacial ages (Turekian, K.K., ed.), 355-379, Yale University Press, New Haven, 1971.

Bloom, A., Atlas of sea-level curves, IGCP Project 61), 1977.

Bloom, A., Late Quaternary sea level change on South Pacific coasts: a study in tectonic diversity, Earth rheology, isostasy and eustasy (Morner, N. A. ed.), 505-516, John Wiley & sons, Chichester, 1980.

Bloom, A., W.S. Broecker, J.M.A. Chappell, R. K. Matthews and K.J. Mesolella, Quaternary sea level fluctuations on a tectonic coast: New ^{230}Th/^{234}U dates from the Huon Peninsula, New Guinea, Quat. Res., 4, 185-205, 1974.

Bloom, A., C. Jouannic, F.W. Taylor, Preliminary radiometric ages from the uplifted Quaternary coral reefs of Efate, Geology of Efate and Offshore Islands (Ash, R.P., J.N. Carney, and A. MacFarlane), 47-49, Regional Report, New Hebrides Condominium Geological Survey, 1978.

Broecker, W.S., D.L. Thurber, J Goddard, R.K. Matthews, and K.J. Mesolella, Milankovitch hypothesis supported by precise dating of coral reefs and deep-sea sediments, Science, 159, 297-300, 1968.

Buddenmeier, R.W., S.V. Smith, and R.A. Kinzie, Holocene windward reef-flat history, Enewetak atoll, Geol. Soc. Amer. Bull., 86, 1581-1584, 1975.

Chai, B.H.T., Structure and tectonic evolution of Taiwan, Amer. Jour. Sci., 272, 389-422, 1972.

Chappell, J., Geology of coral terraces Huon Peninsula, New Guinea: a study of Quaternary tectonic movements and sea-level changes, Geol. Soc. Amer. Bull., 85, 553-570, 1974.

Chappell, J. and H.A. Polach, Holocene sea-level change and coral-reef growth at Huon Peninsula, Papua New Guinea, Geol. Soc. Amer. Bull. 87, 235-240, 1976.

Chappell, J. and H.H. Veeh, Late Quaternary tectonic movements and sea-level changes at Timor and Atauro Island, Geol. Soc. Amer. Bull., 89, 356-368, 1978.

Clark, L.A., W.E. Farrell, and W.E. Peltier, Global changs in postglacial sea level: a numerical calculation, Quat. Res., 9, 265-287, 1978.

Commission of Quaternary Shorelines, INQUA, World shoreline map, Pacific-Indian sector, 1979.

Cooper, A.K., M.S. Marlow, and D.W. Scholl, Mesozoic magnetic lineations in the Bering Sea marginal basin, Jour. Geophys. Res., 81, 1916-1934, 1976.

Curray, J.R., F. Shepard, and H.H. Veeh, Late Quaternary sea-level studies in Micronesia: CARMARSEL expedition, Geol. Soc. Amer. Bull., 81, 1865-1880, 1970.

Dewey, J.F. and J.M. Bird, Mountain belts and the new global tectonics, Jour. Geophys. Res., 75, 2625-2647, 1970.

Easton, W.H. and E.A. Olson, Radiocarbon profile of Hanauma Reef, Oahu, Hawaii, Geol. Soc. Amer. Bull., 87, 711-719, 1976.

Easton, W.H. and T.L. Ku, Holocenen sea-level changes in Palau, West Caroline Islands, Quat. Res., 14, 199-209, 1980.

Hashimoto, W., Problems on the Tainan formation and related formations, brought about by ^{14}C dating, Acta Geol. Taiwan., 15, 51-62, 1972.

Jouannic, C., F.W. Taylor, A. Bloom, and M. Bernat, Late Quaternary uplift history from emerged reef terraces on Santo and Malekula Islands, central New Hebrides island arc, UN ESCAP, CCOP SOPAC Tech. Bull., 3, 91-108, 1980.

Kaizuka, S., T. Matsuda, M. Nogami, and N. Yonekura, Quaternary tectonic and recent seismic crustal movements in the Arauco Peninsula and its environs, central Chile, Geogr. Rep. Tokyo Metropol. Univ., 8, 1-49, 1973.

Kanamori, H., Quantification of earthquakes, Nature, 271, 411-414, 1978.

Kanaya, A., Holocene marine terraces and crustal movement of the Muroto Peninsula, Japan, Geogr. Rev. Japan, 51, 451-463, 1978.

Karig, D.E., Plate convergence between the Philippines and the Ryukyu Islands, Mar. Geol., 14, 153-168, 1973.

Kelleher, J. and W. McCann, Buoyant zones, great earthquakes and unstable boundaries of subduction, Jour. Geophys. Res., 81, 4885-4896, 1976.

Konishi, K., A. Omura, and T. Kimura, 234U-230Th dating of some late Quaternary coralline limestones from southern Taiwan (Formosa), Geol. Palaeont. Southeast Asia, 5, 211-224, 1968.

Konishi, K., A. Omura, and O. Nakamichi, Radiometrical coral ages and sea level records from the later Quaternary reef complexes of the Ryukyu Islands, Proc. Second Intern. Coral Reef Sympo., 2, 596-613, Brisbane, 1974.

Ku, T.L., M.A. Kimmel, W.H. Easton, T.J. O'Neil, Eustatic sea level 120,000 years ago on Oahu, Hawaii, Science, 183, 959-962, 1974.

Labeyrie, J., C. Lalou, and C. Delibrias, Etudes des transgressions marines sur l'atoll de Mururoa par la datation des differents niveaux de corail, Cahiers du Pacifiques, 13, 59-68, 1969.

Lalou, G., J. Labeyrie, and G. Delibrias, Datation des calcaires corralliens de l'atoll de Mururoa (Archipel des Tuamotu) de l'epoque actuelle jesqu'a 500,000 ans, C. R. Aca. Sci., Paris, 263, 1946-1949, 1966.

Lin, C.C., Holocene Geology of Taiwan, Acta Geol. Taiwan, 13, 83-126, 1969.

Machida, H., Tephrochronology of coastal terraces and their tectonic deformation i South Kanto Jour. Geogr. Tokyo, 82, 53-76, 1973.

Machida, H., Paleistocene sea level of South Kanto, Japan, analysed by tephrochronology, Quaternary Studies (Suggate, R.P. and M.M. Cresswell, eds.) 215-222, Roy. Soc. NZ. Wellington, 1975.

Matsuda, T., Estimation of future destructive earthquakesw from active faults on land in Japan, Jour. Phys. Earth, 25, Suppl., S. 251-S 260, 1977.

Matsuda, T., Y. Ota, M. Ando, and N. Yonekura, Fault mechanism and recurrence time of major earthquakes in southern Kanto district, Japan, as deduced from coastal terrace data, Geol. Soc. Amer. Bull., 89, 1610-1618, 1978.

Mesollella, K.J., R.K. Matthews, W. Broecker, and D.L. Thurber, The astronomical theory of climatic change: Barbados data, Jour. Geol., 77, 250-274, 1969.

Molnar, P. and T. Atwater, Inter-arc spreading and cordilleran tetonics as alternates related to the age of subducted oceanic lithosphere, Earth Planet. Sci. Lett., 41, 330-340, 1978.

Nakata, T., T. Takahashi, and M. Koba, Holocene-emerged coral reefs and sealevel changes in the Ryukyu Islands, Geogr. Rev. Japan, 51, 87-108, 1978.

Ota, Y., Late Quaternary vertical movements in Japan estimated from deformed shoreliens, Quaternary Studies (Suggate, R.P. and M.M. Cresswell, eds.), 231-239, Roy. Soc. NZ, Wellington, 1975.

Ota, Y. and Y. Naruse, Marine terraces in Japan, Kagaku, 47, 281-292, 1977.

Ota, Y., Quaternary studies in New Zealand with special reference to the study of sea-level records, Daiyonki-Kenkyu (The Quat. Res.), 15, 141-155, 1977.

Ota, Y., H. Machida, N. Hori, K. Konishi, and A. Omura, Holocene raised coral reefs of Kikai-jima (Ryukyu Islands) -an approach to Holocene sea level study-, Geogr. Rev. Japan, 51, 109-130, 1978.

Ota, Y. and T. Yoshikawa, Regional characteristics and their geodynamic implications of late Quaternary tectonic movement deduced from deformed former shorelines in Japan, Jour. Phys. Earth, 26, Suppl., S 379-S 389, 1978.

Ota, Y. and N. Hori, Late Quaternary tectonic movement of the Ryukyu Islands, Japan, Daiyonki-Kenkyu (The Quat. Res.), 18, 221-240, 1980.

Peng, T.H., Y.H. Li, and F.T. Wu, Tectonic uplift rates of the Taiwan Island since the early Holocene, Mem. Geol. Soc. China, 2, 57-69, 1977.

Pirazzoli, P.A., High stands of Holocene sea levels in the Northwest Pacific, Quat. Res., 10, 1-29, 1978.

Plafker, G., Alaskan earthquake of 1964 and Chilean earthquake of 1960: implications for arc tectonics, Jour. Geophys. Res., 77, 901-925, 1972.

Plafker, G. and J.C. Savage, Mechanism of the Chilean earthquakes of May 21 and 22, 1960, Geol. Soc. Amer. Bull., 81, 1001-1031, 1970.

Ruff, L. and H. Kanamori, Seismcity and the subduction process, Phys. Earth Planet. Interiors, 23, 240-252, 1980.

Schofield, J.C., Late Holocene sea level, Gilvert and Ellice Islands, West central Pacific ocean, N.Z. Jour. Geol. Geophys., 20, 503-529, 1977.

Schackleton, N.J. and N.D. Opdyke, Oxgen isotope and palaeomagnetic stratigraphy of Equatorial Pacific cores V-28-238: oxgen isotope temperatures and ice volume on 10^5 years and 10^6 years scale, Quat. Res., 3, 39-55, 1973.

Szabo, B.J. and L.M. Gard, Age of the South Bight II marine transgression at Amchitka Island, Aleutians, Geology, 3, 457-459, 1975.

Taylor, F.W. and A.L. Bloom, Coral reefs on tecttonic blocks, Tonga island arc, Proc. Third Intern. Coral Reef Sympo., 275-281, Miami, 1977.

Taylor, F.W., B.L. Isacks, C. Jouannic, A.L. Bloom and J. Dubois, Coseismic and Quaternary vertical tectonic movements, Santo and Malekula Islands, New Hebrides island arc, Jour. Geophys. Res., 85, 5367-5381, 1980.

Thurber, D.L., W.S. Broecker, R.L. Blanchard, and H.A. Potratz, Uranium-series ages of Pacific atoll coral, Science, 149, 55-58, 1965.

Tracey, J.I. Jr., S.O. Schlanger, J.T. Stark, D.B. Doan, and H.G. May, General geology of Guam, U.S. Geol. Suvey Prof. Paper, 403-A, 104 p., 1964.

Uyeda, S. and H. Kanamori, Back-arc opening and the mode of subduciton, Jour. Geophys. Res., 84, 1049-1061, 1979.

Veeh, H.H., Th^{230}/U^{238} and U^{234}/U^{238} ages of Pleistocene high sea level stand, Jour. Geophys. Res., 71, 3379-3386, 1966.

Veeh, H.H. and J. Chappell, Astronomical theory of climatic change: support from New Guinea, Science, 168, 862-865, 1970.

Walcott, R.I., Past sea levels, eustasy and deformation of the Earth. Quat. res., 2, 1-14, 1972.

Wellman, H.W., Tilted marine beach ridges at Cape Turakirae, NZ, Jour. Geosci. Osaka City Univ., 10, 123-129, 1967.

Wu, F.T., Recent tectonic of Taiwan, Jour. Phys. Earth, 26, Suppl. S 265-S 299, 1978.

Yonekura, N., Quaternary tectonic movements in the outer arc of Southwest Japan with special reference to seismic crustal deformations, Bull. Dept. Geogr. Univ. Tokyo, 7, 19-71, 1975.

Yonekura, N., Late Quaternary sea level changes and crustal movements in the Pacific region, Gekkan Chikyu (Monthly the Earth), 11, 822-829, 1979.

Yonekura, N. and K. Shimazaki, Uplifted marine terraces and seismic crustal deformation in arc-trenches systems: a role of imbricated thrust faulting EOS 61, 1111, 1980.

Yoshikawa, T., S. Kaizuka, and Y. Ota, Crustal movement in the late Quaternary revealed with coastal terraces on the southeast coast of Shikoku, southwestern Japan, Jour. Geodet. Soc. Japan, 10, 116-122, 1964.

Yoshikawa, T., Y. Ota, N. Yonekura, A. Okada, and N. Iso, Marine terraces and their tectonic deforamtion on the northeast coast of the North Island, New Zealand, Geogr. Rev. Japan, 53, 238-262, 1980.

BASIN FORMATION; THE MASS ANOMALY AT THE WEST PACIFIC MARGIN

R.C.Bostrom[1], K.K.Saar[2], D.A.Terry[3]

[1]University of Washington AK-50, Seattle, WA 98105

[2]Dept. of Geology, Texas A.& M. Univ., College Station, TX 77843

[3]University of Washington AK-50, Seattle, WA 98105

Abstract. The formation of extensional basins at the west Pacific margin takes place in a zone of lithosphere convergence. An additional factor requiring explanation is that contrary to what is expected in a zone of subduction and downflow under convection, the west Pacific is the site of a mass excess marked by a major geoidal high. We have examined the west Pacific region in the light of newly published solutions of the Geos 3 data including low and high order components expressed in the altimetric geoid. We have avoided an assumption that the low-order principal high is the expression of core-mantle topography. The phase relation is that the high-order components, associated with tectonic arcs and foundering lithosphere slab, are spatially related to the crest and west flank of the low order high which characterizes the whole of the western Pacific. In a model of the flow in the sub-Pacific mantle, trans-Pacific westward drift slackens upon encountering the lithospheric upper mantle of Australasia and east Asia. The west Pacific low-order high is the expression of the lag in assimilation in the lower mantle of material added to this region in excess of equilibrium. Within the region excess dense, cool lithosphere is prone to founder, so that subduction zones are clustered in this region. Tectonic arcs are convex towards the ocean and intersect at sharp angles because the back-arc region is the site of material cumulatively added to the asthenosphere, resulting in outward spreading and "overflow tectonics" of a sort resembling that postulated by Ida. The most plausible reason for the accumulation of material in the west Pacific is that passage of the tidal bulge favors the development of the west limb of the East Pacific Rise, hence surface-west flow in the upper mantle.

Introduction

Plate tectonics holds that the west Pacific margin is the site of convergence between the Pacific plate, the Eurasian plate and the Indo-Australian plate. Yet much evidence of the sort adduced by Karig (1971 a,b) suggests that the intervening marginal basins are the site of sea-floor extension. Newly available gravimetric data, in particular the Geos-3 derived altimetric geoid, suggest that the principal mass anomalies of the west Pacific margin are not associated with the downgoing lithosphere slab but with the embedding mantle including the back-arc basin.

In what follows, the gravimetric data have been reviewed, in particular the advantage of using the geoid in preference to gravity data in the examination of such large (i.e. long wave-length) tectonic features as the belt of subduction zones at the west Pacific margin. A flow model consistent with the observed restraints suggests that trans-Pacific drift of the Pacific plate and upper mantle material slackens upon its encounter with the less mobile, lithospheric upper mantle beneath Asia and Australasia. Matter accumulation ensues, because downward assimilation of excess material is impeded by phase transition and viscosity increase. Arcuate features such as the island arc and tectonic trench form around the periphery of the back-arc spreading region.

Mass Anomaly at the West Pacific Margin

In this section we have examined the spatial relation of the large-scale tectonic features of the western Pacific to the mass anomalies described by the gravity field.

Surface Gravity Data

The island arcs of the west Pacific and Indonesia have been the site of gravity observations at sea commencing with those of Vening Meinesz in 1926 (Vening Meinesz, Umbgrove and Kuenen, 1934). The observations of Meinesz and successors (Worzel, 1965) were absolute measurements using pendulums. They first revealed the existence of the intense negative anomalies bordering island arcs on the ocean side. The development of the floating gravimeter(Graf and Schulze, 1961 ; La Coste and Harrison, 1961) has made available an immense number of observations. The values obtained are relative and must be tied to port bases. A comprehensive free-air map of the western Pacific has been prepared by Watts, Bodine and Bowin (1978), having a contour interval of 25 mgals.

Orbital Data

In the last decade, increasingly detailed maps of the geoid have been prepared using satellite orbital data combined with surface gravity, see for example Gaposchkin, 1974, 1979. An important new body of data has recently become available in the shape of the GEM series of potential solutions of values obtained by Geos 3, a satellite equipped with a radar sea-surface altimeter (Stanley, 1979).

Utilization of Gravity Data

The tectonic features of the western Pacific such as the Philippine Basin, formed by back-arc spreading (Scott and Kroenke, 1980), individually have dimensions of thousands of kilometers. Additionally, they are linked in a tectonic zone or belt 11,000 kms in length, embracing the whole of the west Pacific margin. To interpret a feature of such dimensions in terms of gravity data, it is necessary to admit spherical harmonic components down to order three or preferably two. In the latter case it is necessary to employ a geoid referred to the hydrostatic rather than best-fitting geoid. As noted by Chapman (1979) and Bowin (1980), for a given wave-length the gravimetric representation of the field is attenuated by a factor $|k|$, where k is the wavenumber, relative to the geoidal representation. Additionally, with respect to most observations, the long wave-length components of the gravity field have not been corrected for the indirect effect (Chapman and Bodine, 1979). The latter results from the fact that sea-surface gravity observations are made at the elevation of the geoid, not that of the reference ellipsoid. In a region of severe geoidal undulations such as the western Pacific, uncorrected gravity observations thus omit the gravity information fundamentally sought. For these reasons, it is preferable or essential to base interpretation of the large-scale west Pacific features upon the form of the geoid, retaining the low-order harmonic coefficients.

At the other extreme, until recently the geoidal image of the Earth has not contained high-order coefficients, because these are rendered unobtainable by orbital height restrictions; hence the geoid has contained insufficient detail to describe geological features. The restrictions described have been in large part overcome by the device of mounting upon Geos 3 a radar altimeter, making it possible to observe in detail the height of the sea surface, a close approximation to the geoid (Stanley, 1979). Examination of the results of the Geos 3 project (Chapman and Talwani, 1979) indicated that in comparison with the $1^o \times 1^o$ gravimetric geoid, the Geos 3 altimetric geoid contained offsets and long wave-length discrepancies. These authors have shown that below a certain wave-length, about 600 kms, the gravimeter-determined field possesses a better signal to noise ratio. Rapp (1979 a) has used the coefficients of the GEM 9 potential solution to remove from the altimetric geoid the effect of orbital error and bias.

Having in mind the restrictions and advantages noted, in our analysis we have utilized principally the Rapp geoid (Rapp, 1979 b). Its accuracy is attested by its close correspondence with topographic detail, itself constituting a hydrostatic anomaly. We have additionally utilized surface gravity information, principally that represented in the free-air map of Watts, Bodine and Bowin (1978), to examine geologic detail, such as that at the east margin of the Caroline Basin.

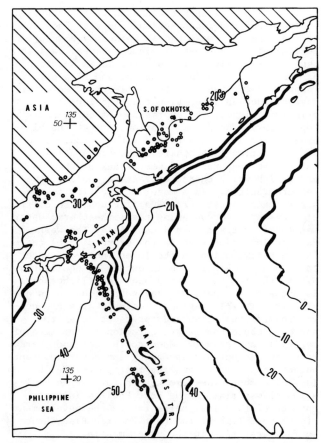

Fig. 1a. North flank of west Pacific geoidal high. Altimetric geoid prepared by Rapp (1979b) using Geos 3 data. Contour interval, 10 ms. Referred to best-fitting ellipsoid, flattening 1/298.256. Small circles: deep epicenters (depth more than 300 kms).

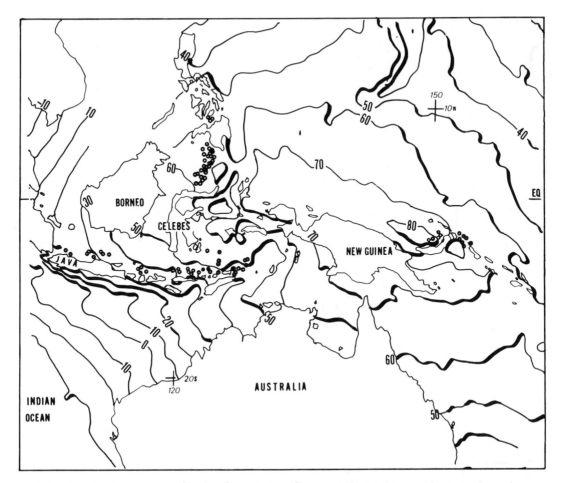

Fig. 1b. Center & west flank of west Pacific geoidal high; geoid as in Fig. 1a.

Location of the Mass Anomalies

Referred to the hydrostatic ellipsoid (Gaposchkin, 1974), the entire west Equatorial Pacific is the site of a positive geoidal anomaly. Reference of the geoid to the best-fitting ellipsoid, thus artificially masking the excess flattening of the Earth (Goldreich and Toomre, 1969), removes the second-degree harmonic component and confirms (Gaposchkin, 1979) that the western half of the Pacific Ocean is the site of a major geoidal high. The eastern Pacific is the site of a mass deficiency. The culmination of the high is at the western Equatorial margin of the Pacific, over northern New Guinea. Low and high order components of the geoid are so related that the Pacific anomaly is constituted of a gradual build-up proceeding westward from the central Equatorial Pacific to its culmination, followed by sharp westward decline into the Indian Ocean low. Referred to the hydrostatic or natural geoid, the Indian Ocean low is of less amplitude and area than the west Pacific high. In the vertical dimension, the low order harmonic components may be related to shallow mass anomalies or to anomalies located as deep as the Earth's core (Bowin, 1980). Crough and Jurdy (1980) in connection with another investigation have computed the portion of the geoidal anomaly due to foundered lithosphere slab. When the effect of this is removed (their figure 3), the long wave-length component is found still to remain, having an amplitude in excess of 80 ms. The expression of slabs is local. In this paper, we do not make the assumption that the mass anomaly corresponding to the long wave-length anomaly is located at the core-mantle boundary, but leave open the option that the source of the anomaly is in the mantle, upper or lower.

Introducing the high-order geoidal components visible on the altimetric geoid (Rapp, 1979b) the northerly flank of the west Pacific high is observed to extend in a spur whose axis, (fig.1a) defined by the 30-meter contour, passes from the eastern part of the Philippine Sea, crosses the surface trace of the foundering slab as indicated by epicenters, and passes into the eastern part of the Sea of Japan.

The southeasterly flank of the geoid high (Fig. 1 b,c) overlies New Guinea, the Melanesian Borderland and the Fiji Plateau. The apex

BASIN FORMATION   53

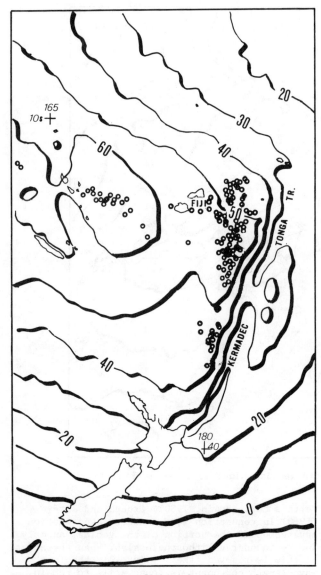

Fig. 1c. South-east flank of west Pacific geoidal high; altimetric geoid as in Fig. 1a.

of the high is not invariably related to the geographic location of the foundering slab. Removal of the geoidal contribution of the foundered slab (Crough and Jurdy, 1980) leaves intact the principal part of the background anomaly, here 60ms in amplitude. In the Papua New Guinea region, the apex of the high is in the back-arc region lying north of New Guinea, whereas subduction appears to take place southwards.

The western flank of the primary geoidal high (fig. 1b) is located over east-central Indonesia, namely the Philippines, Celebes and Borneo. As in the case of the northerly spur, the tectonic trenches of the region are associated with a furrow in the altimetric geoid. Beyond the furrow marking the tectonic trench bordering Sumatra and Java, the geoidal high declines westward into the Indian Ocean low.

Interpretation of the Mass Anomalies

Based on his prolonged researches, Vening Meinesz was inclined to interpret the conspicuous elongated negative anomalies bordering island arcs as the expression of mantle convection (Heiskanen and Vening Meinesz, 1958). Later researchers identified these features, in the same vein, as tectogenes, a component of present-day geosynclinal activity (Holmes, 1944). The advent of plate tectonics theory has caused it to be assumed that the elongated Meinesz lows are an expression of a down-going slab or phenomena secondary to it. Seismologic data in the form of first-motion analyses, absorption, and velocity values (Isacks, Oliver and Sykes, 1968) provide strongly suggestive evidence that in many regions a slab lies in the mantle. Volcanic phenomena (Hatherton and Dickinson, 1969) of the island arc accord with the expectation that a foundering slab must experience differentiation accompanied by the production of andesitic magmas.

Nevertheless, interpretations of the gravity field across tectonic trenches and island arcs have encountered difficulty in reconciling observed data with the expression of models based on our conception of the geometry of a foundering slab. Employing gravity profiles Griggs (1972) was able to reconcile seismologic data with restricted cases of the gravity field, composed of contributions by a down-going slab, a trench, and regional compensation effects. Across the back-arc basin region, however, his model drops to zero and negative values, directly opposite in sign to that now revealed by the orbital geoid. Watts, Talwani and Cochran (1976) have isolated the gravity highs oceanward of the deep-sea trenches of the north-west Pacific from features of longer wave-length upon which they are superimposed. The local highs are attributed to the effect of elastic crustal flexure as the crust enters the region of subduction (Watts and Talwani, 1974). This explanation, however, does not allow for the positive anomaly invariably found landward of the tectonic trench, and which seems to form an integral part of the gravity field. The altimetric geoid shows that the highs oceanward of the tectonic arc and within it are part of a single spur (fig. 1a) extending north from the major west Pacific high; the axis lies on the basinward, inner flank; the oceanward high in reality is composed of the east flank of the spur, defined by a sulcus or furrow overlying the Marianas sea floor trench.

Worzel (1976) has modelled the gravity expression of subduction zones. He has included in his computation the following factors: sea floor topography, the density of sediments, the oceanic crust, the upper and lower layer of the lithosphere and the asthenosphere; and the inversion of minerals during subduction. The gravity expression of Worzel's models closely fits the observed anomaly in the region fronting the island arc. In this region the structure and densities in the model are closely controlled by seismic refraction data. However on the arc and back-arc flank, the anomaly based on the form of a downgoing slab widely departs from what is observed. There, the model anomaly consists of a low of more than one hundred milligals. Worzel points out that to accord with a commonly accepted geometric model of subduction, a high density mass must also be present, overcoming the low so greatly as to produce the high invariably observed in such regions.

Lerch et al. (1974), fig. 2, display the geoid devoid of the high order components contained in Rapp's solution. The axis of the high is located over the back-arc regions, rather than over the island chains themselves. Thus for instance although the foundering slab as depicted by epicenters dips steeply, lying below the extreme eastern part of the Philippine Sea, the geoidal axis marking the center of the mass anomaly passes through Parece Vela, the central part of the Philippine Sea. The geoid and free-air gravity anomaly corrected for the indirect effect display the location of the principal mass anomaly supported in a region, regardless of whether it is in the form of a subcrust density anomaly or one represented by a difference in water depth. The investigations of Bracey and Ogden (1972), Karig (1971 b) and others show that the contemporary addition of material is taking place at the zone of lithosphere foundering in the extreme east of the Philippine Sea; the geoidal anomaly suggests that this process has in the past so operated as to accumulate excess material in what is now the mantle below the central Philippine Sea; the site of foundering, hence the eastern boundary of the Sea is migrating eastward, extending the area of the back-arc basin.

In the Indonesian region, the relation of present subduction to the principal mass anomaly is analogous: the Sumatra-Java arcs and flanking trenches are peripheral to a large mass anomaly, residual even when the expression of the downgoing slab (Crough and Jurdy, 1980) is removed, centered beneath the Celebes Sea. Similarly, the southeast flank of the west Pacific geoidal high, encompassing Melanesia, describes a mass anomaly flanked by trenches and tectonic arcs. The down-going slab only accounts for a fraction of the main mass anomaly. In terms of the anomalous mass, in these regions the back-arc region is the site of the principal accumulation. Tentatively, we interpret this to indicate that although at present the addition of mass takes place principally in the form of downgoing slab plus more dense material peripheral to these regions, this process takes place progressively further from the center of the already-accumulated excess mass as time goes on. If this is correct the arcuate form of many island chains and peripheral trenches is the result of outward spreading of the back-arc region, the site of a net excess of material accumulating since subduction began in these regions. The back-arc regions, already known to be the site of sea-floor spreading (Karig, 1971, a,b; Uyeda and Kanamori, 1979) expand in area, pushing outward the peripheral features in a form of the "overflow tectonics" suggested by Ida (1978).

The significance of the west Pacific geoidal high, pointed out by Kaula (1970, 1972) and the greatest challenge to the interpreter, is that it implies that excess mass is being added to this whole region by an unidentified agency. A deficiency of the type caused by foundering under convection, drawing replacement material into the region, would be the site of a low. What is the agency, acting contrary to the relationship in convection, which sweeps material into the west Pacific region to the extent that a gross excess exists, lagging its downward assimilation?

Furthermore, Forsyth and Uyeda (1975) have shown that the foundering slab is the site of the most intense stress concentration in the mantle, to the extent that it may be a major driving agency in plate tectonics. Why do subduction zones cluster on the west flank and apical region of the west Pacific high? In this connection, it may be noteworthy that of the Earth's three primary geoidal highs, referred to the complete or hydrostatic geoid, two, those centered on the western Equatorial Pacific and South America, are the site of intense subduction. The third, centered on west Africa, is the site of Paleozoic subduction.

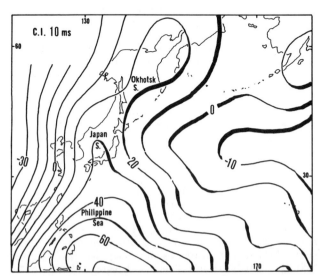

Fig. 2 North flank of geoidal high, GEM 6 geoid, flattening 1/298.2 (Lerch, Wagner, Richardson and Brownd, 1974).

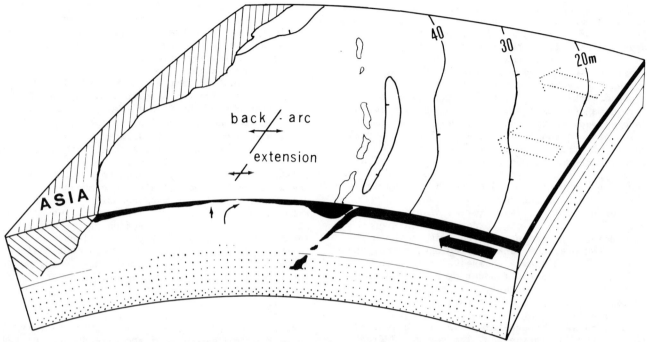

Fig. 3. Flow processes and asthenosphere accumulation at the west Pacific-east Asia margin, at early stage. The essential condition is that downward assimilation of material in the asthenosphere lags its influx from the east, creating a mass excess and geoidal high; a negative anomaly would be present if assimilation led influx, causing material to be drawn in from the surrounds. Foundering of slab having acquired negative buoyancy appears to be initiated by its encounter with region of accumulation.

A Flow Model

A flow model has been attempted (fig. 3) which complies with the restraints imposed by gravimetric data, and is constrained also by tectonic observations. The model invoked calls for the assumption that the trans-Pacific drift of the thickening Pacific plate apparent from plate tectonics is accompanied by drift also of material of the asthenosphere.

Seismic data (Sipkin and Jordan, 1975; Jordan, 1975) have shown that the sub-continental lithosphere is 200 kms or more in thickness, and the underlying asthenosphere correspondingly thin. In contrast, the Pacific plate is 75 kms in thickness and overlies thick asthenosphere. In these circumstances trans-Pacific drift must be impeded or halted as it encounters continental material orders of magnitude more viscous, comprising Australasia and east Asia. The observed long wave-length geoidal high is the expression of the lag in the assimilation of incoming material in the mantle.

Provided only that trans-Pacific drift of material takes place, the steadily westward increasing values of the geoid entail corresponding increase of pressure at depth. The pressure acts upon material already at the point of transition. In addition to the effect of phase densification thus caused, the cooling lithosphere contributes its share to the long wave-length anomaly.

In these terms, the major west Pacific geoidal high is viewed as constituted
1) in the open-ocean region, of phase-densified material, much of it in the form of thickened lithosphere, in excess of hydrostatic equilibrium, and
2) in the apical and west-flanking region of the major high, of foundered slab and phase-densified material in the upper part of the lower mantle, implanted there by slab foundering.

Belt of Foundering

In the model described, cooled dense lithosphere already having negative buoyancy eventually must founder. Foundering is initiated when the west-drifting lithosphere encounters the region where drift is impeded or prevented. Contemporaneously, this is the belt of back-arc spreading adjacent to Asia and Australasia. Unlike material descending in the cool slab, asthenosphere material accumulating in this region is already at ambient temperature. Heat of transformation is not absorbed in bringing material up to ambient temperature, but instead raises the temperature above ambient.

Opinion is divided as to the role of entrained water in the belt of foundering. Most

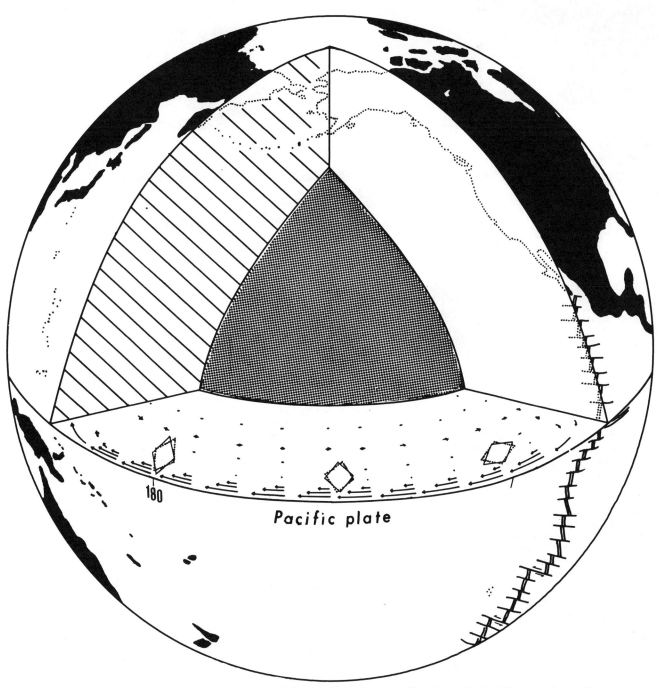

Fig. 4a. A causative agency consistent with the flow pattern in the sub-Pacific mantle. The rhombic specimen volume element describes the deformation under twice-daily passage of the 50 cm tidal bulge. The element deforms from prolate to oblate (solid line), incurring self-cancelling strain. The phase lag in the tides however tends to rotate the element (dotted line) always in the same direction (Bostrom, 1981a). The latter action induces rotational flow dimensionally identical to that in convection, with which it can interact. The form of the tidal flow is shown by streamlines. Although flow is concentrated in asthenosphere, as much material flows slowly east in deep mantle.

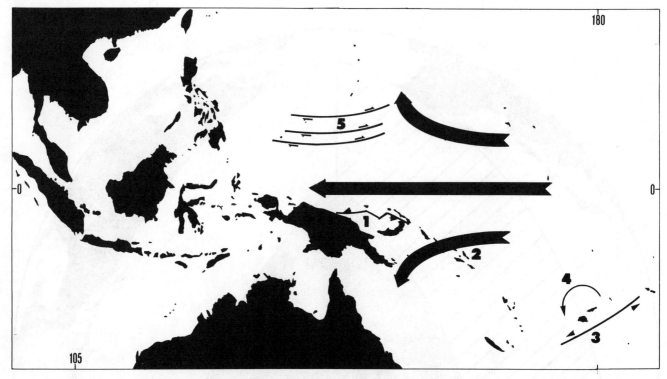

Fig. 4b. Geotectonic features of the western Equatorial Pacific. 1,2,3,4 are elements of the Melanesia zone of sinistral shear. 1, Bismarck Sea tectonic boundary (Johnson, Mutter and Argulus, 1979); 2, Solomon Islands site of sinistral shear between Indo-Australian and Pacific lithosphere plates (Coleman and Packman, 1976); 3, rotation of Fiji Platform (Coleman and Packman, 1976; Harper, 1975); 4, sinistral strike-slip zone of Hess and Maxwell, 1953; 5, dextral shear in Palau, Yap, Mariana arcs (Wu, 1978; Bracey and Ogden, 1972). The large arrows indicate the form of the flow contributed by the system shown in fig. 4a, namely surface west at the Equator, decreasing rapidly with latitude increase, and acquiring sinistral component in southern hemisphere, dextral component in northern.

petrologic constructs however suppose that heat is released rather than absorbed during phase compression. Let us suppose using for example Ringwood's model (Ringwood, 1973) that the P/T gradient of the transitions is 30 bars/^{o}C, positive, and that the transition entails 9% phase densification. The Clapeyron relation is

$$L = dP/dT \times T(v_1 - v_2)$$

in which L is the heat of transition, v the specific volume. Using the values named, the transition per cm^2 per year to account for an increased heat-flow of 1.5 hfu as observed over back-arc basins (Sclater, 1972) is approximately 15 gms. To this must be added (Griggs, 1972) as much or more energy release as material descends under gravity. The excess in heat flow is of the same magnitude as that observed to be associated with mantle convection at a principal site of mantle upwelling, the sea floor ridge system. Thus significantly, the region which gravity suggests is the site of most intense phase densification is the locus of diapirism, sea floor spreading and enhanced heat flow, the phenomena shown by Karig (1974) to characterize the near back-arc basins.

Causative Agency

The flow model described is consistent with observation, but hard to explain in terms of its primary requirement, a driving agency promoting flow into a region already having a mass excess. The purpose of this paper is to identify the flow pattern rather than to account for it, but a suggested pattern must be physically plausible. As noted by McKenzie, Roberts and Weiss (1974), the only agencies having an energy supply sufficient to drive flow in the viscous mantle are buoyancy forces (causing thermal convection) and tidal dissipation. In respect to the first, it has proven hard to model convection having an unbroken flow cell extending one third way around the globe, across the Pacific Ocean.

A flow model driven by buoyancy and passage of the 50 cm tidal wave or bulge has been described elsewhere (Bostrom, 1978, 1981a), fig. 4. The agency accords with peculiarities of the tectonics of the west Pacific margin. These include the region of extraordinary sinistral shear, Carey's Melanesian Megashear (Johnson, 1979; Carey, 1970; Coleman and Packham, 1976). This extends for 6000 kms, forming the northern boundary of the active region between western New Guinea and the Tonga-Kermadec trench. Sinistral rotation is pervasive within the region named, (fig. 4b), typified by the counterclockwise wrench zone originally identified by Hess and Maxwell (1953), extending from the south end of the New Hebrides to the north end of the Tongas. The celebrated but enigmatic "Fiji spiral nebula" (Harper, 1975; Coleman and Packham, 1976) centered upon the Fiji Platform, an area the size of India, is a product of counterclockwise rotation. The Fiji Platform developed as an arc until bodily left-hand rotation became dominant some time prior to 5 million years ago.

At the Equator, the geoidal high, deep seismicity and volcanicity of the Pacific margin form the core of the Indonesian island arc system (fig. 5). Analysis has shown that until the end of the Mesozoic this region formed an unbroken projection, Sundaland, of the Asian continent (Dickinson, 1973; Kropotkin and Krakvarstova, 1965). Hilde, Uyeda and Kroenke (1977) have shown that a re-organization of the tectonic pattern took place in this region during the early Tertiary, with change from a transform regime to one of converging plate boundary in the western Pacific. Rona and Richardson (1978) have shown that a global re-organization of plate motion took place during Eocene time. The re-organization consisted of re-orientation of plate motions with large N-S components into large E-W components, with the initiation of increased E-W spreading. At the west Pacific margin, re-organization involved the break-up of Sundaland and the development in this region of the flow system seen today.

Powell and Johnson (1980) point out that to allow the northward passage of India, Sundaland must during the early Tertiary have lain at least 550 kms and probably much further east of its present position. The region is separated from Australia by an analog or extension of the sinistral shear zone bordering all of Melanesia on the north, earlier described. Powell and Johnson furthermore point out that seismic first-motion analyses show that the northern, Asia boundary of Sundaland is the site of dextral slip, whilst the southern, Australia boundary is the site of sinistral slip. These factors together with the unbroken extension into Sundaland of the west Pacific geoidal high suggest that westerly, trans-Pacific Equatorial flow is invading the entire region, carrying it towards the Indian Ocean low, figure 5.

In the portion of the west Pacific margin north of the Equator during Tertiary times, whilst subduction zones having a N-S alignment extended (Hilde, Uyeda and Kroenke, 1977), the right-lateral component of rotational motion likewise increased. Mirroring the sinistral shear prevalent south of the Equator, and according with the system of zonal mantle flow described by Crook (1980), Bracey and Ogden (1972) have shown that the Yap, Palau, and Marianas arcs (fig. 4b) are displaced by dextral shear.

Conclusions

The data cited suggest the following conclusions as to the sub-Pacific mantle flow; its relation to the formation of marginal basins; and its origin.

Sub-Pacific Flow Pattern

Material of the Pacific lithosphere and asthenosphere is affected by trans-Pacific drift, towards the western margin. The flow slackens towards the western half of the ocean basin. This is because the flow impinges on the less mobile, lithospheric upper mantle present beneath Australasia and east Asia. Flow deceleration is associated, as Kaula (1972) has pointed out, with material accumulation, hence with the development of the west Pacific geoidal high. Directly, the latter results from the fact that downward assimilation of material lags its inflow from the east.

Back-arc basin formation

The excess of material at the west Pacific margin required by the geoidal high is the reason for the burgeoning development of back-arc basins (Uyeda and Kanamori, 1979), so conspicuously absent at the east margin. The Pacific geoidal high is composed of two principal contributions. The first is the long wavelength component, which we attribute to phase compression including lithosphere thickening.

The second contribution is that by foundered slab in zones of subduction. The latter cluster where the thick, dense, cool lithosphere is most prone to founder, in a belt adjacent to Australasia and east Asia. Thus subduction zones are those of concentrated addition of material to the lower mantle. The material overlying subduction zones accumulates, lagging its assimilation at depth, causing the back-arc basin to bulge outward and assume its rounded form. The peripheral arcuate features tend to intersect at sharp angles because they form around individual spreading centers. To judge by the intensively investigated Philippine basin, the site of contemporaneous addition of material is the basin periphery immediately behind the arc.

Driving Agency

The flow system described requires an agency capable of driving material of the upper mantle into a region, the western Pacific, already the site of a gross hydrostatic excess. The only

Fig. 5. Altimetric geoid projected on National Geographic Society Physical Globe (Grosvenor, 1971). SE Asia is at top left, Australia bottom center. The seam marks the Equator. The history and tectonics of this region (Hilde, Uyeda and Kroenke, 1977) and the form of the geoid suggest that the Indonesian region is being invaded by the intense Equatorial trans-Pacific mantle flow, causing its disruption and westward push towards the Indian Ocean, left. The westward crowding of material towards the Pacific margin is indicated by the form of the geoid to have been deflected upon encountering the Australasian continental massif. The Melanesian Borderland, extending for 6000 kms ESE from New Guinea through the Solomons to Fiji, is the site of the extraordinary left-lateral wrenching identified by Carey (1970), Coleman and Packham (1976) and others. To have produced this, Equatorial material flowed westward, towards present-day Indonesia.

agency having the requisite energy and mechanical resources (Bostrom, 1981a) appears to be passage of the M_2 tidal bulge. The latter is prone to cause development of the trans-Pacific west limb of the East Pacific Rise, associated with westward flow in the upper mantle, at the expense of the attenuated east limb. The latter is being overridden and eliminated by the extending west limb of the mid-Atlantic spreading center (Bostrom, 1981 b).

References

Bostrom, R.C., Motion of the Pacific plate and formation of marginal basins; asymmetric flow induction, in Geodynamics of the Western Pacific, Center for Academic Publications Japan, Tokyo, Japan, 103-122, 1979.

Bostrom, R.C., Lithosphere creep, J. Phys. Earth, 29, in the press, 1981, a.

Bostrom, R.C., Formation of the cordillera; flow processes in the sub-Pacific mantle, Pacific Geol., in the press, 1981, b.

Bowin, C.O., Why the Earth's greatest gravity anomaly is so negative, EOS Trans. AGU, 61, 209, 1980.

Bracey, D.R., and T.A. Ogden, Southern Mariana arc; geophysical observations and hypothesis of evolution, Geol.Soc.Amer.Bull., 83, 1509-1522, 1972.

Carey, S.W., Australia, New Guinea and Melanesia in the current revolution in concepts of the Earth, Search, 1, 178-189, 1970.

Chapman, M.E., Techniques for interpretation of geoid anomalies, J.Geophys.Res., 84, 3793-3801, 1979.

Chapman, M.E., and J.H. Bodine, Considerations of the indirect effect in marine gravity modelling, J.Geophys.Res., 84, 3889-3892, 1979.

Chapman, M.E., and M. Talwani, Comparison of gravimetric geoids with Geos 3 altimetric geoid, J. Geophys.Res., 84, 3803-3816, 1979.

Coleman, P.J., and G.H. Packham, The Melanesian Borderlands and Indian-Pacific Plates' Boundary, Earth-Sci. Revs., 12, 197-233, 1976.

Crook, K.A.W., The origin of west Pacific geosynclines by zonal spreading, Tectonophysics, 63, 235-259, 1980.

Crough, S.T., and D.M. Jurdy, Subducted lithosphere, hotspots and the geoid, Earth Plan. Sci.Letts, 48, 15-22, 1980.

Dickinson, W.R., Reconstruction of past arc-trench systems in the western Pacific, in The Western Pacific, Ed. P.J. Coleman; Crane Russack & Co. Inc., N.Y., 473-512, 1973.

Forsyth, D.W., and S. Uyeda, On the relative importance of the driving force of plate motion, Geophys. J. Roy. Astron. Soc., 43, 163-200, 1975.

Gaposchkin, E.M., Earth's gravity field to the 18th degree, J.Geophys.Res., 79, 5377-5411, 1974.

Gaposchkin, E.M., Global gravity field to degree & order 30 from Geos 3 satellite altimetry, Smithsonian Astrophysical Observatory,, Ppt. Ser. No. 1092, 1979.

Goldreich, P., and A. Toomre, Some remarks on polar wandering, J.Geophys.Res., 74, 2555-2567, 1969.

Graf, A., and R. Schulze, Improvements on the sea gravimeter Gss2, J. Geophys. Res., 66, 1813-1821, 1961.

Griggs, D.T., The sinking lithosphere and the focal mechanism of deep earthquakes, in The Nature of the Solid Earth, Ed. E.C. Robertson, 361-384, McGraw-Hill, N.Y., 1972.

Grosvenor, G.M., (Ed.), Physical Globe, 1/31,363,200, National Geographic Society, Washington, D.C., 1971.

Harper, J.F., Subduction-zone vortices, Bull. Austral. Soc. Explor. Geophys. 6, 79-80, 1975.

Hatherton, T., and W.R. Dickinson, The relation between andesitic vulcanism and seismicity in island arcs, J.Geophys.Res., 74, 5301-5310, 1969.

Heiskanen, W.A., and F.A. Vening Meinesz, The Earth and its Gravity Field, McGraw-Hill, NY, 1958.

Hess, H.H., and J.C. Maxwell, Major structural features of the south-west Pacific: a preliminary interpretation of H.O. 5484, bathymetric chart, New Guinea to New Zealand, Proc. 7th Pacific Sci. Congr., II, 14-17, 1953.

Hilde, T.W.C., S. Uyeda, and L. Kroenke, Evolution of the western Pacific and its margin, Tectonophysics, 38,, 145-165, 1977.

Holmes, A., Principles of Physical Geology, 532 p., Nelson and Sons Ltd., London, 1944.

Ida, Y., Oceanic crust in the dynamics of plate motion and back-arc spreading, J.Phys.Earth, 28 Suppl., S 55 - S 67, 1978.

Isacks, B.J., J. Oliver, and L.R. Sykes, Seismology and the new global tectonics, J.Geophys. Res., 73, 5855-5899, 1968.

Johnson, R.W., Geotectonics and vulcanism in Papua New Guinea; a review of the late Cenozoic, BMR J. Geol. Geophys. 4,, 181-207, 1979.

Johnson, R.W., J.C. Mutter and R.J. Argulus, Origin of the Willaumez-Manus Rise, Papua New Guinea, Earth Sci. Letts., 44, 247-260, 1979.

Jordan, T.H., The continental tectosphere, Revs. Geophys.Sp.Phys., 13, 1-12, 1975.

Karig, D.E., Origin and development of marginal basins in the western Pacific, J.Geophys. Res., 76, 2542-2561, 1971 a.

Karig, D.E., Structural history of the Mariana island arc system, Geol.Soc.Amer.Bull., 82, 323-344, 1971 b.

Karig, D.E., Evolution of arc systems in the western Pacific, Ann. Revs. Earth Plan. Sci., 2, 51-75, 1974.

Kaula, W.M., Earth's gravity field; relation to global tectonics, Science, no. 3849, 982-985, 1970.

Kaula, W.M., Global gravity and tectonics, in The Nature of the Solid Earth, (Ed. E.C. Robertson), 385-405, McGraw-Hill, N.Y., 1972.

Kropotkin, P.N., and K. Krakvarstova, Geol. structure of the circum-Pacific mobile belt, Trudi Geol. Sci., 134, 366, Akad. Nauk, USSR, 1965.

LaCoste, L.J.B., and J.C.Harrison, Some theoretical considerations in the measurement of gravity at sea, Geophys.J.Roy.Astron.Soc. 5, 89-103, 1961.

Lerch, F.J., C.A.Wagner, J.A.Richardson amd J.E. Brownd, Goddard Earth Models (5 & 6), Rep. X-921-74-145, Goddard Space Flight Center, Greenbelt, Maryland, 1974.

McKenzie, D.P., J.M.Roberts and N.O.Weiss, Convection in the Earth's mantle; towards a numerical simulation, J.Fluid Mechs., 62, 465-538, 1974.

Powell, C.McA., & B.D.Johnson, Constraints on the Cenozoic position of Sundaland, Tectonophysics, 63, , 91-109, 1980.

Rapp, R.H., Geos 3 data processing for the recovery of geoid undulations and gravity anomalies, J.Geophys.Res., 84, , 3784-3792, 1979 a.

Rapp, R.H., Global anomaly and undulation recovery using Geos 3 altimetric data, Reps.Dept. Geod. Sci., no. 285, Ohio State Univ. Res. Fdtn., Columbus, Ohio, 49 p., 1979 b.

Ringwood, A.E., Phase transformations and their bearing on the dynamics of the mantle, Fortschr. Mineral., 50, 113-139, 1973.

Rona, P.R., and E.S.Richardson, Early Cenozoic global plate reorganization, Earth Plan. Sci.Lrs., 40, 1-11, 1978.

Sclater, J.G., Heat flow and elevation of the marginal basins of the western Pacific, J.Geophys.Res., 77, 5705-5719, 1972.

Scott, R., and L.Kroenke, Evolution of back-arc spreading and arc volcanism in the Philippine Sea; interpretationof Leg 59 DSDP results, in: The Tectonic and Geologic Evolution of South-East Asian Seas and Islands, Monogr. 23, AGU, Wash., D.C., 283-291, 1980.

Sipkin, S.A., and T.H.Jordan, Lateral heterogeneity of the upper mantle determined from the travel times of ScSn J. Geophys. Res., 80, 1474-1484, 1975.

Stanley, H.R., The Geos 3 project, J. Geophys. Res., 84, 3779-3783, 1979.

Uyeda, S., and H. Kanamori, Back-arc opening and the mode of subduction, J.Geophys.Res., 84, 1049-1061, 1979.

Vening Meinesz, F.A., J.M.F.Umbgrove and Ph. A. Kuenen, Gravity Expeditions at Sea, 1923-1932, Vol. II, Pblctn. Neth.Geod.Comm., Delft, 1934.

Watts, A.B., and M. Talwani. Gravity anomalies seaward of deep-sea trenches and their tectonic implications, Geophys.J.Roy.Astron. Soc., 36, 57-90, 1974.

Watts, A.B., M.Talwani, and J.R.Cochran, Gravity field of the NW Pacific Ocean Basin and its Margin; Monogr. 19, AGU, Wash., D.C., 17-34, 1976.

Watts, A.B., J.H.Bodine and C.O.Bowin, Free-air gravity field, in: A Geophysical Atlas of the East and Southeast Asian Seas, Nat. Sci. Fdtn., Washington, D.C., 1978.

Worzel, J.L., Pendulum Gravity Measurements at Sea, 1936-1959, 422 p., J.Wiley & Sons, 1965.

Worzel, J.L., Gravity investigations of the subduction zone, in The Geophysics of the Pacific Ocean Basin and its Margin, Eds. G.H.Sutton, M.H.Manghnani, R.Moberley. Geophys. Monogr. 19, AGU, Washington, D.C., 1-15, 1976.

Wu, F.T., Benioff zones, absolute motion and interarc basin, J.Phys.Earth, 26 Suppl., S 39 - S 54, 1978.

DEPTH ANOMALIES OVER MESOZOIC CRUST IN THE WESTERN PACIFIC

Jacqueline Mammerickx

Scripps Institution of Oceanography, La Jolla, CA 92093 (U.S.A.)

Abstract. Erratic relations between crustal age and depth are so common in the part of the Pacific underlain by Mesozoic crust on which M-sequence magnetic anomalies are developed that there is little likelihood of finding significant areas of "normal" crust on which to base an empirical curve. Departures from the published empirical curves of several hundred meters are documented in three regions: the basin between Shatksky Rise and the Mid-Pacific Mountains, the Central Pacific Basin and the Nauru Basin. At least 4 causes for depth anomalies are identified: uparching around volcanic edifices, volcanic precursor uplift (either active or dead), uplift accompanying intraplate deep-sea volcanism and uparching on the outer swell of trenches. All these have deformed Pacific Mesozoic crust.

Introduction

An empirical relation between the depth of the sea floor - or more precisely the depth of the top of the oceanic crust beneath the sea floor - and the age of that crust is well established, at least for crust younger than about 80 m.y. and can be plausibly explained by thermal contraction of the lithosphere. For crust older than 80 m.y., the data do not fit along the Depth = Age$^{1/2}$ curve which describes younger crust, and Parsons and Sclater (1977) proposed that older crust subsides asymptotically toward some constant depth. One of the areas they chose to show the relationships among heat flow, crustal age and depth is the part of the Pacific between Shatsky Rise and the Mid-Pacific Mountains where the M-sequence of magnetic anomalies is well mapped. Because of the widespread occurrence of significant regional and local departures from the best-fit Age curve in areas of young crust, and because of the known complexities of post M-sequence history in the Western Pacific, such as seamount volcanism, mid-plate flood basalt flows and sill injections (Larson and Schlanger, 1981), it is important to reexamine the age-depth data base used by Parsons and Sclater.

Methods

Three areas in the Western Pacific where M-sequence anomalies have been mapped were chosen for study (Figs. 1 & 2). For each area, I constructed a revised bathymetric chart (Figs. 3a, 4a and 5a) with depths shown by contours in uncorrected meters, using a velocity of sound of 1500 m/second. Depths were not corrected for variable velocities of sound in sea water, because the newest tables of corrections were not available at the time the charts were drawn. The new tables (Carter, 1980) show maximum corrections of +65 m at a depth of 6200 m diminishing to 0 m at 4000 m, a discrepancy not considered significant in this paper.

Bathymetric data come from the data banks of the Scripps Institution of Oceanography and the U.S. National Geophysical and Solar Terrestrial Data Center. A few soundings from older Navy sounding charts were used to substantiate the existence of the Surveyor fracture zone in figure 3a. Track charts showing data control are shown in Figs. 3b, 4b, 5b. On each bathymetric chart, isochrons were drawn using the magnetic anomaly identifications in the literature. Numeric ages have been assigned to the geomagnetic time scale using van Hinte's Cretaceous and Jurassic time scales. (1976a, 1976b)

I. The Shatsky–Mid-Pacific Basin

Morphology

The Basin located between the Shatsky Rise to the northwest, the Emperor-Hawaiian Ridge to the east and the Mid-Pacific mountains to the south (figure 1.) shows northwest-trending magnetic lineations discovered by Hayes and Pitman (1970), identified by Larson and Chase (1972) as Mesozoic in age and more completely mapped by Hilde (1973). These authors recognized two closely parallel branches of the Mendocino fracture zone, here designated Mendocino N and Mendocino S. The revised bathymetric chart (figure 3a) is substantially different from the Chase and others' (1971) chart because a higher grade of bathymetric data was used and because the magnetic lineations and their offsets were used to guide the contouring. Some of the randomly distributed "Mapmaker" seamounts of the old chart now fall on the trace of the magnetically and bathymetrically defined Mendocino N fracture zone and are construed to

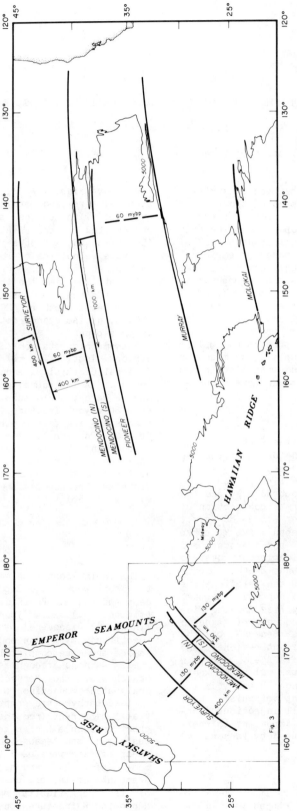

Fig. 1. Regional map showing the basin between Shatsky Rise, the Emperor Hawaiian Ridge and the Mid-Pacific Mountains. 60 and 130 mybp isochrons (dashed lines) show variable offsets through time along the Surveyor and Mendocino fracture zones.

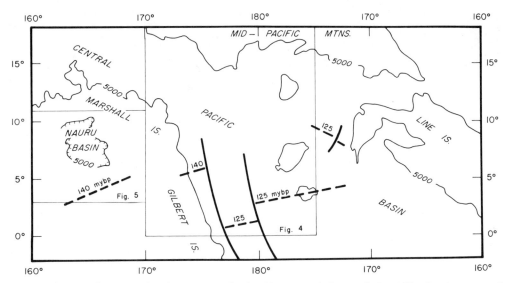

Fig. 2. Regional map showing the location of the Nauru and Central Pacific Basins, south of the Mid-Pacific Mountains.

belong to continuous ridges of this feature. The Mendocino S fracture zone, well defined by a magnetic anomaly offset, is poorly defined in the topography. Still another elongated ridge parallels the Mendocino N fracture zone, 400 kilometers to the north, and is here correlated with the Surveyor fracture zone. Figure 1 shows that the Mendocino fracture zone is also a double feature on younger crust and that the Surveyor fracture zone is located 400 km north of its northern branch. West of the Hawaiian ridge Hilde (1973), Hilde et al. (1976) and Larson and Hilde (1975) show offset of the M4 to M10 magnetic anomalies sequence and none for the sequence M11 to M22. A ridge and trough topography is developed along the Surveyor fracture zone even where there is no magnetic anomalies offset (figure 3a). Schouten and White (1980) have discussed the existence of variable or even zero offset fracture zones in the Atlantic and proposed that fracture zones are zones of decoupling between adjacent spreading centers, decoupling which exists and persists even when there is no offset in the magnetic anomalies. Figure 1 shows that the 0 offset of the Surveyor fracture zone has grown to 400 km between 130 and 60 mybp, while the 320 km offset of M-anomalies of the two branches of the Mendocino fracture zone has grown to 1000 km in the same time interval. The extremely complex history of plate boundaries is such that in the even more recent history (between 35 and 19 mybp) the Surveyor fracture zone offset is again reduced to zero by a series of jumps in the ridge crest (Shih and Molnar, 1975). Alternatively, the identification of M-sequence magnetic anomalies across the Surveyor fracture zone may be in error.

Assuming that the magnetic anomalies are correctly identified, the expected progressive deepening of the crust with increasing age does not happen in this area. West of the Emperor seamounts (figure 3a), and north of the Surveyor fracture zone, the seafloor deepens progressively to 6200 m over the 129 mybp isochron and then shoals on progressively older crust to 5700 m on crust 138 mypb old. Between the Surveyor and Mendocino N fracture zone, the sea floor shoals from 5900 m over the 130 mybp isochron to a minimum of 5700 m, on the 138 mybp isochron. South of the Mendocino N fracture zone the sea floor shoals westward to 5900 m on the 135 mybp isochron.

The three shoals thus described form a broad and subdued southeasterly ridge, $\pm$ 500 m above the surrounding basins, forming an angle to the crustal age trends and recognizable in the fracture zone areas, although complicated by the narrow ridge-and-trough fracture zone topography. This broad shoal, roughly defined by the 5900 m contour, constitutes a regional departure from the best-fit model for depth-vs-age and is a depth anomaly.

The Sediment Cover

Thick sediment accumulations may create broad topographic swells, but the recent isopach map of (Ludwig and Houtz, 1979) shows that sediments are only about 200 m thick over the whole area under discussion. They thicken to about 300 m in the little basin at 33°N and 165° while the basin at 25°N and 170°E has a sediment cover of less than 100 m. My own study of seismic reflection profiles in the area confirms that shoaling of the topography over the depth anomaly is not due to an excess sediment accumulation but is a feature of the crust below.

Two Deep Sea Drilling Sites are located in the area: site 46 (Fischer et al, 1971) and site 307 (Larson and Moberly, 1975). Only 9 m of the sedi-

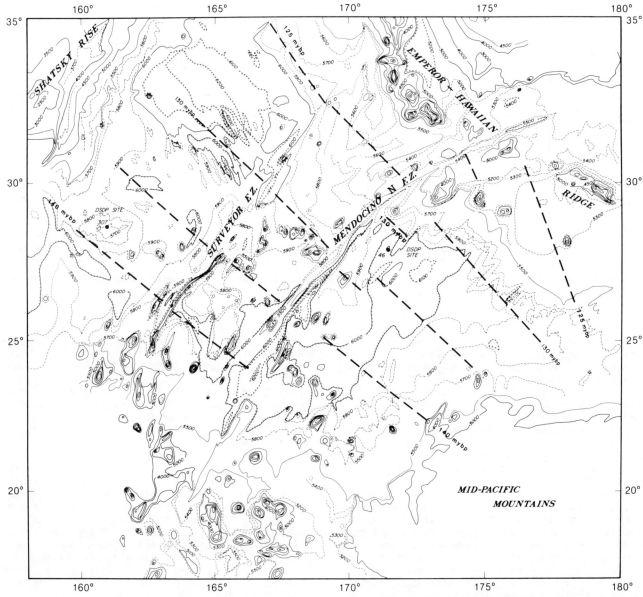

Fig. 3a. Detailed bathymetric chart of the basin between Shatsky Rise and the Mid-Pacific Mountains. Features to notice are: the 0 offset Surveyor fracture zone, the > 500 m depths range of several isochrons and DSDP site 307 at the northwest end of a subdued rige which is a depth anomaly.

ment cover were recovered on site 46 while site 307 reached basalt below 298 m of sediments. The purpose of drilling site 307 on anomaly M 21 was to determine the biostratigraphic age of that anomaly. The age of the oldest fossils recovered are discussed by Larson and Moberly (1975) who assign a Berriasian age to the basement: the age of the site is close enough to the expected one to preclude an episode of more recent flood volcanism as a cause of the depth anomaly.

Depth-Versus Age Correlation

The data for this basin are graphically summarized in figure 6, which illustrates the relationship of depth with age for crust older than 110 mybp. The dashed line is the equation proposed by Parsons and Sclater (1977) for younger ocean crust: $d(t) = 2500 + 350\ t^{1/2}$ m; the heavy solid line is the equation proposed by the same authors

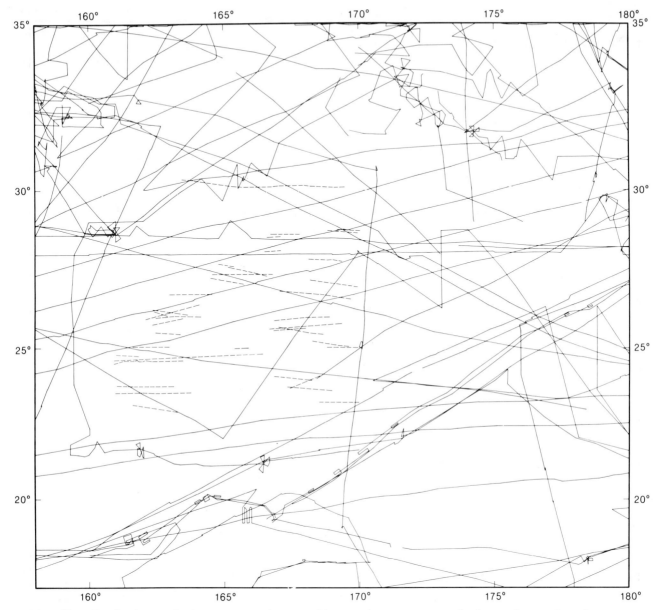

Fig. 3b. Track chart of the most precise soundings and most accurately located tracks for area 3a.

for older ocean crust: $d(t) = 6400 - 3200 \exp(-t/62.8)$ m. The vertical lines represent the ranges of crustal depths along isochrons of various ages in several areas: A, north of the Mendocino N fracture zone; B, south of the Mendocino N fracture zone). These depths have been corrected for isostasy by adding 70% of the sediment thickness derived from the Deep Sea drilling site values and the Ludwig and Houtz (1979) isopach map. A thin solid line in Fig. 6 connects the median values of the depth ranges in area A for isochrons 125, 130, 138, 140 and 145 mybp. The curve so defined is a sinuous one, plunging below Parsons and Sclater (1977) curve for older ocean at the 130 and 140 mybp isochron and then arising above it.

All B values are depth ranges for isochrons located southeast of the Mendocino N fracture zone. Although no curve has been drawn connecting the median value of these depth ranges, the median crustal depths of the crust show a similar pattern. For all these values depth ranges are considerably greater than the ±100 m accepted by Parsons and Sclater.

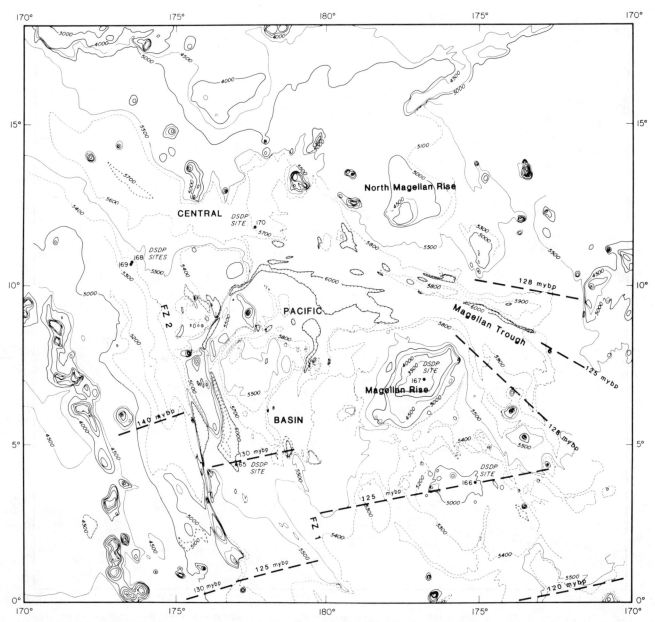

Fig. 4a. Detailed bathymetric chart of the Central Pacific Basin. Features to notice are the Magellan through, deeper than its older flanks, the depths range of several isochrons > 500 m and the location of DSDP site 166 on a depth anomaly similar to the one at DSDP site 307.

II The Central Pacific Basin

Morphology and Crustal Age

The Central Pacific Basin is located between the Marshall and Gilbert Is chains to the west, the Line Is to the east and the Mid-Pacific mountains to the north (figure 2). Its salient topographic features, magnetic lineations and DSDP sites are shown on figure 4a. Larson et al. (1972) discovered a set of west-trending lineations in the Central Pacific Basin which were later dated from Late Jurassic through Early Cretaceous (Winterer et al., 1973 and Larson, 1976). Tamaki and others (1979) interpreted the known set of symmetric fanning lineations north of the Magellan Rise as an spreading ridge abondoned 125 m.y. ago. Mammerickx and Smith (1979) showed that the North Magellan Rise is symmetrically opposite the Magellan Rise, across the west-trending Magellan trough, the site of the abandoned spreading center, and speculated that the two plateaus were

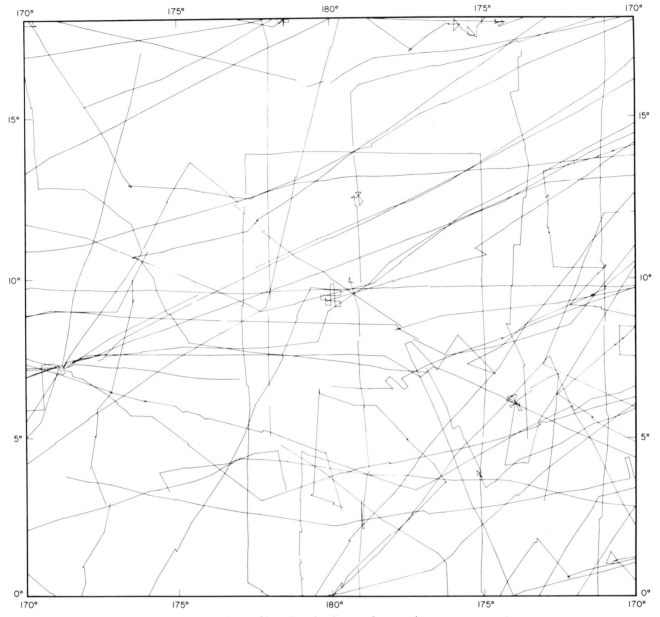

Fig. 4b. Track chart of area 4a.

the dislocated parts of a much larger older feature.

Depth Versus Age Correlation

In this area, well-identified magnetic lineations cross a complex topographic setting, and the area thus provides values for the depth of Mesozoic crust.

Depth ranges along isochrons in this area are plotted on figure 6. Values for the isochrons south of the Magellan Rise and east of fracture zone 1 (figure 4a) are labelled C. The depth corrected for isostasy at site 166, where basalt was reached and which served to date the crust between anomalies M-7 and 8, is also shown. The median depth of the 125 mybp isochron is 540 m shallower than the expected depth, showing clearly that DSDP Site 166, with a sediment cover only 310 m thick is located on a depth anomaly very much like DSDP Site 307. Schlanger et al. (1981) suggest that this area underwent uplift due to midplate volcanism between 90 and 110 mybp, but the crustal age determined at this site is compatible with the age determined by the magnetic lineations, and there is no evidence for more recent midplate volcanism.

West of fracture zone 2, the Marshall-Gilbert

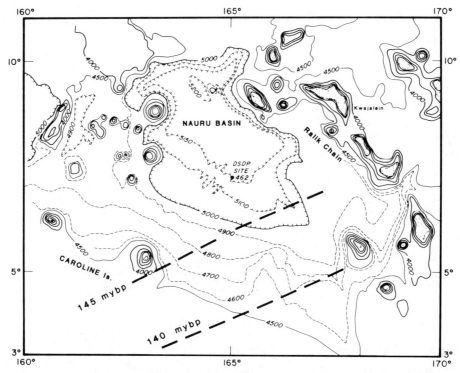

Fig. 5a. Naura Basin and location of DSDP site 462. Notice depths range of 140 and 145 mypb isochrons.

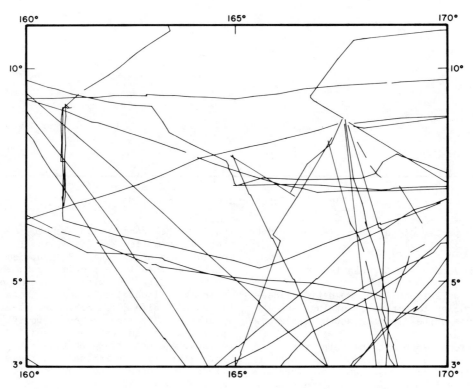

Fig. 5b. Track chart of area 5a.

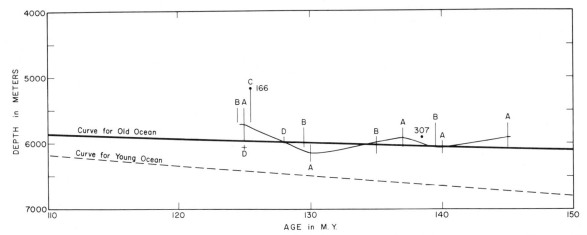

Fig. 6. Heavy solid line is Parsons and Sclater (1977) equation for older ocean crust $d(t) = 6400 - 3200 \exp(-t/62.8)$ m and dashed line is equation for younger oceanic crust $d(t) = 2500 + 350\sqrt{t}$ m. A, B, C, and D are depths ranges along isochrons in 4 areas discussed in the text. The thin solid line joins mean depths for various isochrons in the region north of the Mendocine N fracture zone. (A) Depths of basalt, corrected for isostatic effect are indicated at DSDP sites 166 and 307.

Islands chain is superimposed on a much broader swell. Watts (1976) has extensively discussed similar features around the Hawaiian Ridge Seamount chain. The point to be made here is that the isochrons east of the Gilbert Islands, well developed between the islands and fracture zone 2, do not deepen progressively with age but each one of them slopes down from the swell of the Gilbert-Marshall islands.

North of the Magellan trough, two values for crustal ages 125 and 128 mybp are reported on figure 6 with the letter D. The observation to be made here is that the younger crust is deeper than the older.

III. The Nauru Basin

The Nauru basin (figure 2 and 5a) shows magnetic lineations (Larson, 1976) that belong to the same set discussed in the preceding paragraphs. Leg 61 drilled one hole, 1068 m deep, in the basin (Larson and Schlanger, 1981). The purpose of locating site 462 on magnetic anomaly M 26 (148 my old) was to recover Early Cretaceous and Late Jurassic sediments. Unexpectedly, drilling encountered a volcanic complex of mid-Cretaceous age, more than 500 m thick presumably overlying the Jurassic basement. The volcanic chronology of site 462 is discussed in Larson and Schlanger (1981). The volcanic sequence is complex and it appears that several periods of post ridge crest volcanism took place.

From this drill hole critical observations are made on further complexities in the regional evolutions of Mesozoic crust. Sediments and sedimentary rocks were recovered down to 561 m below the sea-floor and the age of the oldest sediments overlying basalt in that location was Cenomanian (100 mybp). The isostatically corrected depth for the basalt underlying the Cenomanian sediments is $5190 + (0.7 \times 561)$ m = 5582 m. The expected depth for crust of that age, using Parsons and Sclater curve for older ocean, would be 5749 m. Thus the basalt sills and flows encountered at site 462 are 48 my younger than the magnetic lineations would have us expect, and 167 m shallower than expected if they had followed the simple thermal contraction history of aging ridge crust.

In their analysis of these results from site 462, Schlanger and Premoli-Silva (1981) argue that the drill site is anomalously shallow for its age because Mid-Cretaceous thermal rejuvenation associated with large scale emplacement of sills and flows interrupted normal subsidence, uplifting the area to abnormally shallow depths, from which it has since subsided.

In summary the drilling results of site 462 indicate that thick volcanic flow complexes were emplaced over the original oceanic crust tens of millions of years later. These flows are at depths shallower than the expected ones.

Discussion

The relationships between depth and crustal age of three basins in the Mesozoic northwest Pacific show so many complexities and anomalies that one can question the availability of preserved areas of Mesozoic ocean floor sufficiently "normal" or "accident free" to provide a safe basis for model building.

M-sequence isochrons in these basins have depth anomalies considerably larger than 100+ m. Many circumstances can alter the expected behavior of the plates; they are summarized as follows:

1. <u>Arching of the crust around volcanic edi-</u>

fices. A simple look at a bathymetric map of the Pacific shows a large development of seamount chains and plateaus west of Hawaii. Detailed studies of the best known seamount chain, the Hawaiian-Emperor Ridge, shows that the Hawaiian seamounts are surrounded by a moat and an arch: these combined features make up the Hawaiian swell, 1200 km wide (Watts, 1976). Although the Hawaiian seamounts are volcanic constructions, the lower and broader parts of the swell on which the volcanoes are built is made of elevated oceanic crust, not significantly altered by recent volcanism. The origin of the arching of the crust (or the depth anomaly) around the Hawaiian ridge and other seamount chains is much debated (Menard, Watts, Detrick and Crough, 1978) and will not be discussed here. The observation is that volcanic edifices are surrounded by a pedestal of elevated but othrwise normal oceanic crust. The younger part of the Hawaiian Ridge is developed across crust generated during the magnetically quiet time between 80 and 110 mybp; thus no magnetic lineations are observed on the swell, but the older part of the ridge is developed over older mesozoic crust and the swell associated with the chain has elevated nearby M-sequence isochrons well above their expected depths. A similar arching of the crust around volcanic edifices may account for the shoaling of isochrons around the Mid-Pacific mountains and the Marshall-Gilbert chain.

2. Intraplate deep-sea volcanism. Drilling in the Nauru Basin (Larson and Schlanger (1981) and Schlanger and Premoli-Silva (1981)) has brought to light the existence in basins of flows and sills much younger than the age presumed from the magnetic lineations. According to these authors, Cretaceous mid-plate volcanism accompanied by thermally induced uplift produced depths shallower than the expected ones in the Nauru Basin and perhaps over a wide area of the western Pacific.

3. Depth anomalies in areas not affected by mid-plate volcanism. Much of the oceanic basement is created by processes of ridge crest volcanism or "original crestal volcanism". Occasionally more recent episodes of mid-plate volcanism alter this original setting. The two types of depth anomalies discussed above are related to this mid-plate volcanism. Mammerickx (1981) has discussed the possible significance of depth anomalies developed on areas of "original crestal volcanism", undisturbed by mid-plate volcanism. They may be active (precursor domes associated with impending spreading) or dead features once the sites of potential spreading volcanism, where volcanism did not take place but for which the shoaling has been retained as a topographic anomaly. Two depths anomalies described in this paper fit these criteria: 1) the subdued ridge trending SE cutting across the Shatsky-Mid-Pacific Basin (fig. 3a) and 2) the relatively shallow dome south of Magellan Rise at DSDP Site 166. (fig. 4a) In the absence of signs of impending spreading in this area of the western Pacific these two depth anomalies would be fossil features.

4. Trench outer-swell. The detailed bathymetry of the region of the Mesozoic "Japan" lineations (Uyeda et al. 1962 and Hilde 1973) has not been discussed in this paper, but it is evident from the study of existing Bathymetric charts (Mammerickx et al. 1977) that the magnetic lineations bend over the outer swell of the Bonin-Japan-Kuril trenches, and therefore that they too are involved in relatively recent uplift altering the "normal" crustal age depth.

In summary, at least four processes have altered the expected depth of Mesozoic crust and at this point it seems difficult to locate an area in the Western Pacific that is not anomalous in depth.

Acknowledgements. I wish to thank T. W. C. Hilde and S. O. Schlanger for their helpful comments and critiques. This work was supported by grants from the National Science Foundation (OCE 80-18169) and the U.S. Office of Naval Research (N00014-75-C-0152).

References

Chase, T. E., H. W. Menard, and J. Mammerickx, Topography of the north Pacific, Institute of Marine Resources Technical Report series, TR-17, 1971.

Carter, D. J. T., Echo-sounding corrections tables. Formerly Matthews' Tables, NP-139, Hydrographic department, Ministry of Defense, Taunton, England, 1980.

Detrick, R. S. and S. T. Crough, Island subsidence, hot spot, and lithosphoric thinning, Journal of geophysical Research, 83, 1236-1244, 1978.

Fischer, A. G., et al., Initial reports of the Deep Sea Drilling Project, v. 6, Washington and U. S. government printing office, 1329 p., 1971.

Hayes, D. E. and W. C. Pitman, Magnetic lineations in the North Pacific Geol. Soc. America, Mem. 126, 291-314, 1970.

Hilde, T. W. C., Mesozoic sea-floor spreading in the North Pacific, D. Sc. thesis, 84 p., Univ. of Tokyo, Tokyo, Japan, 1973.

Hilde, T. W. C., N. Isezaki and J. M. Wageman, Mesozoic sea-floor spreading in the North Pacific in The geophysics of the Pacific Ocean Basin and its Margin, 19, 205-228, 1976.

Larson, R. L., Late Jurassic and Early Cretaceous evolution of the Western Pacific, J. Geomag. Geoelectr., 28, 219-236, 1976.

Larson, R. L. and C. G. Chase, Late Mesozoic evolution of the Western Pacific Ocean, Geol. Soc. Am. Bull., 83, 3627-3643, 1972.

Larson, R. L., S. M. Smith and C. G. Chase, Magnetic lineations of early Cretaceous age in the western equatorial Pacific Ocean, Earth and Planet. Sci. Letters., 15, 315-319, 1972.

Larson, R. L., R. Moberly et al., Initial Reports of the Deep Sea Drilling Project, v. 32, Washington, U.S. Government Printing Office, 980 p., 1975.

Larson, R. L. and T. W. C. Hilde, A revised time scale of Magnetic Reversals for the early Cretaeous and Late Jurassic, J. Geophys. Res., 80, 2586-2594, 1975.

Larson, R. L. and S. O. Schlanger, Geological evolution of the Nauru Basin, and regional implications in Initial Reports of the Deep Sea Drilling Project, 61, Washington (U.S. Government Printing Office, 1981).

Ludwig, W. J. and R. E. Houtz, Isopach map of sediments in the Pacific ocean basin and marginal sea basins, 2 sheets, Amer. Assoc. Petr. Geologists., Tulsa, Oklahoma, 1979.

Mammerickx, J. and S. M. Smith, North Magellan Plateau, a possible symmetric twin to the Magellan Plateau, Abstract, in EOS, 46, 888, 1979.

Mammerickx, J., Depth anomalies in the Pacific: active, fossil and precursor, Earth and Plan. Sc. Letters, in press, 1981.

Mammerickx, J., R. L. Fisher, F. Emmel, and S. Smith, Bathymetry of the East and Southest Asian Seas, Geol. Soc. America, Map and chart series MC-17, 1977.

Menard, H. W., Depth anomalies and the bobbing motion of drifting islands, J. Geophys. Res., 78, 5128-5137, 1973.

Parsons, B. and J. G. Sclater, An analysis of the variation of ocean floor bathymetry and heat flow with age, J. Geophys. Res., 82, 803-827, 1977.

Schlanger, S. O., H. C. Jenkyns and I. Premoli-Silva, Volcanism and vertical tectonics in the Pacific Basin related to global Cretaceous Transgressions, Earth and Planet. Sci. Letters., 52, 435-449, 1981.

Schlanger and Premoli-Silva, Tectonic, volcanic and paleogeographic implications of redeposited reef faunas of Late Cretaceous and Tertiary age from the Nauru Basin and the Line Islands. Initial reports of the Deep Sea Drilling Project 61, in press, (1981).

Schouten, H. and R. S. White, Zero-offset fracture zones, Geology 8, 175-179, 1980.

Shih, J. and P. Molnar, Analysis and implications of the sequence of ridge jumps that eliminated the Surveyor transform fault, J. Geophys. Res., v. 80, 4815-4822, 1975.

Tamaki, K., M. Joshima and R. L. Larson, Remanent Early Cretaceous sreading center in the Central Pacific Basin, J. Geophys. Res., 84, 4501-4510, 1979.

Uyeda, S., M. Yasui, K. Horai and T. Yabe, Report on geomagnetic survey during JEDS -4 cruise, Oceanogr. Mag., 13, 167-183, 1962.

Van Hinte, J. E., A Jurassic time scale, Am Assoc. Pet. geol. Bull., 60, 489-497, 1976a.

Van Hinte, J. E., A Cretaceous time scale, Am. Assoc. Pet. Geol. Bull., 60, 498-516, 1976b.

Watts, A. B., Gravity and Bathymetry in the Central Pacific Ocean, J. Geophys. Res., 81, 1533-1553, 1976.

Winterer, E. L., J. I. Ewing et al., Initial reports of the Deep Sea Drilling Project, 17, 930 p., 1973.

TRENCH DEPTH: VARIATION AND SIGNIFICANCE*

Thomas W.C. Hilde

Departments of Oceanography and Geophysics, Geodynamics Research Program,
Texas A&M University, College Station, Texas 77843

Seiya Uyeda

Earthquake Research Institute, The University of Tokyo, Tokyo, Japan

Abstract. Trench depths vary systematically with age and therefore mass of the subducting plate for all Circum-Pacific trenches; the greater the age of subducting plate, the greater the trench depth. This relationship holds for plate ages greater than 80 Ma which supports the thermal models for oceanic lithosphere that predict continuous thickening with age. Subduction rates are also systematically greater with greater trench depth, and therefore mass of subducting plate, suggesting that negative buoyancy is a significant driving force for plate motion. A simple bending plate model is presented to account for the above observations. Backarc region trenches display greater depth for a given age of subducting plate than the Pacific basin perimeter trenches. This is in part due to the increased depth of the backarc basins, which we conclude results from compensation of non-equilibrated portions of subducted lithosphere residing in the asthenosphere beneath the backarc regions.

Introduction

Trench depths around the Pacific are highly variable and depth gradients exist along individual trenches. Because so many trench-arc-backarc features are apparently related to the convergence and subduction process in some systematic way (Uyeda, 1982), trench depth may also be an indicator of important features or processes at subduction zones. Until very recently (Hilde and Uyeda, 1981, and Grellet and Dubois, 1982), the possible significance of trench depth variations seem to have been largely overlooked. Working independently, these studies have shown that trench depths vary systematically according to age of the subducting plate and rate of subduction; the greater the age and the downgoing rate of the subducting plate, the greater the trench depth.

We have also observed a clear difference in maximum trench depth and age of subducting plate for those trenches around the perimeter of the Pacific basin and those trenches located in backarc regions. Trench depths for backarc region trenches are greater for a given age of subducting plate than for Pacific perimeter trenches.

In this paper we show trench depth-age of subducting plate correlations for maximum trench depth compared between trenches, and along five individual trenches for which there is sufficient age data, and examine the relevance of these correlations for the evolution of oceanic lithosphere, subduction processes and plate velocities. We also offer an explanation for the anomalously deep backarc regions.

Maximum Trench Depth-Age of Subducting
Plate Relationship

To see if age of the subducting slab and trench depth are related we first examined this relationship between trenches (Fig. 1 and Table 1). To do this we chose a common single point for each trench which might be expected to reflect any systematic age-depth relationship. This is the point of maximum depth. Maximum depths for the Pacific trenches are rather well-known (Fisher and Hess, 1963; and Robert Fisher, personal communication) so we have reliable, narrowly constrained data for at least one parameter. Except for a very few trench regions, sediment thickness on the subducting slab at the trench seldom varies by more than 200 m (Ludwig and Houtz, 1979). This variation would produce less than 200 m difference in depth. We have therefore not corrected the depths for sediment thickness or loading. The ages of subducting slabs at trench depth maxima are, however, somewhat less well-constrained. The uncertainty in the age for each trench depth

---

*Texas A&M University Geodynamics Research Program Contribution No. 28.

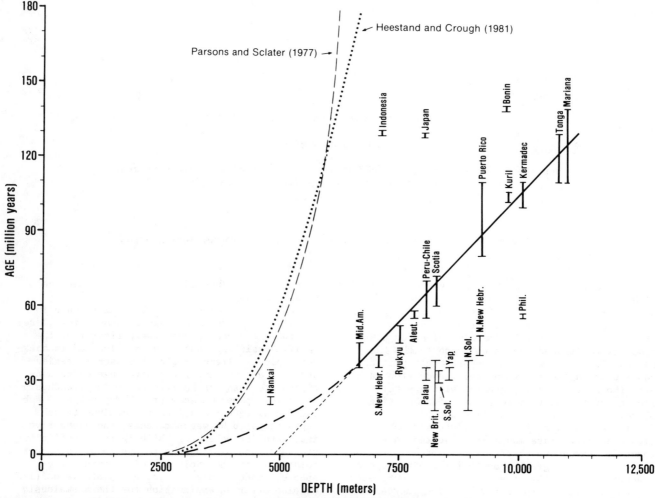

Fig. 1. Maximum trench depth-age of subducting plate relationship. Thick and thin bars represent the likely range of ages for plates subducting at the Pacific basin perimeter and in backarc regions respectively. See Table I for detailed maximum trench depth-age data. The thick straight line is the best fit, referred to as the standard maximum trench depth-age curve (for the broken and dotted lines extended to zero age, see text). Curves labelled Parsons and Sclater (1977) and Heestand and Crough (1981) are proposed depth-age relationships for normal ocean basins. These curves and the standard maximum trench depth-age curve are also shown for reference in figures 2-5.

maximum is given in Table 1, and reflected in Fig. 1 by the vertical lengths of the bars on the plot. Ages of the subducting slabs in all of the trench depth-age plots presented in this paper are based primarily on identified magnetic lineations; AAPG Circum-Pacific tectonic maps (1981), various other seafloor magnetics studies, and DSDP basement ages where magnetic lineations have not been identified. Interpolation was in many cases required to determine the age of the subducting plate precisely at the trench, but age data is now so ubiquitous that we consider the ages we have assigned to be sufficiently accurate to clearly reveal any age-depth relationship that might exist.

Fig. 1 clearly demonstrates the systematic increase of maximum trench depth with increasing age of subducting lithosphere. This relationship holds beyond the age of 80 - 100 Ma where normal ocean basin depths start to deviate from the simple $t^{\frac{1}{2}}$ curve (Parsons and Sclater, 1977). If we take the Middle America, Aleutian, Peru-Chile, Scotia, Puerto Rico, Kuril, Kermadec, Tonga and Mariana trenches, the maximum depth-age relationship is essentially linear with t. For comparison, the ocean basin depth-age relationships postulated by Parsons and Sclater (1977) and by Heestand and Crough (1981) are also shown in Fig. 1. The later authors propose that the $t^{\frac{1}{2}}$ relationship holds for seafloor older than 80 - 100 Ma for areas away from hot spots.

The apparent linear relationship for subducting

Table I

Maximum Trench Depth Data

Trench	Depth			Age		Basin Depth seaward of outer High (Corr. meters)	ΔD Basin/Trench	Source
	Corr. Depth Meters	Position	Source	Age -m.y.	Source			
Mariana	10,980	142°18'E	Hilde/Lee	110-140	----	4710	6270	Mammerickx et al.
Tonga	10,800	23°S	Fisher/Hess	110-130	----	5555	5245	Circum-Pacific
Philippine	10,055	10°N	"	55-57	Hilde/Lee	5950	4105	Mammerickx et al.
Kermadec	10,047	32°S	"	100-110	BMR	5555	4492	Circum-Pacific
Kuril	9,750	45°N	"	102-106	Hilde et al.	5515	4235	"
Bonin	9,695	29°N	Fisher/Hess (Japanese chart)	139-141	"	6155	3540	Mammerickx et al.
Puerto Rico	9,200	67.4°W	Fisher/Hess	80-110	Pitman et al.	5770	3430	Chase/Holcombe
N. New Hebrides	9,165	12°S	"	40-48	Circum-Pacific	4515	4650	Circum-Pacific
North Solomons	8,940	6.5°S	"	18-38	BMR	4515	4425	Mammerickx et al.
Yap	8,527	10.5°N	"	30-35	Circum-Pacific	----	----	----
South Solomons	8,310	162°W	"	29-34	BMR	4515	3795	Circum-Pacific
Scotia	8,264	~60°S	"	60-72	Pitman et al.	4455	3809	"
New Britain	8,245	152°W	"	18-38	BMR	4515	3730	"
Peru-Chile	8,055	23.5°S	"	55-70	Circum-Pacific	4150	3905	Mammerickx/Smith
Palau	8,050	7.5°N	"	30-35	"	----	----	----
Japan	8,000	36.5°N	Japanese chart	128-130	Hilde et al.	6050	1950	Mammerickx et al.
Aleutian	7,800	174°E	Fisher/Hess	55-58	Peter et al.	4985	2815	"
Ryukyu	7,507	24.5°N	"	45-52	Hilde/Lee	5535	1972	Mammerickx et al.
Indonesia	7,125	110°E	Fisher/Hess (Hamilton)	129-131	Circum-Pacific	5650	1475	"
S. New Hebrides	7,070	21°S	Fisher/Hess	30-40	"	4515	2555	Circum-Pacific
Middle America	6,662	94°W	"	35-45	"	4100	2562	Fisher
Nankai	4,800	134°E	Hilde et al.	20-23	"	4610	190	Mammerickx et al.

EXPLANATION: Depth Sources Hilde/Lee = unpublished 1973 R/V CHIU LIEN survey; Fisher and Hess = The Sea, 1963, plus Fisher (personal comm.); Hamilton = Geol. Survey Prof. Paper 1078, 1979; Japanese Chart = Bathymetric Chart Adjac. Seas Nippon, Mar. Saf. Agency, 1966; Hilde et al. = Deep-Sea Res., 1969. Age ---- = interpolation this paper; Hilde/Lee = unpublished W. Phil. Basin; BMR = Bureau of Mineral Resources Earth Sci. Atlas Australia, Plate Tectonics, H.M. McCracken, 1979; Pitman et al. = Age of the Ocean Basins map, GSA, 1974; Circum-Pacific = Plate-Tectonic map of the Circum-Pacific, AAPG,1981; Peter et al. = The Sea, v. 4-II, 1970; Hilde et al. = Tectonophysics, v. 38, 1977. $\frac{\Delta D}{}$ ---- = not used; Mammerickx et al. = Bathy. East and Southeast Asian Seas, GSA, 1976 and Topography South Pacific, Univ. Calif., 1975; Mammerickx/Smith = Bathy. Southeast Pacific, GSA, 1978; Circum-Pacific = AAPG, 1981; Chase/Holcombe = Geologic-Tectonic Map Caribbean, USGS, 1980; Fisher = GSA Bul. v. 72, 1961.

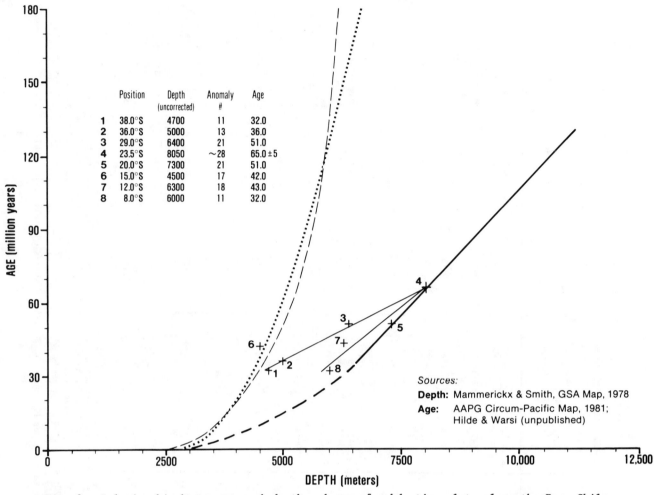

Fig. 2. Relationship between trench depth and age of subducting plate along the Peru-Chile trench.

plate age and maximum depth shown in Fig. 1 is expressed as $D_{max} = 4886 + 48.56\ t$, and hereafter referred to as the standard maximum trench depth-age curve. Its validity for t outside of the range from 35 - 45 Ma to 110 - 140 Ma, corresponding to the age of lithosphere subducting at the Middle America and Mariana trenches, is not supported by data. Nevertheless, projection of the slope intersects with the depth axis at 4886 m for t=0. It is considered more likely, however, that the extension of this relationship in the small t range would swing upwards, as shown by the broken curve, to join the normal ocean depth at t=0. In fact, at the Chile margin triple junction where the active Chile Ridge collides with the Chile trench, the depth is about 2500 m (Herron et al., 1981).

Another interesting feature to be noticed in Fig. 1 is the distinct tendency for lithosphere being subducted in backarc regions to be systematically deeper than for normal ocean lithosphere of similar age that is being subducted at the Pacific perimeter trenches. Thin line-weight bars are used for the backarc trenches to distinguish them from the Pacific perimeter trenches. Subducting plates of some of the backarc trenches, for instance Palau, Yap, New Britain, Solomon and New Hebrides trenches, are not in backarc basins, *sensu stricto*, when present polarities and only active subduction zones are considered. However, all these plates are believed to have formed in backarc environments in the past.

It has already been pointed out (e.g. Sclater et al., 1976; Watanabe et al., 1977) that backarc basins are deeper than normal ocean basins of similar age. The observed difference in maximum depth between backarc trenches and Pacific perimeter trenches appears to be in harmony with the difference in basin depths, and may be due to a common cause as will be discussed later.

A few trenches apparently have anomalously shallow maximum depths; namely the Indonesian, Japan and Bonin trenches, and the Nankai trough. However, plausible explanations exist for each

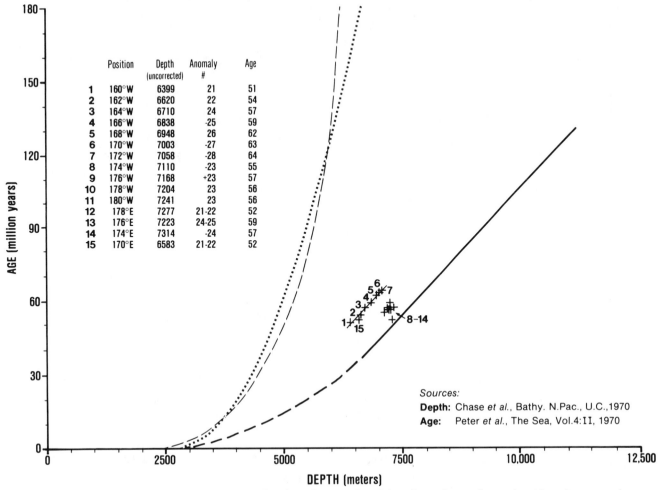

Fig. 3. Relationship between trench depth and age of subducting plate along the Aleutian trench.

of these departures. These will be discussed in the following section.

Trench Depth-Age of Subducting Plate Relationship for Individual Trenches

The trench depth age correlation can be examined along individual trenches for which there is sufficient age data. Five trenches qualify; the Peru-Chile, Aleutian, Indonesian, Japan and Bonin trenches.

Fig. 2 shows age of subducting plate-depth correlations along the Peru-Chile trench. Although this plot does not exactly fit that for the maximum depth defined in Fig. 1, the general trends are remarkably consistent. The trench depth increases from 38°S to 23.5°S where the age of the subducting plate is greatest and then decreases northward as the age of the subducting plate becomes younger. At 15°S where the Nazca Ridge is colliding with the trench, the depth is anomalously shallow as might be expected. The same tendency can be observed at the north and south ends of this trench where large topographic features give rise to anomalously shallow depths. The data from these areas have been excluded. It should be noted that the data for the northern part of the Peru-Chile trench fall very close to the standard maximum trench depth-age curve, while the southern part of the trench appears to increasingly depart from this trend towards the south (points 4 - 1, Fig. 2). This may be due to a corresponding southward increase in sediment volume in the trench axis and on the overriding plate (Hayes, 1974) as we will discuss later.

Isochrons of the slab going into the Aleutian trench are only slightly oblique to the trench axis (Peter et al., 1970). This should and does produce a cluster of points (Figure 3) even though the data are taken along a great length of the trench; depth varies little because age of the subducting plate varies little. Systematic deepening of the trench from 160°W to 172°W is, however, still quite remarkable, reflecting the small age gradient of the subducting plate along this portion of the trench (points 1 - 7).

TRENCH DEPTH

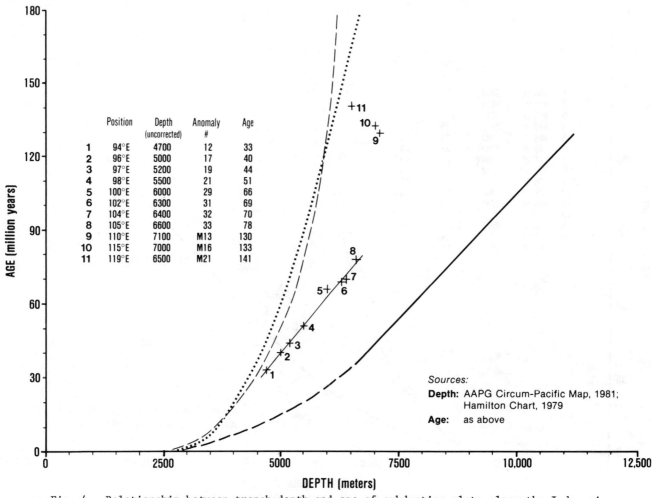

Fig. 4. Relationship between trench depth and age of subducting plate along the Indonesian trench.

From 174°W to 174°E, the Aleutian trench and the isochrons of the subducting plate are essentially parallel. Here the data points, including the deepest point, cluster. There seems to be no obvious reason for this region to be deepest. However, it may be worth noting that numerous north-south fracture zones intersect with the trench along this segment. Possibly they disrupt the slab (Kraus Jacob, personal communication), reducing its elastic strength and allowing parts of the slab to sink more easily. On the other hand, the data points for this region do fall very near the standard maximum trench depth-age curve (Fig. 1), while points 1 - 7 and 15 appear to be too shallow for the plate age. Here, as in the case of the Peru-Chile trench, it appears that where sediments are lacking, either on the oceanic plate or as a significant volume of the overriding plate (points 8 - 14), the age-depth relationship is consistent with the standard curve. A considerable volume of sediment makes up the leading edge of the overriding plate along the eastern Aleutian trench (von Huene, 1972) and trench depth for this region is too shallow.

The Indonesian trench deepens from west to east, corresponding to an increase in age of the subducting slab (Fig. 4). The deepest point of the Indonesian trench (110°E) is therefore not far from where Australia is colliding with this trench system. This, however, is not the oldest part of the plate being subducted. It is still closer to Australia, but is much too shallow for its age (points 9 - 11). The thick crustal material of Australia is made up of less dense rocks than the oceanic portion of the Indo-Australian plate to the west and may be expected to contribute a positive buoyancy force on the subduction process. This is apparently causing the oldest portion of the subducting oceanic plate to be shallower than expected for the age of the slab being subducted. It is being held up by its attachment to the more buoyant Australia. Points 1 - 8 clearly describe increasing depth with increasing age but are all too shal-

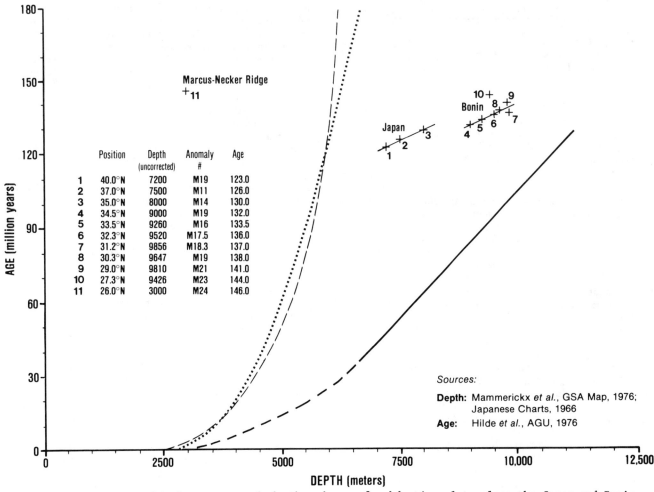

Fig. 5. Relationship between trench depth and age of subducting plate along the Japan and Bonin trenches.

low compared to the standard maximum trench depth-age curve. The extensive accretionary complex of the overriding plate (Hamilton, 1977; Karig et al., 1980; Moore et al., 1980) again correlates with this departure as does the distribution of accreted sediments with similar departures along portions of the Aleutian trench, and probably the southern Peru-Chile trench. Additionally, the combined effects of Ninety East Ridge on the west and Australia on the east, plus other large topographic features on the Indian plate in between may all contribute to making the entire trench anomalously shallow.

The depth-age correlations for both the Japan and Bonin trenches (Fig. 5), show that for these trenches depth increases with increasing age of the subducting slab more rapidly than for the standard maximum trench depth-age curve. The too shallow maximum depth of the Bonin trench may be explained in a similar way as the shallow maximum depth of the Indonesian trench. This trench deepens from north to south, consistent with a N-S increase in age of the subducting slab. However, the large shallow topographic feature (Marcus-Necker Ridge) which is colliding with the trench at ~26°N (point 11) may also be a greater thickness of material that is less dense than normal ocean crust and produce the positive buoyancy effect.

The too shallow maximum depth of the Japan trench can not be explained as above since no large, low density mass is colliding with the trench. However, the depth maximum is near the south end of the Japan trench near the Japan, Bonin trench, Nankai trough triple junction. While ~130 Ma old oceanic lithosphere is subducting into the Japan trench at its deepest point, the very young lithosphere of the Shikoku Basin (~20 Ma) is subducting into the same mantle region, from the south. Recent detailed investigations on the hypocentral distribution of this area (e.g. Shimazaki et al., 1982) reveal that the subducted Pacific and Shikoku basin plates are contorted in a complex configuration. Re-

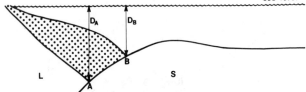

Fig. 6. Schematic cross section of a trench, showing that a thick sediment wedge on the landward plate masks the "real" position and depth of the trench axis.

cently, it has also been suggested that the subducted slab beneath Japan must be held upwards by more steeply dipping portions of the slab at the Kuril and Izu-Bonin trenches, if the area of the Pacific plate is to be conserved after subduction (Yoshio Fukao, personal communication).

Actually, all depths along the Japan-Bonin trench plot much shallower (Fig. 5) than the standard maximum trench depths. Apparently the entire slab is being held up at anomalously shallow trench depths by the above or other effects. It is interesting to note the displacement in the age-depth trend of the Japan-Bonin trench across the intersection of the two trenches (between points 3 and 4 in Fig. 5). This suggests that the complexities beneath Japan produce a greater buoyancy than those affecting the Pacific plate being subducted along the Bonin trench.

Overriding Plate Sedimentary Wedge Effects on Trench Depth

Earlier we stated that the effect on trench depth of sediments overlying the subducting plate is generally insignificant for the regions of maximum trench depth. However, when there is a thick wedge of sediments, either terrigenous or pelagic, under the landward slope of a trench, its effect on trench depth may be highly significant. Fig. 6 shows a schematic cross section of such a case. First, the load of the wedge may act to depress the subducting plate and increase trench depth. But, the more serious effect is apparently the seaward displacement and therefore decrease in the depth of the trench axis. If the "real" trench axis is to be defined as the apex formed by the landward plate L and subducting plate S, it should be at A and the trench depth should be D_A, whereas the actual axis is at B with the depth D_B when the shoreward slope contains large amounts of sediment. Without a thorough knowledge of the structure of the trench forearc, it may be difficult to define A. However, where there is a large volume of forearc sediments such as along the eastern part of the Aleutian trench or along the Indonesian trench, the difference between D_A and D_B is considered quite large. We suspect that the apparent too shallow depths (departure from the standard maximum trench depth-age relationship) along these trench axes (Fig. 3 and 4) and possibly the Japan trench may be explained in this way. The anomalously shallow maximum depth of the Nankai trough (Fig. 1) may also be explained by this effect. It may also be noted that the slightly different trends to the south and north of 23.5°S along the Peru-Chile trench (Fig. 2) may be accounted for by the same effect, since the sediment wedge is suspected to be more fully developed southward along that trench.

Considering the rather consistent association of the departure from the standard maximum trench depth-age relationship with regions of known large forearc sedimentary complexes, we suggest that similar departures for other trench regions may be used to infer the existence and size of forearc sedimentary complexes.

Trench-Basin Depth Differential

Trenches are deeper than the adjacent ocean floor because the oceanic plate is bent downwards when subducted. We have attempted to isolate the contribution to trench depth due solely to subduction of the slab from the depth contributed by normal "floating" compensation of the lithosphere as in the ocean basins by plotting the difference (ΔD) in trench and adjacent basin depths against age of the downgoing plate (Fig. 7). Several interesting features can be observed from the plot of ΔD. First, ΔD seems to increase with the age of subducting plate. Excluding the anomalously deep backarc region trenches and the anomalously shallow Indonesian, Japan and Bonin trenches, the $\Delta D - t$ correlation can be regarded as "roughly"

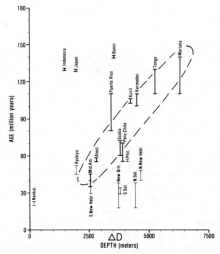

Fig. 7. Relationship between trench-basin depth differential (ΔD) and age of subducting plate. Thick and thin bars are for the Pacific perimeter and backarc trenches, as in Figure 1. Broken oval indicates probable correlation for "well-behaving" trenches.

linear. Such a result might be expected from Fig. 1 in which the difference between trench maximum depth and ocean depth curves increases with age. However, the ΔD - t correlation (Fig. 7) is more scattered than the standard maximum trench depth-age correlation (Fig. 1). This is an intriguing fact but apparently results from the ocean depths corresponding to the adjacent region of each trench maximum depth not being consistent with the idealized depth-age curves as shown in Fig. 8.

Some of this ΔD scatter, from one trench to another, may be accounted for by either or both the effects of thick sediment wedges under the landward slopes and buoyant features colliding with the trenches. Among backarc trenches, the Ryukyu, Philippine and South New Hebrides trenches show ΔD values that are close to the trend of ΔD for the Pacific perimeter trenches, suggesting that their greater maximum depths can be explained simply by the greater depth of the subducting backarc basins. Therefore, these three trenches can be regarded as "well-behaving" as far as ΔD is concerned. However, ΔD for the New Britain, Solomon and North New Hebrides trenches are still anomalously large. Adjacent basin depths for these trenches are only slightly deeper than for normal ocean basin depths of the same age (Fig. 8). One characteristic feature of these trenches is that they form a trench system where the seafloor of former backarc regions is subducting toward the Pacific due to a change in subduction polarity following a major collision in the relatively recent past (Karig, 1972; Karig and Mammerickx, 1972). Possibly the large ΔD for these trenches relates to their Pacificward subduction, for some reason which is not clear at this stage.

Discussion

Trench depth-age relationship

As shown in the previous section, trench depth is closely correlated with age of the subducting plate and maximum trench depth, D_{max}, can be expressed as $D_{max} = 4886 + 48.56\ t$ for 30 Ma < t < 130 Ma. Similar depth-age correlations can be observed along individual trenches, although the quantitative relationship is affected by local and regional factors. It was also found that the depth difference between the trench and the adjacent ocean basin (ΔD) "roughly" correlates with age of the subducting plate. We explore the possible significance of these correlations in the following discussion.

The existence of a systematic relationship between age and depth of the world's ocean basins has been established in previous studies (Langseth et al., 1966; McKenzie, 1967; Turcotte and Oxburgh, 1967; Parsons and Sclater, 1977; and others). Two categories of thermal models have been developed to explain this relationship. One category of models is the cooling half-space models (Turcotte and Oxburgh, 1967; Parker and Oldenburg, 1973; Yoshii, 1973; and Davis and Lister, 1974) in which oceanic lithosphere thickens and ocean depth increases continuously with age as $t^{1/2}$. The other category of models is the cooling plate models (Langseth et al., 1966; McKenzie, 1967), in which both plate thickness and ocean depth approaches certain values as age increases. In general agreement with earlier studies, Parsons and Sclater (1977) found that seafloor older than 80 Ma exponentially approaches an equilibrium depth of about 6400 m. Various additional models have been proposed to explain how the lithosphere matures to a certain thickness beyond which it no longer thickens with age (Crough, 1975; Richter and Parsons, 1975; Schubert et al., 1976; Yoshii et al., 1976; Forsyth, 1977; and Schubert, 1978).

A recent study by Heestand and Crough (1981) suggests that the departure of depth from the half-space model for older seafloor is found only where the oceanic lithosphere has been affected by hot spots. They conclude that seafloor that hasn't been reheated continues to deepen with age beyond 80 Ma and that the half-space model correctly defines the seafloor age-depth relationship, with depth (meters) = $2700 + 295\ t(Ma)^{1/2}$. Both of the depth-age relationships postulated

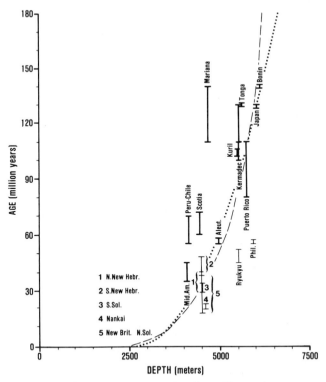

Fig. 8. Ocean basin depths adjacent to trench maximum depth for different trenches. Note that the depths scatter considerably about the ocean basin depth-age curves proposed by Parsons and Sclater, 1977 (broken curve) and Heestand and Crough, 1978 (dotted curve).

by Parsons and Sclater (1977) and Heestand and Crough (1981) are shown in Figs. 1 - 5.

The idea that trenches are the topographic expressions of high density masses sinking into the mantle is well established. Additionally, the topography of trenches and many other trench-arc-backarc features have recently been explained by various kinematic (i.e. Uyeda and Kanamori, 1979), dynamic (i.e. Watts and Talwani, 1975; Bodine and Watts, 1979; Hager, in press; Jones, in press; McAdoo, 1981) and thermo-mechanical models (i.e. McKenzie, 1969; Oxburgh and Turcotte, 1970; Hsui and Toksoz, 1979; Carlson, preprint). Davies (1981) has suggested that depths of both the trenches and backarc basins are due to compensation of subducted lithosphere, and Bostrom (this volume) has concluded that the anomalous depths of backarc regions and associated geoidal high are due to a "bunching up" of lithosphere in the western Pacific. However, the significance of variations in trench depth has only recently been addressed (Hilde and Uyeda, 1981; and Grellet and Dubois, 1982).

The total maximum trench depth, D_{max}, may be expressed as the sum of the seafloor depth D_s and ΔD (Fig. 9).

$$D_{max} = D_s + \Delta D \quad (1)$$

As long as subducting lithosphere is regarded as elastic, the shape of trenches can be simulated by that of a bent elastic plate (e.g. Turcotte and Schubert, 1982). Following Davies (1981), we assume, for simplicity, that the lithosphere is fractured at the trench axis and that stress can be transmitted across the subduction fault but that no significant moment is transmitted. The load is assumed to act at the trench axis. The problem is considered as a two-dimensional one.

The downward displacement w of lithosphere fractured at x=0 is, according to the bending plate theory,

$$w = \Delta D \cdot \exp(-x/\alpha) \cdot \cos(x/\alpha), \quad x \geq 0 \quad (2)$$

where

$$\Delta D = \frac{V_o \alpha^3}{4R} \quad (3)$$

$$R \equiv \frac{Eh^3}{12(1-\nu^2)} \quad : \text{flexural rigidity} \quad (4)$$

$$\alpha \equiv \left\{\frac{4R}{(\rho_a - \rho_w)g}\right\}^{1/4} \quad : \text{flexural parameter} \quad (5)$$

and V_o = load at x=0 per unit length of trench; h = thickness of slab, E = Young's modulus, ν = Poisson's ratio; ρ_a = density of asthenosphere; ρ_w = density of water and g = acceleration of gravity.

From (3) - (5),

$$\Delta D = \frac{V_o}{4} \cdot A \cdot h^{-3/4} \quad (6)$$

where $A = \left\{\frac{4}{(\rho_a - \rho_w)g}\right\}^{3/4} \cdot \left\{\frac{12(1-\nu^2)}{E}\right\}^{1/4}$

Equation (6) indicates, other quantities being equal, that the trench depth should be less for thicker and therefore older subducting plates. This, however, is exactly the opposite of our observation. In the case of trenches, the load applied at x=0 may be considered to consist of two main parts. One is the load from above, i.e. the load of the overriding plate and the other is the pull from below, i.e. the pull by the subducted slab. Since, for the moment, we are interested in trench depth variation with age of the subducting plate, we consider only the latter load. Load by the overriding plate may contribute to the depth of individual trenches but should not be related to age of the subducting plate.

Again following Davies (1981), the pull by the slab may be expressed as

$$V_o = a_T \cdot M \cdot g \cdot \sin^2\delta \quad (7)$$

where M = mass of the slab; δ = dip of the slab and a_T = the efficiency with which the weight of the slab is transmitted.

The mass, M, of subducted slab per unit length of trench may be expressed, though crudely, as

$$M \sim \rho_s \cdot h \cdot L \quad (8)$$

where ρ_s = density of slab, h = thickness of slab, and L = length of slab. The length of slab responsible for the pull is difficult to assess, but if we can replace it by the length of the Wadati-Benioff zone it may, according to Molnar et al., (1979), be expressed as

$$L \sim t \cdot V \quad (9)$$

where t = age of subducting slab, and V = the

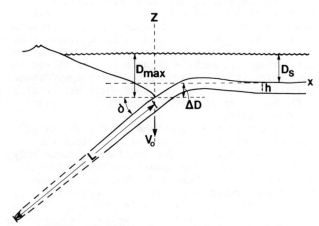

Fig. 9. Bending plate model (see text for details).

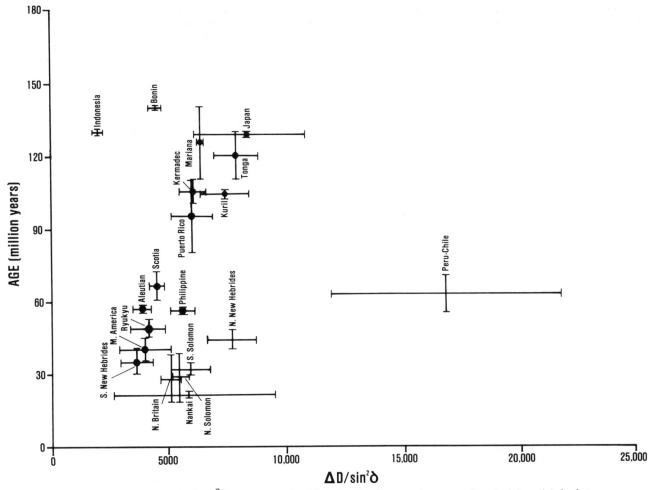

Fig. 10. Plot of $\Delta D/\sin^2\delta$ vs age. "Well-behaving" trenches are denoted by thick dots.

rate of subduction. Moreover, Carlson et al. (in preparation) show that

$$V \sim t^{\frac{1}{2}} \qquad (10)$$

Molnar et al. (1979) and Carlson et al. (in preparation) show that relations (9) and (10) can be explained by the plate thickening model i.e.

$$h \sim t^{\frac{1}{2}} \qquad (11)$$

By inserting (7), (8), (9), (10) and (11) into (6), one obtains

$$\Delta D \simeq t^n \sin^2\delta, \ (n \sim 1.63) \qquad (12)$$

Relation (12) is more strongly time-dependent than the linear relation. Moreover, there is, to some degree, a tendency that the dip of the Wadati-Benioff zone is steeper for older slabs. Therefore, (12) is probably more strongly t-dependent. Fig. 10 shows the $\Delta D/\sin^2\delta$ plot for the trenches using δ-values given in Uyeda and Kanamori (1979), and several other references, namely Seno and Kurita (1979) for the Philippine trench; Jordan (1977) for the Kuril trench; Yoshii (1982) for the Bonin, Japan and Ryukyu trenches; Engdahl (1977) for the Aleutian trench; and K. Siono (1982) for Nankai trough. The δ-value is a poorly constrained quantity because the Wadati-Benioff zones are not well defined for all subduction zones and deep seismic zones are not generally straight as shown in Fig. 9. We have used the dip for depths greater than about 100 km, when the Wadati-Benioff zone is deep enough, and we have attached $\pm 5°$ uncertainty to all of the δ-values. The resultant ranges in $\Delta D/\sin^2\delta$ are reflected by the horizontal bars in Fig. 10. Naturally, the bars are longer for smaller δ-values for the same ΔD.

Despite the inaccuracy of the data, some interesting features are observed in Fig. 10. As in Fig. 7, the Ryukyu, Philippine and South New Hebrides trenches are "well-behaving" trenches. Namely, their greater depths are the result of the deeper floor of the backarc basins. The Peru-Chile trench, however, no longer behaves. The great departure of the Peru-Chile trench results from its small δ-value ($30° \pm 5°$). This probably means that, in effect, the Peru-Chile trench is too deep for the age and dip of the subducting plate. We suspect that the thick actively overriding continental mass may be acting as an additional load that has not been taken into account in our model. The same may be true for the Japan trench ($\delta = 30° \pm 5°$) and Nankai trough ($\delta = 10° \pm 5°$). In all these cases, the mode of subduction, is the Chilean-type of Uyeda and Kanamori (1979) and the overriding plate has a thick continental mass. In the cases of the Japan trench and the Nankai trough, however, the deepening effect due to the overriding plate load may be compensated, at least partially, by the masking effect of a large sedimentary wedge (Fig. 6). In the Chilean case, no significant sedimentary wedge exists. Although it may be fortuitous, the Japan trench plots close to the trend of "well-behaving" trenches in Fig. 10.

The Melanesian trenches still show significant departures from the trend of the "well-behaving" trenches. Excluding these too deep trenches and the too shallow Indonesian and Bonin trenches, other trenches appear to show a rather regular increase in $\Delta D / \sin^2 \delta$ as the model predicts. However, with the considerable scatter of data and the approximate nature of the relationships used in deriving relation (12), further analysis may not be justified.

Undoubtedly, many more local factors are influencing trench depth than those considered in our model. To sort out the influence on trench depth by all the possible factors is beyond the scope of this paper.

Backarc region trenches

In all cases, except Nankai trough, the maximum trench depth-age of subducting plates in backarc regions plot deeper than for the Pacific basin perimeter trenches (Fig. 1). Still, an increasing depth with increasing age trend can be defined. It would therefore appear that in addition to the effect of the mass of the lithosphere being subducted there is some additional common parameter (force) that is causing them to be systematically deeper than Pacific perimeter trenches.

As we noted earlier many backarc basins are anomalously deep when compared to ocean basin age-depth values. Based on Geoid studies, Bostrom (this volume) has suggested mass anomalies in the mantle beneath the Western Pacific basin due to "lithosphere accumulation". Davies (1981) also has concluded that subducted slabs constitute mass anomalies within the mantle. Davies further proposes that this anomalous mass is compensated in part by stresses transmitted up the slab as mentioned earlier, and through the overlying mantle wedge, causing a broad shallow topographic depression in the backarc region. A key point of these and other geoid, gravity and topography studies of backarc regions is the need to have larger mass anomalies than can be reasonably estimated from subducted slabs as defined by the Wadati-Benioff zones. Given continuous subduction of the Pacific plate along the western Pacific margin, at present rates since the change in Pacific plate motion at $\sim$42 Ma (Hilde et al., 1977), $\sim$3500 km of slab has been subducted. Although most of this length is not observed seismically, we suggest that it has not yet completely equilibrated with the surrounding mantle and is the source of the deep mass anomaly, not only in the backarc regions that are close to the Pacific plate subduction zones, but including beneath such distant regions as the West Philippine Basin.

If we denote the quantities for backarc region trenches with a prime in Eq. (1), we get,

$$D'_{max} = D'_s + \Delta D' \qquad (13)$$

Then, for the same t, our observation says

$$D'_{max} - D_{max} = (D'_s - D_s) + (\Delta D' - \Delta D) > 0 \qquad (14)$$

Therefore, if the difference in maximum trench depths between Pacific perimeter trenches and backarc region trenches is caused simply by the difference in the starting seafloor depths, one would expect

$$\Delta D' - \Delta D = 0 \qquad (15)$$

Within the crude accuracy of the present data, the South New Hebrides, Philippine and Ryukyu trenches may be such cases, because ΔD for these trenches plot rather close to ΔD of Pacific perimeter trenches (Fig. 7).

However, as mentioned already, for the Melanesian trenches, i.e. North New Hebrides, North and South Solomon and New Britain trenches, the situation is different and

$$D'_{max} - D_{max} \simeq \Delta D' - \Delta D \qquad (16)$$

This apparently indicates that additional dynamic conditions cause the Melanesian trenches to be anomalously deep. This is a puzzling problem, but it may be pointed out that at these trenches, unlike other trenches, former backarc basin lithosphere is subducting beneath the Pacific Ocean, due to a polarity reversal of subduction.

Conclusion

The most obvious result of the trench depth-age correlations is that trench depth increases almost linearly with age of the subducting plate for all ages of oceanic lithosphere, including lithosphere older than 80 Ma. This strongly supports the thermal models which have the oceanic lithosphere thickening continuously with age, and indicates that trench depth is primarily determined by mass of the downgoing slab. In fact, it can be shown that the simple bending plate model for the case of loading being due to the subducted slab predicts that trench depth increases with age of subducting plate, despite older and thicker plates having greater flexural rigidity.

Although the increasing depth-increasing age trend apparently holds for all trenches, different quantitative relationships exist from trench to trench. The departures from the standard linear relationship of maximum depth by the Indonesian, Japan and Bonin trenches and by individual depths along other trenches all appear to be related either to low density, thick crustal blocks approaching the subduction zones or to the masking effect of real trench depth by thick sedimentary wedges on the overriding plates.

It has also been observed that the trenches in backarc regions have greater depth than the trenches of the Pacific perimeter. The main reason for this difference is probably because the backarc basins are systematically deeper for their age. We believe that deep non-equilibrated subducted portions of the Pacific plate lie beneath the backarc basin regions of the western Pacific. The amount of non-equilibrated slab exceeds that observed by seismicity and constitutes an excess mass in the mantle that is gravitationally compensated, thus causing backarc region basin and trench lithosphere to be at greater depths for age of the lithosphere than in normal ocean basins.

Finally, the correlation between age of subducting slab and trench depth (this paper), and rate of subduction and trench depth (Grellet and Dubois, 1982) lead to what we believe is an obvious conclusion. That is, that mass of an oceanic plate increases continuously with age, and rate of subduction is determined by mass of the subducting plate; the greater the age of the subducting plate the faster it sinks. Subduction is therefore, as Forsyth and Uyeda (1975) suggested, a significant driving force for plate motion.

Acknowledgements. Comments and suggestions from R. L. Carlson, W. B. Watts and G. Davies are greatly appreciated. We also thank Chao-Shing Lee, Waris Warsi and Kristy Barnett for their help with the computations and preparation of figures and manuscript. This research was supported in part by the Office of Naval Research, contract #N00014-80-C-00113.

References

Anonymous, Bathymetric chart of the adjacent seas of Nippon, Maritime Safety Agency, Tokyo, sheets 1 and 2, 1966.
Bodine, J.H., and A.B. Watts, On lithosphere flexure seaward of the Bonin and Mariana trenches, Earth Planet Sci. Lett., 43, 132-148, 1979.
Bostrom, R.C., K.K. Saar, and D.A. Terry, Basin Formation; the mass anomaly at the Western Pacific margin, this volume.
Carlson, R.L., T.W.C. Hilde, and S. Uyeda, On the relationship between the rate of subduction and the age of the subducted lithosphere (in preparation), 1982.
Carlson, R.L., Plate motions, boundary forces, and horizontal temperature gradients: Implications for the driving mechanism (preprint), in press, Convergence and Subduction, Tectonophysics Spec. Issue, edited by T.W.C. Hilde and S. Uyeda, 1982.
Chase, C.G., Subduction, the geoid, and lower mantle convection, Nature, 282, 464-468, 1979.
Chase, J.E., and T.L. Holcombe, Geologic-tectonic map of the Caribbean region, United States Geol. Survey, Misc. Investigations Series, Map I-1100, 1980.
Circum-Pacific Council for Energy and Mineral Resources, Plate-tectonic map of the circum-Pacific region, Amer. Assoc. Petroleum Geol., 5 sheets, 1981.
Crough, S.T., Thermal model of oceanic lithosphere, Nature, 256, 388-390, 1975.
Davies, G.F., Regional compensation of subducted lithosphere: effects on geoid, gravity and topography from preliminary model, Earth Planet. Sci. Lett., 54, 431-441, 1981.
Davis, E.E., and C.R.B. Lister, Fundamentals of ridge crust topography, Earth Planet. Sci. Lett., 21, 405-413, 1974.
Engdahl, E.R., Seismicity and plate subduction in the Central Aleutians, in Island Arcs, Deep Sea Trenches, and Back-arc Basins, AGU, M. Ewing Series 1, edited by M. Talwani and W. Pitman III, 259-271, 1977.
Fisher, R.L., Middle America trench: topography and structure, Geol. Soc. America Bull., 72, 703-720, 1961.
Fisher, R.L., and H.H. Hess, Trenches, in The Sea, 3, edited by M.N. Hill, 411-436, 1963
Forsyth, D.W., The evolution of the upper mantle beneath mid-ocean ridges, Tectonophysics, 38, 89-118, 1977.
Forsyth, D., and S. Uyeda, On the relative importance of the driving forces of plate motion, Geophys. J. Roy. Astr. Soc., 43, 163-200, 1975.
Grellet, C., and J. Dubois, The depth of trenches as a function of the subduction rate and age of the lithosphere, Tectonophysics, 82, 45-56, 1982.
Hager, B.H., R.J. O'Connell, and A. Raefsky, Global mantle flow and comparative subductology, in press, Convergence and Subduction, Tectonophysics Spec. Issue, edited by T.W.C. Hilde and S. Uyeda, 1982.
Hamilton, W., Subduction in the Indonesian region, in, Island Arcs, Deep Sea Trenches and Back-arc Basins, AGU, M. Ewing Series 1, edited by M. Talwani and W. Pitman, III, 15-31, 1977.
Hamilton, W., Tectonics of the Indonesia Region, Geological Survey Professional Paper 1078, U.S.

Gov. Printing Off., Washington, D.C., 345 p., 1979.

Hayes, D., Continental margin of Western South America, in, The Geology of Continental Margins, Springer-Verlag, edited by C.A. Burk and C.L. Drake, 581-590, 1974.

Heestand, R.L., and S.T. Crough, The effects of hot spots on the oceanic age-depth relation, J. Geophys. Res., 86, 6107-6114, 1981.

Herron, E.M., S.C. Cande, and B.R. Hall, An active spreading center collides with a subduction zone: A geophysical survey of the Chile Margin triple junction, Geol. Soc. Amer. Mem., 154, 683-702, 1981.

Hilde, T.W.C., J.M. Wageman, and W.T. Hammond, The structure of Tosa terrace and Nankai trough off southeastern Japan, Deep-Sea Res., 16, 67-75, 1969.

Hilde, T.W.C., N. Isezaki, and J.M. Wageman, Mesozoic sea-floor spreading in the north Pacific, The Geophysics of the Pacific Ocean Basin and its Margins, AGU, Geophysical Mon., 19, 205-226, 1976.

Hilde, T.W.C., S. Uyeda, and L. Kroenke, Evolution of the Western Pacific and its margin, Tectonophysics, 38, 145-165, 1977.

Hilde, T.W.C., and S. Uyeda, Trench depth vs. age of subducting plate, read before OJI Seminar on Accretion Tectonics, Tomakomai, Japan, 1981.

Hilde, T.W.C., and C.S. Lee, Age and evolution of the west Philippine basin: a new interpretation, (in preparation).

Hsui, A., and N. Toksoz, The evolution of thermal structure beneath a subduction zone, Tectonophysics, 60, 43-60, 1979.

Jones, G.M., Isostatic geoid anomalies over trenches and island arcs, in press, Convergence and Subduction, Tectonophysics Spec. Issue, edited by T.W.C. Hilde and S. Uyeda, 1982.

Jordan, T.H., Lithospheric slab penetration into the lower mantle beneath the Sea of Okhotsk, J. Geophys. Res., 43, 473-496, 1977.

Karig, D.E., Remnant arcs, Geol. Soc. America Bull., 83, 1057-1068, 1972.

Karig, D.E., and J. Mammerickx, Tectonic framework of the New Hebrides island arc, Marine Geology, 12, 187-205, 1972.

Karig, D.E., G.F. Moore, J.R. Curray, and M.B. Lawrence, Morphology and shallow structure of the lower trench slope off Nias Island, Sunda Arc, in, The Tectonic and Geological Evolution of Southeast Asian Seas and Islands, AGU, Geophysical Mon., 23, edited by D.E. Hayes, 179-208, 1972.

Langseth, M.G., X. Le Pichon, and M. Ewing, Crustal structure of mid-ocean ridges, 5, Heat flow through the Atlantic Ocean floor and convection currents, J. Geophys. Res., 71, 5321-5355, 1966.

Ludwig, W.J., and R.E. Houtz, Isopach map of sediments in the Pacific Ocean basin and marginal sea basins, AAPG, (2 sheets), 1979.

Mammerickx, J., S.M. Smith, I.L. Taylor and T.E. Chase, Topography of the south Pacific, Univ. Calif., Inst. Mar. Resource Tech. Report Series TR-56, 1975.

Mammerickx, J., R.L. Fisher, F.J. Emmel, and S.M. Smith, Bathymetry of the east and southeast Asian seas, Geol. Soc. America, Map and Chart Series MC-17, 1977.

Mammerickx, J. S.M. Smith, Bathymetry of the southeast Pacific, Geol. Soc. America, Map and Chart MC-26, 1978.

McAdoo, D.C., Geoid anomalies in the vicinity of subduction zones, J. Geophys. Res., 86, 6073-6090, 1981.

McCracken, H.M., Plate tectonics: Australian region, Bureau of Mineral Resources, BMR Earth Science Atlas of Australia, 1979.

McKenzie, D.P., Some remarks on heat flow and gravity anomalies, J. Geophys. Res., 72, 6261-6273, 1967.

McKenzie, D.P., Speculations on the consequences and causes of plate motions, Geophys. J. Roy. Astr. Soc., 18, 1-32, 1969.

Molnar, P., O. Friedman, and J.D.F. Shih, Lengths of intermediate and deep seismic zones and temperatures in downgoing slabs of lithosphere, Geophys. J. Roy. Astr. Soc., 56, 41-54, 1979.

Moore, G.F., J.R. Curray, D.G. Moore, and D.E. Karig, Variations in the geologic structure along the Sunda fore arc, Northeastern Indian Ocean, in, The Tectonic and Geological Evolution of Southeast Asian Seas and Island, AGU, Geophysical Mon., 23, edited by D.E. Hayes, 145-160, 1980.

Oxburgh, E.R., and D.L. Turcotte, Thermal structure of island arcs, Geol. Soc. America Bull., 81, 1665-1688, 1970.

Parker, R.L., and D.W. Oldenburg, Thermal model of ocean ridges, Nat. Phys. Sci., 242, 137-139, 1973.

Parsons, B., and J.G. Sclater, An analysis of the variation of ocean floor bathymetry and heat flow with age, J. Geophys. Res., 82, 803-827, 1977.

Peter, G., B.H. Erickson, and P.J. Grim, Magnetic structure of the Aleutian trench and northeast Pacific basin, in, The Sea, 4-II, edited by A.E. Maxwell, 191-222, 1970.

Pitman, W.C., III, R.L. Larson, and E.M. Herron, Age of the ocean basins determined from magnetic anomaly lineations, Geol. Soc. America, Map and Chart Series, MC-6.

Schubert, G., C. Froidevaux, and D.A. Yuen, Oceanic Lithosphere and asthenosphere: thermal and mechanical structure, J. Geophys. Res., 81, 3525-3540, 1976.

Sclater, J.C., D. Karig, L.A. Lawyer and K. Louden, Heat flow, depth and crustal thickness of the marginal basins of the south Philippine Sea, J. Geophys. Res., 81, 309-318, 1976.

Seno, T., and K. Kurita, Focal mechanisms and tectonics in the Taiwan-Philippine region, in, Geodynamics of the Western Pacific, edited by S. Uyeda, R. Murphy and K. Kobayashi, 249-264, 1979.

Shimazaki, K., K. Nakamura and T. Yoshii, Complicated pattern of the seismicity beneath the metropolitan area of Japan: Proposed explanation by the interactions among the superficial Eurasian plate and the subducted Philippine Sea and Pacific slabs, (abstract), Intern. Symp. Math. Geophys., Bonas, France, in press in Terra Cognita, 1982.

Siono, K., Frontal shape of the seismic slab by normal subduction of dead spreading axis and a preliminary application to southwest Japan, J. Geosci., Osaka City Univ., 25, Art. 2, 19-33, 1982.

Turcotte, D.L., and E.R. Oxburgh, Finite amplitude convection cells and continental drift, J. Fluid Mech., 38, 29-42, 1967.

Turcotte, D.L., and G. Schubert, Geodynamics: Application of Continuum Physics to Geological Problems, John Wiley and Sons, 450 p.m 1982.

Uyeda, S., Subduction zones: an introduction to comparative subductology, Tectonophysics, 81, 133-159, 1982.

Uyeda, S., and H. Kanamori, Back-arc opening and the mode of subduction, J. Geophys. Res., 84, 1049-1061, 1979.

Von Huene, R., Structure of the continental margin and tectonism at the Eastern Aleutian trench, Geol. Soc. America Bull., 83, 3613-3626, 1972.

Watanabe, T., M.G. Langseth, and R.N. Anderson, Heat flow in back-arc basins of the western Pacific, in, Island Arcs, Deep Sea Trenches and Back-arc Basins, AGU, M. Ewing Series 1, edited by M. Talwani and W. Pitman, III, 419-428, 1977.

Watts, A.B., and M. Talwani, Gravity effect on downgoing lithospheric slabs beneath island arcs, Geol. Soc. America Bull., 86, 1-4, 1975.

Yoshii, T., Upper mantle structure beneath the North Pacific and the marginal seas, J. Phys. Earth, 21, 313-328, 1973.

Yoshii, T., Cross sections of geophysical data around the Japanese Islands, this volume.

Yoshii, T., Y. Kono and K. Ito, Thickening of the oceanic lithosphere, in, The Geophysics of the Pacific Ocean Basin and its Margins, AGU, Geophysical Mon., 19, edited by G. Sutton, M. Manghnani, R. Moberly and E. McAfee, 423-430, 1976.

ENTRAPMENT ORIGIN OF MARGINAL SEAS

Zvi Ben-Avraham[1] and Seiya Uyeda[2]

[1]Department of Geophysics, Stanford University, Standford, CA, 94305

[2]Department of Geophysics, Texas A&M University, College Station, TX 77843*

Abstract. The entrapment of a marginal part of a pre-existing ocean, by the formation of an island arc, is one of the important mechanisms of formation of marginal basins. There are two basic ways by which such entrapment can occur. One is by the collision of oceanic plateaus with the subduction zones which could entrap a piece of the large ocean basin by shifting the position of the subduction zone. The other is by transformation of a transform fault to a subduction system. This event is likely to occur when there is a change in the direction of motion of a plate. Under certain conditions a combination of collision of plateaus and the transformation of transform fault may occur, or they may occur in sequence. Some portions of at least four marginal basins have probably formed by the entrapment of old sea floor. They include the Amerasian Basin in the Arctic Ocean, the Bering Sea, the shallow portion of the Okhotsk Sea, and the West Philippine Basin. It is possible that some of the oceanic plateaus in these basins were not formed in place but came to their present position with the pre-existing ocean floor.

Introduction

Several theories have been put forward to explain the mechanism of formation of marginal basins (Figure 1). The entrapment of a marginal part of a pre-existing ocean by the formation of an island-arc has been suggested for the West Philippine Basin (Uyeda and Ben-Avraham, 1972) and the Bering Sea (Scholl et al., 1975). Back arc spreading caused by, or related to, subduction was proposed by Karig (1971a) for most marginal basins. Various models have been proposed for back arc spreading, including extensive magma intrusion (Hasebe et al., 1970), mantle wedge flow induced by subducting slab (McKenzie, 1969; Sleep and Toksoz, 1971) and sea-floor spreading caused by injection of

*On leave from Earthquake Research Institute, University of Tokyo, Tokyo, Japan 113

heated materials from the core-mantle boundary (Artyushkov, 1979). Opening related to a "leaky" transform fault was suggested for the Andaman Sea (Curray et al., 1979). Another mechanism was proposed by Uyeda and Miyashiro (1974) for the Japan Sea, Uyeda (1977) for the Gulf of California, and Eguchi et al. (1979) for the Andaman Sea. This is the initiation of back arc spreading related to a subduction of a spreading ridge. An explanation not related to plate tectonics is the formation of marginal basins by oceanization of continental crust (Beloussov and Ruditch, 1961).

Although most of the marginal basins are bordered by subduction zones on their oceanic side, the formation of a back arc basin does not always accompany the subduction process. The most notable example is the absence of marginal basins back of the Peru-Chile trench. In fact the entire eastern Pacific is lacking marginal basins related to present subduction along the eastern Pacific. Geological studies on land, however, suggest that such basins did possibly exist during the early Mesozoic and Paleozoic along various segments of the eastern Pacific margin (e.g. Churkin, 1974, Dalziel et al., 1974). Moreover, active opening does not appear to be occurring in some of the arcs with back arc basins (Uyeda and Kanamori, 1979; Uyeda, 1979). For example, the Sea of Japan was probably formed by extension in the past, but there is no active opening at present. Thus, there must be something in the present plate tectonics setting which causes this non-uniformity in the distribution and activity of marginal basins. Uyeda and Kanamori (1979) pointed out that the difference in the degree of coupling between the subducting and overriding plates may account for the difference in the tectonics of back arc regions: when the coupling is strong, the back arc region is dominated by a compressive stress regime, whereas when the coupling is weak, the stress in the back arc region tends to be extensional and spreading can occur. Taking the names of the most typical cases, they called the former case the "Chilean-

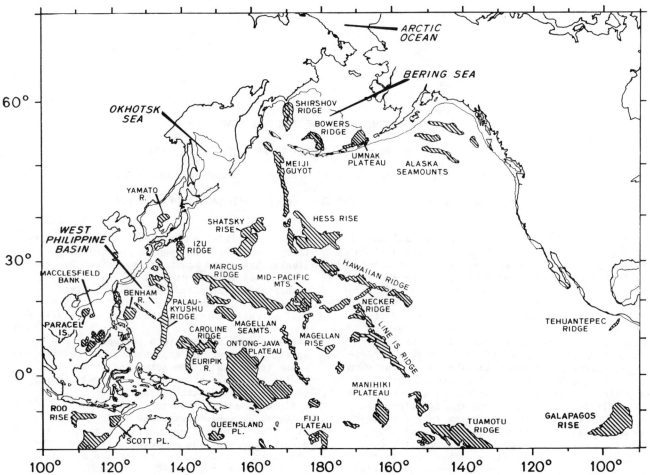

Fig. 1. Distribution of plateaus in the North Pacific. They exist in the main Pacific Basin as well as inside marginal basins. Hatched areas show outlines for plateaus. Included are rises which have been thought of as extinct arcs, ancient spreading ridges, detached and submerged continental fragments, anomalous volcanic piles, or uplifted normal oceanic crust. Marginal basins with names are those which contain portions formed by entrapment of a marginal part of a pre-existing ocean.

type" mode of subduction and the latter the "Mariana-type" mode of subduction.

Several hypotheses have been proposed to explain the cause of different modes of subduction. One hypothesis describes the different modes of subduction as different evolutionary stages of a single process (Kanamori, 1971; Uyeda and Kanamori, 1979). According to this model the nature of the contact between the upper and lower plates changes from tight coupling, like in South America, to decoupling, like in the Marianas, with time. The decoupling results in a oceanward retreat of the trench and back arc opening. A second hypothesis suggests that the age of the subducting plate controls the formation of marginal basins (Molnar and Atwater, 1978). Where the subduction plate is young, therefore hot and light, tight coupling will occur, and when it is old, cold and dense, decoupling will occur and a marginal basin may be formed. A third hypothesis suggests that the motion of the landward plate control the formation of marginal basins (Chase, 1978; Uyeda and Kanamori, 1979). According to this model, when the landward plate moves toward the trench, a tight coupling will occur and when it moves away from the trench a marginal basin will be formed. These models seem to provide adequate explanations to the major aspects of the present distribution of marginal basins, but not all of them. It appears difficult to prove the first hypothesis as the evidence for its validity are hidden within the geologic record of present and ancient active margins, although some attempts are being made along this line (e.g. Kobayashi, 1982). The second hypothesis seems reasonable. However, if it is the sole cause, it does not explain why the Sea of Japan for example has stopped spreading (Uyeda, 1979). The third hypothesis appears to be supported by the present day absolute plate motion. However this is somewhat of an ad hoc explanation and it is hard to prove its validity in the past, because the motions of landward plates relative to trench lines in the geologic past are only poorly constrained.

There appears to be a marked difference in stress regime between subduction zones with active back arc spreading (Mariana-type) and those without active back arc spreading (Chilean-type). More than 90 percent of global seismic energy is released in subduction zones without active back arc basins (Uyeda and Kanamori, 1979). Although an explicit relationship between seismic energy release and tectonic stress would be difficult to provide (Ruff and Kanamori, 1980), it may be reasonable to state that the Chilean-type subduction zones are associated with high stress regimes while the Mariana-type subduction zones are characterized by low stress regimes. This concept is important in dealing with the formation of marginal seas.

Much of the Pacific margin is composed of allochthonous terranes (Coney et al., 1980). The collision of these terranes with continental margins is a major factor in the process of mountain building in the Cordillera systems, and continental growth (Ben-Avraham et al., 1981). In this paper, the importance of collisional tectonics to the formation of marginal basins through entrapment is discussed. An important element in the model is the stress regime during the collision. Entrapment by mechanisms other than collision are also probably important for the formation of marginal basins.

Models of Entrapment

Entrapment of a part of pre-existing ocean by the formation of island-arc can be accomplished in two basic ways. One way is by collision of oceanic plateaus with subduction zones. Oceanic plateaus, ranging in size from 1000 km^2 down to a few square kilometers are embedded in the normal oceanic crust and move with it. They are found in all parts of the world ocean. Their origin ranges from detached and submerged continental fragments to piles of basalt which were formed on the sea floor. In most cases they are isostatically compensated. This indicates a light buoyant root beneath them. As a result, collision of oceanic plateaus with subduction zones produces profound effects including reduced seismicity, shifts of volcanic activity, and shifting of plate boundaries so that new subduction zones form elsewhere (Vogt et al., 1976; Nur and Ben-Avraham, 1981; Ben-Avraham et al., 1981). In fact it has been suggested that the collision of an oceanic plateau with a subduction zone will produce the same effects as continental collision, even if the plateau is made of volcanic material (Ben-Avraham et al., 1981). Because of its thickened crust, the plateau has a light root and probably an anomalous upper mantle. Therefore it is more buoyant than the adjacent oceanic crust, and upon collision it will behave like a continent and be accreted rather than subducted.

Indirect geological evidence indicates that plateaus similar to those that exist in ocean basins today also existed in ancient ocean basins. These ancient plateaus can now be recognized only by their remnants which have been incorporated into continental masses in the form of allochthonous terranes whose stratigraphy and paleomagnetism indicate distant origins (Coney et al., 1980; Ben-Avraham et al., 1981). The nature, history, and character of allochthonous terranes along the Pacific margin are best understood in the northern Cordillera of western North America, particulary in Southern Alaska and British Columbia. Amongst the allochthonous terranes in this region which are candidates for possible ancient oceanic plateaus now incorporated into the continental framework, Wrangelia (Jones et al., 1977) and Cache Creek (Monger, 1977) terranes are the best known examples.

The transformation of oceanic plateaus into allochthonous terranes may take different forms depending on, among other things, whether the tectonic stress is high or low. As stated in the previous section, seismicity implies that the stresses are high along subduction zones without active back-arc spreading. When a collision takes place in a high stress regime, the colliding mass will undergo extensive deformation during the accretion, but the configuration of the plate boundary will be changed only by a modest oceanward shifting of the trench, after which normal subduction will resume (Figure 2). In contrast, the plate boundary geometry may change significantly upon plateau collision in a low stress regime, and may produce a marginal sea by entrapment (Figure 2).

Dickinson (1978) discusses, in concept, two main kinds of crustal collision. In one kind, the active continental margin draws a passive block against it from a more seaward position. After crustal collision, subduction jumps to the opposite side of the accreted block, as depicted above. Dickinson (1978) indicated that microcontinental blocks, dormant island arcs, seamount chains, and aseismic oceanic ridges can all be accreted in this way. In the other kind, an active island arc collides with a passive continental margin. If the overall pattern of plate kinematics induces subduction to resume after such an arc-continent collision, the subduction flips to the opposite side of the accreted arc structure, reversing polarity, and an active continental margin results. We can envision a third type of crustal collision in which an active island arc collides with an active continental margin with opposite polar-

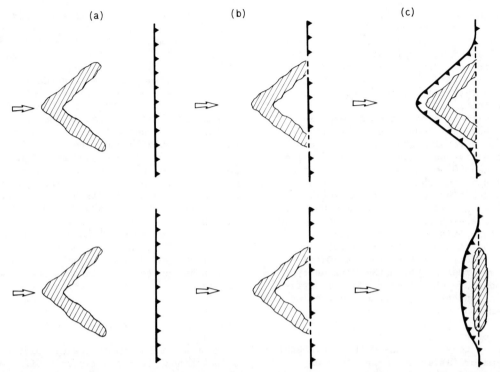

Fig. 2. Collision of an oceanic plateau with a plate boundary. Top: low stress regime (a) Oceanic plateau approaches a subduction zone. (b) The plateau collides with the subduction zones. (c) A new plate boundary forms owing to the presence of the plateau, and a marginal sea is formed by entrapment of old oceanic crust. Bottom: high stress regime: (a), (b) same as in Top. (c) The plateau is accreted to the plate margin and a small change occurs in the configuration of the plate boundary.

ity. In this type, if there is no spreading ridge in between, the relative motion between the two crustal units could be faster than in the other two types. Since it is not possible for a single plate to move in opposite directions simultaneously, the convergence at one subduction zone in such a case must be due to overriding of the upper plate on the fleeing oceanic plate. Examples of such cases are not well known, except the case which will be suggested later for Bowers Ridge and the Bering Sea margin in the past. The Molucca Sea may represent the last stage of this case (Silver and Moore, 1978).

Another mode of entrapment was suggested for the Western Philippine Sea by Uyeda and Ben-Avraham (1972), and for the Bering Sea by Hilde et al. (1977). According to this hypothesis, under certain conditions, entrapment of old ocean basin can occur by transformation of a transform fault to a subduction system. This event is likely to occur when there is a change in the direction of motion of a plate. Conceivably, subduction begins because the sea-floor on one side of the fault can be much younger and therefore hotter and less dense than that on the other side. As a result of the change in the direction of plate motion, the trend of the former transform fault is no longer parallel to the direction of relative plate motion and the side with the older sea-floor will be subducted beneath the other side (Figure 3). The transform fault then becomes a trench-island arc system. Under certain circumstances a combination of collision and the transformation of transform fault to a subduction system can occur.

In the following sections several examples of marginal basin information around the Pacific margin through entrapment are reviewed.

Arctic Basin

Geological studies on land have shown that the northern margin of the Pacific basin is composed of numerous allochthonous terranes of oceanic and continental origin which were successively accreted to the larger continental plates (Jones et al., 1977; Monger and Price, 1979; Fujita, 1978). By analyzing the past movement of some of these terranes Churkin and Trexler (1980) offered the following model for the origin of the Arctic Ocean Basin involving plate accretion and the entrapment of a part of the proto-

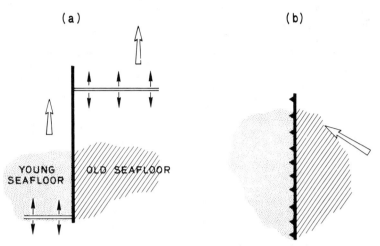

Fig. 3. Entrapment mechanism associated with a change in the direction of plate motion. (a) Active transform fault strikes in the direction of relative plate motion. (b) When a change in the direction of motion of one plate takes place a transformation of the transform fault to a subduction system will occur.

Pacific (Kula?) plate (Figure 4). In early Mesozoic time the major continental plates of the northern hemisphere were all part of the megacontinent Laurasia. As the northern Atlantic opened in the Jurassic, the Siberian and Russian platforms, which had been sutured together in the late Paleozoic, moved eastward around the Arctic while the North American plate moved westward. The moving continents, thus half closed off the ocean that separated Alaska-Chukotka from Siberia. This ocean was part of the Kula plate which during the early Jurassic, extended from the north Pacific into the Arctic. The continuation of the circumpolar drift of continents in combination with northward motion of the Kula and Kolyma plates resulted in collison and suturing of Kolyma and Siberia in the Cretaceous. Multiple collisions and accretion of allochthonous terranes against the Pacific margin of Alaska-Chukotka resulted in orogenic deformation. The collision between the Kula and Eurasia plates is probably represented by the Mendeleyev-Alpha fold belt (Herron et al., 1974). The collision and suturing of Alaska-Chukotka with the Eurasia plate isolated part of the Kula plate, creating a nucleus for the Arctic basin (Churkin and Trexler, 1980).

The Arctic Ocean is composed of two major basins, the Eurasian and the Amerasian, which are separated by the Lomonosov Ridge. The ridge is probably a continental fragment (Wilson, 1963). The above discussion of entrapment origin refers to the Amerasian Basin, of probable Mesozoic age. In terms of plate tectonics, the Eurasian Basin is simply an extension of the North Atlantic, which opened by sea-floor spreading during the past 63 m.y. (Herron et al., 1974).

A previous plate tectonic model for the evolution of the Amerasian Basin has been offered by Herron et al. (1974). They proposed that the basin was formed during the Jurassic magnetic quiet period (180 to 150 m.y.B.P.) by rifting of Kolyma away from North America. This model now seems rather unlikely in view of paleomagnetic data from Siberia. These data indicate that Kolyma has come from the south (McElhinny, 1973) and collided with the Siberian plate in the middle Mesozoic time. The entrapment model is thus a more likely mechanism for the formation of a portion of the Arctic basin.

Bering Sea

The Bering Sea is thought to be a marginal sea formed by the Aleutian Arc, which trapped a portion of the Kula plate (Cooper et al., 1976). Prior to the formation of the Aleutian Arc in late Mesozoic or early Tertiary time, subduction probably took place along the present-day continental margin of the Bering Sea (Cooper et al., 1976). As the arc formed, subduction shifted to the Aleutian Trench some time prior to the change in motion of the Pacific plate 42 m.y. ago (Dalrymple et al., 1977). Hilde et al. (1977) explained the formation of the present Aleutian arc by assuming the existence of a former transform fault connecting segments of the Kula-Farallon ridge, which was turned into a subduction zone. The formation of the arc can also be explained as triggered by the collision of an oceanic plateau with the Mesozoic subduction zone, as follows.

At present three large oceanic plateaus and ridges - the Umnak Plateau, the Bowers Ridge and the Shirshov Ridge, exist in the Bering Sea. Refraction data over the Bowers Ridge (Ludwig et al., 1971) and the Umnak Plateau (Cooper et al., 1980) indicate that a thickened welt of crustal material is present beneath both

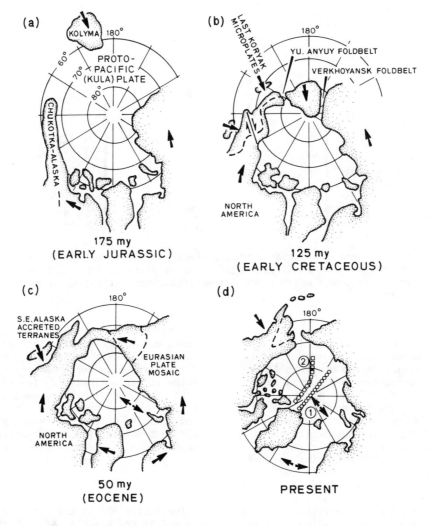

Fig. 4. Paleogeographic reconstructions of the Arctic showing northward drift, collision and accretion (after Churkin and Trexler, 1980).

features. The Bowers Ridge with altered andesitic rocks, a positive magnetic anomaly over its crest, and a sediment wedge on its northern side, is probably an extinct island arc. Newly acquired multichannel seismic profiles (Cooper et al., 1981) clearly reveal a fossil subduction zone on the northern side of the Bowers Ridge. Multichannel seismic profiles show that the Bering Sea margin was also a subduction zone. Thus, there must have been relative motion in the past between the Bowers Ridge and the Bering Sea margin.

It is not clear whether the Umnak Plateau, now situated between the Bering Sea margin and the Aleutian Arc, was formed in situ or not, but it is possible that it came to its present location from elsewhere, like the Bowers Ridge. Thus, a possible scenario is that before formation of the Aleutian Arc, the proto-Bowers Ridge and Umank Plateau moved into their present Bering Sea positions (Ben-Avraham and Cooper, 1981). The collision of the Umnak plateau with the then convergent Bering Sea margin may have caused subduction to terminate and jump southward, forming the Aleutian Arc. Similarly, the Shirshov Ridge, separating the Aleutian and Komandorsky Basins, could have been formed along a large transform fault which was active during the northward motion of the Kula plate (Figure 5) or by rifting away from Kamchatka

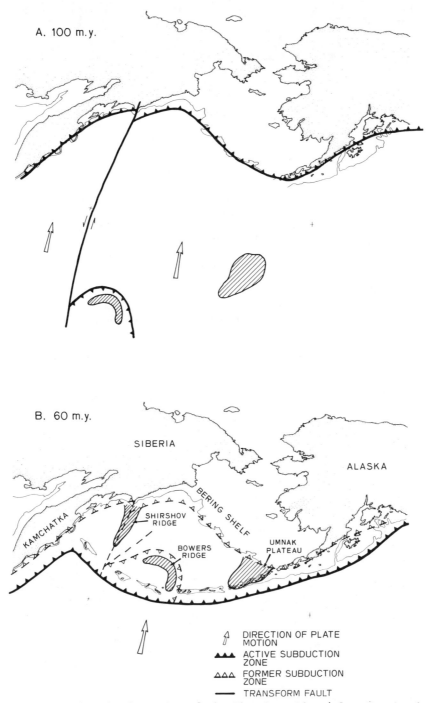

Fig. 5. Conceptual model for the formation of the Aleutian ridge (after Ben-Avraham and Cooper, 1981). Bowers ridge and Umnak plateau are thought to have come from the south with the Kula plate and Shirshov ridge to have formed in place. A - Late Mesozoic; B - Early Tertiary.

(Ben-Avraham and Cooper, 1981). Both mechanisms can explain why the Komandorsky Basin contains less sediments, has higher heat flow and thus is probably younger than the Aleutian Basin.

Okhotsk Sea

Most of the Sea of Okhotsk is underlain by continental crust (Burk and Gnibidenko, 1977;

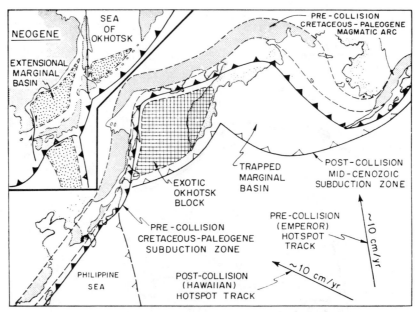

Fig. 6. Sketch map showing inferred evolution of subduction systems before and after accretion of the Okhotsk block by crustal collision followed by a subduction jump (after Dickinson, 1978).

Savostin et al., 1981). Despite this, a calc-alkaline volcano-plutonic belt lay west and north of the present Sea of Okhotsk during late Cretaceous and early Cenozoic time. This has led Den and Hotta (1973) and Dickinson (1978) to propose that a microcontinental Okhotsk block of unknown origin collided with the eastern margin of Eurasia during early Cenozoic time and caused the subduction zone to move into a new easterly location along the Kurile Ridge (Figure 6). The continental fragment, or possibly island arcs (Savostin et al., 1981), lodged in the Okhotsk Sea is now submerged; hence a new marginal sea formed. After the new subduction zone was established back arc spreading may have occurred to form the Okhotsk abyssal plain (or the Kurile Basin).

Geological studies in the northwest Pacific margin have indicated that northeast Siberia evolved through the collision of several terranes (Fujita, 1978), similar to the process which occurred in northwest America (Coney et al., 1980). Possibly there was a different stress regime along the northwest Pacific margin in the Mesozoic than that in the Cenozoic. During the Mesozoic, microcontinents, such as Kolyma, collided with the continental margin and were accreted to the continent, while the collision of the Okhotsk microcontinent with the continental margin during the Cenozoic resulted in shifting of the subduction zone to the oceanic side of the microcontinent to form a marginal sea.

West Philippine Basin

The Philippine Sea is composed of two distinct units; the West Philippine Basin with WNW trending magnetic anomalies and the eastern Philippine Sea with north-south trending magnetic anomalies. Ben-Avraham et al. (1972) found symmetrical magnetic anomalies across the Central Basin fault (the Philippine Ridge) in the West Philippine Basin. On the basis of this and other geophysical data they showed that the Central Basin fault is actually an extinct spreading center. Based on this observation, Uyeda and Ben-Avraham (1972) have proposed that the Philippine Ridge was connected by a large transform fault to the Kula-Pacific Ridge and was migrating northward. Because of this transform fault, the Philippine Ridge was left behind while the other ridges descended beneath Japan. When the direction of motion of the Pacific plate changed at about 42 m.y.B.P. (Dalrymple et al., 1977), subduction started along the former transform fault because the sea-floor west of the fault was much younger and therefore hotter and less dense than that to the east. The transform fault then became a subduction zone (now the Kyushu-Palau Ridge) and the Philippine Ridge became extinct (Figure 7). Thus, the part of the Kula plate west of the transform fault was trapped to become a marginal sea, the West Philippine Basin. After this, Karig's model might explain how the present state in the Phillippine Sea was reached (Karig, 1971b): first the development of the Parece Vela Basin by splitting the Kyushu-Palau Arc into two parts, a process that probably started in the Early Miocene as dated by JOIDES holes No. 53 and 54 (Fischer et al., 1971). Then, the Mariana Trough developed by splitting the West Mariana Ridge in two and causing the eastward bulge of the Mariana Ridge.

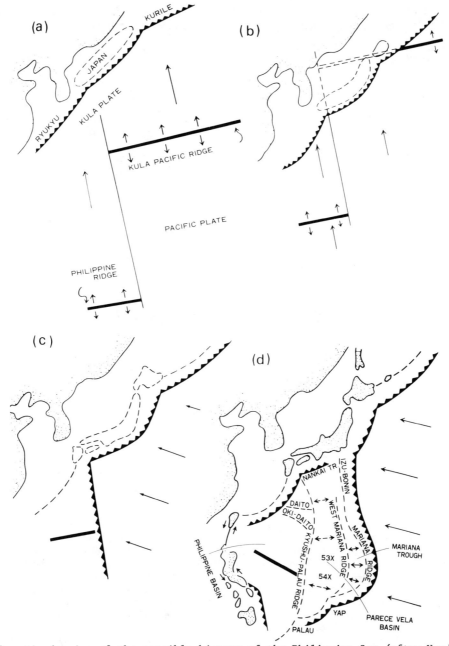

Fig. 7. Schematic drawing of the possible history of the Philippine Sea (after Uyeda and Ben-Avraham, 1972). (a) Possible plate motions and arrangement at about 100 m.y.B.P. (b) As the Kula-Pacific ridge descends beneath Japan, the Philippine Ridge lags behind. (c) About 40 m.y.B.P., the direction of motion of the Pacific plate changes to west-northwest and the transform fault becomes a subduction zone, after which back arc spreading starts. (d) Present state. Extentional opening first started to form the Parece Vela basin and then the Mariana trough. 53x and 54x indicate Deep Sea Drilling Project holes.

There are some problems in the scenario outlined above. One is the original position of the assumed transform fault and the other is the question of why the Kula plate west of the transform fault did not change its motion in harmony with the Pacific plate at 42 m.y.B.P. Both these problems may be related to the collision of the proto Izu-Bonin ridge against Honshu in pre-Tertiary time.

Uyeda and Ben-Avraham (1972) assumed that the transform fault in question was originally at about the present position of the Kyushu-Palau

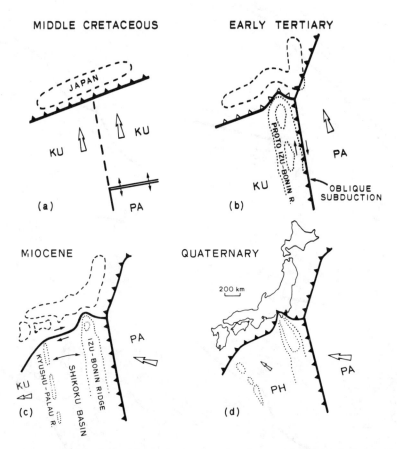

Fig. 8. Plate motion and the formation of the Izu-Bonin arc (modified from Matsuda, 1978). Middle Cretaceous (a) to Quaternary (d), KU - Kula plate; PA - Pacific plate; PH - Philippine Sea plate.

ridge relative to Honshu (Figure 7). This was clearly not right. Because there has been some subduction of the West Philippine Basin into the Ryukyu Trench, the original position of the transform fault should have been more east. The question is how far east. According to Matsuda (1978), the sharp bending of terranes in Central Honshu north of the Izu collision zone took place in pre-Miocene time or earlier (early Tertiary or even late Cretaceous) and the Miocene volcanic front of northeast Honshu extends smoothly southward to the eastern margin of the Izu-Bonin ridge (Figure 8). Matsuda postulated that along the transform fault, there was a line of uplifted features, similar to those along the present 90° East Ridge in the Indian Ocean, to cause the sharp bending of terranes in the early Tertiary time. He called this the Proto Izu-Bonin ridge. From these considerations Matsuda concluded that the Izu-Bonin arc and its ancestral ridge were originally located at essentially their present position relative to Honshu and have stayed there to the present. Thus, when the Pacific plate motion changed from NNW to WNW at about 42 m.y.B.P., the subduction zone was formed south of Central Honshu and not south of Kyushu. This implies that the subsequent episodes of back arc spreading have occurred westward and driven the remnant arcs (Kyushu-Palau and West Mariana Ridges) westward also, keeping the relative position of the trench to Honshu essentially unchanged (Figure 8). This scenario fits better with the idea suggested by Chase (1978) and Uyeda and Kanamori (1979) that trench lines tend to be immobile at the time of back arc spreading.

Now, let us consider about the second problem. It is sometimes questioned why the western part of the Kula plate did not change its motion along with the Pacific plate at 42 m.y.B.P. It is at the present stage not easy to solve this problem in an unequivocal manner as we do not know the dynamics of plate sufficiently. It may, however, be suggested that the collision of the Proto Izu-Bonin ridge with Honshu tended to block further subduction of this part of the Kula plate just as Umnak plateau might have blocked the Bering Sea margin subduction. If this was the case, the transform fault with the Proto Izu-Bonin ridge would have been the log-

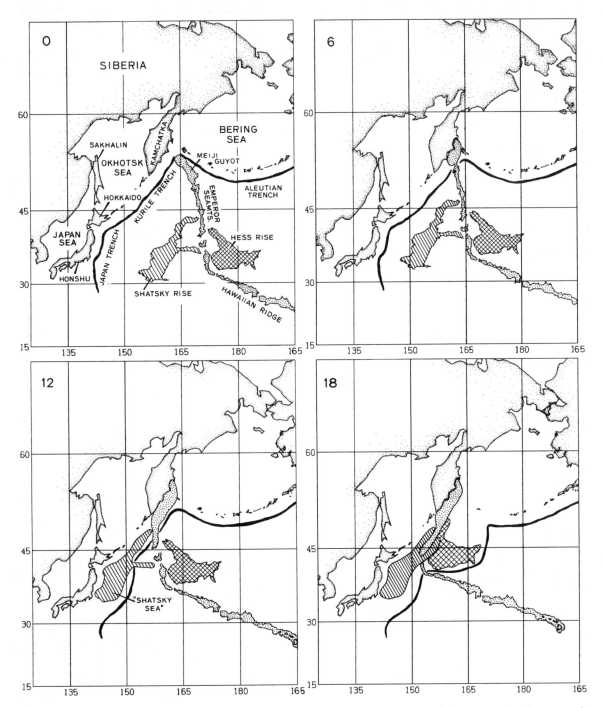

Fig. 9. Sketches of possible future events in the northwest Pacific (after Ben-Avraham et al., 1981) using the present day plate motion parameters of Minster and Jordan (1978).
(A) Present day configuration of Shatsky rise, Hess rise, Emperor seamounts and Hawaiian ridge.
(B) 6 m.y. from now all the plateaus will be to the northwest of their present position. Meiji Guyot, following collision with the subduction zone, will become part of the Kamchatka Margin.
(C) 12 my.y. from now Shatsky rise will be in full collision with north Honshu, Hokkaido and the Kuriles. At this stage the trench might move to the oceanic side of the plateau and a new marginal sea, the "Shatsky Sea" will be formed.
(D) 18 m.y. from now Shatsky rise, Hess rise and the Emperor ridge will all be part of the Eurasia plate as collision takes place and new plate boundaries are formed in this region.

ical line for the new subduction. Both the buoyancy of the ridge and the relative youthfulness of the sea-floor west of the transform fault mentioned above have made the Pacific plate subduct. It is not certain if the collision of the Proto Izu-Bonin ridge exerted any influence on the motion of the Pacific plate, but it apparently did on the Kula plate.

Conclusions

Active back arc basins occur in a very few locations on earth. There must be, therefore, a very unique setting in these places which allow back arc spreading to occur. Some marginal seas were obviously formed by rifting in the past. How spreading initiated in these seas and what made it stop are among the major unsolved problems in geodynamics. For example, in the Shikoku Basin spreading started about 30 m.y.B.P. and ended 20 m.y.B.P. (Kobayashi and Nakada, 1978). Several attempts to relate the initiation of spreading in these areas to plate motions have been made (e.g. Jurdy, 1979) but it is not clear why spreading stopped.

In most of the attempts to explain the origin of marginal basins so far, the irregularities of the sea floor have been ignored. In fact, beside mid-oceanic ridges, oceanic plateaus of various origins exist on the seafloor and are moving with it to the subduction zones. The collision of plateaus with subduction zones could entrap a piece of the ocean basin by shifting the position of the subduction zones. This appears to have occurred in low stress regimes. The collision of plateaus with subduciton zones under high stress regimes will result in accretion of the plateau to the continents without large shifting of the subduction zones.

Another mode of entrapment is by transformation of a transform fault to a subduction system. This event is likely to occur in response to a change in the direction of plate motion. Under certain conditions collision of oceanic plateaus may induce this mode of entrapment. The direction of plate motion may be changed as a result of collision of large oceanic plateaus with subduction zones (Engebretson and Ben-Avraham, 1981), and then the transformation of transform faults to subduction systems will occur.

It has been shown in this paper that some portions of at least four marginal seas have probably been formed by entrapment of old sea floor: the Amerasian Basin in the Arctic Ocean, the Bering Sea, the shallow portion of the Okhotsk Sea, and the West Philippine Basin. When more data are available it may turn out that more basins were initially formed by entrapment. There is some evidence to suggest that at least in some marginal seas back arc spreading was inititated following the entrapment. Examples are the Kurile Basin, the Shikoku and Parece Vela Basins and the Mariana Trough.

A small number of plateaus are presently being consumed at subduction zones with profound geological effects, including reduced seismicity and shifts of volcanic activity (Vogt et al., 1976). Furthermore, the distribution of oceanic plateaus (Figure 1) suggests that additional collisions will occur in the future, e.g. Shatsky Rise with Japan (Figure 9). If we extend the motion of the plates into the future, it is inevitable that new marginal seas, such as the "Shatsky Sea", will be formed in the same way some of them were formed in the past.

The entrapment origin of marginal seas may provide some clue for understanding the origin of the enigmatic plateaus which exist today within these seas. These include the Chukchi Cap in the Arctic Ocean, the Bowers Ridge, Shirshov Ridge and Umnak Plateau in the Bering Sea, the Daito Ridge, Oki-Daito Ridge and Benham Rise in the West Philippine Basin, and many others. It is possible that some of these features were not formed in place, but came from somewhere else.

Acknowledgments. This work was supported by a grant from the Geophysics Program, Division of Earth Sciences, U.S. National Science Foundation. Texas A&M Geodynamics Research Program Contribution #24.

References

Artyushkov, E.V., "Geodynamics", Nauka, Moscow, 328 pp., (in Russian), 1979.

Beloussov, V.V. and E.M. Ruditch, Island arcs in the development of the earth's structure, J. Geol., 69, 647-658, 1961.

Ben-Avraham, Z., C. Bowin, and J. Segawa, An extinct spreading center in the Philippine Sea, Nature, 240, 453-455, 1972.

Ben-Avraham, Z. and A.K. Cooper, Early evolution of the Bering Sea by collision of oceanic rises and north Pacific subduction zones, Geol. Soc. Am. Bull., 92, 485-495, 1981.

Ben-Avraham, Z., A. Nur, D. Jones and A. Cox, Continental accretion and orogeny: from oceanic plateaus to allochthonous terranes, Science, 213, 47-54, 1981.

Burk, C.A. and H.S. Gnibidenko, The structure and age of acoustic basement in the Okhotsk Sea, in "Island Arcs, Deep Sea Trenches, and Back-Arc Basins", Manik Talwani and W.C. Pitman III (eds.), 451-462, Am. Geophys. Union Maurice Ewing Ser. 1, 1977.

Chase, C.G., Extension behind island arcs and motion relative to hotspots, J. Geophys. Res., 83, 5385-5387, 1978.

Churkin, M. Jr., Paleozoic marginal ocean basin-volcanic arc systems in the cordilleran foldbelt, in: "Modern and Ancient Geosynclinal Sedimentation," R.H. Dott Jr. and R.H. Shaver (eds.), Soc. Econom. Paleontologists and Mineralogists, Special Pub. 19, 174-192, 1974.

Churkin, M. Jr., and J.H. Trexler, Circum-Arctic plate accretion-isolating part of a

Pacific plate to form the nucleus of the Arctic Basin, Earth Planet. Sci. Lett., 48, 356-362, 1980.

Coney, P.J., D.L. Jones, and J.W.H. Monger, Cordilleran suspect terranes, Nature, 288, 329-333, 1980.

Cooper, A.K., M.S. Marlow, and D.W. Scholl, Mesozoic magnetic lineations in the Bering Sea marginal basin, J. Geophys. Res., 81, 1916-1934, 1976.

Cooper, A.K., M.S. Marlow and Z. Ben-Avraham, Multichannel seismic evidence for the origin of Bowers Ridge, Geol. Soc. Am. Bull., 92, 474-484, 1981.

Cooper, A.K., D.W. Scholl, T.L. Vallier and E.W. Scott, Resource report for the deep-water areas of proposed OCS lease sale #70, St. George Basin, Alaska, USGS Open-File Report 80-246, 90 p., 1980.

Curray, J.R., D.G. Moore, L.A. Lawver, F.J. Emmel, R.W. Raitt, M. Henry and R. Kieckhefer, Tectonics of the Andaman Sea and Burma, in: "Geological and Geophysical Investigations of Margins", Watkins et al., (eds.), Assoc. Geol. Mem. 29, 189-197, 1979.

Dalrymple, G.B., D.A. Clague and M.A. Lanphere, Revised age for Midway volcano, Hawaiian Volcanic chain, Earth Planet. Sci. Lett., 37, 107-116, 1977.

Dalziel, I.W.D., M.J. deWit and K.F. Palmer, Fossil marginal basin in southern Andes, Nature, 250, 291-294, 1974.

Den, N. and H. Hotta, Seismic refraction and reflection evidence supporting plate tectonics in Hokkaido, Pap. Meteorol. Geophys., 24, 31-54, 1973.

Dickinson, W.R., Plate tectonic evolution of north Pacific rim, J. Phys. Earth, 26, Suppl. S1S19, 1978.

Engebretson, D.C. and Z. Ben-Avraham, Collisional events and the direction of relative motion between western North America and the Pacific Basin plates since the Jurassic, Geol. Soc. Amer., Abstracts with Programs, 13, 55, 1981.

Eguchi, T., S. Uyeda and T. Maki, Seismotectonics and tectonic history of the Andaman Sea, Tectonophys., 57, 35-51, 1979.

Fischer, A.G., B.C. Heezen et al., Initial Reports of the Deep Sea Drilling Project, Vol. VI. (U.S. Government Printing Office, Washington, D.C.), 1329 pp., 1971.

Fujita, K., Pre-Cenzoic tectonic evolution of northeast Siberia, J. Geol., 86, 159-172, 1978.

Hasebe, K., N. Fujii and S. Uyeda, Thermal process under island arcs, Tectonophys., 10, 335-355, 1970.

Herron, E.M., J.F. Dewey, and W.C. Pitman III, Plate tectonic model for the evolution of the Arctic, Geology, 2, 377-380, 1974.

Hilde, T.W.C., S. Uyeda and L. Kroenke, Evolution of the Western Pacific and its margin, Tectonophys., 38, 145-165, 1977.

Jones, D.L., N.J. Silberling and J.W. Hillhouse, Wrangellia - a displaced terrane in northwestern North America, Canad. J. Earth Sci., 14, 2565-2577, 1977.

Jurdy, D.M., Relative plate motions and the formation of marginal basins, J. Geophys. Res., 84, 6796-6802, 1979.

Kanamori, H., Great earthquakes at island arcs and the lithosphere, Tectonophys., 12, 187-198, 1971.

Karig, D.E., Origin and development of marginal basins in the western Pacific, J. Geophys. Res., 76, 2542-2561, 1971a.

Karig, D.E., Structural history of the Mariana island-arc system, Geol. Soc. Amer. Bull. 82, 323-344, 1971b.

Kobayashi, K., Cycles of subduction and Cenozoic arc activity in the northwestern Pacific margin, (This volume).

Kobayashi, K., and M. Nakada, Magnetic anomalies and tectonic evolution of the Shikoku interarc basin, J. Phys. Earth, 26, Suppl., S391-S402, 1978.

Ludwig, W.J., S. Murauchi, N. Den, M. Ewing, H. Hotta, R.E. Houtz, T. Yoshii, T. Asanuma, K. Hagiwara, T. Sato and S. Ando, Structure of Bowers Ridge, Bering Sea, J. Geophys. Res., 76, 6350-6366, 1971.

Matsuda, T., Collision of the Izu-Bonin Arc with Central Honshu: Cenozoic tectonics of the Fossa Magna, Japan, J. Phys. Earth, 26, Suppl., S409-S421, 1978.

McElhinny, M.W., "Paleomagnetism and Plate Tectonics", Cambridge Univ. Press, 358 pp., 1973.

McKenzie, D.P., Speculation on the consequences and causes of plate motions, Geophys. J. Roy. Astron. Soc., 18, 1-32, 1969.

Minster, J.B. and T.H. Jordan, Present-day plate motions, J. Geophys. Res., 83, 5331-5354, 1978.

Molnar, P., and T. Atwater, Interarc spreading and cordilleran tectonics as alternates related to the age of subducted oceanic lithosphere, Earth Planet. Sci. Lett., 41, 330-340, 1978.

Monger, J.W.H., Upper Paleozoic rocks of the western Canadian Cordillera and their bearing on Cordilleran evolution, Canad. J. Earth Sci., 14, 1832-1859, 1977.

Monger, J.W.H. and R. A. Price, Geodynamic evolution of the Candian Cordillera - progress and problems, Canad. J. Earth Sci., 16, 770-791, 1979.

Nur, A. and Z. Ben-Avraham, Volcanic gaps and the consumption of aseismic ridges in South America, in "Nazca Plate: Crustal Formation and Andean Convergence", Geol. Soc. Am. Memoir 154, 729-740, 1981.

Ruff, L. and H. Kanamori, Seismicity and the subduction process, Phys. Earth Planet. Int., 23, 240-252, 1980.

Scholl, D.W., E.C. Buffington, and M.S. Marlow, Plate tectonics and the structural evolution of the Aleutian-Bering Sea region, in: The

Geophysics and Geology of the Bering Sea Region, Geol. Soc. of Amer. Special Paper 151, 1-31, 1975.

Silver, E.A. and J.C. Moore, The Molucca Sea collision zone, Indonesia, J. Geophys. Res., 83, 1581-1691, 1978.

Sleep, N. and M.N. Toksoz, Evolution of marginal basins, Nature, 233, 548-550, 1971.

Savostin, L., L. Zonenshain and Baranov, Geology and plate tectonics of the Sea of Okhotsk (This volume).

Uyeda, S., Some basic problems in the trench-arc-back arc systems, in "Island Arcs, Deep Sea Trenches and Back-Arc Basins", M. Talwani and W.C. Pitman III (eds.), M. Ewing Series 1, American Geophys. Un., 1-14, 1977.

Uyeda, S., Subduction zones: facts, ideas, and speculation, Oceanus, 22, 52-62, 1979.

Uyeda, S., and Z. Ben-Avraham, Origin and development of the Philippine Sea, Nature, 240, 176-178, 1972.

Uyeda, S., and H. Kanamori, Back-arc opening and the mode of subduction, J. Geophys. Res., 84, 1049-1061, 1979.

Uyeda, S. and A. Miyashiro, Plate tectonics and the Japanese Islands: A synthesis, Geol. Soc. of Amer. Bull., 85, 1159-1170, 1974.

Vogt, P.R., A. Lowrie, D.R. Bracey, and R.N. Hey, Subduction of aseismic oceanic ridges: effects on shape, seismicity and other characteristics of consuming plate boundaries, Geol. Soc. of Amer. Special Paper, 172, 59 p., 1976.

Wilson, J.T., Hypothesis of Earth's behaviour, Nature, 198, 925-929, 1963.

EXPLOSION SEISMOLOGICAL EXPERIMENTS ON LONG-RANGE
PROFILES IN THE NORTHWESTERN PACIFIC AND THE MARIANAS SEA

Toshi Asada
Institute of Research and Development, Tokai University,
Hiratsuka, Japan

Hideki Shimamura
Geophysical Institute, Hokkaido University, Sapporo, Japan

Shuzo Asano
Earthquake Research Institute, Tokyo University,
Bunkyo-ku, Tokyo, Japan

Kazuo Kobayashi and Yoshibumi Tomoda
Ocean Reserach Institute, Tokyo University,
Nakano-ku, Tokyo, Japan

Abstract. The structure of the oceanic lithosphere in the Northwestern Pacific was investigated by long-range explosion seismological experiments using ocean bottom seismometers (OBS). Results obtained in 1971, 1973, 1974, 1977 and 1978 are summarized. A distinct velocity anisotropy with magnitude of several percent was found in the Northwestern Pacific. The azimuth of the maximum velocity (8.6 km/sec) is approximately north-south. The anisotropy seems to exist through the entire depth extending from the Mohorovicic discontinuity to the zone below the low-velocity layer. The lithosphere beneath the Marianas Sea seems to be different from that beneath the Northwestern Pacific. Attenuation of seismic waves in the lithosphere is large in the Marianas Sea compared with that in the Northwestern Pacific.

Introduction

The investigation of the seismic structure of the upper mantle, especially the study of vertical distributions of seismic wave velocities near the bottom of the plate, is one of the most important problems in modern geophysics. No accurate determination of the thickness of the plate has yet been made. It is not even known even whether or not the seismic velocity near the bottom of the plate is discontinuous all over the world. Strictly speaking, we do not know if we can define the thickness of the plate.

The importance of studying the upper mantle structure was recognized by geophysicists in the 60's, and seismological experiments on profiles as long as 2,000 km were carried out on the continents. In particular, the experiments using atomic explosions in Nevada and the experiments in the Early Rise projects supplied us with valuable information about the upper mantle beneath the North American continent (Green and Hales, 1968; Hales, 1972; Lewis and Mayer, 1968; Mereu and Hunter, 1969; Mansfield and Evernden, 1966). Recently in Europe, experiments on long-range profiles have been carried out by cooperation among scientists in several countries (Group Grands Profils Sismiques and German Research Group for Explosion Seismology, 1972; Hirn et al., 1973; Bamford et al., 1976). Similar experiments have also been done in the USSR (Ryaboy, 1977). Long-range experiments in an ocean with OBS' swere originally made by a Japanese group (Asada and Shimamura, 1979).

The results obtained in such experiments on the continents can be summarized as follows. There is a low-velocity layer beneath the continents, the depth of which ranges from 40 to 100 km depending on locality. In some areas the low-velocity layer is very thin or lacking. In many cases there is a high-velocity layer with Vp from 8.4 to 8.6 km/sec between the Mohorovicic discontinuity and the low-velocity layer. These facts have been revealed by methods with higher resolution such as explosion seismology.

The structure of the oceanic lithosphere and the velocity distriabution under it were studied only by the method using seismic surface waves before 1970 (Kanamori and Press, 1970). The dependence of the structure on laclity has not been precisely investigated. However, the structure of the oceanic lithosphere and the low-velocity layer and their dependence on

TABLE 1. List of the experiments on long-range profiles

Date	Region	Distance	Sites (OBS)	No. of shots
July 1973	Marianas Sea	60-1120 km	5 (11)	3
Oct. 1974	N W Pacific	800-1240 km	6 (11)	2
Jan. 1976	Marianas Sea	50-1820 km	12 (27)	5
July 1977	N W Pacific	550-1150 km	5 (8)	3
July 1978	N W Pacific	130-1200 km	5 (10)	8

locality are important for the future development of plate tectonics (Yoshii et al., 1976).

There are two methods for studying the structure of the lithosphere along a long profile. One is to carry out an eaxperiment using ocean bottom seismometers (OBS) and shots under the water. The other is to fire many underwater shots along along profile at sea and receive the resulting waves with seismometers installed on land. Hales et al. (1970) set up a profile in the Gulf of Mexico and fired 23 shots which were recorded with a seismic array installed in Mexico. Steinmetz et al. (1977) carried out such an experiment in the Atlantic. These experiments gave interesting results. In both cases high-velocity layers with a velocity of about 8.6 km/sec were found. Hales et al. (1970) obtained a time-distance diagram awith a clear offsets in travel time lines around a distance of about 1,200km.

It seem to us that the OBS is better than the hydrophone for recording waves from distant shots, because the records obtianed by hydrophones, especially those operated at the sea surface, are noisier than those obtained by OBS at the ocean bottom. The authors designed and developed ocean bottom seismometers for recording seismic waves from distant origins, first at the Geophysical Institute of the University of Tokyo and later at the Geophysical Institute of Hokkaido University (Shimamura and Asada, 1974; Yamada et al., 1976). The present paper summarizes the results of our long-range experiments in the Northwestern Pacific.

Ocean bottom seismometers used for these experiments

In 1969 we deployed our first OBS under the sea off the Izu Islands. At that time the available information about noises as well as signals and their spectra at the bottom of the sea was quite inadequate. We therefore designed an OBS using a continuous recorder with three channels of different levels. As we adopted a continuous recorder, various types of noises were recorded, some having very short duration. Shimamura and Asada (1970) also found noises which looked like very near microearthquakes.

As various types of noises may be generated in the sea, an event recorder may not be useful for recording natural earthquakes, because it may often be mistriggered. We also found that some of the natural earthquakes had very sharp pulse-like first arrivals (Asada and Shimamura, 1976), which may not be adequate to activate a trigger.

A cylindrical pressure case, 90 cm long and 19 cm in diameter, is used for our OBS. It is made of aluminum and can be used to a depth of 8,000 m. The system for deployment and recovery is of the tethered type. The noise due to the vibration of the nylon rope is insulated by a heavy anchor and chains. It is possible for our OBS to detect signals as small as 10^{-6} cm/sec. Noises generated by a micromotor and the other mechanical system driven by it are also successfully cut off so as not to disturb the geophone.

Generally speaking, the ocean bottom is very quiet. Especially in the frequency band ranging from 5 to 7_6 Hz, the ground noise is less than 1 to 2×10^{-6} cm/sec. Nevertheless, in the lower frequency range, the amplitude of the noise becomes larger (Asada and Shimamura, 1976). Due to these circumstances, seismic waves generated by a shot using a small amount of explosive which has a higher predominant frequency can be detected by an OBS at a much greater distance than on land.

We obtained clear records of seismic waves from 1 kg shots at a distance of 90 km in the Marianas Sea. We also obtained a clear Pn from shots of explosives of several kilograms at a distance of about 400 km (Asada and Shimamura, 1976). Since our purpose was to study the structures of the oceanic lithosphere and their dependence on locality, it was necessary to conduct long-range experiments in various sea areas. The experiments conducted are listed in Table I, and the results obtained in the respective areas are described later.

The recovery rate of tethered OBS's is about 90 percent in the south seas of outside of the Kuroshio Current area. In the sea near Japan, however, the recovery rate is smaller, mainly because of the Kuroshio Current.

We usually deploy two OBS's at one site, at an interval of about 200 m. A better signal-to-noise ratio can be obtained by stacking two records if the noises for both records are about the same in amplitude. Alternatively we sometimes deploy an OBS with a different characteristic response.

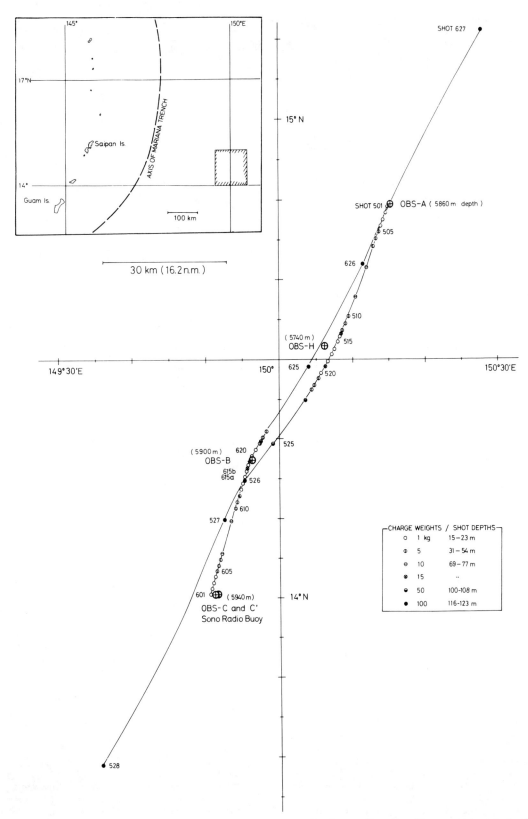

Fig. 1 A. Locations of OBS sites and shots in the Marianas Sea for the experiment in 1971.

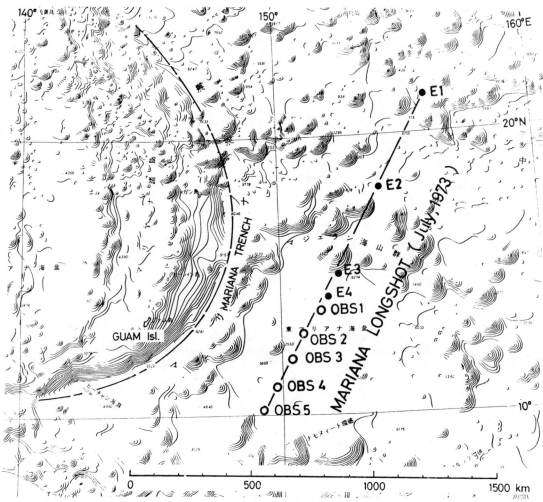

Fig. 1 B. Locations of OBS sites and shots in the Marianas Sea for the experiment in 1973. Black circles indicate the location of shots. The charge sizes are 1.5 tons for E1, 1.5 tons for E2, 0.45 tons for E3, and 0.15 tons for E4.

Experiment in the Marianas Sea

In 1971 and 1973, profiles of 146 km and 1400 km length, respectively, were set for long-range experiments in the Marianas Sea. This area was selected for the first experiments because the bottom topography is fairly flat and wide enough for a long profile, and also because the Marianas Sea is calm compared with the seas near Japan. The locations of the OBS sites and the shots are given in Figure 1 A and B. The record section for the experiment in 1973 is shown in Figure 2. As shown in this record section, the travel time line derived from the first arrivals is simple. The velocity is 8.2 km/sec in the range from 0 to 500 km. No other lines with different velocities are found. The maximum amount of explosives for one shot was 1.5 tons.

In 1976 we carried out an experiment in the same area in which the length of the profile reached 1,800 km. The Hawaii Institute of Geophysics and Sakhalin Complex Scientific Research Institute, USSR, joined in the experiment, bringing the total number of sites to 27 and the total number of OBS's to 41. The location of the OBS sites and the shots are shown in Figure 3 (OBS 21 LONGSHOT-3). We already had obtained a successful result in 1974 in an area to the east of the Ogasawara Islands; i.e., a time-distance line with a velocity of about 8.6 km/sec in the range from 500 to 1,200 km with an offset at about 1,000 km. We therefore expected clear evidence for a high-velocity layer and a low-velocity layer below the high-velocity layer in this area also.

The record section obtained in this experiment is shown in Figure 4 A and B. The length of the profile is as long as 1,800 km. It is impossible to find distinct first arrivals in the range over 500 km. Nor are there any later phases having

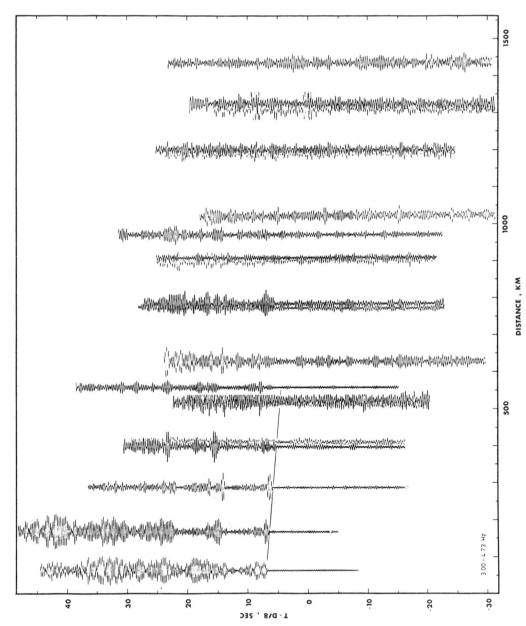

Fig. 2 Record section for the experiment in the Marianas Sea in 1973.

EXPLOSION SEISMOLOGICAL EXPERIMENTS 109

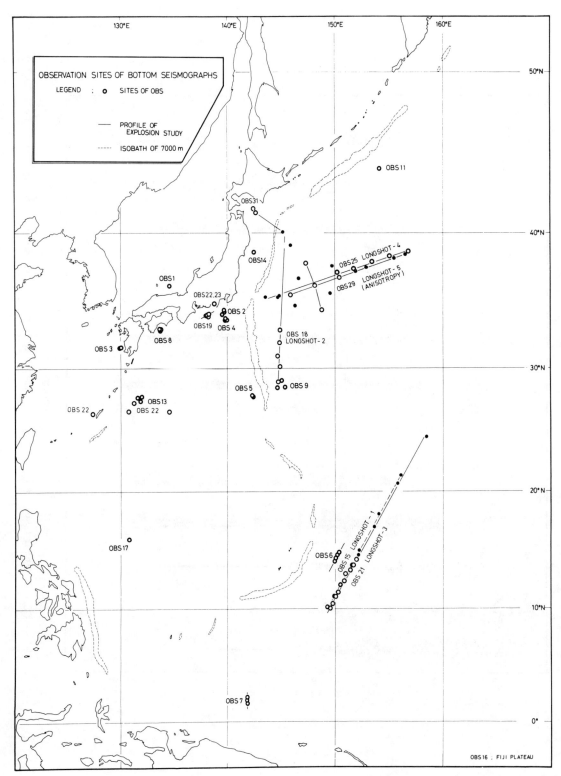

Fig. 3. Locations of OBS sites and shots during the period from 1970 to 1979. The locations of OBS sites and shots for the experiment in the Marianas Sea in 1976 are shown as OBS 21 LONGSHOT-3.

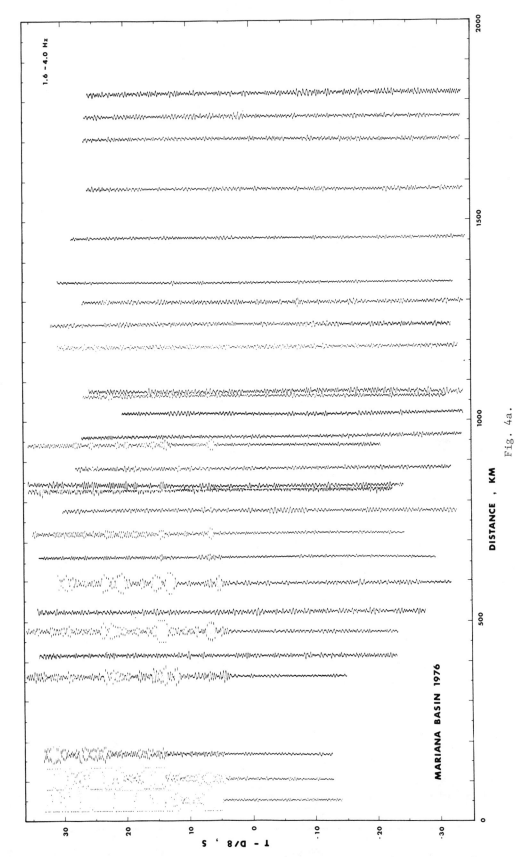

Fig. 4a.

Fig. 4 A and B. Record sections for the experiment in the Marianas Sea in 1976. A: 1.6-4.0 Hz. B: 3.2-6.4 Hz.

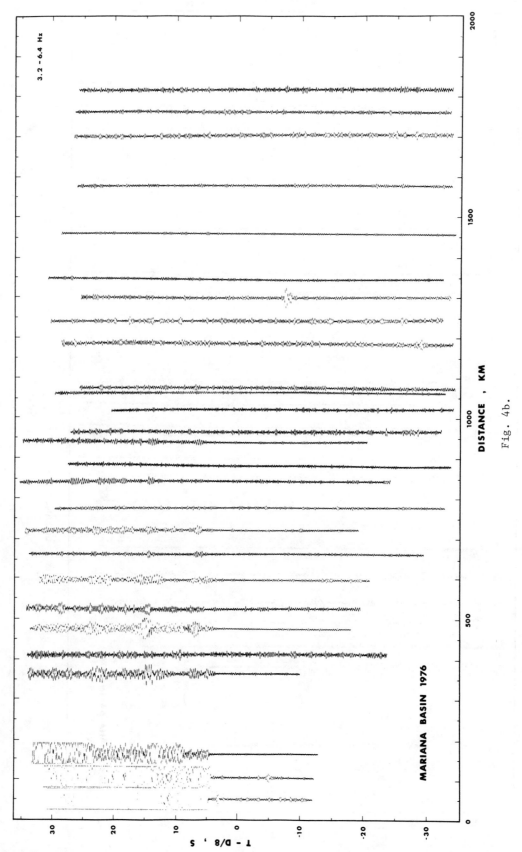

Fig. 4b.

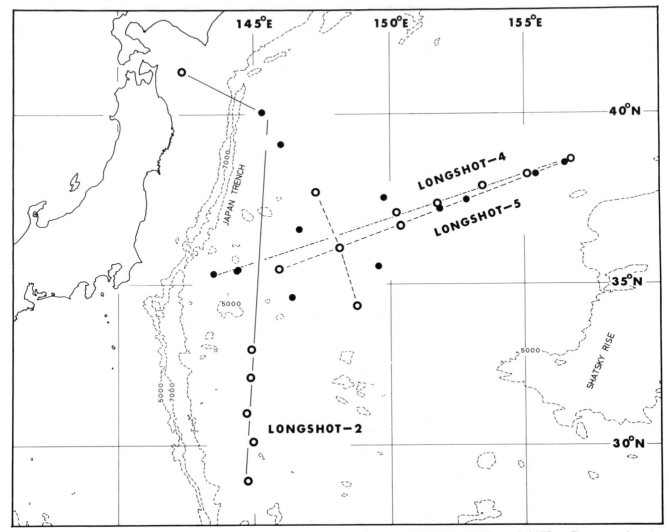

Fig. 5. Locations of OBS sites and shots for the experiments in 1974, 1977, and 1978. The experiment in 1974 is shown as LONGSHOT 2. LONGSHOT 4 was carried out in 1977, and LONGSHOT 5 in 1978.

good correlation with each other. These results were far from what we had expected. It seems to us that the upper mantle beneath the Marianas Sea is not as transparent to seismic waves as that along the profile in the area to the east of the Ogasawara Islands and along the profile roughly perpendicular to it in the Northwestern Pacific. The results obtained may be attributed to one of the following reasons: (a) the attenuation itself is large; (b) the noise level at the bottom is high; (c) both attenuation and noise are large; or (d) there is a thick low velocity layer with such a velocity distribution that causes a very wide shadow zone.

Experiment in the area to the east of the Ogasawara (Bonin) Islands

In 1974 we conducted an experiment along a profile extending north to south in the area to the east of the Ogasawara (Bonin) Islands. The locations of OBS's and the shots are given in Figure 5 (LONGSHOT-2). The number of shots was limited due to an insufficient ship time. We emphasized the range around 1,000 km in planning the experiment, and accordingly omitted observations in the range from zero to several hundred kilometers.

This experiment was successful: clear first arrivals were recorded at the sites over 1,000 km and even at a site 1,235 km from the shot. The outline of this experiment is described in Asada and Shimamura (1976). A record section processed at the Geophysical Institute, Hokkaido University, is shown in Figure 6. The travel time lines in Figure 6 show a velocity of 8.6 km/sec and an offset around 1,000 km. In the first arrivals at sites over 1,000 km, lower frequency is predominant, and in those at sites less than 1,000 km higher frequency is predominant.

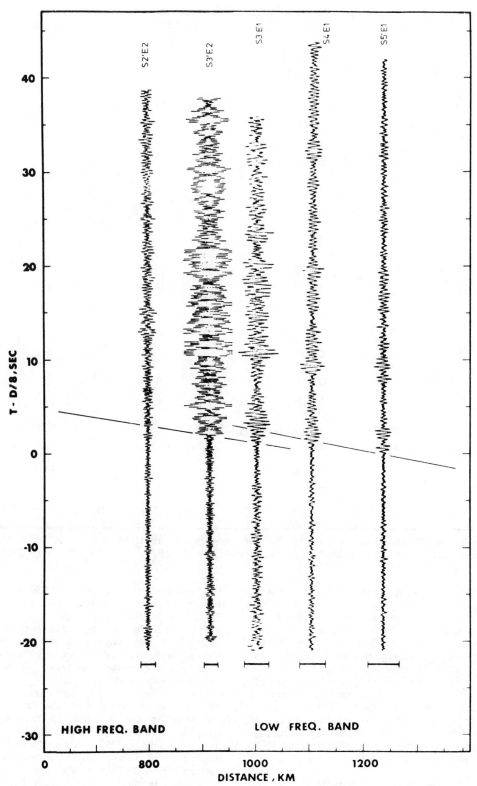

Fig. 6. Record section for the experiment on the profile to the east of the Ogasawara Islands in 1974. Bars indicate the amplitude for 10×10^{-6} cm/sec. The spacing of OBS sites was about 60 nautical miles. However, because of rough surface condition of the sea, shots had to be fired at sites about 60 nautical miles apart, resulting in only six traces.

As the number of seismograms in Figure 6 is only five, the conclusion that the apparent velocity (Vapp) is 8.6 km/sec and the time-distance lines have an offset around 1,000 km may not be convincing enough. However, this conclusion is supported by an analysis of thirty or more records of natural earthquakes which indicate that the Vapp is about 8.6 km/sec on the average (Shimamura and Asada, 1976). Some of the time-distance lines of the waves of natural earthquakes have an offset similar to that of the waves from artificial origins.

Unfortunately, no observations were made in the range less than 800 km. If we assume a velocity of 8.2 km/sec with an intercept time of 7.0 sec for the range less than 600 km in the profile, the structure given in Table 2 is obtained in which a structure is assumed for the crust. It should be noted that, in the record section in Figure 6, most of the first arrivals would be readable even if the amplitude of the noise was much larger. This feature of the record section is quite different from that of the record sections for the experiment in the Marianas Sea (Figure 4).

Experiment in the Northwestern Pacific in 1977

A long-range experiment with OBS's was carried out in 1977 on a profile perpendicular to the north-to-south profile to the east of the Ogasawara Islands. To avoid the Shatzky Rise, however, the angle between the two profiles was not exactly $90°$. The OBS sites and the shots are shown in Figure 5 (LONGSHOT-4). Two profiles were set nearly perpendicular to each other in order to obtain information about the existence of velocity anisotropy for the entire depth of the lithosphere. Due to the limited ship time, only five OBS sites were used, with two OBS's deployed at each site. Two shots of 0.5 tons of explosives were fired at a single site in order to increase the signal-to-noise ratio by stacking two records. We also prepared one 5-ton shot, but it was unsuccessful.

The signal-to-noise ratio in this experiment was surprisingly good, and fairly clear first arrivals were recorded even at the most distant site. Processing by stacking records or other means was not necessary. The record section for this experiment is given in Figure 7. A time-distance line with a velocity of 8.04 km/sec can be drawn on this record section in the range from 650 km to 1150 km. The possibility for other solutions may remain, but the analysis of the records shown in Figure 9 resulted in the same velocity, namely 8 km/sec on the average.

The noise level at the bottom of the sea to the east of the Ogasawara Islands and in the Northwestern Pacific is very low: less than 1 to 2×10^{-6} cm/sec in the frequency range from 5 to 7 Hz. This may be one of the reasons why clear signals were obtained at distant places in both experiments. At the bottom of the Marianas Sea, the level of noise is about 10 db higher than at the bottom of the other two areas. The spectrum of the noise is also different (Yamada and Inatani, in preparation). As already mentioned, however, the first arrivals in the record sections in Figures 6 and 7 were so strong that they could be identified, even if the noise level were higher. The above discussion leads to the conclusion that the signal levels at sites more than several hundred kilometers distant in the Marianas Sea are below the noise levels, not only due to the high level of the noise but also due to the small amplitude of the signals.

Tentative conclusions from the experimental results are as follows:
(1) A layer with a velocity of 8.6 km/sec was found in the area to the east of the Ogasawara Islands. However, the velocity was 8 km/sec along the profile which is not only perpendicular to but also intersects the profile along which the velocity was found to be 8.6 km/sec. Namely, we obtained two values for the velocity of P waves in one area.
(2) The offset in the time-distance lines was found only in the area to the east of the Ogasawara Islands.
(3) The attenuation is dependent on locality. The attenuation in the Marianas Sea is larger than that in the other two areas. Alternatively, beneath the Marianas Sea there may exist a thick low-velocity layer which causes a very wide shadow zone.
(4) If the structure of the oceanic lithosphere is indeed dependent on locality, all the results may be explained. But as the two profiles intersect each other, such a conclusion is not justified, and another interpretation is necessary. We therefore postulate the existence of a velocity anisotropy, or the dependence of the apparent velocity upon the azimuth.

TABLE 2. Structure of lithosphere derived from the data obtained in the experiement on the profile to the east of the Ogasawara Islands in 1974

Velocity	Depth
1.5 km/s	
2.2	0.0
2.2	0.3
4.33	0.31
4.33	2.33
7.05	2.34
7.05	9.85
8.20	9.86
8.20	60.00
8.55	60.10
8.60	80.00
8.40	80.10
8.40	120.00
8.60	120.10
8.80 km/s	300.00 km

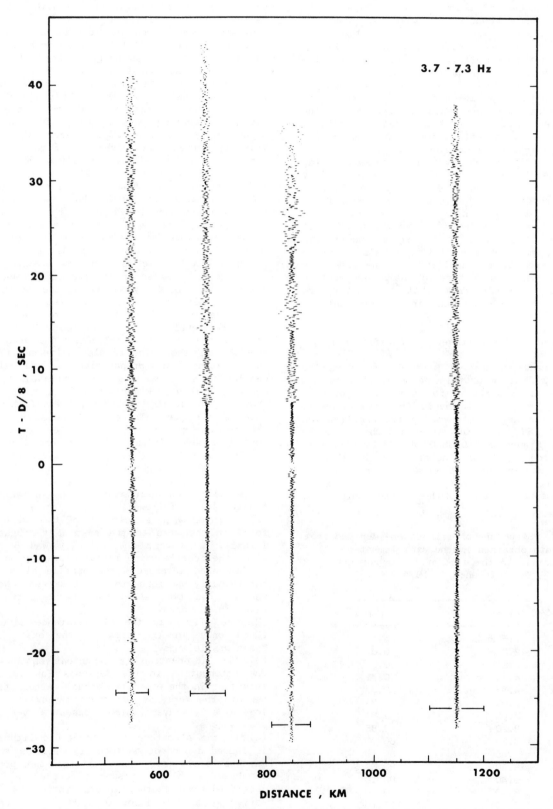

Fig. 7. Record section for the experiment on the profile from east to west in the Northwestern Pacific in 1977.

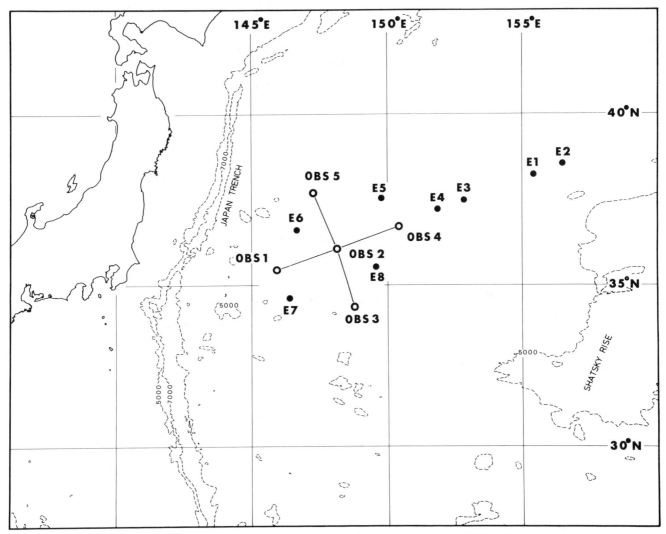

Fig. 8. Location of OBS sites and shots in the experiment in 1978. E means shot.

Experiment in the Northwestern Pacific in 1978

In order to study velocity anisotropy extending over the entire lithosphere, we planned the arrangement of the OBS sites and shots as shown in Figure 8. As we had already conducted two experiments along the profiles perpendicular to and crossing over each other, we had to choose the area shown in Figure 8 where the Kuroshio Current was meandering. The speed of the Kuroshio Current is usually 4 to 5 knots. If the Kuroshio Current was not flowing around the OBS sites during the time for shooting and recovery, conditions would have been satisfactory; however, this was not the case for most of the OBS's. Only the OBS at site 5 recorded seismic waves from the shots successfully. At the other sites we obtained only records with poor signal-to-noise ratio, probably because the anchor could not stand the strong tension of the rope due to the Kuroshio Current. Accordingly, in the present paper the time-distance relations were calculated based on the shot times and the arrival times at site 5. The crustal structure was assumed based on the results of the experiments with hydrophones for calculating intercept times. The crustal structures are almost the same throughout the Northwestern Pacific as that described later, and we may adopt one typical intercept time for calculating time-distance relations.

As we have successful results from only one OBS in the experiment in 1978, it is necessary to know the thickness of the crust at each shot point in order to obtain a time-distance line for Pn. According to the studies on crustal structures using hydrophones in the Northwestern Pacific, the structures were found to be surprisingly similar in the experiments along the seven reversed profiles (Ludwig et al., 1971; Murauchi

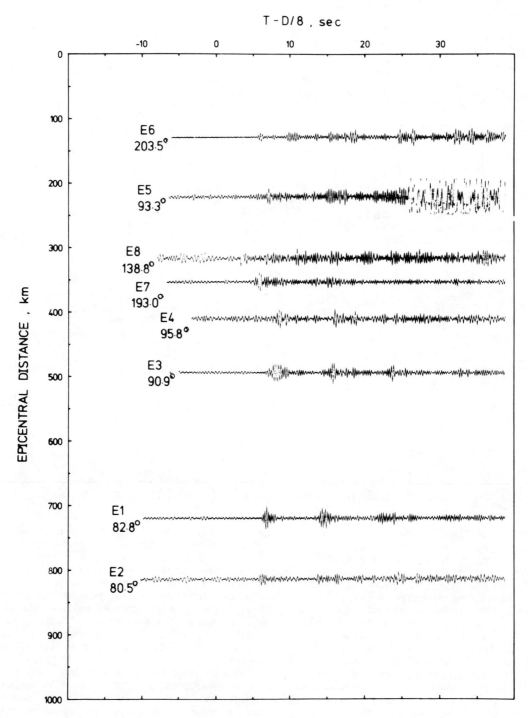

Fig. 9. Record section for the experiment in 1978.

TABLE 3. Structure of lithosphere derived from the data obtained in
the experiment on the profile to the east of the Ogasawara Islands in 1974

Explosion	Distance	Azimuth	TT - D/8	Calculated Velocities	Shot
E-8	317.4km	138.8°	3.12s	8.57 8.58 8.61 km/s	250kg
E-7	354.1 km	193.0°	5.19s	8.06 8.09 8.12 km/s	250kg
E-6	130.1km	203.5°	5.87s	7.88 7.96 8.03 km.s	250kg
E-5	222.4kg	93.3°	6.61s	7.72 7.77 7.81 km/s	250kg
E-4	410.4km	95.8°	6.28s	7.90 7.92 7.95 km/s	300kg
E-3	493.7km	90.9°	6.05s	7.94 7.97 7.99 km/s	500kg
E-2	814.9km	80.5°	6.23s	7.95 7.96 7.98 km/s	600kg
E-1	719.4km	82.8°	6.06s	7.96 7.98 7.99 km/s	500kg

Intercept Time 5.62 5.78 5.94   s

et al., 1973; Den et al., 1974). The time term of the Mohorovicic discontinuity in these profiles is 1.85 sec $\pm$ 0.16 sec on the average. Assuming the water depth is 5,670 m, the intercept time in the time-distance diagrams is 5.78 sec $\pm$ 0.16 sec.

The record section obtained in this experiment with OBS 5 is shown in Figure 9. It is quite obvious from this record section that no one time-distance line can be fitted for the first arrivals. This is particularly true for the range shorter than 400 km. For the range over 400 km, it is possible to fit a line with a velocity of about 8.22 km/sec to the first arrivals. The scattering of the first arrivals in the range shorter than 400 km may be attributable to a velocity anisotropy. The azimuths from the OBS 5 to the shots are within $\pm$ 8° for the shots E 1, 2, 3 and 4, or the shots more distant than 400 km. Therefore, the effect of the anisotropy may be neglected as far as travel times of seismic waves from the shots E 1, 2, 3 and 4 are concerned. The arrival times of P waves from these shots are much later than those in the experiment in 1974. The differences are 2 to 3 seconds in the range from 600 km to 800 km. These values are considered to be large; that is, such differences in time as 2 to 3 seconds for the same distance but for different azimuths are seldom found in the usual explosion seismological experiments. In Table 3 travel-times and distances from shots to OBS 5 are listed.

It takes different times for waves from the shots E 5, 6, 7 and 8 to arrive at OBS 5, although the distances are not much different. For instance, the reduced travel times from E 8 and E 5 differ by 3 seconds, whereas the difference in distance is 100 km. Such a difference in the reduced travel time at two sites close to each other is difficult to interpret based on the usual homogeneous structure. The amplitudes of waves seems to be dependent on the location of the shots. For instance, the waves from E 4 and 5 are small, but the waves from E 7 and 8 are large in amplitude in spite of the approximately equal distance.

The profile on which OBS 5, E 1, 2 and 3 are situated is a reversed profile of the experiment in 1977. The travel times for both profiles can be fitted on one line in the range from 500 km to 1,150 km. There is no dipping sturucture along the east-to-west profile.

On the other hand, the travel times along these profiles are quite different from the travel times along the profile in the 1974 experiment which is perpendicular to the present profile. Also the difference in travel times increases as the distance from the origin increases. We may naturally conclude that the difference between the travel times along the intersecting north-to-south and east-to-west profiles are not due to the structure in the shallow parts of the oceanic lithosphere such as the crust or uppermost mantle, but are due to the structure in the deeper part. The present authors would like to note again that there is an offset in the time-distance lines on the north-to-south profile to the east of the Ogasawara Islands at a distance of about 1,000 km. The difference in the velocities in the two profiles is found in the distance range over 1,000 km, as well as in the distance range of several hundred kilometers. Therefore, the deeper part here refers to the lower lithosphere and the asthenosphere. In other words, we may conclude that there is a velocity anisotropy extending down to the layer below the low-velocity zone.

Conclusions

Together with the tentative conclusions already mentioned, the following conclusions are obtained if we take into account the results obtained in the experiment in the Northwestern Pacific in 1978.

(1) The attenuation of seismic waves is large in the lithosphere or in the low-velocity zone beneath the Marianas Sea. That is, there is a very thick low-velocity layer with a peculiar velocity distribution, which causes a very wide shadow zone. The noise level at the bottom of

the Marianas Sea is high compared with those in the other areas.

(2) There is a velocity anisotropy whose magnitude is several percent in the Northwestern Pacific including the area to the east of the Ogasawara Islands. This anisotropy extends to a depth below the low-velocity layer (Asada and Shimamura, 1976).

(3) Along the azimuth with the minimum velocity, the offset of the time distance lines is not obvious.

(4) Along the azimuth with the maximum velocity, there is an offset of time-distance lines as shown in Figure 6.

(5) The azimuth with the maximum velocity is approximately north-to-south.

(6) In the area of the experiments in the North Western Pacific, the attenuation seems to be small compared with that in the Marianas Sea, even along the profile with smaller velocity for the distance range of over several hundred kilometers.

(7) The results obtained in the explosion seismology experiments are also supported by the results from natural earthquakes.

References

Asada, T., and H. Shimamura, Observation of earthquakes and explosions at the bottom of the Western Pacific: Structure of oceanic lithosphere revealed by longshot experiment, The Geophysics of the Pacifc Ocean Basin and Its Margin, editors, G.H. Sutton, M.H. Manghnani, R. Moberly and E.U. McAfee, A.G.U. Monograph 19, 135-154, 1976.

Asada, T., and H. Shimamura, Long-range refraction experiments in deep ocean, Tectonophysics, 56, 67-82, 1979.

Bamford, B., S. Faber, B. Jacob, W. Kaminski, K. Nunn, C. Prodehl, K. Fuchs, R. King, and P. Willmore, A lithospheric seismic profile in Britain I, Geophys. J. R. astro. soc., 44, 145-160, 1976.

Den, N., W.J. Ludwig, S. Murauchi, J.I. Ewing, H. Hotta, N.T. Edger, T. Yoshii, T. Asanuma, K. Hagiwara, T. Sato, and S. Ando, Seismic refraction measurements in the Northwest Pacific Basin, Jour. Geohys. Res., 74(6), 1421-1434, 1969.

Green, R.W.E. and A.L. Hales, The travel times of P waves to 30° in the central United States and upper mantle structure, Bull. Seism. Soc. Am., 58, 267-289, 1968.

Groupe Grands Profils Sismiques and German Research Group for Explosion Seismology, A long-range seismic profile in France from Bretagne to the Provence, Ann. Geophys. 28, 2, 247-256, 1972.

Hales, A.L., The travel times of P seismic waves and their relevance to the upper mantle velocity distribution, The Upper Mantle, A.R. Ritsema (ed.), Tectonophysics, 13, 447-482, 1972.

Hales, A.L., C.E. Halsley, and J.B. Nation, P travel-times for an oceanic path, J. Geophys. Res. 75, 7362-7381, 1970.

Hirn, A., L. Steinmetz, R. Kind, and K. Fuchs, Long range profiles in western Europe: II Fine structure of the lower lithosphere in France, Zeitschr für Geophys., 39, 363-384, 1973.

Kanamori, H., and F. Press, How thick is the lithosphere?, Nature, 226, 331, 1970.

Lewis, B.T.R., and R.P. Meyer, A seismic investigation of the upper mantle to the west of Lake Superior, Bull. Seism. Soc. Am., 58, 565-596, 1968.

Ludwig, W.L., J.I. Ewing, M. Ewing, K. Murauchi, N. Den. S. Asano, H. Hotta, M. Hayakawa, T. Asanuma, K. Ichikawa and I. Noguchi, Sediments and structure of the Japan trench, Jour. Geophys. Res., 71(8), 2121-2137, 1966.

Mansfield, R.H., and J.F. Evernden, Long range seismic data from Lake Superior seismic experiment, 1963-1964, The Earth beneath the Continent, edited by J.S. Steinhart and T.J. Smith, A.G.U. Monograph 10, 249-269, 1966.

Mereu, R.F., and J.A. Hunter, Crustal and upper mantle structure under the Canadian shield, Bull. Seism. Soc. Amer., 59, 147-165, 1969.

Murauchi, S., N. Den, S. Asano, H. Hotta, T. Yoshii, T. Asanuma, K. Hagiwara, K. Ichikawa, T. Edgar, and R.E. Houtz, Crustal structure of the Philippine Sea, Jour. Geophys. Res., 73(10), 3143-3171, 1968.

Ryaboy, V.Z., Study of the structure of the lower lithosphere by explosion seismology in the USSR, Journ. of Geophysics, 43, 593-610, 1977.

Shimamura, H., T. Asada, and K. Takano, Measurements of microearthquakes at sea bottom, La Mer (Bull. Soc. Fr. Japan. Oceanogr.), 8, 6-12, 1970.

Shimamura, H., and T. Asada, A cassette recorder for an ocean bottom seismometer, Geophys. Bull. Hokkaido Univ., 32, 17-24, 1974.

Shimamura, H., and T. Asada, Apparent velocity measurements on an oceanic lithosphere, Physics of the Earth and Planetary Interiors, 13, 15-22, 1976.

Shimamura, H., T. Asada, and M. Kumazawa, High shear velocity layer in the upper mantle of the Western Pacific, Nature, 269, 680-682, 1977.

Steinmetz, L., R.B. Whitmarsh, and V. S. Moreira, Upper mantle structure beneath the Mid-Atlantic ridge north of the Azores based on observations of compressional waves, Geophys. J. R. astro. Soc., 44, 145-160, 1977.

Yamada, T., H. Shimamura, and T. Asada, Resolving power of OBS recording system using compact cassette magnetic tape in determining time difference between signals in separate channels, Zisin (Journ. Seism. Soc. Japan) 29, 289-297, 1976.

Yoshii, T., Y. Kono, and K. Ito, Thickening of the oceanic lithosphere, The Geophysics of the Pacific Ocean Basin and Its Margin, editors, G.H. Sutton, M.H. Manghnani, R. Moberly, and E.U. McAfee, A.G.U. Monograph 19, 423-432, 1976.

VELOCITY ANISOTROPY EXTENDING OVER THE
ENTIRE DEPTH OF THE OCEANIC LITHOSPHERE

Hideki Shimamura
Geophysical Institute, Hokkaido University,
Sapporo, Japan

and

Toshi Asada
Institute of Research and Development,
Tokai University, Hiratsuka, Japan

Abstract. In 1974, 1977, and 1978, long-range explosion experiments were carried out in the Northwestern Pacific by Asada, Shimamura, Asano, Kobayashi, and Tomoda (1981). In these experiments the present authors and their colleagues determined the existence of a velocity anisotropy extending down to a layer below the low-velocity layer first found in the experiment in 1974. In the present paper apparent velocities of waves along the profiles are obtained for 36 earthquakes, and data from 5 explosions in 1977 and 1978 are added. More detailed information about anisotropy extending over the entire lithosphere was obtained. The azimuth of the maximum velocity is $155(335)^\circ$ and perpendicular to the magnetic lineation in the Northwestern Pacific, although the azimuth of the present sea floor motion is 305°. The mean velocity is 8.15 km/sec. The magnitude of the anisotropy is $\pm$ 0.55 km/sec or 13 percent.

Introduction

A large amount of data about the crustal structure in the oceans was accumulated by the middle of the 1960's. Hess (1964) made a study of the dependence of Vpn on azimuth, and concluded that a velocity anisotropy existed at the upper most mantle and that the azimuth of the maximum velocity was parallel to a fracture zone. Later, experiments particularly designed for the study of velocity anisotropy were carried out by Morris et al. (1969) and Raitt et al. (1969), and the following conclusions were obtained. The magnitude of velocity anisotropy is several percent in most cases, and the azimuth of the maximum velocity is parallel to the spreading direction. However, these experiments were carried out only in the Eastern Pacific, and the lengths of the profiles were all less than 100 km, as hydrophones were used for detecting signals. Because of these short profiles, anisotropies only in the uppermost mantle were detected, and no information was obtained for the deeper part of the oceanic lithosphere. In experiments using hydrophones the range of profiles cannot be over 200 km due to poor signal-to-noise ratio. Another defect of using hydrophones is the difficulty of maintaining the positions of shooting ships and sonobuoys for the extended period of time which is indispensable for precise refraction experiments. Moreover, in analysing data from most explosion experiments with hydrophones, diagrams using reduced-travel-times have not been adopted, probably due to the large scattering of data plots. However, reduced-travel-time curves are necessary in order to obtain sufficient accuracy for the study of anisotropy.

Recently many continental explosion seismological experiments have been carried out in Europe and in the USSR. The existence of anisotropy in the upper part of the continental lithosphere has been revealed by the explosion seismological method as well as by the analysis of surface waves, and papers including reviews have been published already (e.g. Bamford and Crampin, 1977). Most of these papers describe the anisotropy just below the Mohorovicic discontinuity. Only Hirn discussed the anisotropy in the lower lithosphere (1977). According to him, all data can be interpreted consistently, if the anisotropy is assumed to extend down to depths where the lithosphere-asthenosphere transition is supposed to be. However, he could not obtain any relation between the azimuth of the maximum velocity and tectonics in the area.

As far as the oceanic lithosphere is concerned, many problems remain to be solved in spite of its importance in tectonics. First of all, a strict definition of a lithosphere has not yet been proposed. This is because of the lack of knowledge about its structure, thickness, and the nature of its bottom. For instance, it has not been revealed even whether the bottom of a lithosphere is a discontinuity or not.

In order to study these problems, we have carried out long-range experiments in the Northwestern Pacific and the Marianas Sea (Asada et al., 1981). The results obtained in these experiments lead to the conclusion that velocity anisotropy extends to below the low-velocity layer. However, these results are obtained only for limited azimuths. Large numbers of explosions and sufficient ship time are necessary in order to obtain detailed information about velocity anisotropy in the oceanic lithosphere. In order to study the structure of the oceanic lithosphere and asthenosphere, it is indispensable to obtain seismograms at epicentral distances over 1,000 km.

It is helpful for obtaining results with good accuracy to use seismic waves from controlled sources. However, it is difficult to get clear first arrivals of waves from artificial origins at large distances. Carrying out a long-range explosion experiment is expensive and unbelievably time-consuming. It is better to use natural earthquakes as wave sources for obtaining clear records at large distances over 1,000 km.

The profiles of the experiments in 1974 and 1977 were laid on the deep sea basin, where the bottom topography was flat and regarded as a typical oceanic plate (Asada et al., 1981). The principal purpose of these experiments was to obtain records of seismic waves from artificial origins, but the arrays were designed also for measuring apparent velocities of seismic waves from earthquakes located on the land side of the Japan trench.

Many earthquakes are occurring in the sea to the east of Honshu, Japan. Most of their foci are located by the J.M.A. (Japan Meteorological Agency), the I.S.C., and universities having their own dense networks for observing microearthquakes. The network of Tohoku University is highly sensitive and located in the Tohoku (northeast) area which is approximately parallel to the Japan trench, making it possible to locate the foci of earthquakes under the sea with good accuracy.

As the seismic arrays in 1974 (OBS 18 LONG-SHOT-2) and in 1977 (OBS 25 LONG SHOT-4) consisted of OBS's deployed along one line in each case (Asada et al., 1981), we can get the values of Vapp (apparent velocity) along the profiles. The error in determining the location of foci of natural earthquakes is as large as 10 km when they are close to the trench. In order to obtain a precise value of Vapp using data from earthquakes or to get an accuracy of at least 0.1 km/sec, the intervals of the OBS sites must be large enough, because determination of the location of an OBS has an error as large as about 0.4 km, although the timekeeping accuracy is good enough with a DR recorder as well as a crystal clock. Too large spacing of OBS sites is undesirable, due to the possible inhomogeneity in the oceanic lithosphere. In the Northwestern Pacific Basin, bottom topography is flat and featureless. It was possible to lay an array of 10 OBS's at 5 sites with spacing of 100 km. The authors obtained Vapp by dividing phase difference by distance and correcting the value, taking the value of the azimuth from the array to the epicenter into account.

Our OBS's recorded many earthquakes during the period of our experiments. Most of them occurred in the area between the trench axis and the coast of North Honshu Island. An OBS deployed at a site in the ocean basin several hundred kilometers away from the trench recorded more than one earthquake each hour on the average. The total number of Vapp determinations for P waves was 20 during the experiment in 1974 and 16 in 1977.

Even an OBS deployed at a site in the oceanic basin 1,000 km away from the trench axis recorded many earthquakes, with clear first arrivals and saturated trace amplitude. But many of these earthquakes were poorly recorded by the network on land, and the foci were not located. The J. M. A. network locates less than 7 to 8 earthquakes during a month in the area along the Kuril trench near Hokkaido, The number of earthquakes which can be used for obtaining Vapp with good accuracy will increase if the locations of foci are determined by an OBS network placed on the landward side of the trench.

Results

Values of Vapp for P and S waves obtained through the north-south OBS array in the experi-

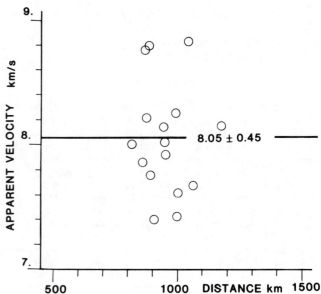

Fig. 1. Apparent velocities for P waves obtained in the 1977 experiment and epicentral distances.

ment in 1974 are given in Shimamura and Asada (1976) and Shimamura et al. (1977). The relation between epicentral distance and Vapp is shown in Figure 3 in the paper cited above. The mean value of Vapp is 8.64 km/sec, while the velocity obtained in the explosion seismological experiment in 1974 is about 8.6 km/sec (Shimamura and Asada, 1976). The mean value for Vapp along the same profile for S waves is 4.88 km/sec.

Along the east-west profile in the explosion experiment in 1977, the velocity was about 8.02 km/sec, as shown in Figure 7 in the paper by Asada et al. (1981). The mean value of Vapp of waves from earthquakes was 8.05 km/sec for P waves and 4.53 km/sec for S waves. The diagram of Vapp and distance is shown in Figure 1. The scattering of Vapp is larger in the experiment in 1977 than in 1974. It should be noted that the epicentral distances of earthquakes which are used for obtaining Vapp range from 650 to 1,800 km as shown in Figure 3 in the paper by Shimamura and Asada (1976) and Figure 1 in the present paper, and the areas for the two experiments overlapped.

In the experiment in 1974 (Asada and Shimamura, 1976; Asada et al., 1981), we found an offset of the travel-time curve at the epicentral distance of about 1,000 km. In addition, four of the travel-time curves of natural earthquakes which are used for calculating Vapp had offsets around 600 to 800 km (Shimamura and Asada, 1976). The P wave velocities in the ranges both nearer and more distant from the epicentral distance of the offset are the same. This makes it clear that the maximum depth of the ray path is below the low-velocity layer when the epicentral distances are more than 800 to 1,000 km.

In some cases we must take account of the effect of the horizontal component of the refraction of seismic waves coming into the descending oceanic lithosphere from the continental lithosphere; in other words, we must trace

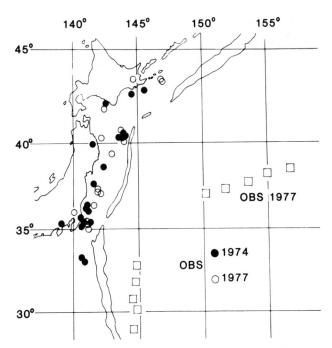

Fig. 3. Location of epicenters of earthquakes used for the diagram in Figure 2.

the ray path, when the ray passes from the continental to the oceanic lithosphere. In calculating Vapp, if the foci are situated within the oceanic lithosphere or at its surface, even if it is descending, it is not necessary to make any correction for refraction. Even when the foci are situated in the shallow part of the shelf inside the trench, the correction is not necessary when they are on the extension of a profile, and the profile is perpendicular to the trench axis.

Otherwise, the problem of ray tracing from a focus in the continental lithosphere to the oceanic lithosphere can be solved under several assumptions, as we have limited knowledge about the shape of the boundary, velocities in both lithospheres, and the direction of the maximum velocity, if an anisotropy exists there. It is not possible to make realistic estimates of these parameters and even more difficult to calculate realistic ray paths. In any case, however, the total amount of the error in calculating Vapp seldom exceeds 1 to 2 percent, when we neglect the effect of the horizontal component of the refraction of the waves passing from the continental to the oceanic lithosphere.

In the present paper, it is assumed that earthquakes occur in the descending oceanic lithosphere. Another assumption is that foci are deep enough and close to the boundary between the continental and the oceanic lithosphere, even if they are in the continental lithosphere.

The depth of the boundary of the descending lithosphere is about 10 to 20 km at a distance of

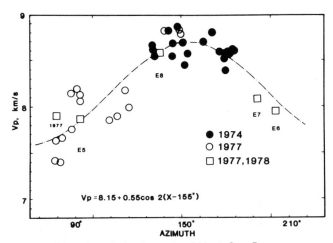

Fig. 2. Seismic velocities for P waves versus azimuth.

100 km from the trench axis and 30 to 40 km at 150 km. Accordingly, if the depth of the foci of earthquakes occurring in the area between the sea of Honshu and the Japan trench is about 30 to 40 km or deeper, the above assumptions are satisfied. Furthermore, even if a focus is shallower (the epicentral distance to OBS sites is usually over several hundred kilometers and much longer than that from the focus to the boundary), the second assumption is approximately satisfied. In the present paper, earthquakes poorly located and/or having poor first arrivals are omitted. As stated before, there are many cases where the accuracy of the location of foci is poor, even through the trace amplitudes of seismograms recorded with OBS's are saturated. Earthquakes deeper than 60 km are also omitted, otherwise Vapp at a very deep layer may be obtained for the same epicentral distance and the same azimuth, due to the nature of Vapp to represent the value for the deepest point of a ray path.

The angle of azimuth is dependent on the location of the focus and the array. In Figure 2 a diagram of Vapp is given with azimuth as abscissa. Solid circles are for the data obtained in 1974 and open circles are for 1977 data. Squares are for the results of explosion experiments in 1977 and 1978.

It is very clear that there is a systematic relation between Vapp and azimuth. The apparent velocity is a maximum at about $155(335)°$ in azimuth. The epicenters of earthquakes used in Figure 2 are shown in Figure 3. The direction of the maximum Vapp, i.e. $155(335)°$ in azimuth, is almost perpendicular to the magnetic lineation in this area obtained by Hilde et al. (1978) as given in Figure 4. The direction does not agree with the direction of the present sea floor movement. The direction of the present sea floor movement is about 305 degrees, $30°$ different from the direction of the maximum Vapp.

Conclusion. The merit of using earthquakes as wave sources is the possibility of obtaining large numbers of events with sufficient signal strength or the possibility of obtaining Vapp at large epicentral distances and for a wide range of azimuth due to the distribution of foci along the Kuril, Japan, and Izu-Mariana trenches. The values of Vapp in Figure 2 are obtained using travel-time curves with epicentral distances ranging from 140 to 1,800 km (Asada et al., 1981; Shimamura and Asada, 1976; Shimamura et al., 1977). Some travel-time curves of natural earthquakes have offsets at distances of several hundred to thousand kilometers (Shimamura and Asada, 1976). Considering these two facts, it is clear that the velocity anisotropy found in the Northwestern Pacific extends down to the layer below the low-velocity layer as described by Asada and Shimamura (1976).

The conclusions of the present paper are as follows:
(1) The mean velocity of P waves in the lithosphere in the Northwestern Pacific is 8.15 km/sec. The magnitude of anisotropy is $\pm$ 0.55 km/sec or 13 percent. The azimuth of the direction along which the velocity is maximum is about $155(335)°$.
(2) The velocity anisotropy is extending to the depth below the low-velocity layer found in the experiment in 1974. The velocities in the layers above and below the low-velocity layer seem to be the same.
(3) The azimuth of the direction of maximum velocity is about $155(335)°$. This is perpendicular to the magnetic lineations in the area where the anisotropy is found (Hilde et al., 1976). The magnetic lineations are not perpendicular to the direction of the present motion of the Pacific plate. The difference in angle is $30°$.

Acknowledgement

The authors would like to express their sincere thanks to Dr. J. Ito, Mr. I. Furuya, Dr. M. Matsu'ura, Dr. K. Suyehiro, and Dr. T. Yamada for their help in the OBS operation. The authors

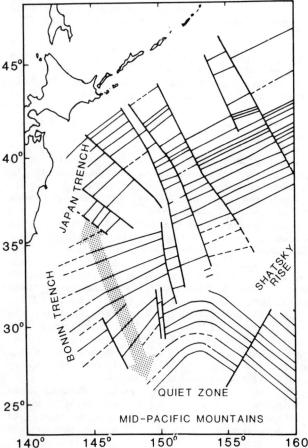

Fig. 4. Magnetic lineations (after Hilde et al., 1976) and the direction of the maximum velocity.

also wish to acknowledge the help provided by Dr. S. Nagumo and Dr. J. Kasahara. We are grateful to Dr. K. Kobayashi, Dr. Y. Tomoda, and Dr. S. Asano for their help throughout the observation with OBS's, and Dr. S. Karato, Dr. Y. Ida and Dr. M. Kumazawa for their helpful discussions. The excellent work of the captains and crews of R/V Hakuhomaru, R/V Boseimaru, and Nihonmaru is greatly appreciated. The present study was carried out as a part of the Geodynamics Project of Japan. The financial support of the Ministry of Education, Science and Culture is acknowledged.

References

Asada, T., and H. Shimamura, Obseravation of earthquakes and explosions at the bottom of the Western Pacific: Structure of oceanic lithosphere revealed by long shot experiment, The Geophysics of the Pacific Ocean Basin and Its Margin, editors, G.H. Sutton, M.H. Manghnani, R. Moberly and E. McAfee, AGU Monograph 19, 135-154, 1976.

Asada, T., H. Shimamura, S. Asano, K. Kobayashi and Y. Tomoda, Explosion seismological experiments on long-range profiles in the Northwestern Pacific and the Marianas Sea (this volume).

Bamford, D. and S. Crampin, Seismic anisotropy -the state of the art, Geophys. J. R. astr. Soc. 49, 1-8, 1977.

Hess, H.H., Seismic anisotropy of the uppermost mantle under oceans, Nature, 203, 629-631, 1964.

Hilde, T.W.C., N. Isezaki and J.M. Wageman, Mesozoic sea-floor spreading in the North Pacific, The Geophysics of the Pacific Ocean Basin and its Margin, editors, G.H. Sutton, M.H. Manghnani, R. Moverly and E.U. McAfee, AGU Monograph, 19, 205-226, 1976.

Hirn, A., Anisotropy in the continental upper mantle: Possible evidence from explosion seismology, Geophys. J. R. astr. Soc., 49, 49-58, 1977.

Morris, G.B., R.W. Raitt, and G.G. Shor, Velocity anisotropy and delay-time maps of the mantle near Hawaii, J. Geophys. Res., 74, 4300-4316, 1969.

Raitt, R.W., G.G. Shor, T.J.G. Francis, and G. B. Morris Anisotropy of the Pacific upper mantle, J. Geophys. Res., 74, 3095-3109, 1977.

Shimamura, H., and T. Asada, Apparent velocity measurements on an oceanic lithosphre, Phys. Earth and Planet. Interiors, 13, 15-22, 1976.

Shimamura, H., T. Asada , and M. Kumazawa, High shear velocity layer in the upper mantle of the Western Pacific, Nature, 269, (5630), 680-682, 1977.

SEISMIC CRUSTAL STRUCTURE AND THE ELASTIC PROPERTIES OF ROCKS
RECOVERED BY DRILLING IN THE PHILIPPINE SEA

R.L. Carlson

Department of Geophysics, Texas A&M University, College Station, Texas 77843

R.H. Wilkens

Department of Geological Sciences, University of Washington, Seattle, Washington 98195

Abstract. Extensive deep drilling in the Philippine Sea has provided a variety of rock samples for laboratory studies of compressional and shear-wave velocities at in-situ confining pressures. Included are volcaniclastic sediments, breccias, basalts and gabbroic rocks. Because these lithologies correspond to the probable compositions of all major units of the oceanic crust, their seismic properties can be directly compared with the results of numerous seismic refraction surveys in the region. The properties of the gabbroic rocks compare well with layer 3 velocities (6.5-7 km/sec), and the basalt velocities generally correspond with the refraction velocities for layer 2 (4.5-5.3 km/sec). Velocities in the upper part of the oceanic crust (layer 1) in this region range from 2 to 4 km/sec. The laboratory velocities of vitric tuffs, breccias and vesicular basalts all fall within this range, suggesting that layer 1 in some areas of the Philippine Sea may include a significant volcanic component.

Introduction

Laboratory measurements of compressional and shear-wave velocities in rocks at in-situ confining pressures are commonly used in conjunction with seismic refraction data to identify likely lithologic constituents corresponding to observed velocity structures. Indirect comparisions between the velocity structures of ophiolites, inferred from laboratory data, and average oceanic crustal structure, for example, have been remarkably successful (Christensen, 1978; Salisbury and Christensen, 1978). Direct comparisions between the properties of rocks and refraction data from the same region have not been made because samples spanning the range of appropriate lithologies are rarely available in areas where comprehensive seismic surveys have been conducted. An exception is the region of the Philippine Sea, which is perhaps the world's most comprehensively studied marginal basin complex. An extensive seismic refraction survey by Murauchi et al. (1968) is amply supplemented by other studies of the crustal structure (Gaskell et al., 1958; Henry et al., 1975; Mrozowski and Hayes, 1979), and the region has been intensely studied by participants in the Deep Sea Drilling Project. Christensen et al. (1974, 1975, 1980) and Carlson et al. (1980) have reported the acoustic properties of numerous samples recovered by drilling. The sample suite includes gabbros, metagabbros, basalts, basaltic breccias and sedimentary rocks, and thus spans the probable lithologies from layer 1 through layer 3 of the oceanic crust. The principal objective of this report is to compare the regional velocity structure with the laboratory data.

Regional Setting

Major physiographic features and the locations of seismic surveys and drilling sites are shown in Figure 1. Bounded on all sides by active subduction complexes, the Philippine Sea region is roughly bisected by the inactive Palau-Kyushu Ridge. To the west of the ridge lies the broad floor of the West Philippine Basin; to the east are the inactive Parece-Vela Basin and the presently active Mariana Trough, which are separated by the inactive West Mariana Ridge. The Mariana Trough is bounded on the east by the presently active subduction complex of the Mariana Arc.

Recent magnetic studies indicate that the West Mariana Basin formed by spreading on the Central Basin Ridge, which apparently ceased about 42 m.y.b.p. (Watts et al., 1977; Louden, 1976). At or about the time that spreading stopped in the

---

Texas A&M University Geodynamics Research Program, Contribution no. 15.

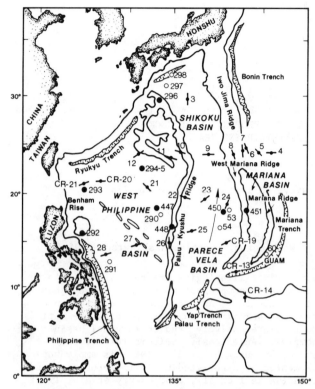

Fig. 1. Location map of the Philippine Sea including DSDP sites and seismic refraction stations. Drilling sites indicated by solid circles are those for which data is included in this study. Refraction station numbers are from original sources (see text).

West Philippine Basin, active subduction of the Pacific Plate began along its eastern margin (Uyeda and Ben-Avraham, 1972). This event may have coincided with or resulted from a general reorganization of plate motions at approximately 40 m.y.b.p. (Hilde et al., 1977; Jurdy, 1979). The Palau-Kyushu Ridge was thus constructed by arc volcanism associated with the early stages of subduction. Subsequently, the volcanic arc split and the Parece-Vela Basin formed by back-arc spreading between the active arc to the east and the Palau-Kyushu Ridge, which was left as an inactive, remnant arc to the west. This process has since been repeated to produce the active Mariana Trough and West Mariana Ridge (Mrozowski and Hayes, 1979; Karig, 1975).

The distribution of sediments in the south Philippine Sea is primarily the result of this complicated tectonic history. The active and inactive arcs have extensive volcaniclastic aprons built on their western slopes, while the remnant arcs have no volcaniclastics to the east. Karig and Moore (1975) have suggested that this results from the cessation of volcanism in the remnant arc at the time that it splits from the frontal (active) arc. The new zone of extension prevents a volcaniclastic buildup in the western half of the newly-forming basin. Clastic aprons on the eastern boundaries of the West Philippine Basin, Parece-Vela Basin and Mariana Trough may reach 2.0 km in thickness (Karig, 1975). Western borders generally exhibit thin sedimentation, although the Benham Rise is capped by a thick, non-volcanic sedimentary sequence.

This study includes samples from eight of the numerous Deep Sea Drilling sites in the region. Of those eight, seven lie between 15 and 25° North latitude. The easternmost (Site 451) is located on the West Mariana Ridge, and the westernmost (Site 292) lies on the Benham Rise. The sites for which we have data are thus distributed in such a way as to form a rough transect of the Philippine sea (Figure 1). The basalts and gabbroic rocks are from the three sites in the West Philippine Basin and from Site 292. The tuffs and breccias are from sites 447, 448, 450 and 451 in the eastern half of the Philippine Sea.

Lithologic Descriptions

Among the vitric tuffs for DSDP Sites 447A, 448, 450, and 451, grain size varies from very fine to sand sized. A majority of the samples contain basaltic clasts, euhedral volcanogenic plagioclase laths are ubiquitous, and authigenic clays, formed by the degradation of volcanic glass shards are present in varying amounts. Many of the samples are well sorted and clasts are well rounded, suggesting transport, probably in the form of turbidity flows. Graded bedding is evident in some cases. Those samples lacking basalt clasts are finer grained and may represent wild-borne ash deposits. Also included are a sample of lapilli tuff (31-296-56-6) and a nannofossil chalk (59-448-35-1).

The breccias recovered from Site 448A consist of angular to subrounded grains of highly-altered aphyric and vitrophyric, vesicular basalt in a matrix of small clasts and zeolite cement. The breccia from hole 447A is composed of angular fragments of olivine-plagioclase basalt with intersertal texture, in a calcite, quartz, and zeolite vein matrix.

In general, the basalts may be described as aphyric to vitrophyric. The groundmass commonly contains plagioclase, pyroxene, olivine and/or glass, and exhibits intersertal to ophitic textures. Sample 447A-14-1 contains abundant plagioclase phenocrysts. The rocks from Site 292 are vesicular and slightly altered.

The seven gabbros and metagabbros from Site 293 have been described previously by Fountain et al. (1975). Four are gabbros or anorthositic gabbros having cumulate textures, and two of the three metagabbros show evidence of cataclasis or granulation.

Laboratory Procedures

The degree of water saturation is known to have a marked effect on compressional-wave velocities in rocks (Wyllie et al. 1958; Nur and Simmons, 1969; Christensen, 1970). Consequently, the samples used in this study were stored in water to maintain water saturation until the measurements were completed. The samples were cut in the form of right-circular cylinders, 2.54 cm in diameter and several cm in length. The ends of the cores were lapped smooth, mutually parallel, and perpendicular to the cylinder axes.

Wet-bulk densities of the igneous samples were estimated from the weights and computed volumes of the cores: sediment densities were determined by immersion using a Jolly Balance.

Prior to the pressure runs, the cylinders were wrapped in 100-mesh screen and jacketed with copper foil to prevent the pressure medium from entering the rock. The screen maintains low pore pressures by providing voids into which pore water can escape at elevated confining pressures. Velocities were measured by the pulse-transmission method as described by Birch (1960).

The velocities and wet-bulk densities of the sedimentary and igneous rocks are listed in Tables 1 and 2, respectively. Bulk densities and compressional and shear-wave velocities have been corrected for pressure and used to compute adiabatic elastic constants. Included in Table 3 are ratios of compressional and shear-wave velocities (V_p/V_s), Poisson's ratio (σ), seismic parameters (ϕ), bulk moduli (k), compressibilities (β), shear moduli (μ), Young's moduli (E), and Lame's constant (λ).

Results

Velocity-density relations for this suite of rocks are illustrated in Figure 2. The dashed lines enclose data for other DSDP basalts, taken from Christensen and Salisbury (1972, 1973), Christensen (1973), and Carlson and Christensen (1977). By comparison, the basalts from the Philippine Sea fall well within the range of compressional velocity data previously reported, while shear-wave velocities appear to be slightly higher than average. Though the vitric tuffs and breccias generally have lower densities and velocities than do basalts, these data appear to lie on the same trend. This result is perhaps not surprising in view of the fact that basaltic clasts are the predominant constitutents of these rocks. One exception is the breccia from Site 447, which is composed of olivine-plagioclase basalt fragments with abundant quartz. By virtue of its very low Poisson's ratio (0.08) the quartz in this sample accounts for high shear-wave velocity in relation to compressional-wave velocity, and for the low value of Poisson's ratio (see Table 2). The compressional-wave velocities of the gabbroic rocks are substantially higher than the velocities in basalts of comparable density.

Data pertaining to the vitric tuffs includes ten pairs of compressional-wave velocities measured parallel and perpendicular to bedding in separate samples from adjacent parts of the drill core. These pairs of data points are connected in Figure 2. In five cases, no significant difference in velocity is observed. In the remaining five cases, the sample which has the higher velocity also has a higher value of density. Hence, the apparent velocity anisotropy of some vitric tuffs is probably related to differences in density between samples from the same depth interval and not to texture or preferred mineral orientation.

Christensen (1972) has shown that V_p/V_s ratios of oceanic serpentinites are significantly lower than those of oceanic basalts or gabbros. The laboratory data compared with refraction data for both V_p and V_s thus provides information on the abundance of serpentinites in the oceanic crust. Similarly, Christensen et al. (1973) have suggested that the V_p-V_s relations for some deep sea sediments differ from V_p-V_s relations for oceanic basalts. The V_p-V_s relations for the samples from the Philippine Sea are shown in Figure 3. Again, the dashed line encloses data from studies of DSDP basalts cited above. The basalt data shows a strong linear trend with ratios generally between 1.8 and 2.0. The quartz-rich breccia has an unusually low Vp-Vs ratio, while the other breccia in which Vs was measured has a value on the basalt trend. The sediments have scattered values generally above 2.1 and though basalts from other regions fall in this range, the ratios for basaltic rocks from the Philippine Sea are significantly lower. Hence, if these basalt, breccia and vitric tuff samples are representative, it may be possible to distinguish between these lithologies from seismic data if and when shear-wave velocities can be determined for the upper levels of the oceanic crust.

Correlation of Laboratory and Refraction Data

A comparison of marine seismic and laboratory data for the Philippine Sea region is illustrated in Figure 4. Compressional-wave velocities reported by Gaskell et al. (1958), Murauchi et al. (1968), Henry et al. (1975), and Mrozowski and Hayes (1979) are summarized in the histogram at the bottom of the figure, and the laboratory data is presented in the form of a plot of velocity versus wet-bulk density. The velocities used to construct Figure 4 were measured at confining pressures which approximate expected in-situ pressures for the various lithologies as indicated.

Seismic velocities observed in the Philippine Sea show good correspondence with the velocity distribution for normal ocean crust. The upper

TABLE 1. Compressional (P) and Shear (S) Wave Velocities of Sediments

Sample		Bulk Density (g/cc)	Mode	Velocity (km/sec) vs. Pressure					
				0.1 kb	0.2 kb	0.4 kb	0.6 kb	0.8 kb	1.0 kb
59-447A-9-1 (97-99)	H*	1.68	P	2.48	2.49	2.51	2.54	2.57	2.61
59-447A-9-1 (100-101)	V*	1.74	P	2.55	2.56	2.58	2.62	2.67	2.70
			S	1.03	1.05	1.08	1.11	1.14	1.18
mean		1.71	P	2.52	2.53	2.55	2.58	2.62	2.66
59-447A-12-3 (4-6)	H	2.00	P	3.38	3.41	3.42	3.42	3.44	3.47
59-447A-12-3 (0-4)	V	1.97	P	3.19	3.21	3.23	3.24	3.26	3.28
			S	1.59	1.60	1.63	1.64	1.66	1.66
mean		1.99	P	3.29	3.31	3.33	3.33	3.35	3.38
59-448-35-1 (71-74) (nanno chalk)	V	1.73	P	2.26	2.27	2.30	2.34	2.36	2.39
			S	0.95	0.97	1.00	1.01	1.02	1.03
59-450-25-1 (28-30)	H	1.56	P	1.97	1.99	2.03	2.07	2.10	2.12
			S	0.92	0.94	0.96	0.97	0.98	0.99
59-450-25-1 (32-35)	V	1.57	P	1.92	1.94	1.98	2.02	2.06	2.09
mean		1.57	P	1.95	1.97	2.01	2.05	2.08	2.11
59-450-27-1 (57-59)	H	1.87	P	2.85	2.87	2.91	2.92	2.96	2.98
59-451-37-1 (35-40)	H	1.92	P	2.83	2.87	2.95	3.02	3.08	3.14
59-451-37-1 (35-40)	V	1.92	P	2.85	2.89	3.01	3.12	3.17	3.18
mean		1.92	P	2.84	2.88	2.98	3.07	3.13	3.16
59-451-46-3 (6-8)	H	2.02	P	3.03	3.12	3.18	3.19	3.21	3.23
59-451-46-3 (10-13)	V	1.96	P	2.97	3.01	3.04	3.08	3.23	3.28
mean		1.99	P	3.00	3.06	3.11	3.13	3.22	3.26
59-451-31-1 (111-115)	V	2.16	P	3.25	3.31	3.37	3.40	3.42	3.42
59-451-54-4 (81-87)	H	2.04	P	3.09	3.13	3.22	3.27	3.29	3.30
			S	1.25	1.30	1.35	1.38	1.41	1.42
59-451-54-4 (81-87)	V	1.96	P	2.76	2.84	2.99	3.09	3.16	3.20
mean		2.00	P	2.93	2.99	3.11	3.18	3.23	3.25
59-451-60-2 (101-104)	V	2.13	P	3.20	3.23	3.28	3.30	3.31	3.35
59-451-64-3 (30-32)	H	1.87	P	2.75	2.76	2.80	2.86	2.91	2.93
59-451-66-4 (142-145)	V	2.09	P	3.08	3.37	3.51	3.52	3.53	3.56
59-451-70-1 (27-31)	H	1.94	P	2.75	2.80	2.88	2.95	3.00	3.03
59-451-70-1 (32-35)	V	1.94	P	2.80	2.86	2.94	3.00	3.14	3.20
mean		1.94	P	2.78	2.83	2.91	2.98	3.07	3.12
59-451-79-1 (42-48)	H	1.95	P	2.64	2.67	2.72	2.76	2.81	2.83
59-451-79-1 (42-48)	V	1.98	P	2.90	2.94	2.99	3.00	3.00	3.00
mean		1.97	P	2.77	2.81	2.86	2.88	2.90	2.92
59-451-86-1 (86-88)	H	2.06	P	2.62	2.67	2.72	2.74	2.78	2.82
			S	1.16	1.18	1.21	1.23	1.24	1.26
59-451-86-1 (71-74)	V	2.10	P	2.60	2.66	2.76	2.83	2.88	2.92
mean		2.08	P	2.61	2.67	2.74	2.79	2.83	2.87
59-451-93-2 (124-126)	H	2.13	P	2.92	3.02	3.06	3.10	3.20	3.23
59-451-93-2 (127-131)	V	2.10	P	2.83	2.90	2.95	3.00	3.20	3.30
mean		2.12	P	2.88	2.96	3.01	3.05	3.20	3.27

H* indicates propagation parallel to bedding.
V* indicates propagation perpendicular to bedding.

TABLE 2. Compressional (P) and Shear (S) Wave Velocities of Basalts and Breccias

Sample	Bulk Density (g/cc)	Mode	Velocity (km/sec) vs. Pressure							
			0.2 kb	0.4 kb	0.6 kb	0.8 kb	1.0 kb	2.0 kb	4.0 kb	6.0 kb
31-292-41-2 (37-40) (basalt)	2.567	P	4.63	4.71	4.78	4.84	4.89	5.06	5.28	----
		S	2.41	2.45	2.48	2.50	2.53	2.62	2.69	----
31-292-41-5 (40-43) (basalt)	2.611	P	4.88	4.93	4.97	5.01	5.04	5.21	5.39	----
		S	2.48	2.52	2.56	2.59	2.62	2.71	2.80	----
31-292-42-5 (29-32) (basalt)	2.607	P	5.04	5.09	5.13	5.16	5.19	5.29	5.41	----
		S	2.64	2.66	2.68	2.70	2.72	2.77	2.83	----
31-292-43-4 (40-43) (basalt)	2.675	P	4.98	5.04	5.09	5.13	5.16	5.26	2.36	----
		S	2.66	2.69	2.71	2.73	2.74	2.79	2.84	----
31-292-44-4 (48-51) (basalt)	2.688	P	5.09	5.16	5.21	5.24	5.27	5.34	5.45	----
		S	2.73	2.75	2.77	2.78	2.79	2.84	2.89	----
31-292-46 (cc) (basalt)	2.792	P	5.33	5.39	5.43	5.47	5.50	5.61	5.73	----
31-294-7-1 (116-119) (altered basalt)	2.462	P	4.63	4.69	4.73	4.75	4.78	4.88	5.04	5.14
		S	2.30	2.35	2.38	2.42	2.45	2.55	2.66	----
31-296-56-6 (10-13) (lapilli tuff)	1.985	P	3.93	3.95	3.96	3.97	3.98	4.00	4.02	----
59-447A-14-1 (74-76) (basalt)	2.735	P	5.58	5.66	5.70	5.73	5.76	5.88	5.99	6.04
		S	2.92	2.95	2.97	2.99	3.01	3.08	3.20	3.25
59-447A-25-2 (24-26) (basalt)	2.831	P	6.00	6.09	6.14	6.18	6.22	6.33	6.45	6.56
		S	3.26	3.29	3.31	3.33	3.34	3.36	3.37	3.37
59-447A-26-4 (30-33)	2.684	P	5.05	5.15	5.20	5.24	5.28	5.40	5.58	5.71
		S	2.58	2.63	2.66	2.69	2.71	2.78	2.86	2.91
59-447A-29-1 (93-96) (breccia)	2.506	P	4.31	4.41	4.50	4.57	4.62	4.84	5.13	5.32
		S	2.51	2.56	2.60	2.64	2.67	2.77	2.90	2.97
59-448-48-3 (57-59) (vesic. basalt)	2.229	P	3.70	3.75	3.77	----	----	----	----	----
		S	1.84	1.92	1.96	----	----	----	----	----
59-448-48-3 (59-61) (vesic. basalt)	2.196	P	3.84	3.88	3.90	----	----	----	----	----
		S	1.93	1.96	1.98	----	----	----	----	----
59-448-48-3 (61-64) (vesic. basalt)	2.230	P	3.60	3.64	3.66	----	----	----	----	----
		S	1.81	1.84	1.86	----	----	----	----	----
59-448-56-4 (124-127) (breccia)	2.087	P	3.25	3.32	3.37	3.41	3.44	3.57	3.75	3.87
59-448-56-4 (127-130) (breccia)	2.076	P	3.18	3.22	3.25	3.27	3.29	3.39	3.57	3.72
59-448-56-4 (134-137) (breccia)	2.064	P	3.65	3.69	3.73	3.75	3.76	3.81	3.90	4.12
59-448A-35-2 (81-84) (breccia)	1.98*	P	2.83	2.86	2.89	2.89	2.92	----	----	----
		S	1.38	1.42	1.46	1.51	1.55	----	----	----
59-448A-57-1 (16-18) (basalt)	2.597	P	4.40	4.45	4.49	4.52	4.53	4.60	4.73	4.84
		S	2.47	2.50	2.52	2.53	2.54	2.56	2.58	2.59

*Reflects 5% correction for chipped sample cylinder.

TABLE 3. Elastic Constants

Sample	Pressure (kbar)	Vp/Vs	σ	φ (km/sec)^2	K (Mb)	B (Mb^{-1})	μ (Mb)	E (Mb)	λ (Mb)
59-447A-9-1 (100-101) (vitric tuff)	0.1	2.48	0.40	5.08	0.09	11.29	0.02	0.05	0.08
	0.2	2.44	0.40	5.08	0.09	11.30	0.02	0.05	0.08
	0.4	2.39	0.39	5.09	0.09	11.25	0.02	0.06	0.08
	1.0	2.29	0.38	5.40	0.09	10.54	0.02	0.07	0.08
59-447A-12-3 (0-4) (vitric tuff)	0.1	2.01	0.33	6.80	0.13	7.46	0.05	0.13	0.10
	0.2	2.01	0.33	6.88	0.14	7.36	0.05	0.13	0.10
	0.4	1.98	0.33	6.88	0.14	7.36	0.05	0.14	0.10
	1.0	1.98	0.33	7.05	0.14	7.15	0.05	0.14	0.10
59-448-35-1 (71-74) (nanno chalk)	0.1	2.38	0.39	3.89	0.07	14.80	0.02	0.04	0.06
	0.2	2.34	0.39	3.89	0.07	14.81	0.02	0.05	0.06
	0.4	2.30	0.38	3.94	0.07	14.58	0.02	0.05	0.06
	1.0	2.32	0.39	4.26	0.07	13.39	0.02	0.05	0.06
59-450-25-1 (28-30) (vitric tuff)	0.1	2.14	0.36	2.75	0.04	23.27	0.01	0.04	0.03
	0.2	2.12	0.36	2.77	0.04	23.01	0.01	0.04	0.03
	0.4	2.11	0.36	2.88	0.05	22.10	0.01	0.04	0.04
	1.0	2.14	0.36	3.15	0.05	19.97	0.02	0.04	0.04
59-451-54-4 (81-87) (vitric tuff)	0.1	2.47	0.40	7.46	0.15	6.57	0.03	0.09	0.13
	0.2	2.41	0.40	7.54	0.15	6.50	0.03	0.10	0.13
	0.4	2.39	0.39	7.93	0.16	6.17	0.04	0.10	0.14
	1.0	2.32	0.39	8.17	0.17	5.96	0.04	0.11	0.14
59-451-86-1 (86-88) (vitric tuff)	0.1	2.26	0.38	5.07	0.10	9.57	0.03	0.08	0.09
	0.2	2.26	0.38	5.27	0.11	9.20	0.03	0.08	0.09
	0.4	2.25	0.38	5.43	0.11	8.90	0.03	0.08	0.09
	1.0	2.24	0.38	5.80	0.12	8.30	0.03	0.09	0.10
31-292-41-2 (37-40) (basalt)	0.4	1.92	0.32	14.16	0.36	2.75	0.15	0.40	0.26
	1.0	1.94	0.32	15.38	0.40	2.53	0.16	0.43	0.29
	2.0	1.94	0.32	16.46	0.42	2.36	0.18	0.46	0.31
	4.0	1.96	0.32	18.14	0.47	2.13	0.19	0.49	0.35
31-292-41-5 (40-43) (basalt)	0.4	1.96	0.32	15.80	0.41	2.42	0.17	0.44	0.30
	1.0	1.93	0.32	16.24	0.43	2.35	0.18	0.47	0.31
	2.0	1.92	0.31	17.24	0.45	2.21	0.19	0.50	0.32
	4.0	1.93	0.32	18.53	0.49	2.05	0.21	0.54	0.35
31-292-42-5 (29-32) (basalt)	0.4	1.91	0.31	16.46	0.43	2.33	0.18	0.48	0.31
	1.0	1.91	0.31	17.00	0.44	2.25	0.19	0.51	0.32
	2.0	1.91	0.31	17.67	0.46	2.16	0.20	0.53	0.33
	4.0	1.91	0.31	18.54	0.49	2.05	0.21	0.55	0.35
31-292-43-4 (40-43) (basalt)	0.4	1.87	0.30	15.74	0.42	2.37	0.19	0.50	0.29
	1.0	1.88	0.30	16.62	0.45	2.24	0.20	0.52	0.31
	2.0	1.89	0.30	17.25	0.46	2.16	0.21	0.54	0.32
	4.0	1.89	0.30	17.83	0.48	2.08	0.22	0.56	0.34
31-292-44-4 (48-51) (basalt)	0.4	1.88	0.30	16.54	0.44	2.25	0.20	0.53	0.31
	1.0	1.88	0.30	17.30	0.47	2.15	0.21	0.55	0.33
	2.0	1.88	0.30	17.74	0.48	2.09	0.22	0.57	0.33
	4.0	1.88	0.30	18.44	0.50	2.00	0.23	0.59	0.35

TABLE 3 (cont.)

Sample	Pressure (kbar)	Vp/Vs	σ	φ (km/sec)^2	K (Mb)	B (Mb^{-1})	μ (Mb)	E (Mb)	λ (Mb)
31-292-7-1 (116-119) (altered basalt)	0.4	1.99	0.33	14.62	0.36	2.77	0.14	0.36	0.27
	1.0	1.95	0.32	14.82	0.37	2.73	0.15	0.39	0.27
	2.0	1.91	0.31	15.09	0.37	2.68	0.16	0.42	0.27
	4.0	1.89	0.31	15.86	0.39	2.53	0.17	0.46	0.28
59-447A-14-1 (74-76) (basalt)	0.4	1.92	0.31	20.42	0.56	1.79	0.24	0.63	0.40
	1.0	1.91	0.31	21.07	0.58	1.73	0.25	0.65	0.41
	2.0	1.91	0.31	21.88	0.60	1.66	0.26	0.68	0.43
	6.0	1.86	0.31	22.25	0.62	1.62	0.29	0.75	0.41
59-447A-25-2 (24-26) (basalt)	0.4	1.85	0.29	22.65	0.64	1.56	0.31	0.79	0.44
	1.0	1.86	0.30	23.79	0.67	1.48	0.32	0.82	0.46
	2.0	1.88	0.30	24.97	0.71	1.41	0.32	0.83	0.50
	6.0	1.95	0.32	27.75	0.79	1.26	0.32	0.85	0.58
59-447A-26-4 (30-33) (basalt)	0.4	1.96	0.32	17.29	0.46	2.16	0.19	0.49	0.34
	1.0	1.95	0.32	18.06	0.49	2.06	0.20	0.52	0.35
	2.0	1.94	0.32	18.81	0.51	1.98	0.21	0.55	0.37
	6.0	1.96	0.32	21.16	0.57	1.74	0.23	0.60	0.42
59-447A-29-1 (93-96) (breccia)	0.4	1.72	0.25	10.70	0.27	3.72	0.16	0.41	0.16
	1.0	1.73	0.25	11.81	0.30	3.36	0.18	0.45	0.18
	2.0	1.75	0.26	13.14	0.33	3.01	0.19	0.48	0.20
	6.0	1.79	0.27	16.36	0.42	2.40	0.22	0.56	0.27
59-448-48-3 (57-59) (vesic. basalt)	0.4	1.95	0.32	9.14	0.20	4.90	0.08	0.22	0.15
59-448-48-3 (59-61) (vesic. basalt)	0.4	1.98	0.33	9.92	0.22	4.57	0.08	0.22	0.16
59-448-48-3 (61-64) (vesic. basalt)	0.4	1.98	0.33	8.72	0.19	5.13	0.08	0.20	0.14
59-448A-35-2 (81-84) (breccia)	0.4	2.01	0.34	5.48	0.10	9.62	0.04	0.10	0.08
	1.0	1.88	0.30	5.29	0.10	9.91	0.05	0.12	0.07
59-448A-57-1 (16-18) (basalt)	0.4	1.78	0.27	11.55	0.30	3.33	0.16	0.41	0.19
	1.0	1.78	0.27	11.89	0.31	3.22	0.17	0.43	0.20
	2.0	1.80	0.28	12.36	0.32	3.09	0.17	0.44	0.21
	6.0	1.87	0.30	14.29	0.38	2.64	0.18	0.46	0.26

levels of the crust (layer 1) are represented by a continuous distribution of velocities without a peak or well-defined modal value between 2.0 and 4.2 km/sec. Igneous basement (layer 2) velocities are grouped between 4.5 and 5.3 km/sec, and another well-developed maximum corresponding to the lower oceanic crust (layer 3) is centered on 6.7 km/sec.

The distribution of laboratory data, illustrated in Figure 4, correlates remarkably well with the velocities detected in seismic surveys. Fountain et al. (1975) have pointed out that the gabbros and metagabbros from the West Philippine Basin have both the expected composition and seismic velocities appropriate for the oceanic layer. Though they extend to just over 6 km/sec, the velocities in most of the massive basalts recovered by deep drilling in the Philippine Sea correlate well with the 4.5 to 5.3 km/sec peak in the survey data. The fact that some of the laboratory velocities for basalts are significantly higher than the layer 2 velocities may be related to the occurrences of large-scale cracks and pores in the basalt formation. Hyndman and

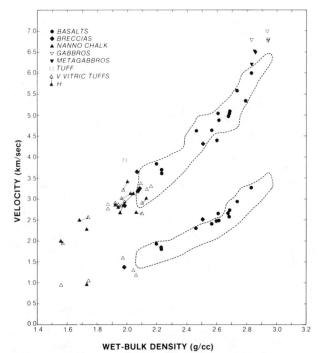

Fig. 2. Compressional and shear-wave velocity versus wet-bulk density of rocks from the floor of the Philippine Sea. Dashed lines enclose fields of data for other deep-sea basalts (see text for sources). Lines connecting data points indicate velocities measured parallel and perpendicular to bedding in separate samples from the same depth interval.

Drury (1976) have pointed out that the upper part of layer two may be fractured or shattered, with the result that the formation velocity can be significantly lower than the velocity of the solid rock studied in the laboratory. Conversely, sea-floor weathering processes which have reduced the velocities and densities of many of these samples may also serve to fill voids and cracks in the formation so that the properties of the altered basalt samples are perhaps more representative of in-situ conditions. All of the remaining lithologies included in the study exhibit velocities between 2.0 and 4.0 km/sec: Velocities in the vitric tuffs extend from 2.0 to 3.5 km/sec, and velocities in the breccias range from 2.7 to 3.6 km/sec. It should be emphasized that all of the gabbro and basalt samples included in this study are from the West Philippine Basin, which is thought to consist of "normal" oceanic lithosphere (Uyeda and Ben-Avraham, 1972; Hilde et al., 1977). If so, the observed correlation between the properties of these basement rocks and seismic data is perhaps to be expected, and these results support the widely-held view that the lower crust beneath ocean basins is gabbroic in composition. We tentatively extend these correlations to the Parece-Vela and Mariana Basins because crust produced by back-arc spreading should be similar to normal oceanic crust in structure and composition, and because the velocity structures in these regions are similar (see Murauchi et al., 1968).

Summary

A principal objective in conducting laboratory studies of acoustic velocities in rocks at elevated confining pressures is the correlation of lithologies with subsurface velocities. The correspondence between rock properties and seismic data from the Philippine Sea is excellent. Gabbroic rocks are known to have velocities appropriate to oceanic layer 3 (e.g. Christensen and Salisbury, 1975). Fountain et al. (1975) have pointed out that the gabbros and metagabbros recovered by drilling in the West Philippine Basin are likely constituents of the lower crust in the Philippine Sea region. Similarly, the correspondence of basalt velocities to layer 2 velocities in the region is

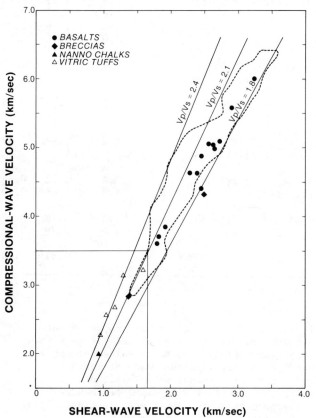

Fig. 3. Compressional versus shear-wave velocity at 0.2 kbar in rocks from the Philippine Sea. Dashed line encloses field of data for deep-sea basalts.

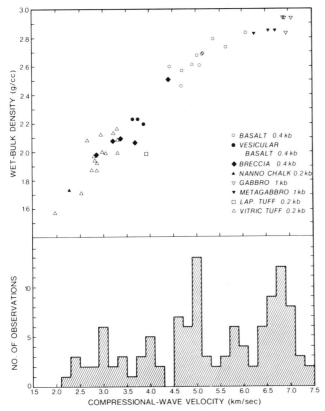

Fig. 4. Refraction data and laboratory velocities for various lithologies from the floor of the Philippine Sea. Laboratory data plotted by wet-bulk density versus compressional-wave velocity measured at pressure indicated. Symbols denote lithology. See text for sources of refraction data.

consistent with the observed correlation between oceanic basalt velocities and the range of layer 2 velocities observed world-wide (Christensen and Salisbury, 1975).

The most significant finding of this study pertains to the nature of the upper portions of the crust in the Philippine Sea. Lithologies not included in the study, such as mudstones and chalks, are likely to be the dominant rock types in the central portions of the basins at large distances from the numerous ridges which have been the sources of other types of sediments. Near the ridges, however, layers having velocities in the 2 to 4 km/sec range cannot be conclusively identified as exclusively sedimentary in nature. That these materials are thought to have accumulated on the flanks of active volcanic arcs (Karig, 1975) suggests that they should consist of mixed sedimentary and volcanic rock types. The fact that vesicular basalts, volcanic breccias and vitric tuffs from the region all have acoustic velocities in the appropriate range is consistent with this view.

Acknowlegments. We wish to thank N.I. Christensen and R.B. Scott for their many helpful suggestions while we were preparing the manuscript. R.P. Moore assisted in compiling the data and revising figures. This study was supported by Office of Naval Research Contract N00014-80-C-0113 and N00014-75-C-0502, and by NSF grant OCE 7817919.

References

Birch, F., The velocity of compressional waves in rocks to 10 kilobars, 1, J. Geophys. Res., 65, 1083-1102, 1960.

Carlson, R.L. and Christensen, N.I., Velocities, densities and elastic moduli of basalt and trachytic tuff, Deep Sea Drilling Project, Leg 39, in Supko, P., Perch-Nielsen, K. et al., Init. Repts. DSDP, 16, 647-649, 1977.

Carlson, R.L., Christensen, N.I., and Wilkens, R.H., Acoustic properties of volcaniclastic sediments recovered from the floor of the Philippine Sea, DSDP Leg 59, in Kroenke, L. and Scott, R., et al., Init. Repts. DSDP, 59, 519-522, 1980.

Christensen, N.I., Compressional wave velocities in basalts from the Juan de Fuca ridge, J. Geophys. Res., 75, 2773-2775, 1970.

Christensen, N.I., The abundance of serpentinites in the oceanic crust, J. Geol., 80, 709-719, 1972.

Christensen, N.I., Compressional and shear wave velocities in basaltic rocks, DSDP Leg 16, in van Andel, T.H., Health, G.R., et al., Init. Repts. DSDP, 16, 647-649, 1973.

Christensen, N.I., Ophiolites, seismic velocities and oceanic crustal structure, Tectonophysics, 47, 131-157, 1978.

Christensen, N.I., Carlson, R.L., Salisbury, M.H. and Fountain, D.M., Elastic wave velocities in volcanic and plutonic rocks recovered on DSDP Leg 31, in Karig, D.E., Ingle, J.C., et al., Init. Repts. DSDP, 31, 607-609, 1975.

Christensen, N.I., Fountain, D.M. and Stewart, R.J., Oceanic crustal basement: a comparison of seismic properties of DSDP basalts and consolidated sediments, Mar. Geol., 15, 215-216, 1973.

Christensen, N.I. and Salisbury, M.H., Seafloor spreading, progressive alteration of layer 2 basalts, and associated changes in seismic velocities, Earth Planet. Sci. Lett., 15, 367-375, 1972.

Christensen, N.I. and Salisbury, M.H., Velocities, elastic moduli and weathering-age relations for Pacific layer 2 basalts, Earth Planet. Sci. Lett., 19, 461-470, 1973.

Christensen, N.I. and Salisbury, M.H., Structure and constitution of the lower oceanic crust, Rev. Geophys. Space Phys., 13, 57-86, 1975.

Christensen, N.I., Salisbury, M.H., Fountain, D.M. and Carlson, R.L., Velocities of compressional and shear waves in DSDP Leg 27 basalt, in Veevers, J.J., Heirtzler, J.R., et al., Init. Repts. DSDP, 27, 445-449, 1974.

Christensen, N.I., Wilkens, R.H., Blair, S.C., and Carlson, R.L., Seismic velocities, densities, and elastic constants of volcanic breccias and basalt from Deep Sea Drilling Project Leg 59, in Kroenke, L., Scott, R. et al., Init. Repts. DSDP, 59, 515-518, 1980.

Fountain, D.M., Carlson, R.L., Salisbury, M.H. and Christensen, N.I., Possible lower crustal rocks recovered on Leg 31 by deep sea drilling in the Philippine Sea, Mar. Geol., 19, M75-M80, 1975.

Gaskell, T.F., Hill, M.N. and Swallow, J.C., Seismic measurements made by H.M.S. Challenger in the Atlantic, Pacific, and Indian oceans and in the Mediterranean Sea, Phil. Trans. Roy. Soc. London, A., 251, 23-58, 1958.

Henry, M., Karig, D.E. and Shor, G.G., Two seismic refraction profiles in the west Philippine Sea, in Karig, D.E., Ingle, J.C., et al., Init. Repts. DSDP, 31, 611-614, 1975.

Hilde, T.W.C., Uyeda, S. and Kroenke, L., Evolution of the western Pacific and its margin, Tectonophysics, 38, 145-165, 1977.

Hyndman, R.D. and Drury, M.J., The physical properties of oceanic basement rocks from deep-drilling on the mid-Atlantic ridge, J. Geophys. Res., 81, 4042-4052, 1976.

Jurdy, D.M., Relative plate motions and the formation of marginal basins, J. Geophys. Res., 84, 6796-6802, 1979.

Karig, D., Basin genesis in the Philippine Sea, in Karig, D., Ingle, J., et al., Init. Repts. DSDP, 31, 857-879, 1975.

Karig, D. and Moore, G.F., Tectonically controlled sedimentation in marginal basins, Earth Planet. Sci. Lett., 26, 233-238, 1975.

Louden, K., Magnetic anomalies in the west Philippine basin, AGU Monogr. 19, Int. Woollard Symp., Washington, D.C., pp 253-267, 1976.

Mrozowski, D.L. and Hayes, D.E., The evolution of the Parece Vela Basin, eastern Philippine Sea, Earth Planet. Sci. Lett., 46, 49-67, 1979.

Murauchi, S., Den, N., Asano, S., Hotta, H., Yoshii, T., Asanuma, T., Hagiwara, K., Ichikawa, K., Sato, R., Ludwig, W.S., Esing, J.I., Edgar, N.T. and Houtz, R.E., Crustal structure of the Philippine Sea, J. Geophys. Res., 73, 3143-3171, 1968.

Nur, A. and Simmons, G., The effect of saturation on velocity in low porosity rocks, Earth Planet. Sci. Lett., 7, 183-193, 1969.

Salisbury, M.H. and Christensen, N.I., The seismic velocity structure of a traverse through the Bay of Islands ophiolite complex, Newfoundland, an exposure of oceanic crust and upper mantle, J. Geophys. Res., 83, 805-817, 1978.

Uyeda, S., and Ben-Avraham, Z., Origin and development of the Philippine Sea, Nature Phys. Sci., 240, 176-178, 1972.

Watts, A.B., Weissel, J.K. and Larson, R.L., Seafloor spreading in marginal basins of the western Pacific, in S. Uyeda (editor) Subduction Zones, Mid-Ocean Ridges, Oceanic Trenches and Geodynamics, Tectonophysics, 37, 161-181, 1977.

Wyllie, M.R.J., Gregory, A.R. and Gardner, G.H.F., An experimental investigation of factors affecting elastic wave velocities in porous media, Geophysics, 23, 459-483, 1958.

SUMMARY OF GEOCHRONOLOGICAL STUDIES OF SUBMARINE ROCKS FROM THE WESTERN PACIFIC OCEAN

M. Ozima, I. Kaneoka, K. Saito*, M. Honda, M. Yanagisawa and Y. Takigami

Geophysical Institute, University of Tokyo, Tokyo 113, Japan

Abstract. High quality $^{40}Ar-^{39}Ar$ and K-Ar radiometric ages of DSDP drilled rocks and some dredged rocks from the western Pacific area are summarized. Although the radiometric ages are generally in good accordance with the oldest fossil ages in the overlying sediments, they are generally considerably younger than the age of the ocean basement estimated from the magnetic anomaly pattern.

During the last IPOD missions in the western Pacific, deep-drilling of ocean basement rocks was successfully carried out at several sites (Leg 57, 58, 59, 60, 60). These rocks were the first samples ever obtained directly from the ocean basement. Absolute ages of these rocks should impose a crucial constraint in understanding the origin of the western Pacific ocean crust. Hitherto, ages of the ocean basement have been inferred from a conventional K-Ar dating applied to sporadically dredged submarine rocks, mostly from the tops of seamounts. Hence, the latter age information suffers from several serious drawbacks; firstly, conventional K-Ar dating does not generally offer an objective criteria with which to judge the reliability of the result. Therefore, we can hardly know whether the obtained K-Ar age represents the true age of the sample or not. Secondly, it is often difficult to conclude whether dredged rocks are truly from the seamount or if they were transported there from elsewhere, for example, by ice rafting. Lately, the $^{40}Ar-^{39}Ar$ stepheating dating technique has turned out to be very successful in yielding reliable radiometric ages of submarine rocks. We therefore decided to use the $^{40}Ar-^{39}Ar$ technique to date IPOD samples from the western Pacific area. The purpose of this report is to summarize the $^{40}Ar-^{39}Ar$ ages of the western Pacific IPOD samples. We also summarize radiometric ages obtained for dredged samples in the western Pacific area and in the Japan Sea for comparison. All the age data have been determined at the Geophysical Institute, University of Tokyo during the last several years and some of the results are reported for the first time in this summary.

The $^{40}Ar-^{39}Ar$ stepheating dating method, which was independently proposed by Sigurgeirsson (1962) and Merrihue and Turner (1966) has been quite successfully applied to meteorite studies by a number of investigators. Ozima and Saito (1973) pointed out the potential advantage of the $^{40}Ar-^{39}Ar$ method over the conventional K-Ar dating method for dating submarine rocks and showed that the method yields reliable ages of some submarine rocks. Lately, Ozima et al. (1979) further demonstrated, from a $^{40}Ar-^{39}Ar$ study on some artificially disturbed rocks, that the method could even recover useful age information from some altered samples. This latter point is particulary important in dating submarine rocks, since most submarine rocks are altered due to sea water weathering. For the experimental details of the $^{40}Ar-^{39}Ar$ dating technique used in our laboratory, readers can refer to Ozima et al. (1977).

Although IPOD drilling was primarily intended to penetrate the ocean basement, some of the drilled rocks are sills. All the samples from seamounts were dredged from near their tops. Hence, the radiometric ages of the latter samples would generally represent the later stage of the volcanism.

In Table 1 we summarized radiometric ages of submarine rocks from the western Pacific. Included are all the IPOD drilled rocks and dredged rocks west of the mid-Pacific seamounts and from the Japan Sea. In the same table, we have included for comparison the oldest fossil age in the overlying sediments and in the dredge hauls, and the age estimated from magnetic anomaly lineations. For other notations, readers may refer to the key to Table 1. Geographic location for the radiometric ages are also shown in Figure 1. Although the absolute age data gives a most crucial constraint in understanding the geological structures of the oceanic crust, this can be fruitfully accomplished only when other data concerning the petrology, geochemistry, and so on are synthetically considered. Hence, in this short report, we are content to make a few general comments on the age results.

We first note that the radiometric ages are

---

*Now at Department of Earth Sciences, Yamagata University, Yamagata 990, Japan

Table 1. Radiometric ages obtained for rocks from the western Pacific Ocean crust.

A. DSDP-IPOD drilled cores

SAMPLE	rock	radiometric[1] age (Ma)	Quality	Method	magnetic anomaly[4] age	oldest[5] fossil age
Leg 17, site 169 (10°40.2'N, 173°3.0'E)	diabase sill	90.0 ± 4	a	$^{40}Ar/^{39}Ar$	Jurassic	Albian
Leg 17, site 170 (11°48.0'N, 177°37.0'E)	basalt	97 ± 3	a	$^{40}Ar/^{39}Ar$	Jurassic	Albian
Leg 31, site 292 (15°49.11'N, 124°39.05'E)	basalt	49 ± 2	a	$^{40}Ar/^{39}Ar$		l. Eocene
Leg 31, site 293 (20°21.25'N, 124°05.65'E)	gabbro	42	b	$^{40}Ar/^{39}Ar$		
Leg 31, site 294 (22°34.74'N, 131°23.13'E)	alkali basalt	49 ± 2	a	$^{40}Ar/^{39}Ar$		
Leg 31, site 296 (29°20.41'N, 133°31.52'E)	lapilli tuff	48	b	$^{40}Ar/^{39}Ar$		l. Oligocene
Leg 57, site 439 (40°37.61'N, 143°18.63'E)	andesite boulder rhyolite boulder dacite boulder	21 20 22.7 ± 1.4	b b a	$^{40}Ar/^{39}Ar$ $^{40}Ar/^{39}Ar$ $^{40}Ar/^{39}Ar$		
Leg 58, site 443 (29°19.64'N, 137°26.36'E)	basalt basalt dolerite	16 8 11	c c c	$^{40}Ar/^{39}Ar$ $^{40}Ar/^{39}Ar$ $^{40}Ar/^{39}Ar$	M6(20∼26 Ma)	m. Miocene
Leg 58, site 445 (25°31.36'N, 133°12.49'E)	andesite boulder in conglomerate sandstone	59 ± 3	a	$^{40}Ar/^{39}Ar$		m. Eocene
Leg 58, site 446A (24°24.04'N, 132°46.49'E)	dolerite sill gabbro sill basalt sill basalt sill	57 ± 1 56 ± 3 54 ± 1 36	a a a c	$^{40}Ar/^{39}Ar$ $^{40}Ar/^{39}Ar$ $^{40}Ar/^{39}Ar$ $^{40}Ar/^{39}Ar$		e. Eocene
Leg 60, site 458 (17°51.85'N, 146°56.06'E)	basalt basalt	34 19 ± 0.2	b a	$^{40}Ar/^{39}Ar$ $^{40}Ar/^{39}Ar$		e. Oligocene
Leg 60, site 459 (17°51.75'N, 147°18.09'E)	basalt	36	c	$^{40}Ar/^{39}Ar$		m. Eocene
Leg 61, site 462 (7°14.92'N, 165°01.89'E)	basalt sill	(120)			M26(∼155 Ma)	Cenomanian
Leg 61, site 462A (7°14.92'N, 165°01.89'E)	basalt sill	110 ± 3	a	$^{40}Ar/^{39}Ar$	M26(∼155 Ma)	Cenomanian

Table 1. (continued) Radiometric ages obtained for rocks from the western Pacific Ocean crust.

SAMPLE	rock	radiometric[1] age (Ma)	Quality	Method	magnetic anomaly[4] age	oldest[5] fossil age
B. Dredged rocks from seamounts						
Renard guyot (17.8°N, 176.1°E)	basalt	89 ± 10	a	$^{40}Ar/^{39}Ar$	Cretaceous	l. Eocene
Wilde guyot (21.2°N, 163.4°E)	andesite	86 ± 2	a	$^{40}Ar/^{39}Ar$	Jurassic	m. Eocene
Lamont guyot (21.5°N, 159.6°E)	oliv. theralite	87 ± 4	a	$^{40}Ar/^{39}Ar$	Jurassic	Eocene
Scripps guyot (23.7°N, 159.5°E)	basalt	98 ± 3	a	$^{40}Ar/^{39}Ar$	Jurassic	m. Eocene
Makarov guyot (29.5°N, 153.4°E)	mugearite	94 ± 1	a	$^{40}Ar/^{39}Ar$	Jurassic	Cenom.-Senon.
Seiko guyot (34.3°N, 143.9°E)	mugearite	102	b	$^{40}Ar/^{39}Ar$	Jurassic	Cretaceous
Seamount D1 (24°12'N, 152°18'E)	altered basalt	78 ± 2	a	$^{40}Ar/^{39}Ar$	> M25 (153 Ma)	
Golden Dragon guyot (21°22'N, 153°17'E)	basalt	95	b	$^{40}Ar/^{39}Ar$	> M25 (153 Ma)	
Seamount D4 (19°31'N, 153°37'E)	basalt	74	b	$^{40}Ar/^{39}Ar$	> M25 (153 Ma)	
Ryofu Seamount (38.0°N, 146.0°E)	basalt	72 ± 1	a	K-Ar	Cretaceous	
Erimo Seamount (40.9°N, 144.8°E)	basalt	80	b	K-Ar	Cretaceous	
Daini-Kashima Seamount (36.1°N, 143.5°E)	baslat	80	c	K-Ar	Cretaceous	
Shatsky Rise (37.1°N, 163.4°E)	basalt	50	c	K-Ar	Jurassic	
JAPAN SEA						
Yamato Bank (39.5°N, 135.3°E)	basalt	22	c	K-Ar		

Table 1. (continued)

Hakusanse (38.6°N, 137.1°E)	andesite	8	c	K-Ar
Meiyo Seamount (39.6°N, 137.7°E)	andesite	13	c	K-Ar
Matsu Seamount (39.5°N, 138.2°E)	andesite	4	c	K-Ar
Nishi-Takuyo Bank (40.0°N, 135.2°E)	granodiorite	225	a	K-Ar/Rb-Sr

Key to Table 1

1. Radiometric age : Decay constants used are as follows.
   For $^{40}K/^{40}Ar$; $\lambda_e = 0.581 \times 10^{-10}$ yr^{-1}, $\lambda_\beta = 4.962 \times 10^{-10}$ yr^{-1}
   $^{40}K/K = 1.167 \times 10^{-4}$
   For $^{87}Rb/^{87}Sr$; $\lambda = 1.42 \times 10^{-11}$ yr^{-1}

2. Quality : Radiometric ages are ranked as a, b, c, according to their quality.
   a : (i) an $^{40}Ar/^{39}Ar$ age which has both a well defined isochron and plateau age, (ii) a concordant K-Ar age determined on several different minerals separated from the rock, (iii) a concordant age determined by K-Ar and Rb-Sr methods.
   b : (i) an $^{40}Ar/^{39}Ar$ age which is determined for an approximate isochron, but does not form an age plateau. (ii) a roughly concordant K-Ar age on several rock or separated mineral samples.
   c : (i) a single K-Ar age. (ii) an $^{40}Ar/^{39}Ar$ age obtained only from the total fusion age, but for which neither an isochron or age plateau is defined.

3. Method : $^{40}Ar/^{39}Ar$: $^{40}Ar-^{39}Ar$ step-heating dating method. Seven temperature fractions were usually obtained and the age information was determined from an Ar-39Ar isochron, an apparent age spectrum and a total fusion age. Rb/Sr : Rubidium-Strontium dating method. The age was determined from an isochron on the separated minerals. K-Ar : K-Ar dating method.

4. Magnetic anomaly age
   Except for those indicated by the magnetic anomaly number, all the basement ages were taken from a compilation by Heezen and Fornari (1975). For Leg 58, site 443, Kobayashi and Nakada (1979), for Leg 61, site 462 and 462A, Cande et al. (1976), for SM D1, Golden SM, SM D4, Larson (written communication, 1968).

5. Oldest fossil age
   This gives the oldest fossil age found in the overlying sediment (drilled core) and in the same dredging haul (dredged sample). For IPOD samples, data are taken from Initial Report DSDP Leg 17, 31, 57, 58, 60 and 61. For the Pacific seamounts, Heezen and Fornari (1975).

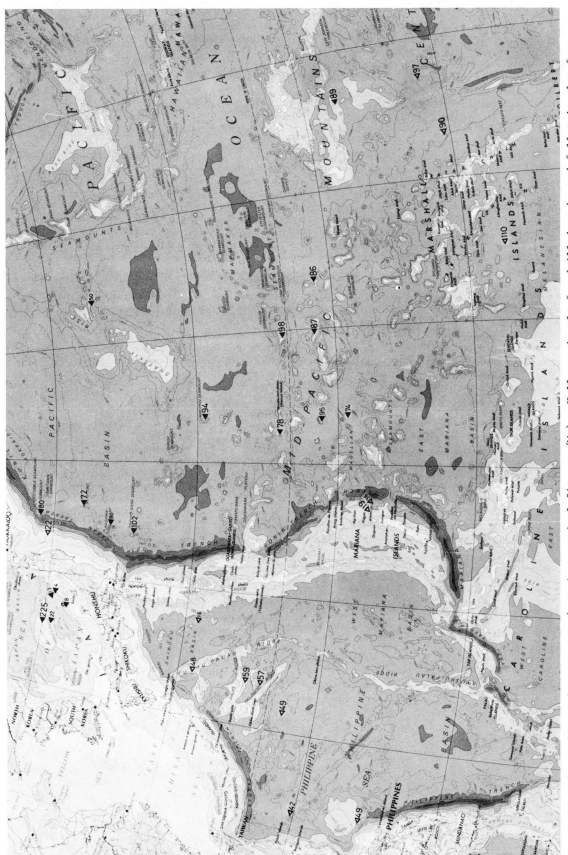

Fig. 1. Radiometric ages are shown in numerical figures (Ma). Hollow triangle for dredged rocks and full triangles for drilled rocks. Size of the numerical figures indicates the quality of the age, the largest figures for quality a (the most reliable), medium for b and the smallest for c (see a key for Table 1 for the explanation, a,b,c).

generally in good accordance with the oldest fossil ages in the overlying sediments or in the dredge hauls. However, we have found no rocks older than Cretaceous, in contrast to the supposed antiquity of the western Pacific ocean; that is, mostly of Jurassic age. This discrepancy may be explained in that either (i) the drilled submarine rocks represents only the veneer of recurrent volcanism in the ocean crust, but not the main oceanic crust, or (ii) the estimated magnetic age is in error. In the case of Leg 17 and Leg 61, neither possibility can not be totally ruled out and the answer will have to wait unitl future drilling samples the deeper oceanic crust.

Most of the Pacific seamounts in Table 1 are located on the supposed Jurassic ocean crust (Heezen and Fornari, 1975). However, all the radiometric ages of these samples are Cretaceous. Although adjustments of the age of the western Pacific lineations have been made since Heezen and Fornari (1975), they are relatively minor for the comparisons we are making. The age results suggest that the seamounts were not formed at the oceanic ridge, but were formed in the inter-plate region long after the oceanic crust was created. They may have been formed by hot spots (Morgan, 1972).

All the samples from the Japan Sea were dated by the K-Ar method and in one case by the Rb-Sr method. Although these ages are of rather poor quality, one might notice that they are much younger than the western Pacific samples. This indicates a relatively younger origin for the Japan Sea. Granodiorite from the Nishi-Takuyo Tai shows a very old age. Judging from the petrology and high $(^{87}Sr/^{86}Sr)_0$ (0.7043, (Ueno, 1969)) the Nishi-Takuyo Tai is a remnant land mass related to the adjacent Honshu Island where similar granodiorites of the same age are widely spread.

References

Cande, S.C., and D.V. Kent, Constraints imposed by the shape of marine magnetic anomalies on the magnetic source, J. Geophys. Res., 81, 4157, 1976.

Heezen, B.C., and D.J. Fornari, Geological map of the Pacific Ocean, Initial Report of the Deep Sea Drilling Project, XXX, 1975.

Kobayashi, K., and M. Nakada, Magnetic anomalies and tectonic evolution of the Shikoku Inter-Arc basin, in Adv. Earth Planet. Sci., ed. by S. Uyeda, R.W. Murphy, and K. Kobayashi, Vol. 6, 391, Center Acad. Publ. Japan, 1979.

Merrihue, C., and G. Turner, Potassium-argon dating by activation with fast neutrons, J. Geophys. Res., 71, 2852, 1966.

Morgan, W.J., Plate motions and deep mantle convection, Hess Mem. Vol. Mem. Geol. Soc. Am., 2, 1972.

Ozima, M., and K. Saito, $^{40}Ar/^{39}Ar$ stepwise degassing experiments on some submarine rocks, Earth Planet. Sci. Lett., 20, 77, 1973.

Ozima, M., M. Honda, and K. Saito, Sea water weathering effect on K-Ar age of submarine basalts, Geochim. Cosmochim. Acta, 41, 453, 1977.

Ozima, M., I. Kaneoka, and M. Yanagisawa, Temperature and pressure effects on $^{40}Ar/^{39}Ar$ systematics. Earth Planet. Sci. Lett., 42, 463, 1979.

Sigurgeirsson, T., Age dating of young basalts with the potassium argon method, Rep. Phys. Lab. Univ. Iceland, 9p, 1962.

Ueno, N., Rb-Sr isotopic studies on granitic rocks and metamorphic rocks in the Ryoke, Abukuma and Hida metamorphic belts, Central Japan, Ph.D. Thesis, University of Tokyo, 1968.

PHENOCRYST ASSEMBLAGES AND H_2O CONTENT IN CIRCUM-PACIFIC ARC MAGMAS

M. Sakuyama

Geological Institute, Faculty of Science, University of Tokyo, Hongo, Tokyo 113, JAPAN

Abstract. Volcanoes of the circum-Pacific convergent plate boundaries are classified into three types according to the crystallization sequence of hornblende, biotite and quartz; type I, quartz → hornblende → biotite; type II, hornblende → quartz → biotite; and type III, hornblende → biotite → quartz. The H_2O content of magma probably increases in the order type I - type II - type III.

The distribution of the three types of volcanoes is different in different arcs. In the Northeastern Japan and Izu-Mariana arcs, the type I volcanoes are predominant (assigned to rank A arcs). In the Lesser Antilles, New Zealand, Ryukyu-Kyushu and Aleutian, type II volcanoes are dominant (rank B arcs). The Cascades, Central America, Andes, Melanesia-Solomon, Indonesia, Philippines and Kurile-Kamchatka are characterized by the type III volcanoes (rank C arcs). The H_2O content in the magmas characterizing each arc seems to increase from the rank A to rank C arcs. The K_2O content also increases in the same way. It is likely that the degree of partial melting of mantle peridotite decreases from the rank A toward the rank C arcs. The degree of partial melting is discussed with respect to the activity of back-arc regions and the average thicknesses of the crust.

The systematic variation of the assemblages of phenocrystic hornblende, biotite and quartz, indicates that arc magmas are not saturated with H_2O. It is concluded from this and the solubility of H_2O in the magma that the maximum H_2O contents in the parental basaltic magmas of arcs are at most 2.5%, thus the partial melting of the peridotite beneath arcs are nearly under dry conditions.

Introduction

Magmatic activity in convergent plate boundaries is most likely linked with the descending movement of the oceanic plate beneath the other oceanic or continental plate, based on such observations as the close correlation between the distribution and chemical compositions of the volcanic rocks and the Benioff zone (Kuno, 1959; Coats, 1962; Dickinson and Hatherton, 1967; Sugimura, 1968; Ringwood, 1974). The detailed mechanisms of magma generation related to the descending plate movement, however, are under active debate at present (Green, 1980). For example, which is the main source of arc magmas, the eclogitic descending slab or the peridotitic mantle wedge above the descending slab? Does the generated magma assend directly to the Earth's surface or reach the surface only after interaction with rocks of the mantle or the crust, and after crystal fractionation of various phases? What determines the distribution of the volcanic centers (Carr, et al., 1973; Marsh, 1970)? To resolve these problems, our studies should be in two directions. One is the extensive study of individual arcs from geophysical, geochemical, petrological and geological aspects. The other is the comparative study of various arcs.

An early comparative study of arc volcanic rocks was made by Kuno (1966). He clarified the basic chemical characteristics of volcanic rocks in several arcs and then classified the volcanic rocks of the circum-Pacific convergent plate boundaries into three series; tholeiite series, high-alumina basalt series and alkali olivine basalt series, based on the total alkali - SiO_2 relations. According to him, the chemistry of volcanic rocks changes from the tholeiite series through the high-alumina basalt series to the alkali olivine basalt series toward the back arc direction in Northeastern Japan and Kamchatka. In the Ryukyu-Kyushu, Aleutian, New Zealand and Indonesian Islands, and the Cascades, the lateral change is from high-alumina basalt series to alkali olivine basalt series without the tholeiite series. On the other hand, only tholeiite series occurs in the Izu-Mariana, Kurile, South Sandwich and Tonga-Kermadec Islands.

Baker (1968) concluded that the Lesser Antilles is more mature arc than the South Sandwich Arc from the comparison of their volumetric ratios of andesites. Miyashiro (1974) extended this to all circum-Pacific convergent plate boundaries. He classified arc volcanic rocks into three categories; calc-alkalic, tholeiitic and alkalic series, and compared the volume ratios with the average crustal thickness and closing rate of the plates in each arc. He concluded that rocks of calc-alkalic series with high SiO_2 contents predominate in mature arcs with greater crustal thickness, where FeO*/MgO ratios of volcanic rocks tend to be small. His discussions were extended to the problem of magma genesis in the peridotitic mantle. However, most of the arc volcanic rocks would be the products of low pressure fractionation, as is discussed later. It would be, thus, difficult to have an insight into magma genetic problems using histograms of SiO_2 contents and FeO*/MgO ratios which are strongly affected by the fractionation processes. It is also possible to conclude from his studies that more differentiated magmas with lower temperatures tend to extrude in more mature arcs. Sugisaki (1976) undertook a similar compilation, with special reference to the closing rates of the plates, and reached the conclusion that higher closing rates favored dominant andesite production with lower K_2O and Na_2O contents. Mainly from geophysical data, Uyeda and Kanamori (1979) classified arcs into two types, Chilean and Mariana types, and discussed the difference of volcanic rocks between them.

It has long been suggested by many workers that H_2O may play an important role in the generation and differentiation of magmas at convergent plate boundaries (e.g., Yoder, 1969; McBirney, 1969; Wyllie, 1973; Ringwood, 1974). In spite of its significance, the estimation of H_2O content in the magmas is difficult due to the degassing associated with volcanic eruptions. Recently, some estimates of H_2O content in intermediate to acidic calcalkalic magmas were made based on the reproduction of the natural crystallization sequence by melting experiments with known amounts of H_2O (Eggler, 1972; Eggler and Burnham, 1973; Maaløe and Wyllie, 1975; Sekine, et al., 1979), EPMA analyses of glass inclusions within phenocrysts (Anderson, 1973; 1979), and thermodynamic considerations (Ewart, et al., 1975; Luhr and Carmichael, 1980; Nicholls, 1980). According to these observations, arc magmas seem to be undersaturated with H_2O through a large part of the differentiation path.

Sakuyama (1979b) studied the change of the crystallization sequence of phenocrysts (quartz, hornblende and biotite) in accordance with the variation of H_2O content in magmas. Based on the characteristic phenocryst assemblages associated with individual volcanoes, the H_2O content in Quaternary magmas of the Northeastern Japan was suggested to increase toward the back-arc side. Absolute values of H_2O content in the magmas can not be estimated by this method, but the relative H_2O abundance can be qualitatively deduced by such simple criteria as phenocryst assemblages.

In this paper, phenocryst assemblages of volcanic rocks of the circum-Pacific convergent plate boundaries will be summarized, with special reference to the relationship of hornblende, biotite and quartz. The petrologic character of the arc-systems are then compared with one another. Some of the major chemical compositional features are also discussed. Ewart (1976) made statistical analyses of the compiled modal and chemical data, and then clarified the general porphyritic nature (plagioclase phyric) of arc volcanic rocks and its large effect on the chemical compositions. However, he did not consider the coexisting relationships among phenocrysts in respective rocks.

Phenocryst Assemblages and H_2O Contents in Magmas

For the following reasons, it is reasonable to consider that most of arc magmas are the products of low pressure crystallization differentiation (< 10 kb), probably from more mafic magmas. Arc volcanic rocks are generally porphyritic (Ewart, 1976); the compositional variations of phenocrysts show good relationships with the rock compositions (Ewart et al., 1973; Stern, 1979; Sakuyama, 1981); coherent and continuous chemical variations are generally observed in the wide chemical range from basalt to dacite (Hawkesworth and Powell, 1980; Katsui et al., 1978); and Mg-rich olivines coexist with An-rich plagioclase phenocrysts in many arc basalts (e.g., Kawano et al., 1961; Ewart et al., 1977; Lopez-Escobar, 1977; Arculus, 1978; Stern, 1979), indicating that they crystallized at shallow depth and at less than 10 kb of pressure. Therefore, phenocrysts are the crystals which precipitated from magmatic melts having slow cooling rates, maintaining nearly equilibrium conditions with the surrounding melt, in shallow (< 10 kb) magma chambers. The following is a discussion of the experimental basis on the relationships between phenocryst assemblages of hornblende, biotite and quartz and H_2O content in magmas.

From the hydrous experimental work on calc-alkalic andesite to dacite compositions under 2 to 10 kb, it is evident that the upper temperature stability limits (=liquidus temperature) of hornblende and biotite are about 950°C and 850°C respectively in calc-alkalic andesite and dacite compositions, and are essentially independent of 1) variations in pressure (Eggler, 1972), 2) the anhydrous bulk chemical composition within the range of calc-alkalic andesite - dacite (Piwinskii and Wyllie, 1968), and 3) the H_2O content in the magma (Eggler, 1972; Eggler and Burnham, 1973; Maaløe and Wyllie, 1977). Of course, the liquidus temperatures of hornblende and biotite are very sensitive against variations of H_2O content under strongly H_2O undersaturated conditions. According to Eggler and Burnham (1973), the liquidus temperature of hornblende in the Mt. Hood andesite melt is suddenly depressed with the H_2O content less than 3.5 %. It would be the same for biotite. In the following discussions, H_2O content in the magmatic melts is assumed to be enough to stabilize hornblende and biotite. The validity of this assumption is partly supported by the natural occurrence of hornblende and biotite described later.

Accordingly, the beginning of crystallization of hornblende or biotite from the calc-alkalic andesitic to dacitic magma can be considered to be largely dependent on the magmatic temperature.

The temperatures of magmas with hornblende phenocrysts, below about 950°C, are lower than those without hornblende, and the temperatures of biotite-bearing magmas, below about 850°C, are lower than those without biotite.

The above deductions from experimental studies are supported by the petrographic observations on natural rocks. The Wo content in orthopyroxene phenocrysts coexisting with hornblende is less than 3.0 mole %, which is generally lower than that in orthopyroxene phenocrysts in hornblende-free rocks (Ewart et al., 1975; Stewart, 1975; Oshima, 1975; Togashi, 1977, Sakuyama, 1979a, b; Cameron et al., 1980; Sakuyama, 1981). These orthopyroxenes coexist with augite phenocrysts and their Mg/Fe ratios are nearly constant (Mg/Mg+Fe atomic ratios between 0.7 and 0.5), regardless of the presence or absence of hornblende. The Wo content in the orthopyroxene phenocrysts thus corresponds to the magmatic temperature, the lower Wo mole % in hornblende-bearing rocks indicating a lower magmatic temperature. The constancy of Wo mole % in orthopyroxene phenocrysts dividing those in hornblende-bearing and hornblende-free rocks, about 3.0 mole %, in various volcanoes with different chemical trends (Sakuyama, 1978b, 1981) also indicates that the liquidus temperature of hornblende is independent of anhydrous bulk chemistry. Using geothermometers, Luhr and Carmichael (1980) and Sakuyama (1981) actually showed that hornblende crystallized from natural andesitic magma at a temperature of less than 950°C. According to Sakuyama (1979a), the Wo content in orthopyroxene phenocrysts in biotite-bearing rocks from the Shirouma-Oike volcano is less than about 2.0 mole %. Ewart et al. (1975) and Cameron et al. (1980) described similar relationships. Although accurate estimates are difficult because of the absence of coexisting augites in the above rocks, these natural observations suggest that the temperature of biotite-bearing magmas are lower than for biotite-free magmas and that the liquidus temperature of biotite is lower than that of hornblende.

On the other hand, the liquidus temperature of quartz drastically decreases with increasing H_2O content in magmas (Whitney, 1975; Wyllie, 1977). The liquidus temperature of quartz intersects with those of hornblende and biotite in the $T - X_{H_2O}$ space. Of course, the liquidus temperature of quartz would be somewhat affected by the anhydrous melt composition compared with those of hornblende and biotite. According to Piwinskii and Wyllie (1968), the liquidus temperature of quartz under H_2O-saturated conditions at 2 kb is nearly constant (about 750 ± 10°C) over a wide range of anhydrous bulk compositions; SiO_2: 58.0 - 69.4, Al_2O_3: 14.8 - 18.7, FeO^*: 2.5 - 5.8, MgO: 1.6 - 3.7, CaO: 3.3 - 7.3, Na_2O: 3.3 - 4.0, K_2O: 0.97 - 3.81. These compositional variations appear to have no effect on the liquidus temperature of quartz, at least under H_2O-saturated conditions. Although the effect of K_2O on the liquidus temperature of quartz is probably large, being next in importance to that of H_2O (e.g., Kushiro, 1975), the K_2O content in natural arc volcanic rocks (with SiO_2 contents of 60 - 65 %) range from about 1 % up to 4 %, which is within the variation of the starting materials of Piwinskii and Wyllie (1968). According to Kushiro (1975), the liquidus boundary between silica mineral and protoenstatite in the MgO - SiO_2 system at 1 atm shifts from about 56 mole % of SiO_2 to about 80 mole % of SiO_2 by adding about 5 wt % of K_2O. But the addition of only 1 wt % H_2O shifts the liquidus boundary between quartz and enstatite to about 90 mole % of SiO_2 at 15 kb (Nakamura, Y., personal communication). This indicates that H_2O is more effective than K_2O on the shift of the liquidus boundary between protoenstatite (or enstatite) and silica mineral. Based on these observations, the variation of K_2O content from 1 to 4 % no doubt affects the liquidus temperature of quartz somewhat, but it would be more effectively depressed by H_2O. Moreover, the estimated H_2O content in natural andesitic or dacitic magmas is generally greater than the K_2O content (Eggler, 1972; Ewart et al., 1975; Anderson, 1973; Luhr and Carmichael, 1980; Sekine et al., 1979). Therefore, it is possible to qualitatively discuss the relative H_2O content by comparing the liquidus temperature of quartz with those of hornblende and biotite in magmas of similar anhydrous compositions.

Fig. 1 shows the schematic liquidus relation of quartz, hornblende and biotite. The phase relations of pyroxenes and plagioclase are omitted from this figure. The horizontal axis represents the approximate bulk H_2O content in the rock - H_2O system. The absolute values have no definite meaning. Three thick arrows (I, II and III) on the figure represent the assumed differentiation courses. For course I, quartz begins to crystal-

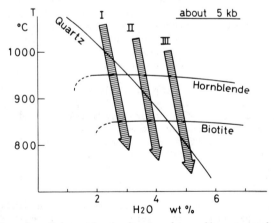

Fig. 1. Schematic isobaric phase diagram and three assumed differantiation courses. Liquidus temperatures of quartz, hornblende and biotite are indicated; those of pyroxenes and plagioclase are omitted. H_2O wt % of the horizontal axis indicates approximate values.

lize after the appearance of pyroxenes and plagioclase, and prior to the appearance of hornblende. Hornblende begins to crystallize below its liquidus temperature. With a further decreasing temperature, to below the biotite liquidus, biotite crystallizes. On the other hand, for course III, hornblende is the first crystallizing phase among the three; then biotite begins to crystallize. Quartz appears at the lowest magmatic temperature. Along course II, quartz begins to crystallize at temperatures between the hornblende and biotite liquidi. Which differentiation course magmas take among three is determinded by the relative H_2O content in the magma; course I represents the differentiation trend of magmas with the lowest H_2O content and course III represents those with the highest H_2O content.

The crystallization sequences along respective differentiation courses can be identified by the phenocryst assemblages in the actual rocks as shown in Fig. 2. Plus signs indicate the phases whose presence or absence is not important in the present discussion. Circles, open or closed, indicate the presence of those phases as phenocrysts. Four different stages of crystallization are shown in the figure. The magmatic temperature of these four stages decreases toward the right. Solid circles in Fig. 2 are the phenocryst assemblages which characterize the respective differentiation course. Course I is characterized by the assemblage of quartz - pyroxenes without hornblende and biotite. This assemblage does not appear in the other two courses. On the other hand, course III is characterized by the assemblage of hornblende - biotite- pyroxenes without quartz. Two assemblages are necessary to characterize course II, hornblende - pyroxenes without quartz and biotite, and hornblende - quartz - pyroxenes without biotite. If there is a group of rocks with various phenocryst assemblages which are possibly co-magmatic, that is, considered to be the products of a continuous crystallization differentiation process from a common primary magma, it is possible to classify all of them into one of the three courses by studying the coexisting relations among hornblende, biotite and quartz.

In Northeastern Japan, whole rock bulk chemical trends of volcanic rocks, including major and trace elements, are generally different in different volcanoes (Kawano et al., 1964; Katsui et al., 1978). The difference is especially clear in the content of incompatible elements. But in one volcano, there is a coherency among the chemical compositions of the various volcanic rocks making smooth chemical variation trends. In some volcanoes two groups of rocks are present in a single cone; pigeonitic rock series and hypersthenic rock series (Kuno, 1950; Kawano et al., 1964; Masuda and Aoki, 1978). Two distinct chemical trends are shown which have been considered to represent two different differentiation trends. However, based on the detailed petrological studies on volcanic rocks of the Myoko and Kurohime volcanoes, central

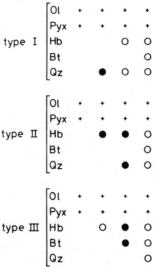

Fig. 2. Change of phenocryst assemblages with decreasing temperature in three types of differentiation. Circles, solid and open, represent the presence of the phases. Assemblages shown by solid circles are the characteristic ones of each type. Plus signes represent the phases which are not important in the present discussion. Ol=olivine, Pyx=pyroxenes, Hb=hornblende, Bt=biotite, Qz=quartz. Four different stages are indicated; magmatic temperature decreases toward the right.

Japan, Sakuyama(1981) showed that the chemical trend indicated by the hypersthenic rock series is not a differentiation trend but a mixing trend, and thus the apparent two differentiation trends can be reasonably reduced to a single trend. Therefore, it would be reasonable to assume that all the volcanic rocks consisting single stratovolcano are co-magmatic. This is also supported by the general constancy of the Sr-isotope ratio of the various volcanic rocks in a single volcano (Katsui et al., 1978).

Of course, there are some cases for which the judgement is impossible. The first case is where all of the three phases (hornblende, biotite and quartz) are absent as phenocrysts in a group of rocks. The second case is where only rocks with the assemblage of quartz and hornblende without biotite are observed but other assemblages are absent. In this case, it is impossible to assign the group of rocks to course I or II, but it is evident that they are not classified into course III (Fig. 2). The third case is where only rocks with the phenocryst assemblage of hornblende without quartz and biotite are observed. In this case, the assignment of the group of rocks to course II or III is impossible, but it is evident that they can not be classified into course I. The fourth case is where quartz, hornblende and biotite are observed always together. An assignment is also impossible in this case.

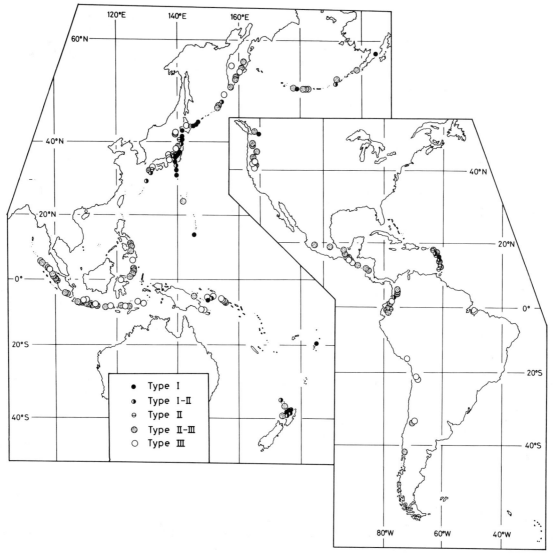

Fig. 3. Volcano types of the circum-Pacific convergent plate boundaries. See text for explanation of types. References cited are listed according to respective arc.

Phenocryst assemblages of
circum-Pacific arc volcanic rocks

Published petrographic data of Quaternary volcanic rocks on the circum-Pacific convergent plate boundaries are compiled with reference to the coexisting relations among quartz, hornblende and biotite phenocrysts. Volcanoes or eruptive centers are divided into following five types; type I, type II, type III, type I-II, and type II-III. Type I, II and III volcanoes are characterized by the magmas whose fractionation courses are assigned to the previously described courses I, II and III respectively. As previously discussed, when the distinction between type I and II, or between type II and III, is impossible, they are classified as type I-II or type II-III respectively. In some rare cases, incompatible assemblages of quartz - pyroxenes (without hornblende and biotite) and hornblende - pyroxenes (without quartz and biotite) are observed in a single volcano. The magma characterizing this volcano is regarded as a transitional between that of type I and II, and is classified as type I-II. Similarly, the coexistence of the assemblages of hornblende - quartz - pyroxenes (without biotite) and hornblende - biotite - pyroxenes (without quartz) is classified as type II-III. The distribution of volcano types is shown in Fig. 3. References cited are listed separately for each arc at the end of the paper.

It is obvious from Fig. 3 that there is a systematic difference in the phenocryst assemblages among arcs. The type II-III and type III vol-

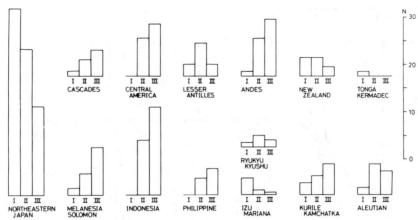

Fig. 4. Frequency distribution diagram of volcano types. Types I-II and II-III add 0.5 units to types I and II, and to types II and III respectively.

canoes are predominant in the Cascades, Central American, Andes, Melanesia-Solomon, Indonesian, Philippine and Kurile-Kamchatka arcs, with these arcs having very few type I and I-II volcanoes. In contrast, type I and I-II volcanoes are predominant in Northeastern Japan and Izu-Mariana arcs. The Lesser Antilles, New Zealand, Ryukyu-Kyushu and Aleutian arcs are dominated by the type II volcanoes, and seem to show an intermediate character between above two groups.

The number of volcanoes of types I, II and III in each arc is shown in Fig. 4. For types I-II and II-III, half unit is added to both types I and II, and to both types II and III respectively. From this figure, the above mentioned differences become clearer.

The arcs are divided into three ranks, A, B and C, based on the position of the peak of the frequency distributions. The arcs of ranks A, B and C are defined by the predominance of the types I, II and III volcanoes respectively. As previously discussed, the H_2O content in magmas increase from type I to type III. The systematic difference on the frequency distributions, thus, corresponds to the difference of H_2O content in the arc magmas, increasing from rank A to C. Of course, it is apparent from the scarcity of data that the frequency distributions of volcano types shown in Fig. 4 and the assignment of rank to the arcs are not well-founded statistically. However, the position of the peak on the histogram would be unchanged with increasing data points. The lateral changes of the chemical compositions of volcanic rocks have been pointed out in earlier studies (e.g., Kuno, 1966; Dickinson and Hatherton, 1967). Sakuyama (1979b) clarified the lateral variation of the phenocryst assemblages across Northeastern Japan, and the similar variations can be expected in other arcs. But the distribution of arc volcanoes is generally concentrated near the volcanic front (Sugimura, 1968). Therefore, although the histograms of Fig. 4 include the lateral variation effects, they probably reflect the frequency distributions of volcano-type near the volcanic front.

In some arcs longitudinal variation of the phenocryst assemblages is observed along the arcs. For example, volcanoes of the types I and I-II seem to be predominant in the Southern Kuriles, but in the Northern Kuriles and Kamchatka those of the types II-III and III seem to be predominant. Other arcs may display similar variations. These variations are not indicated in Fig. 4 because the data are insufficient to clarify such trends.

H_2O contents and major chemistry of arc magmas

The rank of arc, the range and the average of the K_2O contents at 58 - 62 wt % SiO_2 of various volcanic rocks, and those of Al_2O_3/CaO ratio of augite - hypersthene andesites are shown in Fig. 5. In Fig. 6, the average K_2O contents and Al_2O_3/CaO ratios are plotted against the rank. These figures indicate good correlations among three variables; the K_2O content of the andesites and the Al_2O_3/CaO ratios of augite - hypersthene andesites increase with increasing average H_2O content in the arc magmas.

According to Sakuyama (1979b), H_2O can be regarded as one of the incompatible oxide components like K_2O in the Quaternary magmas of the Northeastern Japan, and its content increases away from the volcanic front as shown in Fig. 3. In other words, magmas enriched in K_2O tend also to be enriched in H_2O. The positive correlation between K_2O content and the rank of the arcs shown in Figs. 5 and 6 means that the above trend is also present in various arcs of the circum-Pacific belt. It suggests that the same mechanism causes both the intra-arc lateral variations and the inter-arc variations in the magma compositions.

Sakuyama (1980) suggested that the Al_2O_3/CaO wt. ratio of the groundmass of augite - hypersthene andesite can be used as a possible qualitative indicator of H_2O content in a magma. More than 90 % of the C.I.P.W. norm of the whole rock bulk of augite - hypersthene andesites can be approximated by quartz, diopside, hypersthene, albite, anorthite and orthoclase (e.g., Kawano et al., 1964; Ewart, 1976). K_2O is expressed as the Or component associated with Al_2O_3 and SiO_2 in the C.I.P.W. norm calculation. If K_2O is neglected because of the negligible amount in the phenocrysts, the number of components can be reduced to five; quartz, diopside, hypersthene, albite and anorthite. The most important point is that the whole rock bulk chemistries of augite - hypersthene andesites are satisfactorily approximated by the four component systems such as quartz, diopside, hypersthene and plagioclase (albite and anorthite). On the other hand, phenocrysts of augite - hypersthene andesites are augite, hypersthene and plagioclase with or without olivine and magnetits. Therefore, the groundmasses of augite - hypersthene andesites are very close to the <u>univariant cotectic line</u> at constant pressure in the quartz - augite - hypersthene - plagioclase four component system.

The effect of H_2O on the univariant cotectic line can be qualitatively presumed. Yoder (1965) demonstrated that the increase of P_{H_2O} or H_2O content in the melt at constant pressure causes the reduction of the anorthite phase volume in the anorthite - diopside system, and thus the eutectic point become enriched in the anorthite component. According to Kushiro (1969), the liquidus boundary between diopside and enstatite in the diopside - enstatite - quartz system shifts from the diopside apex under wet conditions as compared with dry conditions. According to these experimental re-

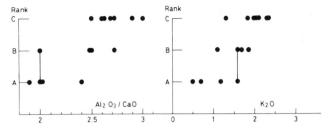

Fig. 6. Correlation among rank of arcs and averages of Al_2O_3/CaO ratios and K_2O content.

sults, the univariant cotectic line made by the augite - hypersthene - plagioclase - melt would be expected to shift from the augite apex toward the hypersthene - plagioclase side with increasing H_2O. As an easy expression of this shift, Sakuyama (1980) plotted the Al_2O_3/CaO wt % ratios of the calculated groundmass or the whole rock compositions of the aphyric rocks against the SiO_2 content. As is expected from the phenocryst assemblages of the associated dacitic rocks, in the Osore-yama volcano which is assigned to the type I volcanoes (on the volcanic front) the Al_2O_3/CaO ratios of the groundmass of augite - hypersthene andesites are about 2.0 at SiO_2 of about 60 % (Togashi, 1977). In the Myoko volcano, assigned to the type II volcanoes (on the back-arc side) the ratios are 3 to 4 at 60 % of SiO_2. The Al_2O_3/CaO ratios of the volcanic rocks of the augite - hypersthene assemblages (with or without olivine) of the Northeastern Japan (Kawano, et al., 1964) are plotted against SiO_2 wt % in Fig. 7. The Nasu zone represents a volcanic chain on the volcanic front side, the Chokai zone a volcanic chain on back arc side. Although the plotted data are somewhat ambiguous because of the presence of abundant phenocrysts, the Al_2O_3/CaO ratios are systematically higher in the Chokai zone than in the Nasu zone. This is also consistent with the tendency of increasing H_2O content in magmas toward the back arc side as shown by Sakuyama (1979 b) on the acidic volcanic rocks.

The variations of Al_2O_3/CaO ratios of augite - hypersthene andesites shown in Figs. 5 and 6, therefore, probably reflect those of the H_2O content in the augite - hypersthene andesite magmas. Magmas, on which classification of the arcs into three ranks was based, are the products of the so-called later stage of differentiation and thus are relatively lower in temperature so as to include quartz, biotite and hornblende. The correlation shown in Fig. 6 suggests that the variation of H_2O content in the late differentiates has the same tendency as the variation in the associated augite - hypersthene andesite magmas without hornblende, biotite or quartz representing the so-called middle stage of the differentiation. This is consistent with the initial assumption that all the andesites and dacites are the continuous differentiation products derived from more basic magmas. In this context, higher Sr isotope

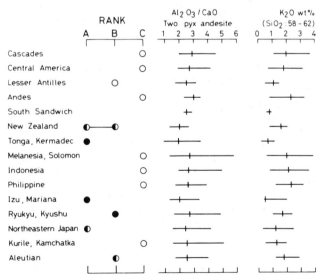

Fig. 5. Arc rank according to the Al_2O_3/CaO ratios of augite - hypersthene andesite, and K_2O content of the rocks. Averages and ranges are indicated. <u>Solid circles</u> indicate active island arcs; <u>half solid circles</u> indicate inactive island arcs; <u>open circles</u> indicate continental arcs (Uyeda and Kanamori, 1979).

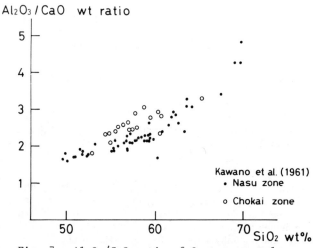

Fig. 7. Al_2O_3/CaO ratio of Quaternary volcanic rocks of the augite - hypersthene assemblage (data from Kawano et al., 1961). The Nasu zone is along the volcanic front; the Chokai zone is on the back-arc side.

ratios of acidic rocks of some arcs, for example the Andes (Francis et al., 1980), New Zealand (Ewart et al., 1977) and Indonesia (Whitford, 1975), might be caused by the selective contamination of Sr (Briqueu and Lancelot, 1979; Francis et al., 1980).

Petrogenetic considerations

If arc magmas are saturated with H_2O, the previously shown variation of phenocryst assemblages can not be observed. Therefore, most of the arc magmas are not saturated with H_2O. This is consistent with the observation that most of the arc volcanic rocks are plagioclase phyric in the wide compositional range from basalt to dacite. According to the H_2O saturated experiments on basalts (e.g., Yoder and Tilley, 1962), andesites (e.g., Eggler and Burnham, 1973) and dacites (e.g., Piwinskii and Wyllie, 1968), the liquidus temperature of plagioclase is lower than that of hornblende. If all the plagioclase phyric rocks are also hornblende phyric, there is a possibility of H_2O saturated conditions. But this is not the case.

The upper limit of H_2O content of the arc magmas can be estimated by the following considerations, if the magmas are not saturated with H_2O. As previously shown, a large part of the arc magmas are affected by low pressure fractionation, say at most 10 kb in pressure. The solubility limit of H_2O in the andesitic melt is about 10 % at 10 kb (Hamilton et al., 1964; Sakuyama and Kushiro, 1979). This value would not change greatly in dacitic or rhyolitic melts which contain hornblende, biotite or quartz phenocrysts. If these dacitic to rhyolitic melts are derived from basaltic parental magmas, and if the H_2O/K_2O ratio is nearly constant during crystallization differentiation, the estimate of the upper limit of the H_2O content in the parental basaltic magmas is possible by use of K_2O as a key oxide. The K_2O content of arc dacites or rhyolites is at most about 4 % (Fig. 5). On the other hand, the basaltic rocks accompanied with K_2O enriched dacite contains at most about 1 % K_2O. If dacite magma with 4 % K_2O is just saturated with H_2O, its H_2O content is at most 10 %, with a H_2O/K_2O ratio of about 2.5. From this ratio, the basalt magma with 1 % K_2O contains at most 2.5 % H_2O. In this context, volcanic rocks on rank C arcs must be derived from basaltic magma which contains at most 2.5 % of H_2O; those on rank A or B arcs must be derived from further H_2O depleted basaltic magmas. Such a small value of the maximum H_2O content in the arc parental magmas shows that they are essentially dry. If peridotitic mantle is assumed as a source material, the parental basaltic magmas beneath arcs should be produced by the dry partial melting of the peridotitic mantle. For example, basaltic magma with 2.5 % of H_2O, probably an alkali olivine basalt magma, is assumed to be made by 20 % melting of mantle peridotite, an unreasonably high degree of partial melting for the alkali olivine basalt. In this case, the initial H_2O content in the mantle is estimated at most to be only 0.5 %. This indicates that the temperature necessary to produce parental basaltic magmas from a peridotitic upper mantle beneath arcs must be close to or exceed the dry peridotite solidus (Mysen and Kushiro, 1977).

The dry solidus of the peridotite is so high (Kushiro et al., 1968; Jaques and Green, 1980) that the calculated geotherms for the upper mantle beneath arcs (Hasebe, et al., 1970; Toksöz, et al., 1971) can not reach it. It is, thus, impossible to produce the parental basaltic magmas beneath arcs on the steady geotherm. One of the possible mechanisms is a diapiric uprise and an associated adiabatic decompression of the upper mantle (Marsh, 1979).

As previously suggested, the global chemical variations among arcs seem to be caused by the same mechanism which causes the lateral variations in a single arc. The most plausible explanation for such a global chemical variation would be the difference in the degree of partial melting of the upper mantle peridotite if the inter- and intra-arc systematic heterogeneity of the mantle composition with respect to incompatible elements, including H_2O, is assumed to be absent. The degree of partial melting may increase from rank C arcs to those of rank A. The difference in degree of partial melting can be translated into the distance from the solidus of the peridotite in the P - T space. Therefore, there are two independent valuables, temperature and pressure, which can change the degree of melting. It is difficult to recognize them separately at present.

Uyeda and Kanamori (1979) divided convergent plate boundaries into the three types based on the nature of back-arc opening and the mode of subduction; 1) Continental arcs without back-arc basins (called continental arcs in this paper), 2) Island arcs with inactive back-arc basins, where back-arc opening is not in progress at present (called inactive island arcs), and 3) Island arcs with active back-arc basins, where back-arc opening is presently in progress (called active island arcs).

The Cascades, Central American, Peru-Chile and Indonesian arcs were designated as continental arcs. The South Sandwich, Tonga-Kermadec, Mariana and Ryukyu arcs were classified as active island arcs. The Lesser Antilles, New Zealand, Northeastern Japan, Kurile and Aleutian arcs were classified as inactive island arcs.

Solid circles in Fig. 5 represent active island arcs; half solid circles the inactive island arcs; open circles the continental arcs. All the continental arcs are the rank C, and most of the active island arcs are the rank A. The inactive island arcs seem to be of the rank B. As a general trend, H_2O and K_2O contents of arc magmas seem to decrease with increasing activity of the back-arc. This might indicate that the degree of partial melting of the mantle beneath arcs is dependent on the back-arc activities. Back-arc opening would be a phenomenon with a long time span, probably several million years to several tens of million years. The correlation between a phenomena with such a long time scale and the degree of mantle melting beneath arcs may indicate that temperature is the determining factor for degree of melting.

On the other hand, the rank of the arcs also correlates with the average crustal thickness (Fig. 8). Arcs of rank A seem to have the thinner crust and those of rank C seeem to have the thicker crust. In other words, the degree of partial melting of the mantle seems to become large with decreasing crustal thickness. Crustal thickness can also be linked with the classification of arcs by Uyeda and Kanamori (1979), such that, if arcs with back-arc opening are considered to be built on newly developed back-arc oceanic plate, active island arcs are expected to have thinner crusts while crustal thickness may be expected to be large in continental arcs. The Kurile-Kamchatka arc was treated as a single arc in previous discussions, but, as previously described, there is a longitudinal variation along the arc. The southern Kurile arc is characterized by type I volcanoes, but in the northern Kurile-Kamchatka arc type II-III and III volcanoes are predominant (Fig. 3). This may be a reflection of the crustal thickness, which is 10 to 15 km in the southern Kurile and gradually changes to 30 to 40 km beneath Kamchatka. It is also possible to consider that this is due to the presence of the Kurile back-arc basin behind the southern Kuriles. Ui and Aramaki (1979) pointed out the negative correlation between K_2O contents in volcanic rocks and the long-wavelength Bouguer gravity anomaly at the volcanoes in Northeastern Japan, in addition to the positive correlation

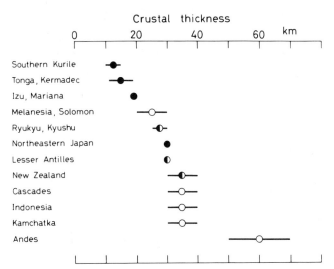

Fig. 8. Rank of arc and average crustal thickness. <u>Solid circles</u> are rank A arcs; <u>half solid circles</u> are rank B arcs; <u>open circles</u> are rank C arcs. The Southern Kurile arc is distinguished from Northern Kurile - Kamchatka region. References for crustal thickness are listed separately.

between K_2O contents and the depth to the Benioff zone. The long-wavelength Bouguer anomaly is approximately in inverse proportion to the crustal thickness. Therefore, their result indicates that the lower K_2O content is associated with the thinner crustal thickness, and that the previous discussions of inter-arc variations are also valid for intra-arc variation.

One of the possible solutions for the role of crustal thickness controlling the degree of partial melting of the upper mantle is that the upward movement of mantle diapirs might stop near the mantle - crust boundary. The degree of melting of the diapirs with the same temperature will be higher when they stop at the shallower depth, that is, at the lower pressure.

Summary

According to the hydrous experimental studies on calc-alkalic andesite to dacite compositions, the crystallization sequence of quartz, hornblende and biotite varies according to H_2O contents in magmas. The crystallization sequence is quartz, hornblende and then biotite with decreasing temperature in magmas with the least H_2O; hornblende, quartz and then biotite in magmas with greater H_2O; and hornblende, biotite and then quartz in magmas with the highest H_2O content. It is, thus, possible to deduce the relative H_2O content in magmas by the analyses of the coexisting relations among quartz, hornblende and biotite.

Each volcano in the circum-Pacific convergent plate boundaries is assigned into one of the above three crystallization sequence, which are termed as types I, II and III respectively. Volcanoes which can not be assigned to type I or II, but are not the type III (according to published assemblages) are classified as type I-II. Similarly, those not of type I are classified as type II-III. The frequency distribution of the volcano types shows systematic differences among the arcs. The Northeastern Japan and Izu-Mariana arcs are characterized by type I volcanoes, which are termed as rank A arcs. In the Lesser Aneilles, New Zealand, Ryukyu-Kyushu and Aleutian arcs, type II volcanoes are predominant. These arcs are termed rank B arcs. Type III volcanoes are predominant in the Cascades, Central American, Andes, Melanesia-Solomon, Indonesian, Philippine and Kurile-Kamchatka arcs, which are termed rank C arcs. These variations probably reflect the difference of average H_2O content in the magmas of these arcs. The K_2O content also increases from the rank A arcs to the rank C arcs. The Al_2O_3/CaO ratio of associated augite - hypersthene andesites, which would be related to H_2O content in magmas, also increases from rank A arcs to rank C arcs. Therefore the similar trend of H_2O variation is present in the augite - hypersthene andesitic magmas. This supports the idea that most arc volcanic rocks are the products of a crystallization differentiation process from more basic magmas.

The systematic variations shown in H_2O and K_2O contents for volcanic rocks among arcs would be due to the different degree of partial melting of mantle peridotite. It seems to be highest beneath rank A arcs and lowest beneath rank C arcs. The degree of melting appears to be higher beneath arcs with more active back-arc regions and smaller average crustal thicknesses.

On the other hand, the variation of H_2O content indicates that most of the arc magmas are not saturated with H_2O. The maximum H_2O content in parental basaltic magmas beneath arcs, estimated by the condition of undersaturation and the solubility data of H_2O, is 2.5 % H_2O. Therefore, the temperature of mantle partial melting, which produces such low H_2O magmas, must be close to that under dry conditions. The estimated geotherm does not exceed the dry solidus of peridotite. Diapiric rise of mantle peridotite is thus suggested as the likely mechanism for magma generation.

Acknowledgements

The manuscript was critically read by Professor I. Kushiro, University of Tokyo. The author is deeply indebted to him for his constructive criticism and improvement of the manuscript.

References

Anderson, A. T., The before - eruption water contents of some high-alumina magmas, Bull. Volcanol., 37, 530-552, 1973.
Anderson, A. T., Water in some hypersthenic magmas, J. Geol., 87, 509-531, 1979.
Arculus, R. J., Mineralogy and petrology of Grenada, lesser Antilles island arc, Contrib. Mineral. Petrol., 65, 413-424, 1978.
Baker, P. E., Comparative volcanology and petrology of the Atlantic island arc, Bull. Volcanol., 32, 189-206, 1968.
Briqueu, L. and Lancelot, J. R., Rb-Sr systematics and crustal contamination models for calc-alkaline igneous rocks, Earth Planet. Sci. Lett., 43, 385-396, 1979.
Cameron, M., Bagby, W. C. and Cameron, K. L., Petrogenesis of voluminous mid-Tertiary ignimbrites of the Sierra Madre Occidental, Chihuahua, Mexico, Contrib. Mineral. Petrol., 74, 271-284, 1980.
Carr, M. J., Stoiber, R. E. and Drake, C. L., Discontinuities in the deep seismic zone under the Japanese Arcs, Geol. Soc. Am. Bull., 74, 271-284, 1973.
Coats, R. R., Magma type and crustal structure in the Aleutian arc, Geophys. Monograph no.6, 92-109, 1962.
Dickinson, W. R. and Hatherton, T., Andesitic volcanism and seismicity around the Pacific, Science, 157, 801-803, 1967.
Eggler, D. H., Water-saturated and undersaturated melting relations in a Paricutin andesite and an estimate of water content in the natural magma, Contrib. Mineral. Petrol., 34, 261-271, 1972.
Eggler, D. H. and Burnham, C. W., Crystallization and fractionation trends in the system andesite - H_2O - CO_2 at pressure to 10 kb, Geol. Soc. Am. Bull., 84, 2517-2532, 1973.
Ewart, A., Mineralogy and chemistry of modern orogenic lavas - some statistics and implications, Earth Planet. Sci. Lett., 31, 417-432, 1976.
Ewart, A., Bryan, W. B. and Gill, J., Mineralogy and geochemistry, and the possible petrogenetic evolution of the volcanic rocks of the Tonga - Kermadec - New Zealand island arc, J. Volcanol. Geotherm. Res., 2, 205-250, 1977.
Ewart, A., Hildreth, W. and Carmichael, I. S. E., Quaternary acid magma in New Zealand, Contrib. Mineral. Petrol., 51, 1-27, 1975.
Francis, P. W., Thorpe, R.S., Moorbath, S., Kretzschmar, G. A. and Mamimill, M., Strontium isotope evidence for crustal contamination of calc-alkaline volcanic rocks from Cerro Galan, northwest Argentina, Earth Planet. Sci. Lett., 48, 257-267, 1980.
Green, T. H., Island arc and continent-building magmatism - A review of petrogenic models based on experimental petrology and geochemistry, Tectonophys., 63, 367-385, 1980.
Hamilton, D. L., Burnham, C. W. and Osborn, E. F., The solubility of water and effect of oxygen fugacity and water content on crystallization in mafic magmas, J. Petrol., 5, 21-39, 1964.
Hasebe, K., Fujii, N. and Uyeda, S., Thermal process under island arcs, Tectonophys., 10, 335-355, 1970.

Hawkesworth, C. J. and Powell, M., Magma genesis in the Lesser Antilles island arc, Earth Planet. Sci. Lett., 51, 297-308, 1980.

Jaques, A. L. and Green, D. H., Anhydrous melting of peridotite at 0 - 15 kb pressure and the genesis of tholeiite basalts, Contrib. Mineral. Petrol., 73, 287-310, 1980.

Katsui, Y., Oba, Y., Ando, S., Nishimura, S., Masuda, Y., Kurasawa, H. and Fujimaki, H., Petrochemistry of the Quaternary volcanic rocks of Hokkaido, north Japan, J. Fac. Sci., Hokkaido Univ., Ser. IV, 18, 449-484, 1978.

Kawano, Y., Yagi, K. and Aoki, K., Petrography and petrochemistry of the volcanic rocks of Quaternary volcanoes of northeastern Japan, Sci. Rep. Tohoku Univ., Ser. III, 7, 1-46, 1961.

Kuno, H., Petrology of Hakone volcano and the adjacent areas, Japan, Geol. Soc. Am. Bull., 61, 957-1020, 1950.

Kuno, H., Origin of Cenozoic petrographic provinces of Japan and surrounding areas, Bull. Volcanol. Ser. II, 20, 37-76, 1959.

Kuno, H., Lateral variation of basalt magma type across continental margins and island arcs, Bull. Volcanol., 29, 195-222, 1966.

Kushiro, I., The system forsterite - diopside - silica with and without water at high pressures, Amer. J. Sci., 267-A, 269-294, 1969.

Kushiro, I., On the nature of silicate melt and its significance in magma genesis: regularities in the shift of the liquidus boundaries involving olivine, pyroxene and silica minerals, Amer. J. Sci., 275, 411-431, 1975.

Kushiro, I., Syono, Y. and Akimoto, S., Melting of a peridotite nodule at high pressures and high water pressures, J. Geophys. Res., 73, 6023-6029, 1968.

Lopez-Escobar, L., Frey, F. A. and Vergara, M., Andesites and high-alumina basalts from the central - south Chile High Andes: geochemical evidence bearing on their petrogenesis, Contrib. Mineral. Petrol., 63, 199-228, 1977.

Luhr, J. F. and Carmichael, I. S. E., The Colima volcanic complex, Mexico. I Post-caldera andesites from Volcan Colima, Contrib. Mineral. Petrol., 71, 343-372, 1980.

Maaløe, S. and Wyllie, P. J., Water content of a granite magma deduced from the sequence of crystallization determined experimentally with water-undersaturated conditions, Contrib. Mineral. Petrol., 52, 175-191, 1975.

Marsh, B. D., Island arc development: some observations, experiments and speculations, J. Geol., 87, 687-713, 1979.

Masuda, Y. and Aoki, K., Trace element variations in the volcanic rocks from the Nasu zone, northeast Japan, Earth Planet. Sci. Lett., 44, 139-149, 1979.

McBirney, A. R., Compositional variations in Cenozoic calc-alkaline suites, in: Proceedings of the Andesite Conference, A. R. McBirney, ed., Oregon Dep. Geol. Mineral. Ind. Bull., 65, 185-189, 1969.

Miyashiro, A., Volcanic rock series in island arcs and active continental margins, Amer. J. Sci., 274, 321-355, 1974.

Mysen, B. O. and Kushiro, I., Compositional variations of coexisting phases with degree of melting of peridotite in the upper mantle, Amer. Mineral., 62, 843-865, 1977.

Nicholls, J., A simple thermodynamic model for estimating the solubility of H_2O in magmas, Contrib. Mineral. Petrol., 74, 211-220, 1980.

Oshima, O., Mineralogical aspects of volcanic eruption (in Japanese with English abstract), Bull. Volcanol. Soc. Japan, 2nd Ser., 20, 275-298, 1975.

Piwinskii, A. J. and Wyllie, P. J., Experimental studies of igneous rock series: a zoned pluton in the Wallowa batholith, Oregon, J. Geol., 76, 205-234, 1968.

Ringwood, A. E., The petrological evolution of island arc systems. J. Geol. Soc. Lond., 130, 183-204, 1974.

Sakuyama, M., Evidence of magma mixing: petrological study of Shirouma-Oike calc-alkaline andesite volcano, Japan, J. Volcano. Geotherm. Res., 5, 197-208, 1979a.

Sakuyama, M., Lateral variations of H_2O contents in Quaternary magmas of northeastern Japan, Earth Planet. Sci. Lett., 43, 103-111, 1979b.

Sakuyama, M. and Kushiro, I., Vesiculation of hydrous andesitic melt and transport of alkalies by separated vapor phase, Contrib. Mineral. Petrol., 71, 61-66, 1979.

Sakuyama, M., Magma mixing and crystallization sequence of magmas beneath Shirouma-Oike, Myoko and Kurohime volcanoes and their bearing on the petrography of Quaternary volcanic rocks of northeastern Japan, Unpublished Ph. D. Thesis: Univ. of Tokyo, 187pp, 1980.

Sakuyama, M., Petrological study of the Myoko and Kurohime volcanoes, Japan: crystallization sequence and evidence for magma mixing, J. Petrol., 22, 553-583, 1981.

Sekine, T., Katsura, T. and Aramaki, S., Water saturated phase relations of some andesites with application to the estimation of the initial temperature and water pressure at the time of eruption, Geochim. Cosmochim. Acta, 43, 1367-1376, 1979.

Stern, R. J., On the origin of andesite in the northern Mariana island arc: implications from Agrigan, Contrib. Mineral. Petrol., 68, 207-219, 1979.

Stewart, D. C., Crystal clots in calc-alkaline andesites as breakdown products of high-Al amphiboles, Contrib. Mineral. Petrol., 53, 195-204, 1975.

Sugimura, A., Spacial relations of basaltic magmas in island arcs. in: Basalts: The Poldervaart Treatise on Rocks of Basaltic Composition ed., by H. H. Hess and the late A. Poldervaart, vol. 2, 537-571, 1968.

Sugisaki, R., Chemical characteristics of volcanic rocks: relation to plate movements, Lithos, 9, 17-30, 1976.

Togashi, S., Petrology of Osore-yama volcano,

Japan (in Japanese), J. Japan. Assoc. Min. Petrol. Econ. Geol., 72, 45-60, 1977.

Toksöz, M. N., Minear, J. W. and Julian, B. R., Temperature field and geophysical effects of a downgoing slab, J. Geophys. Res., 76, 1113-1138, 1971.

Ui, T. and Aramaki, S., Relationship between chemical composition of Japanese island-arc volcanic rocks and gravimetric data, Tectonophys., 45, 249-259, 1978.

Uyeda, S. and Kanamori, H. Back-arc opening and the mode of subduction, J. Geophys. Res., 84, 1049-1061, 1979.

Whitford, D. J., Strontium isotopic studies of the volcanic rocks of the Sunda arc, Indonesia and their petrogenetic implications. Geochim. Cosmochim. Acta, 39, 1287-1302, 1975.

Whitney, J. A., The effects of pressure, temperature and X_{H_2O} on phase assemblage in four synthetic rock compositions, J. Geol., 83, 1-31, 1975.

Wyllie, P. J., Experimental petrology and global tectonics - a review, Tectonophys., 17, 189-209, 1973.

Wyllie, P. J., Crustal anatexis: an experimental review, Tectonophys., 43, 41-71, 1977.

Yoder, H. S., Diopside - anorthite - water at five and ten kilobars and its bearing on explosive volcanism, Carnegie Inst. Wash. Yearb., 64, 82-89, 1965.

Yoder, H. S., Calcalkalic andesites: experimental data bearing on the origin of their assumed characteristics, in: Proceedings of the Andesite Conference, A. R. McBirney, ed., Oregon Dep. Geol. Mineral. Ind. Bull., 65, 77-89, 1969.

Yoder, H. S. and Tilley, C. E., Origin of basalt magmas : an experimental study of natural and synthetic rock systems, J. Petrol., 3, 342-532, 1962.

References for crustal thickness data

Erlich, E. N. and Gorshkov, G. S., Quaternary volcanism and tectonics in Kamchatka, Bull. Volcanol., 42, 1-298, 1979.

Gorshkov, G. S., Geophysics and petrochemistry of andesite volcanism of the circum-Pacific belt, in: Proceedings of the Andesite Conference, A. R. McBirney ed., Oregon Dep. Geol. Mineral. Ind. Bull., 65, 91-98, 1969.

Hamilton, W., Tectonics of the Indonesian region, U. S. Geol. Surv. Prof. Paper 1078, 345pp, 1979.

James, D. E., Andean crustal and upper mantle structure, J. Geophys. Res., 76, 3246-3271, 1971.

Johnson, R. W., Distribution and major - element chemistry of late Cainozoic volcanoes at the southern margin of the Bismark Sea, PNG, BMR Geol. Geophys. Rep., 188, 157pp, 1977.

Karig, D. E., Ridges and trenches of the Tonga - Kermadec island arc systems, J. Geophys. Res., 75, 239-254, 1970.

Murauchi, S., Den, N., Asano, S., Hotta, H., Yoshii, T., Asanuma, T., Hagiwara, K., Ichikawa, K., Sato, T., Ludwig, W. J., Ewing, J. I., Edgar, N. T. and Houtz, R. E., Crustal structure of the Philippine Sea, J. Geophys. Res., 73, 3143-3171, 1968.

Pakiser, L. C. and Zeitz, I., Transcontinent crustal and upper mantle structure, Rev. Geophys. Space Phys., 3, 505-520, 1965.

Reilly, W. I., Gravity and crustal thickness in New Zealand, N. Z. J. Geol. Geophys., 5, 228-233, 1962.

Research Group For Explosion Seismology, Regionality of the upper mantle around northeastern Japan as derived from explosion seismic observations and its seismological implications. Tectonophys., 37, 117-130.

Thompson, A. A. and Evison, F. F., Thickness of the Earth's crust in New Zealand, N. Z. J. Geol. Geophys., 5, 29-45, 1962.

Vajk, R., Correlation of gravity anomalies at sea for submarine topography, J. Geophys. Res., 69, 3837-3844, 1964.

Westbrook, G. K., The structure of the crust and upper mantle in the region Barbados and the Lesser Antilles, Geophys. J. Roy. Astron. Soc., 43, 201-242, 1975.

Wiebenga, W. A., Crustal structure of the New Britain - New Ireland region, in: The Western Pacific: island arcs, marginal seas, geochemistry, P. J. Coleman, ed., 163-177, 1973.

Regional references

CASCADES

Anderson, C. A., Volcanic history of the Clear Lake area, California, Geol. Soc. Am. Bull., 47, 629-664, 1936.

Anderson, C. A., Hat Creek lava flow, Amer. J. Sci. 238, 477-492, 1940.

Condie, K. C and Swenson, D. H., Compositional variation in three Cascade stratovolcanoes; Jefferson, Rainier and Shasta, Bull. Volcanol., 37, 205-230, 1974.

Coombs, H. A., Mt. Baker, a Cascade volcano, Geol. Soc. Am. Bull., 50, 1493-1510, 1939.

Coombs, H. A. and Howard, A. D., Catalogue of the active volcanoes of the world including solfatara fields; Part IX. United States of America, Internat. Volcanol. Assoc., 68pp, 1960.

Fiske, R. S., Hopson, C. A. and Waters, A. C., Geology of Mount Rainier National Park, Washington, U. S. Geol. Survey Prof. Paper, 444, 1-93, 1963.

Greene, R. C., Petrography and petrology of volcanic rocks in the Mount Jefferson area High Cascade Range, Oregon, U. S. Geol. Surv. Bull., 1251-G, 48pp, 1968.

Higgins, M. W., Petrology of Newberry volcano, central Oregon, Geol. Soc. Am. Bull., 84, 455-488, 1973.

Mertzman, Jr., S. A., The petrology and geochemistry of the Medicine Lake volcano, California, Contrib. Mineral. Petrol., 62, 221-247, 1977.

Powers, H. A., The lavas of the Modoc Lava Bed quadrangle, California, Am. Mineral., 17, 253-294, 1932.

Sheppard, R. A., Petrology of a late Quaternary potassium-rich andesite flow from Mt. Adams, Washington, U. S. Geol. Surv. Prof. Paper 575-C, 55-59, 1967.

Smith, A. L. and Carmichael, I. S. E., Quaternary lavas from the southern Cascades, western U.S.A. Contrib. Mineral. Petrol., 19, 212-238, 1968.

Tabor, R. W, and Crowder, D. F., On batholiths and volcanoes - Intrusion and eruption of late Cenozoic magmas in the Glacier Peak area North Cascade, Washington, U. S. Geol. Surv. Prof. Paper, 604, 67pp, 1969.

Verhoogen, J., Mt. St. Helens: a recent Cascade volcano, California Univ. Pubs. Geol. Sci., 24, 263-302, 1937.

Williams, H., Geology of Lassen Volcanic National Park, California, Calif. Univ. Dept. Geol. Sci. Bull., 21, 195-385, 1932.

Williams, H., Mount Shasta, a Cascade volcano, J. Geol., 40, 417-429, 1932.

Williams, H., Mount Thielsen, a dissected Cascade volcano, Univ. Calif. Publ. Bull. Dept. Geol. Sci., 23, 195-213, 1933.

Williams, H., The geology of Crater Lake National Park, Oregon, with a reconnaissance of the Cascade Range southward to Mount Shasta, Carnegie Inst. Washington Publ., 540, 162pp, 1942.

Wise, W. S., Geology and petrology of the Mt. Hood area: a study of High Cascade volcanism, Geol. Soc. Am. Bull., 80, 969-1006, 1969.

CENTRAL AMERICA

Carr, M. J., Rose, W. I. and Mayfield, D. G., Potassium content of lavas and depth to the seismic zone in Central America, J. Volcanol. Geotherm. Res., 5, 387-401, 1979.

Fairbrothers, G. E., Carr, M. J. and Mayfield, D. G., Temporal magmatic variation at Boqueron Volcano, El Salvador, Contrib. Mineral. Petrol., 67, 1-9, 1978.

McBirney, A. R. and Williams, H., Volcanic history of Nicaragua, Univ. Calif. Publ. Geol. Sci., 55, 1-65, 1965.

Mooser, G., Meyer-Abich, H. and McBirney, A. R., Catalogue of active volcanoes of the world including solfatara fields, Part VI, Central America, Internat. Volcanol. Assoc., 146pp, 1958.

Rose, W. I. Jr., Santiaguito volcanic dome, Guatemala, Geol. Soc. Am. Bull., 83, 1413-1434, 1972.

Rose, W. I. Jr., Woodruff, L. G. and Bonis, S. B., Magma composition changes during the 1974 eruption of Volcán Fuego: result of vertical variations of H_2O during shallow intertelluric crystal fractionation, Trans. Am. Geophys. Union, 57, 346, 1976.

Rose, W. I. Jr., Grant, N. K., Hahn, G. A., Lange, I. M., Powel, J. L., Easter, J. and DeGraff, J. M., The evolution of Santa Maria volcano, Guatemala, J. Geol., 85, 63-87, 1977.

Rose, W. I. Jr., Anderson, A. T. Jr., Bonis, S. and Woodruff, L. G., The October 1974 basaltic tephra from Fuego volcano, Guatemala: description and history of the magma body, J. Volcanol. Geotherm. Res., 4, 3-53, 1978.

Ui, T., Recent volcanism in the Masaya - Granada area, Nicaragua, Bull. Volcanol., 36, 174-190, 1972.

Wilcox, R., Petrology of the Paricutin volcano, Mexico, U. S. Geol. Surv. Bull., 965-C, 281-353, 1954.

Williams, H., Volcanic history of the Guatemalan Highlands, Calif. Univ. Pubs. Geol. Sci., 38, 1-86, 1960.

Williams, H., McBirney, A. R. and Dengo, G., Geologic reconnaissence of southern Guatemala, Calif. Univ. Publ. Geol. Sci., 50, 1-56, 1964.

Williams, H. and McBirney, A. R., Volcanic history of Honduras, Calif. Univ. Publ. Geol. Sci., 85, 1-101, 1969.

Woodruff, L. G., Rose, W. I. Jr. and Rigot, W., Contrasting fractionation pattern for sequential magmas from two calc-alkaline volcanoes in Central America, J. Volcanol. Geotherm. Res., 6, 217-240, 1979.

LESSER ANTILLES

Arculus, R. J., Geology and geochemistry of the alkali basalt - andesite association of Grenada, Lesser Antilles island arc, Geol. Soc. Am. Bull. 87, 612-624, 1976.

Arculus, R. J., Mineralogy and petrology of Grenada, Lesser Antilles island arc, Contrib. Mineral. Petrol., 65, 413-424, 1978.

Baker, P. E., Petrology of Mt. Misery volcano, St. Kitts, West Indies, Lithos, 1, 124-150, 1968.

Brown, G. M., Holland, J. G., Siggurdsson, H., Tomblin, J. F. and Arculus, R. J., Geochemistry of the Lesser Antilles volcanic island arc, Geochim. Cosmochim. Acta, 41, 785-801, 1977.

MacGregor, A. G., The volcanic history and petrology of Montserrat, with observations of Mt. Pelée, in Martinique, Phil. Trans. Roy. Soc. Ser. B, 229, 1-90, 1938.

Rea, W. J., The volcanic geology and petrology of Montserrat, West Indies, J. Geol. Soc., 130, 341-366, 1974.

Robson, G. R. and Tomblin, J. F., Catalogue of the active volcanoes of the world including solfatara fields, Part XX West Indies. Internat. Assoc. Volcanol., 56pp, 1966.

ANDES

Casertano, L., Catalogue of the active volcanoes of the world including solfatara fields, Part XV. Chilean continent, Internat. Assoc. Volcanol., 55pp, 1963.

Francis, P. W., Roobol, M. J., Walker, G. P. L., Cobbold, P. R. and Coward, M., The San Pedro and San Pabro volcanoes of northern Chile and their hot avalanche deposits, Geol. Rundschau, 63, 357-388, 1974.

Hantke, G. and Parodi, I. A., Catalogue of the active volcanoes of the world including solfatara fields, Part XIX Colombia, Ecuador and Peru, Internat. Assoc. Volcanol., 73pp, 1966.

Hörmann, P. K., Pichler, H. and Zeil, W., New data

on the young volcanism in the Puna of NW - Argentina, Geol. Rundschau, 62, 397-418, 1973.

Lopez-Escobar, L., Frey, F. A. and Vergara, M., Andesites and high-alumina basalts from the Central - South Chile High Andes: geochemical evidence bearing on their petrogenesis, Contrib. Mineral. Petrol., 63, 199-228, 1977.

Pichler, H. and Zeil, W., Die quartäre "Andesite" - Formation in der Hochkordollere Nord - Chiles, Geol. Rundsch., 58, 866-903, 1969.

Pichler, H. and Zeil, W., The Cenozoic rhyolite - andesite association of the Chilean Andes, Bull. Volcanol., 35, 424-452, 1972.

Zeil, W. and Pichler, H., Die känozoische Rhyolith - Formation im mittleren Abschmitt der Anden, Geol. Rundsch., 57, 48-81, 1967.

SOUTH SANDWICH

Baker, P. E., Comparative volcanology and petrology of the Atlantic island arcs, Bull. Volcanol., 32, 189-206, 1968.

Baker, P. E., Recent volcanism and magmatic variation in the Scotia Arc, in: Antarctic Geology and Geophysics, R. J. Aide ed., 57pp, 1971.

Baker, P. E., Davies, T. G. and Roobol, M. J., Volcanic activity at Deception Island in 1967 and 1969, Nature, 224, 553-560, 1969.

Berninghausen, W. H. and Newman van Padang, M., Catalogue of the active volcanoes of the world including solfatara fields. Part X, Antarctica, Internat. Volcanol. Assoc., 32pp, 1960.

Gass, I. G., Harris, P. G. and Holdgate, M. W., Pumice eruption in the area of the South Sandwich Islands, Geol. Mag., 100, 321-330, 1963.

Weaver, S. D., Saunders, A. D., Pankhurst, R. J. and Tarney, J., A geochemical study of magmatism associated with the initial stages of back-arc spreading. The Quaternary volcanics of Bransfield Strait, from South Shetland Islands, Contrib. Mineral. Petrol., 68, 151-169, 1979.

NEW ZEALAND

Clark, R. H., Petrology of volcanic rocks of Tongariro subdivision. Appendix 2 in the Geology of Tongariro Subdivision, N. Z. Geol. Suv. Bull., 40, 107-123, 1960.

Cole, J. W., Petrology of the basic rocks of the Tarawera volcanic complex, N. Z. J. Geol. Geophys., 13, 925-936, 1970.

Cole, J. W., Petrography of the rhyolitic lavas of Tarawera volcanic complex, N. Z. J. Geol. Geophys., 13, 903-924, 1970.

Cole, J. W., Structure and eruptive history of the Tarawera volcanic complex, N. Z. J. Geol. Geophys., 13, 879-903, 1970.

Cole, J. W., High alumina basalts of Taupo volcanic zone, New Zealand, Lithos, 6, 53-64, 1973.

Cole, J. W. and Nairn, I. A., Catalogue of the active volcanoes of the world including solfatara fields. Part XXII. New Zealand, Internat. Volcanol. Assoc., 156pp, 1975.

Cole, J. W. and Teoh, L. H., Petrography, mineralogy and chemistry of Pureora andesite volcano, North Island, New Zealand, N. Z. J. Geol. Geophys., 18, 259-272, 1975.

Black, P. M., Observations on White Island volcano, New Zealand, Bull. Volcanol. 34, 158-167, 1970.

Ewart, A., The petrography of the Central North Island rhyolitic lavas. Part I - Correlations between the phenocryst assemblages, N. Z. J. Geol. Geophys., 10, 182-197, 1967.

Ewart, A., The petrography of the Central North Island rhyolitic lavas. Part 2 - Regional petrography including notes on associated ash flow pumice deposits, N. Z. J. Geol. Geophys., 11, 478-545, 1968.

Ewart, A., Brothers, R. N. and Mateen, A., An outline of the geology and geochemistry, and the possible petrogenetic evolution of the volcanic rocks of the Tonga - Kermadec - New Zealand island arc, J. Volcanol. Geotherm. Res., 2, 205-250, 1977.

Ewart, A., Green, D. C., Carmichael, I. S. E. and Brown, F. H., Voluminous low temperature rhyolitic magmas in New Zealand, Contrib. Mineral. Petrol., 33, 128-144, 1971.

Gow, A. J., Petrographic and petrochemical studies of the Mt. Egmont andesites, N. Z. J. Geol. Geophys., 11, 166-190, 1968.

Searle, E. J., The petrology of the Auckland basalts, N. Z. J. Geol. Geophys., 4, 165-204, 1961.

Steiner, A., Petrogenetic implications of the 1954 Ngauruhoe lava and its xenoliths, N. Z. J. Geol. Geophys., 1, 325-363, 1958.

TONGA - KERMADEC

Brodie, J. W., Notes on the volcanic activity at Fonualei, Tonga, N. Z. J. Geol. Geophys., 13, 30-38, 1970.

Brothers, R. N. and Martin, K. R., The geology of Macauley Island, Kermadec group, southwest Pacific, Bull. Volcanol., 34, 330-346, 1970.

Brothers, R. N. and Searle, E. J., The geology of Raoul Island, Kermadec group, southwest Pacific, Bull. Volcanol., 34, 7-37, 1970.

Bryan, W. B., Stice, G. D. and Ewart, A., Geology, petrology and geochemistry of the volcanic islands of Tonga, J. Geophys. Res., 77, 1566-1585, 1972.

Ewart, A., A petrological study of the younger Tongan andesites and dacites, and the olivine tholeiites of Niua Foóu island, S.W. Pacific, Contrib. Mineral. Petrol., 58, 1-21, 1976.

Ewart, A., Brothers, R. N. and Mateen, A., An outline of the geology and geochemistry, and the possible petrogenetic evolution of the volcanic rocks of the Tonga - Kermadec - New Zealand island arc, J. Volcanol. Geotherm. Res., 2, 205-250, 1977.

Ewart, A., Bryan, W. B. and Gill, J., Mineralogy and geochemistry of the younger volcanic islands of Tonga, S.W. Pacific, J. Petrol., 14, 429-465, 1973.

Reay, A., Rooke, J. M., Wallace, R. C. and Whelan, P., Lavas from Niua Foóu island resemble ocean-floor basalts, Geology, 2, 605-606, 1974.

Richard, J. J., Catalogue of the active volcanoes of the world including solfatara fields. Part XIII, Kermadec, Tonga and Samoa, Internat. Assoc. Volcanol., 38pp, 1962.

MELANESIA

Blake, D. H., Post Miocene volcanoes in Bougainville Island, Territory of Papua and New Guinea, Bull. Volcanol., 32, 121-138, 1968.

Blake, D. H. and Bleeker, P., Volcanoes of the Cape Hoskins area, New Britain, Territory of Papua and New Guinea, Bull. Volcanol., 34, 385-405, 1970.

Blake, D. H. and Ewart, A., Petrology and geochemistry of the Cape Hoskins volcanoes, New Britain, Papua New Guinea, J. Geol. Soc. Aust., 21, 319-331, 1974.

Blake, D. H. and Miezitis, Y., Geology of Bougainville and Buka Islands New Guinea, B. M. R. Geol. Geophys. Aust. Bull., 93, 56pp, 1967

Colley, H. and Warden, A. J., Petrology of the New Hebrides, Geol. Soc. Am. Bull., 85, 1635-1646, 1974.

Fisher, N. H., Catalogue of the active volcanoes of the world including solfatara fields Part V. Melanesia, Internat. Volcanol. Assoc., 105pp, 1957.

Heming, R. F., Geology and petrology of Rabaul caldera, Papua New Guinea, Geol. Soc. Am. Bull., 85, 1253-1264, 1974.

Heming, R. F. and Carmichael, I. S. E., High temperature pumice flows from the Rabaul caldera, Papua New Guinea, Contrib. Mineral. Petrol., 38, 1-20, 1973.

Johnson, R. W., Distribution and major - element chemistry of late Cainozoic volcanoes at the southern margin of the Bismark Sea, Papua New Guinea, B. M. R. Aust. Res., 188, 157pp, 1970.

Lowder, G. G., The volcanoes and caldera of Talasea, New Britain: Mineralogy, Contrib. Mineral. Petrol., 26, 324-340, 1970.

Lowder, G. G. and Carmichael, I. S. E., The volcanoes and caldera of Talasea, New Britain: Geology and petrology, Bull. Geol. Soc. Am., 81, 17-38, 1970.

Taylor, G. A., The 1951 eruption of Mount Lamington, Papua, B. M. R. Geol. Geophys. Aust. Bull., 38, 117pp, 1958.

Warden, A. J., Evolution of Aoba caldera volcano, New Hebrides, Bull. Volcanol., 34, 107-140, 1970.

INDONESIA

Neumann Van Padang, M., Catalogue of the active volcanoes of the world including solfatara fields Part I. Indonesia, Internat. Volcanol. Assoc., 271pp, 1951.

Whitford, D. J. and Jezek, P. A., Origin of late-Cenozoic lavas from the Banda arc, Indonesia: trace element and Sr. isotope evidence, Contrib. Mineral. Petrol., 68, 141-150, 1979.

Whitford, D. J., Nicholls, I. A. and Taylor, S. R., Spacial variations in the geochemistry of Quaternary lavas across the Sunda arc in Java and Bali, Contrib. Mineral. Petrol., 70, 341-356, 1979.

PHILIPPINES

Moore, J. G. and Melson, W. G., Nuées Ardentes of the 1968 eruption of Mayon volcano, Philippines, Bull. Volcanol., 33, 600-620, 1969.

Neumann Van Padang, M., Catalogue of the active volcanoes of the world including solfatara fields Part II, Philippine islands and Cochin China, Internat. Volcanol. Assoc., 49pp, 1953.

IZU - MARIANA

Isshiki, N., Explanatory text of the geological map of Japan, Hachijo-jima (in Japanese), Geol. Surv. Japan, 58pp, 1959.

Isshiki, N., Explanatory text of the geological map of Japan, Miyake-jima (in Japanese), Geol. Surv. Japan, 82pp, 1960.

Kuno, H., Catalogue of the active volcanoes of the world including solfatara fields Part XI Japan, Taiwan and Marianas, Internat. Volcanol. Assoc., 332pp, 1962.

Schidt, R. G., Geology of Saipan, Mariana islands, Chapter B. Petrology of the volcanic rocks, U. S. Geol. Surv. Prof. Paper, 280-B, 127pp, 1957.

Stern, R. J., Agrigan: an introduction to the geology of an active volcano in the northern Mariana island arc, Bull. Volcanol., 41, 43-55, 1978.

Stern, R. J., On the origin of andesite in the northern Mariana island arc: implication from Agrigan, Contrib. Mineral. Petrol., 68, 207-219, 1979.

Taniguchi, H., Volcanic geology of Kozu-shima, Japan (in Japanese), Bull. Volcanol. Soc. Japan Ser. II, 22, 133-147, 1977.

RYUKYU ISLANDS AND KYUSHU

Kuno, H., Catalogue of the active volcanoes of the world including solfatara fields Part XI. Japan, Taiwan and Marianas, Internat. Volcanol. Assoc., 332pp, 1962.

Lipman, P. W., Mineral and chemical variations within an ash-flow sheet from Aso caldera, southwestern Japan, Contrib. Mineral. Petrol., 16, 300-327, 1967.

Ono, K., Explanatory text of the geological map of Japan, Kuju (in Japanese), Geol. Surv. Japan, 106pp, 1963.

Ono, K., Matsumoto, Y., Miyahisa, M., Teraoka, Y. and Kambe, N., Geology of the Taketa district: Quadrangle series scale 1:50,000. Geol. Surv. Japan, 145pp, 1977.

NORTHEASTERN JAPAN

Isshiki, N., Matsui, K. and Ono, K., Selected bibliography of Japanese volcanoes, Geol. Surv. Japan, 73pp, 1968.

KURILE - KAMCHATKA

Gorshkov, G. S., Catalogue of the active volcanoes of the world including solfatara fields Part VII

Kurile Islands, Internat. Volcanol. Assoc., 99pp 1958.

Vlodavetz, V. I. and Piip, B. I., Catalogue of the active volcanoes of the world including solfatara fields Part VIII. Kamchatka and continental areas of Asia, Internat. Volcanol. Assoc., 110pp, 1959.

ALEUTIAN

Barth, T. F. W., Geology and petrology of the Pribilof islands, Alaska, U. S. Geol. Surv. Bull., 1028-F, 101-160, 1956.

Byers, F. M., Geology of Umnak and Bogoslof Islands, Alaska, U. S. Geol. Surv. Bull., 1028-L, 267pp, 1959.

Byers, F. M., Petrology of three volcanic suites, Umnak and Bogoslof Islands, Aleutian Islands, Alaska, Geol. Soc. Am. Bull., 72, 93-128, 1961.

Coats, R. R., Geology of northern Adak island, U. S. Geol. Surv. Bull., 1028-C, 45-66, 1947.

Coats, R. R., Magmatic differentiation in Tertiary and Quaternary volcanic rocks from Adak and Kanaga islands, Aleutian Islands, Alaska, Geol. Soc. Am. Bull., 63, 485-514, 1952.

Coats, R. R., Geology of northern Kanaga island, U. S. Geol. Surv. Bull., 1028-D, 69-80, 1956.

Coats, R. R., Geologic reconnaissance of Gareloi island, Aleutian islands, Alaska, U. S. Geol. Surv. Bull., 1028-J, 249-256, 1959.

Coats, R. R., Magma type and crustal structure in the Aleutian arc, Geophys. Monograph, no.6, 92-109, 1962.

Coats, R. R., Nelson, W. H., Lewis, R. O. and Powers, H. A., Geologic reconnaissance of Kiska island, Aleutian islands, Alaska, U. S. Geol. Surv. Bull., 1028-R, 563-581, 1961.

Drewes, H., Fraser, G. D., Snyder, G. L. and Barnett, H. F., Geology of Unalaska island and adjacent Insular Shelf, Aleutian islands, Alaska, U. S. Geol. Surv. Bull., 1028-S, 583, 1961.

Fenner, C. N., The Katmai magmatic province, J. Geol., 34, 673-772, 1926.

Kennedy, G. C. and Waldron, H. H., Geology of Pavlof volcano and vicinity Alaska, U. S. Geol. Surv. Bull., 1028-A, 1-19, 1955.

Marsh, B. D., Some Aleutian andesites: their nature and source, J. Geol., 84, 27-45, 1976.

Nelson, W. H., Geology of Segula, Davidof and Khvostof islands, Alaska, U. S. Geol. Surv. Bull., 1028-K, 257, 1959.

Simons, F. S. and Mathewson, D. E., Geology of Great Sitkin island, Alaska, U. S. Geol. Surv. Bull., 1028-B, 21-43, 1955.

Snyder, G. L., Geology of Little Sitkin island, Alaska, U. S. Geol. Surv. Bull., 1028-H, 1959.

Waldron, H. H., Geology reconnaissance of Frosty Peak volcano and vicinity, Alaska, U. S. Geol. Surv. Bull., 1028-T, 677-707, 1961.

PETROLOGY AND GEOCHEMISTRY OF OPHIOLITIC AND ASSOCIATED VOLCANIC ROCKS
ON THE TALAUD ISLANDS, MOLUCCA SEA COLLISION ZONE, NORTHEAST INDONESIA*

C. A. Evans, J. W. Hawkins, and G. F. Moore

Geological Research Division, A-015 Scripps Institution of Oceanography La Jolla, California 92093

Abstract. The Talaud Islands lie within the Molucca Sea collision complex between the converging Halmahera and Sangihe arcs. Much of the islands are a tectonic melange which includes blocks of ophiolite. It is interpreted as slivers of oceanic crust and trench sediments in an uplifted forearc terrane. The ophiolitic blocks preserve a disrupted but complete oceanic crustal section. The sequence includes basalt, diabase, microgabbro, layered gabbro, and serpentinized peridotites.

Basalt pillows are closely associated with Eocene pelagic sediments. Basalts, diabases and microgabbros comprises plagioclase laths (An_{50-60}), augite and magnetite. They display Ti and Fe enrichments and Ni and Cr depletions with increasing fractionation. Abundances of K,Ba,Rb,Sr,Ti,Zr are relatively low.

Layered cumulate gabbros are dominated by olivine-clinopyroxene- plagioclase and orthopyroxene-clinopyroxene-plagioclase assemblages. Layering is defined by the alignment of plagioclase laths and the mafic minerals. Clinopyroxene is often interstitial. Mineral compositions vary in different assemblages. Changes in both the mineral compositions and bulk chemistry of the gabbros reflect crystallization trends of the basaltic liquid.

The peridotites are serpentinized lherzolites and harzburgites and are interpreted as refractory mantle residua.

Miocene volcanic flows (basaltic andesites) are found on the Talaud Islands but are structurally separate from the ophiolitic rocks. They display geochemical trends which are different from the basalts and diabases and are presumed to be island arc derived. They are not considered part of the ophiolite.

Whole rock and mineral chemistry supports the hypothesis that the basalts and gabbros represent oceanic layers 2 and 3 generated at a mid-ocean ridge or back-arc basin. Trace element trends and major element modelling of the basalts suggest that crystal fractionation of a basaltic magma in a shallow magma chamber may account for the primary chemical differences and magma evolution. Subsequent crystal accumulation within the magma chamber yielded the gabbros. This is consistent with their proposed origin at a spreading oceanic ridge.

Introduction

The origin and mechanism of emplacement of ophiolites into orogenic zones has been the subject of considerable recent study. Many models for ophiolite emplacement associated with collisions between island arcs and other "microplates" have been proposed (e.g., Dewey, 1976), but none have been based on observations from modern tectonic settings. An area of active collision tectonics and proposed ophiolite emplacement is the Molucca Sea collision zone of northeastern Indonesia (Fig. 1; Silver and Moore, 1978). The "Molucca Sea Plate" is presently being subducted along its eastern margin beneath the Halmahera arc system and along its western margin beneath the Sangihe arc system (Hatherton & Dickinson 1969; Fitch 1970; Cardwell et al., 1980). The Halmahera and Sangihe arcs are thus colliding (Murphy, 1973; Katili, 1978), and the Molucca Sea represents the collision zone (Silver and Moore, 1978; Hamilton, 1977, 1979). The southern Molucca Sea collision zone is a broad, highly deformed ridge and is composed of very thick, low-density material of low seismic velocity (McCaffrey et al., 1980). The Talaud islands are the emergent portion of a block that forms the northern boundary of the active collision zone. North of Talaud, the collision is apparently complete and convergence has ceased (Hamilton, 1979; Cardwell et al., 1980).

The Talaud islands were visited by Dutch geologists in the 1920's (Rothaan, 1925; van Bemmelen, 1949) and were mapped on a regional scale in 1976 (Sukamto et al, 1980). This mapping indicated the existence of several bodies of ophiolitic rocks. During July and August of 1979, Moore and D. Kadarisman of the Indonesian Geological

---

*Contribution of the Scripps Institution of Oceanography, new series

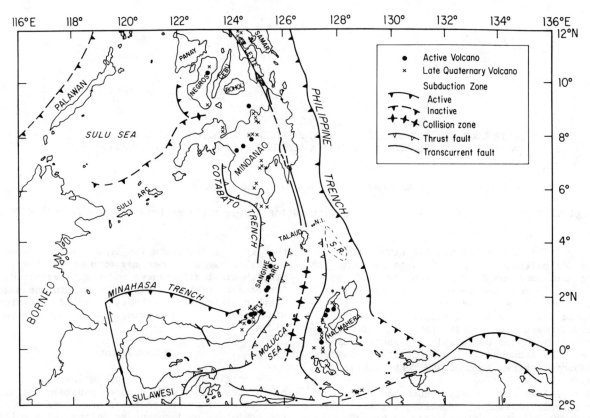

Figure 1. Regional tectonic map of eastern Indonesia, modified from Hamilton(1979) and Silver(1980). "N.I." is an abbreviation for Nanusa Islands and "S.R." for Snellius Ridge.

Research and Development Center carried out a mapping and sampling program on the Talaud Islands. Our program was designed to determine the structural setting of the ophiolitic rocks and to collect representative samples for petrologic and geochemical analyses. This paper reports the results of our sample analysis. A more complete description of the structure and stratigraphy of Talaud is presented in Moore et al. (in press).

Regional Geology

The Talaud Islands comprises five major rock units (Figs. 2 and 3): Pleistocene coralline limestone, mid-Miocene to Pliocene marine sedimentary rocks, andesitic volcanic rocks, tectonic melanges, and ophiolites. These rock units occur in approximately north-south-trending belts commonly separated by faults (Fig. 2).

Tectonic Melanges. Tectonic melanges on Talaud are mappable bodies that display a characteristic internal fabric dominated by penetrative mesoscopic shear fractures and containing inclusions of all sizes immersed in a pervasively sheared, fine-grained matrix. They contain blocks of igneous and sedimentary rocks that range in size from a few centimeters to hundreds of meters.

Igneous rocks are most abundant and are fragments of a dismembered ophiolite (see below). Sedimentary rocks are shale, recrystallized limestone and greywacke that are strongly sheared and boudinaged. Bedding is totally disrupted.

Ophiolites. Three of the largest tectonic inclusions in the melange are mappable bodies of ophiolitic rocks that range in size up to 5 km in width. These ophiolites preserve a disrupted but complete ophiolite sequence. Serpentinized peridotite (lherzolite and harzburgite), cumulate gabbro, fine grained massive gabbros, diabases, pillow basalts and cherts are all present, but no complete, intact ophiolite sequence has been observed on Talaud. Pillow structures are abundant in basalt fragments found in streams, although they are rare in hillside exposures. The interpillow matrix is commonly chlorite, but pink limestone is also found. Bedded cherts and limestones are closely associated with the pillow basalts, but no depositional contacts have been observed. A few Eocene Radiolarians were found in the cherts and red shales (Moore et al, in press). Most of the peridotites are serpentinized (Soeria Atmadja & Sukamto, 1979), and there is deformation on the scale of outcrops and thin sections.

The ophiolites are generally east-dipping slabs

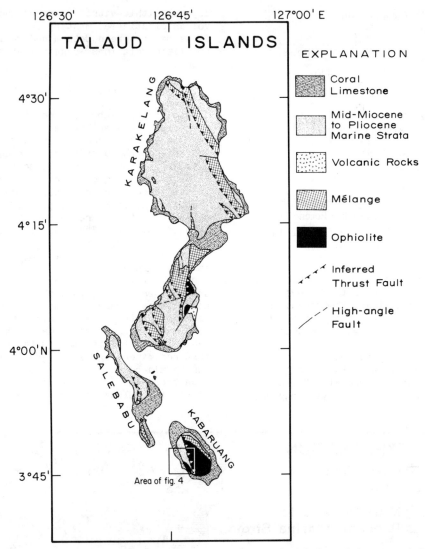

Figure 2. Geologic sketch map of the Talaud Islands, updated from Sukamto and Suwarno (1976).

with north-south striking and steeply east dipping contacts. In southwestern Kabaruang Island (Fig. 4), a vertical sequence through the gabbro-basalt transition was sampled. The lowest structural unit is layered gabbro. At the outcrop scale, layering is defined by the alignment of mafic minerals. Fragments of amphibolitized diabase are intimately mixed with the gabbro near the top of the section. This may represent intrusive relations between gabbros and diabase dikes. A 30-50m thick section of pillow basalt overlies the gabbro-diabase unit. The basalt is in turn overlain by mid-Miocene shales.

The ophiolite, melange and sediments are imbricately thrust together. In one river valley, the sediments are overthrust by a steeply dipping melange zone 75-100m thick. The melange is structurally overlain by massive peridotite. This melange-peridotite unit appears to have a very limited lateral extent. It is overlain depositionally by mid-Miocene strata which are folded into open west-verging folds (Fig. 4). The Neogene strata are overthrust by a block of massive peridotite along central and western Kabaruang (Figs. 2 and 4).

In central Karakelang there are two smaller bodies of ophiolite. Both are east-dipping slabs with high-angle contacts on all sides. One body is massive peridotite and the other is layered gabbro and basalt.

Neogene Strata. Mid-Miocene to Pliocene deep water sedimentary rocks are widely exposed on the Talaud Islands (Fig. 3). The strata occur in elongate, north-trending belts and are separated by melange zones. The strata are moderately to strongly deformed but continuity is generally well preserved except adjacent to melange zones. The trend of major folds is parallel to the

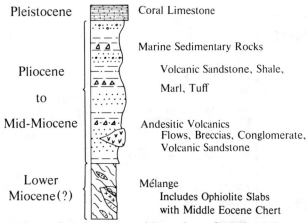

Figure 3. Stratigraphic column for the Talaud Islands (from Moore et al, in press).

outcrop belts and the folds have wavelengths of 10-200 meters. Fold vergence is to the west.

Volcanic detritus is dominant throughout the section. The Neogene strata are apparently both in depositional contact with, and overthrust by the melanges.

Miocene Volcanic Rocks. Interbedded with the sandstones on the east coast of Karakelang Island are exposures of volcanic rocks (Fig. 2) which include flows of basaltic andesite vitrophyres, crystal-lithic-vitric breccias and poorly sorted volcaniclastic sedimentary rocks. The age of the flows has not been determined radiometrically, but field relations suggest that they are mid-Miocene.

Petrographic Descriptions

Basalts and Diabases. The pillow basalts within the melange are spilitized but their variolitic texture is still recognizable. Some plagioclase and olivine phenocrysts are present although the olivine is pseudomorphed by serpentine. Intersertal augite is common. The diabases are generally less altered than the basalts and are predominantly ophitic rocks of plagioclase, augite, and minor magnetite. One sample, T-90, contains abundant olivine. The radiating plagioclase laths have compositions of An_{50-65}. Plagioclase phenocrysts may be as calcic as An_{70}, and are often strongly zoned.

Most of these rocks show evidence of some shearing and have metamorphic mineral assemblages. They are amphibolitized to varying degrees. Retrograde greenschist facies assemblages are present as veins of chlorite, quartz, and epidote. Despite this alteration, their textures, mineralogies and geochemistry are similar to mid-ocean ridge basalts (MORB).

Gabbros. The Talaud gabbros occur as blocks within the melange and as structurally intact

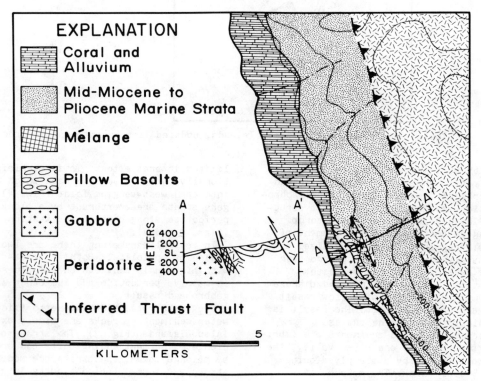

Figure 4. Geologic map of southwest Kabaruang Island showing the field relations of the ophiolite slices. See Fig. 2 for location.

Table 1. Talaud whole rock chemical compositions (wt. %)[a].
a. Basalts, diabases and arc volcanic rocks.

	T-90	T-55	T-14a	T-58	T-30	T-65	T-77	T-77$_{glass}$[b]
SiO_2	46.80	48.28	52.17	49.35	48.79	48.58	54.91	62.39
TiO_2	1.28	.89	1.16	1.16	1.92	2.04	.72	.66
Al_2O_3	15.14	15.49	14.50	16.57	14.00	14.91	15.41	16.66
Fe_2O_3	3.82	3.44	3.48	3.62	4.14	6.01	3.49	
FeO	6.88	5.91	5.79	4.01	6.37	6.10	3.61	3.22[c]
MnO	.28	.15	.17	.27	.27	.30	.12	.07
MgO	13.81	9.51	8.50	6.11	6.55	5.46	6.54	1.61
CaO	10.31	11.27	9.11	12.59	9.97	11.53	9.41	4.53
Na_2O	1.63	2.51	3.02	2.27	3.22	2.30	3.03	4.51
K_2O	.20	.25	.19	.08	.06	.32	1.29	2.85
P_2O_5	.15	.10	.09	.12	.21	.25	.21	.33
H_2O+	1.91	1.11	1.74	1.29	1.69	1.33	.68	
H_2O-	1.38	.53	.80	2.31	1.80	1.95	.72	
SUM	100.37	99.08	100.72	99.75	99.15	101.18	100.14	96.83
FeO*/MgO	.73	.98	1.05	1.19	1.56	2.11	1.03	2.00
Ba(ppm)	39	28	23	12	46	55	124	
Rb	1	1	2			1	22	
Sr	259	234	88	133	219	251	588	
Cr	481	256	229	295	122	104	95	
Ni	423	205	102	157	69	68	32	
V	189	174	240	259	352	192	235	
Zr	105	74	70	75	148	157	110	

[a] Major and trace element chemistry by XRF, except Cr which was done by AA.
[b] Analysis done by microprobe, an average of 5 points
[c] Total Fe as FeO

T-90   Olivine diabase
T-55   Plagioclase-augite-magnetite diabase
T-14a Alterered vitrophyre with minor plagioclase phenocrysts
T-58   Diabase
T-30   Variolitic vitrophyre with plagioclase and serpentinized olivine phenocrysts
T-65   Altered diabase
T-77   Porphyritic hypersthene-augite-plagioclase vitrophyre (about 50% phenocrysts)
T-77$_{glass}$ Groundmass of T-77

sequences associated with the diabases. Included are microgabbros which are coarser grained equivalents of the diabases, and layered cumulate gabbros such that a continuum of textures (vitrophyre through cumulate gabbro) has been sampled.

Cumulate gabbros are layered on both a centimeter and outcrop scale. Layering is defined by alignment of plagioclase laths and varying abundances of the mafic minerals. The gabbros comprise a variety of lithologies which include CPX-OL-PL, CPX-OPX-PL, CPX-OPX-OL-PL, and CPX-AMPH-PL assemblages. Opaque minerals are predominantly magnetite, although Fe-Ni sulfides are seen in one sample (T-85). Chromites are absent. Irregular patches of diopside often forms interstitial patches poikilitically including both plagioclase and olivine. Plagioclase and clinopyroxene may be zoned but zoning is more subtle than in the finer grained gabbros and diabases. The gabbros display the following ranges in mineral compositions: plagioclase, An_{50-80}; olivine, Fo_{84-86}; orthopyroxene, En_{75-85}; and clinopyroxene, $En_{45-52}Wo_{40-50}Fs_{6-11}$. Mineral analyses are by electron microprobe. Representative analyses are given in Table 2.

The gabbros show evidence of some shear deformation. Bent twin lamellae and cleavage planes in plagioclase and pyroxenes and minor pyroxene granulation are sometimes present but are not pervasive. Other alteration includes replacement of mafic minerals by chlorite and green amphiboles, sericitization of plagioclase, and cross-cutting veins of chlorite, epidote, prehnite and uralite.

TABLE 1b. Talaud Gabbros and Peridotites

	T-44	T-94	T-52	T-53	T-82	T-85	T-56	T-18	T-19	T-93
SiO_2	50.16	48.37	49.85	51.85	48.63	43.69	42.07	38.11	37.02	41.14
TiO_2	1.84	2.51	1.39	.32	.20	.08	.01	.02	.01	.04
Al_2O_3	15.00	15.81	13.69	15.96	14.89	17.93	.85	1.11	1.02	1.69
Fe_2O_3	4.31	5.52	3.20	1.10	1.27	1.77	6.79	6.48	6.50	6.51
FeO	6.22	6.94	8.24	4.21	4.02	3.79	2.27	1.76	1.59	2.13
MnO	.86	.17	.22	.13	.10	.09	.11	.13	.12	.13
MgO	7.62	5.52	7.53	9.55	13.28	15.23	39.24	35.48	36.78	38.14
CaO	6.37	7.52	11.62	13.22	13.81	11.04	.57	1.73	1.66	1.40
Na_2O	3.12	3.49	2.06	2.01	1.31	.90	.13	.15	.13	.14
K_2O	.15	.68	.11	.17	.09	---	.08	.09	.07	.05
P_2O_5	.23	.40	.55	.03	.07	.04	.02	.003	.01	.006
H_2O^+	3.17	3.15	.70	.88	1.30	3.17	7.33	12.72	9.04	9.40
H_2O^-	1.17	1.08	.41	.50	.03	.82	1.53	1.95	2.27	1.46
SUM	100.22	101.10	99.24	99.80	99.55	98.55	101.01	99.73	101.43	102.24
Ba(ppm)	400	84	--	23	--	--	--	--	--	--
Rb	--	7	--	--	--	1	--	--	--	--
Sr	179	59	135	148	112	90	2	8	--	--
Cr	269	80	46	197	800	537	1832	1768	1792	1410
Ni	84	142	69	109	294	560	2184	1993	1795	2012
V	198	166	275	163	110	47	44	49	47	52
Zr	119	172	41	12	13	9	6	4	4	6

T-44 Clinopyroxene-plagioclase-magnetite microgabbro
T-94 "
T-52 Sheared and granulated olivine-bearing, two pyroxene gabbro
T-53 Two pyroxene cumulate gabbro
T-82 Cumulate olivine-clinopyroxene-plagioclase gabbro
T-85 Cumulate olivine-clinopyroxene-plagioclase gabbro, with Fe-Ni sulfides
T-56 Serpentinized spinel lherzolite
T-18 Serpentinized OPX-poor lherzolite
T-19 Serpentinized harzburgite
T-93 Granulated and serpentinized spinel lherzolite

TABLE 2. Representative Mineral Analyses
(Analyzed by Electron Microprobe)
a. Olivine

	T-56	T-19	T-82	T-85
MgO	49.83	49.60	44.64	46.24
FeO	8.05	9.40	15.25	13.77
Al_2O_3	---	---	---	---
CaO	.14	.04	.07	.04
TiO_2	.02	---	---	.01
Cr_2O_3	.06	---	.08	.02
MnO	.08	.05	.28	.11
NiO	.40	.30	.18	.10
SiO_2	41.25	40.56	39.34	39.35
SUM	99.85	99.95	99.86	99.63
Fo	91.6	90.4	83.9	85.7

TABLE 2b. Orthopyroxenes

	T-56 opx	T-19 opx	T-85 opx	T-53 opx	T-77 opx
MgO	33.45	33.65	31.88	27.83	28.11
FeO	5.17	6.22	9.57	13.07	11.96
CaO	1.49	.98	1.00	1.91	1.70
Na_2O	---	---	---	---	---
TiO_2	.06	.04	.22	.28	.26
Cr_2O_3	.71	.58	.30	.09	.02
MnO	.14	.09	.13	.13	.37
NiO	.08	.03	.08	.06	.06
Al_2O_3	2.48	3.06	1.40	1.28	1.20
SiO_2	55.99	55.44	54.59	54.25	54.52
SUM	99.63	100.07	99.59	98.90	98.22
En	89.4	88.9	83.9	76.2	78.0
Fs	7.8	9.2	14.2	20.0	18.6
Wo	2.9	1.9	1.9	3.8	3.4

TABLE 2c. Clinopyroxenes

	T-56 cpx	T-93 cpx	T-82 cpx	T-77 cpx
MgO	18.25	16.78	16.83	15.94
FeO	2.75	2.15	4.56	7.99
CaO	20.44	22.87	21.30	20.16
Na_2O	.15	.30	.39	.37
TiO_2	.09	.22	.51	.63
Cr_2O_3	1.46	.93	.69	.06
MnO	.19	.15	.14	.27
NiO	---	.09	.07	.01
Al_2O_3	3.73	4.42	2.82	2.00
SiO_2	52.70	52.17	52.24	51.85
SUM	99.78	100.11	99.60	99.30
En	52.9	48.7	48.5	45.7
Fs	4.5	3.5	7.4	12.8
Wo	42.6	47.8	44.1	41.5

Peridotites. The ultramafic rocks found on Talaud are highly serpentinized lherzolites and harzburgites. They are similar in texture and composition to peridotites from the metamorphosed ultramafic section of other ophiolite suites (for example, Coombs et al, 1976; England and Davies, 1973) and peridotites from oceanic fracture zones and trenches (Bonatti, 1971; Hawkins et al, 1979). The peridotites are thought to be depleted upper mantle material which immediately underlies crustal gabbros. Olivine (Fo_{90-92}), which comprises over 85% of the rock, has been recrystallized and subsequently partly altered to serpentine and magnetite. Orthopyroxene (En_{88-90}), Cr-diopside and chromite are the other primary minerals. Mineral compositions are very constant within samples; chromites display the greatest range in compositions between samples (Cr/Cr+Al = .2-.5).

Orthopyroxenes form large rounded grains (up to 1 cm) with strained cleavage and exsolution lamellae. Clinopyroxene generally occurs as smaller grains but several crystals are often clumped together. Spinels may be symplectically associated with pyroxene crystals in complex and irregular patterns. Additional petrographic descriptions are given in Soeria Atmadja and Sukamto (1979).

Miocene Volcanic Flows. Volcanic flows (T-26, T-77) and associated volcaniclastic sediments are found on the east side of Karekelang island (Sukamto, 1980) (Fig. 1). These rocks are basaltic andesite-andesite vitrophyres and are not considered part of the ophiolite (Moore et al, in press). The vitrophyres are quite fresh and contain approximately 50% phenocryst material by volume. The euhedral phenocrysts include zoned plagioclase (An_{52-63}), hypersthene (En_{75-85}), and clinopyroxene ($En_{48}Wo_{40}Fs_{12}$). Some hypersthene crystals have clinopyroxene rims. The groundmass is a brown glass with high SiO_2, K_2O, Na_2O and microlites of clinopyroxene, plagioclase, and opaque minerals. Whole rock mineral and glass analyses are given in tables 1 and 2.

The field association with the volcaniclastic sediments of Miocene age, coupled with a distinct geochemistry (to be discussed later) suggest that these are island arc volcanic rocks. A definitive arc source for these rocks is unknown, but possible volcanism associated with eastward subduction north of Halmahera is suggested (Moore et al, in press). Another possibility is juxtapositon of the volcanic rocks on eastern Talaud and Nanusa against the rest of the Talaud block by strike-slip movement along the Philippine Fault Zone. The fault zone is a major tectonic feature that traverses the Philippine Islands from Luzon Southward to Mindanao (Allen, 1962; Ranneft et al., 1960). A strand of the fault zone has been traced south from Mindanao to the Talaud Islands (fig. 1) on seismic profiles in the northern Molucca Sea (G. Moore & E. Silver, work in progress, 1981). The volcanic rocks of eastern

TABLE 2d. Chromite

	T-56	T-19	T-93
MgO	12.70	15.18	17.53
FeO	18.49	17.73	12.57
Cr_2O_3	36.65	24.90	20.79
Al_2O_3	31.49	41.34	47.95
CaO	.02	.03	.02
MnO	.33	.18	.16
TiO_2	.09	.04	.11
NiO	.05	.15	.30
SiO_2	.01	.34	---
SUM	99.91	99.88	99.44
α	43.2	27.9	22.5
β	55.4	69.0	77.4
γ	1.4	3.1	.1

α = 100 Cr/(Cr + Al + Fe''')
β = 100 Al/(Cr + Al + Fe''')
γ = 100 Fe'''/Cr + Al + Fe''')

TABLE 2e. Plagioclase

	T-82	T-82	T-53	T-77
K_2O	.03	.02	.03	.36
Na_2O	3.93	2.21	4.32	4.96
CaO	13.57	16.26	12.59	11.56
FeO	.29	.30	.24	1.04
Al_2O_3	30.53	32.94	29.88	28.49
SiO_2	51.27	47.79	52.65	52.99
SUM	99.79	99.74	99.72	99.40
An	65.5	80.2	61.5	52.0
Ab	34.4	19.7	38.2	44.4
Or	.1	.1	.3	3.6

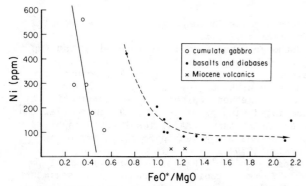

Figure 5. Ni (ppm) vs. FeO*/MgO for Talaud ophiolitic rocks and Miocene volcanic flows. The line through the gabbros is a linear regression and the curve drawn through the diabases and basalts is an eye-ball fit.

Talaud could have been moved several hundred kilometers from the south.

Geochemistry

The basaltic volcanic rocks on Talaud are geochemically similar to typical oceanic layer 2 basalts generated at mid-ocean ridges. They are relatively depleted in magmaphilic elements such as Ti,Zr,Rb,Sr,K,Ba (table 1). The low abundances of these elements separate these rocks from alkali basalts (Engel et al.,1965).

Within the suite of rocks which is structurally dismembered, more primitive liquid compositions may be defined by low FeO*/MgO ratios and high concentrations of Ni and Cr (> 200 ppm)

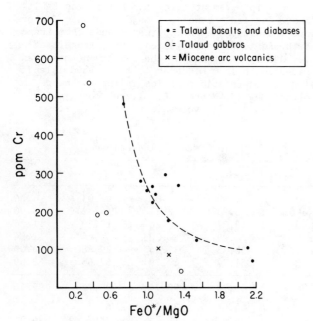

Figure 6. Cr (ppm) vs. FeO*/MgO for the igneous rocks on Talaud.

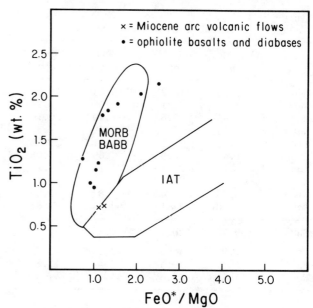

Figure 7. TiO_2 vs. FeO*/MgO for the extrusive rocks on Talaud. The island arc tholeiite (IAT) field is defined by rocks from the Marianas arc. Both mid-ocean ridge basalt (MORB)-back arc basin (BABB) and IAT fields are taken from Hawkins (1980). The Talaud ophiolitic rocks fall within the MORB-BABB field.

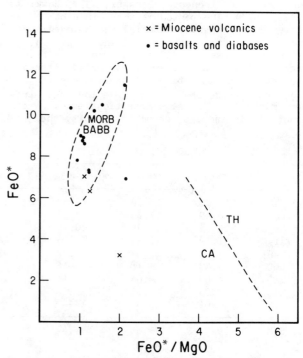

Figure 8. FeO* vs. FeO*/MgO. Most basalts and diabases are within the MORB-BABB field. The Miocene volcanic rocks show uncertain affinities in this plot. CA (calc-alkaline) and TH (tholeiitic) fields are those defined by Miyashiro (1974).

(Miyashiro, 1974; Hawkins 1980). Using these criteria for recognizing more evolved rocks, trends with increasing fractionation can be seen. Ni and Cr are rapidly depleted from the liquid (figs. 5 and 6) suggesting the early crystallization of olivine and clinopyroxene was important in the evolution of the volcanic rocks. This is supported by the abundances of olivine and clinopyroxene in some of the cumulate gabbros. TiO_2 and FeO* enrichments are positively correlated with FeO* (increasing fractionation) as seen in figs. 7 and 8. The rocks fall into MORB fields (Miyashiro, 1974; Hawkins, 1980). For comparison, the Miocene volcanic rocks from Talaud and selected analyses from Halmahera and Sulawesi (the closest known Miocene arc volcanic sources) are also plotted.

The different volcanic rocks on Talaud are clearly segregated using the trace element criteria of Pearce and Cann (1973). One may employ

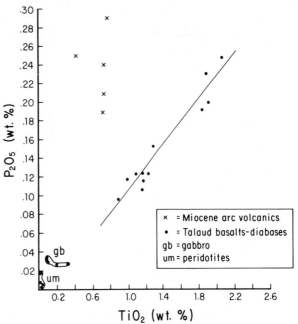

Figure 10. P_2O_5 vs. TiO_2. A linear regression curve is drawn through the basalts and diabases. As in Fig. 9, the Miocene arc volcanics (including the analyses from Sulawesi and Halmahera) have a very different trajectory from the ophiolitic basalts and diabases.

elements which are enriched in the melt with fractionation, yet relatively insensitive to later alteration such as Ti, Zr, P, Sr to separate MORB (and back-arc basin basalts) from island arc tholeiites (IAT). The positive correlations of P_2O_5 and Zr with TiO_2 as displayed by Talaud

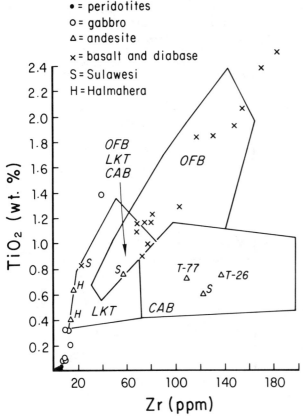

Figure 9. Pearce and Cann (1973) plot of TiO_2 vs. Zr. OFB = ocean floor basalt, LKT = low K tholeiite, CAB = calc-alkaline basalt. "S" and "H" denote rocks from Sulawesi and Halmahera, respectively. All other analyses are Talaud rocks. The Miocene volcanic rocks (T-77 and T-26) display a trend similar to other arc volcanic rocks and are more clearly separated from the ophiolitic basalts and diabases which are similar to MORB.

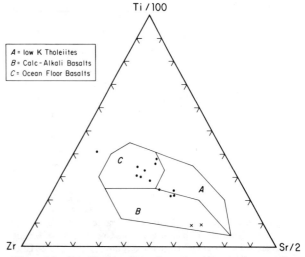

Fig. 11. Ti-Zr-Sr plot for the extrusive rocks on Talaud. Fields are defined by Pearce and Cann (1973). Dots are basalts and diabase, "x" are Miocene volcanic rocks.

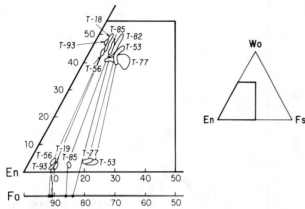

Figure 12. Pyroxene plot of pyroxenes analyzed by microprobe in Talaud gabbros and peridotites. T-18, T-19, T-56, T-93 are peridotites. T-77 is a Miocene vitrophyre. T-53, T-82, T-85 are cumulate gabbros.

basalts and diabases (Figs. 9 and 10) supports a MOR origin. These trends contrast sharply with the arrays of the Miocene volcanics (T-26, T-77) and the arc volcanics from Halmahera and Sulawesi. The arc volcanics have a low concentration of TiO2 which remains relatively unchanged with magma evolution (Miyashiro, 1974). Yet these rocks are enriched in other magmaphilic elements such as Zr, P. The basalts and diabases likewise plot separately on a Ti-Zr-Sr plot (fig. 11), the MORB field being characterized by low Zr and Sr contents (Pearce and Cann, 1973).

The Talaud gabbroic rocks may be separated into two groups: (1) fine grained PL-CPX gabbros which are geochemically similar to the diabases or (2) coarsely crystalline cumulates which are extremely depleted in Ti, P, K and other magmaphilic elements (Table 1, figs. 9 and 10). These cumulate gabbros are similar (petrographically and geochemically) to gabbros recovered from trenches (Hawkins et al, 1979 and unpublished data), from DSDP holes (Hodges and Papike, 1976, and others) and from cumulate sections of other ophiolites (Hawkins and Evans, 1980 and others).

Representative mineral analyses from some of the gabbros and peridotites are given in table 2. Mineral compositions of the gabbros are distinct from similar minerals from the peridotites. A spectrum of pyroxene compositions (fig. 12) is seen. The pyroxene compositions (En content, Cr/Cr+Al) show a positive correlation with the Fo content of the olivines, the An content of the plagioclase, and whole rock Ni and Cr abundances.

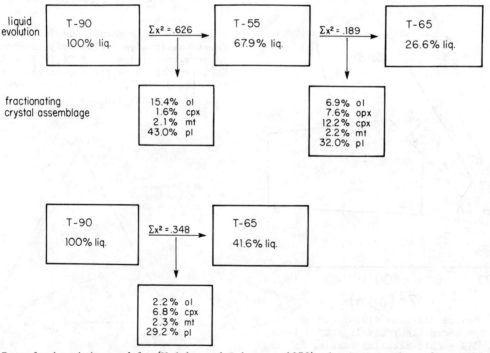

Figure 13. Petrologic mixing models (Wright and Doherty, 1970) showing results of possible crystal fractionation combinations using major element compositions of given parent and derivative liquids. The mineral assemblages are the crystallizing phases. Mineral compositions used in the modelling are microprobe analyses of minerals in the cumulate gabbros. The model reproduces a parent magma (e.g. T-90) by fitting amount of given phenocrysts and a daughter liquid (e.g. T-55). x^2 is the sum of the squares of the differences of the oxides between the modelled and given parent compositions. Both a two stage (T-90 to T-55 to T-65) and a single step (T-90 to T-65) evolutions are given.

These correlations within the cumulate gabbros may define crystallization trends within batches of basaltic magma. Thus, the most olivine-rich gabbro (T-85) has the most forsteritic olivines, high Ni contents (attributable to partitioning of Ni into both olivine and the Fe-Ni sulfides), high Cr in both pyroxenes and the bulk rock, and the most calcic plagioclases (An_{70-80}). It is presumed to occur near the base of the gabbroic section. Upsection, the gabbros (e.g. T-53) are more Fe enriched (whole rock, pyroxene, and olivine), may lack olivine, have lower bulk rock and mineral abundances of Ni and Cr (reflecting liquid depletion of these elements), and have less calcic plagioclase (An 55-65). These differences are graphically displayed in Figs. 5, 6 and 12, and can be seen in bulk rock compositions (Table 1) and mineral compositions (Table 2).

The mineral constituents of serpentinized and sheared lherzolites and harzburgites are more refractory. These rocks likewise have correlations between the Mg content of the silicates and the Cr_2O_3 content and Cr/Cr+Al of the whole rock, the pyroxenes and the chromites. The mineral compositions and element ratios within these rocks are typical of peridotites from other ophiolite suites. We explain the range of compositions representing heterogeneity in a depleted mantle either due to differences in original composition or to subsequent melting history.

Discussion

The rapid depletion of Ni and Cr with increasing fractionation (Fe enrichment) of the Talaud basalts and diabases has suggested that crystal fractionation, especially olivine and clinopyroxene was important in the evolution of the basaltic magmas. The presence of cumulate gabbros with these phases, and the parallel FeO enrichment and Ni and Cr depletion in both whole rock and mineral phases of the gabbros upsection reinforces this hypothesis (Miyashiro et al., 1970). Also, the cumulate gabbros with the most evolved compositions are structurally associated with basalts and diabases. We therefore believe that the primary chemical differences between the basalts and diabases of the Talaud ophiolite may be a result of substantial fractional crystallization within a shallow magma chamber and which was accompanied by crystal accumulation to yield the gabbros. Possible parent-derivative liquid-crystal cumulate relationships were tested with computer modelling (Wright and Doherty, 1970). The program models possible crystal fractionation combinations using the major element compositions of given parent and daughter liquids and the fractionating minerals. The mineral compositions used, with the exception of magnetite (Deer, Howie, and Zussman, 1966), are from microprobe analyses on the cumulate gabbros. The given parent liquid is reproduced by fitting amounts of the given phenocrysts and derivative liquid. Genetic relationships between the basalts and diabases are

TABLE 3. Trace element abundances, both predicted and observed, in Talaud volcanic rocks calculated using modelling solutions given in Fig. 13. Bulk distribution coefficients (D) were calculated using equations and mineral-liquid partition coefficients given in literature (Gast, 1968; Allegre and Minster, 1977; and Hart and Brooks, 1974).

C_o = abundance of given element in the parent liquid (in ppm)

C_{pred} = predicted concentration of daughter liquid as calculated using the modelling solutions.

C_{obs} = actual concentration of the element measured in the rock.

T-90 to T-55				
	D	C_o	C_{pred}	C_{obs}
Sr	.832	259	274	234
Ba	.134	39	54	28
Ni	4.64	423	195	205
Cr	2.28	481	340	256

T-55 to T-65				
	D	C_o	C_{pred}	C_{obs}
Sr	1.095	234	221	251
Ba	.176	28	56	55
Ni	3.66	205	78	68
Cr	3.93	256	92	113

T-90 to T-65				
	D	C_o	C_{pred}	C_{obs}
Sr	.964	259	265	251
Ba	.154	39	77	55
Ni	5.85	423	110	68
Cr	3.56	481	193	113

allowed. Fig. 13 summarizes the major element modelling results. To reinforce confidence in the allowed models, predicted trace element abundances were calculated (Table 3) using mineral-liquid partition coefficients from various sources (Gast, 1968; Allegre et al, 1977; Hart and Brooks, 1974). Discrepencies between predicted and observed trace element data can probably be explained by uncertainties in crystal-liquid partition coefficients, and post eruption alteration of the rock (particularly Ba and Sr). For each parent-daughter pair, the resulting "gabbro" composition from crystal accumulation is given (Table 4). Such modelling results are consistent with the proposed mid-ocean ridge origin.

Having established that the Talaud ophiolite blocks can be best explained as typical oceanic crust, one may contrast rocks from similar set-

TABLE 4. Hypothetical cumulate gabbro compositions created by crystal accumulations of minerals given in the solutions of fractional crystallization modeling (Fig. 13)

	a	b	c	d
SiO_2	41.03	49.67	46.15	44.70
TiO_2	.01	.004	.007	.11
Al_2O_3	13.50	16.34	15.18	14.95
Fe_2O_3	4.55	2.48	3.40	2.70
FeO	9.07	5.67	7.20	7.15
MnO	.05	.11	.08	.04
MgO	23.14	11.90	16.91	19.61
CaO	7.67	11.42	9.83	9.13
Na_2O	.90	1.34	1.72	1.63
K_2O	.04	.04	.04	.02
P_2O_5	.09	.18	.14	.01

a. Gabbro from T-90 evolving to T-55
b. T-55 evolving to T-65
c. T-90 evolving to T-55, then to T-65
d. T-90 evolving directly to T-65

tings. The Talaud islands are thought to be an uplifted forearc terrane (Moore et al., in press). The ophiolite blocks are tectonic slivers within a melange complex of accreted oceanic and trench sediments (Moore et al., in press). An analagous environment would be the lower trench slope of an arc-trench system like the Mariana Islands. In fact, "ophiolites" have been sampled from the Mariana trench slope (Hawkins et al, 1979, and unpublished data). The rock types and their chemistries are similar to Talaud samples (Hawkins et al, 1979; Bloomer, unpublished data). Pelagic sediments, basalts (both oceanic and arc tholeiitic), diabases, gabbros, and serpentinized peridotites have been retrieved. Sediments include serpentine muds and melange-like sediments (sheared, scaly clays) both of which include clasts of mafic and ultramafic rocks (Evans and Hawkins, 1979). Peridotites (serpentinized harzburgites) and associated serpentine muds were dredged within 800m of the surface on the trench-slope break, supporting a hypothesis that there may be considerable thrusting and imbrication of oceanic lithosphere in the forearc region (Evans and Hawkins, 1979).

Thus, we interpret the Talaud ophiolites as fragments of Eocene oceanic crust and upper mantle which were emplaced into a west facing forearc terrain prior to the early Miocene. The melanges represent oceanic and trench sediments that were also accreted at the base of the trench slope. The structures and rock types found on Talaud are consistent with present day trench morphologies, the processes proposed to explain these morphologies (Karig and Sharman, 1975), and rock types recovered from trenches. The Neogene strata are probably sediments that were deposited on top of the ophiolites and melanges in basins on the lower trench slope (Moore and Karig, 1976).

During the collision of the Talaud block with the Sangihe arc, the Talaud block has ridden up onto the Sangihe arc apron and has been lifted above sea level. This subaerial exposure of a collision zone provides a comparative setting to study ophiolite origin and emplacement in subduction zones.

Acknowledgements. We would like to acknowledge the efforts by H. M. S. Hartono, Director of the Indonesian Geological Research and Development Center, and Ismail Usna and Rab Sukamto of GRDC for their help in arranging the field work on Talaud. Dadang Kadarisman of the GRDC participated in the field program. S. Bloomer, J. Melchior, R. La Borde and R. Fujita gave tremendous help and advice on the geochemical analyses. This research was supported by NSF Grants EAR 78-22750 and EAR 80-07429.

References

Allegre, C. J., M. Treuil, J-F Minster, B. Minster, F. Albarede, Systematic use of trace elements in igneous processes, Part I: Fractional crystallization processes in volcanic suites, Contrib. Min. and Petrol., 60, 57-75, 1977.

Allen, C. R., Circum-Pacific faulting in the Philippines-Taiwan region, J. Geophys. Res., 67, 4795-4812, 1962.

Arai, S., Dunite-harzburgite-chromite complexes as refractory residue in the Sangun-Yamaguchi zone, Western Japan: J. Petro., 21, 141-165, 1980.

Cardwell, R. K., B. L. Isacks, and D. E. Karig, The spatial distribution of earthquakes, focal mechanism solutions and subducted lithosphere in the Philippine and northeast Indonesian Islands: in: Hayes, D. E. (ed.), Geologic/Tectonic evolution of Southeast Asia, Am. Geophys. Union Mono, 1980.

Coleman, R. G., Ophiolites, Springer-Verlag, New York, 119 p., 1977.

Coombs, D. S., C. A. Lanis, R. J. Norris, J. M. Sinton, D. J. Borns, and D. Craw, The Dun mountain ophiolite belt, New Zealand, its tectonic setting, constitution, and origin, with special reference to the southern portion: Am. J. Sci., 276, 561-603, 1976.

Cowan, D. S., Deformation and metamorphism of the Franciscan subduction zone complex northwest of Pacheco Pass, California, Geol. Soc. Amer. Bull., 89, 1623-1634, 1974.

Deer, W., R. Howie, and J. Zussman, An introduction to the rock forming minerals, Longman Group Ltd., London, 528 p, 1966.

Dewey, J. F., Ophiolite obduction, Tectonophysics, 31, 93-120, 1976.

Dewey, J. F., Suture zone complexities: a review, Tectonophysics, 40, 53-67, 1977.

Dixon, T. H. and R. Batiza, Petrology and Chemistry of recent lavas in the Northern Marianas: implications for the origin of island arc basalts, Contrib. Mineral. Petrol. 70, 167-181, 1979.

Engel, A. E. J., C. G. Engel, and R. G. Havens, Chemical characteristics of oceanic basalts and the upper mantle, Geol. Soc. Am. Bull., 76, 719-734, 1965.

England, R. N., and H. L. Davies, Mineralogy of ultramafic cumulates and tectonites from Eastern Papua: EPSL, 17, 416-425, 1973.

Evans, C., and J. Hawkins, Mariana arc-trench system, petrology of "seamounts" on the trench-slope break: EOS, 60, 968, 1979.

Fitch, T. J., Earthquake mechanisms and island arc tectonics in the Indonesian-Philippine region, Bull. Seismol. Soc. Am., 60, 565-591, 1970.

Flower, M. F. J., P. T. Robinson, H. U. Schmicke, and W. Ohnmacht, Petrology and Geochemistry of igneous rocks, DSDP leg 37, in: Initial reports of the Deep Sea Drilling Project, 37, U.S. Govt. Printing Office, Washington, D.C., 1976.

Gast, P. W., Trace element fractionation and the origin of tholeiitic and alkaline magma types, Geochim. Cosmochim. Acta, 32, 1057-1086, 1968.

Hamilton, W., Subduction in the Indonesian region: AGU Maurice Ewing Series, 1, 15-31, 1977.

Hamilton, W., Tectonics of the Indonesian region, USGS Prof. Paper, 1078, 345 p, 1979.

Hamlyn, P. R., and E. Bonatti, Petrology of mantle derived ultramafics from the Owens Fracture zone, NW Indian ocean: implications for the nature of the oceanic upper mantle, EPSL, 48, 65-79, 1980.

Hart, S. R., and C. Brooks, Clinopyroxene-matrix partitioning of K, Rb, Cs, Sr, Ba, Geochim. Cosmochim. Acta, 32, 1057-1086, 1974.

Hatherton, T. and W. R. Dickinson, The relationship between andesitic volcanism and seismicity in Indonesia, the Lesser Antilles, and other island arcs, J. Geophys. Res., 74, 5301-5310, 1969.

Hawkins, J. W., Petrology of back-arc basins and island arcs: Their possible role in the origin of ophiolites, Proc. Int. Symposium on Ophiolites, Cyprus, 1979, 1980.

Hawkins, J. W., and R. Batiza, Petrology and geochemistry of an ophiolite complex: Zambales Range, Luzon: EOS, 58, 1244, 1977a.

Hawkins, J. W., S. Bloomer, C. Evans, and J. Melchior, Mariana arc-trench system: petrology of the inner trench wall, EOS, 60, 968, 1979.

Hodges, F. N., and J. J. Papike, DSDP site 334: Magmatic cumulates from oceanic layer 3, J. Geophys. Res., 81, 4135-4151, 1976.

Hsu, K. J., Principles of melanges and their bearing on the Franciscan-Knoxville paradox, Geol. Soc. Am. Bull., 79, 1063-1074, 1968.

Jackson, E. D., H. W. Green, and E. M. Moores, The Vourinos ophiolite, Greece: Cyclic units of lineated cumulates overlying harzburgite tectonite, Geol. Soc. Am. Bull., 86, 390-398, 1975.

Katili, J. A., Past and present geotectonic position of Sulawesi, Indonesia, Tectonophysics, 45, 289-322, 1978.

McCaffrey, R., E. A. Silver, and R. W. Raitt, Crustal structure of the Molucca Sea collision zone Indonesia: in: Hayes, D.E. (ed.), Geologic/Tectonic evolution of Southeast Asia, Am. Geophys. Union Monograph, 1980.

Miyashiro, A., Volcanic rock series in island arcs and active continental margins, Am. J. Sci., 274, 321-355, 1974.

Miyashiro, A., F. Shido, F., and M. Ewing, Crystallization and differentiation trends in abyssal tholeiites and gabbros from mid-oceanic ridges, EPSL, 7, 361-365, 1970.

Moore, G. F., C. A. Evans, Hawkins, and D. Kadarisman, Geology of the Talaud Islands, Molucca Sea collision zone, NE Indonesia, J. Structural Geol., in press.

Moore, G. F., and D. E. Karig, Development of sedimentary basins on the lower trench slope, Geology, 4, 693-697, 1976.

Moore, G. F., and D. E. Karig, Structural geology of Nias, Implications for subduction zone tectonics, Am. J. Sci., 280, 193-223, 1980.

Murphy, R. W., Diversity of island arcs: Japan, Philippines, northern Moluccas, Australian Petrol. Explor. Assoc. J., 11, 19-25, 1973.

Pearce, J. A., and J. R. Cann, Tectonic setting of basic volcanic rocks using trace element analyses, EPSL, 19, 290-300, 1973.

Ranneft, J. S. M., R. M. Hopkins Jr., A. J. Froelich, and J. W. Gwinn, Reconnaissance geology and iol possibilities of Mindanao, Am. Assoc. Geologists Bull., 44, 529-568, 1960.

Roeder, D., Philippine arc system--collision of a flipped subduction zone?, Geology, 5, 203-206, 1977.

Rothaan, H. Ph., Geologische en petrographische schets der Talaud en Nanusa eilanden, Jaarb. v/h. Mijnwezen Ned. Indie, ver. II, 174-220, 1925.

Silver, E. A., and J. C. Moore, The Molucca Sea collision zone, Indonesia, J. Geophys. Res., 83, 1681-1691, 1978.

Sinton, J. M., Equilibration history of the basal alpine type peridotite, Red Mountain, New Zealand, J. Petrol., 18, 216-246, 1977.

Soeria Atmadja, R., and R. Sukamto, Ophiolite rock association on Talaud Islands, East Indonesia, Bull. Geol. Res. Dev. Centre, 1, 17-35.

Stern, R. J., On the origin of andesite in the northern Mariana island arc: Implications from Agrigan, Contrib. Mineral. Petrol., 68, 207-219, 1979.

Sukamto, R., Tectonic significance of melange on the Talaud Islands, northeastern Indonesia, in: Hartono, H. M. S. and Barber, A. J., (eds.), Geology and tectonics of eastern Indonesia, 1980.

Sukamto, R., and H. Suwarno, Melange di daerah Kepulauan Talaud: presented at meeting of Indonesian Association of Geologists, December, 1976 (Direktorat Geologi, Bandung), 1976.

Sukamto, R., N. Suwarna, J. Yusup, and M.

Monoarfa, Geologic map of Talaud Islands, 1:250,000: Geol. Res. Dev. Centre, 1980.

Van Bemmelen, R. W., The Geology of Indonesia, Printing Office, The Hague, 793 p., 1949.

Wright, T. L., and P. C. Doherty, A linear programming and least squares computer method for solving petrologic mixing problems, Geol. Soc. Am. Bull., 81, 1995-2008, 1970.

MAGMATIC EVOLUTION OF ISLAND ARCS IN THE PHILIPPINE SEA

Robert B. Scott[1]

Department of Geology, Texas A&M University, College Station, TX 77843

Abstract. Magmatic arc evolution of both remnant and modern arcs in the South Philippine Sea follows temporal petrologic patterns generally attributed to magmatic arcs, but does not follow generally accepted spatial petrologic patterns. Arc-tholeiitic basalts from the 42- to 32-m.y.-old Palau-Kyushu arc contain plagioclase, augite, bronzite, pigeonite and olivine phenocrysts whereas the calc-alkalic basalts from the 20- to 10-m.y.-old West Mariana arc contain plagioclase, augite, titanomagnetite, orthopyroxene and olivine phenocrysts. Major- and trace-element abundances of lavas from the Palau-Kyushu and West Mariana arcs record an evolution from arc-tholeiitic to calc-alkalic affinities with time: the low-field-strength (Th, Pb, Rb, Ba, K, Sr) to high-field-strength (Ti, Zr, Y, Nb, Ta, P) element ratios increase from older to younger remnant arcs. However, this trend does not extend uniformly from West Mariana calc-alkalic basalts to Pliocene to Recent calc-alkalic basalts of the Mariana arc: although Ba/Zr ratios do increase with time, Sr/Zr ratios do not. The $^{87}Sr/^{86}Sr$ ratios become appreciably more radiogenic with time, both within each arc and between arcs. No evidence exists to suggest that an early tholeiitic period of volcanism, similar to that of the Palau-Kyushu arc, was repeated after sundering of the Palau-Kyushu arc during growth of the West Mariana arc. No progressive change in chemistry occurs in contemporaneous basaltic lavas from four localities from the trench side to the backarc side of the Palau-Kyushu arc. Also there is an absence of evidence of arc-tholeiitic affinities on the trench side of the calc-alkalic lavas on either the West Mariana or the modern Mariana arc. Large differences in Ti/Zr ratios rule out a comagmatic origin of Palau-Kyushu lavas and low abundances of rare-earth and high-field-strength elements make extensive crystal fractionation during magma ascent unlikely.

Introduction

Careful scrutiny of the temporal and spatial relations of magma genesis in magmatic arcs

[1]Present address: U.S. Geological Survey, MS 954 Denver Federal Center, Denver, CO 80225

[Arculus and Johnson, 1978; Arculus, 1976; Johnson, 1977] has emphatically demonstrated that exceptions to generalized models of magmatic evolution of arc-trench systems [Kuno, 1950; Kuno, 1959; Kuno, 1966; Sugimura, 1960; Rittmann, 1953] are common. These exceptions need to be emphasized in research efforts to avoid possibly misleading, over-simplified models, such as proposed by Jakes and White [1969 and 1972] and Jakes and Gill [1970]. The sequence of remnant arcs [Scott and Kroenke, 1980] and the modern arc [Hussong, Uyeda et al., 1982] that comprises the Mariana arc-backarc basin complex in the South Philippine Sea (Figure 1) progressively formed from west to east [Karig, 1975], preserving a relatively unaltered and accessible record of magmatic evolution. The goals of this paper are to define the chemical and petrological character of magmatic evolution in the Palau-Kyushu and West Mariana remnant arcs, and to compare this character to drilled and exposed rocks of the modern Mariana arc, and finally to compare the entire complex to the generalized model of magmatic arc evolution.

Magmatic arcs have been classified into two types [Uyeda and Kanamori, 1979]; those on the eastern sides of ocean basins (Chilean type) have continental continuity, accretionary wedges and abundant, high magnitude, shallow, interplate earthquakes but no backarc basins. In contrast, those of the western sides of ocean basins (Mariana type) have island arcs and backarc basins and, at best, very few interplate earthquakes and accretionary wedges. Clearly, any magmatic evolution scheme at convergent plate boundaries must include explanations of both arc types. The backarc basin type makes a particularly valuable subject of study because in the process of sundering, they leave a progression of remnant arcs behind the active eastward-moving frontal arc complex. The petrological and chemical character of early magmatic stages are preserved in these remnant arcs, unaffected by subsequent burial, tectonism and hydrothermal metamorphism commonly associated with mature magmatic arcs. The classical magmatic arc evolution model makes no distinction between the Mariana and Chilean types.

In view of this background, critical questions

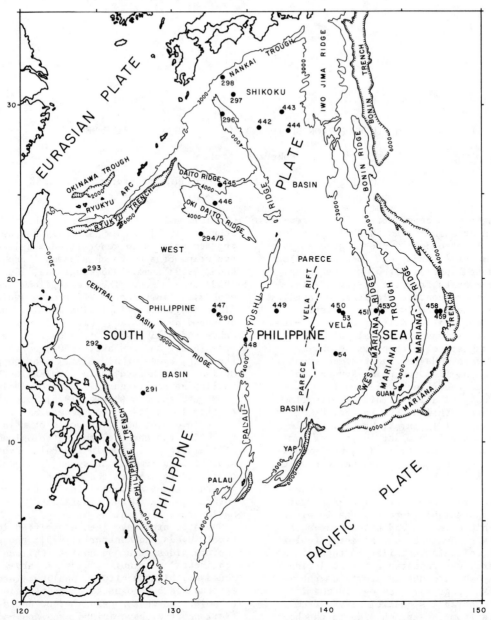

Fig. 1. Map of the Philippine Sea showing the major bathymetric features in meters. The Nankai-Ryukyu-Philippine trench system forms the boundary where the Philippine plate is being consumed and, in turn, east of the Bonin-Mariana trench system, the Pacific plate is being consumed. Land is drawn in bold lines, submarine ridges are shown in lighter lines, trenches and troughs are drawn with closure contours and the Parece Vela Rift [Mrozowski and Hayes, 1979] is drawn as a discontinuous line. The elongate feature in the central part of the West Philippine Basin, originally called the Central Basin Fault [Hess, 1948], has been renamed the Central Basin Ridge to avoid misleading genetic connotations. DSDP drill sites for Legs 6,31,58,59 and selected sites for Leg 60 are shown by solid dots. Dredge sites 1396 and 1397 were collected 137 and 86 km north of Site 448, respectively, by the International Working Group on the IGCP Project "Ophiolites" [1977].

need answering: Does the primative tholeiitic stage of arc volcanism repeat itself each time an arc is sundered and a new magmatic arc grows behind the frontal arc complex on backarc basement [Scott and Kroenke, 1981], or does the magmatic evolution proceed with no break to the calc-alkalic stage? Are all the magmatic zones represented in the Mariana type; that is, does an arc-tholeiitic zone exist on the trench side of a calc-alkalic zone? Do trace-element and isotopic

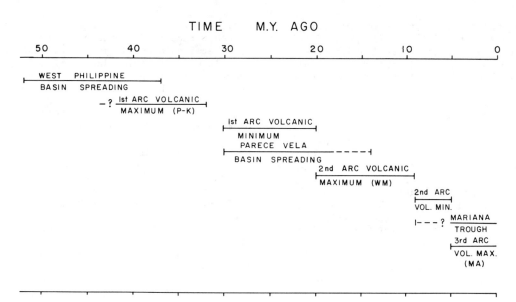

Fig. 2. Chronology of the Philippine plate evolution [Scott and Kroenke, 1980]. The duration of arc volcanic maxima, arc volcanic minima and marginal basin spreading are compared. Note the correlation between periods of arc volcanic minima and periods of early backarc spreading. Palau-Kyushu Ridge is abbreviated to P-K, West Mariana Ridge to WM and Mariana arc to MA.

ratios fit the temporal and spatial evolutionary patterns predicted by major-element characteristics? Do logical petrologic explanations exist for both the generalized evolutionary patterns and departures from those patterns?

Tectonic and Volcanic Evolution Model

As a result of the investigations associated with the Deep Sea Drilling Project (DSDP) Legs 59 [Kroenke, Scott et al., 1978; Scott and Kroenke, 1981; Scott et al., 1981] and 60 [Hussong, Uyeda et al., 1978; Hussong, Uyeda et al., 1982] and regional western Pacific tectonic evolution [Hilde et al., 1977], new insight has been provided into the progressive tectonic evolution of the South Philippine Sea (Figure 1). New data suggest that the timing of arc volcanism and backarc-basin spreading is more complex than Karig [1972; 1974; 1975] visualized when he postulated that maxima in arc volcanism coincide with maxima in backarc spreading, based upon data resulting from investigations of DSDP Legs 6 and 31. Now, additional data from Legs 59 and 60 suggest that initial periods of backarc spreading coincide with lulls in arc volcanism and that peaks in arc volcanism persist after backarc spreading has ceased (Figure 2). As previously recognized [Karig, 1975], remnant arc precursors of the modern Mariana arc and intervening backarc basins progressively developed from west to east to form the Palau-Kyushu Ridge, the Parece Vela Basin, the West Mariana Ridge, and finally the modern Mariana Trough and Mariana Ridge. Scott and Kroenke [1980 and 1981] have concluded that whereas each remnant arc west of the current Mariana arc contains only a segment of the history of Philippine Sea arc volcanism, the forearc region of the present-day arc [Ingle, 1975; Cloud et al., 1965; Schmidt, 1957; Stark, 1963; Tracey et al., 1964; Hussong, Uyeda et al., 1982] contains the entire arc complex evolution from more than 40 m.y. ago to present. The eastern portions of the old magmatic arcs of the Palau-Kyushu and West Mariana arcs were incorporated in the present forearc during sundering of those magmatic arcs. Using this philosophy, the following sequence of events has been proposed (Figure 2): 1) Abundant volcanism on the Palau-Kyushu Ridge occurred from at least 42 to 32 m.y. ago, and is recorded in the modern forearc complex, in the arc-derived volcaniclastic debris drilled in the West Philippine Basin adjacent to the Palau-Kyushu Ridge, and in arc-basement rocks drilled on the ridge itself. 2) Arc volcanism decreased about 30 m.y. ago, seemingly coincident with sundering of the arc along its magmatic axis and onset of seafloor spreading forming the Parece Vela Basin. 3) Parece Vela Basin spreading continued from 30 m.y. ago until early to middle Miocene, 18 to 14 m.y. ago. 4) The lull in arc volcanism ended about 20 m.y. ago. 5) Forearc and basin sediments once again recorded evidence of arc volcanism, this time from West Mariana arc activity, which persisted from 20 m.y. ago to 11-9 m.y. ago. 6) Arc volcanism decreased shortly thereafter, marking a second volcanic lull, coincident with sundering of the arc along its magmatic axis and onset of backarc spreading. 7) Modern Mariana arc volcanism began about 5 m.y. ago and continues at present, forming the third peak in volcanic activity. 8) This latest spreading episode is presently occurring in the Mariana Trough. The specific goal of this paper will be to place the chemical and petrologic character of South Philippine Sea arc volcanism into this new temporal and spatial tectonic scheme.

Previous Petrologic Work

Very little previous work exists on the petrology of the Palau-Kyushu and West Mariana arcs. The most extensive petrologic studies of Palau-Kyushu igneous rocks were done by the International Working Group on the IGCP Project "Ophiolites" [1977] using suites of samples collected during cruise 17 of the R/V DIMITRY MENDELEEV from dredge site 1397 and 1396 close to DSDP drill Site 448. These rocks include vesicular subalkalic basalts and basaltic andesites, two-pyroxene gabbros and low-grade pumpellyite-, prehnite- and greenschist-facies hydrothermal metamorphic rocks. Chemically these rocks are typical of arc tholeiites with high FeO/(FeO/MgO) ratios (0.78) and low Ni, Cr, Sr and Ba contents (40, 300, 300, 50 ppm, respectively); the low Ba and Sr and high iron to magnesium ratios rule out the possibility that these may be calc-alkalic series differentiates even though the Al values (19%) are high. Ingle, Karig et al. [1975] drilled highly vesicular clasts typical of arc volcanism in volcaniclastic debris on the Palau-Kyushu Ridge at DSDP Site 296 on Leg 31, some 1500 km north of DSDP Site 448; the petrographic character of this material is essentially identical to that encountered at Site 448. Also abundant volcanism has been recorded on Palau Island in the Ngeremlengui, Aimeliik and Babelthuap Formations [Ingle, 1975], and Mizuno et al. [1975] and Shiki et al. [1974] report 50 m.y. old intermediate composition plutonic rocks dredged from the northern end of the Palau-Kyushu Ridge-Oki Daito Ridge complex. Vesicular dacite has been dredged from the West Mariana Ridge and is considered by Karig and Glassley [1970] to be evidence of young, shallow arc volcanism.

Studies of volcanic rocks exposed on Guam, Saipan and Tinian, part of the forearc of the southern Mariana Ridge, and the active magmatic arc farther to the north have also provided insight into the magmatic evolution of the Mariana complex. Guam contains the most complete sections of the uplifted portion of the present-day forearc region; evidence for two episodes of volcanic activity [summarized by Ingle, 1975] is present on Guam. The first is an older, tholeiitic episode that lasted from at least 42 to about 34 m.y. ago. During this period, pillow lavas and volcaniclastic breccias of the Alutom Formation and the overlying tholeiitic pillow basalts, flows and flow breccias of the lower Facpi volcanic member of the Umatac Formation were deposited. Only sporadic volcanism was recorded in this lower member between 30 and 25 m.y. ago. After a lull in arc volcanism, a second period of volcanic activity occurred on Guam between 16 and 13 m.y. ago. This period of volcanism has calc-alkalic affinities and is recorded in the upper Facpi volcanic member and the overlying basalt and andesite flows of the Dandan flow member, both of the Umatac Formation [Stark, 1963; Tracey et al., 1964]. No record of more recent arc volcanism exists on Guam.

Saipan has a similar record of activity [Schmidt, 1957; Cloud et al., 1956]; the first volcanic period lasted from about 42 to 41 m.y. ago during which andesite and rare basaltic flows, conglomerates, pyroclastic rocks and tuffs of the Densiyama and Hagman Formations accumulated. Then between 17 and 14 m.y. ago, the tuffaceous facies of the Tagpochau Limestone and the flows of the Fina-sisu Formation represent the second volcanic period. Only the older suite occurs on Tinian where 42 to 37 m.y. old pyroclastic deposits have been found. Thus, the periods of major arc volcanism preserved in the present-day forearc of the Mariana arc complex match the proposed periods of volcanic maxima given in Figure 2. Only the Pliocene to Recent calc-alkalic islands of the more northern part of the modern arc are now active [Stern, 1979].

Analytical Methods

Major- and trace-element analyses of rocks from the Palau-Kyushu Ridge and West Mariana Ridge have been made by several members of the scientific party of DSDP Leg 59 and by collaborators using various analytical techniques in different laboratories. Suites of samples from the Palau-Kyushu Ridge drilled at Site 448 were analyzed by Scott [1981] using atomic-absorption spectrophotometry, instrumental neutron-activation analysis, colorimetry and x-ray fluorescence at Texas A&M University and using electron-microprobe analysis at the University of Texas at Austin. X-ray fluorescence data were collected at the University of Birmingham on representative samples from all igneous units collected during Leg 59 in collaboration with shipboard scientists [Mattey et al., 1981]. Further studies at the Institut de Physique du Globe, Paris [Wood et al., 1981] combine the Birmingham x-ray fluorescence data with epithermal neutron-activation data. Ishii [1981] concentrated on the pyroxene chemistry using microprobe techniques but also contributed major element tables of microprobe, wet chemical and x-ray fluorescence data on whole rocks from both remnant arcs. The nature of altered rocks from the Palau-Kyushu Ridge was chemically and isotopically explored by Aldrich et al. [1981] using atomic-absorption spectrophotometry at Texas A&M University and mass spectrometry at the University of Alberta; Hajash also investigated remnant arc alteration using atomic-absorption spectrophotometry and x-ray diffraction methods at Texas A&M University. The Sr isotopic composition and further chemical studies of rocks collected on Legs 59 and 60 were accomplished by Armstrong and Nixon [1981] using mass-spectrometry and x-ray fluorescence at the University of British Columbia. Analytical techniques and the bulk of data are given in references above; only selected data critical to significant petrological conclusions will be used in the following discussion.

Petrologic Character of Volcanic Rocks of the Palau-Kyushu and West Mariana Ridges

Nearly 600 m of arc basement were drilled at DSDP Site 448 on the Palau-Kyushu Ridge and slightly more than 860 m of similar material were encountered at Site 451 on the West Mariana Ridge. In the 914-m Hole 448A, this arc basement consists of interbedded highly vesicular pillow lavas, flows and related volcaniclastic debris that were intruded by sills and dikes in the lower half of the sequence. In the 930.5-m Hole 451 neither flows nor intrusive bodies were encountered in the monotonous volcaniclastic breccia except in the last few meters where an altered igneous body was recovered in the last core.

Petrography

Petrographically the rocks on the Palau-Kyushu Ridge are similar to arc tholeiites; they contain common phenocrysts of labradioritic plagioclase, augite, bronzite and pigeonite with only rare pseudomorphic olivine. One coarse-grained clast in the volcaniclastic breccias consists of hornblende, plagioclase, quartz, magnetite and ilmenite, a mineralogy that suggests the presence of plutonic hornblende diorite. The 26 igneous units [Kroenke, Scott et al., 1981, Chapter 3 for stratigraphic details] may be separated into three petrographic groups based upon their phenocryst assemblages: from oldest to youngest, 1) clinopyroxene basalts (augite- and plagioclase-bearing), 2) pigeonite basalts (pigeonite-, bronzite-, augite-, and plagioclase-bearing), and 3) olivine basalts (olivine-, bronzite-, augite-, and plagioclase-bearing). The olivine-bearing group near the top of the section shows evidence of intraunit olivine fractionation indicated by more abundant olivine pseudomorphs near the base of each unit. Microprobe analyses of plagioclase and mafic phenocrysts from the glassy margins of flows and pillow lavas sampled from Holes 448 and 448A are typical of the compositions of mineralogies of arc-tholeiitic basalts [Scott, 1981]. These results are consistent with those of Ishii [1981] for the tholeiitic basalts of the Palau-Kyushu Ridge.

Petrographic examination also indicates that these volcanic and volcaniclastic units were affected by low-temperature hydrothermal metamorphism [Aldrich et al., 1981; Hajash, 1981] indicated by a vertical mineralogic zonation of clays, zeolites and amphiboles. However, relatively fresh glassy rims of pillows in the upper portion of the sequence are preserved and seem to have been unaffected by the alteration.

Because the same style of hydrothermal alteration that affected the Palau-Kyushu Ridge also affected the volcaniclastic breccia of the West Mariana Ridge at Site 451, and because no fresh pillow lavas or flows were cored, the petrologic character of the West Mariana Ridge is based upon the petrography and abundance of relatively immobile elements of altered clasts within these breccias. Phenocryst assemblages in these highly vescicular clasts commonly contain andesine plagioclase, augite, orthopyroxene and titanomagnetite; these characteristics are those of calc-alkalic volcanic associations [Holmes, 1921]. Ishii [1981] also reports the phenocryst chemistry of these calc-alkalic basalts and andesite.

Chemical Character of Arc Volcanism

Major Elements. Major-element analyses of partially crystalline and holocrystalline rocks are misleading because even low-temperature hydrothermal alteration greatly affects major-element abundances along crsytal boundary paths of aqueous ion exchange [Scott and Hajash, 1976]; only glass rims of flows and pillows are relatively impermeable to this alteration. Comparison of atomic absorption and x-ray fluorescence whole-rock analyses with electron-microprobe glass analyses of the highly vesicular pillow lavas and flows found on the Palau-Kyushu Ridge demonstrate this effect (Figure 3). Clearly, the scatter in whole-rock compositions, which lie ambiguously between a tholeiitic trend (Skaergaard) and a calc-alkalic trend (Asama and Amagi), is absent in glass analyses; these glass analyses trend away from the Skaergaard line toward alkali enrichment within each flow unit as expected for minor, shallow crystal fractionation of basalts. Although these glass analyses displayed in the AFM diagram are not as definitive as desirable for an explicit classification, the glass analyses plotted on the alkali-silica diagram of Kuno [1965] (Figure 4a) and the SiO_2-FeO^*/MgO diagram of Miyashiro [1974] (Figure 4b) leave no doubt that the major elements fit a tholeiitic composition. Silica contents of all analyzed samples from Site 448, corrected for volatiles added during alteration, fall between 48 and 60% SiO_2. Of the 24 igneous units analyzed at Site 448, 14 have average ranges below 53% SiO_2 and can be termed tholeiitic basalts; 10 units have average ranges between 53 and 55% SiO_2 and can be termed tholeiitic basaltic andesites. One clast with 60% SiO_2 with an andesitic composition has plutonic textures and therefore is a diorite.

Trace Elements. The 9 clasts analyzed from Site 451 on the West Mariana Ridge [Mattey et al., 1981] are highly altered and major-element plots are not meaningful for purposes of classi-

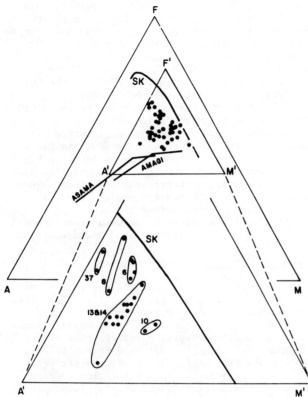

Fig. 3. AFM diagram ($[Na_2O + K_2O]-FeO^*-MgO$ weight % where FeO^* is total Fe expressed as FeO) of analyses of rocks from Site 448 on the Palau-Kyushu Ridge. The upper triangular plot shows whole-rock atomic-absorption and x-ray fluorescence analyses of representative samples of major volcanic units. The trend lines for the tholeiitic Skaergaard (SK) and the calc-alkalic Asama and Amagi volcanoes of Japan are shown for comparison. The small triangular area (A'-F'-M') in the upper diagram is expanded and projected downward with apex coordinates A'=45% A, 40% F, 15% M; F'=5% A, 80% F, 15% M; M'= 5% A, 40% F, 55% M. The expanded portion of the AFM diagram shows glass electron-microprobe analyses from six volcanic units that have fresh glass margins.

fication; however, trace-element analyses do provide definative criteria for classification of igneous rocks both on the Palau-Kyushu and West Mariana Ridges. Table 1 summarizes a comparison of selected trace- and major-element criteria for distinguishing among tholeiitic mid-ocean-ridge basalts, arc-tholeiitic basalts and cal-alkalic basalts. Also the average chemical character of the rocks analyzed from the two drill sites on the Palau-Kyushu and West Mariana Ridges is given. Samples analyzed from Site 448 on the Palau-Kyushu Ridge have an average Sr content of 169 ppm, an average Ba content of 49 ppm and an average Ce/Y ratio of 0.9; these definitely are indicative of arc-tholeiitic affinities [Jakes and Gill, 1970; Gill, 1970]. This conclusion is further substantiated by the condrite normalized rare-earth element (REE) pattern; the basaltic units fall within the range stipulated by Jakes and Gill [1970] to be representative of arc tholeiites (Figure 5). Both clasts have other trace element characteristics of arc tholeiites, and therefore their higher abundances of REE probably represent the effects of crystal fractionation of tholeiitic magmas rather than the presence of calc-alkalic magmas.

As discussed earlier, the 9 whole-rock, x-ray fluorescence analyses of clasts from Site 451 [Mattey et al., 1981] on the West Mariana Ridge do not distinguish between arc-tholeiitic or calc-alkalic affinities using major-element criteria; there has been too much mobilization of Si, Fe, Mg, Na and K during secondary alteration [Hajash, 1981]. Therefore, the criteria used for Site 448 in Figures 3 and 4 cannot be employed. Only the high Al_2O_3 value is supportive of a calc-alkalic character. Trace-element abundances, however, strongly support a calc-alkalic affinity (Table 1) with high Ba (184 ppm) and Sr (512 ppm) contents and light REE/ heavy REE ratios less than 1. Of these 9 analyses, 8 have basaltic SiO_2 contents; only 1 is andesitic with 57% SiO_2.

Wood et al. [1979a] distinguish between volcanic series found in different tectonic environments using a plot of relatively immobile trace elements; for example, Th, Hf and Ta abundances plotted on a triangular diagram can distinguish

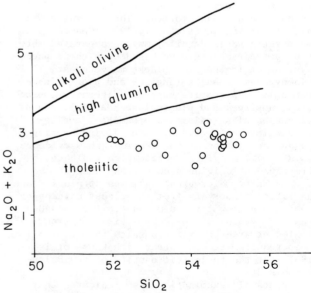

Fig. 4a. $SiO_2-(Na_2O + K_2O)$ weight % diagram [Kuno, 1965] of fresh glass analyses from the Palau-Kyushu Ridge at Site 448.

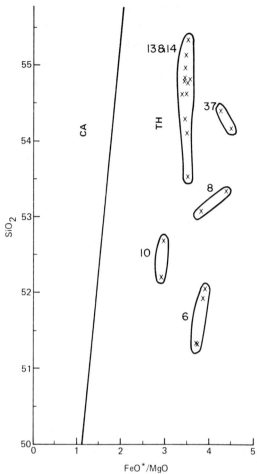

Fig. 4b. SiO$_2$-(FeO*/MgO) diagram [Miyashiro, 1974] of fresh glass analyses from the Palau-Kyushu Ridge at Site 448. Unit numbering after Scott [1981].

between calc-alkalic and tholeiitic orogenic lavas and also among normal (N-type) mid-ocean ridge basalts (MORB), plume (E-type) MORB and oceanic island tholeiites (Figure 6). Because Th is more incompatible than Hf, the more primitive arc tholeiites should have lower Th/Hf ratios. Ta is found to be very low in abundance in orogenic magmas relative to MORB magmas [Jordon and Trevil, 1978; Sun et al., 1979; Wood et al., 1979a]; it seems probable that Ta is withheld from orogenic magmas during melting in a residual phase [Wood et al., 1979b]. Experimental work by Helman and Green [1979] indicates that rutile and sphene have considerable stability in hydrous systems; because Ta is strongly partitioned into Ti phases, it is reasonable to explain the low Ta abundances of orogenic magmas as the effect of residual Ti phases during hydrous melting [Wood et al., 1981]. In any case, the work reported here makes a clear distinction among Palau-Kyushu Ridge arc-tholeiitic basalts, West Mariana Ridge calc-alkalic basalts and andesites and various MORB types.

Petrology

Crystal Fractionation. A fundamental question in the investigation of any volcanic assemblage is whether the units are comagmatic; that is, can the chemical variations be linked by crystal fractionation, or must they have originated as chemically distinct batches of magma, related instead to differences in degrees of partial melting and/or differences in source rock chemistry. A normative olivine-diopside-quartz triangular diagram of glass chemistry from flows drilled on the Palau-Kyushu Ridge [Scott, 1981] indicates that the uppermost units that consist of olivine basalts and pigeonite basalts have glass chemistries that trend from the observed invarient point along the two-pyroxene cotectic (Figure 7) for oceanic tholeiites. The olivine-basalt glasses lie closer to the invarient point. Initially, these relationships seem to relate the volcanic variations to a comagmatic suite created by crystal fractionation involving olivine resorption of relatively primitive magmas at the invarient point and crystallization of orthopyroxene and clinopyroxene to produce more evolved magmas with increasing normative quartz content. Also the decrease of both Co and Cr with increasing normative quartz seems to support this contention because both Co and Cr are strongly partitioned into pyroxenes, but several difficulties negate this line of reasoning. First, the olivine basalts are the most evolved with the highest Fe/Mg ratios, not the least evolved. Second, the pigeonite basalts are slightly more evolved and the 3-phase clinopyroxene basalts are the most primitive with the lowest Fe/Mg ratios. These phenocryst assemblages and degrees of "evolution" suggest that the order of crystallization for a comagmatic series would have been from clinopyroxene basalt to pigeonite basalt to olivine basalt, the reverse of that based on the normative diagram. Trends in Co and Cr abundance may well be related to pyroxene fractionation, but they obviously do not prove a comagmatic relation.

Trace-element ratios produce an even more convincing argument that these magmas cannot have a comagmatic history. In particular, the ratios of the immobile, high-field-strength elements, for example the Ti/Zr ratio, have a wide range of values (Figure 8). The young, olivine basalt units (6 and 8) from Site 448 have a ratio of about 200 whereas the underlying pigeonite basalt units only have a ratio of about 100 and the clinopyroxene basalt units have a wide range of ratios between 196 and 122. Because no Ti- or Zr-bearing phases are found in the basalts from Site 448, and because the differences in the small partition coefficients for these elements in existing crystalline phases are too minute to effectively change the ratios during fractional

TABLE 1. Trace- and Major-Element Character of Orogenic and Oceanic Basalts

Elements[1]	MORB[2]	CA[3]	AT[4]	Site 448	Site 451
Fe/Mg	1.11	1.4	<2.0	4.6	2.0
Al_2O_3%	16.6	17.7	15.6	13.8	18.2
K_2O%	0.07	1.25	0.43	0.45	0.73
Ba ppm	10	300	50	49	184
Sr ppm	110	400	200	169	512
Ni ppm	150	20	20	14	8
Cr ppm	400	50	20	25	33 (4)[6]
LREE/HREE	<1	~10	≤1	<1[5]	1.9 (1.3)[7]
Hf/Th[8]		<2.5	>2.5	5.7	1.3

1) The Fe/Mg and K data for Site 448 are from microprobe analyses of glasses [Scott, 1981]; Al and trace-element data are from x-ray fluorescence data [Mattey et al., 1981]; chondrite normalized REE data are from instrumental neutron-activation analyses [Scott, 1981]; Site 451 data are from Mattey et al. [1981] with exceptions 7 and 8 listed below.
2) MORB = mid-ocean ridge basalts [Johnson, 1979; Bass et al., 1973; Charmichael et al., 1974].
3) CA = calc-alkalic series differentiates [Jakes and White, 1972; Taylor, 1969].
4) AT = arc-tholeiitic series differentiates [Jakes and Gill, 1970].
5) LREE/HREE (Light REE/Heavy REE) = La/Lu
6) Cr = 33 ppm, average of 9 analyses, one with 174 ppm; excluding 174 ppm, Cr = 4 ppm.
7) 1.9 = Ce/Y ratio; 1.3 = La/Tb ratio [Wood et al., 1981].
8) Hf/Th data [Wood et al., 1981].

crystallization, these magmas cannot be comagmatic. If, on the other hand, the magmatic system existed under more hydrous conditions that enhanced the stability of Ti-bearing phases [Helman and Green, 1979], then the more evolved magmas may have appreciably lower Ti/Zr ratios. Such seems to be the case for the clasts analyzed from Site 451 on the West Mariana Ridge because these rocks are titanomagnetite-bearing. Ishii [1981] has emphasized the probability that hydrous magmatic conditions have more greatly affected this stage of magmatic evolution; using pyroxene geothermometry, he has shown that the crystallization temperature for the tholeiitic Palau-

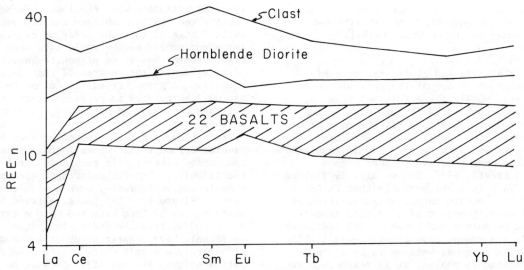

Fig. 5. Chondrite normalized rare-earth element (REE) patterns for twenty two analyses of igneous units from the Palau-Kyushu Ridge. Twenty-two basaltic units fall within the narrow range reported for arc tholeiites by Jakes and Gill [1970]. The hornblende diorite clast and another clast have higher overall REE contents but have other trace-element characteristics of arc tholeiites; these clasts may attribute their higher REE abundances to crystal fractionation.

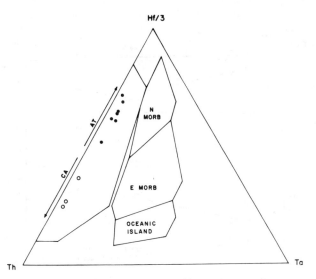

Fig. 6. Th-Hf/3-Ta triangular plot to distinguish among rocks of calc-alkalic, arc-tholeiitic, ocean island tholeiites and various mid-ocean ridge basalt affinities. Data are from Wood et al. [1981]. After Wood et al. [1979a]. Palau-Kyushu rocks = filled circles; West Mariana rocks = open circles.

Kyushu Ridge basalts is between 1130°C and 1075°C but that for the calc-alkalic West Mariana Ridge basalts is only 1055°C. He attributes this difference to a reduction of the pyroxene liquidus temperature by the presece of water. Therefore, the spread of values for the Ti/Zr ratios between 146 and 54 for the West Mariana Ridge clasts does not rule out a comagmatic relation, particularly since the lowest Ti/Zr ratio is found in an andesite. However, other ratios (for example, the Ce/Y, Y/Zr and Ce/Zr ratios) show similar ranges [Mattey et al., 1981] that cannot be explained by a hydrous effect. It is concluded that the igneous rocks of neither remnant arc can be considered comagmatic.

Evolution in Time and Space. By yet unresolved processes, the abundance of low-field-strength elements generally increases relative to high-field-strength elements in time, that is from the Eocene-Oligocene Palau-Kyushu basaltic volcanism of arc-tholeiitic character, to the Miocene West Mariana Ridge volcanism of less pronounced calc-alkalic character, and to the Plio-Pleistocene volcanism of the northern Mariana Islands of somewhat more pronounced calc-alkalic character (Figure 9). Barium shows a progressive increase with time and space from the old Palau-Kyushu lavas (50 ppm), though the West Mariana lavas (200 ppm), to the modern Mariana arc lavas (350 ppm). Strontium shows a similar trend but no increase between West Mariana and modern Mariana volcanism. Although data are lacking from the West Mariana Ridge because the volcaniclastic breccias were too highly altered to conduct meaningful Sr isotopic analyses, there is a distinct increase in the radiogenic Sr component from the Palau-Kyushu lavas to the modern Mariana lavas [Armstrong and Nixon, 1981]; also they show a significant increase in more radiogenic magmas with time on the Palau-Kyushu Ridge (0.70333 to 0.70356). Zirconium shows no significant variation but TiO_2 values show an appreciable decrease in the two arcs with calc-alkalic affinites, probably the result of removal of Ti in shallow-level titanomagnetite fractionation. Aluminum oxide follows the expected progressive trend from arc-tholeiitic affinities to highly aluminous calc-alkalic rocks, but calc-alkalic basalts from the younger ridges are surprisingly Ca-rich. The FeO/MgO ratios follow the expected decrease from arc-tholeiitic basalts to calc-alkalic basalts. Silica values are included to indicate the narrow range of basalt compositions selected.

Although the chemical progression generally fits the classical trend in time, it apparently does not fit the trend in space because there is no record of tholeiitic volcanism between the West Mariana calc-alkalic lavas (Site 451) and its trench or between the Mariana arc (active northern islands) and its trench. Magmatic evolution in space can better be pictured by attempting to reconstruct the relative positions of Site 448, island exposures on Guam, Saipan and Tinian, and the forearc Sites 458 and 459 at the time of arc volcanism between 42 and 32 m.y. ago on the Palau-Kyushu Ridge (Figure 10).

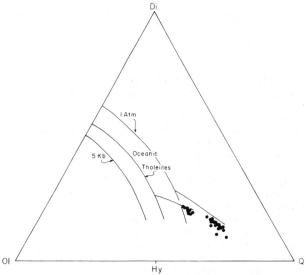

Fig. 7. Normative olivine(Ol)-diopside (Di)-quartz(Q) diagram [Shibata, 1976] showing compositions of glasses for olivine- and orthopyroxene-bearing basaltic units from Site 448. Note that the compositions fall close to the nattural oceanic tholeiite diopside-hypersthene(Hy) cotectic.

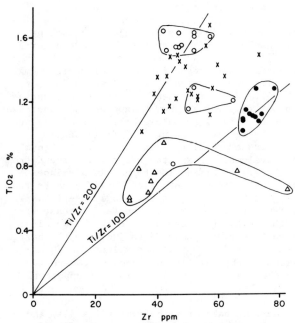

Fig. 8. TiO$_2$-Zr diagram for basalts from Site 448 on the Palau-Kyushu Ridge and Site 451 on the West Mariana Ridge. Open circles are olivine basalts; filled circles are pigeonite basalts; crosses are clinopyroxene basalts; all from Site 448. Open triangles are from Site 451. Data from Mattey et al. [1981].

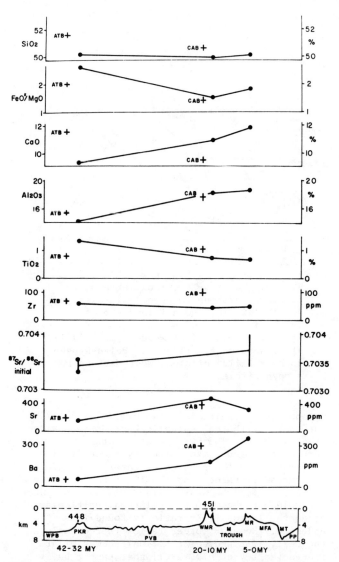

Fig. 9. Comparison of the averaged chemistry of arc-tholeiitic basalts and calc-alkalic basalts relative to time and space. The 18° parallel profile shows the positions of the West Philippine Basin (WPB), the Palau-Kyushu Ridge (PKR), the Parece Vela Basin (PVB), the West Mariana Ridge (WMR), the Mariana Trough (M), the Mariana Ridge (MR), the Mariana Trench (MT), the Pacific plate (PP) and two drill sites for Leg 59, 448 and 451. Crosses mark the average abundances of arc-tholeiitic basalts (ATB) and calc-alkalic basalts (CAB). Abundances for CAB and ATB were obtained from Jakes and White [1972], Jakes and Gill [1970] and Taylor [1969]. Data points for the two remnant arcs and the modern arc are shown as solid circles: Ba from Mattey et al. [1981] and P. Fryer, personal comm. in Mattey et al.; Sr isotopic ratios from Armstrong and Nixon [1981] and Meijer [1976]; Sr, Zr, TiO$_2$, Al$_2$O$_3$, CaO, FeO/MgO, and SiO$_2$ from Mattey et al. [1981], Scott [1981] and Stern [1979]. The period of activity is shown below each arc.

Some difficulties related to the differences in age of magmatic sites exist: Site 448 rocks are about 43 m.y. old [Sutter and Snee, 1981]; Guam, Saipan, and Tinian lavas range from about 42 to 32 m.y. old; and Sites 458 and 459 are thought to be early Oligocene(?) and pre-middle Eocene [Meijer et al., 1982]. In spite of these minor differences in age and limited definitive analyses on the island exposures, the uniformity of the major- and trace-element abundances from these arc-tholeiitic basalts is impressive. Just how closely spaced these four localities were during Palau-Kyushu arc activity cannot be determined. It can be proposed with reasonable certainty that they were somewhat closer together than depicted in Figure 10, which was drawn by replacing the modern magmatic arc with the old Palau-Kyushu remnant arc. The abnormally wide modern Mariana forearc probably can be attributed to magmatic additions to the existing forearc region during reconstruction of the arc following arc sundering. If this is the case, then the present-day forearc contains anywhere from 50 to 100 km of magmatic arc additions intermixed with the original forearc. Whatever the case, no evidence of significant lateral change in the chemistry of these arc-tholeiitic rocks exists. Before this conclusion is accepted without reservation, however, definitive trace-element studies

Site 448					Guam		Site 458					Site 459				
LFS		HFS		MaE			LFS		HFS		MaE		LFS		HFS		MaE
Ba	53	Zr	59	Al_2O_3	14.2	SiO_2 49.4	Ba	29	Zr	71	Al_2O_3	16.9	Ba	28	Zr	52	Al_2O_3 15.1
Sr	167	Nb	1	CaO	9.3	Al_2O_3 15.1	Sr	135	Nb	3	CaO	6.6	Sr	111	Nb	5	CaO 11.3
Rb	10	Y	32	FeO/MgO	2.6	FeO/MgO 0.8	Rb	8	Y	21	FeO/MgO	2.4	Rb	14	Y	15	FeO/MgO 1.5
K_2O	0.4	P_2O_5	0.17	SiO_2	50.2	TiO_2 0.5	K_2O	0.43	P_2O_5	0.10	SiO_2	51.8	K_2O	0.4	P_2O_5	0.11	SiO_2 52.4
$^{87}Sr/^{86}Sr$	0.70356	TiO_2	1.34			CaO 10.0	$^{87}Sr/^{86}Sr$	0.70384	TiO_2	0.91			$^{87}Sr/^{86}Sr$	0.703666	TiO_2	0.59	
"	0.70333																

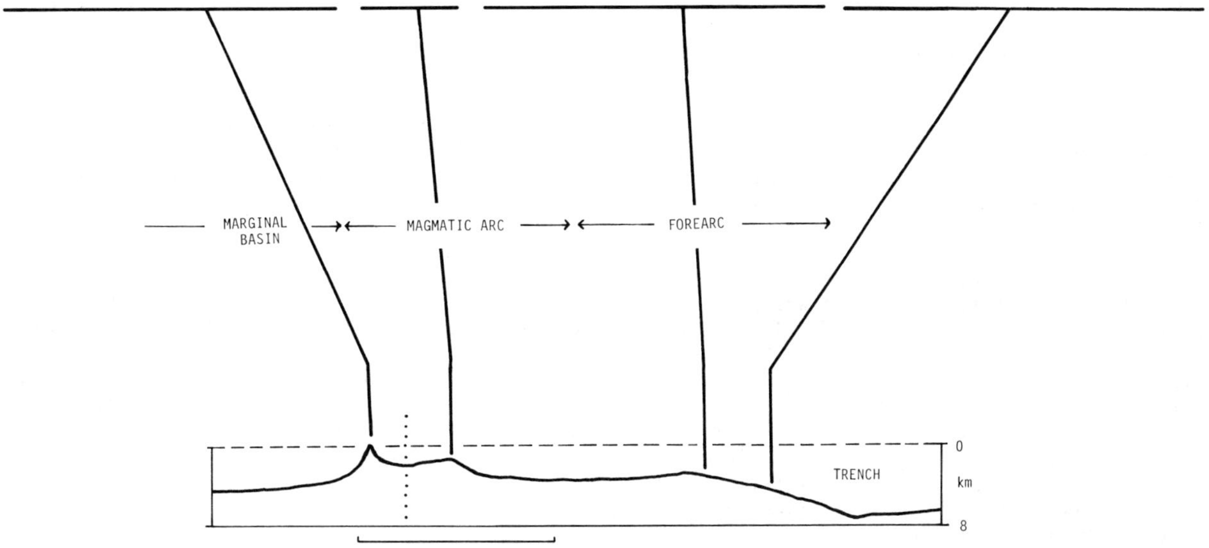

Fig. 10. Comparison of the chemistry of arc-tholeiitic basalts of the first volcanic maximum (42-32 m.y. ago); samples are from the major portion of the magmatic arc of the Palau-Kyushu arc and the presumed forearc at that time. Selected low-field-strength and high-field-strength elements, major elements and initial Sr isotopic ratios are given where available. Data for Site 448 are from Scott [1981], Mattey et al. [1981] and Armstrong and Nixon [1981]; data for Guam (including Saipan and Tinian) are from Schmidt [1957], Stark [1963] and Cloud et al. [1956]; data for Sites 458 and 459 are from Armstrong and Nixon [1981].

of exposures of arc-tholeiitic and calc-alkalic affinities on the present-day islands should be made.

Perhaps during initial backarc spreading in the Parece Vela Basin both the apparent lull in arc volcanism and the apparent lack of an arc-tholeiitic stage of volcanism on the West Mariana Ridge can be accounted for if these missing arc-tholeiitic magmas are mixed with backarc basin magmas during early backarc basin spreading. Although Mattey et al. [1981] and Zakariadze et al. [1981] find no evidence of an arc-tholeiitic influence on the chemistry of basin basalts from DSDP Site 449 in the Parece Vela Basin, the rate of magma genesis is roughly 15 times greater in the backarc basins than in arcs; thus, the small chemical distinctions between arc-tholeiitic basalts and backarc basin or mid-ocean ridge basalts are probably immeasurable when greatly diluted by large volumes of backarc basin basalts. Either better sampling close to backarc basin margins or more sensitive geochemical criteria must be obtained to test this possibility.

Origin of Trace-Element Character. An important charcteristic of orogenic magmas is their high ratio of low-field-strength elements (Th, Pb, Rb, Ba, Sr, K) to high-field-strength elements (Zr, Ti, Ta, P, Nb, Y) relative to basalts derived from mid-ocean ridges [Saunders et al., 1980]. Mattey et al. [1981] have outlined two hypotheses for the origin of island arc tholeiites with these features. One requires that these tholeiites be produced by dehydration of some combination of subducted sediments and altered oceanic crust with selective aqueous transport of low-field-strength ions from these sources into the overlying mantle wedge [Hawkesworth et al., 1977; Saunders et al., 1980] where hydrous partial melting of that wedge occurs [Ringwood, 1977]. But presumably these magmas must undergo extensive olivine fractional crystallization to create the Fe-rich, Mg-, Cr-, and Ni-depleted compositions that characterize arc-tholeiitic basalts [Nichols and Ringwood, 1973; Nichols, 1974]. Although extensive olivine

TABLE 2. Ti, Zr, and Ta Abundances

Elements[1]	Palau-Kyushu Arc-Tholeiitic Basalts							West Mariana Calc-Alkalic Basalts		
Ti %	0.92	0.69	0.49	0.67	0.57	0.75	0.86	0.35	0.46	0.39
Zr ppm	47	50	45	74	40	39	60	31	66	83
Ta ppm	0.03	0.04	0.04	0.05	0.02	0.01	0.04	0.02	0.07	0.10

1) Data from Wood et al. [1981].

fractionation will not greatly affect the shape of the rare-earth element pattern of arc-tholeiitic basalts, the low abundance of rare-earth elements and high-field-strength elements, both similar to MORB chemistry, do suggest that these magmas could not have aquired their high Fe/Mg ratios by abundant olivine fractionation. If the Palau-Kyushu arc-tholeiitic basalts are not co-magmatic, for reasons stated earlier, it does not mean that the possibility of any crystal fractionation must be excluded on this basis. It does mean that the degree of fractional crystallization is severely limited by the rare-earth and high-field-strength abundance constraints and cannot be used to explain chemical differences between magma batches with greatly different Ti/Zr, Ce/Y, Y/Zr or Ce/Zr ratios.

The second hypothesis of Mattey et al. [1981] is considered by them to be more likely: moderate- to high-degree partial melting of the subducting crustal basaltic slab can produce Fe-rich but Ni- and Cr-poor melts that avoid the necessity of extensive crystal fractionation to explain these characteristics. The abundance of low-field-strength elements presumably have been transported from oceanic sediments and hydrothermally altered oceanic crust by dehydration, and the REE and other trace elements retain MORB characteristics. They further contend that low abundances of Ti, Zr, and Ta [Wood et al., 1981] may be related to residual phases such as rutile and zircon that retain these elements during hydrous partial melting. If this is the case, then Ta values and Ti values should be related by the degree of retention of these phases. However there seems to be no correlation between Ta and TiO_2 contents in the Palau-Kyushu arc-tholeiitic basalts or in the West Mariana calc-alkalic basalts and andesite (Table 2). This implies that the control of Ta abundances in these magmas is not related to stabilization of Ti phases during partial melting. A strong correlation between Zr and Ta for both sets of rocks, however, does require explanation (Figure 11). In particular, the Ta intercept between 20 and 30 ppm Zr is intriguing. The highly linear relationship between Zr and Ta, in itself, is not surprising because of the high correlation between Zr and Nb [Graham and Mason, 1972] and the similarity in geochemical behavior of Ta (Ta^{5+}, 0.72 A radius) and Nb (Nb^{5+}, 0.72 A radius) [Whittaker and Muntus, 1970], but the distinct positive Zr intercept suggests that a more complex relation exists, either during partial melting, or during crystal fractionation. Perhaps the increased stability of Ti-bearing phases containing significant Ta but low Zr concentrations during hydrous melting may cause this behavior; further investigation of Ta behavior in mafic melts is obviously needed. In conclusion, it seems that neither method of deriving orogenic magmas is free from problems and therefore the unique hypothesis, and perhaps the discriminating data are still lacking.

Conclusions

Several important conclusions can be reached from the foregoing discussions: 1) Both major-element and trace-element evidence places severe limitations on the degree of crystal fractiona-

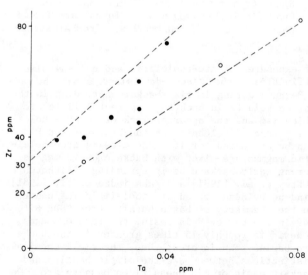

Fig. 11. Zr-Ta plot for arc-tholeiitic basalts (solid circles) from the Palau-Kyushu Ridge and calc-alkalic basalts (open circles) from the West Mariana Ridge. Data from Wood et al. [1981].

tion possible in the sequence of lavas on the Palau-Kyushu Ridge. 2) Similar evidence excludes the possibility of a comagmatic origin of this sequence. 3) The classical evolution trend from arc-tholeiitic affinities to calc-alkalic affinities with time is upheld in general, but small unexplained departures exist. 4) The classical spatial sequence from arc-tholeiitic affinities adjacent to the trench side of the arcs to calc-alkalic affinities farther from the trench does not exist. 5) There is no evidence that after sundering of an arc, the early arc-tholeiitic stage of evolution repeats itself during renewed volcanism along its new magmatic arc; instead, it appears that magmatic evolution passes on to the calc-alkalic stage. 6) Neither selective aqueous transport of low-field-strength ions into the region of hydrous partial melting nor the high-degree of slab partial melting method of derivation of orogenic magmas is free from problems.

Acknowledgements. This manuscript is the Texas A&M Geodynamics Research Project contribution no. 16. The efforts of shipboard colleagues on Leg 59 and the petrologists of Leg 60 of the Deep Sea Drilling Project made this synthesis possible. The understanding, patience and typing of Sally Scott are deeply appreciated. Will Carr has provided particularly critical editorial assistance and Thomas Hilde has been most encouraging during preparation of this manuscript.

References

Aldrich, J.B., T.T. Tieh and R.B. Scott, Alteration of remnant arc debris Site 448, Leg 59, Palau-Kyushu Ridge, Philippine Sea, in L. Kroenke, R.B. Scott et al., Init. Repts. DSDP, 59, pp.737-742, U.S. Government Printing Office, Washington, D.C., 1981.

Arculus, R.J., Geology and geochemistry of the alkali basalt-andesite association of Grenada, Lesser Antilles island arc, Geol. Soc. Am. Bull., 87, 612-624, 1976.

Arculus, R.J. and R.W. Johnson, Criticism of generalized models for the magmatic evolution of arc-trench systems, Earth Planet. Sci. Lett., 39, 118-126, 1978.

Armstrong, R.L. amd G.T. Dixon, Chemical and Sr isotopic composition of igneous rocks from Deep-Sea Drilling Project 59 and 60, in L. Kroenke, R.B. Scott et al., Init. Repts. DSDP, 59, pp.719-727, U.S. Government Printing Office, Washington, D.C., 1981.

Bass, M.N., R. Moberly, J.M. Rhoads et al., Volcanic rocks cored in the central Pacific, Leg 17: Deep Sea Drilling Project, in E.L. Winterer, J.I. Ewing et al., Init. Repts. DSDP, 17, pp. 429-446, U.S. Government Printing Office, Washington, D.C., 1973.

Charmichael, I.S.E., F.J. Turner and J. Verhoogen, Igneous Petrology, pp. 1-739, McGraw-Hill, New York, N.Y., 1974.

Cloud, P.E., Jr., R.G. Schmidt and H.W. Burke, Geology of Saipan, Mariana Islands, Part 1, General geology, U.S. Geol. Surv. Prof. Pap., 280-A, 1-126, 1956.

Gill, J.B., Geochemistry of Viti Levu, Fiji and its evolution as an island arc, Contr. Miner. Petrol., 27, 179-187, 1970.

Graham, A.L. and B. Mason, Niobium in meteorites, Geochim. Cosmochim. Acta, 36, 917-922, 1972.

Hajash, A., Altered rocks from Deep Sea Drilling Project Leg 59, in L. Kroenke, R.B. Scott et al., Init. Repts. DSDP, 59, pp. 736-739, U.S. Government Printing Office, Washington, D.C., 1981.

Hawkesworth, C.J., R.K. O'Nions and R.J. Pankhurst et al., A geochemical study of island arc and back-arc tholeiites from the Scotia Sea, Earth Planet. Sci. Lett., 36, 253-269, 1977.

Helman, P.L. and T.H. Green, The role of sphene as an accessory phase in the high pressure partial melting of hydrous mafic compositions, Earth Planet. Sci. Lett., 42, 191-201, 1979.

Hess, H.H., Major structural features of the western North Pacific, an interpretation of H.O. 5485, bathymetric chart, Korea to New Guinea, Geol. Soc. Am. Bull., 59, 417-466, 1948.

Hilde, T.W.C., S. Uyeda and L. Kroenke, Evolution of the Western Pacific and its margins, Tectonophys., 38, 145-165, 1977.

Holmes, A., Petrographic Methods and Calculations with Some Examples of Results Achieved, pp. 1-515, Murby, London, 1923.

Hussong, D.M., S. Uyeda et al., Leg 60 ends in Guam, Geotimes, 23, 19-22, 1978.

Hussong, D.M., S. Uyeda et al., Init. Repts. DSDP, 60, pp. 1-929, U.S. Governemnt Printing Office, Washington, D.C., 1982.

Ingle, J.C., Jr., Summary of late Paleogene-Neogene insular stratigraphy, paleo-bathymetry, and correlations, Philippine Sea and Sea of Japan region, in J.C. Ingle, Jr., D.E. Karig, et al., Init. Repts. DSDP, 31, pp. 837-855, U.S. Government Printing Office, Washington, D.C., 1975.

Ingle, J.C., Jr., D.E. Karig, et al., Init. Repts. DSDP, 31, pp. 1-927, U.S. Government Printing Office, Washington, D.C., 1975.

International Working Group on the IGCP Project "Ophiolites", Initial report of the geological study of oceanic crust of the Philippine sea floor, N. Bogdanov, editor, Bollettino del Grupo di Lavoro aulle Ofioliti Mediterranee, 2, pp. 137-168, Bologna, Estratto da Ofioliti, 1977.

Ishii, T., Pyroxene geothermometry of basalts and an andesite from the Palau-Kyushu and West Mariana Ridges, Deep Sea Drilling Project Leg 59, in L. Kroenke, R.B. Scott et al., Init. Repts. DSDP, 59, pp. 693-718, U.S. Government Printing Office, Washington, D.C., 1981.

Jakes, P. and A.J.R. White, Structure of the Melanesian arcs and correlation with distribution of magma types, Tectonophys., 8, 222-236, 1969.

Jakes, P. and A.J.R. White, Major and trace element abundance in volcanic rocks of orogenic areas, Geol. Soc. Am. Bull., 83, 29-40, 1972.

Jakes, P. and J.B. Gill, Rare earth elements and the island arc tholeiiic series, Earth Planet. Sci. Lett., 9, 17-28, 1970.

Johnson, J.R., Transitional basalts and tholeiites from the East Pacific Rise, 9°N, J. Geophys. Res., 84, 1635-1651, 1979.

Johnson, R.W., Distribution and major-element chemistry of late Cainozoic volcanoes at the southern margin of the Bismark Sea, Papua New Guinea, Bureau Miner. Res. Aust. Rep., 188, 1-169, 1977.

Jordon, J.L. and M. Revil, Utilisation des propriétés des éléments fortement hygromagmatophiles por l'étude de la composition chimique et de l'hétérogenéité du manteau, Bull. Geol. Soc. France, 19, 1197-1205, 1978.

Karig, D.E., Remnant arcs, Geol. Soc. Am. Bull., 83, 1057-1068, 1972.

Karig, D.E., Evolution of arc systems in the western Pacific, Ann. Rev. Earth Planet. Sci., 2, 51-75, 1974.

Karig, D.E., Basin genesis in the Philippine Sea, in D.E. Karig, J.C. Ingle, Jr. et al., Init. Repts. DSDP, 31, pp. 857-879, U.S. Government Printing Office, Washington, D.C., 1975.

Karig, D.E. and W.E. Glassley, Dacite and related sediments from the West Mariana Ridge, Philippine Sea, Geol. Soc. Am. Bull., 81, 2143-2146, 1970.

Kroenke, L., R.B. Scott et al., In the Philippine Sea--old questions answered and new questions asked, Geotimes, 23, 20-23, 1978.

Kroenke, L., R.B. Scott et al., Init. Repts. DSDP, 59, pp. 1-820, U.S. Government Printing Office, Washington, D.C., 1981.

Kuno, H., Petrology of Hakone volcano and the adjacent areas, Japan, Geol. Soc. Am. Bull., 61, 957-1020, 1950.

Kuno, H., Origin of Cenozoic petrographic provinces of Japan and surrounding areas, Bull. Volcanol., 20, 37-76, 1959.

Kuno, H., Fractionation trends of basalt magmas in lava flows, J. Petrology, 6, 302-321, 1965.

Kuno, H., Lateral variation of basalt magma across continental margins and island arcs, Bull. Volcanol., 29, 195-222, 1966.

Mattey, D.P., N.G. Marsh and J. Tarney, The geochemistry, mineralogy, and petrology of basalts from the West Philippine and Parece Vela Basins and from the Palau-Kyushu and West Mariana Ridges, in L. Kroenke, R.B. Scott et al., Init. Repts. DSDP, 59, pp. 753-800, U.S. Government Printing Office, Washington, D.C., 1981.

Meijer, A., Pb and Sr isotopic data bearing on the origin of volcanic rocks from the Mariana island-arc system, Geol. Soc. Am. Bull., 87, 1358-1369, 1976.

Meijer, A., E.Y. Anthony and M. Reagan, Petrology of volcanic rocks from the fore-arc sites, in D.M. Hussong, S. Uyeda et al., Init. Repts. DSDP, 60, 709-729, U.S. Government Printing Office, Washington, D.C., 1982.

Miyashiro, A., Volcanic rock series in island arcs and active continental margins, Am. J. Sci., 274, 321-355, 1974.

Mizuno, A., Y. Okuda, K. Tamaki, Y. Kinoshita, M. Yuasa, M. Nakajima, F. Munakami, S. Terashima and K. Ishibashi, Marine geology and geological history of the Daito Ridge area, northwestern Philippine Sea, Marine Science, 7, 484-491, 1975.

Mrozowski, C.L. and D.E. Hayes, The evolution of the Parece Vela Basin, eastern Philippine Sea, Earth Planet. Sci. Lett., 46, 49-67, 1979.

Nichols, I.A., Liquids in equilibrium with peridotitic mineral assemblages at high water pressures, Contr. Mineral Petrol., 45, 289-316, 1974.

Nichols, I.A. and A.E. Ringwood, Effect of water on olivine stability in tholeiites and the production of silica-saturated magmas in island arcs, J. Geol., 81, 285-300, 1973.

Ringwood, A.E., Petrogenesis in island arc systems, in M. Talwani and W. Pitman III, editors, Island Arcs, Deep Sea Trenches, and Back-Arc Basins, pp. 311-324, Am. Geophys. Union Maurice Ewing Series I, Washington, D.C., 1977.

Rittman, A., Magmatic character and tectonic position of the Indonesian volcanoes, Bull. Volcanol., 14, 45-58, 1953.

Saunders, A.D., J. Tarney and S.D. Weaver, Transverse geochemical variations across the Antarctic Peninsula: implications for the genesis of calc-alkaline magmas, Earth Planet. Sci. Lett., 46, 344-360, 1980.

Schmidt, R.G., Petrology of the volcanic rocks, Part 2, Petrology and soils, Geology of Saipan, Mariana Islands, U.S. Geol. Surv. Prof. Pap., 280-B, 127-175, 1957.

Scott, R.B., Petrology and chemistry of arc tholeiites on the Palau-Kyushu Ridge, Site 448, Leg 59, Deep Sea Drilling Project, in L. Kroenke, R.B. Scott et al., Init. Repts. DSDP, 59, pp. 681-692, U.S. Government Printing Office, Washington, D.C., 1981.

Scott, R.B. and A. Hajash, Initial submarine alteration of basaltic pillow lavas, Am. J. Sci., 276, 480-501, 1976.

Scott, R.B. and L. Kroenke, Evolution of back arc spreading and arc volcanism in the Philippine Sea: Interpretation of Leg 59 DSDP results, in D.E. Hayes, editor, The Tectonic and Geologic Evolution of Southeast Asian Seas and Islands, Amer. Geophys. Union Monogr. 23, 283-291, 1981.

Scott, R.B. and L. Kroenke, Periodicity of remnant arcs and back-arc basins of the South Philippine Sea, in R. von Huene, editor, Oceanologica Acta, SP, 193-202, 1981.

Scott, R.B., L. Kroenke, G. Zakariadze, and A. Sharaskin, Regional synthesis of the results of DSDP Leg 59 in the Philippine Sea, in L.

Kroenke, R.B. Scott et al., Init. Repts. DSDP, 59, pp. 803-815, U.S. Government Printing Office, Washington, D.C., 1981.

Shibata, T., Phenocryst-bulk rock composition relations of abyssal tholeiites and their petrogenic significance, Geochim. Cosmochim. Acta, 40, 1407-1417, 1976.

Shiki, T., H. Aoki, H. Suzuki, M. Masashino and Y. Okuda, Geological and petrographic results of the GDP-8 cruise in the Philippine Sea, Marine Science, 6, 51-55, 1974.

Stark, J.T., Petrology of the volcanic rocks of Guam, U.S. Geol. Surv. Prof. Pap., 403-C, 1-32, 1963.

Stern, R.J., On the origin of andesite in the northern Mariana Island Arc: implications from Agrigan, Contrib. Mineral. Petrol., 68, 207-219, 1979.

Sugimura, A., Zonal arrangement of some geophysical and petrological features in Japan and its environs, J. Fac. Sci. Tokyo Univ., 1112, 133-153, 1960.

Sun, S.S., R.W. Nesbitt and A.Y. Sharaskin, Geochemical characteristics of mid-ocean ridge basalts, Earth Planet. Sci. Lett., 44, 119-138, 1979.

Sutter, J.F. and L.W. Snee, K/Ar and $^{40}Ar/^{39}Ar$ dating of basaltic rocks from Deep Sea Drilling Project Leg 59, in L. Kroenke, R.B. Scott et al., Init. Repts. DSDP, 59, pp. 729-734, U.S. Government Printing Office, Washington, D.C., 1981.

Taylor, S.R., Trace-element chemistry of andesite and associated calc-alkaline rocks, in A.R. McBirney, editor, Proceedings of the Andesite Conference, Bull. 65, Dept. Geology & Minerals Industry, Univ. of Oregon, 43-63, 1969.

Tracey, J.I., Jr., O.S. Seymour, J.T. Stark, D.B. Doan and H.G. May, General Geology of Guam, U.S. Geol. Surv. Prof. Pap., 403-A, 1-104, 1964.

Uyeda, S. and H. Kanamori, Back arc opening and the mode of subduction, J. Geophys. Res., 84, 1049-1061, 1979.

Whittaker, E.M.W. and R. Muntus, Ionic radii for use in geochemistry, Geochim. Cosmochim. Acta, 34, 945-956, 1970.

Wood, D.A., J.L. Jordon and M. Trevil, A re-appraisal of the use of trace-elements to classify and discriminate between magma series in different tectonic settings, Earth Planet. Sci. Lett. 45, 326-336, 1979a.

Wood, D.A., J. Varet, H. Bougault et al., The petrology, geochemistry and mineralogy of North Atlantic basalts: A discussion based on IPOD Leg 49, in B.P. Luyendyk, J.R. Cann et al., Init. Repts. DSDP, 49, pp. 695-755, U.S. Government Printing Office, Washington, D.C., 1979b.

Wood, D.A., D.P. Mattey, J.L. Jordon, N.G. Marsh, J. Tarney and M. Trevil, A geochemical study of 17 selected samples from the basement cores recovered at Sites 447, 448, 450 and 451, Deep Sea Drilling Project 59, in L. Kroenke, R.B. Scott et al., Init. Repts. DSDP, 59, pp. 743-752, U.S. Government Printing Office, Washington, D.C., 1981.

Zakariadze, G.S., L.V. Dmitriev, A.V. Sobolev and N.M. Suschevskaya, Petrology of basalts of Holes 447A, 449, and 450, South Philippine Sea Transect, Deep Sea Drilling Project Leg 59, in L. Kroenke, R.B. Scott et al., Init. Repts. DSDP, 59, pp. 669-680, U.S. Government Printing Office, Washington, D.C., 1981.

GEOLOGY AND PLATE TECTONICS OF THE SEA OF OKHOTSK

L. Savostin, L. Zonenshain, B. Baranov

Institute of Oceanology, Academy of Sciences of the USSR, Moscow, USSR

Abstract. The Sea of Okhotsk was formed in Miocene time at the site of the Late Cretaceous volcanic arc and marginal basin of this region. Two volcanic arcs existed in the Miocene: the Kuril and the Sakhalin. No less than 200 km of oceanic crust have been subducted under the Sakhalin arc in the western part of the Sea of Okhotsk. Considerable parts of the Sea of Okhotsk, corresponding to the Academy of Science and Institute of Oceanology rises, subsided about 1000 m during the subduction process and their sedimentary cover experienced folding, which continues to the present time. At the same time there occurred the opening of the Kuril Basin behind the Kuril arc. At present, the Sea of Okhotsk is a region of the junction between four lithosphere plates: the Pacific, North American, Eurasian and Amurian. The Sea of Okhotsk microplate is located between them and its interaction with these plates has produced the present pattern of features in the region.

Introduction

The Sea of Okhotsk region includes, besides the Sea of Okhotsk proper, a highly complex bordering land area. The Sea of Okhotsk has a trapeziform shape. It is bounded by the Kuril island arc to the south, by Kamchatka to the east, by the Asiatic continental margin to the north and northwest, and by Sakhalin and its structural continuation into Hokkaido on the west. Plate tectonics explains the formation of the Sea of Okhotsk structures by the interaction of the Pacific, Eurasian, Amurian and North American plates.

The Sea of Okhotsk region has been described by many authors. The synopsis by Markov et al. (1967) is still up-to-date. More recent works on Sakhalin have been made by Melankholina (1973), Raznitsyn (1978), Rozhdestvensky (1975); on the Sea of Okhotsk by Gnibidenko (1976); and on Kamchatka by Legler (1977). A great synopsis on the entire region was published by the Sakhalin Research Institute (Structure . . . , 1976). Geophysical research results have been presented by Kosminskaya et al. (Structure . . ., 1964), Kochergin et al. (1970) and by many others. Tectonic generalizations on the marginal seas and island arcs of northeastern Asia exist by Yu. M. Puscharovsky (1972), Gnibidenko (1979) and Rodnikov (1979), and a scheme of the structure and evolution of the region was published by Parfenov et al. (1978).

The Sea of Okhotsk region is underlain by crust of different types. Regions with typically oceanic crust (the Kuril Basin) and suboceanic crust (Derugin Deep) within the Sea of Okhotsk region coexist with regions underlain by continental crust (shelf area). The Sea of Okhotsk is separated from the Pacific by the Kuril-Kamchatka trench along which the subduction of the Pacific plate occurs.

A general mosaic pattern of the Sea of Okhotsk region results from a T-shape junction of features (Fig. 1). The NE Kuril island arc intersects the Sakhalin-Hokkaido fold zone at almost a right angle. The longitudinal Sakhalin structures seem to have no continuation to the north, as if they rest against the edge of the Asia continent. In the vicinity of Penzhinskaya Bay in the northern Sea of Okhotsk, the Cenozoic features of Kamchatka meet at a sharp angle with the Mesozoic features of the Okhotsk-Chukotsk volcanic belt. The Okhotsk basin is framed by structures of different age, the strikes of which follow the outline of the Sea of Okhotsk. For a long time this fact was used to establish the boundaries of the ancient Okhotsk massif which was believed to occupy almost the entire area of the Sea of Okhotsk. Markov et al. (1967), however, have shown that the Okhotsk massif does not exist and the Sea of Okhotsk floor seems to consist of the same rock complexes that outcrop in the Sakhalin and Kamchatka fold systems. Dredging of bedrock from the Sea of Okhotsk floor revealed Late Cretaceous effusives and granitoides (Geodekyan, et al., 1976; Gnibidenko, Ilyev, 1976), supporting this concept.

The Sea of Okhotsk is composed of structural elements of different ages. The late Mesozoic Okhotsk-Chukotsk volcanic belt borders the region on the Asia continent side. The volcanic belt overlaps a folded basement, the age of which ranges from Precambrian (Okhotsk massif) to Mesozoic (Verkhoyansk-Kolyma area). Late Paleozoic and

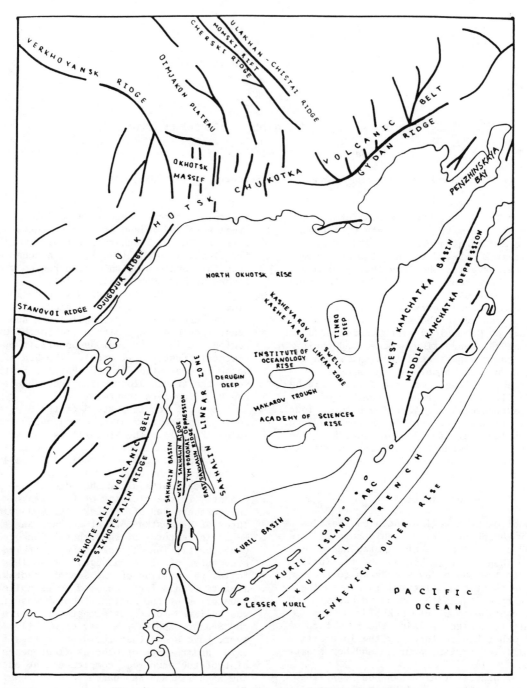

Fig. 1. Topography of the Sea of Okhotsk Region.

Mesozoic fold complexes compose the Sakhalin and Hokkaido regions. The active folding of these complexes took place during the Late Cretaceous and early Tertiary periods. The Kamchatka folded zones also date back to the Late Cretaceous-Tertiary period. Massifs of older rocks, like the Ganalsky outlier, occur in these fold zones.

Presumably, similar prominent features also exist on the Sea of Okhotsk floor. The recent Kuril island arc was formed during Late Oligocene-Early Miocene. The pattern of the Sea of Okhotsk thus is a result of a prolonged evolution, though its main evolutionary stage falls within Late Cretaceous and Cenozoic time.

Kuril Island Arc

Two major ranges make up the Kuril island arc. The first, the Great Kuril range, is a typical volcanic belt originated due to subduction of the Pacific plate. Offshore on the Pacific side, it is paralleled by a nonvolcanic submarine range, the southwestern section of which is the Lesser Kuril island chain. Its northeastern section is completely covered by the sea and is usually called the Vityaz Rise.

Within the Great Kuril volcanic arc there seems to be no rocks older than Late Oligocene-Early Miocene age. Islands of the Great Kuril range display intensive Pleistocene and recent volcanism. Gorshkov (1967) recognized two volcanic zones: the major zone, associated with the axial ridge and a western zone, embracing islands located on the western slope of the island arc, within the back-arc Kuril basin. These zones vary in geographical location and also in concentration of alkalic volcanic rocks which is higher in the western zone. The increase in concentration of alkalics corresponds to the increase in the depth to the Benioff zone (Piskunov, et al., 1979).

Deep seismic sounding data show that the Kuril island arc flanks are underlain by continental crust while its central segment seems to be underlain by oceanic crust. In other words, it was formed on different types of basement. The Lesser Kuril range consists mainly of Late Cretaceous volcanic complexes (Kazakova et al., 1970; Streltsov, 1976) which are remnants of the Late Cretaceous volcanic arc. In some regions, for instance on the Shikotan island (Lesser Kurils), within these complexes there occur allochthonous layered gabbro and associated sheeted rocks (Melankholina, 1978). Results of dredging by Vasilyev and others (1979) show that the basement of the Kuril islands is composed of pre-Late Cretaceous metamorphic sedimentary rocks, granites, granodiorites and granosyenites. Similar rocks were found in numerous nodules in the lavas of the Kuril islands (Federchenko, Rodionova, 1975). Obviously these rocks compose the basement on which the Late Cretaceous arc originated. Late Cretaceous volcanic rocks are also exposed on the adjacent parts of Kamchatka and eastern Hokkaido. The Kuril volcanic arc was formed on these Mesozoic complexes following a long interval of time, the Paleocene and Eocene.

Kuril Trench

The Kuril island arc (Fig. 2) slopes steeply southeast into the Kuril (or Kuril-Kamchatka) trench. A wealth of geological and geophysical data is available. The Kuril trench structure has been analyzed by Udintsev (1955), Kosminskaya, et al., (Structure . . . , 1964), Markov et al., (1967), Kochergin et al., (1970), Snegovskoi (1974), Vasilyev et al., (1979), Zonenshain, Savostin (1978), Zonenshain et al., (1980), and by many others. The width of the steep inner slope of the trench is 70-90 km. The angle of the slope averages 7-10°. The axis of the trench ranges from 7.5 to 9 km in depth (local maximum 10.5 km) and has a pronounced V-shape. The outer slope of the Kuril-Kamchatka trench, similar to other trenches, is more gentle than the inner slope. The seafloor rises through a series of small scarps to the edge of the Zenkevich outer rise. The rise has a mainly flat and even surface at 4500-5000 m depth but includes some seamounts.

Data at hand, in particular seismic reflection data (Fig. 6), allow more detailed descriptions of each of the elements of the Kuril-Kamchatka trench innumerated above.

Inner slope. Cross-sections of the Kuril-Kamchatka trench show two relatively flat terraces on the inner slope bounded by three steep scarps (Fig. 3 and 4). The first terrace has a depth of 3000-3200 m, the second 5800-6300 m. The height of each scarp, thus, is about 3000 m.

The 3000-3200 m deep terrace continues without significant breaks along the entire inner slope of the trench. Its width ranges from 15 to 40 km. Its surface is even and slightly hilly.

Its transition to the middle-slope is often marked by a basement high. The terrace is everywhere covered by sediments. The rough acoustic basement, dissected into blocks and considerably displaced relative to each other, is observed under the sediments. Sedimentary thickness in the depressions separating these blocks reaches 1300-1500 m. The sedimentary layers, disturbed and dissected by faults, can presumably be dated back to late Tertiary or Quaternary, as can the sedimentary cover on the upper scarp.

The upper part of middle scarp usually consists of a series of sharp summits. Each of the summits (200-300 m) crowns an independent block of 2-4 km in width. These blocks can be regarded as slide blocks of the country rocks dropped downward towards the trench.

The deeper terrace, though observed on practically all profiles crossing the Kuril trench, is less pronounced than the shallower one. On the southwestern flank of the trench it has a width of 10-15 km and regularly occurs at depths of 5900 to 6300 m. On the northeastern flank, especially on the inner Kamchatka slope, this terrace is considerably narrower, being only 1-2 km in width.

As a rule, the deeper terrace contains a crest several hundred meters in height, thus forming between it and the wall of the middle scarp a narrow depression filled up by sediments. It is found approximately at the depth to which the Benioff zone projects to the surface.

The lower scarp steeply slopes down towards the trench axis. It contains several small terraces which are either slide-blocks or blocks bounded by faults related to trench tectonics.

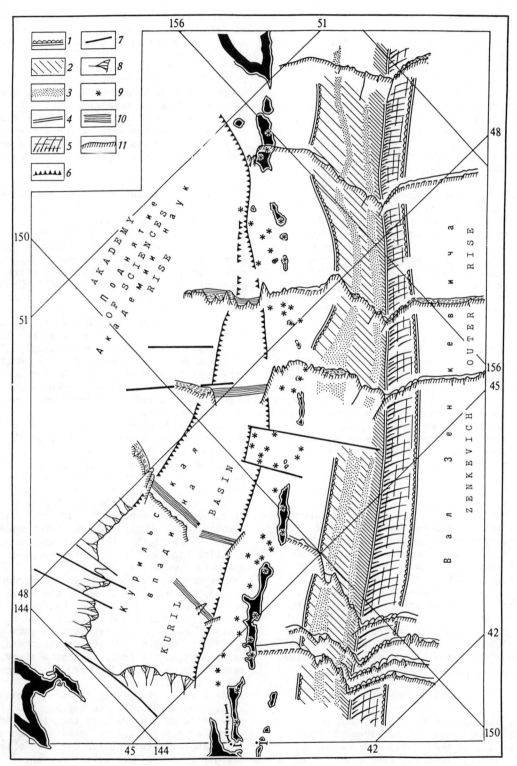

Fig. 2. Main structures of Kuril trench and Kuril basin. 1-upper breaks on the inner and outer slopes of the trench, 2-scarps on the inner slope of the trench, 3-terraces on the inner slope, 4-axis of the trench, 5-outer slope (thin lines show fault directions), 6-fault scarps of the Kuril basin, 7-faults, 8-submarine canyons, 9-volcanoes, 10-sediments, 11-acoustic basement.

An additional terrace is seen on many profiles at 7000-8000 m depth (Fig. 3). It has a gentle and flattened topographic relief and is probably entirely composed of sediments. This lower terrace seems to correspond to accumulations of slide material at the foot of the inner slope.

No major deformations are observed in a thin sedimentary cover on the trench bottom. Moreover, the multi-channel profiling data (Garkalenko et al., 1978) revealed that the oceanic sedimentary layers and underlying basalt basement of the outer slope continue undeformed for 50 km under the inner slope structures.

Similar structures have been revealed for other trenches such as the Japan (Hilde, et al., 1976), Nankai (Ludwig et al., 1973), and Java trenches (Beck and Lehner, 1974).

These facts can only be explained by underthrusting of the oceanic plate (or by overriding of the island arc). It should be noted that the subduction of oceanic basement together with its sedimentary cover under the island arc was theoretically predicted by Lobkovskii and Sorokhtin (1976). According to this model, at certain subduction rates, sediments begin to function as

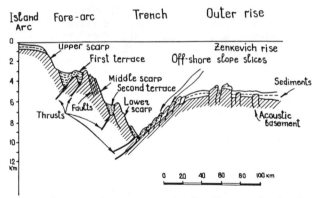

Fig. 4. Section across the Kuril trench showing the main structures.

lubricants. This model explains the undeformed sediments at the trench bottom: sediments undergo deformation only at depth after they are subducted under the island slope. Results of deep drilling in the Japan and Mariana trenches (legs 56,57 and 60 by "Glomar Challenger" - Scientific Party, 1978 a and b) support this model. The obtained data revealed weak manifestation of a "bulldozer" effect - scraping-off of sediments from the oceanic plate in front of the inner slope and imposed great restrictions on accretion models (Karig, 1974; Karig and Sharman, 1975; Seely et al., 1974). The oceanic sediments are apparently involved in the processes occurring at depth within island arcs, thus undergoing dehydration and, at greater depths, melting.

The major morphological characteristic of the Kuril trench inner slope is its strict demarcation between the three scarps, each of which corresponds to the boundary of independent tectonic blocks. These scarps seem to result from the dissection of island-arc basement into individual tectonic slices, possibly overthrust southeastward to the trench due to subduction of the oceanic plate.

<u>Outer slope</u>. The outer slope displays general uniformity along the whole trench length. The trench bottom gently ascends from the trench axis to the Zenkevich outer rise. The acoustic basement and oceanic sediments are generally similar to those recorded on the Zenkevich rise. Small elevations and depressions in the acoustic basement typical of the elevated segments of the rise are also revealed on the slope. A pronounced slice-like structural pattern in the outer slope is clearly traced on all profiles (Fig. 2 - 4). The slope is formed by a descending series (usually 5 - 7) of scarps separately by 5-10 km. Profiles of this step-like pattern show that the bottom of the sedimentary cover and the basement surface at each lower scarp is apparently upthrusted several tens of meters (sometimes up to 100 m) with respect to the basement top of the neighboring seaward block. All these scarps are accompanied by faults dipping towards the trench

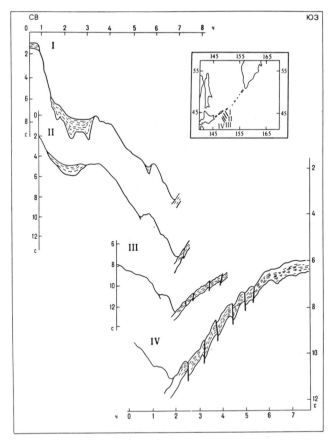

Fig. 3. Seismic reflection profile line drawings across the Kuril trench. Location of the profiles is shown on the inset.

axis, thus producing the effect of a goffered surface or a super ripple, its more steep slopes being oriented towards the rise and its more gentle slopes towards the trench axis. The most obvious explanation of this pattern is that the oceanic plate, consisting of a series of tectonic slices is being subducted or "pushed" under the scarps of the inner slope. It should be noted, that the outer rise is usually marked by normal faults originating under the tension conditions typical of the upper segment of the bending oceanic plate (Ludwig et al., 1966; Hilde and Sharman, 1978).

As a rule, normal faults are common within the rise and upper outer slope. In the vicinity of the trench, however, normal faults change into reverse faults. Consequently the lower outer slope is dominated by compression, not tension. This is believed to be in good agreement with subduction of the oceanic plate in the trench.

Hanks (1971) presents data on earthquakes in the Kuril trench near its outer slope with focal mechanism solutions that suggest compression across the trench strike. He concludes that horizontal compressive stress with a minimum magnitude of 2-4 Kbar diminishes the tension strains which originate at the initial bending of the oceanic plate.

At present, several models are proposed to explain the typical geometry of the outer slope and rise which differ in rheology of the subducting plate: elastic models (Hanks, 1971; Watts and Talwani, 1974; Caldwell et al., 1976; Lobkovskii and Sorokhtin, 1976), elastic-plastic models (McAdoo et al., 1977; Forsyth and Shapple, 1978) and viscous models (DeBremaecker, 1977; Melosh, 1978; Melosh and Raefsky, 1980). All of them explain the observed topography, but result in different stresses in front of the island arc. The problem of stress type in a subduction zone is closely connected with the problem of back-arc basin origin. In case these stresses are so great as proposed by the elastic models (up to 10 Kbar after Watts and Talwani, 1974 and up to 2 Kbar after Lobkovskii and Sorokhtin, 1976), it is difficult to explain the origin of tension conditions and formation of new oceanic crust in back-arc basins. This problem still waits a solution.

Sakhalin - Hokkaido

The present Sakhalin-Hokkaido mountain system is not a single unit from an evolutionary standpoint. Its western zone consists of a great thickness of terrigenous materials ranging from Late Cretaceous to recent in age. It occupies the West-Sakhalin depression in Sakhalin and the Issikari-Rumoi depression in western Hokkaido. Volcanic layers (lavas and tuffs of predominantly calc-alkaline composition) occur within the Miocene terrigeneous deposits. They indicate the existence of a Miocene volcanic island arc in the position of present Sakhalin, and therefore one more subduction zone in addition to that at the Kurils. Folded sedimentary units of the West Sakhalin depression extend westward under the Tatarsky Strait. Melankholina (1973) notes that the folding process seems to still be active. The axial zone of Sakhalin and Hokkaido consists of late Paleozoic and Mesozoic (including Jurassic and Early Cretaceous) fold complexes. Ophiolite sheets and greenschist metamorphic units dominate. The main folding dates back to Middle and Late Cretaceous. The folding was accompanied by formation of paired metamorphic belts: high pressure products in the west and high temperature products in the east (Myashiro, 1972). This suggests an eastward dipping subduction zone off the Asian continent during Early Cretaceous time. Eastern Sakhalin and eastern Hokkaido (Nemuro peninsula) are made up of Late Cretaceous predominantly calc-alkaline and alkaline-basalt volcanic sequences and siliceous terrigeneous accumulations. These deposits underwent folding during uppermost Cretaceous or Early Paleogene time. Thrust features and nappes with westward obduction of ophiolite sheets are typical of the Late Mesozoic pattern of Sakhalin and Hokkaido.

The present structure of Sakhalin was formed only during the Late Pliocene (Geology of the USSR, v. XXXIII). The two longitudinal ridges, the West and East Sakhalin Mountains - separated by Tym-Poronaiskaya depression, were formed during this period. Maximal velocity gradients of uplift are concentrated in the southern mountain part of the island. The Tymski diagonal fault (Galtsev-Bezuk, Solovjov, 1965) marks the boundary between the southern part of the island and its northern low-land part.

Two major fault systems, orthogonal and diagonal, are distinguished in Sakhalin (Galtsev-Bezuk, 1964). The first system includes the longitudinal fault, the second system includes NW and NE striking faults. Strike-slip faults described by Rozhdestvensky (1969, 1972, 1975, 1976) play a significant role in the general fault pattern. They generally have a submeridional strike. Some of them, however, for instance Tym-Poronaisky, Central, etc., curve in plan and so the strike changes from NE to NW. Within the northeast region, steeply dipping normal faults are mapped. They have formed depressions that are filled by Pliocene-Quaternary sedimentary deposits. NW striking segments of these faults form a sequence of closely spaced reverse faults. Young (post-Pliocene) nappes are also mapped there. The amount of overthrusting is undetermined. The overlap zone revealed by wells has the width of 500 m.

The majority of the Sakhalin faults onset in Early Miocene. The main period of displacement along these faults, however, falls within Pliocene-Quaternary time. These displacements continue to the present. Displacement magnitude along strike-slip faults varies from 14 km in northern Sakhalin (Schmidt peninsula) to 25 km in its central part (West Sakhalin mountains).

Kamchatka

Kamchatka has two distinct geologic provinces: western and eastern. Western Kamchatka consists of thick terrigeneous deposits layed down during the Late Cretaceous and entire Cenozoic periods. They fill the West Kamchatka depression. The axis of this depression seems to be located in the adjacent Sea of Okhotsk TINRO Deep (Fig. 1). The basement depression is unknown. It is probably formed by Mesozoic and late Paleozoic fold complexes similar to those in Koryakia, for instance.

The range of young volcanoes of eastern Kamchatka constitutes a continuation of the Kuril arc to the north. The recent volcanism of Kamchatka has lasted since Miocene time. It inherited the previous Paleogene (Oligocene-Eocene) zone of island arc volcanism traced by calc-alkaline series of the East Kamchatka synclinorium. Young volcanic eruptions are traced northward to about the latitude of the Kamchatka peninsula cape where the T-shaped junction of the Aleutian arc and Kamchatka trench occurs. The Paleogene volcanic belt continues through the whole of Kamchatka and runs northward to the Olutorsky zone. The latter, curving, continues into the Shirshov and Bowers Ridges in the Bering Sea. The basement on which the Cenozoic island arcs originated, is exhumed in three regions associated with the axial zone of the Central and East Kamchatka anticlinoria and also with the range of the eastern peninsula. Late Cretaceous fold complexes with small metamorphic massifs (Ganalsky Rise etc.) are represented here. Late Cretaceous deposits, stripped in the anticlinoria, are formed by volcanogenic - siliceous series: the Iruneiskaya suite and its analogues. Many geological investigations show that this sequence, with considerable composition of calc-alkaline volcanics, corresponds to the island arc complex which marks the Late Cretaceous volcanic arc named Iruneiskaya (Markov et al., 1967; Markovskii, Rotman, 1969; Belyi, 1974).

Within the eastern peninsula (like Kronotskii, Kamchatka Cape, etc.) ophiolite sheets outcrop (Markov, 1975). They are presumably remnants of the oceanic bed on which the Iruneiskaya volcanic arc originated. The Mesozoic-Cenozoic evolution of Kamchatka is a successive change of one island arc series to another with considerable reassembly in pattern, embracing presumably the very Late Cretaceous and Early Paleogene. The Late Cretaceous and Paleogene West Kamchatka Basin seems to be an analogue of the present marginal seas located behind the volcanic arc. As a peculiarity of the neotectonic pattern connected with the evolution of the recent volcanic arc of Kamchatka, special notice should be payed to the conditions of tension which govern Central Kamchatka (Erlikh, 1973) and which have resulted in the origin of the Central Kamchatka graben. Legler (1976) stated that besides tension, a significant role is played by sinistral displacements along the northeastern faults bordering the Central Kamchatka graben. A significant role in the Kamchatka structure is also played by sublatitudinal fracture zones (Suprunenko et al., 1973; Dmitriev et al., 1979).

Okhotsk-Chukotsk Belt

The Okhotsk-Chukotsk belt is the major feature of the Okhotsky shore (Fig. 1). Its main structural elements are associated with the Apt-Turonean volcanic series (Belyi, 1974). Volcanic complexes are mainly formed by andesitic-basalt, andesites and ignimbrite fields. Numerous granitoid intrusions are associated with them. The Okhotsk-Chukotsk volcanic belt originated in the location of the early Mesozoic Uda-Murgal island arc (Parfenov et al., 1978), separating the Pacific (Kula plate) and the Verkhoyano-Chukotsk system of microplates. In the Late Jurassic and Late Cretaceous, after a period of intensive folding and numerous microplate collisions in this region, a single continental massif was formed (Fudjita, 1978). This event has considerable influence on both the interacting oceanic and continental lithosphere plates. In Late Neogene the island arc pattern (Uda-Murgal island arc) was changed into an active continental margin (Okhotsk-Chukotsk volcanic belt) similar to the recent Andean margin of South America. Grigorash and Moralev (1974), analyzing changes in volcanic chemistry across the belt using the Dickinson and Hatherton method (1967), established that the Benioff zone associated with the volcanic belt was gently dipping under the Asia continent at an angle of 20-25°.

Sea of Okhotsk

Major structural elements of the Sea of Okhotsk are clearly seen from bathymetric data (Udintsev, 1957). They are (Fig. 1): Kuril Basin, Academy of Science and Institute of Oceanology Rises, Makarov trough, separating them, Derugin and TINRO Deeps, Pri-Sakhalin linear zone, Kashevarov swell (and St. Ion Is.), Kashevarov trough, North Okhotsk Rise, and the vast shelf of the northern Sea of Okhotsk.

<u>Kuril Basin</u>. The Kuril Basin follows the Kuril island arc for 1000 km. Its broadest western segment is about 300 km wide. Eastward, near Kamchatka, it wedges out and continues through Kamchatka as the Central Kamchatka graben. Its average depth is 3300 m. It belongs to the type of marginal basin with moderate heat flow (average 2.3 $\mu cal/cm^2 sec$), no granite layer and suboceanic crust (thickness 10 km, seismic veloc-

ities 6.7 km/sec) (Structure . . . , 1964), and weak positive (60 mgal) free-air gravity anomalies (Kogan, 1975). The magnetic anomalies are long-wave and positive (Shreider, Lukyanov, 1979; Kosygin, 1978). The Basin is bounded by steep 1-2 km high scarps inclined towards the center of the depression at 10°. Thickness of the sedimentary cover (Fig. 6) in the central basin is no less than 3.5 km (Tuezov, et al., 1979). It is composed of thin layered deposits with presumably great concentrations of turbidites. Two deposition complexes are distinguished: an upper reflective sequence and a lower acoustically transparent one (Snegovskoi, 1974). Layers of the upper complex have 800 m thickness and subhorizontal bedding. Seven to ten reflectors of various length are distinguished. Thickness of the lower complex reaches 3.2 km. A pronounced boundary with a velocity of 4.5 km/sec marks the basement of these deposits, which may be the top of a consolidated volcanogenic-sedimentary unit (Popov, Anosov, 1978). Seismic refraction data indicate a 1.5 km thickness for this unit which is underlain by 6.7 km/sec velocity layer. A high velocity layer ($V_p = 7.6$ km/sec) is recorded at the basement of the crust at 11 km depth from the sea surface (Popov, Anosov, 1978). Acoustic basement, when recorded in the reflection data, reveals rough topography with an amplitude up to 1 km. Narrow bodies and seamounts occasionally penetrate the sedimentary cover.

One of these seamounts was examined in detail during the 17th cruise of R/V "Dmitry Mendeleev" (Savostin et al., 1978). It lies on the traverse north of Iturup Is. at 46°N, 147°E. The mountain crowns an elongated ridge and has a cone shape with a flat top (Fig. 5b, c, d). The ridge has a sublongitudinal orientation and is 32 km by 8 km. A 400 γ positive magnetic anomaly is found over the ridge with a horizontal gradient of 70 γ/km. Five measurements of heat flow were made in the vicinity of the cone-like mountain, two of them (st. 1463a and 1463b) showing anomalously high values (to 8 $\mu cal/cm^2 sec$) (Fig. 5a). These are the highest heat flow values in the Sea of Okhotsk. Computations show that comparatively low heat flow, recorded on the slope and foot of the mountain can be due to near-surface distorting factors, while the high values can be caused by heat sources located at shallow depths and of rather recent origin. Anomalously high heat flow (4 $\mu cal/cm^2 sec$) has also been recorded on another seamount in the Sea of Okhotsk (Gorshkov, 1977).

Andesite and andesitic-basalt composition of seamounts (Ostapenko, 1977) located at the foot of the northwestern slope of the Kuril island arc has been determined by dredging. The dredged rocks have higher alkalinity than the andesites of the island arc. Tholeiitic basalts were dredged from one of the seamounts in the Kuril Basin proper (Govorov, personal communication, 1980). These rocks are similar to those formed in active inter-arc basins.

It should be specially noted that in the central basin the sedimentary layers are completely undeformed and horizontal (Fig. 6) with steep dip only at marginal scarps. Deformation of the sedimentary cover is recorded only near the foot of the Kuril island arc. Here, the anticlinal folds have a dip of 20-25° and an amplitude of about 400 m (Snegovskoi, 1974). Folds can be traced through deformations. Characteristic properties of these folds (vergence towards the island arc and insignificant width of the fold belt) allow them to be interpreted as being formed along a fault boundary. A pronounced scarp borders the Kuril Basin on the north. It includes two series of closely spaced faults which appear as normal faults on seismic reflection records (Fig. 6, pr. 3,9). Separate wedges of basement rocks and sediments slope southward along them. A positive magnetic anomaly (400 γ) is associated with the scarp.

Academy of Science Rise. The Academy of Science Rise (Fig. 1) is an uplifted isometric block, 200-250 km across (Fig. 6, pr. 2, 7, 8). In the north and east it is bounded by faults with outcrops of acoustic basement. Deep seismic sounding data show wedging out of a granite layer within the Rise, the crust being composed of two layers: the upper layer with a velocity of 5.3 km/sec and the lower one with a velocity of 6.4-6.6 km/sec. The total thickness of the crust is 20-25 km. The sedimentary layer is very thin, 250 m in average (Udintsev et al., 1976). A gentle arch with a sedimentary cover reduced to 50-100 m is found in the central part of the Rise. Scarps border narrow synclines or graben-synclines with sedimentary layers dipping 5-10°. Thickness of these layers increases up to 500-700 m.

A free-air gravity anomaly (up to 65 mgal), confined to the Academy of Science Rise indicates that it is under isostatic non-equilibrium conditions (Kogan, 1975). Short positive magnetic lineations with magnitudes to 400 γ exist over the Rise. Heat flow values are as high as 2.0-2.3 $\mu cal/cm^2 sec$ (Savostin et al., 1974).

Bedrock dredged from outcropping acoustic basement includes Cretaceous basalts, andesitic basalts, plagio-liparites, gabbro-diorites and granodiorites (absolute age is 95-85 my) (Geodekyan et al., 1976).

Makarov trough. The Makarov trough (Fig. 1) has a west-northwest orientation. Its width is about 100 km. It has a subcontinental type of the crust with 5.8-6.0 km/sec velocities. Seismic reflection data (Fig. 6, pr. 2) show that the Makarov trough represents a region of considerable subsidence, the sedimentary thickness being as much as 1500 m. Acoustic basement is rough and dissected into narrow blocks. Sedimentary layers are folded, forming a system of young folds with the dip of the fold limbs being up to 10°. Inclination of the layers gradually reduces upward the section, thus, revealing a concurrent sedimentation and deformation.

Institute of Oceanology Rise. The Institute

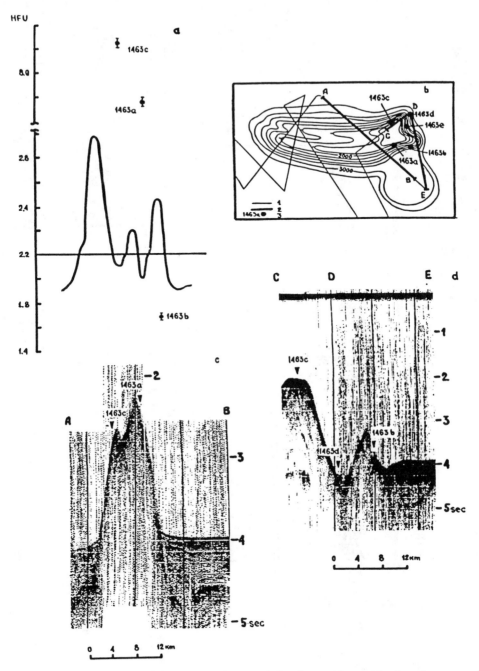

Fig. 5. Seamount in the Kuril basin near Iturup island. a-theoretical and measured heat flow values, b-bathymetry: 1-ship tracks, 2-location of the sections, 3-heat flow stations, c and d-seismic reflections profiles.

of Oceanology Rise (Fig. 1) has an isometric shape. Its western edge is separated from the Derugin Deep by a series of sharp scarps with vertical relief up to 2 km. It has a subcontinental crust with a thickness of 15-20 km. The central part of the Rise (Fig. 6, pr. 6) is underlain by a vast basement high with a flat top surface. Sedimentary thickness does not exceed 250 m and forms a broad gently sloping rise. Almost all flanks of the basement are rugged. Bedrock outcrops are observed along many scarps. Adjacent to the scarp regions, numerous strong deformations like near fault synclines and narrow fold zones (dip 5-7°) occur. Late Cretaceous diorites, granodiorites and plagio-liparites have been dredged from the marginal scarps of the Rise.

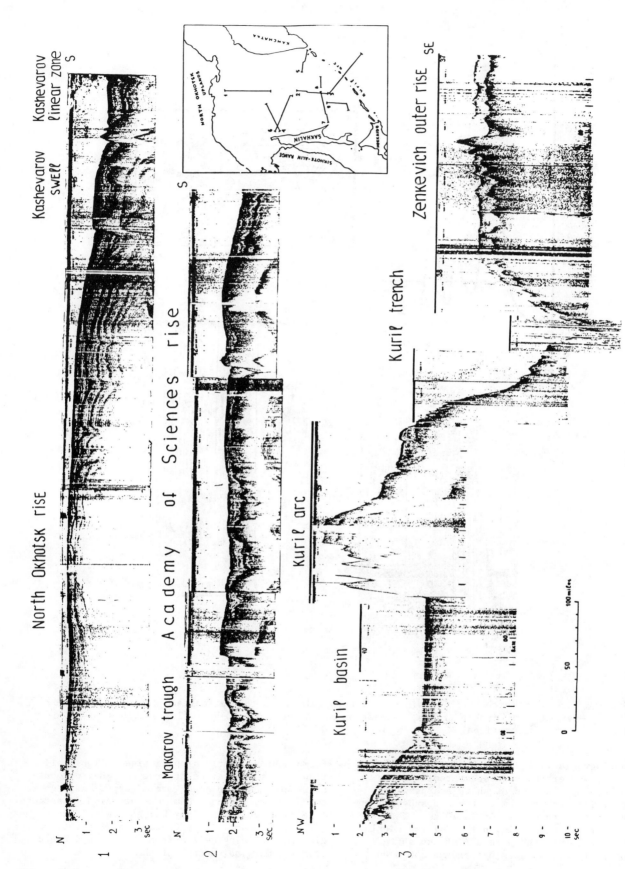

Fig. 6. Seismic reflection profiles in the Sea of Okhotsk. Location of the profiles is shown on the inset.

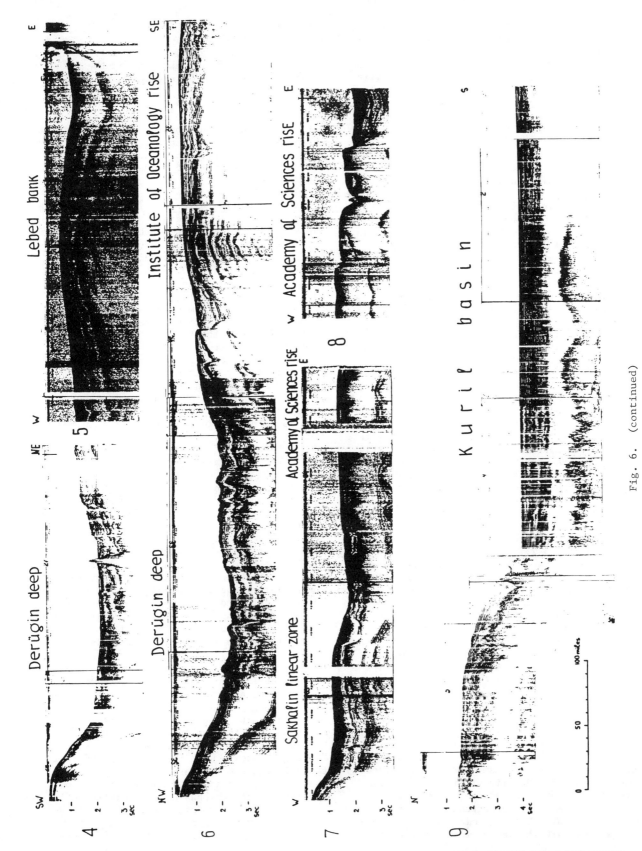

Fig. 6. (continued)

Derugin Deep. The Derugin Deep (Fig. 1) is mainly bounded by the Institute of Oceanology Rise and Sakhalin shelf, and in general has a longitudinal trend. Its maximum depth is 1700 m. Its deep structure is anomalous. Two distinct layers lie beneath the sedimentary cover: the upper layer with a velocity of 6.3-6.4 km/sec and a thickness of 6-7 km, and the lower one with a velocity of 7.2 km/sec and a thickness of 12-16 km. These two layers, after Kosminskaya et al., Structure . . . (1964), comprise the crust. The Moho discontinuity, however, is not distinct here and the crust-mantle transition layer is about 10 km thick.

Seismic profiling data recorded a maximum sedimentary thickness of 1500 km (Fig. 6, pr. 4, 6). In many places, however, the acoustic basement is exposed; for example, on the top of seamounts. The eastern part of Derugin Deep has rugged bottom topography. Sedimentary layers form a system of narrow folds with dips up to 10°. On the whole, fold crests incline towards the Deep center. The sediments also display sinistral strike-slip displacements.

Kashevarov zone. The Kashevarov linear zone (Fig. 1) is a boundary between the northern shallow area of the Sea of Okhotsk and the deeper southern area. It has a clearly articulated linear shape and maximum width of 100 km. The zone strikes in a northwestern direction. Within the Kashevarov zone, several elongated faults dissect its acoustic basement into narrow uplifted and subsided blocks. Sedimentary thicknesses are greatest in the depressions, reaching 1500 m (Fig. 6, pr. 1). To the northwest the Kashevarov zone becomes significantly narrower and in the vicinity of Kashevarov bank it continues as a single fault. It separates the Kashevarov swell from the Derugin Deep. This zone, especially its central broad part, has heat flow values of 1.5-1.6 $\mu cal/cm^2 sec$. However, in its narrow western and southeastern parts values of 2.5-2.7 $\mu cal/cm^2 sec$ are found.

The Kashevarov swell lies north of the linear zone (Fig. 6, pr. 1) and has a continental crust. The northwestern Kashevarov swell is a large uplifted block with numerous bedrock outcrops (greenschists have been dredged). These greenschists originated as a result of metamorphism of basic rocks similar to that of the tholeiitic basalts of the oceanic floor (Baranov et al., 1979). The sedimentary cover is either thin (no more than 0.5 km) or absent. Rough magnetic and gravity fields and isolated anomalously high heat flow (4.0 $\mu cal/cm^2 sec$) are typical of the Kashevarov swell. The swell can apparently be interpreted as a fracture zone similar to the Kashevarov linear zone to the south.

TINRO Deep. The TINRO Deep (Fig. 1) and Lebed trough follow the western Kamchatka coast. They are poorly studied. Seismic profiling has not detected the acoustic basement. A horizontal reflector at a depth of 250 m cannot be interpreted as basement. Deep seismic sounding determined a sedimentary thickness of 6.5 km (Geodekyan et al., 1978). Magnetic anomalies are slight and positive.

Sakhalin linear zone. The Sakhalin linear zone (Fig. 1) borders Sakhalin and the Sakhalin shelf features from the central part of the Sea of Okhotsk. Seismic reflection profiles (Fig. 6, pr. 7) show it as a flexure, complicated by rises and depressions of the basement and also by a series of faults. Dislocation of the sedimentary cover is local and confined to the faults only. In its northern part, an opaque area appears in the cover. At some places, narrow uplifts of acoustic basement exist that can be interpreted as dykes.

Basic tectonic features. In general, two major tectonic regions can be distinguished in the Sea of Okhotsk: the northern area, including the North Okhotsk shelf and North Okhotsk Rise, and the southern area, embracing the greater portion of the Sea of Okhotsk region (Fig. 7).

The tectonics of the northern part can be characterized as passive. It includes an undeformed surface sedimentary sequence of several kilometers. The continuations of the West Sakhalin and West Kamchatka depressions are apparently buried under the shelf. It is inferred that the sedimentary cover there includes depositions since Late Cretaceous.

TINRO Deep, with some reservations, appears to be limited to the same region. Although its structure is poorly known, we can consider it to be a large depression, probably of late Mesozoic age.

The boundary of the northern region follows a large east-southeast fault (Kashevarov linear zone). The fault can be distinguished by a sharp change from the undeformed North-Okhotsk shelf into the rugged topography of the southern Sea of Okhotsk. It can also be recognized by regional magnetic lineations and high heat flow. This fault is presumably an eastern continuation of the Mongolo-Okhotsk lineament.

The southern region, in contrast to the northern one, has been very active with intensive young, late Cenozoic tectonic movements. This fact is supported by the entire collection of geological-geophysical data: rugged topography, a sometimes linear magnetic field, considerable gradients in the gravity anomalies, isostatic non-compensation, high heat flow, and almost ubiquitous traces of young strong deformations.

The Kuril Basin sharply contrasts with the other features within the active southern region. It is a deeply subsided feature with no "granite" layer under it. Nevertheless, there is no significant gravity anomaly which might be expected because of the near-surface position of heavy "basalt" masses. With the very high heat flow, typical of the Kuril Basin, we have evidence for the existence of a heated mantle with a lower density beneath the basin. On seismic reflection profiles the Kuril Basin has a well-pronounced shape of a graben bounded by normal faults. It has absolutely horizontal bedding of undeformed

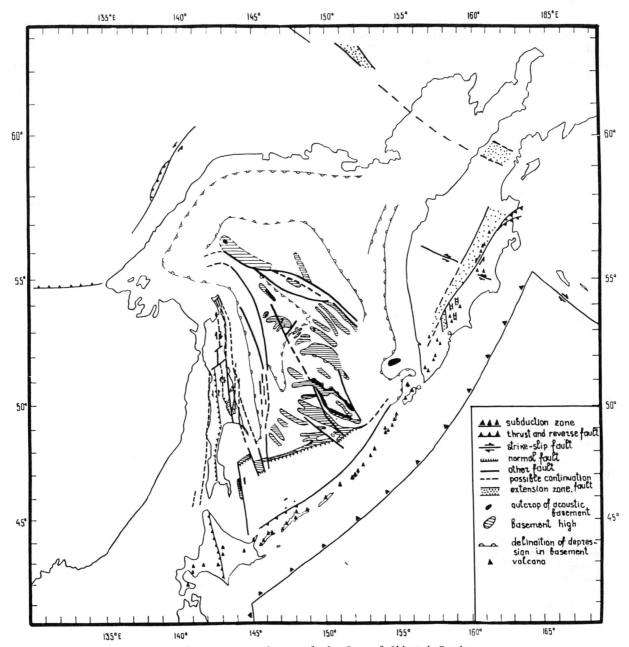

Fig. 7. Structural map of the Sea of Okhotsk Region.

sedimentary layers within the interior. A set of faults with southward inclination to the centre of the Basin define the northern boundary. The Kuril Basin was formed, no doubt, under extensional conditions. Although some profiles of the Basin show a rugged basement topography, this in no way affects the structure of the upper, thick sedimentary cover (Fig. 6, pr. 9). Consequently, sedimentation was accompanied by subsidence and the sediments buried a dissected basement. All of the geological-geophysical data indicate recent formation of the Kuril Basin as a result of extension.

The rest of the active southern region, including the Academy of Science and Institute of Oceanology Rises and Derugin Deep can, in general, be similarily described.

First, the sedimentary cover is extremely thin; its average thickness being 200-300 m and in some places the basement is bare of sediments (Fig. 8). Only in the Derugin Deep does the sedimentary thickness reach 500-1000 m. To our great regret, we have no data on the age of the sediments there. Assuming an average rate of sedimentation of 5 cm/1000 y, one can calculate the age of the sedimentary basement equal to 10 my (Miocene). Similar to the Anivsky Bay (Krasnyi et al., 1975), the sedimentary basement presumably coincides with the basement of the Marujamskaya suite (Late Miocene-Pliocene). Up to Late Miocene the central Sea of Okhotsk area, including Academy of Sciences and Institute of Oceanology Rises, was uplifted, no doubt above sea level. Thus, we can conclude that it has undergone a minimum subsidence of 1000 m during the recent geological episode.

Secondly, the sedimentary cover is deformed in practically all places in the central Sea of Okhotsk. Seismic reflection profiles clearly show numerous dislocations. We can determine several types of deformations: gentle domed swells located within the central parts of the Academy of Science and Institute of Oceanology Rises; narrow asymmetric graben-synclines developed along the faults that mark the swell boundaries; systems of small folds; and zones of strong folding (like in Derugin Deep). The dip of beds is usually no more than several degrees, reaching $10°$ only near faults. The deformation of the sedimentary layers is gradually reduced upward in the section. So, we can conclude that sedimentation and deformation proceeded simultaneously.

On the whole, this greatly deformed sedimentary cover of the central Sea of Okhotsk sharply contrasts to the horizontal non-dislocated sedimentary layers within the Kuril Basin. We can thus conclude that a compression regime dominates at present in the central Sea of Okhotsk region, but not in the Kuril Basin.

The Sakhalin linear zone is a peculiar feature within the central Sea of Okhotsk. It is accompanied by a magnetic lineation and high heat flow. It can be assumed to be one of the submeridional faults making up the series bordering Sakhalin and separating the Okhotsk microplate from the Amurean plate. Young magmatic emplacements in this zone are possible.

The basement of the Central Sea of Okhotsk is comprised predominantly of Late Cretaceous volcanics and granitoids (absolute age 80-90 my). It also seems to be formed of some old greenschists, i.e. by the same rocks which are developed in the Sakhalin and Kamchatka regions. The conclusion is apparently true that there is no massif under the Sea of Okhotsk and that Mesozoic features bordering the Sea of Okhotsk continue into this region.

Bedrock outcrops on the floor of the Sea of Okhotsk are grouped into a submeridional belt which crosses the basic structural strikes of deformations in the sedimentary cover (Fig. 8). At the same time this belt bounds the Derugin Deep on the east and is parallel to the strike of the Sakhalin. It can be supposed that this belt reflects a more ancient structural pattern which is unconformably overlain by younger deposits with subsequent deformations. Major morphological elements of the central Sea of Okhotsk, the belt of eastern rises and Derugin Deep in the west, are remnants of fossil features, presumably of Late Cretaceous age.

The Sea of Okhotsk is a complex geodynamic system. It combines a large zone of extension, which not so long ago was prevailing in the Kuril Basin, and a large zone of compression embracing the central Sea of Okhotsk. Another compression zone, confined mainly to the inner trench slope, can also be distinguished. The entire region, thus, consists of alternating compressional and extensional zones.

Formation of the pattern of the Sea of Okhotsk occurred mainly during Late Cretaceous and Cenozoic time. Three main stages can be distinguished: Late Cretaceous, Paleogene and recent (Neogene-Quaternary). Changes from one stage to another were marked by considerable reassembly in the pattern. Formation of volcanic arc systems within Kamchatka and the central Sea of Okhotsk and the evolution of the Okhotsk-Chukotsk volcanic belt dates back to Late Cretaceous time. This stage was completed by folding and in some places by intrusion of granitoides. Almost the whole Paleogene, at least until middle Oligocene, was comparatively quiet. The volcanic arc can be reconstructed only within Kamchatka. Late Oligocene marked the onset of the recent pattern of the Sea of Okhotsk. Within this stage an intermediate period can be distinguished; the period before Late Miocene. Before Late Miocene another volcanic arc besides the Kurils existed west of it. It was confined to Sakhalin.

Starting from recent times back to the Late Cretaceous we shall consider the interaction of the lithospheric plates, believed to have controlled the evolution of the Sea of Okhotsk region.

Seismicity and recent plate tectonics

Several discrete seismic belts can be traced within the Sea of Okhotsk region (Fig. 9). The Kuril (Kuril-Kamchatka) seismic belt is the most pronounced. Seismicity of this belt has been investigated in great detail (Stauder and Mualchin, 1976; Fedotov, 1966; Tarakanov et al., 1977; Averyanova, 1975; etc). The Benioff zone dips under the Kuril arc at about $45°$ (Fig. 10). The zone of maximum seismic activity is concentrated in a narrow wedge of the inner trench slope. All the large earthquakes are shallow. Focal mechanisms for large shallow earthquakes with $M > 7$ indicate subhorizontal compression perpendicular to the strike of the Kuril island arc (Balakina et al., 1972; Stauder and Mualchin, 1976; Simbireva et al., 1977). Intermediate and deep earthquakes are concentrated in a belt of 80 km thickness dipping northwestwardly to depths of 600-650 km. The plate tectonic model suggests that they mark the position of the consuming rigid oceanic slab. The distribution of Sn and Sc travel time residuals shows that the downgoing plate can be traced to the depth of 1000 km (Jordan, 1977). On the flanks of the Kuril seismic belt, are events of the strike-slip type

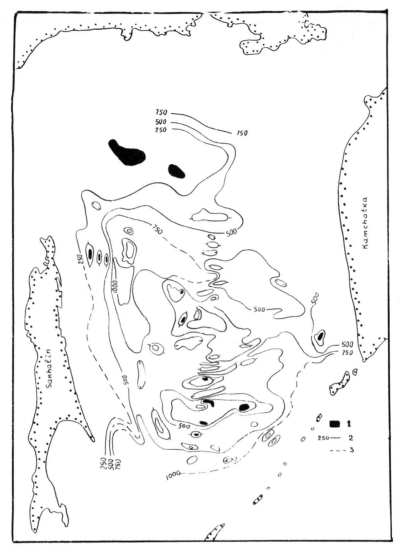

Fig. 8. Thickness of sedimentary cover in meters. 1-outcrops of acoustic basement, 2-isopachs, 3-uncertain interpolation.

resulting presumably from the shear stresses along the strike of the volcanic belt. These shear motions are related to the arcuate geometry of the volcanic belt and oblique subduction. Detailed mechanisms of such displacements were considered by Fitch (1972) and Legler (1976).

Both terminations of the Kuril seismic belt (southwestern and northeastern) are junctions of three seismic belts: the Kuril belt and the two others.

On the southwest, at the latitude of the Sangarsky or Tsugaru Strait (near southernmost Hokkaido), the Kuril belt meets the seismic belt of the Japan trench, and the Sakhalin (Sakhalin-Hokkaido) belt. The seismic pattern of the southern junction is rather complex. An epicentral belt bifurcates westward off the Kuril-Japan trench crossing southern Hokkaido. Available focal mechanisms of the large earthquakes indicate dextral strike-slip latitudinal displacements (Stauder and Mualchin, 1976). This belt is very short. Its length does not exceed 300 km. It joins the meridional Sakhalin seismic belt at the western end of Hokkaido.

Earthquakes within the Sakhalin belt are concentrated on both flanks of Sakhalin, though the majority are confined to its western margin. The magnitude of some of them (for instance, Moneronskoye, 5.7.1971) is over 7. Several northwestward and northeastward clusters of epicenters can be distinguished when both strong and weak earthquakes are considered (Fig. 11). The Sakhalin belt presumably continues southward into the northwestern Honshu margin. To the north, the longitudinal seismic zone is traced to the northern termination of Sakhalin. Within Udskaya

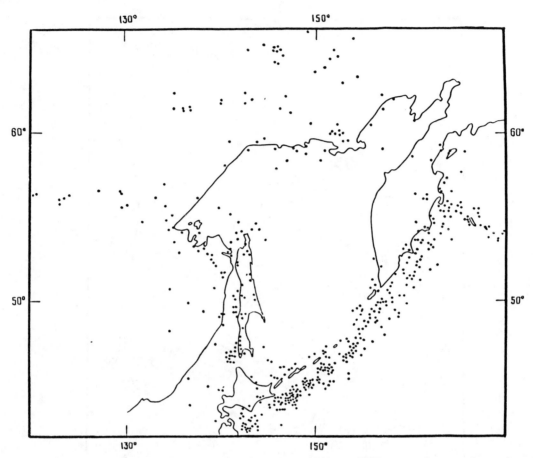

Fig. 9. Seismicity of the Sea of Okhotsk Region from 1913 to 1975. Depths are less than 50 km. Magnitudes for the Kuril arc are 5 to 7, for other parts >4.5 (from the catalogues of the Institute of Physics of the Earth and New catalogue of strong earthquakes..., 1977).

Bay, the belt meets the Baikal-Stanovoi seismic belt and further to the north, the Cherski Ridge seismic belt. Focal mechanisms of the earthquakes of the Sakhalin island arc have been studied by many authors (Balakina, 1976; Borob'eva, 1978); Oskorbin et al., 1968; Chapman and Solomon, 1976 etc.). However all the studies were limited mainly to the solutions of focal mechanisms of the Moneronskoye (05.09.71) and Nogligskoye (02.10.64) earthquakes, whose hypocenters were located in the southern and central part of the island respectively.

The present paper describes four new focal mechanisms (Fig. 12, Table 1) located mainly in the northern part of Sakhalin. All available fault-plane solutions can be divided into the three major groups: the northern group of predominantly normal faults with dextral strike-slip components, the central group of dextral strike-slip motion and the southern group of mainly compression character with small dextral strike-slip components.

In the vicinity of the Kamchatka Cape the Kuril seismic belt meets at a right angle with the Aleutian seismic belt. Near the junction of these two seismic belts, all the focal mechanisms of the Aleutian arc earthquakes have dextral components of displacement along the northwestern planes, which coincides with the strike of the Aleutian arc (Cormier, 1975).

A shallow seismic belt is traced from the Kuril-Aleutian junction northeastward to the Karaginski island. Further, it turns sharply to the northwest and a segment with diffused seismicity crosses Kamchatka structures, through Koni peninsula into the Cherski Ridge (Fig. 1). The segment limited by the trench junction and Karaginski island is characterized by compression mechanisms (Cormier, 1975; Stauder and Mualchin, 1976) or by compression with a dextral component of northeastward displacement (Zobin and Simbereva, 1977; Zobin, 1979). Focal mechanisms for the northwestern seismic zone, which crosses Kamchatka, show subvertical northwestward displacements (Fig. 13).

The two seismic belts meet in the southern part of the Cherski Ridge (near the Kolyma river head). One of them runs eastward to Kamchatka, while the second extends southwestward along the Sea of Okhotsk shore. A single earthquake zone is traced from the Lena river head to the Laptev Sea shelf and further, along the Gakkel Ridge into the Arctic Basin (Sykes, 1965; Lasareva and Misharina, 1965; Chapman and Solomon, 1976; Zonenshain et al., 1978; Savostin and Karasik, 1980). The most prominent topographic features in this region coincide with the seismic belt, for instance, the Cherski Ridge, the mountain chains of which are dissected by narrow graben-like depressions (Grachev et al., 1970, 1971; Grachev, 1979, 1976; Artem'ev, 1972; Naimark, 1976). Some grabens like Momo-Selenyakhsky, Verkhnenersky and others reach 250 km in length and 40 km in width. They are often displaced relative to each other, but together they comprise a pronounced system of the Momsk Rift. Many authors attribute the present Eurasia-North America plate boundary to this zone of high seismicity and tectonic activity (Le Pichon, 1968; Demenitskaya and Karasik, 1969; Churkin, 1972; Chapman and Solomon, 1976; Zonenshain et al., 1978; Savostin and Karasik, 1980). However, these authors disagree on the exact position of this boundary.

An independent zone of shallow seismicity is traced in the rear of the Kuril arc. Earthquake epicenters are confined to the northwestern slope and to the adjacent areas of the basin. They are mainly concentrated behind the arc's southern and northern segments. Focal mechanism solutions (Fig. 12 and 13) show a stress field of subhorizontal compression across the strike of the Kuril basin. Displacements within the earthquake foci located on the slope (08.05.62) and near the foot of the arc (07.02.62) are of the reverse-fault type. A strike-slip component increases with distance from the arc (focal mechanisms in Fig. 13 and 14). In the earthquake foci of the reverse-fault type both nodal planes are approximately parallel to the arc's strike. One of the nodal planes plunges steeply southwestward under the island arc, while the second one has a southeastward plunge. Solovjov et al. (1964) show that the

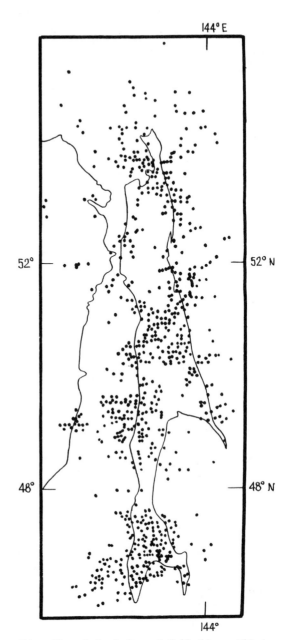

Fig. 11. Seismicity of Sakhalin. All instrumentally determined epicenters are shown from 1963 to 1974 (after Oskorbin, 1977).

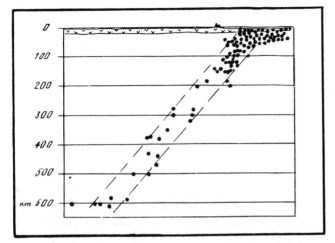

Fig. 10. Section of the Kuril seismofocal zone (after Tarakanov et al., 1977).

spatial distribution of aftershocks accompanying the earthquake of 07.05.62 corresponds to the plane that plunges southeastward under the island arc. Consequently, this is a fault plane with the reverse-fault displacement of the island arc relative to the Kuril basin. Similar displacements can be presumed for other earthquakes near the inner slope. Analogous displacements behind the island arc during Quaternary are varified by the analysis of vertical movements along the Great Kuril arc which show continuous

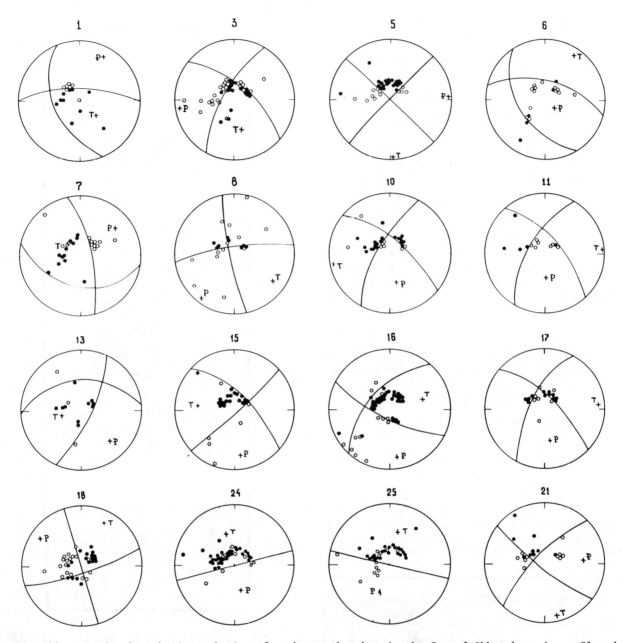

Fig. 12. New focal mechanism solutions for the earthquakes in the Sea of Okhotsk region. Closed dots=compression, open dots=tension (see figure 13, 14 and table 1 for locations).

upward movement of the arc (Grabkov and Pavlov, 1972).

Recent plate boundaries. Seismic belts mark the plate and microplate boundaries of the Sea of Okhotsk region. Chapman and Solomon (1976), using seismic data, examined 5 possible models of plate boundaries within this region. They chose the variant in which the Eurasian/North American plate boundary follows the Sakhalin and Cherski Ridge earthquake belts. The Sea of Okhotsk, thus, was attributed to the North American plate and the Sakhalin earthquakes were considered as the result of the Eurasian and North American plate interaction. We strongly believe that available data suggest a more complex model (Fig. 15) in which the seismic belts reliably define an independent Sea of Okhotsk microplate. This plate embraces almost the entire region of the Sea of Okhotsk, including West Kamchatka. The Sakhalin seismic zone bounds the Sea of Okhotsk microplate in the west and separates it from the Amurian plate. Within Udskaya Bay there is a triple junction between the Sea of Okhotsk microplate and the Amurian and Eurasian plates. The Sea of Okhotsk - Eurasia

Table 1. Parameters for mechanism solutions for shallow earthquakes in the Sea of Okhotsk Region

Date	Position		H	M	Plane 1 Az	Pl	Plane 2 Az	Pl	T_{axis} Az	Pl	P_{axis} Az	Pl	Ref
Northeast Asia													
1. May 20.1963	72.1N	126.7E		50	154	60W	81	64N*	28	3	120	41	7
2. Sep. 9.1968	66.17N	142.13E	39	5.0	174	45W	104	68N*	150	46	44	11	7
3. Jun. 5.1970	63.26N	146.18E		5.4	31	67W	132	62E*	165	37	260	4	7
4. May 18.1971	63.92N	146.1E	0	5.9	43	89W	133	83E*					4
5. Jan.13.1972	61.94N	147.04E	33	5.3	47	84E	140	66W*					4
					45	88W	135	90*	180	2	90	1	7
6. Aug.12.1975	70.76N	127.12E	16	5.1	145	43S	106	54N*	32	7	136	68	7
7. Jan.21.1976	67.73N	140.03E	18	5.0	168	72E	106	34S*	292	55	55	20	7
Sakhalin Is.													
8. Mar. 9.1963	46.16N	141.11E	42	5.0	350.	72W	82	76N*	126	15	216	4	7
9. Oct. 2.1964	51.9N	143.3E	10		102	35S	12	90*					1
10. Dec.24.1967	54.82N	142.55E	3	5.0	28	66NW	310	64NE*	258	2	166	38	7
11. Jan.10.1971	55.01N	142.34E	13	4.9	30	60NW	318	62NE*	86	1	174	45	7
12. Sep. 5.1971	46.4N	141.2E		7.3	179	25W	28	68E*	319	64	108	23	3
					328	54W	348	38E*					4
13. Nov. 4.1973	53.86N	141.54E	13	5.0	25	72E	84	34N*	257	54	136	22	7
Kuril basin													
14. May 7.1962	45.3N	146.7E	25	7.0	40	60NW	200	32SE*	155	73	302	14	5
15. July 1.1964	46.01N	146.96E	20	5.3	312	60NE	48°	84SE	274	27	136	38	7
16. July 25.1968	45.64N	146.8E	19	6.5	297	72SW	41	52NW	72	40	172	12	7
17. July 21.1974	46.1N	145.4E	30	4.9	314	70NE	34	60NW	82	4	173	37	7
18. Jan. 2.1975	46.83N	151.15E	5	6.0	72	78S	343	90*	28	9	298	9	7
Kamchatka													
19. Feb. 3.1957	53.6N	159.1E	30	5.5	129	80SW	34	79NW*	77	21	343	9	2
20. Dec.14.1966	56.7N	161.45E	10	4.6	138	80NE	50	80SW*	273	13	184	2	6
21. Aug.14.1976	56.56N	155.15E	29	4.9	44	70NW	308	76SW*	166	7	84	26	7
Kamchatka north arc-arc junction													
22. Nov.22.1969	57.8	163.5	33	6.3	212	75E	172	19W	319/58		112/29		5
23. June.19.1970	57.45N	163.5E	10	5.2	142	76SW	53	90*	5	10	96	10	6
24. Jan. 21.1976	58.93N	163.57E	7	5.4	76	0	256	90*	346	45	166	45	7
25. Jan.22.1976	58.96N	163.73E	42	5.2	103	0	283	90*	13	45	193	45	7

* fault plane

Reference: 1-Oskorbin et al., 1967; 2-Ale, 1973; 3-Balakina, 1972; 4-Chapman, Solomon, 1976; 5-Stauder, Mualchin, 1976; 6-Zobin, Simbiereva, 1977; 7-this study.

plate boundary coincides with the zone of scattered seismicity on the Sea of Okhotsk shore near the Djugdjur Ridge (Fig. 1). The triple junction between the Sea of Okhotsk microplate and the Eurasian and North American plates falls within the southern Cherski Ridge. The Sea of Okhotsk - North America plate boundary is defined by the northwestern earthquake zone which continues from the Koni peninsula to Karaginsky Island. The seismic zone of the Kuril Basin is the southeastern boundary of the Sea of Okhotsk microplate. It separates the Sea of Okhotsk microplate from the Kuril Island Arc micorplate to the southeast (Zonenshain and Savostin, 1979). The later is contiguous with the Pacific Plate along the Kuril seismic belt. At present the Sea of Okhotsk area consists of the two microplates: the Sea of Okhotsk and the Kuril. They are surrounded by the Pacific, North American, Eurasian and Amurian plates.

The majority of the plate and microplate boundaries within the Sea of Okhotsk region manifest themselves in the form of topographic and tectonic features. The only exception is the northern boundary of the Sea of Okhotsk microplate. It separates the microplate from the North American plate and, being connected with the Karaginsk-Gizhiginskaya extensional zone discordantly crosses the topographic features without significant surface manifestations. This boundary is probably very young, presumably of the Holocene age, with insufficient time for topographic features to be developed. Geological data indicate that formation of the transverse extension zones in northern Kamchatka occurred during the Pliocene-Quaternary time (Dmitriev et al., 1979).

It may be more natural, based on the structure and morphology of the region, to take the Kashevarov linear zone, Fig. 1, (which separates the southern lowered Sea of Okhotsk shelf from its northern shallow segment) as the northern boundary of the Sea of Okhotsk microplate. This zone has rough topography and heat flow anomalies.

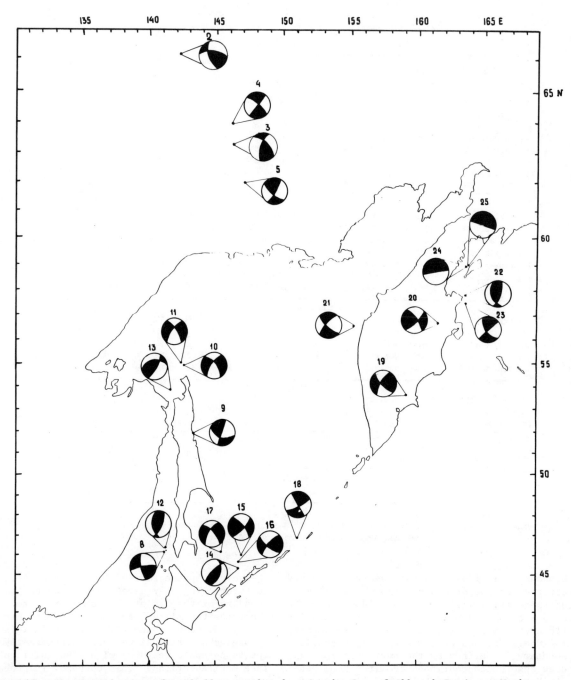

Fig. 13. Focal mechanisms for shallow earthquakes in the Sea of Okhotsk Region. Numbers correspondent to those in Table 1.

However, the seismic network established in the area registers no earthquakes along this zone. One can believe that the Kashevarov zone recently ceased its activity as a boundary and the present boundary was formed a little northward. In case this assumption is true, the northern Sea of Okhotsk was a part of the North American plate until recently.

Instantaneous Kinematics. We have the following data on the nature of the boundaries around the Sea of Okhotsk microplate.

The Kuril island arc microplate is separated from the Pacific plate by a consuming plate boundary which corresponds to the Benioff zone along which northwestward subduction of the Pacific plate occurs. Parameters of the movement of the Pacific plate relative to the Kuril island arc mirocplate were computered by slip vectors within

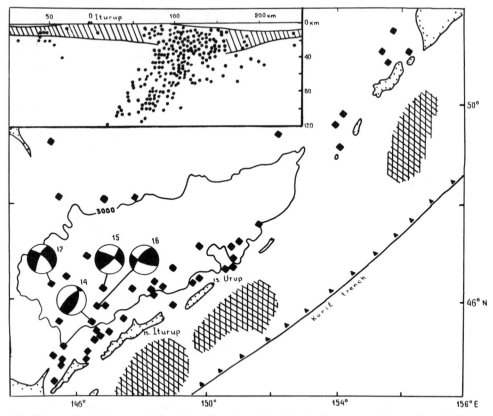

Fig. 14. Shallow seismicity and focal mechanisms in the back-arc region. Hatched areas are of maximum seismic energy release (after Averjanova, 1975). Section shown on the inset after Fedotov, 1965.

the earthquake foci (Stauder and Mualchin, 1976; Ichikawa, 1971; Balakina et al., 1972). The pole of relative rotation has coordinates 66.1°N, 119.2°W with an angular velocity $9 - 10 \times 10^{-7}$ degrees/year.

At present, according to the fault-plane solutions (Fig. 13 and 14), a general relative uplift of the Kuril island range over the Kuril basin is taking place at present. Besides the reverse faults a sinistral component of displacement is recorded.

The movements of the Sea of Okhotsk microplate relative to the Amurian plate are revealed by the earthquakes and recent Sakhalin structures. Focal mechanisms show a change from tension in northern Sakhalin to compression in southern Sakhalin, through the dextral displacements in its central part. In accordance with this fact, northern Sakhalin displays faults with a dextral slip component along them, while the southern part of the island displays young folds and overthrusts. Such a great variety of displacements along the same boundary seems to indicate proximity of the pole of rotation. During the recent episode, for the last 5 my, the displacement is about 25 km (after Rozhdestvensky, 1975), i.e. the rate of motion is 0.5 cm/year.

Verjbitskaya and Savostin (in press) used data on slip vectors along the plate boundaries in the Sea of Okhotsk region and computered the plate motion parameters by closure of motion vectors. The following plates were considered: Eruasian, North American, Amurian and Sea of Okhotsk. Data by Zonenshain et al. (1978) and Savostin (in press) were used to calculate motions for the North American and Eurasian plates. Interaction between the Amurian and Eurasian plates is seen from the data on focal mechanisms for the earthquakes within the Baikal rift zone (Zonenshain and Savostin, 1979). Slip vector azimuths for the Sakhalin earthquakes were used to compute relative motions for the Amurian and Sea of Okhotsk plates (Table 2). To perform the necessary calculations we used the relative motion rate between the Amurian and Sea of Okhotsk plate equal to 0.5 cm/year. This value, however, is the least reliable of all the available data and is probably higher than the real one. The results are shown in Table 3. The kinematic scheme is shown in Fig. 16.

The parameters of plate motion show that along the northern shore of the Sea of Okhotsk, the Sea of Okhotsk plate moves northeastward relative to Eurasia at the rate of about 1 cm/year. Sinistral slip at the rate of 1-1.5 cm/year is believed to be taking place between the

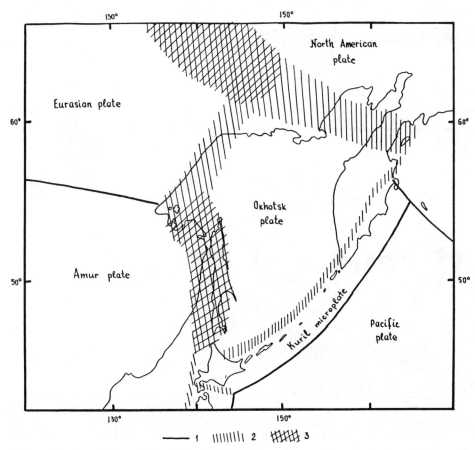

Fig. 15. Lithospheric plates of the Sea of Okhotsk region. 1-certain boundaries, 2-uncertain boundaries, 3-diffuse boundaries.

North American and Sea of Okhotsk plates. The pole of relative rotation for the Amurian and Sea of Okhotsk plates is 52.88°N, 140.11°E, that is, in the Lower Amur river near Sakhalin. It can explain the change in geodynamic conditions observed along the plate boundary. On the whole, if we consider the motion of all plates with respect to Eurasia, the Sea of Okhotsk plate is being gripped by the North American and Amurian plates and undergoing a clockwise rotation round an axis located in the western part of the plate (51.12°N, 143.67°E). The rate of rotation is 7.5×10^{-7} degrees/year. This rotation causes a sinistral displacement of the plate along the northern shore of the Sea of Okhotsk with respect to the Eurasian and North American plates. The Sakhalin structure can be well explained by the interaction of the Amurian and Sea of Okhotsk plates. In the north, where the Sea of Okhotsk plate moves relative to the Amurian plate, there arise normal faults with strike-slip components and sometimes even grabens. The uplift and fold deformations within southern Sakhalin can be explained by plate convergence. Young block displacements and dislocations of the sedimentary cover on the subsided shelf of the Sea of Okhotsk seem to be a result of the general compressional conditions related to the plate interactions in this region.

Origin of the Kuril back-Arc basin. There is no single concept for the origin of the Kuril basin or the other back-arc basins (Uyeda, 1977).

Table 2. Data set used to compute OKH/AMUR pole position

	Lat.	Long.	Azimuth measur.	Azimuth comput.	Difference	Ref.
1.	46.16N	141.11E	80	83.8	-3.80	
2.	46.3	140.9	90	84.9	5.09	V/76
3.	46.4	141.2	89	83	5.9	V/76
4.	46.5	140.9	81	84	-3.72	V/76
5.	46.7	141.0	82	83	-1.89	V/76
6.	51.9	143.3	12	8.4	3.58	O/67
7.	53.95	141.25	-65	-67.7	2.6	
8.	53.99	142.25	-56	-53.43	-2.57	
9.	54.82	142.55	-62	59.87	-2.1	
10.	55.01	142.34	-60	-63.66	3.65	

O/67 - Oskorbin et al., 1967;   V/76 - Vorobjeva, 1976

Back-arc basins with some exceptions (the eastern Bering Sea) (Cooper et al., 1977) are underlain by an oceanic crust which is much younger than the oceanic crust subducting under the island arc. This young crust originates as a result of tensional processes, probably similar to those of the seafloor spreading (Le Pichon et al., 1975; Watts et al., 1977; etc.). Some scientists, however, deny the possibility of tension behind island arcs, for instance, Sorokhtin (personal communication). Several models have been suggested to explain the origin of back-arc basins: (1) Models of mantle diapirs uplifting as a result of mantle heating due to friction of the subducting slab (Karig, 1971a); (2) Models explaining the formation of marginal seas by asthenospheric convective flow. Packham and Folvey (1971) proposed back-arc basin formation within the West Pacific as due to a sharp change of the convective flow system in the asthenosphere after the collision between India and Eurasia. As a result of this collision asthenospheric flow which had formerly been consumed beneath Eurasia began to emerge upward along East Asia giving rise to the formation of oceanic crust of back-arc basins. Nelson and Temple (1972) have assumed the presence of a major single eastward asthenospheric flow, branches of which when they run against sinking plates, emerge to the surface in back-arc basins. Some authors attribute the formation of a new oceanic crust in back-arc basins to small secondary convective cells which appear in back-arc areas due to the turbulence of the asthenosphere due to the affect of the down-going lithospheric slab (McKenzie, 1969; Lobkovskii and Savostin, 1976; Sleep and Toksoz, 1971; Toksoz and Hsui, 1978); (3) Models of back-arc origin due to mid-oceanic ridge subduction, for back-arc basins like the Sea of Japan (Uyeda and Miyashiro, 1974); (4) Models according to which island-arcs are formed where old oceanic lithosphere has subducted mainly due to its own weight. In this case the volcanic arc migrates oceanward and gives rise to the opening of back-arc basins; and (5) Models where an upper continental slab moves away from the subduction zone and a back-arc basin is generated in its place.

Active continental margins may be formed where young oceanic lithosphere is pushed down under a continental plate by a plate driving force (Mol-

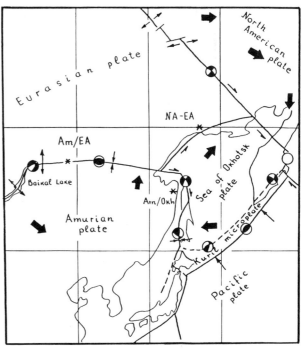

Fig. 16. Plate kinematics of the Sea of Okhotsk Region. Thick arrows show plate motion relative to the Eurasian plate, thin arrows show relative motion of plates along their boundaries. Asterisk are the poles of rotation. Typical mechanism solutions are shown.

Table 3. Parameters of plate and microplate motions in Sea of Okhotsk Region

Plates	Pole Lat.	Long	Rate (10^{-6} deg/year)
Eur/N.Amer	59.46N	141.26E	0.219
Eur/Amur	56.35	118.20	0.142
Amur/Okhot	52.24	140.09	0.831
N.Amer/Okhot	47.09	144.85	0.478
Eur/Okhot	50.57	143.93	-0.694

nar and Atwater, 1978). Where the upper slab moves toward a subduction active continental margins also exist (Wilson and Burke, 1972; Moberly, 1972; Chase, 1978; Jurdy, 1979; Uyeda and Kanamori, 1979; Zonenshain and Savostin; 1979a, 1980). However, it appears that formation of back-arc basins is accompanied by aseismic slip and decoupling of the subducting and overriding plate, while active continental margins only are accompanied by coupling (Kanamori, 1971; 1977).

We believe that the Kuril Basin originated in the Miocene, arising mainly from the last of the innumerated models. Its formation seems to have resulted from the movement of the Okhotsk plate away from the subduction zone. Zonenshain and Savostin (1979a), computing a global closure of movements relative to the western Pacific subduction zone, found that the Sea of Okhotsk plate moved away from the western Pacific subduction zone with a clockwise rotation at the rate of 2.67×10^{-7} degrees/year around a pole of 37.7°N, 168.9°W. Best coincidence of the contour lines of the northern and southern boundaries of the Kuril Basin is obtained by the rotation around a pole of 33.6°N, 156.9°W at the angle of 3°. Both pole positions are very close. The rotation reflects the total movement from Miocene to recent time. At present, however, the Kuril arc is undergoing a steep uplift over the Kuril Basin floor where a compression regime

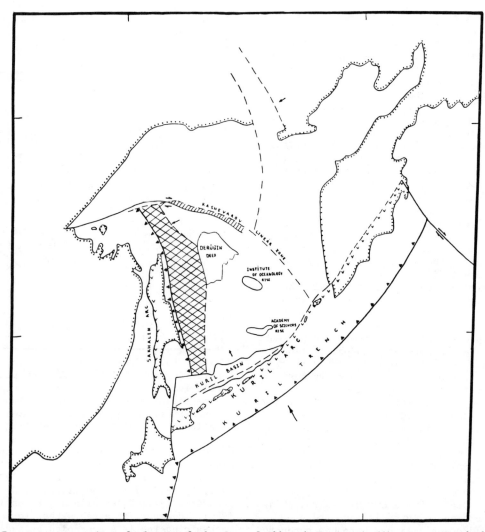

Fig. 17. Reconstruction of plates of the Sea of Okhotsk Region in Miocene. Hatched area is oceanic crust, consumed during Miocene under the Sakhalin arc.

predominates. The same pattern is observed on western Honshu Island, which marks the boundary between the Japan arc and the Sea of Japan (Fukao and Furumoto, 1975). Consequently, the extensional period resulting in the opening of the back-arc basins was completed by recent time and changed into a compressional period. Such changes probably were numerous. Periods of back-arc basin opening recorded in the identified magnetic lineations, for instance, in Sikoku Basin, are comparatively short, 5-7 my (Watts and Weissel, 1975). A change from decoupling to coupling in the Kuril arc subduction zone is in agreement with these changes (K. Sudo, personal communication).

Formation of the recent pattern of the Sea of Okhotsk, however, cannot be realized without analysis of the plate interactions during the entire Neogene-Quaternary period.

History of plate and microplate motions within the Sea of Okhotsk during Neogene-Quaternary Time

On the whole, the recent pattern of plate motions has been set in the Sea of Okhotsk region since the very beginning of Miocene or Late Oligocene time. During this period, however, one significant reorganization occurred. It took place in the Late Miocene. During Pliocene-Quaternary time the plate interaction pattern remained unchanged.

During Miocene time the pattern was slightly different (Fig. 17). Calc-alkaline volcanics, recorded in West Sakhalin, allow reconstruction of the Miocene volcanic arc following Sakhalin in a sub-longitude direction. This Miocene volcanic should have been accompanied by a subduction zone. If we compare data on the petro-

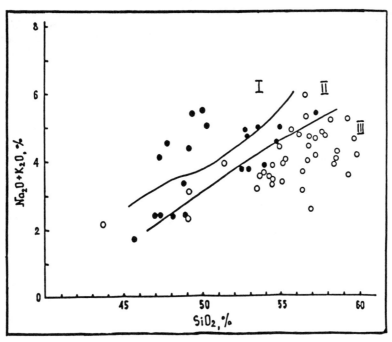

Fig. 18. Diagram showing correlation between K_2O-SiO_2 in Miocene volcanics of Sakhalin. Closed circles are volcanics of West Sakhalin, open circles are volcanics of East Sakhalin. I, II, III are fields of alkaline, calk-alkaline and tholeiitic series.

chemistry of the Miocene Sakhalin volcanism to the MacDonald diagramme, it is clear that alkalinity increases westward; that is, the subduction zone was located somewhere east of Sakhalin (Fig. 18). Considering that the K_2O concentration is a function of depth to the Benioff zone, it may be assumed that the Sakhalin arc was controlled by the Benioff zone dipping under Sakhalin at about 45°. Consequently, it was exposed on the surface not far from the eastern Sakhalin margin, in the vicinity of the Derugin Deep. It is possible to account for the sedimentary prism along the eastern Sakhalin margin if it is a fossil deep trench filled up with sediments. We also can interpret the Derugin Deep as a subsided marginal part of the plate subducted into the trench. The Derugin Deep seems to have an oceanic basement, maybe representing a fossil bed of a Late Cretaceous marginal basin (see below). In this case the Derugin Deep can be regarded as a part of the fossil oceanic plate subducted in the trench.

Miyashiro (1972) established a correlation between the rate of relative displacement of converging plates and chemistry of volcanic series of island arcs. According to the rate of plate convergence, island arcs can be subdivided into three groups. Reduction of rates is marked by a regular increase in alkalinity of volcanics and by a change of andesites to trachiandesites. On Sakhalin, an analogous increase in alkalinity of andesites took place in the Late Miocene (Semenov, 1975). If we compare the Sakhalin curves of alkalinity-silica function for andesites and trachiandesites with curves obtained by Miyashiro (1972) it is clear that the Miocene curves for andesites correspond to the first-group island arcs, while the Pliocene curves for trachiandesites belong to the third group (Fig. 19). Thus, high speed plate convergence (up to 9 cm/year) which occurred within the Sakhalin arc during the Early-Middle Miocene was reduced to 2 cm/year during Late Miocene, and during Pliocene-Quaternary time the subduction ceased. This presumably could occur as a result of blocking of the Sea of Okhotsk by continental massifs. As a result, the sedimentary cover in general, and in particular in the Derugin Deep, underwent deformation.

The consuming plate-boundary marked by the Sakhalin volcanic arc continued to the northern termination of Sakhalin. The triple junction could presumably have been located there in the Miocene (Fig. 17). From the west, it met the Baikal-Stanovoi zone separating the Amurian and Eurasian plates; from the east, the Sea of Okhotsk - North America plate boundary. It probably coincided with the Kashevarov linear zone which separated the uplifted and lowered shelf parts of the Sea of Okhotsk and bounded the region of young deformation of the subsided shelf. Along the Kashevarov zone, one can suppose strike-slip displacements between the Sea of Okhotsk microplate and the North American plate. The pole of

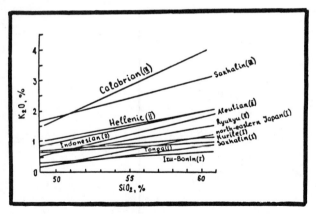

Fig. 19. Relationships between K_2O-SiO_2 in lavas of other island arcs (after Miyashiro, 1972). Data of the Miocene and Pliocene of the Sakhalin are shown as well (after Semenov, 1975). I, II, III-correspond to arcs with different subduction rate (see text for explanation).

relative rotation for these plates, computed by the Kashevarov zone strike, has the coordinates of 68.7°N and 131.4°W.

Formation of the Sakhalin volcanic arc coincided in time with the opening of the Kuril Basin. The plate interaction scheme may have been as follows. The Pacific plate was subducting under the Kuril arc. On its southern flank, the Kuril island-arc microplate was cut by a submeridional Sakhalin plate boundary. A part of the Pacific plate motion was realized in the westward migration of the Sea of Okhotsk microplate. The Sea of Okhotsk microplate began to move away from the Kuril arc causing the opening of the Kuril Basin and at the same time was subducted under Sakhalin resulting in the origin of the Sakhalin volcanic arc. Subduction ceased in Late Miocene, probably owing to the impossibility of the sialic slabs within the inner Sea of Okhotsk to subside in the subduction zone. Collision of these blocks with Sakhalin caused uplift of Sakhalin and deformation of the sedimentary cover, in particular, folding of and formation of the "green tuffs". Simultaneously, the Kuril Basin ceased to open.

History of plate motions within the Sea of Okhotsk region during the Late Cretaceous and early Tertiary

The Late Cretaceous evolution of the Sea of Okhotsk is recorded in bed rocks dredged from its bottom (Geodekyan et al., 1976). These rocks, in general, belong to the calc-alkaline island-arc complex. No doubt, remnants of the Late Cretaceous island arc are buried under the Sea of Okhotsk.

Two major regions of Late Cretaceous magmatism can be distinguished within the Sea of Okhotsk (Fig. 20); the western region within the Sikhote-Alin belt, and eastern region within the East Sakhalin mountains, on the Lesser Kuril islands and on eastern Hokkaido (Nemuro region). These two regions are sharply different; the Sikhote-Alin belt is comprised of a continental volcanic-plutonic association, while the volcanism within the eastern part underwent submarine conditions and was accompanied by the accumulation of chert-graywacke and deep-sea sediments (Melankholina, 1973). Regions of Late Cretaceous magmatism are divided by an amagmatic West Sakhalin depression and by the Late Cretaceous fold belt of Sakhalin and Hokkaido. Magmatic rocks from the Sea of Okhotsk are attributed to the eastern region.

Outcrops of Late Cretaceous magmatic rocks within the eastern region are greatly dispersed. This hinders the reconstruction of a detailed paleogeography. It is important for island arcs to establish their polarity, i.e. to determine the position of their frontal part oriented to the trench, and of their rear occupied by a back-arc basin.

The Kuno diagram shows that volcanics within the Sea of Okhotsk and of the Matakotanskaya and Zelenovskaya suites of the Lesser Kuril islands generally fall within the field of the calc-alkaline series and partially within the tholeiite series (bottom volcanics of the Sea of Okhotsk) (Fig. 21). Rocks from East Sakhalin and the Lesser Kuril suite, localized only to the west of the Lesser Kuril range, belong to the alcalic-basalt series.

It is possible to distinguish two approximately submeridional volcanic zones; an eastern of predominantly calc-alkaline volcanism, and a western zone of basalt-alkaline volcanism. Therefore, it can be concluded that the frontal part of the Late Cretaceous volcanic arc faced to the east. Remnants of the deep-sea trench are probably buried somewhere under the southern continuation of the TINRO Deep and Lebed depression. Derugin Deep can be interpreted as a deep relict basin of a Late Cretaceous marginal sea. Anomalous crustal structure under this Deep supports this inference (see above).

The second volcanic arc, in the eastern region, called Iruneiskaya (Belyi, 1974) is reconstructed within Kamchatka. Belyi believes that the Iruneiskaya arc had a westward facing frontal part. Petrochemical data presented by Markovsky and Rotman (1969) and Khotin (1976), however, contradict this supposition. To the east, within the Kamchatka peninsula, tholeiites are developed. Calc-alkaline rocks dominate in the Valaginsky and Kumrock ridges, while volcanics of basalt-alkaline series dominate to the west, in the Middle Ridge. It can be postulated, therefore, that the Iruneiskaya arc faced east, towards the ocean. The West Kamchatka depression and its continuation buried under TINRO Deep are, in this case, remnants of a Late Cretaceous marginal sea completely filled by sediments.

Using Dickinson and Hatherton's method (1967),

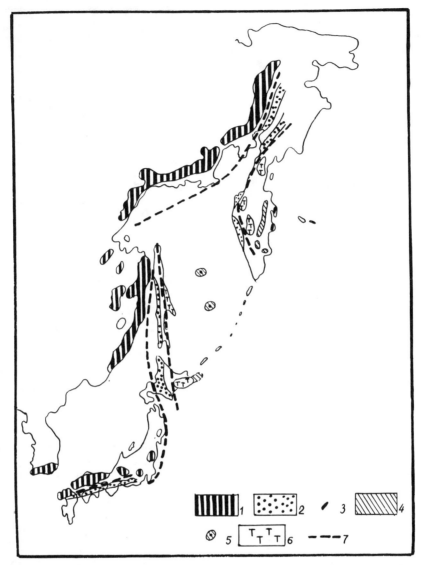

Fig. 20. Distribution of upper Cretaceous (Senonian) rocks in the Sea of Okhotsk Region. 1-continental volcanic-plutonic complexes, 2-marine terrigeneous complexes, 3-6 = volcanic-siliceous-terrigeneous complexes, 3-with tholeiitic volcanism, 4-with calc-alkaline volcanism, 5-dredged samples with alkaline volcanism, 7-boundaries between complexes.

the dip of the fossil Benioff zones can be estimated. It was 30-40° under the Sea of Okhotsk arc, the Benioff zone being projected to the southern continuation of TINRO Deep, and 45-50° under the Iruneiskaya arc, the zone being projected to the region directly to the east of the Kamchatka peninsula.

The Late Cretaceous Sea of Okhotsk volcanic arc and back-arc basin probably underwent varied deformations. Within the East Sakhalin arc there was intensive folding with successive formation of tectonic nappes thrust to the west. Within the Lesser Kuril islands and Nemuro region on Hokkaido, Late Cretaceous deposits are dislocated very little. The floor of the Sea of Okhotsk did not seem to undergo strong deformations either. Otherwise, the relict Derugin Deep could not have existed. The fold area seemed to be concentrated in a narrow belt of West Sakhalin and did not affect the greatest part of the Okhotsk volcanic arc.

Paired metamorphic belts are typical of the Sakhalin linear zone of deformations and its continuation on Hidaka within Hokkaido. They are: the belt of glaucophane metamorphism of high pressures to the west, and the belt of plagiogneissic metamorphism of high temperature to the east (Miyashiro, 1972). A similar polarity of metamorphism is typical of many island arcs, the metamorphic belt of high pressure being confined

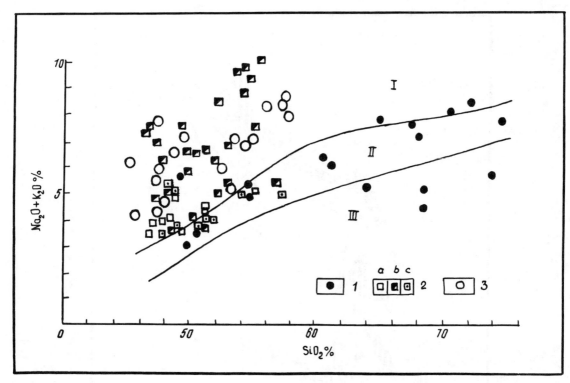

Fig. 21. Relationships between $K_2O+Na_2O-SiO_2$ in lavas of the Sea of Okhotsk Region, I, II, III are the same as in Fig. 18. 1-lavas of the bottom of the Sea of Okhotsk, 2-lavas of Lesser Kuril: a-matakotan, b-malokuril.

to the Benioff zone and the belt of high temperature metamorphism to the rear of the arcs. Consequently, we can speak about the deformations of the pre-Late Cretaceous volcanic arc which is marked by rocks of the Mesozoic Hidaka group within Hokkaido and by the Sakhalin Nabilskaya series. It can be concluded that the frontal part of the arc faced west, the trench was located somewhere under the West Sakhalin depression, and the Benioff zone was inclined eastward. During the Mid-Late Cretaceous there was a large reassembly of the region. An older Mesozoic arc ceased to exist and a new one originated with a Benioff zone plunging to the west. At the same time intensive deformation, accompanied by westward ophiolite obduction, commenced along the old arc, mainly along the previous Benioff zone.

The Late Cretaceous volcanic arc system, including the Okhotsk and Iruneiskaya arcs, was separated from the Asian continental margin by a basin of unknown width and within which the formation of a continental volcanic-plutonic belt was in process. This belt much resembled the recent Andean belt of South America and should have had a trench on its off-shore side. The place of this supposed trench seems to be covered by deposits of the West Sakhalin depression and by Tatarsky Strait waters. A general cross section of the junction area between Asia and the Pacific in the Late Cretaceous contained two Benioff zones with westward subduction.

An independent Late Cretaceous Pre-Okhotsk plate (microplate) can thus be inferred (Fig. 22). It was bordered by an active continental margin of Eurasia and by the Pacific, from which it was separated by the Sea of Okhotsk arc and Iruneiskaya volcanic arc. In the Pacific, the arcs were conjugate to the Kula plate which was moving northward. The Late Cretaceous Pre-Okhotsk plate was considerably larger than the recent Sea of Okhotsk microplate. It should have embraced large areas of oceanic crust which were eventually subducted under the Eurasia active continental margin. Remnants of this crust can be seen in ophiolite sheets within East Sakhalin. They are found in Late Cretaceous nappes. The ophiolite obduction occurred during the very Late Cretaceous - Early Paleogene. This was presumably induced by collision of the Eurasia active continental margin and the Sea of Okhotsk volcanic arc.

During Paleogene the pattern was quite different. The greater part of the region (remnants of oceanic bed excluded) underwent uplift. Accumulation of sediments occurred mainly in front of the fold chains. Carboniferous sediments are distributed almost everywhere. The most striking feature is an almost complete absence of volcanism in the region. In the interval between the Late Cretaceous and Neogene-Quaternary, epochs of intensive volcanism and

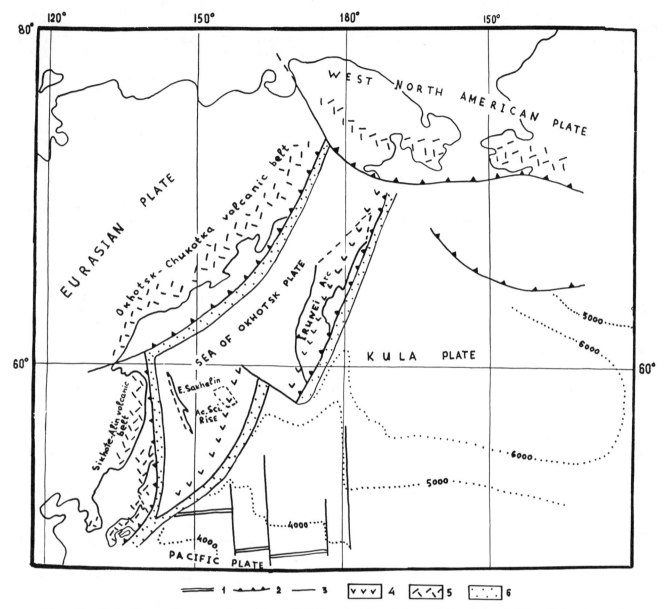

Fig. 22. Late Mesozoic reconstruction of lithospheric plates of the Sea of Okhotsk Region. 1-mid-ocean ridges, 2-trenches, 3-transform faults, 4-island arcs, 5-active continental margins, 6-outer rise. Paleoisobaths are shown by dotted lines (after Savostin et al., 1980).

large changes in tectonic pattern, there was a quiet period which is in disagreement with the notion of a continually active transition zone between Eurasia and the Pacific. The Paleogene quiet period, however, can probably be explained by reduced interaction between the Pacific and Eurasia. During the Late Cretaceous the Kula plate moved towards Eurasia. By Eocene, at least by 50 my ago, the Kula-Pacific Ridge was submerged under the Aleutian trench and the Pacific plate there met Eurasia. In the Early Paleogene the pole of rotation for the Pacific plate relative to the Eurasian was at 1.27°N, 122.42°W and the slip vector for the Pacific plate with respect to Eurasia had a north-northeastern orientation, that is approximately parallel to the Eurasia margin. Thus, there were generally only slip movements between the Eurasian and Pacific plates. The absence of subduction zones and associated volcanic belts can be explained by this supposition. In Mid-Paleogene time, a period of global changes in lithosphere plate movements started. The Pacific plate sharply changed its movement relative to Eurasia. The pole of relative rotation shifted 70° of latitude to the north (67.9°N, 75°W). As a result, convergence commenced between the Pacific and Eurasian plates causing formation of the recent system of consuming plate-boundaries and associated island arcs.

Conclusions

The present Sea of Okhotsk occupies the former site of a Late Cretaceous volcanic arc and back-arc basin system. The Late Cretaceous arc was located at least 1000 km further from the Asian continent than the present Kuril Arc. It was separated from Asia by an active continental margin marked by the Okhotsk-Chukotsk volcanic belt and by an oceanic basin which was subsequently completely subducted beneath the Asian continent.

The recent structural pattern of the Okhotsk region began its formation after an Early Paleogene period of reduced tectonic activity. The present Okhotsk region arc and back-arc system began developing in Mid-Paleogene along with the other present arc systems in the Western Pacific. During the Miocene there existed two parallel volcanic arcs: the Kuril and the Sakhalin. No less than 200 km of the eastern Sea of Okhotsk oceanic crust were consumed in the Sakhalin subduction zone. Remnants of this crust can be seen in Derugin Deep. As a result of the subduction beneath Sakhalin a considerable part of the Sea of Okhotsk underwent $\sim$1000 m subsidence and the sedimentary cover was deformed. Simultaneously, the Kuril basin originated as a result of migration of the Sea of Okhotsk plate away from the Kuril volcanic arc. At present, that tensional regime no longer exists and the Kuril arc is being steeply uplifted over a deep Kuril Basin.

The Sea of Okhotsk contains a complex pattern of recent plate interaction typical of the Western Pacific. Four lithospheric plates meet in the region: the Pacific, North American, Eurasian and Amurian. An independent microplate, the Sea of Okhotsk plate, is locked between them. Its boundaries are marked by seismic belts through Sakhalin, the Kuril back-arc basin, Kamchatka, Penzhinskaya Bay and along the Asian shore. The most pronounced is the Sakhalin belt, the rest showing a great dispersion of seismicity. In other words, the plate boundaries are not yet completely formed and are presumably undergoing the first stages of their formation. On the whole, all plates in the region are converging due to a general compressional regime which governs the region. As a result, the sedimentary cover on the subsided shelf of the Sea of Okhotsk is greatly fractured. The young uplifts and fold system of Sakhalin are caused by convergence of the Sea of Okhotsk and Amurian plates. Since the pole of their rotation is located near the northern termination of Sakhalin the compressional regime governs southern Sakhalin only, while northern Sakhalin is controlled by strike-slip displacements.

REFERENCES

Averjanova, A. N., Deep seismotectonics of island arcs (northwestern Pacific), Nauka, M., 218 p., 1975.

Balakina, L. M., Focal mechanism of Moneronsky earthquake and its strongest aftershocks, Papers by SakhKNIT, 39, 88-93, 1976.

Balakina, L. M., A. V. Vvedenskaya, N. Y. Golubeva, L. A. Misharina and E. I. Shirokova, The Earth's elastic stress field and earthquake focal mechanisms, M., Nauka, p. 192, 1972.

Baranov, B. V., V. V. Gabov and N. M. Suschevskaya, Metamorphic zone of the Kashevarov Bank (Sea of Okhotsk), in complex research of the World Ocean, V. 1, M., Institute of Oceanology, 95-98, 1979.

Beck, R. H. and P. Lehner, Oceans-new frontiers in exploration, AAPG Bull., V. 58/3, 376-395, 1974.

Belyi, B. F., On comparative tectonics of Western Pacific volcanic arcs, Geotectonics, 4, 85-101, 1974.

Caldwell, G. C., W. F. Haxby, D. E. Karig and D. Z. Turcotte, On the applicability of a universal elastic trench profile, Earth Planet. Sci. Lett., 31, 239-246, 1976.

Chapman, M. E., Sean C. Solomon, North American-Eurasian Plate Boundary in Northeast Asia, Geophys. Res., 81, 921-930, 1976.

Chase, C. Y., Extension behind island arcs and motion relative to hot spots, J. Geophys. Res., 83, 5358-5387, 1978.

Cormier, V., Tectonics near the junction of the Aleutian and Kurilo-Kamchatka arcs and the mechanism for middle Tertiary magmatism in the Kamchatka basin, Geol. Soc. Amer. Bull., 86, 446-459, 1975.

Churkin, M., Western boundary of the North American continental plate in Asia, Geol. Soc. Am. Bull., 83, 1027-1972.

Debremaecker, J. C., Is the oceanic lithosphere elastic or viscous?, J. Geophys. Res., 82, 2001-2004, 1977.

Demenitskaya, R. M. and A. M. Karasik, The active rift system of th Arctic Ocean, Tectonophysics, 8, 345, 1969.

Dickinson, W. R. and T. Hatherton, Andesitic volcanism and seismicity around the Pacific, Science, 157, 801-803, 1967.

Dmitriev, V. D., G. P. Dekin and N. T. Demidov, Latitudinal zones of Kamchatka and their seismicity, XIV. Pacific Congress, Section BI, Collection of abstracts, M., 41-42, 1979.

Erlikh, E. N., Recent structure and Quaternary volcanism of the Western Pacific Ring, Nauka, Novosibirsk, p. 243, 1973.

Fedorchenko, V. I., and R. I. Rodionova, Xenoliths of the Kuril islands lavas, Nauka, Novosibirsk, p. 138, 1975.

Fedotov, S. A., Deep structure, upper mantle properties and volcanic activity of the Kuril-Kamchatka island arc after seismic data for 1964, in

Volcanism and the Earth's deep structure, Papers of the II All-Union Volcanological Meeting, Nauka, M., 8-25, 1966.

Fitch, T. I., Plate convergence, transcurrent faults and internal deformation adjacent to Southwest Asia and the Western Pacific, J. Geophys. Res., 77, 4432-4460, 1972.

Forsyth, D. and W. Chapple, A mechanical model of the oceanic lithosphere consistent with seismological constraints, abstract, EOS, 59, p.372, 1978.

Fukae, Y., and M. Furumoto, Mechanism of large earthquakes along the eastern margin of the Japan Sea, Tectonophysics, 25, 247-266, 1975.

Galtsev-Bezuk, S. O., Recent tectonics, Atlas of the Sakhalin island., M., 1967.

Geodekjan, A. A., Yu. P., Neprochnov, I. I. Elnikov, et. al, New data on TINRO deep structure in the sea of Okhotsk. Papers ANSSSR, 243, 2, 449-452, 1978.

Geodetjan, A. A., G. B. Udintsev, B. V. Baranov, et. al., Bed rocks within the central Sea of Okhotsk region, Soviet geology, 6, 12-31, 1976.

Geology of the USSR, v. XXXIII, Sakhalin Geological description. M., Nedra, 431 p., 1970.

Gnibidenko, G. S., Rift system of the Sea of Okhotsk floor, Papers ANSSSR, 229, 1, 163-165, 1976.

Gnibidenko, G. S., Sea floor tectonics of marginal basins of the Far East, Nauka, M., p. 60, 1979.

Gnibidenko, G. S., and A. Ya. Iljev, Composition and seismic wave velocity of the acoustic basement of the central Sea of Okhotsk, Papers ANSSSR, 229, 2, 431-414, 1976.

Gorkalenko, I. A., B. A. Bondarenko, A. V. Zhuravlev, and S. A. Ushakov, New data on the Earth's deep structure within the Kurilo-Kamchatka trench, Papers ANSSSR, 234, 1, 135-137, 1977.

Gorshkov, A. P., Regional heat flow of the Kuril island arc and ways to establish active sea volcanoes, in Geology and geophysics of the floor within the northwestern Pacific, Vladivostok, 75-81, 1977.

Gorshkov, G. S., Volcanism of the Kuril island arc, Nauka, M., p. 286, 1967.

Grachev, A. F., Monsk continental rift (north-eastern region of the USSR). Geophys. methods of the reconnaissance in the Arctic, 8, NIIGA, Leningrad, 1973.

Grachev, A. F., R. M. Demintskaya and A. M. Karasik, Momsk continental rift and its relationship to the mid-oceanic structure of Gakkel Ridge, Geophys. methods of reconnaissance in the Arctic, 6, NIIGA, Leningrad, 1971.

Grabkov, V. K. and Yu. A. Pavlov, Recent movement and crystallic composition of the Earth's crust near the Kuril island arc, Papers ANSSSR, 203, 3, 1972.

Jordan, T. H., Lithospheric slab penetration into the lower mantle beneath the Sea of Okhotsk, J. Geophys. Res., 43, 473-496, 1977.

Jurdy, Donna M., Relative plate motions and the formation of marginal basin, J. Geophys. Res., 84, 6796-6802, 1979.

Hanks, T. C., The Kuril trench Hoffaido Rise system: large shallow earthquakes and simple models of deformation, Geophys. J. Roy. Astron. Soc., 23, 174-189, 1971.

Hilde, T. W. C., N. Iseraki, J. M. Wageman, Mesozoic sea-floor spreading in the north Pacific, in The Geophysics of the Pacific ocean basin and its margin, Am. Geophys. Un., Washington, 205-266, 1976.

Hilde, T. W. C., G. F. Sharman, Fault Structure of the Descending Plate and its Influence on the Subduction Process: EOS, v. 59, no. 12, p. 1182, 1978.

Ishokawa, M., Reanalysis of mechanism of earthquakes which occured in and near Japan and statistical studyies on the nodal plan solutions obtained, 1926-1968, Geophys. Mag. 35, 207-273, 1971.

Kanamori, H., Great earthquakes of island arcs and the lithosphere, Tectonophysics, 12, 187-198, 1971.

Kanamori, H., Seismic and aseismic slip along the subduction zones and their tectonic implications in island arcs, in deep-sea trenches and back arc basins, edited by M. Talwani and W. C. Pitman III, Amer. Geophys. Union, Washington, 163-174, 1977.

Karig, D. E., Evolution of the arc systems in the Western Pacific, Ann. Rev. Earth Planet. Sci., 2 51-75, 1974.

Karig, D. E. and G. F. Sharman III, Subduction and accretion in trenches, Geol. Soc. Am. Bull., 86, 377-389, 1975.

Kazakova, E. N., K. F. Sergeev, and M. I. Streltsov, Stratigraphy of volcanic sedimentary formations of the Lesser Kuril range, in Geology and geophysics of the Pacific belt, Novoaleksandrovsk, Papers by SakhKNII, 25, 1970.

Khotin, M. Yu, Effusive-ciliceous formation of the Kamchatka Cape (structure and position in the series of other Upper Cretaceous formations of Kamchatka), M., Nauka, p. 195, 1976.

Kogan, M. G., Gravity field of the Kuril-Kamchatka arc and its relation to thermal regime of the lithosphere, J. Geophys. Res., 80, 1381-1390, 1975.

Kochergin, E. V., M. L. Krasnyi, P. M. Suchev and I. K, Tuezov, Anomalous geomagnetic field in the northwestern Pacific mobile belt and its connection with the tectonic structure, Geology and geophysics, 12, 1970.

Krasnyi, M. L., Yu, A. Pavlov, S. S. Snegovskoi, B. I. Vasiliev, A. A. Kulikov and V. Yu. Kosygin, Geological strucutre of the Aniva Bay bottom (South Sakhalin) by results of complex geophysical research, DAN, 222, 2, 421-424, 1975.

Legler, V. A., Cenozoic evolution of Kamchatka and plate tectonics, in Plate tectonics (Tectonic processes, their sources of energy and plate dynamics), M. Institute of Oceanology, 137-169, 1977.

Le Pichon, X., J. Francheteau, and C. F. Sharman, Rigid plate accretion in an inter-arc basin: Mariana Trough, Jour. Phys. Earth (Tokyo), 23, 251-156, 1975.

Lobkovskii, L. I., and O. G. Sorokhtin, Condition of sedimentary subduction in trenches, in Plate tectonics (Dynamics of the subduction zone). M., Institute of Oceanology, 89-102, 1976.

Lobkovskii, L. I. and L. A. Savostin, The secondary convection and nature of high heat flow in the marginal seas of the Western Pacific, papers ANSSSR, 258, 5, 1980.

Ludwig, W. J., J. I. Ewing, S. Murauchi, N. Den, S. Asano, H. Hotta, M. Hayakawa, T. Asanuma, K. Ichikawa, and I, Nagurchi, Sediments and structure of the Japan trench J. Geophys. Res., 71, 2121-2137, 1966.

Ludwig, W. J., N. Den and S. Murauchi, Seismic reflection measurements of southwest Japan margin, J. Geophys. Res., 78, 2508-2516, 1973.

Markov, M. S., Metamorphic units and basalt layer of the Earth's crust within island arcs, Nauka, M., p. 274, 1975.

Markov, M. S., V. N. Averjanova, I. P. Kartashov, I. A. Solovjeva and A. S. Shuvae, Mesozoic-Cenozoic history of the Earth's crust within the Sea of Ohkotsk region. M., Nauka, p. 221, 1967.

Markovsky, B. A. and V. K. Rotman, Late Cretaceous geosynclinal volcanic-sedimentary formations of the northwestern Pacific mobile belt, Izv. ANSSSR, geol. ser., 6, 18-34, 1969.

McAdoo, D. C., D. L. Turcotte, and J. G. Caldwell, The analysis of an elastic-perfectly plastic lithosphere entering a trench under both transverse and horizontal bedding, abstract, EOS, 58, 498, 1977.

McKenzie, D. F., Speculations on the consequences and causes of plate motions, Geophys. J. Roy. Astron. Soc., 18, 1-32, 1969.

Melankholina, E. N., West Sakhalin depression and its analogues in the Pacific belt. M., Nauka, 1973.

Melankholina, E. N, Gabbroides and parallel dykes in Shikotan island structure (The Lesser Kuril range), Geotectonics, 128-136, 1978.

Melosh, H. J., Dynamic support of the outer rise, Geophys. Res. Lett., 5, 321-324, 1978.

Melosh, H. J. and A. Raefsky, The dynamical origin of subduction zone topography, Geophys. J. R. Astron. Soc., 60, 333-354, 1980.

Miyashiro, A., Metamorphism and related magmatism in plate tectonics, Am. J. Sci., 272, 629-656, 1972.

Molnar, P. and T. Atwater, Interarc spreading and Cordillerian tectonics as alternates related to the age of subducted oceanic lithosphere, Earth Planet. Sci. Lett., 41, 330-340, 1978.

Moerly, R., Origin of lithosphere behind island arcs with reference to the Western Pacific, Geol. Soc. Amer. Mem., 132, 35-55, 1972.

Naimark, A. A., Recent tectonics of Momsk region (north-eastern USSR), papers ANSSSR, 229, 1, 1976.

Nelson, T. H. and P. G. Temple, Main stream mantle convection: a geological analysis of plate motion, Bull. Amer. Assoc. Petrol. Geol., 56, 226-246, 1972.

New catalogue of strong earthquakes within the USSR region. M., Nauka, 534, 1977.

Oskorbin, L. S., A. A. Poplavskaya and V. I. Zanukov, Nogliksokeye earthquake on October 2, 1964. Sakh-KNII, Yuzhno-Sakhalinsk, p. 86, 1967.

Oskorbin, L. S., Seismicity of the Sakhalin island, in Seismic regions of the Kuril Islands, Primorja and Priamurja ed. by S. L. Solovjev, Vladivostok, 1977.

Ostapenko, V. F., Some aspects of recent history of Pre-Kuril region of the Sea of Okhotsk related to the research of sea volcanoes of the region, in Papers of SakhKNII, 48, 1976.

Packham, G. H. and D. A. Falvey, An hypothesis for the formation of marginal zone in the western Pacific, Tectonophysics, 11, 79-109, 1971.

Parfenov, L. M., I. P. Voinova, B. A. Nataljin and D. F. Semenov, Geodynamics of the northeastern Asia in Mesozoic and Cenozoid time and nature of the volcanic belt, J. Geophys. Earth, Suppl., 26, 503-525, 1978.

Piskunov, B. I., A. I. Abdurakhmanov and Kim Chun Jn., Correlation depth and lcoation of magmatic chambers of the Kuril volcanoes, Papers ANSSSR, 244, 4, 1979.

Popov, A. A. and G. I. Anosov, New data on the Earth's crust structure in the Kuril Basin, Papers ANSSSR, 240, 166-168, 1978.

Puscharovsky, Yu. M., Introduction in the geotectonics of the Pacific region, M., Nauka, 1978.

Raznitsyn, Yu. N., Serpentinite melange and olistostrome of the south-eastern region of the East Sakhalin mountains, Geotectonics, 2, 96-108, 1978.

Rodnikov, A. G., Island arcs of the Western Pacific, M., Nauka, p. 149, 1979.

Rozhdestvensky, V. S., Strike-slips within the Tym-Poronaisky fracture zone in Sakhalin, Papers ANSSSR, 230, 678-680, 1976.

Rozhdestvensky, V. S., Faults within the East Ridge of Schmidt peninsula in Sakhalin, Geol. and Geophysics, 10, 1972.

Rozhdestvensky, V. S., Faults of the East Sakhalin mountains, Papers ANSSSR, 187, 1, 1969.

Rozhdestvensky, V. S., Faults within the Northeastern Sakhalin, Geotectonics, 2, 85-97.

Savostin, L. A., B. V. Varanov, and L. P. Zonenshain, On possible origin of sea mountains of the Kuril Basin within the Sea of Okhotsk, Papers ANSSSR, 242, 676-679, 1978.

Savostin, L. A., A. F. Beresnev and G. B. Udintsev, New data on heat flow through the Sea of Okhotsk floor, Papers ANSSSR, 215, 846-849, 1974.

Savostin, L. A. and A. M. Karasik, Recent plate tectonics of the Arctic Basin and of northeastern Asia, Tectonophysics, 74, 111-145, 1981.

Scientific Party, Transects begun near the Japan trench, Report on DSDP Glomar Challenger Log 56, Geotimes, March 22-26, 1978a.

Scientific Party, Japan trench transects, Report on DSDP Glomar Challenger Log 57, Geotimes, April, 16-21, 1978a.

Seely, D., P. Vail and G. Walton, Trench slope model, in Continental margins, edited by C. Burckand, C. Drake, 249-261, 1974.

Semenov, D. F., Neogene magmatic formation of the South Sakhalin, Khabarovsk, 207 p. 1975.

Shreider, A. A. and S. Lukjanov, Strucutre of the magnetic field of the Sea of Okhotsk, in complex research of the World ocean, M. Institute of Oceanology, M., 1979.

Simbireva, I. G., S. A. Fedotov and V. A. Feofilatov, Geodynamics of te Kuril-Kamchatka arc according to seismic data, in Volcanism and geodynamics, Moscow, Nauka, 91-102, 1977.

Sleep, N. H. and M. N. Toksoz, Evolution of marginal basins, Nature, 33, 548-550, 1971.

Snegovskoi, S. S., Reflection research and tectonics of the southern Sea of Okhotsk region and adjacent Pacific margin. Novosibirsk, Nauka, p. 87, 1974.

Solovjev, S. L., L. M. Poplavskaya, M. P. Zaraiskii, West Iturup earthquake on May 7-8, 1962, Geology and geophysics, 7, 55-62, 1964.

Stauder, W. and L. Mualchin, Fault motion in the large earthquakes of the Kuril-Kamchatka arc and of the Kuril-Hoffaido Corner, Geophys. Res., 81, 297-308, 1976.

Streltsov, M. I., Dislocation of the southern Kuril island arc. M., Nauka, 131 p. 1976.

Structure of the Earth's crust within the transition zone from the Asian continent to the Pacific, M., Nauka, 1964.

Structure of the Earth's crust and upper mantle within the transition zone from the asian continent to the Pacific, Nauka, Novosibirsk, 367, p. 1976.

Suprunenko, O. I., T. A. Andieva and P. N. Safronov, Kronotsko-Krutogorovskaya zone of sublatitude Kamchatak faults, Papers ANSSSR, 209, 1398-1340, 1973.

Sykes, L. K., The seismicity of the Arctic, Bull. Seismol. Soc. Am., 55, 501-518, 1965.

Tarakanov, R. Z., Kim Chun Un. and R. I. Sukhomlinova, Regularities in spatial distribution of the hypocentres of Kuril-Kamchatka and Japan regions and their connection with the peculiar features of geophysical fields, in Geophysical researches in the transitional zones from the Asian continent to the Pacific, Nauka, M., 67-77, 1977.

Toksoz, M. N. and A. T. Hsui, Numerical studies of back-arc convection and the formation of marginal basins, Tectonophysics, 50, 177-196, 1978.

Tuezov, I. K., M. L. Krasnyi and S. S. Snegovskoi, Sedimentary deposits of the southearn Sea of Okhotsk region, Papers ANSSSR, 244, 1206-1210, 1979.

Udintsev, G. B., The Sea of Okhotsk floor topography, Papers of the Institute of Oceanology, ANSSSR, is. 22, 3-76, 1957.

Udintsev, G. B., Bottom topography of the Kuril-Kamchatks Basin, in Research of the Kuril-Kamchatka Basin, M., Izd-vo ANSSSR, Papers of the Institute of Oceanology, 12, 1955.

Udintsev, G. B., A. F. Beresven, A. A. Geodekjan, E. G. Mirlin, L. S. Savostin, A. A. Shreider, B. V. Baranov, and A. V. Belyaev, Preliminary results of geologo-geophysical research in the Sea of Okhotsk and the Northwestern Pacific, R/V "Vityaz", in Geologo-geophysical research in the transition zone from the Asian continent to the Pacific, M., Sovetskoyo radio, 19-29, 1976.

Uyeda, S., Some basic problems in the trench-arc-back-arc systems, in Island arcs, Deep sea trenches and Back-arc basins, Maurice Ewing Ser., I, edited by M. Talwani and W. C. Pitman III, 1-14, Amer. Geophys. Union, Washington, D.C., 1977.

Uyeda S. and A. Miyarshiro, Plate tectonics and Japanese islands: a synthesis, Geol. Soc. Am. Bull., 88, 1159-1170, 1974.

Uyeda, S. and H. Kanamori, Back-arc opening and the mode of subduction, Geoph. Res., 84, 1049-1061, 1979.

Vasiljev, B. I., G. Zhiltsov, and A. A. Suvorov, Geological structure of the southwestern arc-trench Kuril system, Nauka, M., 105 p. 1979.

Verjpinskaya, A. and L. A. Savostin, Instantanwous Kinematic of the plates of the Sea of Okhotsk Region, Oceanology, 1980.

Vorobjeva, E. A., Focal mechanism of the great aftershock of the earthquake 5(6) September 1976, in Seismic regions of Sakhalin Island, Vladivostok, 1977.

Watts, A. B. and M. Talwani, Gravity anomalies seaward of deep-sea trenches and their tectonic implications, Geophys. J. Roy. Astron. Soc., 26, 57-90, 1974.

Watts, A. B. and J. K. Weissel, Tectonic history of the Shikoku marginal basin, Earth. Planet. Sci. Lett., 25, 239-250, 1975.

Watts, A. B., J. K, Weissel and R. L. Larson, Sea floor spreading in marginal basin, of the western Pacific, Tectonophysics, 37, 167-181, 1977.

Wilson, J. T. and K. Burke, Two types of mountain building, Nature, 239, 448-449, 1972.

Zobin, V. M., Focal mechanism of strong Kamchatka earthquakes for 1969-1973 and their aftershocks, Volcanism and seismicity, 5, 74-88, 1979.

Zobin, V. M., I, G. Simbiereva, Focal mechanism of earthquakes in Kamchatka-Commandor region and heterogeneity of the active seismic zone, Pure and Appl. Geolphys., 15, 1977.

Zonenshain, L. P., L. M. Natapov, L. A. Savostin and A. P. Stavsky, Recent plate tectonics of North-Eastern Asia in relation to opening of the north Atlantic and Arctic basin, Oceanology, 5, 846-853, 1978.

Zonenshain, L. P. and L. A. Savostin, Introduction to Geodynamics, Nedra, M., p. 311, 1979.

Zonenshain, L. P., B. V. Baranov, L. A. Savostin V. A. Legler, and L. P. Merklin, Deep sea trenches as compressional features, Izv. ANSSSR, geol. ser. 6, 96-108, 1980.

Zonenshain, L. P. and L. A. Savastin, Geodynamics of the Baikal Rift Zone and plate tectonics of Asia, Tectonophysics, 76, 1-45, 1981.

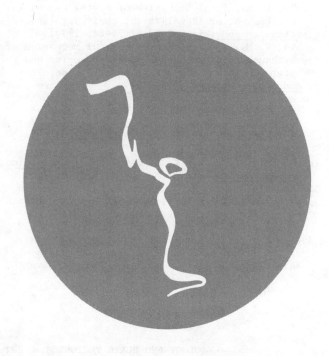

SEISMOFOCAL ZONES AND GEODYNAMICS OF THE KURIL-JAPAN REGION

Roman Z. Tarakanov and Chung Un Kim

Sakhalin Complex Scientific Research Institute, Far East Science Center,
Academy of Science of the USSR, 694050, Novoalexsandrovsk, Sakhalin, USSR

Abstract. Characteristics of the spatial distribution of earthquake hypocenters in the Kuril-Kamchatka and Japan deep seismic zones for 1904-1975 are discussed. Several island arcs forming an intricate tectonic node are observed here. A mosaic distribution of earthquake hypocenters within the focal or Wadati-Benioff zones is observed. It is shown that individual regions of the focal zones display anomalous seismic body wave propagation velocities and focal mechanisms. Within the region, four asthenospheric layers of low velocity are distinguished at depths of 60-80, 110-150, 220-290 and 400-460 km. Areas where these layers are crossed by the seismic focal zones are the most probable regions for the formation of magma. Various aspects of the geodynamics of the region are discussed from this standpoint.

Seismofocal Zones

A summary map of epicenters and a number of vertical cross-sections along and across the Kuril-Japan structures have been compiled to study the spatial distribution of earthquake hypocenters in the Kuril-Kamchatka and Japan Wadati-Benioff zones. Observational data ($M \geq 4$) for 1961-1975 were used for compiling the map and cross-sections for the 0-200 km depth interval. Data for the period of instrumental observations since 1904 were used for foci deeper than 200 km and for the largest earthquakes ($M \geq 7\frac{1}{2}$). The catalogued data includes about 10,000 earthquakes with $M \geq 4$. The sources for these data are listed separately at the end of the references.

The general pattern of the spatial distribution of earthquakes in the Kuril-Kamchatka and Japan zones is shown in Fig. 1. One can see that the level of seismic activity is extremely high and has a complicated distribution in the areas under consideration. There are several island arcs within the region which are interwoven into a complex tectonic node. Miyamura (1969) showed that seismic activity and depth to the foot of a focal zone depend on the age of an island arc. The Japan and Sakhalin-Hokkaido Island Arcs within which the maximum focal depth does not exceed 400 km are classified by him as relatively old ones; while the Kuril-Kamchatka and Izu Bonin arcs are considered by him to be relatively young and tectonically most active island arcs with maximum focal depths reaching 600-650 km.

The areas located between the chain of active volcanoes and the trench axis, where focal zones extend to the sea floor, and the seismic activity reaches the maximum level, are of the greatest interest within the region. As observed in Fig. 1, the epicentral zones of the regions under consideration can be divided into many parts with sharply differing levels of seismic activity.

The main features of the epicenter distribution (Fig. 1) are as follows. 1. The level of seismic activity changes abruptly within the Kuril-Kamchatka and the Japan regions. The zone of earthquake epicenters located between the volcanic chain and the trench is separated from the zone of deep-focus earthquakes by an almost aseismic space. 2. Over the large distance from Northern Kamchatka to Izu-Bonin Islands, there is a parallelism of the main structural elements: the epicentral zones corresponding to different focal depths, chain of active volcanoes, and trench. It is noteworthy that similar structural elements of different ages interpenetrate one another, creating the appearance of a large single zone. 3. In the area where the focal zone crops out, the number of earthquakes having foci in the crust and upper mantle increases remarkably with the age of the island arc. 4. The seismic activity tends to increase for areas with deeper Moho discontinuities. 5. Areas located beneath volcanic zones are characterized by relatively weak seismic activity. 6. In areas adjacent to the trench, on the continental side, almost aseismic portions of "triangular" shape are observed. The Hokkaido "triangle" is the most significant among them.

Fig. 2 shows a generalized vertical section running along the Kuril-Kamchatka and Izu-Bonin Island Arcs. A distinct feature of the section is a very complex shape of the lower edge (foot) of the focal zones. The depth to the foot changes sharply along the strike of the arcs.

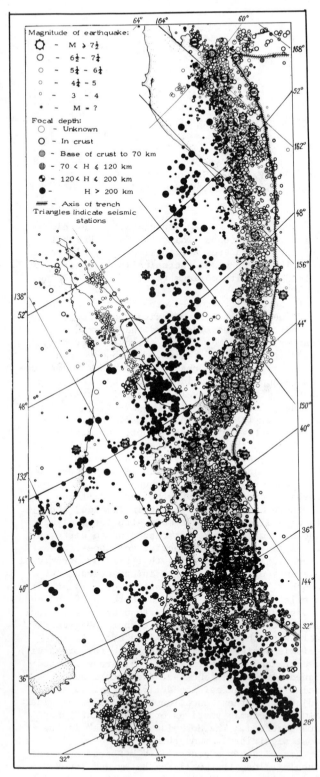

Fig. 1. Earthquake epicenters in the Kuril-Kamchatka and Japan regions and adjacent territories. $H \leq 200$ km for the 1961-1975 period, $H \geq 200$ km for the 1904-1975 period.

For the Kuril-Kamchatka region, the maximum earthquake concentration is observed in the upper part of the focal zone to a depth of about 100 km, and the maximum activity of deep-focus earthquakes is observed at its foot. The minimum value of the energy released during earthquakes is observed in the middle part of the focal zones. A relatively smaller number of earthquakes with foci at depths shallower than 200 km is observed in the Izu-Bonin zone.

The following features of epicenter distribution are remarkable: 1. Depths to the base of the relatively young Kuril-Kamchatka and Izu-Bonin focal zones are minimal (350 and 400 km) in the areas where they cross or intersect the relatively old Japan and Sakhalin-Hokkaido arcs, and increase in depth with distance in both directions from the crossing areas. 2. Shallow and intermediate depth earthquakes disappear as one moves from this area towards the Japan sea, and the hypocenters are localized within relatively narrow belts which plunge gently down to a depth of 600-650 km. 3. On vertical projection, these localized zones (at depths of 300-650 km) form knife-like protrusions which may join near the Khankaisky Massif in Primorie. 4. On the projection along the Kuril-Kamchatka zone, there are characteristic V-shaped clusters of hypocenters from the surface down to 100-200 km. These features, and the aseismic zones shown on the map (Fig. 1) appear to be a reflection of long strike and deep transverse and subtransverse faults as delineated by geological and geophysical data (Ermakov, 1979; Gnibidenko et al., 1980).

Vertical hypocenter sections running across the structures have also been plotted for the Kuril (Fig. 3-a) and Japan (Fig. 3-b) focal zones. The thickness of the focal zones varies in different sections from 60 to 90 km, being 75 km in most cases.

A decrease in dip angle of the focal zones as one moves towards the area of their crossing with the older structures is a characteristic feature of the relatively young Kuril-Kamchatka and Izu-Bonin Island Arcs. The decrease in the dip is accompanied by an increase in the relative number of earthquakes with focal depths shallower than 200 km as compared to the number of deeper earthquakes. The dip angle decreases from 50° to 38° for the Kuril-Kamchatka zone and from 60° to 36° for the Izu-Bonin zone. It is remarkable that in the places where relatively old arcs are crossed with young ones the dip angle has the minimum value for the older arc also.

A double focal plane dipping both under the Asian continent and under the Pacific Ocean has been distinguished for the Japan zone. The dip angles both under the continent and under the ocean are the same within the accuracy limits (Tarakanov et al., 1977). Due to less accuracy in location of the offshore events it is possible that the apparent eastward dip is unreal.

Characteristics of the spatial distribution

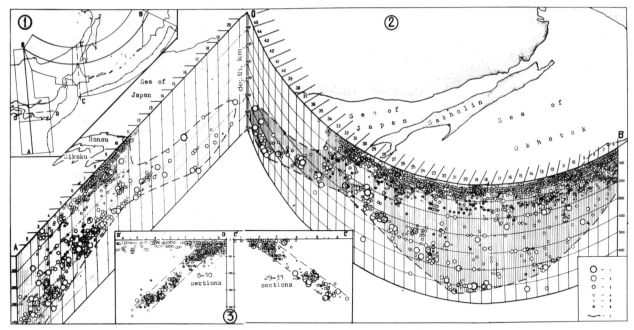

Fig. 2. Generalized vertical sections of the Kuril-Kamchatka and Izu-Bonin zones.
1. Axes of vertical projections: (AO) along the Izu-Bonin zone; (OB) along the Kuril-Kamchatka zone; (C-C') across the Kuril-Kamchatka zone; (D-D') across the Izu-Bonin zone.
2. Projections of earthquake foci on a vertical plane, oriented along axes OA and OB.
3. Projections of earthquake foci on vertical planes, oriented across the structures (along profiles C-C' and D-D'). (1-6) classification of earthquakes according to magnitude: (1) M ≥ 7½; (2) M = 6½-7¼; (3) M = 5¼-6¼; (4) M = 4¼-5; (5) M = 3-4; (6) magnitude unknown; (7) the trench axis.

of hypocenters in the Japan region are dealt with in a number of papers (Wadati et al., 1969; Katsumata and Sykes, 1969; Katsumata, 1970; Utsu, 1968, 1974; Hasegawa et al., 1978). A double-plane structure of the focal zone, with characteristic focal mechanisms for each of the dipping layers spaced at 30-40 km, has been observed for northeastern Japan (Hasegawa et al., 1978). A folded structure of the focal zone has been established beneath the junction area of the northeastern Japan and the Kuril arcs (Moriya, 1978). This deformation of the focal zone may be of a plastic character (Sasatani, 1978). Okada (1978) showed that the roof of the seismic focal layer in the Kanto and Tohoku regions is characterized by a thick asymmetric transitional zone with low velocity.

Anomalies of Seismic Wave Velocities

Seismological and geological-geophysical studies have shown that the upper mantle of the continent-to-ocean transition zone has a complex layered-block structure (Oliver and Isacks, 1967; Utsu, 1967, 1971; Tarakanov and Leviy, 1968; Kebeasy, 1969; Kuzin, 1974; Tarakanov and Kim, 1975, 1979; Novie dannie..., 1978).

In order to account for the numerous observations of anomalies of P and S wave travel-times, and of amplitudes and periods, Utsu (1967, 1971), offered a model of the upper mantle structure beneath northeastern Japan. The main element of the model is an anomalous dipping layer having a thickness of about 100 km and high values of V and Q. Earthquake hypocenters are located in the upper part of this layer. The mantle in contact with this layer is characterized by relatively low values of V and Q. Low V and Q values are more enhanced in this mantle wedge on the the continental side. A similar model of anomalous upper mantle structure was offered by Oliver and Isacks (1967) for the Tonga region.

Following Utsu, Oliver and Isacks, other scientists made attempts to improve the concept of velocity and Q distribution in the upper mantle based on various sets of seismological and geophysical data (Kanamori, 1968; Ishida, 1970; Tada, 1972; Noguchi and Okada, 1976; Hamada, 1977; Yoshii, 1977). The authors obtained some additional data that in general do not contradict the model by Utsu and Oliver and Isacks, but suggest a more complex structure of individual upper mantle blocks.

A complex structure of the focal zone is revealed from propagation velocities of seismic waves. An averaged velocity model for the focal zone of the Kuril-Japan region was constructed

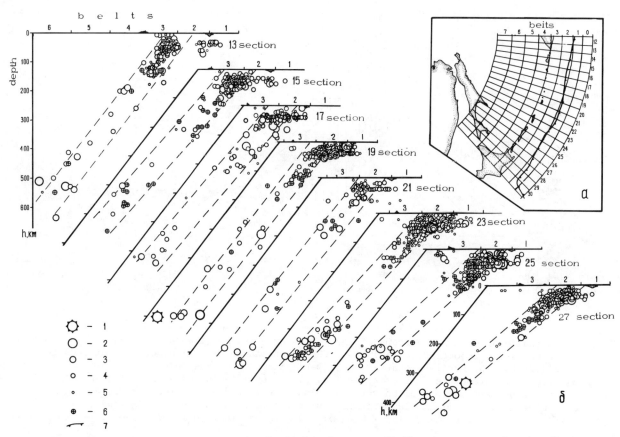

Fig. 3-a. Vertical sections across the Kuril focal zone. (1-7) -the same as in Fig. 2; insert (a) shows the scheme of local coordinates.

based on data from both land and ocean bottom stations (Tarakanov, 1978). A block under the island arc is recognized by anomalously low P-wave velocities both in its continental part ($\delta Vp = 0.3$-0.4 km/s) and in the focal zone proper ($\delta Vp = 0.1$-0.2 km/s) beneath the volcanic range. In the high seismicity part of the focal zone, P-wave velocities are anomalously high both in the zone proper ($\delta Vp = 0.1$-0.2 km/s) and in the block adjacent to it on the Pacific side ($\delta Vp = 0.3$-0.4 km/s). P-wave velocities on the average approach the standard ones at the depths of 450 km in the focal zone, while they are about 0.2 km/s higher in the underlying high velocity block.

Studies to reveal details of the velocity section in the focal zone have also been made. Kuzin (1974) observed an intricate velocity mosaic for a relatively small extent of the focal zone in South Kamchatka. The author showed that a simple concept treating the focal zone at all the depths as having either low or high velocities is not supported by data of detailed observations.

A similar pattern of mosaic P-wave velocity distribution was obtained by Boldirev et al. (1978) for the South Kamchatka region (Fig. 4). Using the method of three-dimensional velocity distribution (Anikonov et al., 1974), the authors analysed anomalies from 1000 earthquakes with focal depths of 0-200 km for 16 Kamchatka stations. Similar to the model by Kuzin, alternate areas of low and high velocity were observed along the strike of the focal zone. It is quite remarkable that an area of low velocities in the upper part of the section (at depths to 80 km) corresponds to high velocity areas at depths of about 150 km beneath the Shipunsky and Kronotsky Peninsulas. A significant anisotropy of the P-wave velocity is also observed. A general tendency of P-wave velocities to increase towards the ocean is observed across the strike of the focal zone.

The P-wave velocity field for the upper mantle has been studied in detail beneath different tectonic zones of Kamchatka (Fedotov and Slavina, 1968; Fedotov et al., 1976). Fedotov et al. showed that the velocity field of longitudinal waves is of a striped mosaic character beneath Kamchatka. Velocities are on the average low and equal to 7.6 km/s for the whole area studied. Highest velocities of 7.8-7.9

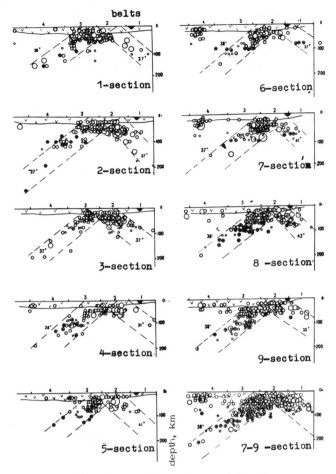

Fig. 3-b. Vertical sections of the Japan focal zone, oriented across the structures. The triangle indicates the trench axis.

km/s are characteristic of the upper mantle beneath the Pacific focal zone on the oceanic side, and beneath the northwestern half of Kamchatka. The eastern volcanic belt is characterized mainly by low velocities of 7.4-7.6 km/s. Anomalous velocities of 7.0-7.2 km/s were also found beneath individual portions of the belt (Slavina and Fedotov, 1974).

Detailed velocity sections have been plotted for the junction zone of the Kuril-Kamchatka and Japan Island Arcs (Tarakanov and Kim, 1979). Significant horizontal velocity heterogeneities are observed along the Hachinohe station (Honshu I.) - Shikotan I. profile extending for about 600 km (Fig. 5). The velocity section along this profile is an intricate mosaic of blocks with relative low and high P-wave velocities. At distances of 100-200 km from the Hachinohe Station, P-wave velocities are anomalously low (7.3 to 7.5 km/s) and vary with depth from the base of the crust down to a depth of 70 km. The adjacent focal zone under the south-western extension of the Kamuikotan-Hidaka belt contrasts greatly to the above area in the distribution character of P-wave velocities. Here velocities are anomalously high within the depth interval of 50-120 km (8.0-8.6 km/s). An area of high P-wave velocity is also distinguished beneath the southeastern extention of the Nemuro Peninsula-Lesser Kuril Isles anticlinal zone (8.2-8.4 km/s).

A velocity section for P-waves has also been plotted along the profile running from the Hachinohe Station (Honsu I.) to the trench (Tarakanov and Kim, 1979). Vp-isolines are oriented mainly in the direction of the focal zone dip, with velocities increasing from 7.4 km/s in the initial part of the profile to 8.2 km/s beneath the area of maximum seismic activity. Velocities decrease to 8.0 km/s near the trench. A complex pattern of P-wave velocity distribution in the focal zone exists in relation to the characteristics of seismic activity. Hypocenters of the largest earthquakes with M=7.8 - 8.4 are observed for regions with relatively high velocity and are attributable to the boundaries of blocks with different velocities or to areas with high velocity gradients. Large earthquakes in the vicinity of the Ishikari synclinorium occur in relatively low velocity (7.5-7.6 km/s) material (Fig. 5).

Focal Mechanisms

Characteristics of the focal mechanisms in the Kuril-Kamchatka and Japan zones have been studied by a number of authors (Averianova, 1968, 1975; Ichikawa, 1971; Abe, 1973; Balakina, 1974; Simbireva et al., 1976; Sasatani, 1978). Although differing in detail, these authors' interpretations reveal a general similarity: near-horizontal compressive stresses oriented across the structures prevail in the regions under discussion.

Studying the dynamic parameters of a number of foci with M>5, Averianova (1968, 1975) obtained an intricate mosaic distribution pattern of the stress field.

Studies on focal mechanism features by Simbireva et al. (1976) are of great interest. The authors performed a reconstruction of the stress field based on the analysis of a number of shear displacements in the foci of earthquakes with magnitudes of 3.5<M<7.0. The regional stress field revealed for large earthquakes (M>7.0) is shown to be complicated by a local field formed by different heterogeneities.

Fig. 6 shows the main features of the local stress field by means of projecting the paths of compression and tension axes along the focal zone of the Kuril-Kamchatka arc. The maximum variability of the local stress field is observed near the Kamchatka peninsulas and bays, as well as in the northern and southern parts of the island arc, i.e. in the terminal parts of the focal zone at depths to 40 km. Much greater stability of

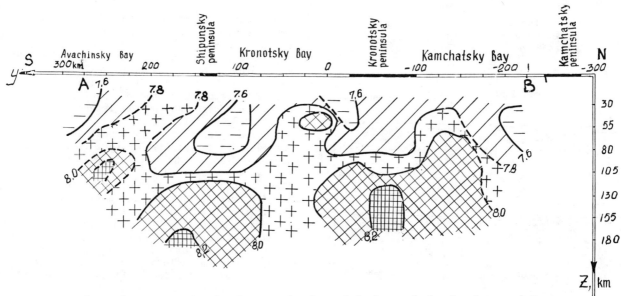

Fig. 4. Velocity section for P waves in the axial plane of the focal zone (after Boldirev et al., 1978)

the local stress field is observed in the depth range of 100-120 km.

Focal zone sections characterized by predominantly horizontal tensions may be of some interest for geodynamics. Among these, a large area of a complex structure beneath the Kuril Isles is especially remarkable (Fig. 6). It is distinctly defined by a strip of maximum gradients in the local stress field. The upper part of this area is at the depth of 100-200 km where an asthenospheric layer of low velocity may occur, and the temperature is close to the melting point of the rocks. This may in general account for some anomalies in the vicinity of Middle Kuril Isles: i.e., the decrease in depth to the roof of the high conductive layer delineated from magnetotelluric sounding data (Lyapishev, 1980), mimimum seismic activity, the existence of practically aseismic lateral (Bussol and Krusenstern) faults, minimum crustal thickness, and active volcanicity (Goryachev, 1966; Tarakanov and Kim, 1980).

It is quite remarkable that the above area of predominant tension (at the depths of about 300 km) is located approximately under the abyssal sea of the Okhotsk Basin. This area is also known to be characterized by anomalously high heat flow, low seismic activity, minimum crustal thickness in the Sea of Okhotsk region, and upper mantle density heterogeneities. Similar to the Middle Kuril Isles region, the above features may be caused by a high degree of partial melting of the upper mantle material.

Asthenospheric Layers

Recently, due to the progress of geodynamic studies, many new data on anomalous properties of the upper mantle have appeard. Scientists began to distinguish various types of asthenospheric layers, which were originally defined as layers with low and variable viscosity, (from 10^{17} to 10^{20} P) (Artyushkov, 1979): the high electric conductivity layers (Marderfeld, 1977; Tuezov, 1975; Artyushkov, 1979; Sychev, 1979), low-V and low-Q layers (Gutenberg, 1959; Fedotov, 1963; Tarakanov and Leviy, 1967; Magnitsky, 1968; Utsu, 1971; Farberov, 1974; Artyushkov, 1979), and anomalously low velocities beneath active volcanic regions (Fedotov and Kuzin, 1963; Fedotov et al., 1976; Novie dannie, 1978). All the above features are distributed within a large depth interval, from the base of the crust to 400 or more kilometers. It is quite remarkable that the differences in depths at which different anomalous properties are manifest decreases gradually toward the tectonically active island arc sections, where these anomalous properties are more distinct. Unfortunately, many authors tend to attribute all the above anomalous properties to the same asthenospheric layers. A layer characterized by low rigidity and low viscosity is usually called asthenosphere. The asthenosphere so understood exists beneath all the Earth regions, including beneath platforms and crystalline shields. Such an asthenospheric layer occurs due to decreased diffusion viscosity of rocks manifest during slow and sufficiently large deformations (Artyushkov, 1979).

In general, the asthnospheric layer should not be identified with low velocity and high electric conductivity layers. In each case, one should specify which of the anomalous properties are expressed in the asthenospheric layer. It is noteworthy that the asthenospheric layer has the properties of high electric conductivity and low

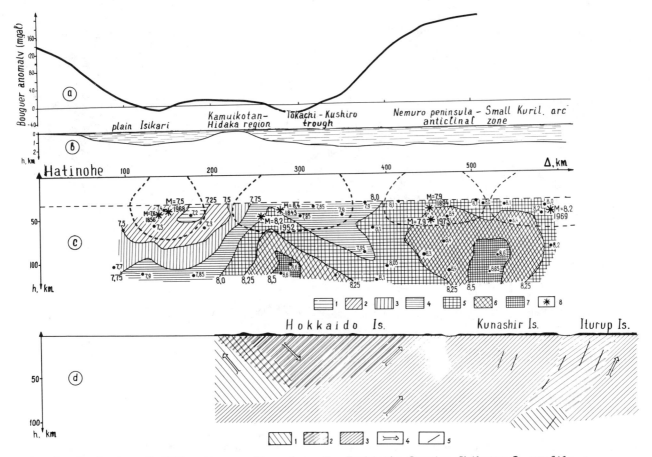

Fig. 5. Variation of different anomalies along the Hachinohe Station-Shikotan I. profile. (a) Bouguer anomalies (Stroyev, 1977); (b) seafloor topography along the profile; (c) V_p velocity fields in the focal zone; (1-7) V_p velocity variation intervals: (1) $V_p < 7.25$ km/s, (2) $V_p=7.25$-7.5 km/s, (3) $V_p=7.5$-7.75 km/s, (4) $V_p=7.75$-8.0 km/s, (5) $V_p=8.0$-8.25 km/s, (6) $V_p=8.25$-8.5 km/s, (7) $V_p > 8.5$ km/s, (8) hypocenters of large earthquakes; (d) vertical projection of predominant seismic dislocations (1-3) of (Averianova, 1968): (1) reversed strike-slip faults type I, (2) strike-slip faults type I, (3) reversed strike-slip faults type II, (4) trends of the movements of the upper "continental" side of the fault, parallel to the island arc, (5) lines of fault sectional planes.

velocity only under critical thermodynamic conditions, i.e. when the temperature at a given depth is at the melting point of the material.

Some authors think that the asthenospheric layer is always characterized by low velocity. Referring to numerous literatures reporting that the low velocity layer is distributed locally (Ryaboi, 1979; Artyushkov, 1979), it may be concluded that asthenospheric layers are also distributed locally (Sychev, 1979).

Seismic wave velocity and viscosity of a material are known to depend mainly on pressure and temperature. Their values increase with pressure increase and decrease with temperature increase. It is assumed that, in a low velocity asthenospheric layer, temperature increase with depth has a greater effect on velocity and viscosity than pressure increase. High electric conductivity of the material can also be accounted for by the temperature effect.

Artyushkov (1979) explains this effect by the abrupt prevalence of ionic conduction over electronic conduction.

It can probably be assumed that the location of a layer having anomalous properties in the upper mantle is determined mainly by the proximity of temperatures to the melting temperature of the mantle material at the depth.

The existence of anomalous properties of asthenospheric layers is also supported by other data (Ritman, 1964; Isaev, 1969; Tarkov, 1970).

It should be noted that the determination of such an important parameter as the depth to a layer having anomalous properties by means of the existing methods is difficult and always implies some uncertainty. For the focal zones of the Pacific seismic belt where anomalous properties may be expressed distinctly, their position can be determined from the magnitude distribution with depth of the largest earth-

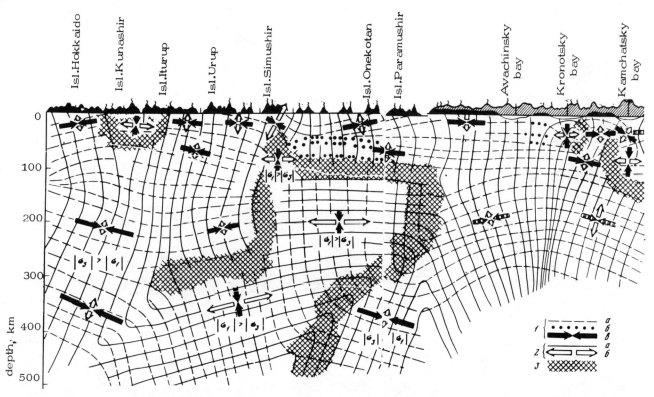

Fig. 6. Projection of trajectories of compression and tension axes in the vertical plane along the Kuril-Kamchatka focal zone (after Simbireva et al., 1976). (1) trajectories of compression axes: (1-a) oriented along and at a small angle to the trend of the focal plane, (1-b) oriented across the focal plane, (1-c) spatial position of the main normal compression axes (δ_3). The length of an arrow corresponds to the quantitative interrelation of tensile and compressive forces. (2-a) trajectories of tension axes oriented along and at a small angle to the trend of the focal plane; (2-b) spatial position of the main normal tension axes (δ_1). (3) zones of maximum gradients of local field of stresses.

quakes for a long period of time. Distribution of $M_{max}(h)$ (where M_{max} is the maximum magnitude of earthquakes at a given depth h) yields reliable qualitative information on the variation of rigidity of focal zones with depth (Fig. 7). Data on the largest earthquakes (since 1896) in the Pacific seismic belt have been used to plot $M_{max}(h)$. A sharply expressed non-uniform trend with depth is the main feature of the $M_{max}(h)$ curve. Distinct minima of the curve are observed at depths of 60-80, 110-150, 220-290 and 400-460 km on a background of general decrease in maximum magnitudes. The first three of the minima practically coincide in depth with low velocity layers revealed by means of velocity sections (Tarakanov and Leviy, 1967). This suggests that the depths corresponding to minima of the $M_{max}(h)$ curve are where asthenospheric layers of low rigidity and velocity intersect the focal zone. A polyasthenospheric model of the upper mantle structure in the Kuril-Japan region has been proposed based on the above distribution of $M_{max}(h)$, and the variation with distance of seismic wave amplitudes and velocities, and the derivation of the empirical travel-time curve (Tarakanov and Leviy, 1968).

It should be noted that such events as earthquakes with magnitude $M \geq 7$ could not be overlooked even at the early period of instrumental observations. It means that distinct extremes of the $M_{max}(h)$ curve reflect the real features of rigidity variation in the focal zones. The sharp boundary of different densities of hypocenters shown at a depth of 60 km in Fig. 7 can be assumed to be the lower boundary of the lithosphere. The above data are testimony to the existence of four low velocity asthenospheric layers in the focal zone. This fact is important for geodynamics.

A Model of Upper Mantle Processes

The major structural elements in the Kuril-Kamchatka and Japan zones (the trench, chains of volcanic edifices, uplifts, and trough axes) and their associated anomaly isolines of geophysical fields are known to have trends that agree well with isodepth trends of the seismic focal zone. It would be quite logical to

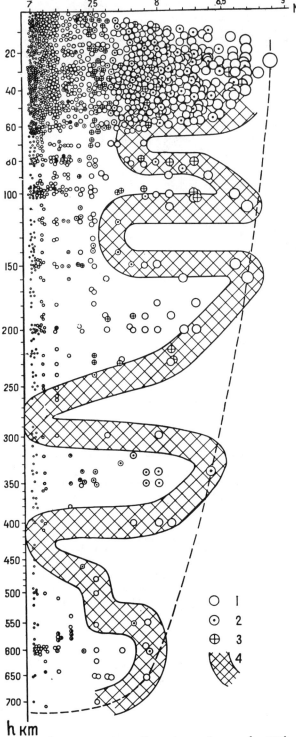

Fig. 7. Variation of maximum observed earthquake magnitudes with depth. (1-3) large earthquakes: (1) in the Circum-Pacific belt other than the Kuril-Japan zone, (2) in the Kuril-Japan zone, (3) in other seismically active regions of the Earth, (4) probable band of values Mmax.

suppose that all the above features are caused by one common cause. Such a cause can be deep differentiation of melts of the mantle material, first portions of which are formed at certain depths within the focal zone.

Vertical "columns" characterized by anomalous absorption of seismic waves, magmatic chambers (at different levels) that screen transverse waves and aseismic sections ("holes") in the focal zone are distinct beneath volcanic regions. High heat flow is also observed there (Wadati et al., 1969; Katsumata, 1970; Utsu, 1974; Faberov, 1974; Tuezov, 1975; Navie dannie, 1978; Sychev, 1979).

These features do not contradict the above assumption of common cause, but authors are usually reluctant to locate the source of primary magma formation within the focal zone because "on the average" it is a compressional zone (Ichikawa, 1971; Balakina, 1974; Avarianova, 1968, 1975; Simbireva et al., 1976; Sasatani, 1978).

As has been shown, the focal zone is very nonuniform in its spatial distribution of hypocenters and velocity structure. It consists of a mosaic of high and low velocity blocks; and tension can prevail in low velocity and low seismic activity (Averianova, 1968; Simbireva et al., 1976). As noted above, low velocity asthenospheric layers are treated by many authors as zones where the mantle material is near melting or partially melted (Faberov, 1974).

Based on the above information, a hypothesis is offered according to which the locations where asthenospheric layers cross the seismic focal zone should be most favorable for the formation of primary melt fluids (Tarakanov and Leviy, 1968). It is known that these intersections occupy a certain position in space, and should manifest themselves one way or another on the paths of differentiates moving towards the Earth's surface. The probable location of formation of primary fluids within the focal zone is illustrated in Fig. 8. A vertical section of the crust (the scale is exaggerated 2.5 times) and upper mantle along the Iturup I. - Terpenia Bay profile (Suvorov, 1975) is shown for comparison of deep and shallow structures. The alternating layers of low and high velocity, and rigidity are depicted according to the polyasthenospheric model (Tarakanov and Leviy, 1968). Why is it that just these regions should be constant energy generators that feed tectonic processes in island arcs? This seems explainable with two assumptions: 1. Weak rigidity (low viscosity) layers are a constant feature of the Earth as a planet, and they serve to weaken the interaction between different layers and to assure isostatic equilibrium, i.e. they are "buffer" protective layers. The existence of these layers is quite probable from a thermodynamic standpoint. They are formed under certain conditions, i.e. the effect of temperature increase with depth prevailing over the effect of pressure increase,

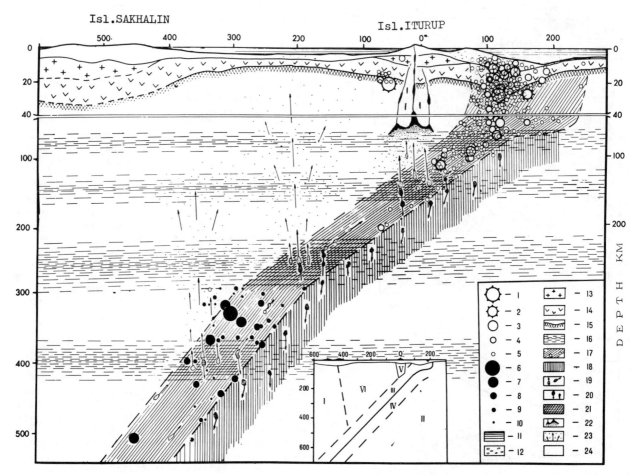

Fig. 8. A structural model and scheme of possible processes in the crust and upper mantle. Earthquake classification according to energy class K, where energy $E=10^k$ joules (1-5), and magnitude (6-10): (1) k=16, (2) K=15, (3) K=14, (4) K=12-13, (5) K=10-11, (6) $M \geq 7\frac{1}{2}$, (7) $M=6\frac{1}{2}-7\frac{1}{4}$, (8) $M=5\frac{1}{4}-6\frac{1}{4}$, (9) $M=4\frac{1}{4}-5$, (10) $M \leq 4$. Crustal structure (11-15) (Suvorov, 1975): (11) water, (12) sedimentary and volcanogenic-sedimentary rocks layer, (13) granitic layer, (14) basaltic layer, (15) Moho boundary. Upper mantle structure and processes (16-24): (16) low velocity asthenospheric layers, (17) focal zone of earthquakes, (18) dipping layer with anomalously high seismic wave velocities, (19) rise of fluids and fusible components, (20) sinking of heavy differentiates, (21) probable primary areas of magma formation, (22) assumed area of magma accumulation, (23) flow of heat and fusible components from possible areas of magma differentiation, (24) high velocity layers. Upper mantle blocks (inserts): (I) continental, (II) oceanic, (III) focal zone of earthquakes, (IV) dipping block with anomalously high seismic wave velocities, (V) discompacted block beneath active volcanicity regions, (VI) high heat flow area.

and the temperature of the layer approximates the melting point of the material. 2. Focal zones are areas of release of stresses that result from interaction of dissimilar crust and mantle blocks of continental and oceanic structures. The maximum tangential stresses theoretically should trend in an angle of 45° when nearly horizontal forces of interaction between these blocks are predominant. This dip angle is most probable for most of the focal zones.

Let us assume that maximum tangential stresses are directed on a medium divided into layers with different rigidity (fluidity) as a result of interaction between continental and oceanic blocks. In this case, energy will accumulate rather rapidly and be released as earthquakes in the more rigid layers. In the more plastic asthenospheric layers (crossing the focal zone), stresses will relax, and the energy will in part be transformed into heat and enhance the plastic properties of the asthenosphere, bringing its material up to the partial melting state. Since the process of interaction between continental and oceanic blocks exists for a long time, it is a constant source of energy for both seismic and magmatic (volcanic) processes.

Judging from the fact that earthquakes are observed in the asthenospheric layers, but seismic waves propagate through them with great attenuation, the material within them might be in a partially melted state. Fusible components that are formed after differentiation in the focal zone, probably as intratelluric solutions (Volokhov, 1979), move upwards to the nearest serious obstacle, for instance the base of the lithosphere, carrying a large amount of heat energy with them. Accumulating in significant amounts and possessing large amounts of heat energy, these fluids form magmatic chambers. On reaching critical pressures, magma can break through upwards to a level where its density is equalized with that of the environment. This process can be repeated many times until magma flows out onto the surface or intrudes into the surrounding rocks (Artyushkov, 1979).

It is assumed that processes of material differentiation that play an important part in the formation and evolution of structures in the formation and evolution of structures in the transitional zone occur in the foci of primary magma formation, as well as on the paths of melts (fusible heat-carrying fluids), especially in magmatic chambers.

Different authors allege quite different probable depths for the foci of magma generation, which vary from 20 to 250 km or deeper (Gorshkov, 1956; Kuno, 1977; Markhinin and Stratula, 1971; Farberov, 1974). Analysis of the available data shows that the above differences in depths of magma formation can be explained by the fact that different data reflect different stages of the magmatic process:

1. Initial partial melting of the upper mantle material in the areas where asthenospheric layers are crossed by the seismic focal zone. They are located at depths of 120-160 km beneath the Kuril-Kamchatka and Japan Island arcs (Kuno, 1966; Tarakanov and Leviy, 1968; Markhinin and Stratula, 1971).

2. Rising heated and gas-enriched fusible components from the primary areas of melt generation. Some discompacted areas in the form of an irregular cylinder or cone appear to be located at depths of 60-120 km and are revealed from seismic data as areas of anomalously high absorption of transverse waves and low body wave velocities (Farberov, 1974).

3. Formation of lens-shaped reservoirs (pockets) of magma prior to its rapid rise and surface extrusion or intrusion into the crust. The depth to these magmatic chambers is 20-60 km (Gorshkov, 1956; Bagin et al., 1971; Fedorchenko and Rodionova, 1975).

The assumption that these are several areas of primary magma generation at different depths within the focal zone is not contradicted by data on the relation between the volcanic rock composition and the depth to the seismic focal zone (Kuno, 1966; Markhinin and Stratula, 1971; Piskunov et al., 1979).

The Possible Role of the Focal Zone
in the Formation and Evolution of
Structures in the Transitional Zone

The anomalous properties of the upper mantle and its structural characteristics both beneath volcanic regions and the whole marginal sea-island arc - trench system are testimony to the leading role of the focal zone in the formation and evolution of the transitional zone features. But scientists do not yet agree about the form and depth at which the "organizing and transforming" role of the focal zone occurs.

Some scientists suppose that focal zones are superdeep faults that serve as dipping channels along which fusible differentiates of basic and ultrabasic magmas (Artyushkov, 1979; Sychev, 1979) and mobile fluids and intratelluric solutions (Volokhov, 1979) are supplied into the upper mantle and crust. Such a view of the role of the focal zone presents serious difficulties in accounting for a number of observed features: low heat flow within a focal zone, weak attenuation of seismic waves, high average seismic activity, and clearly expressed zoning in the marginal sea-island arc - trench system.

The above difficulties can be overcome for the most part if the above proposed scheme of processes is assumed to exist in the upper mantle of the transitional zone. From this standpoint, the focal zone is a constant generator of energy, supporting the processes of partial melting and differentiation of the material in the area where the asthenospheric layers cross with the seismolocal zone.

Based on the above assumptions we can expect the following features above the magma formation areas; high heat flow, anomalously low seismic activity or so-called aseismic "holes" in the focal zone, areas of high attenuation of transverse waves on the paths of fusible differentiates and fluids, lens-shaped magma reservoirs formed at the lithosphere boundary, magma "pockets" at several levels above a magmatic reservoir, intrusive bodies, crustal roots, volcanic edifices (or recent volcanicity). These features beneath volcanic regions are observed with minor variations in different regions of the Pacific belt including the Kuril-Kamchatka and Japan regions.

The primary areas of magma formation shown schematically in Fig. 8 should differ greatly since they can be under different thermodynamic conditions and at different stages of evolution. For example, the mantle material appears to be mostly differentiated in the area where the focal zone crosses the first asthenosphere layer (60-90 km) beneath the Lesser Kuril Isles. This region preserves traces of former volcanic activity: structures peculiar to an island arc, thick crustal roots, and intrusive bodies clearly observed in the magnetic and gravity fields.

Particularly interesting are regions of abyssal depths in inland seas that overly the assumed foci of magma generation at depths of 220-300 km. Here the differentiation of the mantle material does not result in active volcanism. Probably magma does not flow out onto the surface because it is diffused within the two overlying asthenospheric layers (H_1 = 60-90 km and H_2 = 120-160 km). The characteristic features in this region are the anomalously high heat flow, high attenuation of seismic waves and the presence of numerous intrusive bodies. It is possible that this "hot and highly absorbing pillow" in the upper mantle contributes somehow to the preservation of the long term crustal isostatic equilibrium. The downward movement of relatively heavy differentiates should be accompanied by formation of compacted areas (including dipping ones) both in the focal zone proper and in the layer adjacent to it on the Pacific side (Fig. 8), as well as by down-warping of individual parts of the crust and upper mantle.

From this standpoint, the trench can probably be represented as a sinking area that resulted from the subsidence of the high velocity layer adjacent to the focal zone.

Conclusions

The following main conclusions can be reached.

1. Within the regions under discussion, several island arcs entwined into a complex tectonic node, forming a number of features that indicate that the basement of the arcs are of different ages, and that their focal zones are superimposed.

2. The data suggest that the upper mantle in the Kuril-Japan region (especially the seismic focal zone) is of an intricate mosaic structure. This mosaic structure is manifest in the spatial distribution of earthquake hypocenters, anomalous seismic wave velocities, focal mechanisms, and in features of different geophysical fields. It appears to be a typical continent-to-ocean active margin transition zone.

3. In general, the asthenospheric layers may not possess the properties of high electric conductivity and low velocity. These anomalous properties manifest themselves under certain critical thermodynamic conditions, i.e. when the temperature at a given depth approximates the melting point of the material.

4. The regular mutual location of seismic zone isolines, the main structural elements of the Kuril-Kamchatka and Japan Island Arcs (trench, island chains, inland sea, and axes of uplifts and troughs) and their associated geophysical fields appear to have resulted from the processes of primary differentiation of the upper mantle material that occur at different levels (Tarakanov and Leviy, 1968) within the focal zone.

5. The differences in depths of the foci of magma formation (from 20 to 250 km and deeper), postulated by different authors, can probably be explained by the fact that different data reflect individual stages of magma evolution (initial partial melting; rise of fusible components along vertical paths; and formation of reservoirs of fluid magma).

6. A distinct correlation is observed between the spatial distributions of areas of low velocity, seismic faulting that corresponds to the tensional stresses, and low seismic activity.

7. The localization of primary magma formations to the regions where asthenospheric layers cross the seismic focal zone and the localizations of material differentiation and vertical paths of the differentiates, are the key to explaining the properties of the individual upper mantle anomalies and the main structural and evolutionary features of island arc systems.

8. Assuming that the formation and evolution of regions, that are at present at different stages of tectonic evolution, proceeded according to the scheme offered in this paper, the spatial configuration of relict seismic focal zones (Paleozones) and their related structures can probably contribute to the delineation of metallogenic provinces.

Acknowledgements. The authors are grateful to the scientists of the Sakhalin Complex Scientific Research Institute, Far East Science Center, Academy of Sciences of the USSR, for their discussion of the paper and critical remarks, to Drs. I. G. Simbireva and S. A. Boldirev for their permission to use some of their figures, and to Mr. M. S. Fedoryshyn for translating this paper into English.

References

Abe, K., Tsunami and mechanism of great earthquakes, Phys. Earth Planet, Interiors, 7, 143-153, 1973.

Anikonov, Yu.E., N.B. Pivovarova, and L.B. Slavina, Three-dimensional velocity field in the Kamchatka focal zone, Matematicheskie problemi geofiziki, 5, 1, 92-117, Novosibirsk, 1974.

Artyushkov, E.V., Geodinamika, Nauka, Moscow, 1979.

Averianova, V.N., Detalnaya kharakteristika seismicheskikh ochagov Dalnego Vostoka, Nauka, Moscow, 1968.

Averianova, V.N., Glubinnaya seismotektonika ostrovnikh dug, Nauka, Moscow, 1975.

Bagin, V.I., S.Yu. Brodskaya, G.N. Petrova, and D.M. Perchorsky, Depth of volcano foci in the Kuril-Kamchatka Island Arc from data of thermomagnetic studies on volcanic rocks, Izv. AN SSSR, Fizika Zemli, 5, 57-68, 1971.

Balakina, L.M., Earthquake foci and stress field in the earth's crust and upper mantle of the marginal zone of the Pacific Ocean, Geofizika dna Tikhogo okeana, Nauka, Moscow, 1974.

Boldirev, S.A., N.B. Pivovarova, and L.B. Slavina, The three-dimensional velocity field and anisotropy of the Benioff-Zavaritsky zone near the coast of Kamchatka, Novie dannie o stroenii kori i verkhnei mantii Kurilo-Kamchatskogo i Yaponskogo regionov, 100-110, Vladivostok, 1978.

Farberov, A.I., Magmaticheskie ochagi vulkanov vostochnoi Kamchatki po seismologicheskim dannim, Nauka, Novosibirsk, 1974.

Fedorchenko, V.I., and R.I. Rodionova, Ksenoliti v lavakh Kurilskikh ostrovov, Nauka, Novosibirsk, 1975.

Fedotov, S.A., On absorption of transversal waves in the upper mantle and energetic classification of local earthquakes of intermediate depth, Izv. AN SSSR, ser. geofiz., 6, 8, 29-849, 1963.

Fedotov, S.A., and I.P. Kuzin, Velocity section of the upper mantle near South Kuril Islands, Izv. AN SSSR, ser. geofiz., 5, 670-685, 1963.

Fedotov, S.A., and L.B. Slavina, Estimation of longitudinal wave velocity in the upper mantle beneath northwestern Pacific and Kamchatka, Izv. AN SSSR, Fizika Zemli, 2, 70-83, 1968.

Fedotov, S.A., L.B. Slavina, L.S. Shumilina, and A.A. Gusev, Longitudinal wave velocities in the upper mantle beneath Kamchatka, Seismichnost i glubinnoe stroenie Sibiri i Dalnego Vostoka, 180-189, Vladivostok, 1976.

Gnibidenko G.S., T.G. Bikova, O.V. Veselov, V.M. Vorobiev, Ch.U. Kim, and R.Z. Tarakanov, Tectonika Kurilo-Kamchatskogo glubokovodnogo zheloba, Nauka, Moscow, 1980.

Goriachev, A.N., Osnovnie zakonomernosti tectonicheskogo razvitia Kurilo-Kamchatskoi skladchatoi oblasti, Nauka, Moscow, 1966.

Gorshkov, G.S., On the depth of magmatic foci of the Kluchevskoi Volcano, Dold. AN SSSR, 106, 4, 703-705, 1956.

Gutenberg, B., Wave velocities below the Mohorovicic discontinuity, Geophys. J. Roy Astron. Soc., 2, 4, 223-232, 1959.

Hamada, K., Travel-time anomalies of longitudinal seismic waves and the upper mantle structure in Japan, Geofizicheskie issledovania zoni perekhoda ot Aziatskogo kontinenta k Tikhomu okeanu, 40-55, Nauka, Moscow, 1977.

Hasegawa, A., N. Umino, and A. Takagi, Doubleplaned structure of deep seismic zone in northeastern Japan Arc, Novie dannie o stroenii kori i verkhnei mantii Kurilo-Kamchatskogo i Yaponskogo regionov, 68-75, Vladivostok, 1978.

Ichikawa, M., Reanalisis of mechanism of earthquakes which occured in and near Japan and statistical studies on the solutions obtained, 1926-1968, Geophys. Mag., 35, 207-274, 1971.

Isaev, E.N., On viscoces flow of asthnosphere beneath island arcs, Dokl. AN SSSR, 184, 2, 1969.

Ishida, M., Seismicity and travel-time anomaly in and around Japan, Bull. Earthq. Res., Inst. Tokyo Univ., 48, 1023-1051, 1970.

Kanamori, H., Travel-times to Japanese stations from Longshot and their geophysical implications, Bull. Earthq. Res. Inst., Tokyo Univ., 46, 841-856, 1968.

Katsumata, M., Seismicity and some related problems in and near the Japanese islands, Quart. J. Seismol., 35, 75-142, 1970.

Katsumata, M., and L.R. Sykes, Seismicity and tectonics of the western Pacific: Izu - Mariana - Caroline and Ryukyu - Taiwan regions, J. Geophys. Res., 74, 5923-5948, 1969.

Kebeasy, R.M., On the anomaly of travel time of P-waves observed at Japanese stations, Bull. Earthq. Res. Inst., Tokyo Univ., 47, 3, 1969.

Kuno, H., Lateral variation of basalt magma type across continental margins and island area, Bull. Volcanol., 29, 195-222, 1966.

Kuzin, I.P., Focalnaya zona i stroenie verkhnei mantii v rayone vostochnoi Kamchatki, Nauka, Moscow, 1974.

Liapishev, A.M., Deep geoelectric structure of Middle Kuril Isles from Sq-variation data, Glubinnie elektromagnitnie zondirovania Dalnego Vostoka, Vladivostok, 1980.

Magnitsky, V.A., Sloi nizkikh skorostei verkhnei mantii Zemli, Nauka, Moscow, 1968.

Marderfeld, B.E., Beregovoi effect v geomagnitnikh variatsiakh, Nauka, Moscow, 1977.

Markhinin, E.K., and D.S. Stratula, Some petrological geochemical and geophysical aspects of the relation of volcanicity to Earth's interior, Vulkanizm i glubini Zemli, 11-16, Nauka, Moscow, 1971.

Miyamura, S., Seismity of Japan and its neighbourhood, Izv. AN SSSR, Fizika Zemli, 7, 21-50, 1969.

Moriya, T., Folded structure of intermediate depth seismic zone and attenuation of seismic waves beneath the arc junction at southwestern Hokkaido, Novie dannie o stroenii kori i verkhnei mantii Kurilo-Kamchatskogo i Yaponskogo regionov, 59-67, Vladivostok, 1978.

Noguchi, S., and H. Okada, Anomalous seismic wave transmission and the upper mantle structure in and around Hokkaido, Subterranean structure in and around Hokkaido and its tectonic implication, 28-43, Sapporo, 1976.

Novie dannie o stroenii kori i verkhnei mantii Kurilo-Kamchatskogo i Yaponskogo regionov (Proceedings of the third Soviet-Japan Symposium on geodynamics and volcanicity), Col. papers, Vladivostok, 1978.

Okada, H., New evidences of the dipping lithosphere and its structure derived from precursors to ScS phases in Japan, Novie dannie o stroenii kori i verkhnei mantii Kurilo-Kamchatskogo i Yaponskogo regionov, 76-89, Vladivostok, 1978.

Oliver, J., and B. Isacks, Deep earthquake zones, anomalous structure in the upper mantle and lithosphere, J. Geophys. Res., 72, 16, 4259-4275, 1967.

Piskunov, B.N., A.I. Abdurakhmanov, and Ch.U. Kim, "Composition-depth" correlation and position of magmatic foci of Kuril volcanoes, Dokl. AN SSSR, 244, 4, 939-940, 1979.

Riaboi, V.Z., Structura verkhnei mantii territorii SSSR po seismicheskim dannim, Nedra, Moscow, 1979.

Ritman, A., Vulkani i ikh deyateinost, Mir, Moscow, 1964.

Sasatani, T., Mechanisms of mantle earthquakes near the junction of the Kurile and the Northern Honshu Arcs, Novie dannie o stroenii kori i verkhnei mantii Kurilo-Kamchatskogo i Yaponskogo regionov, 90-99, Vladivostok, 1978.

Simbireva, I.G., S.A. Fedotov, and V.D. Feofilaktov, Heterogeneities of the stress field in the Kuril-Kamchatka Arc as derived from seismological data, Geologia i geofizika, 1, 70-86, 1976.

Slvaina, L.B., and S.A. Fedotov, Longitudinal wave velocities in the upper mantle beneath Kamchatka, Seismichnost i seismicheskii prognoz svoistva verkhnei mantii i ikh sviaz s vulkanizmom na Kamchatke, 188-199, Nauka, Novosibirsk, 1974.

Stroyev, P.A., Bouguer anomalies and crustal thickness in the Sea of Japan transitional zone, Geofizicheskie issledovania zoni perekhoda ot Aziatskogo kontinenta k Tikhomu okeanu, 127-136, Nauka, Moscow, 1977.

Suvorov, A.A., Glubinnoe stroenie zemnoi kori Yuzhno-Okhotskogo sektora po seismicheskim dannim, Nauka, Novosibirsk, 1975.

Sychev, P.M., Glubinnie i poverkhnostnie tectonicheskie protsessi severo-zapada Tikhookeanskogo podvizhnogo poyasa, Nauka, Moscow, 1979.

Tada, T., P-wave velocity distribution of the down going slab. Zisin, 25, 310-317, 1972.

Tarakanov, R.Z., and N.V. Leviy, A polyasthenospheric model of the Earth's upper mantle as derived from seismological data, Dokl. AN SSSR, 176, 8, 571-574, 1967.

Tarakanov, R.Z., and N.V. Leviy, A model for the upper mantle with several channels of low velocity and strength, The crust and upper mantle of the Pacific area, AGU, Washington 43-50, 1968.

Tarakanov, R.Z., and Ch.U. Kim, On the anomalous dipping layer adjacent to the focal zone on the Pacific side, Geofizicheskie issledovania zemnoi kori v zone perekhoda ot Aziatskogo materika k Tikhomu okeanu, 30, 87-99, Vladivostok, 1975.

Tarakanov, R.Z., Ch.U. Kim, and R.I. Sukhomlinova, Regularities of spatial distribution of hypo-centers in the Kuril-Kamchatka and Japan regions and their relation to the features of geophysical fields, Geofizicheskie issledovania zoni perekhoda ot Aziatskogo kontinenta k Tikhomu okeanu, 67-77, Nauka, Moscow, 1977.

Tarakanov, R.Z., A model of medium structure in a focal zone and the adjacent mantel for the Kurile-Japan region, Novie dannie o stroenii kori i verkhnei mantii Kurilo-Kamchatskogo i Yaponskogo regionov, 111-126, Vladivostok, 1978.

Tarakanov, R.Z., and Ch.U. Kim, Velocity mosaic in the upper mantle of the Kuril-Japan region, Vulkanologia i seismologia, 1, 82-96, 1979.

Tarakanov, R.Z., and Ch.U. Kim, Seismity features in the Kuril-Kamachatka zone and the problem of distinguishing transverse faults, Seismichnost i mechanismi ochagov zemletriasenii Dalnego Vostoka, Vladivostok, 1980.

Utsu, T., Anomalies in seismic wave velocity and attenuation associated with a deep earthquake zone, J. Fac. Sci. Hokkaido Univ., 7, 3, 1-25, 1967.

Utsu, T., Seismic activity in Hokkaido and its vicinity, Geophys. Bull. Hokkaido Univ., 20, 51-75, 1968.

Utsu, T., Seismological evidence for anomalous structure of island arcs with special reference to the Japanese region, Rev. Geophys. Space Phys., 9, 839-890, 1971.

Utsu, T., Distribution of earthquakes in the Japanese islands and its vicinity, Kagaki, 44, 739-746, 1974.

Volokhov, I.M., Magmi, intratellurichekie rastvori i magmaticheskie formatsii, Nauka, Novosibirsk, 1979.

Wadati, K., T. Hirono, and T. Yumura, On the attenuation of S-waves and the structure of the upper mantle in the region of Japanese island. Pap. Meteorol. Geophys., 20, 49-78, Tokyo, 1969.

Yermakov, B.V., Correlation of faults and endogenic processes on the example of some areas of northwestern Pacific mobile belt, Geodinamicheskie issledovania, 6, 112-126, Moscow, 1979.

Yoshii, T., Crustal and upper mantle structure beneath the Pacific Ocean, Japan Islands, and the Sea of Japan, Geofizicheskie issledovania zoni perekhoda ot Aziatskogo kontinenta k Tikhomu okeanu, 7-13, Moscow, 1977.

Sources of Earthquake Data

Atlas zemletryasenii v SSSR, Nauka, Moskow, 1962.

Cataloque of major Earthquakes which occurred in the vicinity of Japan (1885-1950). Suppl. of Seism. Bull. Centr. Meteorol. Observ. Japan for the Year 1950, Tokyo, 1952.

Cataloque of major earthquakes which occurred in and near Japan (1926-1956). Seism. Bull. Japan Meteorol. Agency, Suppl. voi 1, Tokyo, 1958.

Novii catalog silnih zemletryasenii na territorii SSSR. Nauka, Moskow, 1977.

Seismologicheskii bulleten Dalnego Vostoka (1961-1975) Novoaleksandrovsk, 1962-1976.

The Seismological Bulletin of the Japan Meteorol. Agency (1957-1975), Tokyo, 1957-1976.

HEAT FLOW AND GEODYNAMICS OF THE TRANSITION ZONE FROM ASIA TO THE
NORTH PACIFIC

P.M. Sychev, V.V. Soinov, O.V. Veselov, N.A. Volkova

Sakhalin Complex Research Institute, Far East Science Center of the Academy
of Sciences of the USSR, Novoalexandrovsk, Sakhalin, 694050, USSR

Abstract. In the region under consideration both local and regional anomalies of high heat flow can be distinguished. The latter are restricted predominantly to the depressions of the Japan, Okhotsk and Bering marginal seas. Corrections for sedimentation and topography do not exceed 20-50 percent. Within the regions of high heat flow the corrections are, as a rule, minimal. The contribution of radiogenic heat from the Earth's crust equals 10 to 50 percent of the normal values. Therefore we can suggest that the observed high heat flow is caused mainly by deep-seated sources. Interpretations of the heat flow and other geophysical data suggest that the excessive heat is generated at depths of not more than 30-40 km. The sources of excessive heat are evidently zones of partial melting. The local anomalies of high heat flow are related to zones of partial melting at shallower depths within the Earth's crust. The generation of partial melting zones is explained as resulting from "penetrating convection," i.e. rapid supply of hot material in the liquid phase from the Earth's interior, and its subsequent intrusion into the top layers of the upper mantle by the mechanism of hydro-rupture.

Introduction

For the explanation of processes operative in the upper mantle beneath island arc-trench systems, heat flow data are of great importance. As a rule, anomalously high heat flow is recorded in marginal seas and in some regions of island arcs. Continental or oceanic regions have mainly normal values. Such a pattern of heat flow distribution is clearly distinguished in the northwestern Asia-Pacific transition zone.

For the region being discussed there are many heat flow measurements, and more are routinely being made. This paper is a review of geothermal investigations in the region to date. It also examines the nature of the anomalously high heat flow.

We hope that the conclusions reached in this paper will help to understand the geodynamics of other island arc systems.

Distribution of Heat Flow in the Region

Despite the nonuniform distribution of measurements in the region, they are of sufficient number to recognize the main peculiarities of heat flow distribution. Several heat flow compilations have been made within the region (Veselov et al., 1974a, 1974b, 1978; Watanabe et al., 1977; Smirnov and Sugrobov, 1979, 1980a; Yoshii, 1979). The description of heat flow regularities is based predominantly on published data from these studies and additional measurements made by the authors during 1975-1979 (unpublished data). The mean heat flow value used for discussion of the data is 55-60 mW/m^2 (Lubimova, 1968).

Marginal Seas

Anomalously high heat flow values in marginal seas is a remarkable feature of heat flow distribution in the Asia-North Pacific transition zone (Fig. 1). Similar average values of about 90-95 mW/m^2 are found in the basins of the Japan and Okhotsk marginal seas. The Bering Sea basin is different. Although Aleutian basin heat flow is rather low and variable within the range of 50-70 mW/m^2 (Marshall, 1978), the Comandor basin of the Bering Sea is, like the Japan and Okhotsk Seas, marked by anomalously high heat flow, reaching an average of 135 mW/m^2.

It is noteworthy that high heat flow values have also been recorded in rather shallow basins with thick sedimentary fill (up to 6-8 km). For example, in the Tatar Straits in the northern end of the Japan Sea, the average heat flow value is 110 mW/m^2 while maximum values in the western part of the straits reach only 43 mW/m^2. Within the Deryugin basin heat flow also varies, from 32 to 210 mW/m^2, 82 mW/m^2 being the average.

Slightly lower and variable heat flow is observed in the areas of submarine rises in both the Japan and Okhotsk Seas.

Land

The tendency to low heat flow on submarine rises, as compared to the basins, is also

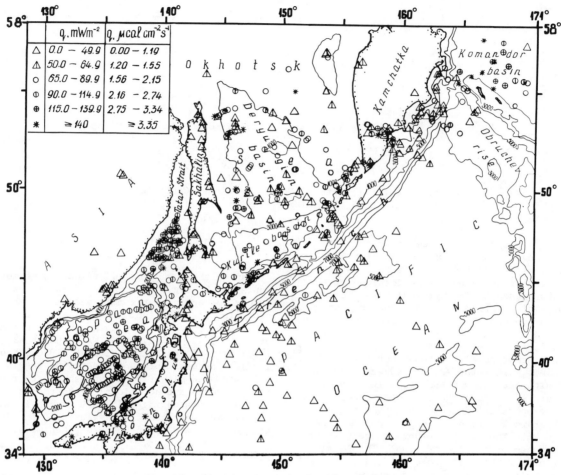

Fig. 1. Heat flow in the Asia-Pacific transition zone

recognized in the transition to land. However, there are some exceptions going onto the island arcs.

The island arcs of the region are not as well studied as the basin areas. Using the few measurements made on and near land in the Aleutian, Comandor and Greater Kurile arcs, a tendency to higher than normal heat flow values is noted. For the Aleutian arc the average is 69 mW/m^2 (Smirnov and Sugrobov, 1979); for the Comandor arc it is 98 mW/m^2 and for the Greater Kurile arc it is 85 mW/m^2. The above averages are very approximate, due to the wide range of heat flow, implying heterogeneity in sources. This conclusion is confirmed by a heat flow distribution pattern in Japan where zones of both high and normal heat flow values are also observed (Yoshii, 1979).

Lower heat flow is observed in Sakhalin, where the average value is 50 mW/m^2. Minimum values are obtained in the north (about 46 mW/m^2) and values closer to the average in the south of the island. However, in the center of the island a value of 75 mW/m^2 has been measured in a borehole.

Heat flow similar to that of Sakhalin is observed in the continental part of the Soviet Far East (44 mW/m^2). Slightly higher heat flow exists in Korea (Mizutani et al., 1970).

It is noteworthy that the transition from areas of high heat flow, characteristic of marginal sea basins, to areas with lower values on land is usually rather sharp and takes place over a distance of a few tens of kilometers.

The Northwestern Pacific Basin

In this part of the Pacific heat flow is, in general, close to the worldwide average value. The lowest but somewhat variable heat flow is found adjacent to the trenches.

Along the oceanward side of the trenches heat flow is rather homogeneous. Within the outer gravity high east of the Kurile-Kamchatka trench (marginal Zenkevich swell), the mean value of heat flow equals 50 mW/m^2, but its distribution is irregular. The mean heat flow values in the southern and northern parts of the area are about 45 mW/m^2, while in the central part it is 58 mW/m^2. A single measurement in the central area gave a value of 91 mW/m^2.

The distribution of heat flow on Obruchev Rise, which is the northern extremity of the submarine Emperor Ridge, is irregular. In this region a wide scatter of heat flow values, from 24.5 to 99 mW/m^2, is observed. In some cases the scatter is evidently due to poor measurements because of incomplete penetration of the probe into sediments (Smirnov and Sugrobov, 1979). However, both the average value (70 mW/m^2) and individual values definitely show that the Obruchev Rise has slightly higher heat flow than the adjacent northwest Pacific Basin.

Corrections

Heat flow measurements may be strongly influenced by factors such as sedimentation, vertical movements, contrasting heat conductivity, topography, deep-seated structures etc.

In the Sea of Okhotsk and on Sakhalin (Volkova, 1978), on the inner slope of the Kamchatka trench and in the northwestern Pacific (Smirnov and Sugrobov, 1980a) the influence of sedimentation was taken into account. A correction for sedimentation on the average amounts to about 10-20 percent, sometimes as much as 30 percent. When the correction for sedimentation is made we get larger heat flow values for marginal seas. Some very low values obtained in the Kurile-Kamchatka trench might be caused not by sedimentation but by landslides. With this in mind, the actual heat flow for the Kurile-Kamchatka trench may be close to the normal value (Smirnov and Sugrobov, 1980a).

The influence of relief and vertical movements estimated for Sakhalin is not large, not more than 10 percent which is within the measurement error (Volkova, 1978).

Finally, the influence of the Earth's crust and upper mantle heterogeneity was evaluated (structural factor). It has been suggested that in such cases for increased crustal thicknesses of 10-20 kilometers heat flow values may be lowered by 10-20 percent, and conversely it can be higher by 5-15 percent in areas of subcontinental crust (Smirnov and Sugrobov, 1980a).

The consideration of different types of corrections shows that they do not greatly change the general pattern of heat flow distributions. Besides, some corrections, for example that for the sedimentation and the decreased crustal thickness in basins, are of opposite sign. In this connection, the map of "background" flow (Smirnov and Sugrobov, 1980a) is very close to the map of measured values.

Radiogenic Heat

Rock radioactivity data allows us to access the contribution of the radiogenic component to the total value of heat flow. The contents of uranium, thorium and potassium, as major heat-generating elements were investigated in rocks of different composition and age from Kamchatka (Puzankov et al., 1977), the Kurile islands (Leonova and Udaltsova, 1975), Primorie (Smyslov et al., 1979) and Sakhalin (Volkova, 1975). Averaging the available data allows construction of a model of radioactive element contents and specific heat generation in the Earth's major crustal layers (Veselov and Volkova, 1981).

Mean values of specific heat generation within sedimentary, "granitic" and "basaltic" layers are 1.22-1.26; 1.54-1.70, 0.41-0.46 W/m^{-3} in continental crust; 1.13, 1.19, 0.41 W/m^{-3} in the continental crust of Sakhalin; 0.35-0.80, 0.54-0.76, 0.18-0.28 W/m^{-3} in the subcontinental crust of the Greater Kuriles; and 0.76-0.80, 0.45-0.97, 0.11-0.54 W/m^{-3} in the oceanic crust. The above estimates are, to a certain degree, tentative as the radioactive element content in the deep Earth's crust layers can be determined only by indirect methods. Using these estimates we may conclude that the contribution of the crustal component in the continental parts of the Far East and Sakhalin may account for 50-70 percent of the total heat flow. In the Kurile island arc and regions with subcontinental crust, the possible value of radiogenic heat decreases to 10 percent. Therefore, radioactivity of the crustal rocks, at least in the marginal seas, appears to be a minor factor contributing to the heat flow.

Temperature Distribution Within the Earth's Crust and Upper Mantle

The data on heat flow distribution, radioactivity and heat conductivity allow estimates to be made of the temperatures within the Earth's crust and upper mantle. Such estimates, (Smirnov et al., 1974; Volkova, 1975; Veselov et al., 1976; etc.), are becoming more refined. With availability of new data, the distribution of temperatures along the profiles of deep seismic soundings in the Sea of Okhotsk and in the Bering Sea was determined (Smirnov and Sugrobov, 1980b; Veselov, unpublished). The main conclusion from the above calculations was that beneath the marginal sea basins isotherms sharply rise. So, the 1100-1200°C isotherms, at which partial melting of matter can begin, are recorded in the Komandor and Bauers basins (Bering Sea) at depths of 10-40 km. In the Kurile basin of the Sea of Okhotsk these isotherms are at depths of 20-40 km (Fig. 2). The isotherms plunge sharply under the Kurile island arc where under the inner trench slope they lie at depths of 100 km or more. The proposed reconstructions are based on the assumption of steady heat flow. In fact it is reasonable to consider that the heat flow field, at least in some areas (Kurile-Kamchatka and Japan island arcs) are unstable (Smirnov and Sugrobov, 1980b) and in some cases the areas of partial melting can occur at the base of the Earth's crust.

Another attempt was made to determine a temperature distribution within the upper mantle on the basis of the known heat flow distribution at the surface (Soinov and Soloviev, 1978). The

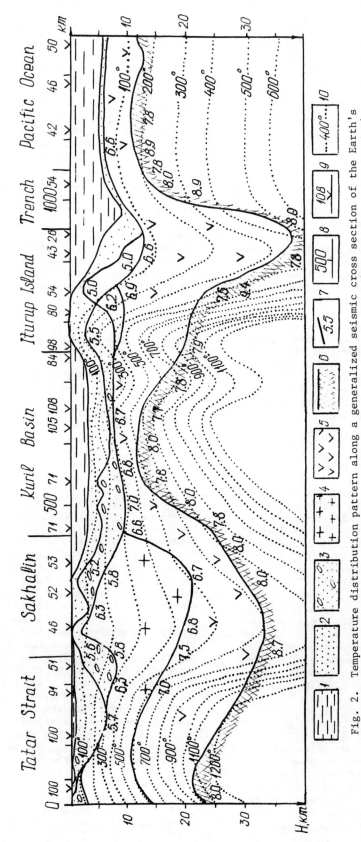

Fig. 2. Temperature distribution pattern along a generalized seismic cross section of the Earth's crust for the Tatar Strait - Sakhalin - Kurile Basin - Pacific profile (compiled by O.V. Veselov). Layers: 1 - water; 2 - sediments; 3 - volcanics and sediments; 4 - "granitic"; 5 - "baslatic"; 6 - subcrustal; 7 - refractors; 8 - km marks; 9 - heat flow stations (in mW/m²); 10 - isotherms (°C)

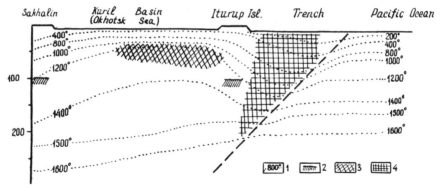

Fig. 3. Steady state model of temperatures in the upper mantle of the Kurile Basin (compiled by V.V. Soinov and V.N. Soloviev). 1 - isotherms (°C); 2 - electro-conductive layer in the upper mantle beneath Sakhalin and Iturup; 3 - area of predicted partial melting; 4 - area of the strongest earthquakes.

required temperatures were obtained using a Dirichlet problem for an area irregular form (Fig. 2). As boundary conditions on the left and right sides, we adopted the "normal" temperatures of the upper mantle of the continent and ocean, respectively. At the top, temperatures at a depth of 30 km, calculated by means of the heat flow distribution, were given. The distribution obtained is shown in Fig. 3. The disturbance of isotherms under the Kurile basin is traced to depths of 150-180 km and the temperature difference between this area and adjacent parts of the mantle is 200-400°. The 1200°C isotherm gently rises from a depth of 100 km beneath Sakhalin up to a depth of 40 km beneath the central Kurile basin and sharply plunges from Iturup Island to 140 km towards the Kurile-Kamchatka trench. Using the calculated temperatures we can distinguish a zone of possible partial melting of the upper mantle at depths of 30 to 75 km, stretching 150-180 km from Iturup towards Sakhalin.

The above calculations are consistent with theoretical arguments which imply that very thick zones of partial melting must have existed for a long period of time (30-60 m.y.) to explain the high heat flow observed in marginal seas (Kutas, 1978). It is possible that such zones occur at depths of 40-200 km.

Sharp submergence of isotherm and temperature gradients in the region between Iturup and the trench axis correspond to maximum seismic activity in this part of the island arc to the depths of 150-200 km (Fig. 3).

As already noted within other arc-backarc regions, local anomalies of high heat flow are observed. For the anomalous region east of central Sakhalin, calculations were made to determine the limiting depth of the upper edge of a thermal source represented by a magma instrusion with initial temperature of about 1200°C. Calculated depths range from 5 km to 24 km (Savostin, 1974; Eremin et al., 1976). Since the calculated depths were the deepest limiting case, the actual depth may be smaller (10-15 km). Similar shallow sources (in the crust) may be postulated for some of the other local heat flow anomalies.

Comparison of Temperature
Distribution With
Other Geophysical Data

The results of calculations of possible temperatures within the Earth's crust and upper mantle in some cases correlate with other geophysical data.

It has been noted that high electric conductivity may be due to the rise of the 1200°C isotherm on the inner side of island arcs (Uyeda and Rikitake, 1970; Rikitake, 1971). Honkura (Honkura, 1974, 1975) proposes a depth interval of 30-80 km for a conductive layer under the Japan Sea. However, under Japan, conductivity at these depths decreases so strongly that this layer becomes obscured. He also shows that conductivity of this layer beneath the Japan and Philippine seas is very high and thus cannot be attributed only to temperature rise; partial melting of the layer by 4-6 percent should be assumed. It is worth adding that high electric conductivity by a rather low degree of partial melting is possible only in cases where the liquid phase of the layer occurs, not in the form of isolated pockets, but distributed within a network of interconnected channels (Shankland and Waff, 1977).

Seismological data on the structure of the upper mantle along the inner side of island arcs are contradictory. The reexamination of surface waves beneath the Japan Sea indicates a low velocity layer from a depth of 40 km to a limiting depth studied of 125 km (Evans et al., 1978). A low velocity layer both for S- and P-waves was earlier proposed for the region as a whole at depths from 60-70 km to 180-200 km (Kanamori, 1970). For comparison, beneath the Solomon Sea where high heat flow is also observed, a low velocity layer is reported at depths of 35-65 km (Sundaralingam, 1978). A low velocity layer at depths of 85-150 km is established from an oceanward side of the Japan trench as well (Asada and Shimamura, 1976). A decrease in velocity and strong attenuation of both P- and S-waves down to depths of 120-170 km has also been observed below the volcanic belts of the Kurile (Fedotov and Kuzin, 1963), Japan (Maki, 1976), and Aleutian (Grow and Qamar, 1973) island arcs.

The zone of low velocity and high attenuation is characteristic, evidently not only to island arcs. Similar phenomena are observed under the Yamato Rise and central Sea of Okhotsk, possibly to the depths of 250-300 km (Barazangi et al., 1975).

Low velocity layers probably also occur at greater depths. Under Sakhalin, particularly, a low velocity layer possibly lies at depths of about 240-300 km (Noguchi and Okada, 1976). The decrease in P-wave velocities with respect to adjacent regions beneath such large structures as the Kurile-Kamchatka, Aleutian and other arcs takes place at depths exceeding 1000 km (Julian and Sengupta, 1973).

Among other characteristics of the upper mantle structure in island arc regions we should mention the occurrence of an obliquely dipping high-velocity layer identified with a subducting lithosphere (Utsu, 1971; Suyehiro and Sacks, 1979).

Studies of explosion seismology show a decrease in velocity of longitudinal waves at the M-discontinuity to 8.0 km/s in the Komandor basin (Shor and Fornari, 1976) and less than 8.0 km/s in the Kurile basin (The Earth's crust structure, 1964). However, normal velocities of 8.1 km/s are recorded at Moho in the Japan Sea (Regionality of crust..., 1979). It is interesting to note that low Moho velocities (7.2-7.6 km/s) are observed in the area of the outer slope of the Japan trench with a gravity maximum adjoining this region to the east (Houtz et al., 1978).

In spite of some decrease in longitudinal wave velocity at the Moho in marginal seas, the transition from the Earth's crust to upper mantle is evidently rather sharp (Structure of the Earth's crust..., 1964; Helmberger, 1968). Another type of velocity transition is observed in areas of island arcs. In some regions, for example in Kamchatka, there is a distinct low velocity layer (V_p = 7.5-7.7 km/s) with a thickness of 75 km at the base of the Earth's crust (Anosov et al, 1978).

It is shown above that heat flow is rather closely correlated with the crustal thickness. The lower the heat flow, the greater is the crustal thickness (Ehara, 1978). In other words, heat flow is high in places where the upper mantle surface lies at shallow depths.

The analysis of gravity fields shows that marginal seas are characterized, as a whole, by positive values of isostatic anomalies, which in the Sea of Okhotsk reaches +20 mgl (Anomalous fields..., 1974). Positive isostatic anomalies are not limited only to the basins, but are more clearly expressed on submarine rises and island arcs, implying that the crustal thickness there is not enough for isostatic equilibrium. To be more exact, the Earth's crust of these regions is uplifted with respect to a normal level. By contrast, trenches are characterized by negative isostatic anomalies that can be caused by excessive downwarping of the Earth's crust.

Discussion

It is noteworthy that calculations of temperatures within the Earth's crust and upper mantle, and electromagnetic and seismological data, all rather unambiguously show that the upper rim of the excessive heat sources in marginal seas lie at depths of about 30-40 km. The lower boundary of the heat sources is not as obvious but may lie at a depth of about 150 km. Judging from electromagnetic data these sources of excessive heat are probably zones of partial melting.

Gravity data, in general, also show zonation of isostatic anomalies similar to the heat flow and other geophysical trends and on the whole support the existence of hot upper mantle zones in marginal seas. It is interesting that the total gravity anomaly may be positive for large topographic features such as submarine rises, island arcs, and for Sakhalin (Sychev, 1973). These features, as has already been mentioned, usually have low velocity and low density "roots," to depths of 150-200 km in the underlying mantle (Anomalous gravity..., 1974; Egorkin et al., 1977). Even though surface heat flow for these thick crustal regions may be low (Fig. 1), it is possible that the gravity anomaly relates to thermal and chemical processes at their "roots."

So, the available data imply that anomalously high heat flow in marginal seas is caused, most probably by the presence of zones of partial melting within the upper mantle. What are the reasons for the generation of these zones?

It is natural to suggest that generation of extensive and thick zones of excessive temperatures is caused by the influx of heat from the Earth's interior. The most common models for their explanation are based on the concept of place tectonics. In terms of this concept a high heat flow in marginal seas is usually related either to generation of heat at the upper edge of subducting lithosphere due to friction (Hasebe et al., 1970), uplift of a thermal diapir (Karig, 1971), induced convection (Toksöz and Bird, 1977) or a combination of some of the above factors. However, there is a great deal of uncertainty about these processes of heat generation and its transport (cf. Uyeda, 1979). The heat flow distribution and geologic-geophysical data for the region under consideration should allow us to reach some preliminary conclusions on these problems.

As shown in Fig. 1, heat flow in the region changes sharply from place to place. Fig. 2 shows a profile across the major structures from the continent to the Pacific. These figures show high heat flow in the narrow zone of the Tater Strait, which is separated from the Kurile basin by a narrow zone of normal or low heat flow in Sakhalin.

Such a sharp alternation of relatively narrow areas of high and low heat flow can hardly be attributed to a mechanism of conductive heat transport from great depths, or to the presence of thermal diapirs or convective currents of a large scale.

On the other hand, ascending movements of convective currents or diapirism should result in uplift of the Earth's crust and tension over the area. But the Tatar Strait, where sediment thicknesses reach 6-8 km, at least for deposits in Tertiary time, has been subjected to downwarping. A considerable thickness of sediments is also observed in the Kurile basin (4-5 km). In both cases the sedimentary deposits lie horizontally and are undeformed. Thus, there is no evidence for compression; the downwarping could imply tension.

Heat flow is closely connected with the age of the marginal seas. Marginal seas are often considered as newly formed basins (Karig, 1971). In fact, this problem is highly complex. Geological and geophysical data allow one to suggest rather old ages for some marginal seas (Sychev and Snegovskoi, 1976; Sychev, 1979). In particular, the age of the oldest sediments in the Kurile Basin is evidently ~100 m.y. However, the underlying second layer in some cases may be newly-formed as a result of sill intrusions. At any rate, the formation of some submarine rises is undoubtedly related to magma intrusion into the Earth's crust after sedimentation.

Some publications (McKenzie and Sclater, 1968; Horai and Uyeda, 1969) have discussed the question concerning the relation between high heat flow and magma injection into the Earth's crust, and sometimes this mechanism has been preferred to explain marginal sea formation and the high heat flow (Pacham and Falvey, 1971). Taking into account the data about the upper mantle structure and the age of marginal seas one doubts the generation of marginal basin crust by this mechanism. However, the idea of heat transport by magma injection deserves further attention.

As to heat generation by viscous friction from subduction of lithosphere into the upper mantle, some authors (Hsui and Toksöz, 1979; etc) favor this process while others (Ito, 1978; etc) do not share this idea. It is difficult to explain the high heat flow zones in Tatar Strait and Deryugin basin by this model, because they lie beyond the projection of the earthquake focal zone to the surface.

Finally, let us discuss the heat flow in the northwestern Pacific. The most interesting feature in this region, as was mentioned above, is the high heat flow on the Obrutchev Rise. The Rise being the northern continuation of the Emperor seamount chain, it is natural to attribute the observed high heat flow to the after effect of the heating of the lithospheric plate passing over a "hot spot." However, quantative estimates show that under conditions of steady heat transport (Birch, 1975), the "hot spot" effect disappears at a distance of about 100 km. The non-stationary relation of lithosphere to the "hot spot" will not produce a long affect on surface heat flow (Severina, 1979). The high heat flow on the Obrutchev Rise is caused, evidently, by more recent tectono-magmatic processes operative in the Earth's crust and upper mantle under the region.

The slight increase in heat flow in the zone of gravity maximum on the oceanic side of the Kurile-Kamchatka and Japan trenches may relate to relatively recent magmatism. It is quite possible that the observed decrease in the seismic wave velocity at Moho is not accidental. To the east of the trench at the latitude of the South Kurile arc basalts were dredged from the slopes of small submarine hills on the marginal swell. Their K-Ar ages are 31.6 and 41.0 m.y. (Vasiliev et al., 1978), i.e. younger than the magnetic predicted age for the basement in this region. These results are far from conclusive.

However, if these data are correct this would imply either stability of the marginal swell structures and a rather long period of development, or in the convergence scheme the arrival at the swell of products of mid-plate magmatism. If the magmatic activity is genetically related to processes operative in the island arc the age of the rocks presents problems for the general scheme of subduction. This problem is currently being studied and will be the subject of future publication.

Hypothetical Model of the Formation
of Partial Melting Zones in the Upper Mantle

The formation of partial melting in the upper mantle under the vast area of the Sea of Okhotsk and beneath local zones (Tatar Strait) is probably closely related with the mechanism of heat transport from the Earth's interior. For the formation of a spatially limited zone of partial melting a rapid supply of heat would be needed. A possible mechanism may be "penetrating" convection (Elder, 1965). This implies the upward flow of hot masses of relatively low density. The rise of hot masses evidently takes place along weakened zones or deep fractures in the upper mantle (Sychev, 1976). However, the rise of hot material from the Earth's interior hardly takes place just below the areas of high heat flow. As it has already been mentioned a considerable decrease in velocity, and probably density, of the upper mantle to depths of 200-250 km is inferred under major, thick structures

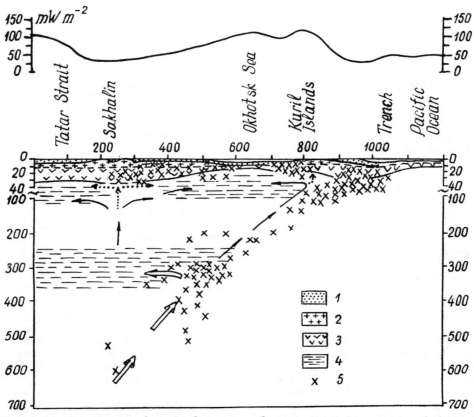

Fig. 4. Hypothetical mechanism for the formation of excessive temperature and partial melting zones. Layers: 1 - sediments; 2 - "granitic"; 3 - "basaltic"; 4 - partial melting zones; 5 - earthquakes. Arrows show the direction of primary and secondary differentiate movement.

of the Earth's crust. Therefore, it is probable that there is a rise of hot material just below the structures characterized by increased crustal thickness related to the supply of igneous material (Sychev, 1976). In this case, to explain the formation of zones of partial melting in the upper mantle beneath negative, or thin structures of the crust, horizontal as well as vertical transport of ascending hot material is required. A hypothetical model of such a process is shown in Fig. 4.

The rise of hot material via vertical and oblique channels is suggested, from depths of 240-350 km or deeper. If this hot material is represented by ultrabasic magma, its ascent will be restricted to density differences with the environment. As an example let us consider ascending magma with a density intermediate between that of the upper mantle and the crust. In this case ascending magma reaching the base of the Earth's crust would envelope it by moving towards the highest elevation of the upper mantle surface. Areas of its highest position are the zones of thinnest crust, i.e. basins of the marginal seas.

The above mechanism for the ascent of hot material is similar to that proposed by Artyushkov (1979) who considers regions of uplifted mantle surface as "traps" for the ascending hot, relatively light material. Evidently, the mechanism of magma rise is more complex. In this connection let us discuss some physical premises used as a basis for this model (Fig. 4).

It is assumed that ruptures in the upper mantle and ascent of liquid material are caused mainly by excessive pressure due to density differences between the ascending fluid material and the environment. The average tensional strength of the Earth's crust and upper mantle is estimated at $1 \times 10^7 N/m^2$ ($1 \times 10^7 Pa$) (Magnitsky, 1965). If we assume that the thickness of the partial melting layer situated in the upper mantle is at 50 km, and the density difference between liquid and solid phases is 0.1 g/cm^3, then the excessive pressure or the tensile stresses on the upper limit of the layer will be approximately 500 bar, ($5 \times 10^7 Pa$), i.e. quite sufficient for the formation of expansion fractures. Liquid material will rise to the level where its density becomes equal to that of the environment. If the influx continues the material will be spread laterally by the mechanism of hydro-rupture. Estimates show (Popov, 1972) that even thin sills (about

10 m) can spread laterally for hundred kilometers. Thus the Earth's crust and upper mantle rocks are not serious obstacles for the movement of liquid material if its density is lower than that of the enclosing rocks. At the same time we may assume that the liquid phase in the zones of partial melting of the upper mantle beneath marginal seas consists of ultrabasic material. In an extreme case the excessive pressures and tensile stresses would become so high that the liquid phase, for example, of basalt may be erupted into the crust and to the Earth's surface.

The differentiation ("separation") of fluid material takes place also during its rise from great depths. Therefore, a small amount of secondary differentiates of basic composition may rise via vertical channels and intrude into the Earth's crust or be accumulated at its base. The main part of ultrabasic melts will be in the upper mantle beneath basins of marginal seas.

Major or thick crustal structures seem to have peculiar "roots" in the upper mantle represented by low velocity and density zones, so that the paths of the ascending ultrabasic melts may be deflected from the vertical direction at great depths. If ultrabasic magmatism in the upper mantle of island arcs is an important factor in this process, we can derive some promising conclusions. One is that in such a case, ascending high-temperature mantle material results from gravity differentiation at great depths and can be considered as a primary thermal source.

Discussions of other aspects and conclusions of the proposed model are not within the scope of the present paper. Some of these aspects have been discussed previously (Sychev, 1973, 1979).

Conclusion

The pattern of heat flow in the northwestern Pacific shows several interesting features. The most important of them is the presence of high heat flow in the marginal seas, particularly in their basins. Estimates of temperatures and comparison of the heat flow with other geophysical results imply that the high heat flow is caused by the presence of high temperature partial melting at depths of 20-40 km. Local anomalies of high heat flow in these regions are associated with thermal sources in the Earth's crust.

High heat flow in marginal seas may be explained in terms of plate tectonics. However, it is obvious that existing models do not reflect all the processes operative in the upper mantle under the island arcs and trenches. The hypothetical model presented here is based on the concept of "penetrating" convection or magmatic injections resulting from the ascent of relatively light and hot material. Such a model considers ultrabasic magmatism, caused by gravity differentiation in the Earth's interior, and stepwise ascent of differentiates towards higher horizons as the driving force and widespread mechanism (Sychev, 1979).

Heat flow is not homogeneous within the oceanic part of the region under consideration. High heat flow over the Obruchev Rise, and possibly over the marginal swells rimming the trench outer sides, implies magmatic activity in these regions. Additional work is certainly needed to understand the thermal structure and geodynamics of the northwestern Pacific region, and particular emphasis is needed on the trench outer swells.

REFERENCES

Gainanov, A. G., Yu. A. Pavlov, P. A. Stroev, P. M. Sychev and I. K. Tuesdov, <u>Anomalous gravity fields of the Far East marginal seas and adjacent part of the Pacific</u>, P. M. Sychev, editor, Novosibirsk, (in Russian), 108 p., 1974.

Anosov, G. I., S. K. Kibbenina, A. A. Popov, K. F. Sergeev, V. K. Utnasin and V. I. Fedorchenko, <u>Deep seismic sounding in Kamchatak</u>, Moskow, (in Russian), 130 p., 1978.

Artushkov, Ye. V., <u>Geodynamics</u>, Moskow, (in Russian), 328 p., 1979.

Asada, T, and H. Shimamura, Observation of earthquakes and explosions at the bottom of the Western Pacific: structure of oceanic lithosphere revealed by longshot experiment, in geophysics of the Pacific Ocean basin and its margin, in <u>Geophys. Monogr. Ser. 19</u>, edited by G. H. Sutton, M. H. Manghami and R. Morberly, AGU, Washington, D.C., 135-153, 1976.

Barazangi, M., W. Pennington and B. Isacks, Global study of seismic wave attenuation in the upper mantle behind island arcs using P-waves, <u>J. Geophys. Res.</u>, 80, 1079-1092, 1975.

Birch, F. S., Conductive heat flow anomalies over a hot spot in a moving medium, <u>J. Geophys. Res.</u>, 80, 4825-4827, 1975.

Egorkin, A. V., V. Z. Ryaboy, L. N. Starobinets and V. S. Druzhinin, Velocity sections of the upper mantle according to the DSS data on land <u>Akad. Nauk SSSR, Ser. Fizika Zemli, 7</u>, 27-41, (in Russian), 1977.

Ehara, S., The heat flow in the Hoffaido-Okhotsk region and its tectonic interpretation, in <u>The structure and geodynamics of the lithosphere of the North-West Pacific according to the geophysical data</u>, edited by S. L. Soloviev, 86-98, (in Russian), 1978.

Elder, J. W., Physical processes in geothermal areas, in <u>Terrestrial Heat Flow</u>, edited by W. H. K. Lee, <u>Geophys. Monogr. 8</u>, Amer. Geophys. Un., 211-239, 1965.

Eremin, G. D., N A Volkova, O. V. Veselov, Interpretation of the local anomaly of the heat flow near the east coast of the Sakhalin Island, in <u>Geotermy (geothermal investigations in the USSR), part II</u>, edited by Ya. B. Smirnov, Moscow, 183-187, (in Russian), 1976.

Evans, J. R., K. Suyehiro and I. S. Sacks, Mantle

structure beneath the Japan Sea, a re-examination, Geophys. Res. Lett., 5, 487-490, 1978.

Fedotov, S. A. and I. P. Kuzin, The velocity section of the upper mantle in the region of the southern Kurile Islands, Izv. Akad. Nauk SSSR, Ser. Geophys., 5, 670-688, (in Russian), 1963.

Grow, J. A., A. Qamar, Seismic-wave attenuation beneath the central Aleutian Arc, Bull. Seismol. Soc. Amer., 63, 2155-2166, 1973.

Hasebe, K., N. Fujii and S. Uyeda, Thermal processes under island arcs, Tectonophysics, 10, 335-355, 1970.

Helmberger, D. V., The crust-mantle transition in the Bering Sea, Bull. Seismol. Soc. Amer., 58, 179-214, 1968.

Honkura, J., Electrical conductivity anomalies beneath the Japan Arc, J. Geophys. Geoelectr., 26, 147-171, 1974.

Honkura, J., Partical melting and electrical conductivity anomalies beneath the Japan and Philippine seas, Phys. Earth. Planet. Inter., 10, 128-134, 1975.

Horai, K., and S. Uyeda, Terrestrial heat flow in volcanic areas, in The Earth's Crust and Upper Mantle, Geophys. Mongr. 13, edited by J. H. Pembroke, Amer. Geophys. Un., 95-109, 1969.

Houtz, R. E., P. Buhl, P. Stoffa, C. Windish, S. Murauchi, Observation of the decrease in upper mantle seismic velocity beneath the Japan-Bonin trench, EOS Trans. Amer. Geophys. Un., 59, p. 321, 1978.

Hsui, A. T. and M. N. Toksoz, The evolution of thermal structures beneath a subduction zone, Tectonophysics, 60, 43-60, 1979.

Ito, K., Ascending flow between the descending lithosphere and the overlying asthenosphere, J. Geophys. Res., B 83, 262-268, 1978.

Julian, B. R. and M. K. Sengupta, Seismic travel time evidence for lateral inhomogeneity in the deep mantle, Nature, 242, 443-447, 1973.

Kanamori, H., Mantle beneath the Japanese arc., Phys. Earth. Planet. Inter., 3, 475-483, 1970.

Karig, D. E., Origin and development of the marginal basins in the Western Pacific, J. Geophys. Res., 76, 2542-2561, 1971.

Kutas, R. I., Heat flow field and the thermal model of the Earth's Crust, Naukova dumka, Kiev, 148 p. (in Russian), 1978.

Leonova, L. L. and N. L. Udaltsova, Geochemistry of the uranium and thorium in the volcanic process on the example of the Kurile-Kamchatka zone, in Novosibirsk, edited by K. N. Rudich, 102 p., (in Russian), 1974.

Lubimova, E. A., Thermics of the Earth and the Moon, edited by G. S. Gorshkov, Moscoe, 278 p., 1968.

Magnizky, V. A., Interior structure and physics of the Earth, Moscow, 380 p., (in Russian), 1968.

Maki, T., P-wave velocity structure in the upper mantle beneath the Japanese Islands, Zisin, 29, 233-245, 1976.

Marshall, V. B., Recent heat flow measurements in the Aleutian Basin, Bering Sea, EOS Trans. Amer. Geophys. Un., 59, 384, 1978.

McKenzie, D. P., and J. G. Sclater, Heat flow inside the island arc of the North-West Pacific, J. Geophys. Res., 73, 3173-3179, 1968.

Mizutani, H., K. Baba, N. Kobajashi, C. C. Change, C. H. Lee, and J. S. Kang, Heat flow in Korea, Tectonophysics, 10, 183-203, 1970.

Noguchi, Sh. and H. O. Ocada, Anomalous seismic wave transmission and the upper mantle structure in and around Hokkaido, in Structure in and around Hokkaido and its tectonic implcation, 28-43 (in Japanese, with English abstract), 1976.

Pacham, G. H. and D. A. Falvey, An hypothesis for the formation of marginal seas in the Western Pacific, Tectonophysics, 11, 79-109, 1971.

Popov, V. S., The estimations of the velocity of the basitic dykes and sill intrusions, Geohimiya, 6, 713-718 (in Ressian), 1972.

Puzankov, Yu. M., V. A. Bobrov and A. D. Duchkov, Radioactive elements and heat flow of the crust of the Kamchatka Peninsula, in Novosibirsk, edited by A. S. Mitropolsky and B.M. Garshin, 125 p., (in Russian), 1977.

Regionality of crust and upper mantle structure around Japan as derived from large explosions at sea, Tectonophysics, 56, 130, 1979.

Rikitake, T., Electric conductivity anomaly in the Earth's crust and mantle, Earth. Sci. Rev., 7, 35-65, 1971.

Savostin, L. S., A. F. Beresnev and G. B. Udintsev, The new data on the heat flow through the bottom of the Okhotsk Sea, Dokl. Akad. Nauk SSSR, 215, 846-849, (in Russian), 1974.

Severina, N. S., The non-stationary termal field in the moving media over the mantle plume, in Experimental and theoretical study of the heat flows, edited by M. P. Volarovich, Moscow, 212-216, (in Russian), 1979.

Shankland, T. J. and H. S. Waff, Partial melting and electrical coductivity anomalies in the upper mantle, J. Geophys. Res., 88, 5409-5417, 1977.

Shor, G. G. and D. J. Fornari, Seismic refraction measurements in the Kamchatka basin, Western Bering Sea, J. Geophys. Res., 81, 5260-5266, 1976.

Smirnov, Ya. B., V. M. Sugrobov and N. G. Sugrobova, The heat flow, hydrothermal activity and dynamics of the evolution of deep zones of the regions of Cenozoic volcanism, in Geodynamics, magma formation and volcanism, edited by M. M. Vasilevsky, Petropavlovsk-Kamchatsky, 175-196, (in Russian), 1974.

Smirnov, Ya. B. and V. M. Sugrobov, The Earth's heat flow in the Kurile-Kamchatka and Aleutian provinces, I. Heat flow and geotectonics, Vulkanologiya i seismologiya, 1, 59-73, (in Russian), 1979.

Smirnov, Ya. B and V. M. Sugrobov, The Earth's heat flow in the Kurile-Kamchatka and Aleutian provinces, II. The map of measured and phone heat flow, Vulkanologiya i seismologiya, 1, 16-31, (in Russian), 1980a.

Smirnov, Ya. B. and V. M. Sugrobov, The Earth's heat flow in the Kurile-Kamchatka and Aleutian provinces. III, The estimations of the deep temperatures and thickness of the lithosphere, Vulkanologiya i seismologiya, 2, 3-18, (in Russian), 1980b.

Smislov, A. A., U. I. Moiseenko and T. Z. Chadovich, The thermal regime and radioactivity of the Earth, Leningrad, 191 p., (in Russian), 1979.

Soinov, V. V. and V. N. Soloviev, Stationary model of the temperatures of the upper mantle of the Okhotsk Sea region, in Geophysical fields of the Asia Pacific transition zone, edited by M. L. Krashy, Yuzhno-Sakhalinsk, 53-56, (in Russian), 1978.

Structure in the Earth's crust in the Asia-Pacific transition region, edited by E. I. Galperin and I. P. Kosminskaya, Moscie, 308 p., (in Russian), 1964.

Suyehiro, K. and I. S. Sacks, P- and S- wave velocity anomalies associated with the subducting lithosphere in the Japan region, Bull. Seismol. Soc. Amer., 69, 97-114, 1979.

Sunduranlingan, K., Uppermost upper mantle beneath Solomon Sea, Search, 9, 155-156, 1978.

Sychev, P. M., Upper mantle structure a nature of deep processes in island arc and trench system, Tectonophysics, 19, 343-359, 1973.

Sychev, P.M., Deep structure and crust formation in the North-West Pacific, in Volcanoes and Tectonophere, edited by H. Aoki and S. Iizuka, Tokyo, Tokai Univ. Press, p. 341-357, 1976.

Sychev, P. M. and S. S. Snegovskoi, Abyssal depression of the Okhotsk, Japan and Bering seas, Pacific Geol., 11, 57-80, 1976.

Sychev, P.M. Deep and surface tectonic processes of the North-West Pacific mobile belt, Moscow, 208 p., (in Russian), 1979.

Toksoz, M. N. and P. Bird, Formation and evolution of marginal basins and continental plateaus, in Island Arcs, Deep Sea Trenches and Back-Arc Basins, Amer. Geophys. Un. Monograph. Ser., Washington, D. C., 379-393, 1977.

Utsu, T., Seismological evidence of anomalous structure of island arcs with special reference to Japanese region, Rev. Geophys. Space Phys., 9, 839-890, 1971.

Uyeda, S. and T. Rikitake, Electrical conductivity anomaly and terrestrial heat flow, J. Geomagn. Geoelectr., 22, 75-90, 1970.

Uyeda, S., Subduction zones: facts, ideas and speculations, Oceanus, 22, 52-62, 1979.

Vasiliev, B. I., M. G. Egorova, E. G. Zhiltsov, D. I. Podzorova, M. F. Skorikova and A. A. Suvorov, The new data on the structure of the Zenkevich swell, Izv. Akad. Nauk SSSR, Ser. Geol., II, 121-142, (in Russian), 1978.

Veselov, O. V., N. A. Volkova, G. D. Eremin, N. A. Kozlov and V. V. Soinov, Study of the heat flow in the North-West Pacific, in Geotermiya, otcheti o geotermicheskih issledovaniyah v SSSR, vip. 1-2. edited by Ya. B. Smirnov, Otcheti za 1971-1972 gg., Moscow, 87-90, (in Russian), 1974a.

Veselov, O. V., N. A. Volkova, G. D. Eremin, N. A. Kozlov, and V. V. Soinov, The measurement of heat flow in the Asia-Pacific transition zone, Dokl. Akad. Nauk SSSR, 217, 897-900, (in Russian), 1974b.

Veselov, O. V., Yu. A. Pavlov, V. V. Soinov, R. Z. Tarakanov, and V. I. Fedorchenko, The Earth's crust and upper mantle of the north-west part of the Asia-Pacific transition zone (structure, heterogeneities, recent processes), The upper Mantle and its heterogeneities. in The structure and upper mantle in the Asia-Pacific transition zone, edited by A. A. Yanshin, Novosibirsk, 249-265, (in Russian), 1976.

Veselov, O. V., N. A. Volkova, V. V. Soinov, and G. D. Eremin, Geothermal study by the SakhKNII in the north-west part of the Pacific mobile belt, in, Structure and geodynamics of the lithosphere of the North-West Pacific according to the geophysical data, edited by S. L. Soloviev, Materiali tretiego Sovetsko-Japonskogo simposiuma v g. Yuzhno-Sakhalinske, 1976, Vladivostok, 99-104, (in Russian), 1978.

Veselov, O. V. and N. A. Volkova, The radioactivity of the rocks of the Okhotsk Sea region, in Geophysical fields of the transition zone of the Pacific type, edited by M. L. Krasny, Vladivostok (in print), 1981.

Volkova, N. A., The temperature distribution in the Earth's crust of the Southern Okhotsk Sea region, in The Earth's crust and upper mantle of the Asiatic part of the Circum Pacific, edited by P. M. Sychev, Yuzhno-Sahkalinsk, 202-212, (in Russian), 1975.

Volkova, M.A., component of the heat flow due to the radioactivity of the crust in the north-west of the Asia-Pacific transition zone, in Geophysical study of the Earth's crust and upper mantle structure in the Asia-Pacific transition zone, edited by I. K. Tuezov, et.al., Vladivostok, 221-226 (in Russian), 1975.

Volkova, N. A., Heat flow and some aspects of the geological history of Sakhalin in Cenozoic era. in Geophysical fields of the Asia-Pacific transition zone, edited by M. L. Krasny, Yuzhno-Sakhalinsk, 57-62, (in Russian), 1978.

Watanabe, T., M. G. Langseth and R. N. Anderson, Heat flow in Back-Arc Basins of the Western Pacific, in Island Arcs, Deep Sea Trenches and Back-Arc Basins, III, edited by M. Talwani and W. C. Pitman, Amer. Geophys. Un. Monogr. Ser., Washington, D. C., 137-161, 1977.

Worzel, J. L., Gravity investigation of the subduction zone, in Geophysic Pacific Ocean Basin and Margin, edited by G. H. Sutton, M. H Manghami, and R. Moberly, Washington D.C., 1-15, 1976.

Yoshii, T., Compilation of geophysical data around the Japanese Islands, Bull. Earthq. Res. Inst. Univ. Tokyo, 54, 75-117 (in Japanese, with English abstract), 1979.

THE TECTONICS OF THE KURIL-KAMCHATKA DEEP-SEA TRENCH

Helios Gnibidenko, Tanya G. Bykova, Oleg V. Veselov,
Vladimir M. Vorobiev and Alexander S. Svarichevsky

Sakhalin Complex Science Research Institute, Novoalexandrovsk, Sakhalin 694050, U.S.S.R.

Abstract. A regional geological and geophysical investigation of the Kuril-Kamchatka deep-sea trench and the zone of its intersection with the Aleutian deep-sea trench was carried out in 1976-1979 by R/V "Pegas", "Otvazhniy", "Morskoi Geophysik" and "Orlik" of the Sakhalin Complex Scientific Research Institute of the Academy of Sciences of the USSR as a part of the "International Geodynamics Project". The data together with those obtained by the US and Japanese researchers on the Aleutian and Japan trenches allowed the authors to establish the general structure of the Kuril-Kamchatka deep-sea trench. The upper crust structure of the Kuril-Kamchatka trench continental slope consists of a system of horst-anticlinorial uplifts of the acoustic basement and partially filled graben-synclinorial troughs stretching in the north-east direction following the general trend of the trench. From dredging data, the acoustic basement rock associations of the horst-anticlinorial uplifts of the trench continental slope are found to be Pre-Neogene complexes of the deformed geosynclinal volcanogenic and sedimentary deposits broken by gabbroids, granodiorites and granitoids. Graben-synclinal troughs are filled with sedimentary-volcanogeneous deposits mainly of the Neogene-Quaternary ages, the thickness of which in some basins exceeds 3000 m. In some cross-sections of the Kuril-Kamchatka deep-sea trench at the base of the continental slope, there are lenticular sedimentary bodies (with thickness up to 1000 m if the seismic P-wave velocity is about 2 km/s) which are characterized by weak discontinuous reflections. Those sedimentary bodies seem to be formed as a result of landslides. Turbidite deposits with thickness up to 1000 m are identified in some regions of the trench. In the Komandor part of the Aleutian trench, there is a thick layer of non-deformed turbidite deposits. The oceanward slope of the trench is composed of a sedimentary layer with a thickness of about 100-300 m, lying on the oceanic crust. According to the dredging data, the acoustic basement of the marginal oceanic swell (Hokkaido Rise) is mainly composed of metamorphosed basalts and seems to include intercalated sedimentary rock. From whole rock K/Ar data, the period of intensive basaltic volcanism on the Hokkaido Rise is estimated to be from Cretaceous to Paleogene. Since the turbidite layers in the Kuril-Kamchatka and Aleutian trenches are not deformed, recent compressional forces which are postulated by "plate tectonics" appear absent. Crustal faults along the Kuril-Kamchatka and Aleutian trenches are distinct in the seismic data and can be mainly characterized as "normal faults", especially on the oceanward side, which points to tensional conditions. Faults that strike transverse to the trenches are established mainly from magnetic data. The magnetic anomaly field is subdivided into regions of anomalies that are subparallel to the trench (southern part) and subtransverse to the trench (northern part). A vast region of oceanic plate adjacent to the central part of the Kuril-Kamchatka trench is characterized by the absence of linear magnetic anomalies. From the above observations one can conclude that the structural evolution and the geodynamic regime of the Kuril-Kamchatka trench are not in agreement with some postulates of plate tectonics, particularly because:

1) Localized regions of compression and dilatation exist within the crust and upper mantle of the Kuril island arc and adjacent deep-sea trench produce a mosaic of "normal" and "reverse" faulting (Averianova, 1975). This fact is not in agreement with the simple compression model for island arcs and trenches due to subduction;

2) The sedimentary turbidite complex within the deep-sea trench adjacent to the Asiatic continental margin is not deformed and the crust in the trench regions is mainly under tension.

Vertical crustal movements are predominant in the lithosphere of the Kuril-Kamchatka deep-sea trench, resulting in the formation of the main trench features, whereas the horizontal crustal movements result in plicate deformations and displacements along the faults.

Introduction

The study of the structure and geological history of geoanticlinal uplifts of island arcs, as-

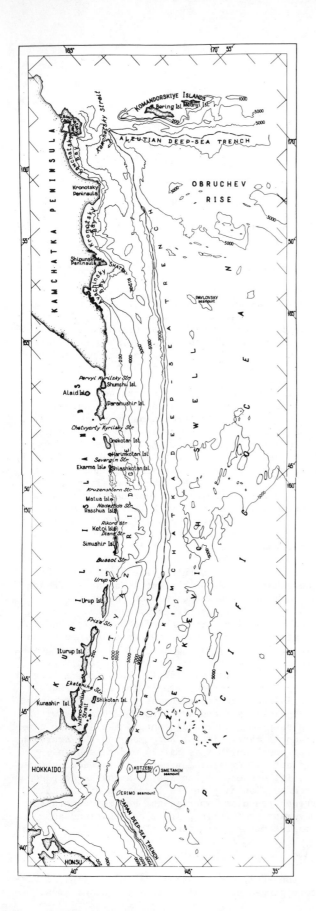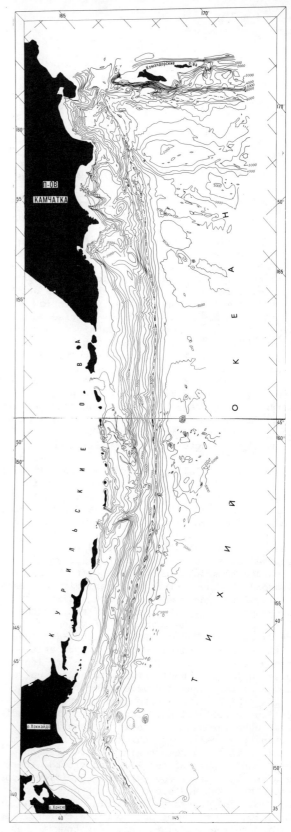

Fig. 1. Place name map (a) and bathymetric map (b) of the Kuril-Kamchatka deep-sea trench.

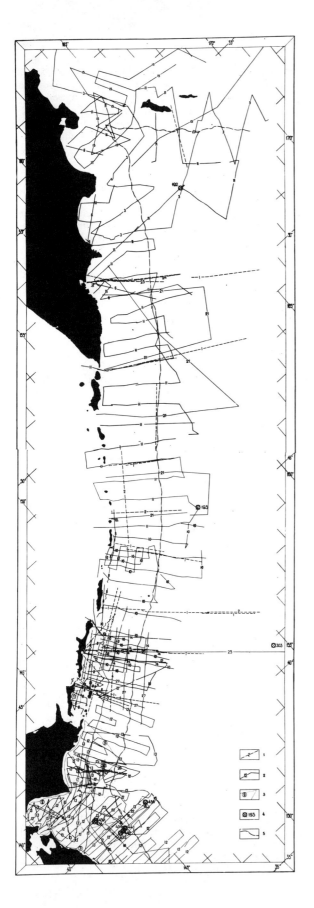

sociated deep-sea trenches and marginal oceanic rises provides a means of evaluating the various geological models of crustal evolution that may apply to the geodynamic processes occurring in the Asia-to-Pacific transition zone (Malahoff, 1970; Matsuda and Uyeda, 1971; Karig and Sharman, 1975; Hilde et al., 1976, 1977). A regional geological and geophysical study of the Kuril-Kamchatka deep-sea trench was carried out by the Sakhalin Complex Scientific Research Institute as a part of "The International Geodynamic Project" mainly during cruises 9 (1976) and 13 and 15 (1977) of R/V "Pegas" and also during cruises of R/V "Otvazhniy", "Orlik" and "Morskoi Geophysik". The region explored covers (Fig. 1) the Kuril-Kamchatka deep-sea trench and its intersections with the Aleutian and Japan deep-sea trenches. The results of these investigations together with the data obtained during the cruises of the US R/V "Bartlett" (1970) in the area of the Aleutian trench and during the cruises of the Japanese R/V "Hakurei Maru" (1976) in the zone of intersection between the Kuril

---

Fig. 2. (Opposite) Ships tracks of seismic reflection and refraction profiles of the Kuril-Kamchatka deep-sea trench.

Explanation of index: (1) seismic refraction profiles, (2) continuous single-channel and multi-channel profiles, (3) areas of detailed single-channel surveys, (figures on profiles correspond to figures in list of data sources), (4) sites of deep-sea drilling "Glomar Challenger" (Initial Reports..., 1973, 1975; Japan..., 1978); (5) deep-sea trench axis.

Sources of data: 1 - Crustal structure..., 1964; 2 - Tulina, 1969; 3 - Scholl et al., 1976; 4 - Suvorov and Zhiltsov, 1972; 5 - profiles of the Sakhalin Complex Sci. Res. Institute, 1973; 6 - profiles of the Pacific Expedition of VNIIMORGEO, 1971 and 1972; 7 - Snegovskoy, 1974; 8 - profiles of the Sakhalin Complex Sci. Res. Institute, 1974; 9 - Sakurai et al., 1975; 10 - profiles of the Sakhalin Complex Sci. Res. Institute, 1976, R/V "Pegas" cruise N 11; 11 - profiles of the Sakhalin Complex Sci. Res. Inst., 1977, R/V "Pegas" cruise N 13 and 15; 12 - Geological Investigation..., 1977; 13 - Murauchi and Asanuma, 1977; 14 - Bondarenko et al., 1977; 15 - Rabinowitz and Cooper, 1977; 16 - Ludwig et al., 1966; 17 - Den et al., 1971; 18 - Den and Hotta, 1973; 19 - Okada et al., 1973; 20 - Minaev et al., 1975; 21 - Udintsev et al., 1976; 22 - profiles of the Inst. of Ocean Investigations of the Tokyo Univ., 1978; 23 - Tulina et al., 1972; 24 - profiles of the Sakhalin Complex Sci. Res. Inst. and Inst. Physics of Earth of the Acad. Sci. USSR, 1972; 25 - profiles of the Sakhalin Complex Sci. Res. Inst., May-June 1978; 26 - profiles of the Sakhalin Complex Sci. Res. Inst., Sept.-Oct. 1978; 27 - profiles of the Pacific Expedition UYZHMORGEO, 1977.

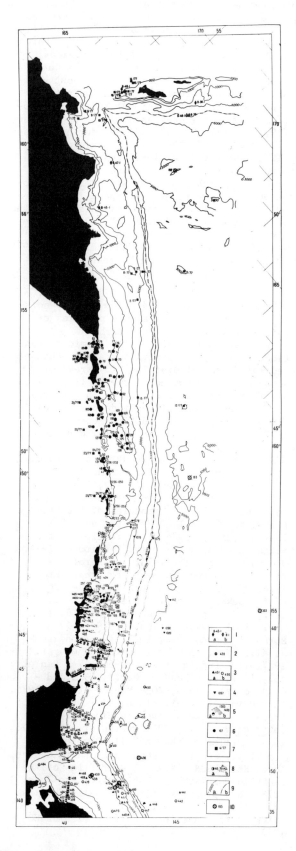

and Japan deep-sea trenches allowed us to clarify the general tectonic system of the Kuril-Kamchatka deep-sea trench and to propose a model of its structural evolution. Along the trench, many seismic, magnetic and gravity observations (Fig. 2) and dredgings (Fig. 3) were made.

Bathymetry

The bathymetry of the trench and adjacent regions (Fig. 1) has been compiled from the data obtained during the R/V "Pegas" cruises and the available navigation chart and published data (Udintsev, 1955; Mikhailov, 1970; Bathymetric Atlas..., 1973; Geological Investigation..., 1977).

The continental slope of the trench is characterized by deep-sea terraces at depths of about 4000-4500 m and 6000-7000 m. In addition, some narrow terraces are found at depths of about 1000-2000 m and 8000-9000 m. The length of the widest (up to 60 km) terrace at 4000-4500 m depth exceeds 500 km. Rises about 200-300 m high are often located on the outer side of the terraces. The terraces in the northern part are considerably narrower (widths rarely exceed 10 km) and not as long as those in the southern part. Inclinations of the continental slope to the thalweg or axis of the trench are generally 3-7° and seldom exceed 10° in the lower part of the slope. Canyons, among which Kushiro, Bussol, Avachinsky, Zhupanovsky, Kronotsky and Kamchatsky are the largest, stretching for distances of 100 km or more. The depth of the canyons often exceeds 500-1000 m. The shelf edge is usually at about 100-160 m below which there occurs a subsided shelf or avanshelf extending to the depth of 500 m. Linearity and rather steep in-

Fig. 3. (Opposite) Sites of dredge stations and cores.
1-8 - Sites and Sites Numbers: 1 - cruise of R/V "Bartlett" in 1970 (U.S. Geol. Surv.; Scholl et al., 1976), a - dredge, b - piston core; 2 - cruise of R/V "Vityaz" in 1974 (Sakhalin Complex Sci. Res. Inst.); 3 - cruise of R/V "Hakurey-maru" in 1976 (Geol. Surv. Japan; Geological Investigation..., 1977); 4 - cruise of R/V "Pegas" in 1976 (Sakhalin Complex Sci. Res. Inst.); 5 - cruise of R/V "Otvazhniy" in 1976 (Sakhalin Complex Sci. Res. Inst.), a - sites, b - area of dredging; 6 - cruise of R/V "Orlik" in 1977 (Sakhalin Complex Sci. Res. Inst.); 7 - cruise of R/V "Pegas" in 1977 (Sakhalin Complex Sci. Res. Inst.); 8 - cruise of R/V "Otvazhniy" in 1973 (Sakhalin Complex Sci. Res. Inst., Vasiliev, 1974), a - sites, b - area of dredging; 9 - isobaths (a) and trench axis (b); 10 - sites of deep drilling by "Glomar Challendger" (Initial Reports..., 1973, 1975; Japan..., 1978).

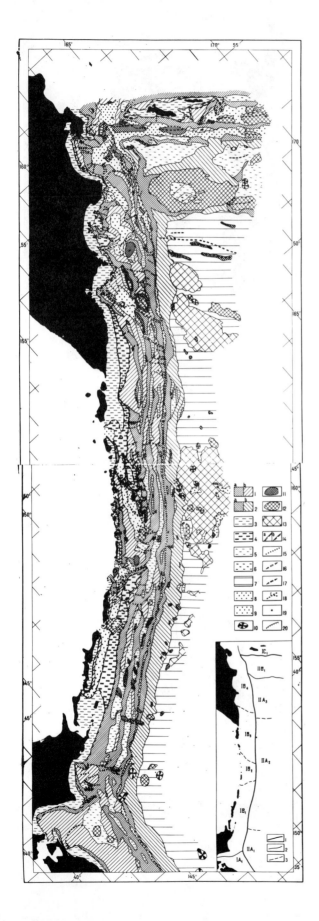

clination, exceeding 25° and in some cases reaching 30°-40°, are characteristic of the slope along the Komandor part of the Aleutian trench. From the Pacific abyssal plain the average inclination of the trench slope is 3-7° and in some places 10°.

The bottom of the Kuril-Kamchatka trench is a flat plain 1-2 km wide, up to 15 km in some places. The bottom plain is about 5-10 km wide in the Japan trench and about 40 km wide in the Komandor part of the Aleutian trench.

The shelves of the Hokkaido-Kuril-Kamchatka-Aleutian region are nearly at the same depth, about 100-160 m, and one can conclude that they subsided as a result of glacioeustatic elevation of sea level up to 100-120 m (Kulakov, 1973) during Holocene time (Fig. 4). The avanshelf de-

---

Fig. 4. (Opposite) Geomorphological map of the Kuril-Kamchatka deep-sea trench.

In the lower corner is a scheme of geomorphological division of the Kuril-Kamchatka deep-sea trench area. Boundaries: 1 - regions, 2 - provinces, 3 - area. I - the Japan-Kuril-Kamchatka-Aleutian region. Provinces: IA - Japan mountain system; IB - Hokkaido-Kuril-Kamchatka mountain system; IC - Aleutian mountain system. Areas: IA_1 - Honshu continental slope; IB_1 - Hokkaido-South Kuril continental slope; IB_2 - Central Kuril continental slope; IB_3 - North Kuril continental slope; IB_4 - Kamchatka continental slope; IC_1 - Komandor continental slope. II - Northwestern Pacific region. Provinces: IIA - marginal oceanic rise; IIB - Emperor seamount. Areas: marginal oceanic rise: southern part - IIA_1; central part - IIA_2 and northern part - IIA_3; Meiji rise - IIB_1.

Symbols: slopes: 1 - continental (a - steep, >8°; b - gentle); 2 - oceanic (a - steep, >8°; b - gentle). Planation surfaces: 3 - abrasional; accumulative-abrasional and accumulative plains of the shelf; 4 - denudational and accumulative-denudational plains of the avanshelf; 5 - accumulative and denudation deep-sea bench; 6 - accumulative-denudation crest surfaces of ridges and rises; 7 - accumulative plains of the marginal oceanic rise; 8 - accumulative plains of the bottom of intermountain depressions and basins; 9 - accumulative plains of the deep-sea trench. Seamounts and rises: 10 - single cones (volcanoes) and massives of joined volcanoes; 11 - separately located seamounts and ridges (steep-sloped); 12 - single rises (gentle-sloped); 13 - crest surfaces of the marginal oceanic rise. Other symbols: 14 - outer edge of the planation surfaces (a - shelf, b - avanshelf); 15 - deep-sea trench axis; 16 - axes of elongated depressions; 17 - ridge axes; 18 - thalwegs of submarine valleys; 19 - summits of seamounts and rises; 20 - scarps.

nudational plains are probably due to crustal subsidence. Both shelves and avanshelves are relict subaerial morphosculptural forms which submerged beneath the sea level.

Geophysical fields

Magnetic field

Two regions are distinguished in the pattern of the magnetic anomaly field of the Kuril-Kamchatka deep-sea trench (Fig. 5): the Northwestern Pacific plate, and the continental slope. The boundary between these two regions is not along the trench axis, but is displaced to the continental slope. It is mainly located at the 4000 to 6000 m deep terraces.

Linear magnetic anomalies with a northeastern trend (70°), which are traced nearly parallel to the Kuril trench up to the Tuskarora transverse fault, exist only near the junction between the Kuril-Kamchatka and Japan trenches. To the southwest, these anomalies cross the Japan trench and have been cited as evidence for oblique underthrusting of the plate (Hilde et al., 1976). To the northeast these linear anomalies are absent ("quiet zone" according to Hilde et al., 1976). The differently oriented trends of the anomaly field in this area are difficult to explain in terms of plate tectonics.

The northwest trending features have been attributed to fracture zones (Hilde et al., 1976) but can be traced far behind the trench axis, into the continental slope.

Gravity field

The gravity field map (Fig. 6) was compiled from the available data (Segawa, Tomoda, 1976; Watts et al., 1976; Geological Investigation..., 1977). It has been suggested (Worzel, 1976), that the negative gravity anomaly at the trenches should be much broader than the observed anomaly to be compatible with subduction of the oceanic plate. The crust adjacent to the trench is not in isostatic compensation (Gainanov et al., 1974). Both subsidence and over-compensation of the crustal block on the continental slope seem to be stipulated by the higher density of the upper mantle beneath the subsiding block (up to 0.2 g/cm^3 according to Gainanov et al., 1974) as compared to the adjacent parts of the upper mantle.

The intensive positive free air anomaly for the southern part of the Kuril geoanticlinal uplift is explained by the crustal higher density (up to 2.9-3.0 g/cm^3 according to Kosygin, Pavlov, 1975), whereas the upper mantle here seems to have a reduced density by 0.06 g/cm^3 (Segawa, Tomoda, 1976). This lower density may reflect a high temperature of the upper mantle beneath this part of the geoanticlinal uplift and possibly some partial melting.

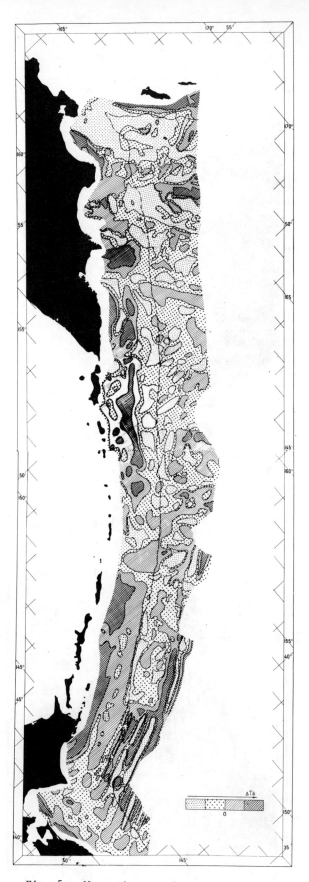

Fig. 5. Magnetic anomalies of the Kuril-Kamchatka deep-sea trench.

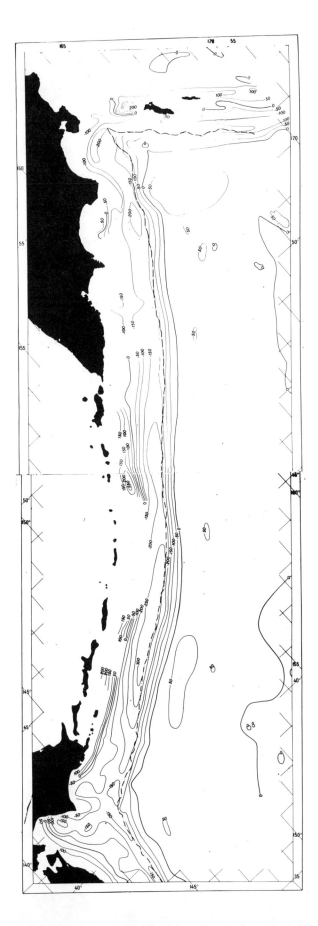

Heat flow

Between the Kuril-Kamchatka trench axis and the volcanic arc, heat flow is low (up to 0.6 HFU). Within the Kuril-South Kamchatka volcanic arc it sharply increases (1.5-2.5 HFU, Fig. 7, 8 and 9). Locally some very high heat flow values (up to 5.5 HFU) are distinguished, which seem to be associated with hydrothermal systems and volcanoes within the western part of the geoanticlinal uplift. Crustal temperature calculation have been carried out along the DSS profiles crossing the deep-sea trench (Fig. 8 and 9).

Tectonics

According to the seismic, deep-sea drilling and dredging data in the Kuril-Kamchatka, Aleutian and Japan trenches and in the adjacent Northwestern Pacific plate, the main structural features and tectonic character of the acoustic basement and the sedimentary cover can be distinguished (Fig. 10).

The acoustic basement

The acoustic basement of the Kuril-Kamchatka arc consists of magmatic and sedimentary rocks, considerably deformed and metamorphosed, with seismic P-wave velocities higher than 3.5-4.0 km/s. Metamorphic rocks with seismic P-wave velocities of about 5 km/s outcrop in the axial part of the Lesser Kuril ridge horst-anticlinorium. Paleogene-Mesozoic to probably the Upper Paleozoic is the stratigraphic range of the geosynclinal deformed complex of the continental slope acoustic basement. Lower Miocene deposits also seem to be constituents of the acoustic basement in the volcanic part of the Kuril geoanticlinal uplift.

The acoustic basement of the Northwestern Pacific plate coincides with the top of the oceanic second layer, which is composed of basalts, and some intercalated sedimentary rocks. Seismic P-wave velocities in basalts (Initial Reports..., 1973, 1975) are about 4.3-5.6 km/s. Layers of siliceous rocks and porcellanites alternating with clays and nannofossil oozes are distinguished in sites 303 and 304 (Fig. 3) of the basement roof. Seismic P-wave velocities in these layers are 3.0-5.0 km/s, due to the siliceous interbeds, and these layers are about 50 m thick.

The sedimentary cover

The sedimentary cover filling the system of graben-synclinal troughs on the continental slope is considerably less deformed than the

---

Fig. 6. (Opposite) Free-air gravity anomaly map (according to Watts et al., 1976 and Geological Investigation..., 1977).

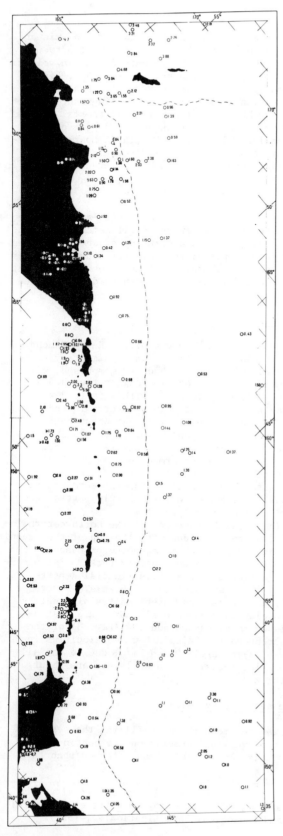

acoustic basement. Seismic P-wave velocities are about 1.6-2.2 km/s and reach 3-3.2 km/s at the sediment base in synclinal troughs.

According to the seismic and deep-sea drilling data (Fig. 2 and 3) the sedimentary cover in the troughs of the continental slope is subdivided into two or three layers, often separated by angular unconformities. The internal structure of the sediments in the troughs usually shows good stratification indicative of turbidite deposition. Structures indicating landslides are also found in the sedimentary cover.

On the oceanic plate, sedimentary cover is represented by a horizontal layering with average thickness of about 250-300 m. In some places the thickness increases up to 400-600 m, reaching 1000 m or more near the northern end of the Emperor seamounts (Obruchev rise). The internal structure of the oceanic plate sediments is conformable with the acoustic basement and is well-stratified. Continuous seismic profiling and deep-sea drilling data (Fig. 2 and 3) show two main layers. The upper layer (Initial Reports..., 1973, 1975) is composed of pelagic, diatomic and radiolarian oozes, clays and argillites with volcanic ash interbeds. Sites 192, 193, 303 and 304 indicate it is Lower Miocene up to Holocene in age. The lower layer within the Obruchev rise area is composed of chalk and argillites. It is Lower Miocene up to Cenomanian-Turonian (lower part of Upper Cretaceous) in age, inclusive (Initial Reports..., 1973).

The sedimentary cover of the oceanic plate, in the area adjacent to the base of the continental slope in regions of the Kuril-Kamchatka, Aleutian and Japan trenches, is overlain by a well-stratified layer with thickness up to 1000 m of probable turbidite origin. Within the trench the turbidite layer has a horizontal structure without perceptible deformation. Within the Kuril-Kamchatka trench the turbidite layer is distinguished only in the Kamchatka part. In some parts of this trench the landslide layer with thickness up to 1000 m is distinguished. From the continuous seismic profiler records it is characterized by poorly correlated or absent internal reflections (Fig. 11 and 12).

Deep-sea trench tectonics

The available data on the structure of the upper part of the crust (Fig. 2, 3, 4 and 5) of the Kuril-Kamchatka deep-sea trench area are summarized in the tectonic map (Fig. 10), a brief analysis of which is given below.

The Kuril-Japan trench junction. The structure of the zone between the Kuril-Kamchatka and Japan trenches is a gradual intersection of structural elements of the Japanese and Hokkaido-Kamchatka

Fig. 7. (Opposite) Heat flow measurements of the Kuril-Kamchatka region. Heat flow unit is 1 mccal·cm^{-2}·c^{-1} = 41.868 mw·m^{-2}.

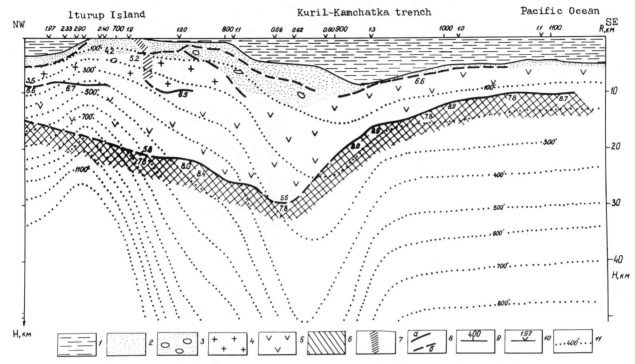

Fig. 8. Distribution of deep temperatures along profile 3b (see location, Fig. 11). 1-6 - crustal and upper mantle layers: 1 - water, 2 - sedimentary, 3 - volcanogenic-sedimentary, 4 - granite-metamorphic, 5 - metabasaltic, 6 - subcrustal; 7 - deep faults; 8 - seismic boundaries: a - certain, b - uncertain; 9 - profile in km; 10 - points of heat flow value; 11 - isotherm.

horst-meganticlinorium (Fig. 10; Honza et al., 1978). Within northeastern Honshu there is the Kitakami horst-anticlinal zone. In the axial part of it there is a miogeanticlinal sequence of Silurian rocks with a thickness of about 350 m. These rocks are overlain by Devonian-Carboniferous molasse and contain andesitic rocks (Geology..., 1977). Folding deformation took place in most parts of the Kitakami horst-anticlinal zone in the Carboniferous-Permian time (Setamai orogenesis). In the Late Paleozoic, the eugeosyncline moved to the northeast where the Late Permian-Triassic-Jurassic Iwadzumi and Tero geosynclinal zones are composed of folded eugeosynclinal formations with a thickness amounting to 5000 m. These formations are overlain with an angular unconformity by the Aptian-Albian Miyako group of shallow-water calcareous sandstones and conglomerates, which in turn are overlain unconformably by Oligocene lagoon-continental deposits (Geology..., 1977). As a whole, the Upper Cretaceous-Paleogene layers dip at low angles eastward towards the Ishikari graben-synclinorium. According to the seismic data (Den and Hotta, 1973), the thickness of the sedimentary cover in the Ishikari depression reaches 4000 m. The sedimentary cover seems to be composed of slightly deformed Neogene diatomic argillites, glauconitic shales and tufogenic sandstones (Geology..., 1977). In the southeast, the Ishikari graben-synclinorium is terminated by the Oyashio horst-anticlinorial uplift. In the axial part of this uplift, according to the deep-sea drilling data (Japan Trench..., 1978), there occur Upper Cretaceous clay aleurolites overlain by Oligocene-Neogene turbidites with a thickness of about 1000 m. The Erimo horst-anticlinal zone stretches along the base of the continental slope in the form of a smooth arc. Site 436 (Japan Trench..., 1978) revealed cherts and argillites of probably Upper Cretaceous age. K-Ar ages of basalts dredged from the Erimo horst near the axial part (Fig. 13) are about 52.8-80.1 my (Ozima et al., 1970). This indicates that deposits older than Upper Cretaceous can be expected within the Erimo zone.

According to the seismic refraction data (Fig. 14) the rocks beneath the sedimentary cover of the Oyashio rise have a velocity of about 4.8 km/s and probably are deformed geosynclinal deposits. The multi-channel seismic reflection data for the Japan trench near its junction with the Kuril-Kamchatka trench do not necessarily indicate subduction. Here the top of the second layer beneath the continental slope of Japan trench is traced for a distance of about 50 km from the trench axis and has been cited as evidence for subduction, but is overlain by the geosynclinal rocks of the continental slope. The eastern part of the Erimo horst-anticlinal zone gradually joins with the

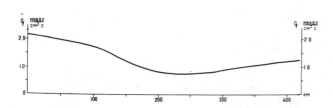

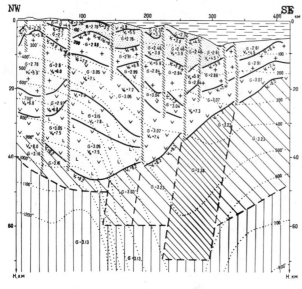

Fig. 9. Distribution of deep temperatures along profile 7-0 (see location, Fig. 11). (Deep seismic sounding..., 1978).

1-6 - crustal and upper mantle layers: 1 - sedimentary, 2 - granite-metamorphic, 3 - metabasaltic, 4 - crustal-mantle mixture, 5 - upper mantle (a - below asthenosphere, b - above asthenosphere), 6 - asthenosphere; 7 - isoterms.

Hidaka-Frontal horst-anticlinorium in which the Hidaka metamorphic complex was formed in Upper Cretaceous-Paleogene time from the eugeosynclinal deposits of the Sorati group. The Sorati group has stratigraphic range from Upper Permian to Lower Cretaceous (Geology..., 1977).

The Kuril continental slope. This slope is separated by the Bussol graben into southwestern and northeastern parts (Fig. 10, also Fig. 1a). In the structure of the southwestern part of the Kuril slope, there are horst-anticlinorial uplifts and graben-synclinorial troughs, the trend of which is parallel with the general trend of the trench. The trend, however, deviates northeastward at the Bussol transverse graben. Ac-

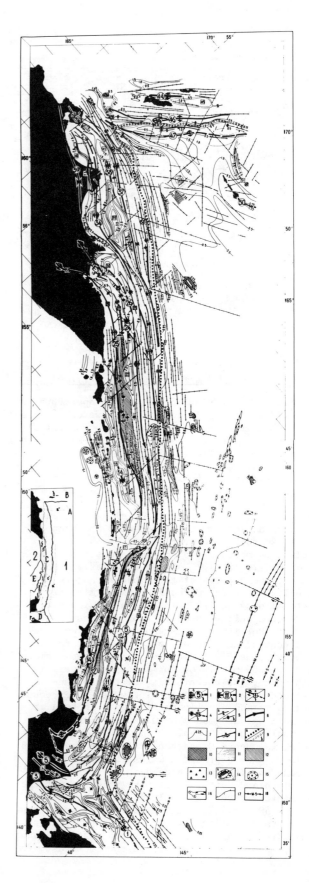

cording to the seismic data, sedimentary cover in the South-Kuril graben-synclinorial trough is well-stratified and seems to be composed of terrigenous and volcanogenic-terrigenous deposits. Their thickness in the Kunashir, Urup and Friza depressions (Fig. 1a) reaches 2500-3000 m (Snegovskoy, 1974; Tuezov et al., 1975). Three layers can be distinguished within the trough's sedimentary complex. The upper layer (about 100 m thick) lies horizontally and unconformably overlaps the intermediate layer. It overlaps the acoustic basement on the edges of the trough. The age of this layer seems to be Quaternary. The intermediate layer is about 1000 m thick. It is wedged out towards the trough edges where it unconformably overlaps the lower layer, especially in the southeastern flange (Fig. 15). The unconformities of the intermediate layer on the Kunashir (Early Miocene), Lovtsovsk (Early Miocene-Middle Miocene) and Alekhinsk (Upper Miocene-Lower Pliocene) suites near the Kunashir Island, and also on the Tebenkovsk (Early Miocene-Lower Pliocene) suites near the Iturup Island allows one to conclude that this layer is Upper Pliocene in age. It can be correlated with the Golovin suite (Middle Pliocene-Upper Pliocene) of Kunashir Island and the Parusny suite (Middle Pliocene-Upper Pliocene) of Iturup Island (Benz et al., 1971). The lower layer, concordantly underlying the intermediate layer in the axial part of the South-Kuril trough, is probably Lower Pliocene-Miocene in age and not likely older than Early Miocene as it lies disconcordantly on the Paleogene-Upper Cretaceous rocks of the Lesser Kuril ridge horst-anticlinorium.

Some large anticlinal and synclinal folds exist

Fig. 10. (Opposite) Tectonic map of the Kuril-Kamchatka deep-sea trench.
Inset. First order structural elements: 1 - the Pacific thalassocraton, 2 - northwestern part of the Pacific folded belt. Second order structural elements: A - geoanticlinal uplift of the Emperor seamounts, B - Aleutian folded-block uplift, C - Kuril-Kamchatka folded-block uplift, D - Japan folded-block uplift, E - Okhotsk Sea deep-sea basin. Third order structural elements: a' - marginal oceanic rise, a" - Obruchev rise, b - Komandor block, c' - Hokkaido-Kamchatka megaanticlinorium, c" - superimposed volcanic arc.
Index x. Fourth order structural elements: 1 - horst-anticlinal uplifts (a - axial zone sedimentary cover less than 250 m or absent, b - buried by sedimentary cover); 2 - graben-synclinorial troughs (a - established, b - inferred. Fifth order structural elements: 3 - horst-anticlinal uplifts (a - established, b - inferred), 4 - graben-synclinal troughs (a - established, b - inferred). Sixth order structural elements: 5 (a - anticlines, b - synclines); 6 - undulation of axial hinge lines, 7 - isopachs of the sedimentary cover, in km, 8 - faults, arrows indicate dip and displacement, 9 - deep-sea trench axis (a - without turbidites, b - with turbidites), 10 - acoustic basement outcrop of the Pacific plate, 11 - deformed geosynclinal Cretaceous-Paleogene-Lower Miocene complex, 12 - deformed geosynclinal Miocene-Pliocene complex, 13 - slump, 14 - volcanic structures (a - outcrops of acoustic basement on summits, b - with sedimentary cover on summits, and unstudied), 15 - probable Quaternary volcanic edifices, 16 - graben-synclinal troughs on marginal oceanic rise, 17 - boundary between marginal oceanic rise and Pacific plate, 18 - magnetic anomalies and numbers (after Hilde et al., 1976).

Index numbers of named structural elements. 1 - Oyashio horst-anticlinorial uplift, 2 - Ishikari basin, 3 - Ishikari graben-synclinorium, 4 - Kamuikotan horst-anticlinal zone, 5 - Hidaka-Frontal horst-anticlinorium, 6 - South Erimo horst-anticlinal zone, 7 - Eastern Erimo horst-anticlinal zone, 8 - Erimo horst-anticlinal uplift, 9 - Tokachi graben-synclinal zone, 10 - Tokachi horst-anticlinal zone, 11 - Outer graben-synclinorial trough, 12 - Lesser Kuril ridge horst-anticlinorium, 13 - Kunashir basin, 14 - Central Kuril graben-megasynclinorium, 15 - South-Kuril graben-synclinorial trough, 16 - Near-Axial horst-anticlinal zone, 17 - Bussol graben, 18 - North-Kuril graben-synclinorial trough, 19 - Paramushir basin, 20 - Chirinkotan horst-anticlinal zone, 21 - Shiashkotan slump, 22 - Orlik basin, 23 - Pegas horst-anticlinorium, 24 - Avacha graben-synclinorial trough, 25 - Dalnyaya graben-synclinal zone, 26 - Avacha horst-anticlinorial uplift, 27 - Povorotnaya horst-anticlinal zone, 28 - South horst-anticlinal zone, 29 - Conform graben-synclinal zone, 30 - Avacha basin, 31 - Ganalsk horst-anticlinal zone, 32 - Nalychevsk horst-anticlinal zone, 33 - Zhupanovsk basin, 34 - Kronotsk horst-anticlinorial uplift, 35 - Kronotsk basin, 36 - Periphery horst-anticlinal zone, 37 - Near-Kamchatka horst-anticlinorial uplift, 38 - Kronotsk graben-synclinorial trough, 39 - East-Kamchatka graben-synclinorium, 40 - Chazhminsk basin, 41 - Kamchatka basin, 42 - Udalyonnoe horst-anticlinal uplift, 43 - Kamchatka Cape horst-anticlinal uplift, 44 - Komandorsk horst uplift, 45 - Bering basin, 46 - South-Komandorsk horst-anticlinorial uplift, 47 - Prodolnoye horst-anticlinorial uplift, 48 - Uglovoye horst-anticlinal uplift, 49 - Obruchev horst uplift, 50 - Emperor seamounts horst uplift (volcanorium), 51 - Orthogonal swell.

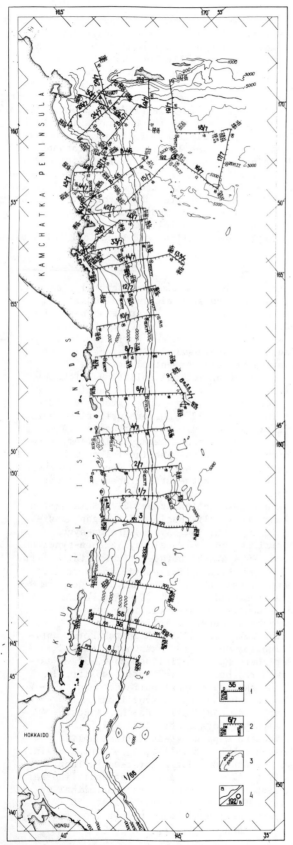

within this trough. Their formation seems to have resulted from the motion along faults penetrating from the basement to the sedimentary cover. The amplitude of these structural elements reaches tens and hundreds of meters with fold widths up to 10 km.

The axial part of the Lesser Kuril ridge horst-anticlinorium is traced from Shikotan Island (Fig. 1a) northeastward to the Bussol graben, for more than 500 km. To the southwest the horst-anticlinorium is traced more than 250 km, from Shikotan Island through the Nemuro peninsula to Hokkaido where, from the available data (Isomi, 1968; Tomoda, 1973; Ogawa and Suyama, 1976), it turns northward to the Kitami-Jamato rise, northeastern of Hokkaido (Fig. 10). According to the seismic data (Fig. 15) two structural-formational complexes can be distinguished in this horst. They are composed of well-pronounced anticlinal structures on the northwestern edge where there is a normal fault which divides the horst and the South-Kuril graben-synclinal trough. The absence of regular seismic reflections is characteristic of the lower structural-formational complex. This complex outcrops in the axial zone of the horst-anticlinorium. The Lesser Kuril islands are within the lower complex area. They are composed of Upper Cretaceous-Paleogene volcanics and volcanogenic-sedimentary deposits, broken by gabbroids, which as a whole form a spilite-diabasic rock association, associated with the mature stage of the geoanticlinal uplift development (Gavrilov and Solovieva, 1973; Sergeev, 1976). The available K-Ar ages (Fig. 13) are within the range of 62-160 my, which points to Upper Cretaceous-Jurassic intrusive activity. From the dredging data (Fig. 3; sites 1094, 1095, 1109, 1110, 1119, 1123 and 1124), the lower structural-formational complex is considered to be composed of greywackes, coaly-micaceous-chloritic shales, andesites, basalts, metadiabases, hornfelses, granodiorite-porphyres, granite-porphyres, quartz porphyres and broken down granites with seismic P-wave velocities ranging 5.0-6.5 km/s. The dredged rocks differ considerably in composition from the rock complex of the Lesser Kuril islands in the occurrence of metamorphosed varieties of rocks and granitoids. Probably, the structural-formational complex outcropping in the central part of the horst-anticlinorium, occurs stratigraphically lower than the complex of volcanogenic-sedimentary rocks and volcanics outcropping in the Lesser Kuril islands. It is, perhaps, of Pre-Campanian age.

The upper structural-formational complex of

Fig. 11. (Opposite) Index map for seismic refraction and continuous seismic reflection profiles shown in this paper. 1 - number of profile, time and length in km; 2 - number of profile, time and length in the hours; 3 - isobaths in m; 4 - (a) axis of the trench, and (b) site of deep-sea drilling.

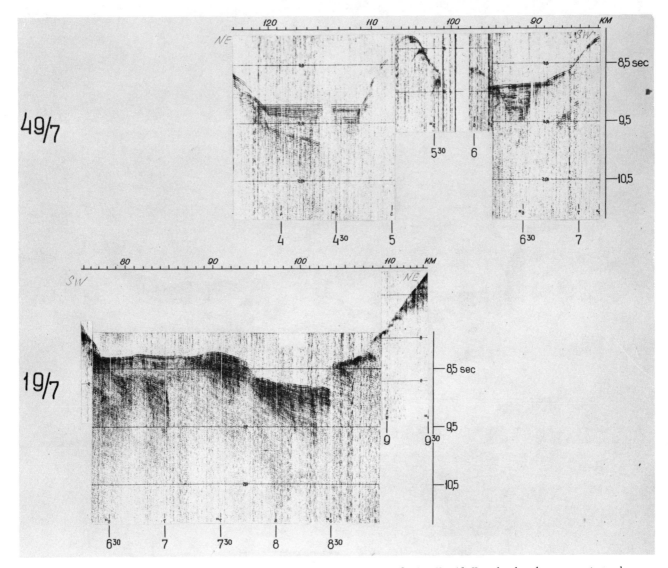

Fig. 12. Continuous seismic profiles for the thalweg of the Kuril-Kamchatka deep-sea trench (see locations, Fig. 11).

the Lesser Kuril ridge horst-anticlinorium is well-stratified (Fig. 15) and can be subdivided into lower and upper layers. The lower layer is characterized by shorter reflections than the upper one. Both layers conformably dip at an angle of 15-25° from the horst-anticlinorium axial zone. The internal structure of both layers is characterized by intensive deformations consisting of faults and relatively large, near-fault anticlinal and synclinal folds. According to the dredging data (Fig. 3; sites 1091, 1093, 1094, 1095, 1101, 1102, 1111, 1119, 1121, 1123, 1125, 1126 and 1128) the upper structural-formational complex is composed of volcanogenic-sedimentary deposits and volcanics and includes greywackes, argillites, tuff diatomites, polymict conglomerates, tuffs of andesite-basalts and lavas of basalt and andesite. The lower layer seems to be characterized by large amounts of volcanics,

whereas the upper one is composed mainly of volcanogenic-sedimentary deposits. Seismic P-wave velocities measured in samples of the upper structural-formational complex range from 2.2 to 3.5 km/s. According to the relics of diatoms, silicoflagellates, spores and pollen, and also from the microfauna relics (Vasiliev, 1974) the age of the upper structural-formational complex of the Lesser Kuril ridge horst-anticlinorium ranges from Oligocene-Miocene up to Pliocene.

The linear horst uplift, separating the Vneshny graben-synclinorial trough from the horst-anticlinorium of the Lesser Kuril ridge, seems to be composed (Fig. 3; site 1091) of metamorphosed greywackes, phyllites, tuffs of medium composition, andesite-porphyrites, rhyolites, and broken down gneissic rocks with seismic P-wave velocities (measured from samples) ranging from 4.0 to 6.0 km/s. The Vneshny graben-synclinorial trough is

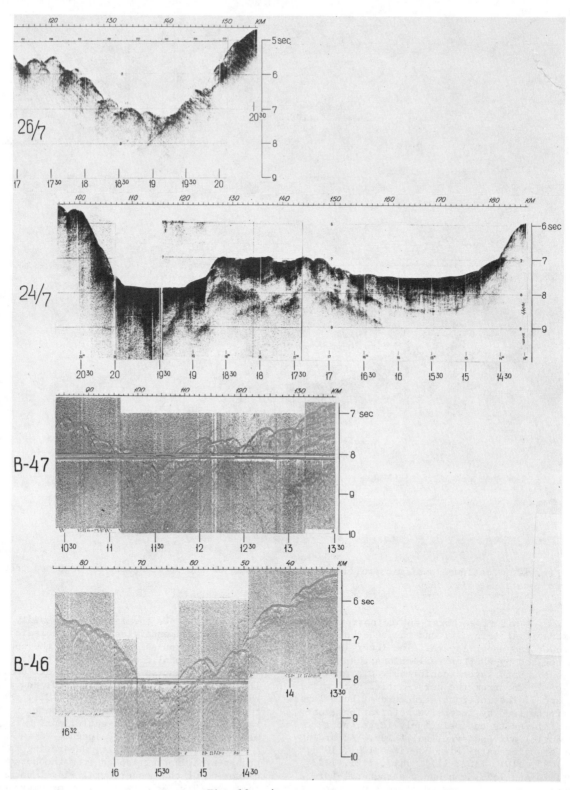

Fig. 12. (continued)

in the northwestern edge of the Hidaka-Frontal horst-anticlinorium and is filled by a layer of well-stratified, probably, turbidite deposits with thickness up to 1.5-2.5 km. The trough's inner structure is a small syncline with dip angles of layers towards the center up to 10°. According to the dredging data (Fig. 3; sites 1104, 1107 and 1108) the sedimentary complex in the trough is

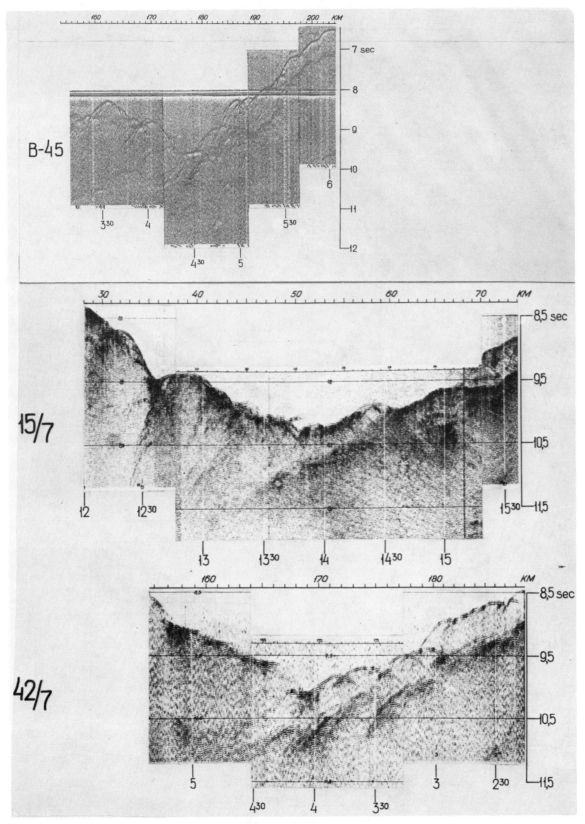

Fig. 12. (continued)

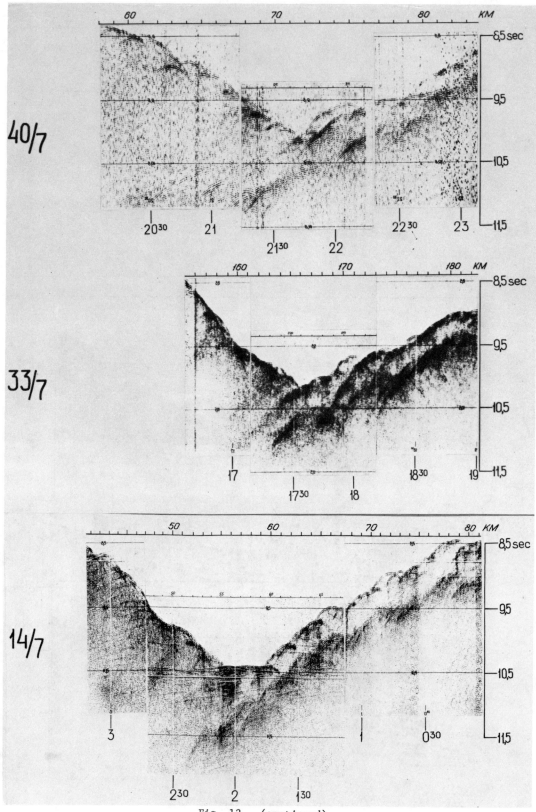

Fig. 12. (continued)

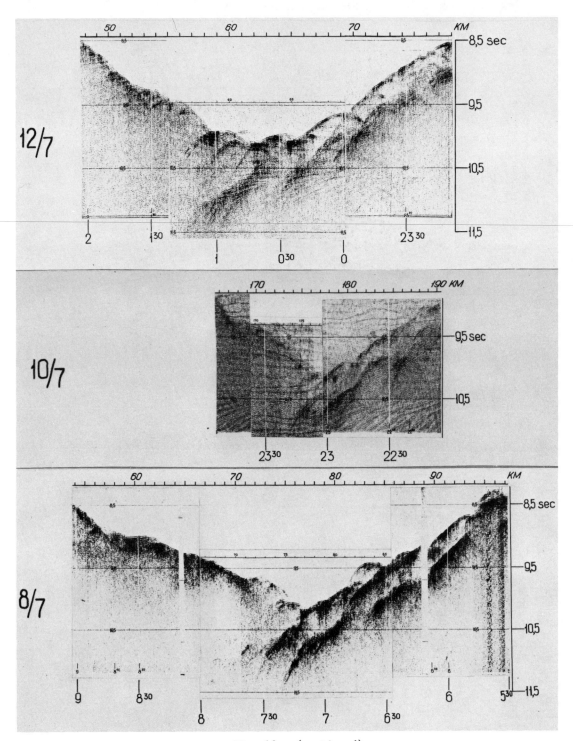

Fig. 12. (continued)

composed of aleurites, sandstones, aleurolites, diatomites, tuffites and tuffs. From analysis of the spores, pollen and diatoms in the samples of aleurites and tuff diatomites taken from sites 1104 and 1132, this sedimentary layer seems to be of Pliocene-Pleistocene ages.

The Hidaka-Frontal horst-anticlinorium stretches along the outer edge of the deep-sea terrace at depths of 3.5-4.5 km, for a distance of more than 1000 km. Dredging in the axial zone of the Hidaka-Frontal horst-anticlinorium (Fig. 3; sites 1104, 1108 and 1132) showed that the acoustic basement here is composed of metamorphosed greywackes, tuffs and tuff breccias of basic and intermediate composition, basalts, andesites, dacites, rhyolites, hornfels, quartz

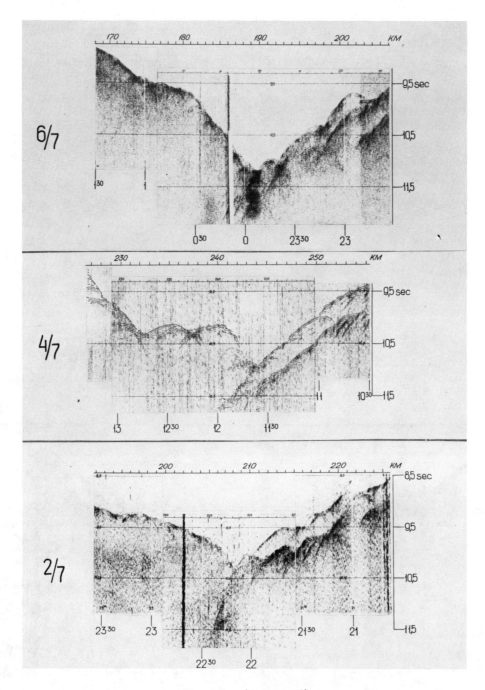

Fig. 12. (continued)

diorites and syenites. Seismic P-wave velocities (measured on samples) ranges between 4.0-6.0 km/s. Dredged rock associations, considerable metamorphism and cataclas indicate the far advanced tectonic development of the Hidaka-Frontal horst-anticlinorium.

The roof of the second oceanic layer is traced by the seismic profiling data (Fig. 12) beneath the Near-axial horst-anticlinal zone (16 in Fig. 10), dipping at angles of 5-15°, for a distance of about 15 km from the trench axis. A complex of deposits of the Near-axial horst-anticlinal zone wedges out at the trench axis.

To the northeast of the Bussol graben, the North-Kuril graben-synclinorium (18 in Fig. 10) stretches for a distance of about 400 km, having a centriclinal closure eastward of Paramushir Island (Fig. 1a). This trough is about 30-50 km

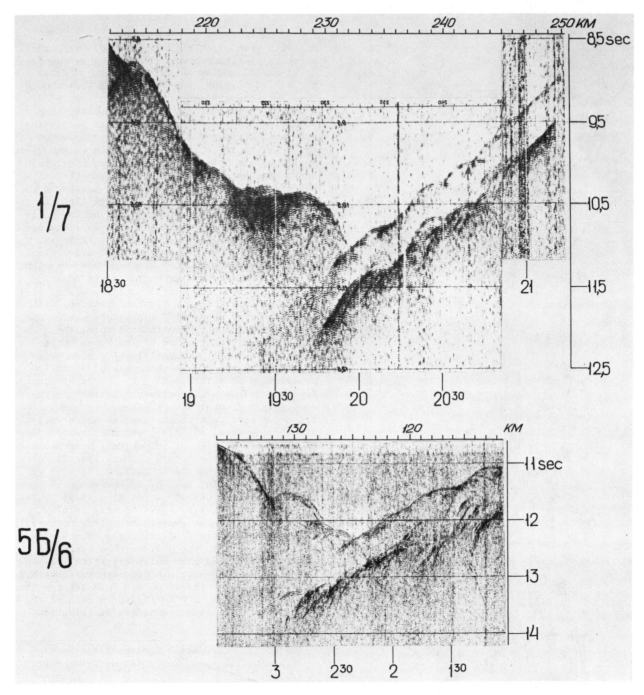

Fig. 12. (continued)

wide. Together with the South Kuril graben-synclinorium it forms the Central-Kuril graben-megasynclinorium (Fig. 10). In the northern part, the North-Kuril graben-synclinorium is partly filled by a well-stratified layer (Fig. 16) of probable turbidite and molasse deposits with a thickness of about 3000 m. This layer has relatively weak folds, but intensive fault deformations. The palynological data show that Upper Miocene deposits seem to occur at the base of the sedimentary complex and therefore the upper part of the section apparently includes not only Upper Miocene, but also Pliocene deposits. The Pegas horst-anticlinorium (23 in Fig. 10) is traced from the Ganalsk horst-anticlinal zone (Kamchatka) to the Bussol transverse graben, within which this synclinorium seems to have an echelon-like relationship with the Lesser Kuril ridge horst-anticlinorium. According to seismic profiling data (Fig. 16), the horst-anticlinorium axial zone is characterized by short reflections. The upper layer is represented by well-stratified

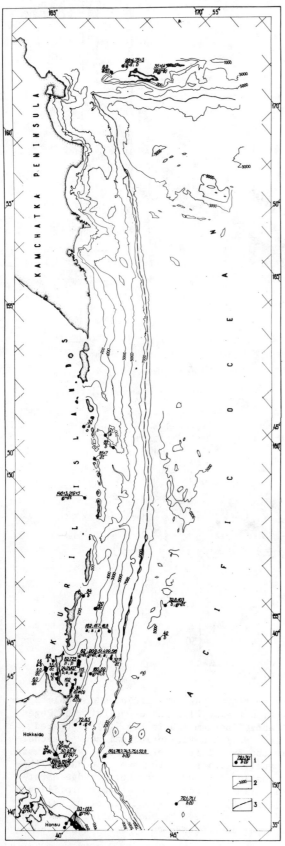

deposits with clearly distinguishable structural elements of higher orders. Dredging in the Pegas horst-anticlinorium axial zone (Fig. 3; sites 61-65, 71-77, 107-109, 115-118 and 123-130) showed that rock associations here are represented by a geoanticlinal volcanogenic-sedimentary complex, involving basalts, andesites, quartz porphyries, and their tuffs and also tuffites, greywackes and pellitomorphic limestones. This volcanogenic-sedimentary complex is broken by diorites, leucocratic and hornblende granites. From the palynological data, the rocks from the northwestern edge of the axial zone (Fig. 3; site 116) are of Middle Miocene age. Farther south of the anticlinorium axial zone (Fig. 3; site 128) the age of basalts is 85 my. From these data, it is suggested that the Pegas horst-anticlinorium axial zone is composed of Upper Cretaceous rocks with Paleogene and Neogene deposits at its edges.

According to the coastal data (Aprelkov, 1971), Oligocene-Lower Miocene flysch formations with thicknesses of about 2500 m outcrops within the South-Kamchatka horst-anticlinal zone, the northeastern continuation of the Pegas horst-anticlinorium. It is overlain by a Miocene andesite formation with a thickness of about 1500 m, which is in turn overlain by the Upper Miocene-Pliocene volcanogenic molasse. Farther northward within Kamchatka, the Pegas horst-anticlinorium continues in the Ganalsk horst-anticlinal zone (Fig. 10, #31), which is composed of an ophiolite complex of intensively deformed and metamorphosed Mesozoic or, perhaps, partly Paleozoic deposits. The oldest radiometric data of the Ganalsk complex the 487 my Rb-Sr age of plagio-granites (Gnibidenko et al., 1974; Dyufur et al., 1977; Tararin, 1977; German, 1978).

The Avacha graben-synclinorium (Fig. 10, #26) extends southeastward of the Pegas horst-anticlinorium. Within the Orlik graben synclinal zone, it is filled with well-stratified sedimentary layers, probably turbidite deposits, with a thickness of more than 2000 m. Fold deformations are observed along the southeastern edge of the trough, normal faults are widely distributed

Fig. 13. (Opposite) Radiometric ages of rocks for the Kuril-Kamchatka deep-sea trench region.

1 - sites (numerator is the age in millions of years, denominator is the name of rock and figure in brackets is reference); 2 - isobaths in meters; 3 - trench axis. References: (1) - Scholl et al., 1976a; (2) - Schmidt, 1978; (3) - Yagi, 1969; (4) - Nozawa, 1975; (5) - Ozima et al., 1970; (6) - Bikerman et al., 1971. Names of rocks: a - andesite, a-b - andesite-basalt, b - basalt, g - gabbro, g-d - gabbro-diabase, gr - granite, grdt - granodiorite, gt-am - garnet-amphibolite, dl - dolerite, dl-m - dolerite-monzonite, dt - diorite, dc - dacite, p - porphyrite, plgr - plagiogranite, s - syenite.

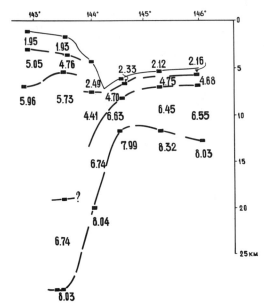

Fig. 14. Section (1/66) obtained by seismic refraction in the northern Japan trench (see location in Fig. 11; after Ludwig et al., 1966).

within this zone (Fig. 17). Clear transgressive onlap of the upper sedimentary layer is observed along the trough's northwestern edge. Along its southeastern edge, there are unconformities between the upper and lower folded layers. Within the central part of the Orlik graben-synclinal zone, the stratified sedimentary complex seems to be of Pliocene-Pleistocene age. Eastward of the Avacha graben-synclinorium (Fig. 10, #24) there is the Avacha horst-anticlinal uplift (Fig. 10, #26), extending to the Nalychev horst-anticlinal zone (Fig. 10, #32) in Kamchatka. There is no data on the composition and age of this horst-anticlinorial uplift in the marine area. Within the Nalychev horst-anticlinal zone, this uplift is composed of cherty-volcanogenic Upper Cretaceous-Paleogene rocks, overlain by a Miocene andesite formation (Geology of the USSR, v. 31, 1964; Aprelkov, 1971).

A large basin (Figs. 10 and 17) is superimposed on the northern part of the Avacha horst-anticlinorial uplift. According to the available data on the adjacent coast (Geology of the USSR, v. 31, 1964), this basin is filled with 1000-2500 m Upper Miocene-Pliocene molasse.

The Kamchatka continental slope. The Kronotsk horst-anticlinorial uplift (Fig. 10, #34) connects with the East-Kamchatka graben-synclinorium (Fig. 10, #39). This graben-synclinorium is composed of intensively deformed Oligocene-Lower Miocene flysch with a thickness of about 8000 m, above which there occurs a less deformed sedimentary-volcanogeneous Middle Miocene-Pleistocene molasse with a thickness of about 3500-4000 m (Geology of the USSR, v. 31, 1964; Gnididenko et al., 1974). According to the dredging data (Fig. 3; sites 2, 3, 6 and 13), the Kronotsk horst-anticlinorial uplift is composed of volcanogenic greywackes, cherty rocks, tuffs, basalts and andesites, probably intruded by gabbroids, biotite-hornblende, quartz diorites, biotite-hornblende and hornblende granites, and hornblende granodiorites. The rocks are metamorphosed to zeolite and partly to the greenschist facies. The axial zone of the Kronotsk horst-anticlinorial uplift involves a complex of rocks outcropping in Shipunsky and Kronotsky peninsulas and also in the central and western parts of the Kamchatsky Mys peninsula (Fig. 1a). This complex (Geology of the USSR, v. 31, 1964; Sadreev and Dolmatov, 1965; Serova, 1966; Borzunova et al., 1969; Shapiro, 1976) is composed of altered basalts, andesite-basalts, andesites and their tuffs and aleurolites and sandstones with a total thickness of about 5000 m, and intruded by diorites, gabbroids and serpentinized peridotites. The lower 1000 m of the section is of Upper Cretaceous age, whereas the upper part is of Paleocene-Lower Miocene age. The entire complex is intensively folded and is broken by numerous faults and thrust-faults.

The Kronotsk graben-synclinorial trough (Figs. 10 and 18), between the Kronotsk and Pri-Kamchatka horst-anticlinorial uplifts, is filled with a layer of well-stratified, probable turbidite deposits, the thickness of which (if the average seismic P-wave velocity is about 2 km/s) exceeds 3500 m. Unconformities are found in the upper layer and at the trough edges. Layers with disturbed internal structure are probably due to land-slides. The age of the sediments is unknown, but considering the structural association of this trough with the Nerpichya basin (west of the Kamchatsky Mys peninsula), which is filled with moderately to slightly deformed Miocene-Pliocene terrigenic molasse, one can infer that the stratigraphic range of the sedimentary complex in the Kronotsk graben-synclinorial trough may be from Middle-Upper Miocene to Holocene.

The Pri-Kamchatka horst-anticlinorial uplift (Fig. 10, #37) consists of exposed acoustic basement from 49°N and more distinctly, from 51°N. According to dredge samples (Fig. 3; sites 4, 8, 9 and 10) the rocks here are represented by volcanogenic greywackes of intermediate and acidic compositions, tuff diatomites, and pelitomorphic clay limestones and basalts, quartz diabases, and biotite-hornblende granites.

From the tuff diatomites of the deepest site (Fig. 3; site 10), the following diatoms are identified (V.S. Pushkar): Melosia albicans; Denticula kamtschatica; Stephonopyxis shenckii and numerous species of the Thalassiosira genus; Th. antiqua; Th. excentrica; Th. gravida f. fossilis; Th. nativa; Th. ocstrupii; Th. zabelinae, which characterize the Early Pliocene. These tuff diatomites dredged from the base of the continental slope seem to be part of the sedimentary cover on the acoustic basement which is older than

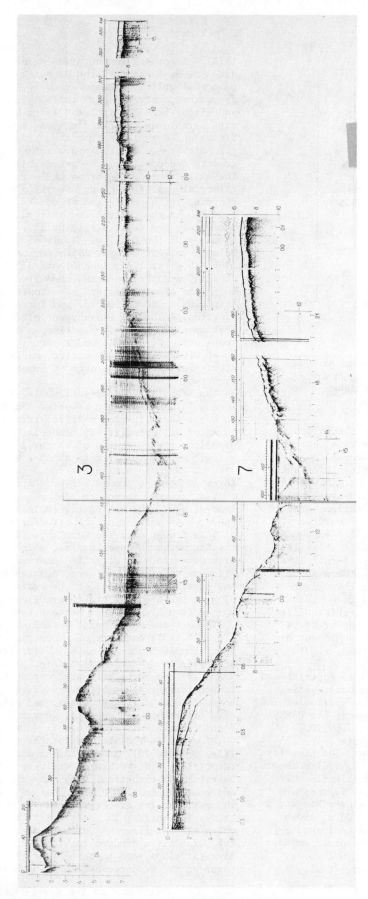

Fig. 15. Continuous seismic profiles across the southern Kuril-Kamchatka trench (see location in Fig. 11).

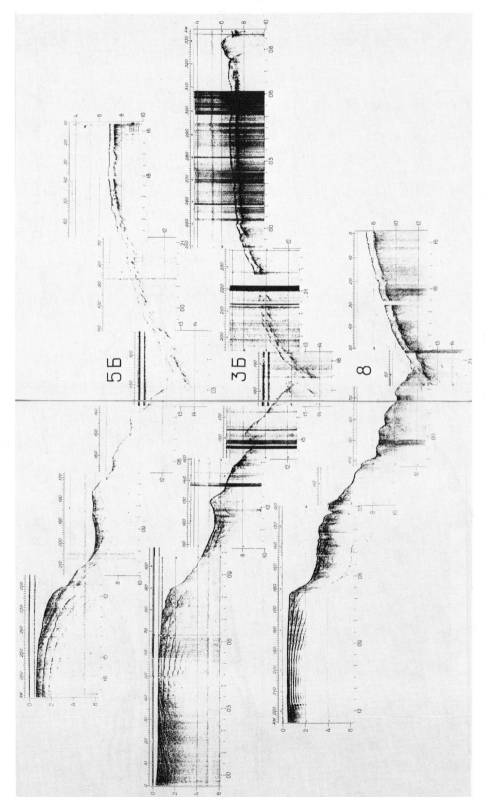

Fig. 15. (continued)

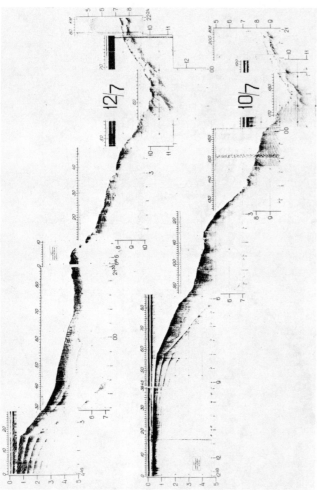

Fig. 16. Continuous seismic profiles across the middle Kuril-Kamchatka trench (see location in Fig. 11).

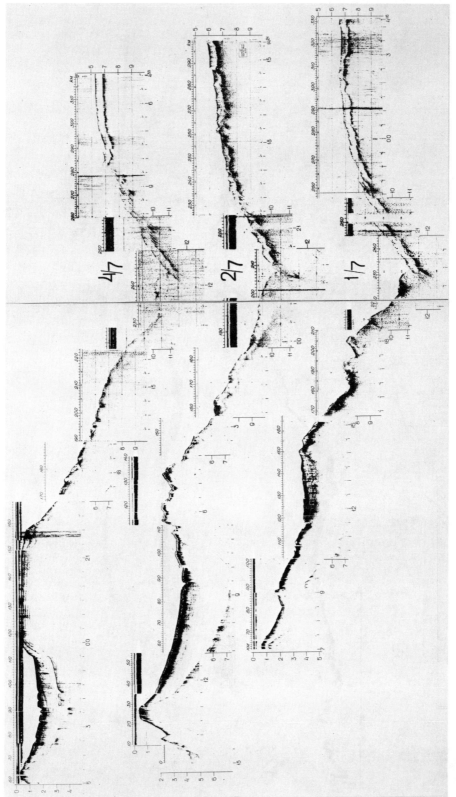

Fig. 16. (continued)

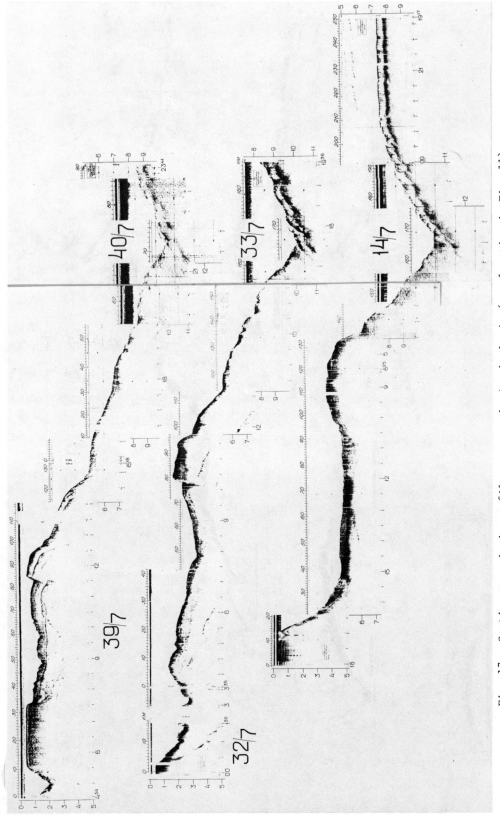

Fig. 17. Continuous seismic profiles across the Avacha basin (see location in Fig. 11).

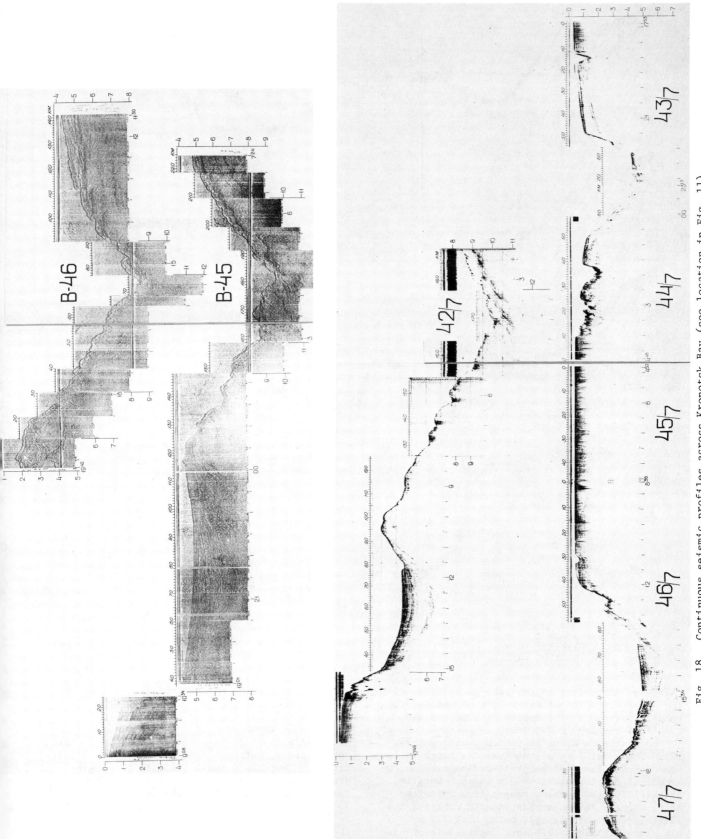

Fig. 18. Continuous seismic profiles across Kronotsk Bay (see location in Fig. 11).

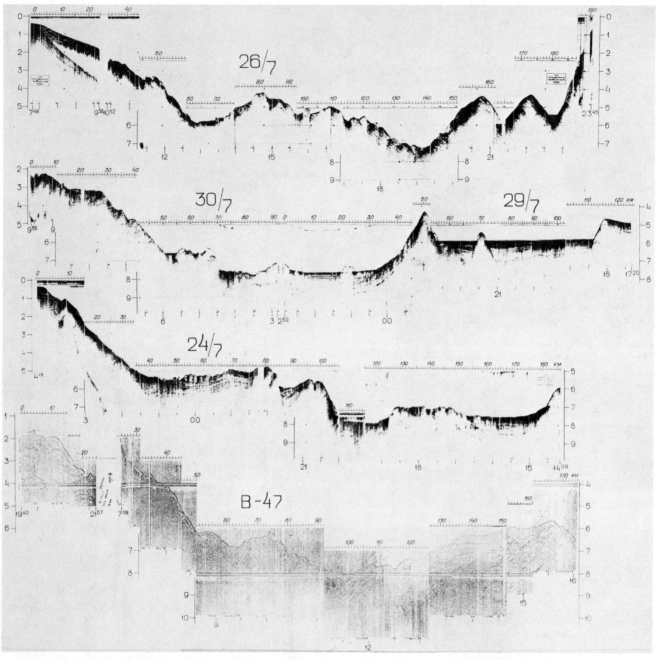

Fig. 19. Continuous seismic profiles across the junction between the Kuril-Kamchatka and Aleutian trenches (see location in Fig. 11).

Pliocene. According to the rock association of the dredged samples at sites 4, 8, 9 and 10, the acoustic basement is composed of a mature geosynclinal complex.

The Kamchatka-Aleutian trench junction (Fig. 19). The axial part of the Komandor block is composed of deformed volcanogenic-sedimentary eugeosynclinal complex about 4000 m thick. Stratigraphically, the complex ranges from Paleogene to Pliocene. The gabbroids, quartz diorites and plagio-granites (K-Ar ages being 35 ± 1.4 my, Oligocene; Schmidt, 1978) are intruded into this complex. According to the dredging data on the northern slope of the Komandor block (Fig. 3; sites 14 and 15), the base of the uplift appears to be composed mainly of basalts, diabases, gabbro-diabases, tuffs of basic and intermediate compositions, metamorphosed into the greenschist facies. There is a small quantity of plagio-granites and felsite-porphyries in the dredged samples. K-Ar ages of the basalts (site 15) are 75 ± 3 my and for the gabbro-diabases (site 15), 96 ± 4 my. From these data one can conclude that the basement of the Komandor block involves a

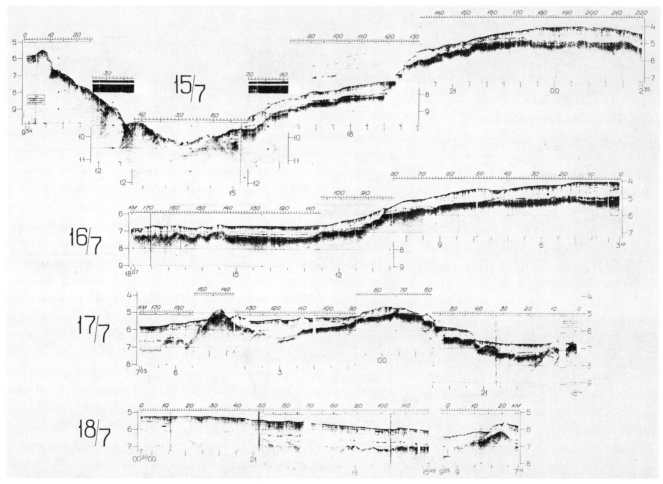

Fig. 20. Continuous seismic profiles through site 192 across the Obruchev rise (see location in Fig. 11).

complex of Upper Cretaceous eugeosynclinal rocks metamorphosed into the greenschist facies. The upper part of the Komandor block was dredged at sites 16, 17, 18, 19 and sites B-49-I (Fig. 3). Rocks dredged from sites 16, 17, 18 and 19 consist of lithoclastic and vitroclastic tuffs and breccias of basic and intermediate compositions and also slightly lithified tuff diatomites. In the tuff diatomites V.S. Pushkar determined: Actinocyclus ingens; Cosmodiscus insignis; Cosmodiscus intersectus; Denticula hustedtii; Denticula lauta; Kisseleviella carina; Mediaria splendida, indicating a Middle Miocene age. Igneous rocks from these sites consist of altered basalts and diabases. From sites B-49-I (Fig. 3) (Scholl et al., 1976) andesites, cherty-diatom shales and argillites were dredged. K-Ar ages of the andesites dredged from this site are 8.3 ± 0.4 my (Middle Miocene), which is well correlated with the diatom flora relics from the sedimentary rocks dredged from the same site, i.e. a complex of Denticula hustedtii-Denticula lauta.

The structure of the northern part of the Emperor seamounts within the junction region between the Aleutian and Kamchatka trenches is a system of uplifts having two structural strikes (Fig. 10): submeridional and northwestern.

The central part of the Obruchev rise (Fig. 1a) is covered by a thick pelagic deposits occurring on the acoustic basement, which according to results of drilling at site 192 (Initial Reports..., 1973) is represented by alkaline basalts. Basalts are overlain by a 1044 m thick sedimentary cover, represented (from below-upwards) by Cretaceous-Paleogene deposits of chalk and argillites with a thickness of 132 m, Lower-Middle Miocene argillites with a thickness of 207 m, Middle-Upper Miocene diatomaceous oozes with a thickness of 155 m, Upper Miocene-Pliocene diatomaceous oozes with a thickness of 410 m and Upper Pliocene-Pleistocene-Holocene aleurite clays with volcanic ash intercalations with a thickness of 140 m (Fig. 20). The formation of the relatively thick sedimentary cover on the Obruchev uplift (Scholl et al., 1977) is considered to be the result of accelerated (35-55 m/my) sedimentation beginning from Upper-Middle Miocene to Holocene inclusive, due to the intensive supply of clay material carried by the Kamchatka current spreading along the Obruchev rise.

The structural pattern of the junction between the Aleutian and Kamchatka trenches (Fig. 10)

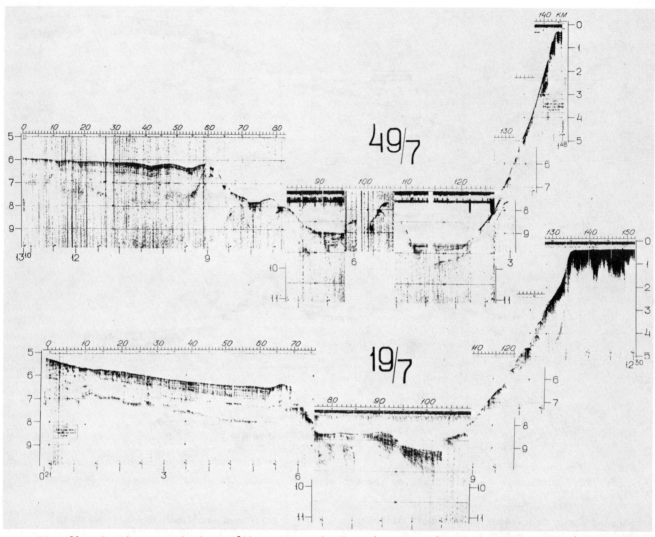

Fig. 21. Continuous seismic profiles across the Komandor part of the Aleutian trench (see location in Fig. 11).

shows a gradual change of structural elements from the Obruchev rise to the Pri-Kamchatka horst-anticlinorial uplift. The Uglovoye and Prodolnoye uplifts (Fig. 10, #48 and 47) are traced across the trench into the base of the Kamchatka continental slope. Thus, if the basement of the main horst-anticlinorial uplifts of the junction between the Aleutian and Kuril-Kamchatka trenches is of upper Cretaceous age, it is reasonable to suggest that the geosynclinal structural pattern within this zone originated in Cretaceous time.

The Aleutian deep-sea trench within the Komandor block area is a graben, formed as a result of normal faulting (Fig. 21), partly filled by turbidites with a thickness of about 1 km. The age of the turbidite deposits in the Aleutian trench is unknown, although the high rate of sedimentation suggested there (more than 1000 m/my; Scholl, 1974) would indicate that they are not older than Pleistocene. A lack of deformation indicates that formation of this graben occurred before deposition of the turbidites.

The oceanward slope. Over its whole length, the oceanward slope of the Kuril-Kamchatka trench is characterized by numerous longitudinal faults dipping towards the trench axis (Fig. 12). These faults appear to be "normal" faults, indicative of a tensional environment. The thickness of the sedimentary layer, from the marginal oceanic rise to the trench slope, usually decreases to less than 250 m (seismic velocity being assumed as 1.7-1.8 km/s). On the oceanic slope this layer fills the grabens and can be traced to the base of the continental slope. Where there are turbidite or land-slide lenses, the oceanic sedimentary layer is buried beneath them. It is significant that in some cases the thickness of the first layer is very small and the acoustic basement (oceanic second layer) outcrops to the sea bottom, usually in the horst swells (Figs. 10 and 12).

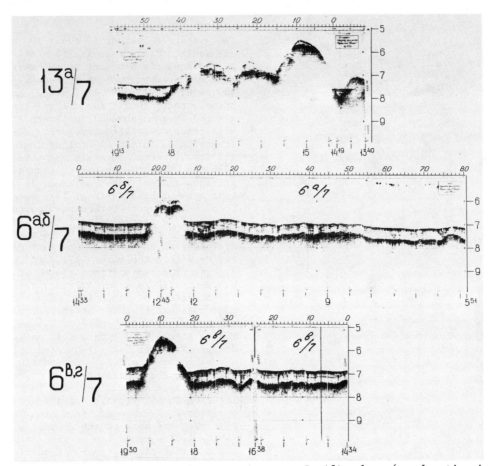

Fig. 22. Continuous seismic profiles of the Northwestern Pacific plate (see location in Fig. 11).

The trench axial zone. The trench axis is curved (Figs. 1 and 10) where the trench intersects with the transverse faults. Since these curved parts of the trench coincide with the turbidite wedges, which began to form in Pre-Quaternary time, it is quite possible that the transverse deformation of the trench occurred as long ago as Pliocene.

The Kuril-Kamchatka trench is not filled with sediments. A large portion of the trench contains only the first layer of the oceanic plate. In places, however, these sediments are overlain by a thin layer (about 100 m) of turbidites, or by relatively thick land-slide lenses. Observations from the "Archimed" bathyscaphe (Houot, 1976) made in 1962 at the depth range of 9500-9545 m on the Iturup Island traverse showed that the trench bottom is flat and covered with yellowish ooze with colonies of sea feathers. Within the Kuril-Kamchatka and Aleutian trench junction, the turbidites thickness increases up to 1000 m. As profile 26/7 (Fig. 19) indicates, near the northeastern end of the Kuril-Kamchatka trench, there are saddle-like uplifts of the basement. The turbidites are south of here. The source of these turbidites seems to be mainly by delivery of sediments from the Kamchatka River, not from the Bering Sea. Farther north of the saddle-like uplifts, at 55°30'N, the Kuril-Kamchatka trench is filled with more than 1500 m of well-stratified turbidites. Their stratigraphic range in the Komandor basin is Upper Miocene-Pleistocene and Holocene (Initial Reports..., 1973).

Tectonics of the Northwestern Pacific plate

The marginal oceanic rise. Within the oceanic plate, the marginal rise is a second order structural element, poorly pronounced in relief. The sediment cover of the marginal oceanic rise consists of an alternation of acoustically "transparent" and "opaque" layers, the number of which may range from 2-3 to 6-8 (Fig. 22). According to the drilling data from adjacent areas (Initial Reports..., 1973, 1975), the upper 70 m of the sediments are composed of diatomaceous, diatomaceous-radiolarian oozes and zeolitic pelagic clays with erratic pebbles and ashes. The underlying 70-90 m is represented by pelagic clays and nannoplanktonic oozes alternating with cherts. The seismic P-wave velocities in this part are 2.7-2.8 km/s. The stratigraphic range of the sedimentary cover is from Valanginian to Holocene.

Fig. 23. Crustal thickness of the Kuril-Kamchatka trench region.

The average thickness is about 300 m and the average sedimentation rate is 2.5 m/my.

The acoustic basement of the marginal oceanic rise was dredged on some seamounts (Fig. 3; sites 453, 456, 1089, 1090, 1112, 12/77 and 3/77). The primary rocks in all the dredges are strongly altered (metamorphosed) and brecciated basalts with trachydoleritic structure. Besides basalts, fragments of tuffs, aleurolites, greywackes, argillites, cherts, hornfels, schists, andesites, diabases, felsite-porphyres, granodiorites, granites and aplites (Vasiliev, 1978) were dredged. K-Ar ages of basalts (Fig. 13), range from 80.1 to 32.6 my (Upper Cretaceous-Oligocene) and a K-Ar ages of a granodiorite (site 1112) is 103 my. Such an age scattering seems to testify to the prolongation of the last stage of magmatic activity within the rise. All dated samples were dredged from the uplifted slopes of the acoustic basement which may be isolated, relatively long-lived centers of volcanism. The acoustic basement adjacent to these horsts may be older than the basement of the uplifts. It is suggested (Ozima et al., 1970) that K-Ar dating of basalts dredged from seamounts indicates only the minimal age of their formation. Ice rafting has been suggested for the occurrance of some of these rocks, such as the granodiorites, but their seafloor location in the vicinity of steep slopes of exposed acoustic basement favor the interpretation that they are part of the oceanic plate.

The sediments of the marginal rise are cut by the vertical dislocations. Acoustic basement is often exposed on the faults in this region. All this testifies to the active vertical movement of crustal blocks within the marginal oceanic rise, producing a system of longitudinal grabens and horsts.

The oceanic plate. The sedimentary cover of the oceanic plate can be characterized by the results obtained at site 303 (Initial Reports..., 1975). The hole was drilled on M-4 magnetic anomaly and reached acoustic basement at a depth of 285 m. This basement is composed of tholeiitic basalts metamorphosed into the zeolitic facies. The seismic P-wave velocity measured in basalts is about 4.5 km/s. The sedimentary cover is composed (from below-upwards) of nannofossil oozes with chert intercalations, having a thickness of about 35 m, above which there are diatomaceous-radiolarian oozes with zeolitic pelagic clays (with intercalations of cherts at its lower part). The thickness of this layer is about 250 m. According to the faunal relics, the stratigraphic range of the sedimentary cover is from Valanginian to Late Pleistocene. Sedimentation rates range from 0.5 to 16 m/my.

The difference between the sedimentary cover on the oceanic plate and that on the marginal rise, is the absence of the upper, relatively thick acoustically opaque layer on the oceanic plate. The opaque layer was probably formed by the occurrence of pyroclastic intercalations.

Deep Structure

Seismic refraction data (Crustal structure..., 1964; Ludwig et al., 1966; Zverev and Merklin, 1966; Tulina, 1969; Den et al., 1971; Okada et al., 1973; Suvorov, 1975; Bulin, 1977), two regions with different types of crust are distinguished: the trench continental slope region and the region of the oceanward slope and adjacent oceanic plate (Fig. 23).

On the continental slope eastward of Iturup Island, the crust has its maximum thickness along the Kuril geoanticlinal uplift. In the south of the Kuril geoanticlinal uplift the upper part of the crust is composed of the sedimentary-volcanogeneous complex with velocities of 4.2-5.6 km/s (Popov et al., 1978) and a thickness of about 3-5 km. This complex is overlain by the sedimentary cover with a thickness up to 3 km and velocities of 1.8-3.6 km/s. The sedimentary-volcanogenic deformed complex is underlain by a layer with velocities of 6.0-6.2 km/s, which seems to correspond to the granite-metamorphic layer. The seismic boundary between these two complexes has not been clearly defined due to insufficient observations. The metabasaltic layer with the velocity of about 6.7 km/s has a thickness of 10-13 km and is separated from the upper mantle by the M discontinuity. The P_m velocity is 7.8-8.4 km/s, but the M boundary is often indistinct. Crustal thickness in the central part of the Kuril geoanticlinal uplift decreases to 10-15 km. There the maximum thickness (15 km) is on the continental slope near the trench and the main part of the crust seems to be represented by the metabasaltic and metadioritic layers. 8.0-8.4 km/s is the characteristic Pm-velocity. Farther to the northeast, the crustal thickness on the continental slope rapidly increases up to 25 km mainly due to the thickening of the metabasaltic layer. Within Avacha Bay it reaches 40-45 km (Fig. 9). As a whole, the increase of crustal thickness on the trench continental slope corresponds to the existence of the outer horst-anticlinorial uplifts.

The oceanic crust consists of three layers. The upper, sedimentary layer (layer 1) has a thickness seldom exceeding 0.5 km. Seismic P-wave velocities within it are about 1.7-2.0 km/s. This layer overlies the sedimentary-volcanogenic layer (layer 2) with velocities of 4.5-5.0 km/s. Layer 2 seems to be composed of alternations of basalts and sedimentary rocks. The thickness of this layer usually ranges from 1 to 2 km. Below layer 2, layer 3 with velocities of 6.4-7.4 km/s and a thickness of 4-8 km, probably consist of gabbro and metamorphosed basalts. The Pm-velocity beneath the oceanic crust in this region ranges from 7.8 to 8.9 km/s.

By the converted wave earthquake method (CWE) (Bulin, 1977), boundaries of wave conversions were distinguished in the sections below the M discontinuity along Kamchatka. These boundaries indicate that there is a considerable stratification in the upper mantle. According to the CWE data, the crustal thickness beneath the Kuril goeanticlinal uplift is about 10 km greater than that obtained by deep seismic sounding. This gives reason to believe that the crust beneath the Kuril uplift is continental rather than subcontinental.

Geological Development

Paleogeography

The data on rock associations of the acoustic basement on the continental slope give reasons to believe that the Hokkaido-Kamchatka horst-meganticlinorium is in a stage of advanced tectonic development. These associations indicate the mature, apparently, orogenic stage of this geosynclinal system.

Since the Hidaka-Frontal (Fig. 10, #5), Avacha (#26), Kronotsk (#34) and, probably, Pri-Kamchatka (#37) horst-anticlinorial uplifts are characterized by a large crustal thickness (about 25-45 km; Figs. 8, 9, 23), which approximates normal continental crustal thickness, it is reasonable to suggest that this zone was formed earlier than the structure-formational zones of the Lesser and Greater Kuril horst-anticlinoria. Structural relations between the Lesser Kuril ridge and Hidaka Frontal horst-anticlinoria allow us to believe that the Paleogene-Miocene structure-formational complex of the Lesser Kuril ridge horst-anticlinorium is superposed on the Pre-Paleogene complex of the Hidaka-Frontal horst-anticlinorial uplift. The late geosynclinal formations of the southeastern part of the Lesser Kuril ridge structure-formational zone were formed in the Pre-Paleogene period on the sialic basement of the northwestern edge of the Hidaka-Frontal structure-formational zone. The geosynclinal development of the Hidaka-Frontal structure-formational zone seems to have been accomplished by deformations in Paleogene and Upper Cretaceous periods (Geology..., 1977), whereas this zone apparently began to form in Lower Mesozoic-Upper Paleozoic time. K-Ar ages of diorite dredged from site 1113 (southeast of the Hidaka-Frontal horst-anticlinorial uplift axial zone; Fig. 13) appear to be 327 m.y. (Upper Carboniferous).

By the time of folding deformations in the Hidaka-Frontal, Avacha, Kronotsk and Pri-Kamchatka structure-formational zones in the Upper Cretaceous, there already existed a mature geoanticlinal uplift at the place of Pegas (Fig. 10, #23); the Lesser Kuril ridge structure-formational zones. During Paleogene its axial zone was uplifted and thus the Paleogene-Miocene structure-formational complex of this zone was superimposed on the sialic basement of the Hidaka-Frontal, Avacha, Kronotsk and Pri-Kamchatka horst-anticlinorial uplifts, which gradually subsided beginning in the Paleogene and involved the edge of the adjacent Pacific plate. It was, therefore, in Paleogene time when the deep-sea trench morphostructure began to form.

The main deformations within the Pegas Lesser Kuril ridge horst-anticlinorial uplift seem to have occurred in Upper Miocene-Lower Pliocene time. These movements, probably, caused not only folding within the horst-anticlinorium, but also uplift to sea level in Pliocene of a large portion of its axial zone followed by formation in the Late Pliocene-Pleistocene of the abrasional platform (shelf).

Since Miocene-Pliocene volcanics and associated deposits of the southern part of the Greater Kuril ridge and Southern Kamchatka form the andesitic rock associations (Piskunov and Gavrilov, 1970; Aprelkov, 1971), which characterizes the mature stage of geoanticlinal uplift development, the formation of the geosynclinal structure-formational zone within this region began in Paleogene and, probably, earlier. From Late Miocene till Holocene time the Greater Kuril ridge geoanticlinal zone has been uplifted (Grabkov and Pavlov, 1972). Intrageoanticlinal ensialic South-Kuril and North-Kuril troughs seem to have been formed as a single zone, receiving sedimentation from Miocene time.

Thus, the main sequence of events within the Kuril-Kamchatka deep-sea trench can be reconstructed as follows:

Late Paleozoic-Early Mesozoic: The ensimatic formation of a geosyncline began at the place of the Hidaka-Frontal, Avacha, Kronotsk and, probably, Pri-Kamchatka structure-formational zones. An embryonic geoanticlinal uplift started in the Early Mesozoic at the place of the Pegas and Lesser Kuril ridge structure-formational zone. Volcanism and condensed sedimentation took place on the Northwestern Pacific plate, and embryonic geoanticlinal uplift started at the place of the Komandor block.

Late Mesozoic-Early Paleogene: Orogenesis occurred within the Hidaka-Frontal, Avacha, Kronotsk and Pri-Kamchatka structure-formational zones. The mature stage of geoanticlinal uplift was reached along the Pegas-Lesser Kuril ridge structure-formational zones. The Hidaka-Frontal, Avacha, Kronotsk and Pri-Kamchatka horst-anticlinorial uplifts subsided and the deep-sea trench formation began. The appearance of the embryonic geoanticlinal uplift within the Great Kuril ridge structure-formational zone possibly started at this time. Formation of the mature geoanticlinal uplift occurred within the Komandor structure-formational zone. Multivent volcanism and acceleration of pelagic sedimentation occurred on the Northwestern Pacific plate. The Obruchev rise was formed.

Late Paleogene-Early Miocene: Orogenesis took place within Pegas-Lesser Kuril ridge structure-formational zone. The mature stage of the geosyncline development was reached in the Greater Kuril ridge structure-formational zone. The crustal block of the Kuril-Kamchatka continental slope subsidence and deep-sea trench formation continued. Deformation followed by subsidence took place within the Komandor block, and the Aleutian trench was formed. There was attenuation of the volcanic activity on the Northwestern Pacific plate and acceleration of pelagic sedimentation.

Late Miocene-Early Pliocene: Subsidence of the Kuril-Kamchatka continental slope and deep-sea trench formation continued. Deformations and uplifts took place within the Lesser Kuril ridge, Pegas and Greater Kuril ridge structure-formational zones, followed by subsidence within the graben-synclinal zones. Subsidence occurred within the Komandor block.

Late Pliocene-Pleistocene: Deformation and uplift took place within the Greater Kuril ridge structure-formational zone. The crustal block of the Kuril-Kamchatka continental slope subsided. The Kuril-Kamchatka deep-sea trench morphostructure developed close to the recent contours. Turbidites were deposited in the Aleutian Trench. Uplift occurred within the Komandor block.

Recent Geodynamic Regime

The large region of the Asia-to-North Pacific transition zone, including the deep basins of the marginal seas, the island arc systems and the deep-sea trenches, is in a geosynclinal state. The Kuril-Kamchatka deep-sea trench and its junctions with the Aleutian and Japan deep-sea trenches, which are at different stages of development, are elements of this geosynclinal system. The Greater Kuril ridge geoanticlinal uplift is eugeoanticlinal, whereas the continental slope and deep-sea trench are generally characterized as miogeosynclinal, with turbidite sedimentation. Miogeoanticlinal regimes characterize the Obruchev and Komandor rises. A Pre-geosynclinal thalassocratonic regime exists for the adjacent part of the Northwestern Pacific plate.

If one attempts to correlate the structural pattern of the Far East seas with the displacement of the Pacific plate, whose north and northwestward motion is postulated since Mesozoic time (Hilde et al., 1976; 1977), it is difficult to make a non-contradicting geodynamics scheme for the Asia-to-North Pacific transition zone.

The development of the structures of the Asia-to-North Pacific transition zone may be explained by thermodynamic conditions and related volume charges of the regions crust and upper mantle. Recent thermodynamic conditions in the crust of the Hokkaido-Kamchatka horst-meganticlinorium are characterized by sharply differentiated temperature and pressure distributions (Figs. 7, 8 and 9). The observed heat flow pattern suggests different temperature and pressure zones across the region. This pattern indicates the presence of metamorphic facies of low and medium pressures and high temperatures for the volcanic region, and facies of high pressures and low temperatures within the continental slope, similar to what was suggested for Japan (Takevchi and Uyeda, 1965). In the zone of high temperature there are conditions for the granulite facies of metamorphism in the lower crust, and at depths of 45-60 km conditions favorable for forming basic magma. Granitization and the formation of acid melts is also possible within the crust at depths of about 20 km.

Within the trench continental slope zone of low temperature conditions, the granulite-ecologite phase transition may occur at depths of about 25 km. A block of ecologitic rocks with a density of about 3.45-3.50 g/cm^3 may exist to depths of 70-80 km in this zone.

With eclogitization of the lower part of the crustal metabasaltic layer (Pingwood and Green, 1972), over a period of time (Kudryavtsev et al, 1969), the Gabbro-eclogite phase transition will result in a decrease in the volume of the upper mantle and a subsidence of the overlying crust. Conversely, effects of high temperature on the basaltic and andesite-basaltic crust lying above zones of partial melting (Gordienlo, 1975) would result in volume increases which would be manifested by uplift. Considerable thermoclastic stresses can be expected between the heated block of lithosphere that is the Greater Kuril ridge geoanticlinal uplift and Eastern Kamchatka zone, and the zone of relatively "cold" lithosphere of the trench continental slope zone. The above mentioned thermoclastic stresses, being caused by a horizontal temperature gradient of 3-5° c/km, would reach 10^7 dn/cm^2 which approximates the strength limits of rocks at depths of 15-30 km. This may also play an important role in the motion between the zones.

These conditions may be responsible for the various periods of zonal uplift, subsidence and faulting between the zones that have characterized the development of the region.

Acknowledgments

We thank Dr. D. Scholl from the Pacific-Arctic Department of the Marine Geology of the U.S. Geological Survey for kindly placing R/V "Bartlett" cruise data at our disposal. We thank Dr. V.S. Puchkar for diatom identifications and Mrs. A.A. Gracheva and Mr. E.S. Ovcharek for the K-Ar dating. The authors express their gratitude to Mrs. Z.S. Polyakova for translation of this paper into English.

REFERENCES

Andieva, T. A., O. I. Suprunenoka and V. N. Shimareav, Magnetic field of the Near-Kamchatka sea areas, Soviet Geology, 3, (in Russian), 119-124, 1977.

Aprelikov, S. E., Tectonics and volcanism history of the South Kamchatka, Geotectonics, 2, (in Russian), 47-61, 1974.

Averianova, V. N., Deep seismotectonic of island arcs: northwestern Pacific, Nauka, Moscow, p. 175 (in Russian), 1975.

Bathymetric Atlas of the North Pacific Ocean, sheets 2010-2012, 2108-2111, 2208-2209, U. S. Naval Oceanographic Office, Washington, 1973.

Bevz, V. E., I. C. Smirnoz and T. P. Koroleva, On the geologic structure of Large Kuril ridge islands, Proc. of the Sakhalin Department of the Geograph. Soc. of the USSR, 2, (in Russian), 83-101, Yuzhno-Sakhalinsky, 1971.

Bikerman, M., M. Minato and M. Hunahashi, K-Ar age of the garnet amphibolite of the Mitsushi district, Hidaka province, Hokkaido, Japan, Chikyu kagaku, Earth Sci., 25, 27-29, 1971.

Bondarenko, B. A., I. A. Garkalenko, A. V. Zhuravlyov, Yu. B. Kazmin, A. A. Kazimirov, V. I. Kovalyou and S. A. Ushakov, New data on the earth crust deep structure of the Kuril-Kamchatka trench, Reports of the Acad. of Sci. of the USSR, 234, (in Russian), 135-137, 1977.

Borzunova, G. P., V. A. Seliverstov, M. Uy. Hotin and M. N. Shaprio, Paleogene of the Kamchatka Cape Peninsula (Eastern Kamchatka), Proc. of the USSR Acad. of Sci., Geol. Ser., 11, (in Russian), 102-109, 1969.

Bulin, N. K., Deep structure of the Kamchatka and the Kuril Islands from seismic data, Soviet Geology, 5, (in Russian), 140-148, 1977.

Crustal structure of the Asia-to-Pacific transition zone, edited by E. G. Galperin and I. P. Kosminskaya, Nauka, Moscow, (in Russian), p. 286, 1964.

Deep seismic sounding of the Kamchatka, edited by A. A. Popov, H. S. Gnibidenko, Fig. 35, Nauka, Moscow, (in Russian), 1978.

Den, N. and H. Hotta, Seismic refraction and reflection evidence supporting plate tectonics in Hokkaido, Papers in Meteorology and Geophys., 24, 31-54, 1973.

Den, N., H. Hotta, S. Asano, T. Yoshii, N. Sakajiri, Yo. Ichinose, M. Motoyama, K. Kakiichi, A. F. Beresnev and A. A. Sagalevitch, Seismic refraction and reflection measurements around Hokkaido, Part 1, Crustal structure of the continental slope off Tokachi, J. Phys. Earth., 19, 329-345, 1971.

Dyufur, M. S., E. M. Ereshko, M. M. Lebedev, I. A. Sivertzeva and A. M. Smirnova, On spore-pollen complexes from metamorphosed deposits of the Kamchatka and age of enclosing them layers, Problems of regional geology, 2, (in Russian), 103-113, 1977.

Gainanov, A. G., Yu. A. Pavlov, P. A. Stroev, P. M. Sychev and I. K. Tuezov, Anomalous gravity fields of Far East marginal seas and adjacent part of the Pacific Ocean, Nauka, Novosibirsk, (in Russian), 58, 1974.

Gavrilow, V. K. and N. A. Solovieva, Volcanogenic-sedimentary formations of geoanticlinal uplifts of the Small and Large Kuril Islands, Nauka, Novosibirsk, 24-79, 1973

Geological Investigation of Japan and Southern Kurile Trench and Slope Areas, GH 76-2 Cruise April-June 1976, edited by E. Honza, Geol. Surv. Japan, Kawasaki-shi, Japan, 10-16, 23-67, 1977.

Geology and Mineral Resources of Japan, v. 1, Geology, edited by K. Tanaka and T. Nozawa, Geol. Surv. Japan, Kawasaki-shi, 39-43, 193, 246-247, 253, 369, Fig. 4-2, 1977.

Geology of the USSR, v. 31, The Kamchatka, Kuril and Komandor Islands, Part 1, edited by G. M. Vlasov, Nedra, Moscoe, (in Russian), 100-101, 188-215, 401, 652-659, 1964.

German, L. L., The most ancient crystalline complexes of the Kamchatka, Nedra, Moscow, (in Russian), p. 117, 1978.

Gnibidenko, H. S., S. Z. Gorbachev, M. M. Levedev and V. l. Marakhanov, Geology and deep structure of Kamchatka Peninsula, Pacific Geol., 7, 1-32, 1974.

Gordienko, V. V., Heat anomalies of geosynclines, Naukova Dumka, Kiev, (in Russian), 11-14, 1975.

Gravkov, V. K., and Yu. A. Pavlov, The recent movements and crustal isostatic state within the Kuril Island Arc area, Reports of the Acad. of Sci. of the USSR, 203, (in Russian), 650-653, 1972.

Hilde, T. W. C., N. Isezaki and J. M. Wageman, Mesozoic Sea-Floor Spreading in the North Pacific, in The Geophysics of the Pacific Ocean Basin and Its Margin, Geophys. Mon. 19. edited by G. H. Satton, M. Manghnani, R. Moberly, AGU, Washington D. C., 205-228, 1976.

Hilde, T. W. C., S. Uyeda and L. Kroenke, Evolution of the western Pacific and its margin, Tectonophysics, 38, 145-165, 1977.

Honza, E., K. Tamaki and F. Murakami, Geological map of the Japan and Kuril Trenches and the Adjacent areas, Geol. Surv. Japan, Kawasaki-shi, 1978.

Houot, G., Twenty years in the bathyscaphe, Hydrometeorological Publisher, Leningrad, (in Russian), pp. 96-97, 1976.

Initial Reports of the Deep Sea Drilling Project, edited by R. Supko, U. S. Government Printing Office, Washington, D. C., 19, 468, 555, 657, 899, 1973.

Initial Reports of the Deep Sea Drilling Project, edited by S. M. White, U. S. Government Printing Office, Washington, D. C., 31, 17-27, 45-55, 1975.

Isaev, E. N. and V. I. Tikhonov, On the relation between tectonics and magnetic field of the Kuril-Kamchatka arc, Reports of the Acad. of Sci, of the USSR, 175 (in Russian), 161-164, 1967.

Isomi, N., Tectonic Map of Japan, Sc. 1:2000000, Geol. Surv. Japan, Kawasaki-shi, 1968.

Japan Trench transected, Geotimes, 23, 16-21, 1978.

Jeffreys, H., Theoretical aspects of continental drift, in Plate Tectonics - Assessments and Reassessments, Edited by Ch. K. Kahle, AAPG, Tulsa, Oklahoma, 395-405, 1975.

Karig, D. E., G. F. Sharman, Subduction and accretion in Trenches, Geol. Soc. Amer. Bull., 86, 377-389, 1975.

Kosygin, V. Yu. and Yu. A. Pavlov, Geologic nature of the anomalous gravitational field of the southern part of the Kuril Island Arc, Reports of the Acad. of Sci. of the USSR, 220, (in Russian), 672-675, 1975.

Kudryavtsev, V. A., V. T. Melamed and V. N. Sharapov, On the dynamics of processes of regional metamorphism and palingenesis, Geology and Geophysics, 6, (in Russian), 37-46, 1969.

Kulakov, A. P., Quaternary coastal lines of the Seas of Okhotsk and Japan, Nauka, Novosibirsk, (in Russian), p. 174, 1973.

Ludwig, W. J., J. I. Ewing, M. Ewing, S. Murauchi, N. Den, S. Asano, H. Hotta, M. Hayakawa, T. Asanuma, K. Ishikawa and I. Noguchi, Sediments and structure of the Japan Trench, J. Geophys. Res., 71, 2121-2137, 1966.

Malahoff, A., Some possible mechanisms of gravity and thrust faults under oceanic trenches, J. Geophys. Res., 75, 1992-2001, 1970.

Matsuda, T. and S. Uyeda, On the Pacific-type orogeny and its model-extension of the Paired Belts Concept and possible origin of marginal seas, Tectonophysics, 11, 5-27, 1971.

Mikhailov, O. V., Some new data on the bottom relief of the Kuril-Kamchatka deep-sea trench, Proc. of the Ocean. Inst. of the Acad. of Sci. of the USSR, Moscow, 86, (in Russian), 72-76, 1970.

Minaev, Yu. N., A. A. Suvorov, B. V. Alexeev and G. I. Anosov, New data on the sedimentary cover structure in the zone of conjugation between the Kuril and Japan deep-sea trenches from the seismic reflection wave data, Proc. of the Sakhalin Complex Sci. Res. Inst., Yuzhno-Sakhalinsk, 37, (in Russian), 223-226, 1975.

Murauchi, S. and T. Asanuma, Seismic reflection profiles in the Western Pacific, 1965-74, Univ. Tokyo Press, 92-100, 142-146, 1977.

Nozawa, T., Radiometric age map of Japan, Granitic rocks, Scale 1:2000000, Map Series, 16-1, Geol. Surv. Japan, 1975.

Ogawa, K. and J. Suyama, Distribution of aeromagnetic anomalies, Hokkaido, Japan and its geologic implication, in Volcanoes and Tectonosphere, edited by H. Aoki, S. Iizuka, Tokai Univ. Press, Tokyo, 207-215, 1976.

Okada, H., S. Suzuki, T. Moriya and S. Asano, Crustal structure in the profile across the southern part of Hokkaido, Japan as derived from explosion seismic observations, J. Phys. Earth, 21, 329-354, 1973.

Ozima, M., I. Kaneoka and S. Aramaki, K-Ar ages of submarine basalts dredged from seamounts in the Western Pacific Area and discussion of oceanic crust, Earth and Planet. Sci. Letters, 8, 237-249, 1970.

Piskunov, B. N. and V. K. Gavrilov, Neogene volcanogenic-sedimentary formation of the Kuril Islands, Reports of the Acad. of Sci. of the USSR, 192, (in Russian), 1111-1113, 1970.

Popov, A. A., G. I. Anosov, V. V. Argentov, S. K. Bikkenina and I. G. Sivkova, Seismic refraction wave investigations in the Far East marine polygens, Geology and Geophysics, 10, (in Russian), 109-118, 1978.

Rabinowitz, P. D. and Q. Cooper, Structure and

sediment distribution in the western Bering Sea, Marine Geology, 24, 309-320, 1977.

Ringwood, A. E., and D. H. Green, Study of phase transitions, in The Earth's Crust and Upper Mantle, Edited by P. J. Hart, Mir, Moscow, (in Russian), 1972.

Sadreev, A. D. and B. K. Dolmatov, New data on the volume and age of effusive-pyroclastic and tuftogenic-sedimentary formations of the Kronotsk Peninsula, Proc. of the USSR Acad. of Sci., Geol. Ser., 7, (in Russian), 122-126, 1965.

Sakurai, M., M. Nagano, T. Nagai, T. Katsura, M. Tozawa and K. Ikeda, Submarine geology off the southern coast of Hoffaido, Rept. Hydograph Res., 10, 1-37, 1975.

Schmidt, O. A., Tectonics of the Komandor Islands and the Aleutian ridge structure, Nauka, Moscow, (in Russian), 1978.

School, D. W., Sedimentary sequences in the North Pacific Trenches, in The Geology of Continental Margins, edited by C. A. Burk, C. L. Drake, p. 499, Springer-Verlag, N.Y., 1974.

School, D. W., J. R. Hein, M. Marlow and E. C. Buffington, Meiji sediment tongue: North Pacific evidence for limited movement between the Pacific and North American plates, Bull. Geol. Soc. Amer., 88, 1567-1576, 1977.

School, D. W., M. S. Marlow, N. S. Macleod and E. C. Buffington, Episodic Aleutian Ridge igneous activity: implications of Miocene and younger submarine volcanism west of Buldir Island, Geol. Soc. Amer. Bull., 87, 547-554, 1976.

Segawa, J. and Y. Tomoda, Gravity measurements near Japan and study of the upper mantle beneath the oceanic trench-marginal sea transition zone, in The Geophysics of the Pacific Ocean Basin and Its Margin, Geophys. Mon. 19, edited by G. H. Satton, M. Manghnani, R. Moberly, AGU, Washington, D. C., 35-52, 1976.

Sergeev, K. F., Tectonics of the Kuril Islands System, Nauka, Moscow, (in Russian), p. 81, 1976.

Serova, M. Ya., Foraminifera of the Paleocene deposits of Eastern Kamchatka, Nauka, Moscow, (in Russian), p. 24, 1966.

Shapiro, M. N., The tectonic development of the Kamchatka eastern boundary, Nauka, Moscow, (in Russian), p. 24, 1966.

Snegovskoy, S. S., Seismic reflection investigations and tectonics of the Okhotsk Sea southern part and the adjacent part of the Pacific Ocean, Nauka, Novosibirsk, (in Russian), p. 65, 1974.

Suvorov, A. A., Crustal deep structure of the South-Okhotsk sector from seismic data, Nauka Novosibirsk, (in Russian), 84-54, 1975.

Suvorov, A. A. and E. G. Zhiltosov, Deep seismic sounding results within the South Kuril islands, Proc. of the Sakhalin Complex Sci, Res. Inst., Yuzhno-Sakhalinski, 26, (in Russian), 74-81, 1972.

Takeuchi, H. and Uyeda S., A possibility of present-day regional metamorphism, Tectonophysics, 2, 59-68, 1965.

Tararin, I. A., Geology and petrography of green schist formations of the Kamchatka Ganalsk ridge, in Mineralogy and petrography of metamorphic and metasomatic rocks of the Far East, edited by S. A. Korenbaum, O. V. Avchenko, Vladivostok, 10-37. 1977.

Tomoda, Y., Maps of Free Air and Bouguer Gravity Anomalies in and around Japan, Sc. 1:3000000, Univ. Tokyo Press, 1973.

Tuezov, I. K., M. L. Kraxny, B. I. Vasiliev, A. A. Kulikov and V. I. Mikhailov, Geological structure of the Kuril island arc southern part, Geology and Geophysics, 12, (in Russian), 63-71, 1975.

Tulina, Uy. V., Detail seismic investigations of the Earth's crust at the southern Kuril Islands, in Structure and Evolution of the Earth's Curst in the Soviet Far East, edited by E. E. Fotiadi, I. K. Tuezov, Nauka, Moscoe, (in Russian), 90-96, 1969.

Tulina, Uy. V., S. M. Zverev and G. A. Krasilschikova, The Earth's crust and upper mantle within the focal zone at Eastern Kamchatka, in Seismic Characteristics of Mohorovicic Discontinuity, edited by N. I. Davydova, Nauka, Moscow, (in Russian), 66-79, 1972.

Udintsev, G. B., The Kuril-Kamchatka basin relief, Proc. of Ocean Inst. of the Acad. of Sci. of the USSR, Moscow, 12, (in Russian), 16-61, 1955.

Udintsev, G. B., A. F. Beresnev, A. A. Geodekyan, E. G. Mirlin, L. A. Savostin, A. A. Shreider, V. V. Baranov and A. V. Belyaev, Preliminary data of geological and geophysical investigations in the Okhotsk Sea and in the north-western part of the Pacific Ocean on a board R/V "Vityaz", in Geological and Geophysical Investigations of the Asia-to-Pacific Transition Zone, edited by B. S. Volvovsky, A. G. Rodnikov, Soviet Radio Publisher, Moscow, (in Russian), 19-29, 1976.

Vasiliev, B. I., On geologic structure of the Small Kuril ridge Pacific shelf, Reports of the Acad. of the USSR, 219, (in Russian), 1437-1440, 1974.

Vasiliev, B. I., M. G. Egorova, E. G. Zhiltsov, D. I. Podzorova, M. F. Skorikova and A. A. Suvorov, New data on the structure of the Zenkevich marginal oceanic rise, Proc. of the USSR Acad. of Sci., Geol. Ser., 11, (in Russian), 127-142, 1978.

Watts, A. B., M. Talwani and J. Cochran, Gravity field of the northwest Pacific Ocean basin and its margin, in The Geophysics of the Pacific Ocean Basin and Its Margin, Geophys. Mon. 19, edited by G. H. Satton, M. Manghnani, R. Moberly, AGU, Washington, D. C., 17-34, 1976.

Worzel, J. L., Gravity investigations of the subduction zone, in The Geophysics of the Pacific Ocean Basin and Its Margin, Geophys. Mon. 19, edited by G. H. Satton, M. Manghnani, R. AGU, Washington, D. C., 1-15, 1976.

Yagi, K., Petrology of the alkalic dolerities of the Nemuro Peninsula, Japan, in Igneous and Metamorphic Geology, Geol. Soc. Amer. Mem. 115, edited by L. Larsen, 103-147, 1969.

Zverev, S. M. and L. R. Merklin, Deep structure of the Kuril-Kamchatka deep-sea trench south-eastern slope, Geotectonics, 5, (in Russian), 58-65, 1966.

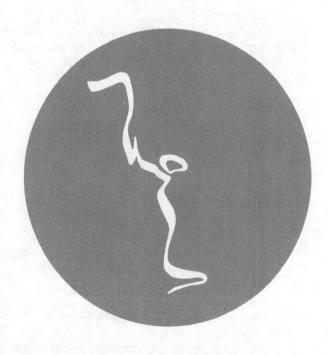

CYCLES OF SUBDUCTION AND CENOZOIC ARC ACTIVITY
IN THE NORTHWESTERN PACIFIC MARGIN

Kazuo Kobayashi

Ocean Research Institute, University of Tokyo
Nakano, Tokyo 164 Japan

Abstract. Volcanic and tectonic events including uplift and subsidence in the island arc as well as opening of back-arc basin are correlatable to various stages of subduction. A large amount of available information obtained from DSDP records and dredge samples in the northwestern Pacific indicates that the uplift of the arc is nonisostatically caused by excess molten magma accumulated beneath it and that subsidence occurs soon after the magma is released. Rates of subsidence of arcs or remnant arcs are much greater than those of the normal oceans at which the basins isostatically subside by the thermal process in the lithosphere. A model of subduction cycle is postulated to elucidate Cenozoic events in the northwestern Pacific margin.

Introduction

Several different features of island arcs are recognized in the northwestern margin of the Pacific Ocean. Each seems to be correlated to a specific configuration of subduction zone [Uyeda, 1982]. The northeastern Honshu, Japan is one of the examples of island arcs. Land area above the sea level has a width of nearly 200 km in the northeastern Honshu. Its length parallel to the trench axis exceeds 600 km. The continental crust beneath the land and slope is about twice as wide as the land itself. East-west cross-sections of topography, crustal structure, gravity anomaly, heat flow and seismicity were compiled by Yoshii [1979], which, therefore, are not repeated. The point worth being emphasized here is the large positive gravity anomaly along the eastern coast and shelf of the arc, which amounts to nearly 200 mgal and has a half-wavelength of about 100 km. This gravity anomaly indicates that the northeastern Honshu arc is not in the isostatic equilibrium. It is in remarkable contrast with the active spreading ridge at which the gravity anomaly is nearly zero. Ida [1978] explicitly pointed out that this positive gravity anomaly is caused and maintained by continuous supply of excess crustal material brought with the subducting oceanic lithosphere, although many authors have implicitly claimed the same idea.

If the island arc is not supported by the elastic strength of the downgoing slab and if it is isostatically adjusted, it would subside by about 2000 m to compensate the positive anomaly of 200 mgal. If such a case happens in the northeastern Honshu, the whole island is submerged. This situation is not fictitious but was realized in the past, e.g. middle Miocene. Correlation of this circumstance to other factors will be discussed later in this paper.

Another example of isostatic subsidence of island arc is that of remnant arc [Karig, 1972]. If a fragment of arc is left behind a spreading inter-arc basin, the remnant arc is no longer kept high enough to be an emerged island and subsides to become a submerged ridge or a chain of seamounts. Kyushu-Palau Ridge trending approximately north to south from Kyushu, Japan to Palau is a typical remnant arc left behind the Shikoku Basin and Parece Vela Basin. Daito Ridge and Benham Rise are similar topographic features in the west Philippine Basin. Since several Deep Sea Drilling Project (DSDP) holes have revealed their subsidence history in relation to the opening of the adjacent basin, the next two sections will be devoted to a review of the DSDP results and consideration of the subsidence process.

The succeeding section will deal with the Cenozoic tectonic history of the southwestern Japan. Tectonic and volcanic events in this region are more or less related to the subduction of oceanic lithosphere at Nankai Trough. Cycles of truncation, rejuvenation and growth of subduction are traced in geological evidence buried under the sea floor or exposed on land. A general scheme is proposed in the next section to summarize the relationship of each stage of subduction with volcanism, crustal stress, and vertical movement in various zones of the island arc. Neogene events in the northeastern Honshu, Japan are elucidated in the last section in a framework of the present subduction model. Some results of DSDP Legs 56-57 are included in the consideration.

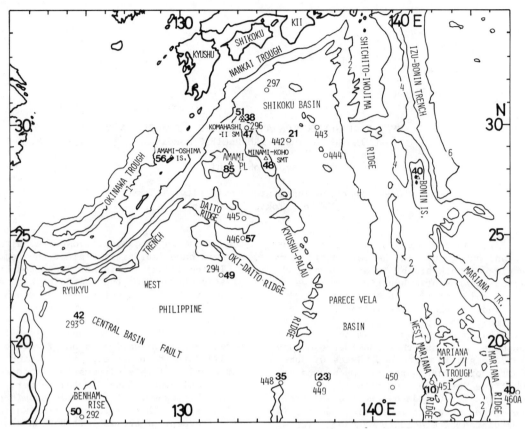

Fig. 1. Outline of topography and basement ages revealed by DSDP holes (circles with site numbers) and dredge hauls (triangles) in the Philippine Sea region. Numbers in brackets denote micropaleontological ages, and the others are ^{40}Ar-^{39}Ar or K-Ar ages in Ma. Sources of data are described in text.

Subsidence of the Kyushu-Palau Ridge
and its synchronism with opening of
the Shikoku Basin and the Parece Vela Basin

Geological history of the Kyushu-Palau Ridge (Fig. 1) was revealed largely by two holes, 296 and 448 of the DSDP. In particular hole 296 (29°20'N, 133°32'E, D=2920 m) provided the evidence and age constraints of subsidence of this now submerged linear topography. Fig. 2 is a summary of age and lithology of hole 296 compiled by Kobayashi and Isezaki [1976] based upon the Initial Report [Karig, Ingle et al., 1975]. ^{40}Ar-^{39}Ar age of a volcaniclastic rock in the bottom of the hole is 47.5 Ma [Ozima et al., 1977]. This age is consistent with a K-Ar age (48.5 Ma) of granodiorite dredged from Minami-Koho seamount about 300 km south of the hole in the Kyushu-Palau Ridge [Mizuno et al., 1977]. Microfossils contained in volcaniclastic sediment layer immediately above the dated rock were identified to be of early Oligocene (approximately 30 Ma BP). A hiatus of about 16 Ma is thus recognized between the rock fragments and overlying sedimentary layer. Some shallow-water fossils found in the sedimentary layer show that this site was emerged and eroded by sea waves until about 30 Ma BP and subsided after that. The uppermost horizon of the volcaniclastic sediment contains microfossils indicating shoal depth and age of about 28 Ma BP so that this site subsided by 650 m/2 Ma=0.33 mm/y, which is nearly one order of magnitude higher than that of isostatic subsidence of the normal ocean floor formed at the mid-ocean ridge (Fig. 3).

Accumulation of fine volcaniclastic materials with thickness of 650 m indicates existence of a nearby source of arc volcanism until about 28 Ma BP. Lithological and paleontological record of hole 296 shows that the supply of volcaniclastics ceased at 28 Ma BP but subsidence continued and water depth at the site increased thereafter. Rate of sedimentation at this site is 325m/Ma at 30∼28 Ma BP, 20 m/Ma at 28∼23 Ma BP, 8 m/Ma at 23∼5 Ma BP and 36 m/Ma after 5 Ma BP. Increase in sedimentation rate after 5 Ma BP is due to volcanic ash blown from the northern Ryukyu Islands in which arc volcanism became active lately.

Similar subsidence of the Kyushu-Palau Ridge

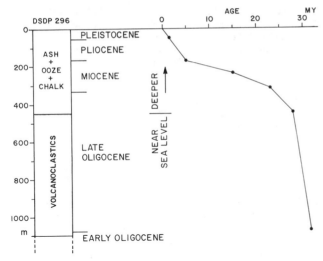

Fig. 2. A diagram showing age and lithology of DSDP hole 296 drilled at the northern portion of Kyushu-Palau Ridge (compiled by Kobayashi and Isezaki [1976] from Karig, Ingle et al. [1975]).

is suggested also by hole 448 (16°21'N, 134°53'E, D=3483 m) of DSDP Leg 59 at the southern portion of the ridge [Kroenke, Scott et al., 1981]. At this site sequences of pillow lavas and sills of island arc tholeiite and volcaniclastic sediment of middle Oligocene were recovered from the bottom 600 m of the core. Occurrence of pillow lavas represents submarine volcanic eruption but nannochalk overlying the volcanics implies hyperbyssal environment at the initial stage. $^{40}Ar-^{39}Ar$ age of the basalt is 32~35 Ma [Sutter & Snee, 1981], which is consistent with the fossil age of the core-bottom sediment. Analysis of benthic foraminifera indicates that the environment was changed to pelagic at about 23 Ma BP.

Rocks dredged from the crestal zone of the Kyushu-Palau Ridge also revealed the origin, ages and subsidence of the ridge. More than 5 cruises of Bosei-maru and Tokaidaigaku-Maru II operated by the Japanese Geodynamics grant were focused to this area as well as the Daito Ridge described in the following section [National Committee for Geodynamics Project of Japan, 1975; 1979a,b; Kobayashi, 1977]. In 1973 many granodioritic rocks coated by ferromanganese oxides were collected by dredge hauls at the slope of Komahashi II seamount (29°56'N, 133°19'E, D=2250 m) located north of hole 296 in the northern part of the Kyushu-Palau Ridge. K-Ar age of one of the rocks is about 38 Ma [Shibata and Okuda, 1975; Shibata et al., 1977] and fission track age of another sample is about 51 Ma [Nishimura, 1975]. From Komahashi seamount (28°05'N, 134°38'E, D=1350 m) south of hole 296, basalts, andesites and coral limestones were dredged. Occurrence of reefal limestones containing late Oligocene foraminifers indicates subsidence of the site since that time.

Results of DSDP holes and dredge hauls seem to show that duration of volcanic activity at the Kyushu-Palau Ridge was probably long from as old as 48 Ma BP till 28 Ma BP and the subsidence started in middle or late Oligocene (about 28~25 Ma BP) roughly simultaneously throughout the whole Kyushu-Palau Ridge.

Identification of magnetic lineations in the Shikoku Basin [Kobayashi and Nakada, 1976; Shih, 1980; Watts and Weissel, 1975] and in the Parece Vela Basin [Mrozowski and Hayes, 1979] together with DSDP holes 442, 443 and 444 in the Shikoku Basin [Klein et al., 1978; Klein, Kobayashi et al., 1980] and holes 449 and 450 [Scott, Kroenke et al., 1980] indicated almost unequivocally that both basins started to open at about 30 Ma BP and ceased to spread at approximately 15 Ma BP in the axial zone of the basins. It is more than coincidental that subsidence of the Kyushu-Palau Ridge bagan soon after the Shikoku Basin and Parece Vela Basin started to open.

Cause of subsidence of the ridge and its synchronism with opening of the inter-arc basin can be explained as follows;
(1) Before the inter-arc basin opened, the Kyushu-Palau Ridge was the western portion of an active island arc together with the present Izu-Bonin-Mariana arc.
(2) The Kyushu-Palau Ridge was underlain with a subduction slab and held nonisostatically by excess materials brought with the oceanic lithosphere. Volcanism and magmatism were active

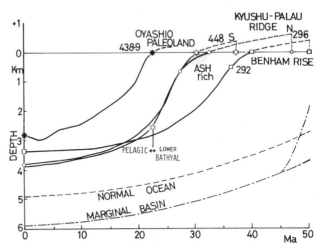

Fig. 3. Subsidence of Kyushu-Palau Ridge (north and south), Benham Rise, and Oyashio Paleoland (fore-arc of the northeastern Honshu, Japan) revealed by DSDP holes 296, 448, 292 and 438-439. Subsidence curves of the normal ocean and marginal basin 50 Ma old are shown for comparison. Note that depth of marginal basin is variable in the initial 3~5 Ma dependent upon configuration of the subduction zone. As to sources for each curve, see the following sections of this paper.

there. A part of the ridge was kept emerged above the sea level.

(3) After the inter-arc basin began opening, the underlying excess materials were spent to form the spreading basin floor. As the Kyushu-Palau Ridge was removed from the downgoing slab by the opening basin, volcanic activity ceased and the ridge isostatically subsided quickly. Gravity anomaly over the ridge is now nearly zero except for small rugged topographies [Watts, 1976].

Subsidence and northward drift of Daito Ridge, Amami Plateau and Benham Rise accompanied by opening of the west Philippine Basin

Daito and Oki-Daito Ridges are also remnant arcs presumably older than the Kyushu-Palau Ridge. The two ridges are submarine topographies with water depths generally greater than 1000 m except for Daito and Oki-Daito Islands which are uplifted atolls. The trend of the ridges is NW-SE and oblique to that of the Kyushu-Palau Ridge. Occurrence of Nummulites boninensis in limestones and sediment samples dredged from the crestal regions of the Daito and Oki-Daito Ridges indicated subsidence of the ridges [Mizuno and Konda, 1977; Shiki et al., 1977], as this giant foraminifera is equivalent to a shoal water fossil of middle Eocene found on land at Hahajima of Bonin Islands. Small pebbles of greenschist, hornblende schist and serpentine were dredged from the Daito Ridge, indicating regional metamorphism related to formation of this ridge. Andesite and diorite were included. In contrast dredge hauls at the Oki-Daito Ridge recovered mostly basalt but no metamorphics.

Two holes were drilled in the Daito Ridge area during Leg 58 of the Glomar Challenger; one hole 445 (D=3377 m) in a small basin located in the western portion of the Daito Ridge and the other (hole 446, D=4952 m) in a deep basin between the Daito and Oki-Daito Ridges. Hole 445 recovered carbonate sediments, mudstones, sandstones and conglomerates. No igneous basement was reached. The bottom of the hole consists of conglomerates with large clastic rocks and irregularly interbedded sandstones of slump origin. Angular shape of clasts implies a nearby source of these rocks. Nummulites boninensis similar to that dredged from the Daito Ridge is abundant. Paleontological age of coexisting nannofossils is middle Eocene (about 48 Ma). At least 23 sills of alkali basalt were recovered from the lowest unit of hole 446. $^{40}Ar-^{39}Ar$ age of basalt is 57 Ma [Ozima et al., 1980], which is older than any other samples recovered from the Kyushu-Palau Ridge. Older ages of igneous rocks have been reported with andesite and granodioritic rocks dredged from the crest (D=1350 m) of Amami Plateau, a topographic high in a triangular area surrounded by the Kyushu-Palau Ridge, Daito Ridge and Ryukyu Trench. Their K-Ar ages are 82∼85 Ma and 75 Ma, respectively [Matsuda et al., 1975]. Nummulites boninensis was recovered also from this area, indicating that it subsided after middle Eocene. Another evidence of subsidence was obtained from DSDP hole 292 at the crest of Benham Rise (D=2943 m) in the southwestern margin of the west Philippine basin. The hole penetrated basaltic basement which was dated to be 50 Ma [Ozima et al., 1977]. Microfossils contained in recovered sediments revealed that subsidence began at about 40 Ma BP. A few million years after that the crest became deeper than 500 m.

Ages of subsidence of these areas seem to be nearly synchronous with opening of the west Philippine basin. Similarity of the overall topography of the central basin fault to midoceanic ridge and apparent symmetry of magnetic anomalies in respect to the fault were pointed out by Ben-Avraham et al. [1972]. Magnetic lineations in the west Philippine basin have been identified by Watts et al. [1977] and Shih[1978]. Andrews [1978] showed by a side-scanning sonar investigation that topographic trend of the basin is essentially parallel to magnetic lineations and the central basin fault took an echelon configuration while the basin was opening. It seems to be reasonably concluded that the basin began opening about 60 Ma BP and almost ceased to spread at about 40 Ma BP. Basaltic rocks recovered from the bottom of DSDP holes 292, 293 and 294-295 of Leg 31 were dated by the $^{40}Ar-^{39}Ar$ method and their ages are represented in Fig. 1 [Ozima et al., 1977]. All of these ages appear to be quite consistent with the opening history of the basin. These results show that the Daito and Oki-Daito Ridges, Amami Plateau and Benham Rise started to sink after the west Philippine Basin began opening or ceased to open.

Correlation of subsidence of the remnant arcs to opening of the basin may depend upon configuration of arc-trench system. No direct evidence is available regarding where the trench and subduction existed while the west Philippine basin was opening. One important result relevant to the ancient geological configuration is the northward drift of the western portion of the Philippine Sea floor. Paleomagnetic measurement of cores from holes 445 and 446 [Kinoshita, 1980] convincingly indicated that the Daito Ridge with an adjacent basin was located at the equatorial region or most probably in the southern hemisphere and that it has drifted northward at a rate faster than 5 cm/y (Fig. 4). Microfossil species [Okada, 1980], pollen [Tokunaga, 1980] and types of clay minerals [Chamley, 1980] contained in each stage of the DSDP cores unanimously showed that the sites were in warmer climate in earlier times. Louden [1977] provided similar paleomagnetic evidence with DSDP holes 292, 290 and 294. He also showed from phase shifting of magnetic anomalies in the west Philippine basin that the basin has drifted northward as a whole and was later rotated clockwise by more than 40 degrees.

On the basis of these paleomagnetic and paleoclimatological evidences as well as configuration

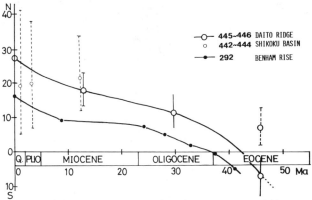

Fig. 4. Paleolatitude of DSDP sites 442 to 446 [Kinoshita, 1980] and 292 [Louden, 1977] compiled by Klein and Kobayashi [1980, 1981].

of ages and rock types of basement, Klein and Kobayashi [1980; 1981] postulated, in the geolgical summary of Leg 58, a model of evolutionary history of this region. As illustrated in Fig. 5, a subduction proceeded southward along an E-W trending arc-trench system in late Cretaceous to early Paleocene period. It seems likely that the northern portion of Kyushu-Palau Ridge composed of granodioritic rocks was a frontal arc, the Daito Ridge and Amami Plateau containing andesite and metamorphics of greenschist facies were the volcanic arc and the Oki-Daito Ridge was an inner arc facing the opening basin. Magmatic activity in these arcs started long before the west Philippine basin began opening. Most of these arcs were emerged above the sea level and no sedimentation occurred. The whole arcs were drifting northward together with most parts of opening west Philippine basin. Investigation of the Shimanto Belt occurring in the southern margin of the southwestern Japan [Kanmera and Sakai, 1975; Taira et al., 1979; Taira and Tashiro, 1981] indicated that a subduction existed nearly northward along the southwestern Japan in Paleogene period. A granitic rock of 49∿56 Ma in ages reported at Amami-Oshima in the Nansei-shoto Islands north of Ryukyu Islands [Shibata and Nozawa, 1967] implies existence of northwestward subduction in the same period. Dual subduction zones were probably existent in the ocean north of the northern Kyushu-Palau Ridge and accelerated the northward drift of the Daito and west Philippine regions.

It was proposed that the southern portion of the present Kyushu-Palau Ridge was a transform fault between the spreading west Philippine basin and the Pacific Ocean floor [Uyeda and Ben-Avraham, 1972; Uyeda and Miyashiro, 1974; Hilde et al., 1977]. It is easily recognized that the northern and southern portions of the Kyushu-Palau Ridge have quite different topography; the southern portion has topography resembling that of the fracture zone with a narrow median trough. Boundary between the northern and southern ridges is situated at the intersection of Oki-Daito Ridge and Kyushu-Palau Ridge.

Around the termination of opening, the west

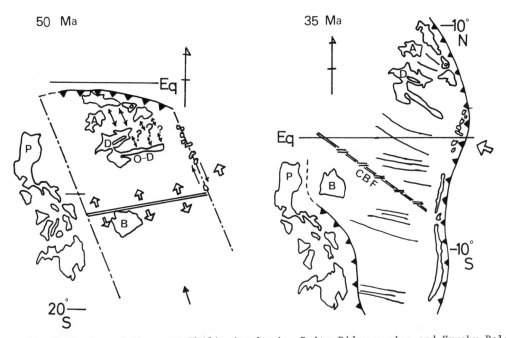

Fig. 5. Evolution of the west Philippine basin, Daito Ridge region and Kyushu-Palau arc [Klein and Kobayashi, 1980; 1981]. Abbreviations: (A) Amami Plateau; (D) Daito Ridge; (O-K) Oki-Daito Ridge; (B) Benham Rise; (P) Philippines; (CBF) central basin fault.

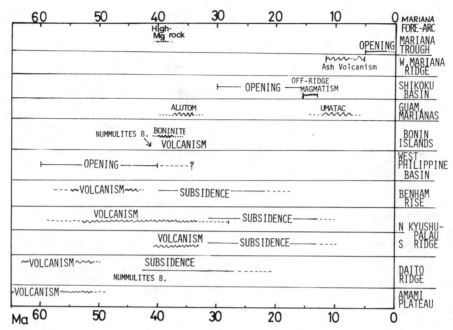

Fig. 6. Volcanic and tectonic events in the Philippine Sea region.

Philippine basin was rotated clockwise by more than 40 degrees. This rotation may be caused by the change of motion of the Pacific floor from northward to northwestward at about 42 Ma BP as shown by the bent of Emperor-Hawaiian chain. After the rotation of the west Philippine basin the trend of northern Kyushu-Palau Ridge became close to that of the southern ridge (transform fault). As direction of relative motion of the Pacific basin was no longer parallel to the southern Kyushu-Palau Ridge, northwestward subduction began along the previous transform fault somewhile after 42 Ma BP.

The whole Kyushu-Palau Ridge became an active island arc, which we call the "Kyushu-Palau arc". Igneous rocks from the Kyushu-Palau Ridge dated as 32 Ma and 37 Ma were formed by this arc volcanism. Bonin Islands were fore-arc of this volcanic islands. Occurrence of Nummulites boninensis in Daito Ridge and Hahajima, Bonin Islands indicates that both Daito Ridge and Bonin Islands were geographically close together. It implies that the basement of Hahajima may be older than this stage of arc volcanism and a part of the previous arc simultaneous to the formation of Daito Ridge.

The subduction of the Pacific basin beneath the Kyushu-Palau arc caused the opening of a back-arc basin, the Shikoku and Parece Vela Basins at about 30 Ma BP. The Kyushu-Palau arc was rifted so that the Kyushu-Palau Ridge and the Izu-Bonin-Marianas arc were separated by the newly-formed basins. The former was left behind the active arc, while the latter continued to be still overlying the subduction zone for a long time after the opening of these basins. The Kyushu-Palau Ridge turned out to be inactive and subsided after about 30 Ma, while in the Guam Island, Marianas (Umatac Formation) thick layers of basalts, andesites and dacites of Miocene in age occur with those of Eocene to Oligocene (Alutom Formation).

Later at about 5 Ma BP the Mariana Trough began opening to separate the west Mariana Ridge and Mariana Islands [Bracey and Ogden, 1972; Karig et al., 1978; Hussong, Uyeda et al., 1982; Bibee et al., 1980]. DSDP hole 451 of Leg 59 on the west Mariana Ridge (18°01'N, 143°17'E, D=2060 m) provided evidence for the following sequence of events [Scott et al., 1980];
1) The latest intense volcanic activity started about 11 Ma BP and continued to accumulate thick (about 850 m) volcaniclastic debris until about 9 Ma BP.
2) After the drastic decay of intense eruptions at about 9 Ma BP, sporadic volcanism continued until about 5 Ma BP.
3) After the Mariana Trough began opening, volcanism at the west Mariana Ridge ceased and subsidence of the ridge proceeded.

This interpretation seems to be slightly inconsistent with the age relation proposed by Rodolfo [1980]. However, K-Ar dating of volcanic rocks in Bonin Islands suggests the existence of a volcanic event at the fore-arc region in 10 to 5 Ma BP. This event may be correlated to the above-mentioned sporadic volcanism prior to the opening of the Mariana Trough, although no back-arc spreading occurred after that behind the Bonin arc-trench system. In Quaternary, volcanic activity is relatively intense in the Izu-Iwojima arc but weak in the Marianas arc. Difference in Quaternary volcanic activity between the two arcs may possibly be due to whether or not the back-arc basin is opening. If the back-arc basin is opening, the excess materials supplied by the downgoing oceanic slab would be consumed for spreading ocean floor but not for arc volcanics, although the arc volcanism usually lasts longer than back-arc opening.

Fig. 6 illustrates the sequence of events in various regions in the Philippine Sea area.

Cenozoic volcanism and tectonism of the southwestern Japan in harmony with subduction at Nankai Trough

The Shikoku Basin is bordered in its north by Nankai Trough at which the Shikoku Basin lithosphere is now sinking into the asthenosphere beneath the southwestern Japan. Studies of seismicity, focal mechanism and seismic wave propagation in the southwestern Japan have demonstated the length, configuration and direction of motion of the downgoing slab [Kanamori, 1971; 1972; 1977; Kanamori and Tsumura, 1971; Fitch and Scholz, 1971; Shiono and Mikumo, 1975; Shiono, 1977; Seno, 1977]. Under the Kii Peninsula the deepest earthquakes related to this slab are only about 70 km deep, while in Kyushu the slab reaches a depth greater than 100 km. In Ryukyu Trench which is presumably a southern extension of Nankai Trough, the downgoing slab is as long as 300 km and its end reaches a depth of about 250 km. From the analysis of the slip vectors of inter-plate earthquakes a pole of the present rotational motion of the Philippine plate relative to the Eurasia plate was given to be at 45.5°N, 150.2°E [Seno, 1977]. According to model RM 1 of the global motions of plates [Minster et al., 1974], the rate of the rotation relative to the Eurasia plate was given to be approximately 1.2°/Ma (clockwise).

Based upon the length of deep-focus earthquake zone, frequency of large thrust earthquakes and magnitude of slips along the thrust faults accompanied by these earthquakes, Kanamori [1972] estimated the duration of the present cycle of subduction to be about 2 Ma. Karig et al. [1975] postulated that the limited occurrence of turbidite in DSDP hole 297 only in cores younger than 5 Ma BP and prior to 3 Ma BP indicates the disappearance of Nankai Trough in a period between 5 and 3 Ma BP. If the trough was deep enough turbidity current from land should have been trapped and would not have reached the Shikoku Basin floor. Therefore, subduction at Nankai Trough was once truncated at about 5 Ma BP and rejuvenated at about 3 Ma BP. It is noteworthy that estimates of the duration of the present subduction cycle from completely independent data are consistent.

As the present subduction does not reach a depth of 100 km, it is quite reasonable that there exist no active volcanoes in the southwestern Honshu except only a few active or Quaternary volcanoes such as Daisen. In contrast the present northeastern Honshu has a distinct characteristic feature of volcanic belt, i.e. volcanoes are aligned west of a line called volcanic front [Sugimura, 1960; Sugimura and Uyeda, 1973]. The front is nearly parallel to the axis of the Japan Trench with a distance of about 300 km. The volcanic front is observed in other island arcs, also. Distance between the front and trench axis varies but is always no less than 200 km. Remarkable is that the front is situated on a line at which the depth of the downgoing slab is about 100 km regardless of the dip angle of subduction. The depth of 100 km is, therefore, considered to be a necessary condition to attain a high temperature sufficient to form molten magma for the source of volcanoes. In central Kyushu and the northern Ryukyus the slab sinks deeply enough to form many active volcanoes such as Aso, Sakurajima and Tokara. No prominent active volcano is seen in the southern Ryukyu arc. This situation may possibly be correlated to existence of the Okinawa Trough, which is presumably a still opening (magma-consuming) back-arc basin [Herman et al., 1979; Lee et al., 1980].

By tracing the older volcanic front, time and configuration of the subduction zone in the past are recognizable to some extent. For instance the volcanic front in early Miocene in the northeastern Honshu was nearly parallel to but about 30 km east of the present front. Nakamura [1969] postulated that this westward shift of the front was due to the eastward drift of northeastern Honshu. However, an alternative interpretation is that the position of Honshu and trench was the same but the dip angle of subduction in early Miocene was steeper than that of the present so that 100 km depth was reached by the slab on a line about 30 km east of the present front.

Cenozoic history of subduction at the Japan Trench will be discussed more in detail in the next section. It must be noted here that the subduction zone has changed its length and dip angle with a period of the order of 1 to 10 Ma. It has even been disrupted and rejuvenated as shown in the Nankai Trough region.

Onset of the previous subduction in the southwestern Japan can be detected from ages of Cenozoic igneous rocks occurring on land overriding the downgoing slab. A number of outcrops of felsic volcanic and plutonic rocks have been reported in the Pacific coast and inland sea district (Setouchi) of Kyushu, Shikoku and Kii peninsula of Honshu [Nozawa, 1968; Oba, 1977]. Most of their ages were re-determined by K-Ar method [Shibata, 1973; Tatsumi, this volume]. Updated ages of these rocks are quite concentrated around 13∼15 Ma BP (Fig. 7). Regional variation in chemical composition of these rocks indicates the existence of a shallow-angle subduction zone at that time [Nakada and Takahashi, 1979].

However, it seems to be curious that the

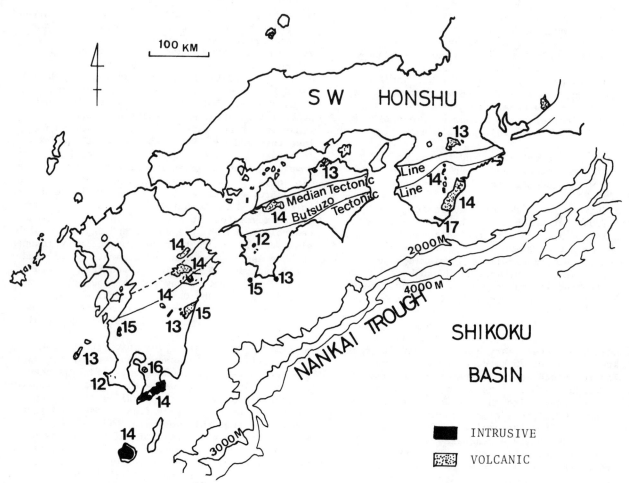

Fig. 7. Ages and geographic distribution of felsic to intermediate igneous rocks in the outer zone and Setouchi province, southwestern Japan. Numbers indicate ages of rocks [Shibata, 1978].

location of coastal rocks is too close to the present axis of the Nankai Trough. The distance between the 'volcanic front' and the axis of trough is only 100 km. The downgoing slab can not reach a depth of 100 km at which temperature becomes high enough to form magma at such a distance. The second question to be solved is why the magmatic activity was so short-lived. In normal volcanic arcs such as northeastern Honshu the arc volcanism lasts for at least a few million years. Three alternative explanations seem to be possible for this short-lived fore-arc volcanism (Fig. 8);

(1) Soon after subduction begins, the asthenosphere beneath the fore-arc toe is not cooled by the cool oceanic lithosphere. The end of the short downgoing slab brings H_2O and CO_2 into the relatively hot asthenosphere. As Kushiro [1974; 1982] has shown by experiment, the melting temperature substantially decreases under the effect of H_2O and CO_2 and Mg-rich magma is formed at a relatively low temperature. If the overlying lithosphere of the fore-arc region happens to be extensional, the magma can rise up to form a high magnesian andesite. The boninite occurring at Bonin Islands [Kuroda and Shiraki, 1975; Kuroda et al., 1979] may have been produced by this process. K-Ar age of boninite was determined by Tsunakawa [1980] to be about 40 Ma, which is quite consistent with the age of the onset of the new northwestward subduction after the Pacific plate changed direction of its motion. The high magnesian rock recovered from DSDP hole 461 [Hussong, Uyeda et al., 1982], which is located in the fore-arc toe of Mariana Trench slope, seems to be also a product of this stage.

In the case when the overriding lithosphere is under a compressional state, magma can not ascend so easily as under the extensional stress and may stay longer in the crust and upper mantle to give rise to crystal differentiation and melting of wall rocks. Igneous rocks thus formed are generally felsic to intermediate.

(2) When an active ridge or very young ocean floor sinks, the asthenosphere beneath the arc

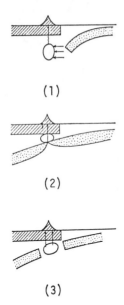

Fig. 8. Three possible explanations for the origin of fore-arc volcanism.

is not cooled but even heated by the downgoing slab. H_2O and CO_2 are supplied by the oceanic slab. Tatsumi (this volume) indicated that 'Sanukitoid' in the Setouchi volcanic belt in 13 Ma BP was formed by subduction of the Shikoku Basin floor while a widespread off-ridge volcanism occurred in the basin around 15 to 13 Ma BP [Klein et al., 1978; Klein, Kobayashi et al., 1979]. During the interval of opening, a part of the present Nankai Trough was acting as a transform fault between the opening basin and the southwestern Japan, since the trend of the trough was nearly parallel to the direction of spreading of the basin. It seems very likely that the Shikoku Basin or even an embryonic portion of the basin began subducting, as the paleomagnetic investigation of DSDP cores shows the northward drift of the basin since its opening. Marshak and Karig [1977] claimed that movement of the triple junction of Nankai Trough, Ryukyu Trench and Izu-Bonin Trench may be responsible for the fore-arc volcanism but the ages are apparently inconsistent.

(3) When a previously long and continuous downgoing slab is disrupted at a place beneath the arc, the asthenosphere in the disrupted gap may be heated to form magma. If it occurs beneath the fore-arc toe, abnormal volcanic activity may proceed. However, as the disruption does not necessarily occur beneath the fore-arc but either side of it, this situation is not to explain the fore-arc volcanism but to elucidate a drastic change in tectonic stress configuration and rifting of the inner arc.

Fig. 9 illustrates a chronological summary of geological events in the southwestern Japan, Nankai Trough and Shikoku Basin. The southwestern Japan arc is divided into three; the inland sea region (Setouchi), the outer zone in its south and the inner zone in its north. It is clearly recognized that uplift of the outer zone is correlatable to the early to matured stages of each subduction cycle, while subsidence of the whole arc occurs in the later stage of each subduction cycle.

Investigation of the ancient shoreline has indicated that the latest tectonic uplift of the southwestern Japan coast began only less than 1 Ma BP [Ota and Yoshikawa, 1978]. It implies that uplift of the land coast starts after the downgoing slab reaches an appropriate length.

Pattern of the stress field is also correlatable to subduction. Kobayashi and Nakamura [1978] showed by the analysis of the alignment of dyke swarms that the direction of the maximum stress was changed from NNW to WWN at about 2 Ma BP. This result implies that the direction of subduction at the Nankai Trough was NNW for a period prior to about 2 Ma BP and thereafter changed to WWN, i.e. the present direction deduced from the slip vector of thrust-type earthquakes. It was also demonstrated that the outer zone was under a compressional stress field throughout these period, while the inner zone and inland sea region were under an extensional stress from about 21 Ma BP and 11 Ma BP. This duration seems to be correlatable to a period for which no sufficiently long subduction slab existed beneath the southwestern Japan arc. The preceeding subduction at the Nankai Trough was nearly completely disrupted at about 21 Ma BP as the Shikoku Basin opened to an appreciable size in a direction roughly prependicular to the subduction.

The downgoing slab which started to sink again was composed of a very young and extraordinary hot lithosphere of the new inter-arc basin. Its negative buoyancy may have been so small that it would not subduct promptly. The dip angle of

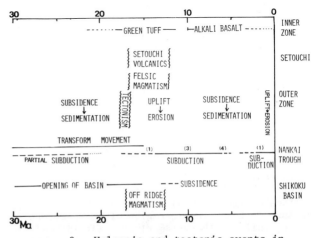

Fig. 9. Volcanic and tectonic events in the Shikoku Basin, Nankai Trough and southwestern Japan.

subduction may have been very shallow. This postulation seems to be consistent with the petrological evidence described above. It seems quite plausible that the molten excess materials transported with the hot oceanic slab were accumulated beneath the arc.

Since the overriding lithosphere was easily heated by the hot slab and became thin and fragile, the stress field on the surface of the overriding arc was changed to be extensional in a way similar to a skin of an expanding rubber ball. I propose that the extensional circumstance in the inner zone of the southwestern Japan between 21 Ma and 11 Ma BP was caused by such a process. After the molten excess mass gave rise to an episodic felsic to intermediate magmatism in the southwestern Japan arc, the sinking slab was no longer hot and magmatism ceased until the slab reached the depth of 100 km to cause the normal island-arc volcanism.

A model of the subduction cycle

Geological evidence given in the preceding sections can be explained by a model in which a subduction is assumed to start, to grow and to finally be disrupted. After disruption the detached slab sinks quickly into asthenosphere and a new subduction is soon rejuvenated. It is an endless cycle. An average period of one cycle seems to be approximately 10 to 20 Ma, although it depends upon the rate of subduction, thermal regime beneath the arc and direction of the relative plate motion.

In the embryonic stage of subduction the downgoing slab is short and shallow, as illustrated in Fig. 10 (1) and Fig. 8 (1). Kanamori [1971] suggested that the present configuration of deep-focus earthquakes beneath Kii peninsula, southwestern Japan, Aleutian and the western portion of Alaska is in this stage. Large thrust earthquakes occur on the contact plane between the continental and oceanic lithospheres. It was mentioned in the preceding section that the ancient 'Kyushu-Palau arc' was also in this stage at about 40 Ma BP. A difference between the ancient Bonin-Mariana system and the present southwestern Japan is that in the former the overriding arc is under an extensional stress, while a highly compressional stress prevails in the latter. No fore-arc volcanism is observed in the outer zone of the present southwestern Japan. However, there exists some evidence of magmatic intrusion in the fore-arc toe of Nankai Trough [Den et al., 1968; Geol. Survey of Japan, 1977]. Heat flow at the axial zone of the Nankai Trough is extraordinarily high in contrast to the generally low heat flow at the trench axis [Watanabe et al., 1970]. There are many thermal springs in the fore-arc region of southwestern Japan, while no thermal springs exist in the oceanward side of the volcanic front of northeastern Japan.

In the second stage next to the embryonic

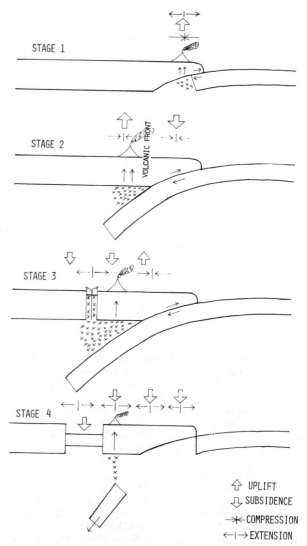

Fig. 10. A model of subduction cycle (modified by the author after Kanamori [1971; 1977] and Kobayashi and Isezaki [1976]).

stage, the subduction zone becomes longer than 200 km. The mantle beneath the fore-arc region is cooled by the downgoing cold oceanic lithosphere so that magmatism does not proceed further. The fore-arc then subsides after the accumnlated magma extrudes. Instead, the leading edge of the downgoing slab is now deeper than 100 km below which temperature is high enough to form molten magma (Fig. 10-(2)). As the downgoing slab horizontally pushes the overriding arc, compression prevails in the arc so that while some of the molten materials are extruded indruded as felsic to intermediate magma, most of them are accumulated deep beneath the arc. The arc is non-isostatically uplifted and heat flow is high. The present northeastern Honshu is in this stage.

The third stage (Fig. 10-(3)) is attained when the molten materials are accumulated beneath the arc too much. The stress field in the overriding arc tends to be extensional when the space above the downgoing slab swells up. The resultant state of stress in the arc is decided by the balance sheet of compression caused by the push of the oceanic lithosphere and extension due to the inner excess mass. If the plate including the overriding arc moves oceanward relative to the oceanic slab, the compression is too large to cause an extensional state in the balance. If, on the other hand, the overriding plate moves away from the oceanic slab, the extensional state is more easily attained with a small amount of underlying excess mass. The eastern margin of the present Pacific is an extremity of the former case, while the western Pacific is more or less close to the latter circumstance. The possible role of the motion of the overriding plate in controlling the stress state in back-arc regions was stressed by Uyeda and Kanamori [1978]. In most places in the northwestern Pacific, the frontal or outer zone of arc is in compression due to the force exerted to the overriding arc by the downgoing slab. In the inner zone of the arc the stress is changed to extensional, as interaction between the downgoing slab and overriding arc is limited to the fore-arc and frontal arc zones [Nakamura and Uyeda, 1980]. Eventually the back-arc zone is rifted and magma ascends to form the back-arc basin. When the rifted zone is narrow, the depth of magma source is so large as the crustal thickness of arc that the composition of rocks is alkali-basalt or arc tholeiite. Water depth of back-arc basin is abnormally shallow, because it is still supported by the subducting slab and excess amss. Gravity is highly positive. After an appropriate width of basin is formed, the magma source becomes shallower and abyssal tholeiite is generated. The inner zone of the arc starts to subside, since the excess mass is consumed to form the back-arc basin. The back-arc basin at first subsides repidly to attain isostasy and then isostatically subsides as the thickness of lithosphere increases with time.

In the last stage of the subduction cycle, the downgoing slab is disrupted and falls down by its own negative buoyancy (Fig. 10-(4) and Fig. 3-(3)). The arc and back-arc basin are generally in extensional stress and subside almost isostatically because no pushing force of the oceanic plate is exerted in this stage. Volcanism in the arc is usually weak except for a possible sporadic activity due to increase in temperature caused by disruption of cold salb. Opening of back-arc basin in about to cease. The Marianas at present is going to be in this stage. There exists a gap of deep-focus earthquakes [Katsumata and Sykes, 1969] indicating the disruption of slab in the Mariana arc. Gravity anomaly at Marianas arc is about 30 mgal, which indicates that isostacy has not been perfectly attained yet.

Kanamori [1971] and Abe [1972] reported occurrence of high-angle normal-fault earthquakes dislocating the oceanic lithosphere off Sanriku, northeastern Japan (1933), at Aleutian (1965) and at Peru (1970). Eissler and Kanamori [1979] showed a similar evidence found at the axial zone at Tonga Trench. A normal fault of about 1600 m at the crest of Daiichi Kashima seamount [Mogi, 1980] may have possibly been formed by this type of normal fault. It seems very likely that the downgoing slab is disrupted by repeated occurrences of such normal faults. The disrupted slab may possibly fall down quickly (perhaps in shorter than 1 Ma) into deep asthenosphere. The excess mass accumalated on the slab is also released into the asthenosphere. Therefore, it seems to be reasonable to assume that opening of the back-arc basin does not proceed any more in this stage.

Soon after the final extinct stage a new subduction is usually rejuvenated. The first stage is restored and a new cycle starts again. As stated in this article, repeated pulses of back-arc opening (so called 'Karig process') as well as episodic tectonic and volcanic activity of island arc can be explained by the present model of subduction cycle, first postulated by Kanamori [1971; 1977], modified by Kobayashi and Isezaki [1976] and further revised in this paper. Niitsuma [1978] successfully explained the sedimentary history of the northeastern Japan on the basis of a similar model of subduction cycle, although his model lacks account of the effect of excess mass treated in the present model.

Cenozoic activity of the northeastern Japan and cycle of subduction

Magmatism in the northeastern Japan was very active in Cretaceous, in early Neogene (23 to 13 Ma BP) and in Quaternary. It was weak and sporadic in Paleogene and in middle to late Neogene. This cycle can easily be explained by the present model of subduction cycle.

Evidence of an event marking the onset of a cycle of subduction in early Neogene is the Oyashio paleoland off the Sanriku coast of the northeastern Japan found by DSDP Leg 57, holes 438, 439 [Von Huene et al., 1981]. It was shown that a landmass composed of presumably Cretaceous sediment was emerged until about 20 Ma BP at a site on the present deep-sea terrace only about 100 km landward of the axis of Japan Trench. At about 23 Ma BP [Yanagisawa et al., 1980] dacitic rocks were erupted from a volcano adjacent to the site and many fragments of rocks were transported and deposited at the site. The landmass started to subside quickly. At present its top is at a depth of 2600 m covered by sediment of 1000 m in thickness and by water of 1600 m in depth (see Fig. 3). I propose that the uplift and volcanism

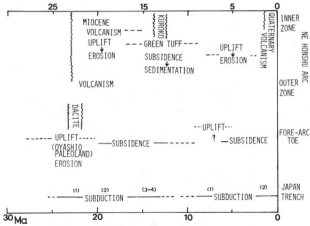

Fig. 11. Volcanic and tectonic events in the northeastern Honshu, Japan.

at the fore-arc toe was a product of the fore-arc magmatism in the embryonic stage of the subduction cycle (Fig. 10-(1)). Soon after this event, subduction grew to the second stage to give rise to arc volcanism in the outer and inner zones of the northeastern Japan (Fig. 11). After about 17 Ma BP extensional field prevailed in the inner zone and most of the island was submerged. At about 13 Ma BP the arc was rifted and the "Kuroko" ore was formed there [Horikoshi, 1975]. As the subduction was disrupted shortly after the disruption of the slab, the rifted trough did not grow up to a back-arc basin.

The next subduction was soon rejuvenated to form the present deep-focus earthquake zone. As length of the slab is nearly 1000 km and rate of the plate motion is about 8 cm/y at the Japan Trench, it takes about 12 Ma to attain the present subduction configuration. Active volcanism in Quaternary and at present indicates that plenty of moten magma is accumulated beneath the inner zone of the arc. As the mechanical interaction between the oceanic slab and overriding arc is still strong, the whole arc is under compression. If the boundary between the oceanic slab and overriding arc is more 'lubricated' to reduce the interaction, the inner zone will turn out to be under extension and eventually the back-arc basin may open in future. Another plausible configuration of the Japan Trench slab in future may be the disruption of downgoing slab without opening of the back-arc basin, as admonished by the high-angle normal faults such as the 1933 earthquake off Sanriku and Daiichi Kashima seamount.

Acknowledgements. This manuscript is a summary of talks partly given at the CCOP-SEATAR Conference held at Bandong, Indonesia, in October 1978 and at two domestic conferences on Geodynamics Project of Japan. I acknowledge the financial support to the project by the Ministry of Education, Science and Culture of Japan.

References

Abe, K., Lithospheric normal faulting beneath the Aleutian Trench, Phys. Earth Planet. Interiors, 5, 190-198, 1972.

Abe, K., Mechanism and tectonic implications of the 1966 and 1970 Peru earthquakes, Phys. Earth Planet. Interiors, 5, 367-379, 1972.

Ando, M., Source mechanisms and tectonic significance of historical earthquakes along the Nankai Trough, Japan, Tectonophysics, 27, 119-140, 1975.

Andrews, J.E., Morphologic evidence for the evolution of the central basin fault spreading center in the west Philippine basin, Abstract of paper presented to GSA Annual meeting, Toronto, Oct. 24, 1978.

Ben-Avraham, Z., C. Bowin, and J. Segawa, An extinct spreading centre in the Philippine Sea, Nature, 240, 453-455, 1972.

Bibee, L.D., G.G. Shor, and R.S. Lu, Inter-arc spreading in the Mariana Trough, Mar. Geol., 35, 183-197, 1980.

Bracey, D.R., and T.A. Ogden, Southern Mariana Arc: Geophysical observations and hypothesis of evolution, Geol. Soc. Amer. Bull. 83, 1509-1522, 1972.

Chamley, H., Clay sedimentation and paleoenvironment in the area of Daito Ridge (northwestern Philippine Sea) since the early Eocene, In Klein, G., K. Kobayashi et al., Initial Rep. DSDP v. 58, Washington, D.C., 669-682, 1980.

Den, N., S. Murauchi, H. Hotta, T. Asanuma, and K. Hagiwara, A seismic refraction exploration of Tosa deep-sea terrace off Shikoku, J. Phys. Earth, 16, 7-10, 1968.

Eissler, H.K., and H. Kanamori, A large normal-fault earthquake at the junction of Louisville Ridge and the Tonga Trench and its geophysical implication, (abstract) EOS 60, 878, 1979.

Fitch, T.J. and C.H. Scholz, A mechanism for underthrusting in southwest Japan: A model of convergent plate interactions, J. Geophys. Res., 76, 7260-7292, 1971.

Geological Survey of Japan, Geological Map around Nankai Trough, Mar. Geol. Map Series 4, 1977.

Herman, B.M., R.N. Anderson, and M. Truchan, Extensional tectonics in the Okinawa Trough, In Watkins, J.S., L. Montadert and P.W. Dickerson (ed.) Geological and Geophysical Investigations of Continental Margins, AAPG Memoir 29, 199-208, 1979.

Hilde, T.W.C., S. Uyeda, and L. Kroenke, Evolution of the western Pacific and its margin, Tectonophysics, 38, 145-165, 1976.

Horikoshi, E., Genesis of Kuroko-stage deposits from the tectonic point of view (in Japanese), Bull. Vol. Soc. Japan, 20, 341-354, 1975.

Hussong, D., S. Uyeda et al., Initial Reports of Deep Sea Drilling Project, vol. 60, Washington, D.C. (U.S. Government Printing Office), 1982.

Ida, Y., Oceanic crust in the dynamics of plate motion and back-arc spreading, J. Phys. Earth, 26, Special issue, 55-67, 1978.

Kanamori, H., Great earthquakes at island arcs and the lithosphere, Tectonophysics, 12, 187-198, 1971.

Kanamori, H., Seismic and aseismic slip along subduction zones and their tectonic implications, In Talwani, M., and W.C. Pitman III (ed.) Island Arcs, Deep Sea Trenches and Back-Arc Basins, AGU ME series 1, 163-174, 1977.

Kanamori, H., and K. Tsumura, Spatial distribution of earthquakes in the Kii peninsula, Japan, south of the median tectonic line, Tectonophysics, 12, 327-342, 1971.

Kanmera, K., and S. Sakai, Which part of the present ocean floor does the production field of the Shimanto Formation correspond to ? GDP Communication Journal on Tectonic Geology, Japan, 3, 55-64, 1975.

Karig, D.E., Origin and development of marginal basins in the western Pacific, J. Geophys. Res., 76, 2542-2561, 1971.

Karig, D.E., Remnant arcs, Geol. Soc. Amer. Bull., 83, 1057-1068, 1972.

Karig, D.E., R.N. Anderson, and L.D. Bibee, Characteristics of back-arc spreading in the Mariana Trough, J. Geophys. Res., 83, 1213-1226, 1978.

Karig, D.E., J.C. Ingle, et al., Initial Reports of Deep Sea Drilling Project, v. 31, Washington, D.C. (U.S. Government Printing Office), 1975.

Katsumata, M., and L. Sykes, Seismicity and tectonics of the western Pacific: Izu-Mariana Caroline and Ryukyu-Taiwan regions, J. Geophys. Res., 74, 5923-5948, 1969.

Kinoshita, H., Paleomagnetism of sediment cores from Deep Sea Drilling Project Leg 58, Philippine Sea, In Klein, G., K. Kobayashi et al., Initial Reports of DSDP v. 58, 765-768, 1980.

Klein, G., and K. Kobayashi, Geological summary of the north Philippine Sea, based on Deep Sea Drilling Project Leg 58 results, In Klein, G., K. Kobayashi et al., Initial Reports of DSDP v. 58, 951-962, 1980.

Klein, G., and K. Kobayashi, Geological summary of the Shikoku Basin and northwestern Philippine Sea, Leg 58, DSDP/IPOD drilling results, Oceanologia Acta, Proc. 26th ICG, 181-192, 1981.

Klein, G., K. Kobayashi, H. Chamley, D.M. Curtis, H.J.B. Dick, D.J. Echols, D.M. Fountain, H. Kinoshita, N.G. Marsh, A. Mizuno, G.V. Nisterenko, H. Okada, J.R. Sloan, D.M. Waples and S.M. White, Off-ridge volcanism and sea-floor spreading in the Shikoku Basin, Nature, 273, 746-748, 1978.

Klein, G., K. Kobayashi et al., Initial Reports of Deep Sea Drilling Project v. 58, Washington, D.C. (U.S. Government Printing Office), 1980.

Kobayashi, K., Marine Geophysics, In Recent Progress of Natural Sciences in Japan, vol. 2, Science Council of Japan, 115-127, 1977.

Kobayashi, K., and N. Isezaki, Magnetic anomalies in the Sea of Japan and the Shikoku Basin: Possible tectonic implications, The Geophysics of Pacific Ocean Basin and Its Margins, Geophys. Monogr. 19, 235-251, Am. Geophys. Union, Washington D.C., 1976.

Kobayashi, K., and M. Nakada, Magnetic anomalies and tectonic evolution of the Shikoku Inter-Arc Basin, J. Phys. Earth, 26, 391-402, 1978.

Kobayashi, Y., and K. Nakamura, Restoration of tectonic stress field of Tertiary southwest Japan by means of dykes, Abst. of Paper International Geodynamics Conference, March 13-17, Tokyo, 86-87, 1978.

Kroenke, L., R. Scott et al., Initial Reports of Deep Sea Drilling Project, v. 59, Washington, D.C. (U.S. Government Printing Office), 1981.

Kuroda, N., and K. Shiraki, Boninite and related rocks of Chichijima, Bonin Islands, Japan, Rep. Fac. Sci. Shizuoka Univ., 10, 145-155, 1975.

Kushiro, I., Melting of hydrous upper mantle and possible generation of andesitic magma: An approach from synthetic systems, Earth Planet. Sci. Lett., 22, 294-299, 1974.

Kushiro, I., Petrology of high-MgO bronzite andesite resembling boninite from site 458 near the Mariana trench, In Hussong, D., S. Uyeda et al., Initial Reports of DSDP v. 60, 731-734, 1982.

Lee, C.-S., G.G. Shor Jr., L.D. Bibee, R.S. Lu, and T.W.C. Hilde, Okinawa Trough: Origin of a back-arc basin, Mar. Geol. 35, 219-251, 1980.

Louden, K.R., Paleomagnetism of DSDP sediments, phase shifting of magnetic anomalies, and rotations of the west Philippine basin, J. Geophys. Res., 82, 2989-3002, 1977.

Marshak, R.S. and D.E. Karig, Tripple junctions as a cause for anomalously near-trench igneous activity between the trench and volcanic arc, Geology, 5, 233-236, 1977.

Matsubara, Y., and T. Seno, Paleogeographic reconstruction of the Philippine Sea at 5 m.y. BP Earth Planet. Sci. Lett., 51, 406-414, 1980.

Matsuda, J., K. Saito, and S. Zasu, K-Ar ages and Sr isotope ratio of the rocks in the manganese nodules obtained from the Amami plateau, western Philippine Sea, Symp. Geol. Problems of the Philippine Sea, Geol. Soc. Japan, 99-101, 1975.

Minster, J.B., and T.H. Jordan, Present day plate motions, J. Geophys. Res., 83, 5331-5354, 1978.

Mizuno, A., and I. Konda, Discovery of Eocene larger foraminiferas from the sea floor of the Oki-Daito Ridge and some related problems, Bull. Geol. Survey Japan, 28, 639-648, 1977.

Mizuno, A., K. Shibata, S. Uchiumi, M. Yuasa, Y. Okuda, M. Nohara and Y. Kinoshita, Granodiorite from the Minami-Koho seamount on the Kyushu-Palau Ridge, and its K-Ar age, Bull. Geol. Survey Japan, 28, 507-511, 1977.

Mizuno, A., Y. Okuda, S. Nagumo, H. Kagami and N. Nasu, Subsidence in the Daito Ridge and associated basins, north Philippine Sea, In Watkins, J.S., L. Montadert, and P.W. Dickerson (ed.), Geological and Geophysical Investigations of Continental Margins, AAPS Memoir 29, 239-244, 1979.

Mogi, A., and K. Nishigawa, Breakdown of a sea-

Mrozowski, C.L. and D.E. Hayes, The evolution of the Parece Vela basin, eastern Philippine Sea, Earth Planet. Sci. Lett., 46, 49-67, 1979.

Nakada, S., and M. Takahashi, Regional variation in chemistry of the Miocene intermediate to felsic magmas in the outer zone and the Setouchi province of southwest Japan, J. Geol. Soc. Japan, 85, 571-582, 1979.

Nakamura, K., and S. Uyeda, Stress gradient in arc-back arc regions and plate subduction, J. Geophys. Res., 85, 6419-6428, 1980.

National Committee for Geodynamics Project of Japan, National Report of Geodynamics Project of Japan, No. 5, 6, 7, 1975; 1979.

Niitsuma, N., Magnetic stratigraphy of the Japanese Neogene and the development of the island arcs of Japan, J. Phys. Earth, 26, Suppl., 367-378, 1978.

Nishimura, S., Fission track age of a granite nucleus in a manganese nodule from Kamahashi-daini seamount, In, Some Geol. Problems of the Philippine Sea Region, Geol. Soc. Japan, 104, 1975.

Nozawa, T., A review of the radiometric ages of the Japanese granitic rocks, Geol. Soc. Malaysia Bull., 9, 91-98, 1977.

Oba, N., Emplacement of granitic rocks in the outer zone of southwestern Japan and geological significance, J. Geol., 85, 383-393, 1977.

Okada, H., Calcareous nannofossils from Deep Sea Drilling Project sites 442 through 446, Philippine Sea, In Klein, G., K. Kobayashi et al., Initial Reports of DSDP v. 58, 549-566, 1980.

Ota, Y., and T. Yoshikawa, Regional characteristics and their geodynamic implications of Late Quaternary tectonic movement deduced from deformed former shorelines in Japan, J. Phys. Earth, 26, Suppl. 376-390, 1978.

Ozima, M., I. Kaneoka, and H. Ujiie, ^{40}Ar-^{39}Ar age of rocks and development mode of the Philippine Sea, Nature, 267, 816-818, 1977.

Ozima, M., Y. Takigami, and I. Kaneoka, ^{40}Ar-^{39}Ar chronological studies on rocks of Deep Sea Drilling Project sites 443, 445 and 446, In, Klein, G., K. Kobayashi et al., Initial Reports of DSDP v. 58, 917-920, 1980.

Rodolfo, K.S., Sedimentological summary: clues to arc volcanism, arc sundering, and back-arc spreading in the sedimentary sequences of Deep Sea Drilling Project Leg. 59, In Kroenke, L., R. Scott et al. Initial Reports of DSDP v. 59, 621-624, 1980.

Scott, R.B., L. Kroenke, G. Zakariadze and A. Sharaskin, Evolution of the south Philippine Sea: Deep Sea Drilling Project Leg 59 results, In Kroenke, L., R. Scott et al. Initial Reports of DSDP v. 59, 803-816, 1980.

Seno, T., The instantaneous rotation vector of the Philippine Sea plate relative to the Eurasian plate, Tectonophysics, 42, 209-226, 1977.

Shibata, K., Contemporanity of Tertiary granites in the outer zone of southwest Japan, J. Geol. Sur. Japan, 29, 551-554, 1978.

Shibata, K., and T. Nozawa, K-Ar ages of granitic rocks from the outer zone of southwest Japan, Geochem. J., 1, 131-137, 1967.

Shibata, K., and Y. Okuda, K-Ar age of a granite fragment dredged from the 2nd Komahashi seamount, Bull. Geol. Survey Japan, 26, 71-72, 1975.

Shibata, K., A. Mizuno, M. Yuasa, S. Uchiumi, and T. Nakagawa, Further K-Ar dating of tonalite dredged from the Komahashi-daini seamount, Bull. Geol. Survey Japan, 28, 503-506, 1977.

Shih, T.-C., Magnetic anomalies in the west Philippine basin and Shikoku Basin, PhD dissertation Univ. of Texas, Galveston, 1978.

Shih, T.-C., Magnetic lineations in the Shikoku Basin, In Klein, G., K. Kobayashi et al., Initial Reports of DSDP v. 58, 783-788, 1980.

Shiki, T., Y. Misawa, I. Konda, and A. Nishimura, Geology and geohistory of the northwestern Philippine Sea, with special reference to the result of the recent Japanese research cruises, Mem. Fac. Sci. Kyoto Univ., ser. Geol. Mineral, 43, 67-78, 1977.

Shiono, K., Focal mechanisms of major earthquakes in southwest Japan and their tectonic significance, J. Phys. Earth, 25, 1-26, 1977.

Shiono, K., and T. Mikumo, Tectonic implications of subcrustal, normal faulting earthquakes in the western Shikoku region, Japan, J. Phys. Earth, 23, 257-278, 1975.

Shiraki, K., Y. Yusa, N. Kuroda and K. Ishioka, Chrome-spinels in some basalts from Guam, Mariana island arc, J. Geol. Soc. Japan, 183, 49-57, 1977.

Shiraki, K., and N. Kuroda, The boninite revisited, J. Geogr., 186, 174-190, 1977.

Sugimura, A., Zonal arrangement of some geophysical and petrological features in Japan and its environs, J. Fac. Sci., Univ. Tokyo sect. 2, 12, 133-153, 1960.

Sugimura, A., and S. Uyeda, Island Arcs, Japan and Its Environs, Elsevier, 247pp. 1973.

Sutter, J.F. and L.W. Snee, K/Ar and ^{40}Ar/^{39}Ar dating of basaltic rocks from Deep Sea Drilling Project heg 59, In Kroenke, L., R. Scott et al., Initial Reports of DSDP v. 59, 729-734, 1980.

Taira, A., Formation of sediment body in the arc-trench system and cyclic subduction model: Earth (Chikyu), 1, 860-868, 1979.

Taira, A., and M. Tashiro (ed.), Geology and paleontology of the Shimanto belt, Rinya-Kosaikai Press, Kochi, 1981.

Tatsumi, Y., Miocene Setouchi volcanic belt, southwest Japan and origin of the Shikoku Inter-arc basin (this volume).

Tokunaga, S., Palynological study of Paleogene sediments from Deep Sea Drilling Project sites 445 and 446, Philippine Sea, In Klein, G., K. Kobayashi et al., Initial Reports of DSDP v. 58, 597-600, 1980.

Tsunakawa, H., K-Ar dating of volcanic rocks in

the Bonin islands (abstract), Bull. Vol. Soc. Japan, 25, 307, 1980.

Uyeda, S., Some basic problems in the trench-arc-back arc system, In, Talwani, M., and W.C. Pitman III (ed.), Island Arcs, Deep Sea Trenches and Back-Arc Basins, AGU ME-1, 1-14, 1977.

Uyeda, S., Subduction zones: An introduction to comparative subductology, Tectonophysics, 81, 133-159, 1982.

Uyeda, S., and Z. Ben-Avraham, Origin and development of the Philippine Sea, Nature, Phys. Sci., 240, 176-178, 1972.

Uyeda, S., and H. Kanamori, Back-arc opening and the mode of subduction, J. Geophys. Res., 84, 1049-1061, 1979.

Uyeda, S., and A. Miyashiro, Plate tectonics and the Japanese islands: a synthesis, Geol. Soc. Amer. Bull., 85, 1159-1170, 1974.

Von Huene, R., M. Langseth, N. Nasu, and H. Okada, Summary, Japan trench transect, In Scientific Party, Initial Reports of DSDP, v. 56-57, Part 1, 473-488, 1980.

Watanabe, T., D. Epp, S. Uyeda, M. Langseth and M. Yasui, Heat flow in the Philippine Sea, Tectonophysics, 10, 205-224, 1970.

Watts, A.B., Gravity field of the northwest Pacific Ocean basin and its margin: Philippine Sea, Geol. Soc. Amer., MC-12, 1976.

Watts, A.B., and J.K. Weissel, Tectonic history of the Shikoku marginal basin, Earth Planet. Sci. Lett., 25, 239-250, 1975.

Watts, A.B., J.K. Weissel, and R. Larson, Sea-floor spreading in marginal basins of the western Pacific, Tectonophysics, 37, 167-181, 1977.

Yanagisawa, M., T. Takigami, M. Ozima and I. Kaneoka, ^{40}Ar-^{39}Ar ages of boulders drilled at site 439, Leg 57, Deep Sea Drilling Project, In, Scientific Party Leg 57, Initial Reports of DSDP v. 56-57 Part 2, 1281-1284, 1980.

Yoshii, T., A detailed cross-section of the deep seismic zone beneath northeastern Honshu, Japan, Tectonophysics, 55, 349-360, 1979.

THE ROLE OF OBLIQUE SUBDUCTION AND STRIKE-SLIP TECTONICS IN THE EVOLUTION OF JAPAN

Asahiko Taira

Department of Geology, Kochi University, Kochi, Japan 780

Yasuji Saito

National Science Museum, Shinjuku, Tokyo, Japan 160

Mitsuo Hashimoto

Department of Earth Science, Ibaraki University, Mito, Japan 310

Abstract. The pre-Neogene geologic framework of Japan consists of the pre-Jurassic terranes (older terranes) and Jurassic to Tertiary subduction terranes, juxtaposed in a complex manner and bounded by strike-slip mobile zones. Within these strike-slip mobile zones, the exotic tectonic blocks of Devonian to Triassic sedimentary and igneous rocks, metamorphic rocks of 200 - 400 m.y., and high- and low-pressure metamorphosed ophiolites have been emplaced with abundant serpentinite forming serpentinite melange. The slicing and lateral migration of subduction terrane and development of strike-slip mobile zone systems were possibly associated with oblique subduction. The Cretaceous systems of the southwest Japan show two strike-slip zones which developed in- and out-side of the frontal arc, forming boundaries to the fore-arc basin and fore-arc shelf basin respectively. The oblique subduction and related strike-slip zones have played a major role in re-orientation of subduction terrane, emplacement of exotic blocks and serpentinite, and development of the fore-arc basin and shelf basin in the geotectonic evolution of Japan.

Introduction

The pre-Neogene geologic framework of Japan has been classified into many "belts" and "tectonic zones (structural belts or tectonic lines)" that are grouped into several major terranes. Although there are differences of opinion among geologists, a general view of the geotectonic division has been achieved as expressed in Tanaka and Nozawa (1977) and the recent 1:1,000,000 geologic map of Japan (Hirokawa, 1978) shown in Figure 1A.

In classic geosynclinal interpretations, the terranes in Japan belong to several geosynclines (see summary by Takai and others, 1963; Minato and others, 1965; Yoshida, 1975; Tanaka and Nozawa, 1977). Among these terranes, the Chichibu belt and its equivalents (Tamba, Mino, and Ashio belts) make up the major framework of the pre-Cretaceous system of southwest Japan and have been interpreted as deposits of the Late Paleozoic Honshu geosyncline (Minato and others, 1965) in which sandstone, shale, chert and limestone were accumulated with contemporary basaltic submarine volcanism.

In northeast Japan, the South Kitakami region, which shows predominantly shallower facies, has been considered as a marginal part of the Honshu geosyncline. The Hidaka geosyncline, which has been considered as a typical "Alpine type" orogenic belt, occupies the major part of Hokkaido (Minato and others, 1965).

The Shimanto and Nemuro belts represents post-Jurassic geosynclines (Harata and others, 1978) and are distributed along the outer margin of the Japanese Islands suggesting that there has been a tectonic polarity toward the outer zone (Matsumoto, 1967).

Pre-Neogene Japan experienced two major orogenic cycles: the Late Paleozoic to Early Mesozoic Honshu orogeny (Minato and others, 1965) and the Middle to Late Mesozoic Sakawa orogeny (Kobayashi, 1941). In Hokkaido, however, the Late Cretaceous to Paleogene Hidaka orogeny was important (Minato and others, 1965).

Within these geosynclines, there are several narrow belts along which "anomalous" rock assemblages (older sedimentary, igneous and metamorphic rocks as well as ultramafics) occur that are sometimes called "tectonic zones (structural belts)". These include the Kurosegawa (Ichikawa and others, 1956), Nagato (Murakami and Nishimura, 1979) and Maizuru (Nakazawa, 1973) tectonic zones.

In geosynclinal interpretations, the "anomalous" older rocks in the tectonic zones have been

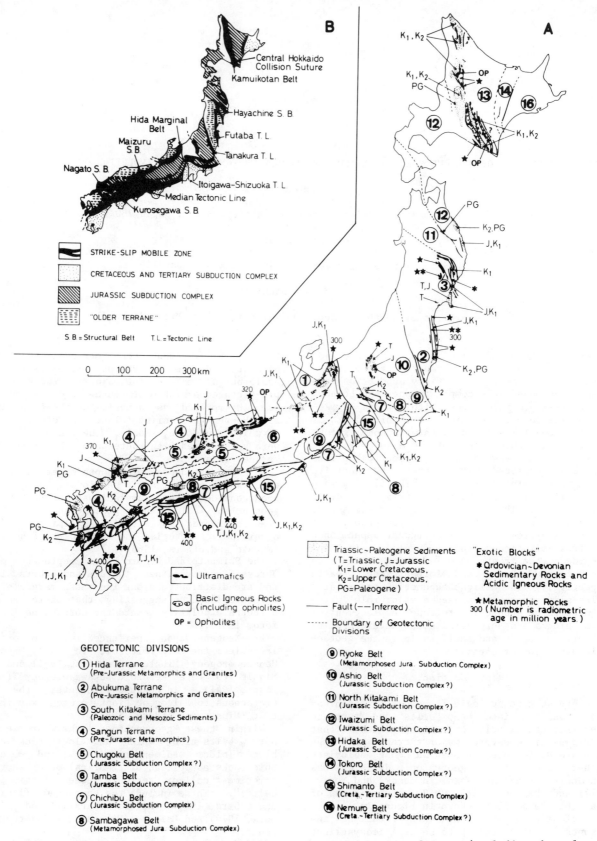

Fig. 1. A: Pre-Neogene geotectonic divisions and tectonic zones of Japan (excluding the volcanic and granitic rocks), B: Interpretation of pre-Neogene geologic framework of Japan.

considered as the uplifted probably pre-Cambrian basement of the Honshu geosynclinal deposits (Minato and others, 1965). Kimura (1973, 1977) suggested that some of these tectonic zones were the site of long existant island chains in the geosyncline. Several authors have concluded that these zones were part of island or remnant arcs (e.g. Ichikawa and others, 1972).

In recent years, models based on plate tectonics have been presented suggesting that the subduction processes played a major role in the evolution of Japan (Matsuda and Uyeda, 1971; Uyeda and Miyashiro, 1974). Taking account the occurrence of Siluro-Devonian acidic volcanic rocks in the Kurosegawa tectonic zone, Suzuki and others (1979) and Hada and others (1979) indicated that was a part of an island arc and the major part of the Chichibu belt was a marginal sea. They suggested that "tectonic erosion" due to subduction modified the old island arc into the present day fragmented appearance.

Horikoshi (1972, 1979) on the other hand suggested that "anomalous" blocks in the Kurosegawa tectonic zone are part of a microcontinent transported to the "Kurosegawa trench". Maruyama(1981) considered that the Kurosegawa tectonic zone represents "serpentinite melange" in which older tectonic blocks of various rock types are mixed with the intruded ultramafics. Kanmera, in Kanmera, Hashimoto, and Matsuda (1980), proposed a subduction model for the evolution of southwest Japan. He suggested that the Chichibu belt includes fragments of a microcontinent transported probably from the Pacific (Nur and Ben-Avraham, 1977, 1978), a similar view previously suggested by Saito and Hashimoto (1982). They indicated that some rocks in the southern Kitakami region, Abukuma massif and Kurosegawa tectonic zone are "exotic" elements of Japan based on sedimentary, paleontological and paleomagnetic evidence. In Hokkaido, the plate tectonic model (Okada, 1974, 1979) requires the collision of two arc-systems at the Kamuikotan "suture zone" during Cretaceous or Paleogene.

We have noticed that not only the ultramafics, tectonic lenses of older rocks and small "anomalous" high-pressure type metamorphic bodies (Hashimoto, 1978) occur along the boundaries of the tectonic divisions, but almost all of the pre-Tertiary shallow marine to non-marine sediments are distributed along these boundaries (see Fig. 1). All of the previously recognized "tectonic zones" coincide with these boundaries. We suspect that these boundaries are strike-slip mobile zones separating the major geologic zones of Japan (Taira and others, 1981). In this paper, an alternative explanation for the pre-Neogene geologic framework of Japan is proposed, in which we stress the role of oblique subduction and strike-slip tectonics at the convergent margins.

Geologic Terranes in Japan

In a simplified view, the pre-Neogene tectonic divisions of Japan can be classified into three terranes and several strike-slip mobile zones (Fig. 1B). The three terranes include:
1) Older terranes,
2) Jurassic subduction complex,
3) Cretaceous to Tertiary subduction complex.

The first group consists of Hida (Kano, 1973, 1975; Hiroi, 1978), Sangun(Hashimoto,1964; Nishimura and others, 1977), Abukuma (Kano and others, 1977; Umemura, 1979; Maruyama, 1979) and South Kitakami(Onuki, 1956,1969; Saito, 1968) terranes. The radiometric dates (Shibata, 1979) and stratigraphic relationships suggest that tha majority of these belts are older than Jurassic and can be classified as "older" terranes. However, there no convincing data exists to show that they once formed a coherent geologic unit. Instead they possibly are parts of exotic terranes and subduction complexes juxtaposed to ancient Asian margin in pre-Jurassic time.

The second group, the Jurassic subduction complex (accretionary and associated slope deposits) includes most of the tectonic divisions of Japan and consisting of the Chichibu, Tamba-Mino,Ashio, North Kitakami, Iwaizumi, Hidaka and Tokoro belts and possibly their metamorphosed counter-parts, the Sambagawa and Ryoke belts(see Fig. 1A). Some of these belts were once considered as Late Paleozoic geosynclinal terrane based mainly on fusulinid fossils. However, recent progress in the study of conodonts (Hayashi, 1968; Koike and others, 1971; Igo, 1979; Suyari and others, 1980; Isozaki and Matsuda, 1980) and radiolarians(Nakaseko and Nishimura, 1979; Ishiga and Imoto, 1980) indicates that they are composed of mixed rock facies in which Carboniferous to Triassic limestones and cherts are embedded mostly in Jurassic terrigenous sediments (Taira and others, 1979b; Yao and others, 1980). The limestones are often associated with basaltic rocks and have been interpreted as seamount complexes with capping calcareous reef material (Ota, 1968; Maruyama and Yamasaki, 1978). Paleomagnetic analysis indicates some of the basalts were formed in the southern hemisphere(Hattori and Hirooka, 1979). The cherts range in age from Permian to Jurassic and often "alternate" with Jurassic sandstone and mudstone. These data indicate that the age of the nonclastic rocks (limestone, chert and possibly basalt) are older than the associated clastic rocks(sandstone and mudstone). These units are often chaotically mixed to various degrees (Tanaka, 1980), suggesting that the belts they make up are subduction terrane in which the oceanic plate material (nonclastics) was mixed with trench-fill clastic sediments. The Sambagawa (high P/T) and Ryoke(low P/T) metamorphic belts have been interpreted as the metamorphosed part of such a subduction terrane (Miyashiro, 1972, 1973).

The third group, the Shimanto belt and probably the Nemuro belt, is a well documented example of Cretaceous and Tertiary subduction complex (Kanmera and Sakai, 1975; Kanmera, 1976; Taira, 1979, 1981; Taira and others, 1979a; Taira and others, 1981; Suzuki and Hada, 1979). Recent studies of the biostratigraphy of the Shimanto belt have revealed tha detailed structure of this complicated belt showing that this terrane is a typical accretionary complex formed in association with subduction of an oceanic plate(Katto and Tashiro, 1978, 1979a, b; Matsumoto, 1980; Noda, 1980; Okamura, 1980; Takayanagi, 1980; Tashiro, 1980). The Shimanto belt is composed of highly deformed flysch and melange. The melange contains blocks of chert and basalt in a sheared shaly matrix. A good example is the Cretaceous melange in Kochi, Shikoku containing Valanginian to Cenomanian radiolarian chert and basalt with inter-pillow nannolimestone (Valanginian age) embedded in an Coniacian to Campanian shaly matrix (Taira and others, 1980). The paleomagnetic measurements suggest that the basalt and interpillow limestone were formed in a low latitude area, some thousands of kilometers from the contemporary magmatic arc of southwest Japan (Taira and others, 1980). In the melange of the Shimanto belt, the age of non-clastic rocks is older than the associated terrigenous clastics. The melange of the Shimanto belt has been interpreted as accreted trench olistostrome deposits, initially derived from the outer wall of the trench where faulting and bending caused the fracturing of the oceanic plate and later tectonized by plate subduction process (Taira and others, 1980).

"Tectonic Zones" as Strike-Slip Mobile Zones

The above terranes are all bounded by tectonic zones which are thought to be strike-slip mobile zones (see Fig. 1). The reasons for this interpretation are:
1) The tectonic zones separate different kinds of terranes. Some of them show an estimated large scale strike-slip displacement reaching up to 400 km displacement (e.g. Tanakura tectonic zone, Otsuki, 1975; Median tectonic line, Ichikawa, 1980; see summary by Otsuki and Ehiro, 1979).
2) They sometimes run in a branching manner, an anastamosing pattern similar to the known strike-slip systems such as in southern California(Crowell, 1952).
3) Within the zones, there are many faults, usually of high angle and grouped in a narrow belt.
4) Highly sheared and serpentinized ultramafics are associated. A similar occurrence of the ultramafics along the strike-slip fault has been reported from the Sumatran subduction margin (Page and others, 1979) and the Solomon "fractured arc" (Hackman, 1973).
5) Within the zones, there are "exotic" tectonic lenses surrounded by serpentinite (serpentinite melange). These "exotic" rocks include the Ordovician to Triassic unmetamorphosed igneous and sedimentary rocks (Adachi and Igo, 1980; Igo and others, 1980; Hamada, 1961; Kobayashi and Hamada, 1974), high grade metamorphic rocks of pre-Jurassic age, frequently 300 - 400 m.y. old (Nozawa, 1977) and dismembered ophiolite complexes (Maruyama, 1978). These rock associations are totally unrelated suggesting that serpentinite melange includes laterally as well as vertically emplaced tectonic blocks.
6) The tectonic zones are accompanied by clastic sedimentary basins. The pattern of sedimentation and geologic structure show great similarity to the previously documented examples of the strike-slip sedimentary basins (see summary by Reading, 1980).

In Kyushu, the Kurosegawa tectonic zone(Fig.1B) which contains "lenses" of Silurian to Triassic rocks and high-grade metamorphic rocks with abundant serpentinite outcrops in the middle, separating the Chichibu belt to the south and Ryoke metamorphic belt to the north (Karakida and others, 1977). Along this zone, the Jurassic to Paleogene elongate sedimentary basins were developed, now faulted into a braided pattern. In the eastern Kyushu, the Kurosegawa tectonic zone branches to form two separate systems, and within this divergent zone enormously thick deposits of Upper Cretaceous sediments (Ohnogawa Group) were accumulated. The 20,000m thick Ohnogawa Group consists of laterally deposited clastic sediments which were fed from a NE direction and now dip toward the same NE direction (Teraoka, 1970). This pattern is quite similar to the previously reported examples of strike-slip basins in California (Ridge basin, Crowell, 1974a, b) and in Norway (Hornelen basin, Steel, 1976; Steel and Gloppen, 1980).

Two branched tectonic zones enclose the Sambagawa and Chichibu belts and separate these from the Ryoke belt to the north and Sambosan (a subbelt of the Chichibu belt) and Shimanto belts to the south in the Island of Shikoku. The northern branch is the famous Median Tectonic Line of Japan. A large lateral as well as vertical displacement has been estimated along this fault (Yabe, 1956, 1963; Ichikawa, 1980): several tens of kilometers left lateral displacement during the Cretaceous and Paleogene and minor right lateral motion in later time have been estimated by structural analysis of the Cretaceous deposits (Miyata and others, 1980) and en echelon folding patterns in the Sambagawa belt (Hara and others, 1977, 1980). Along this fault system, a narrow Upper Cretaceous clastic basin (Izumi basin) was formed (Fig. 2). The Izumi Group shows a structural and sedimentary pattern similar to the Ohnogawa Group. The strata shows some 40,000 m thick lateral accumulation with a migration of depocenter, dipping to the east with en echelon synclinal structures, and the sediment was fed from the east to west direction (Suyari, 1973). This pattern, again, is quite similar to the previously documented examples of strike-slip basins (see Reading, 1980 for summary).

The southern branch, the Kurosegawa tectonic zone, extends into the middle of Shikoku (Suzuki, 1977). It contains lenses of Silurian to Devonian limestone and shale in association with rhyolitic tuffs (Yoshikura and Sato, 1976). It also contains such rocks as metabasalts, barroisite-bearing schist, rodingite, jadeite-glaucophane

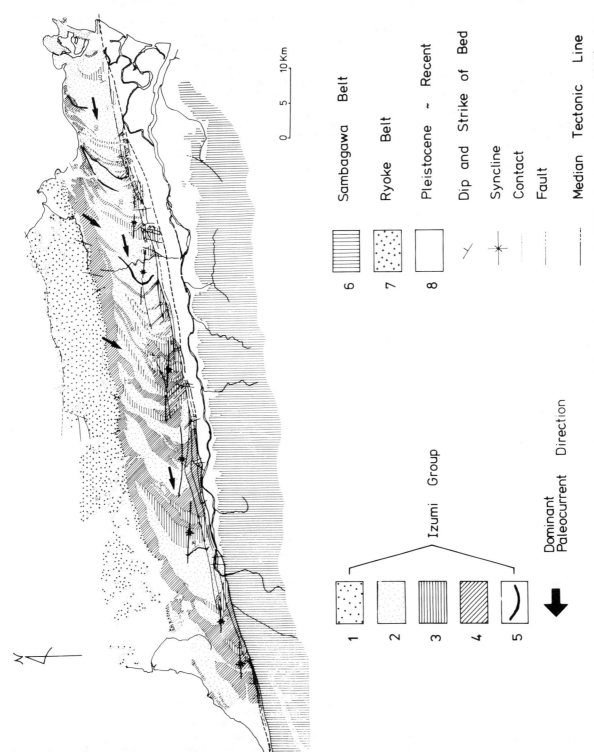

Fig. 2. Geologic map of the Izumi Group in northeastern Shikoku (after Suyari, 1973; Katto and others, 1977). See Fig. 1 for location. 1=Conglomeratic facies of the Izumi Group, 2=Sandy facies of the Izumi Group, 3=Sandstone and mudstone alternation facies of the Izumi Group, 4=Muddy facies of the Izumi Group, 5=Tuff (key bed), 6=Metamorphic rocks of the Sambagawa belt, 7=Metamorphic rocks and granitic rocks of the Ryoke belt, 8=Pleistocene and Recent sediments.

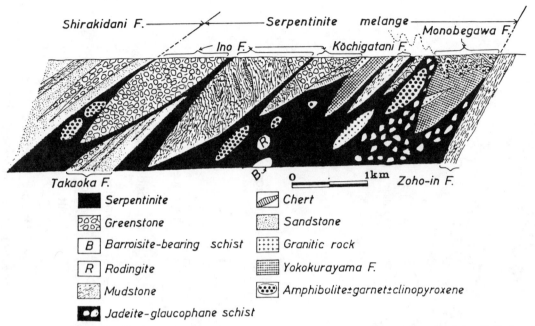

Fig. 3. Schematic cross section of the Kurosegawa tectonic zone at Ino, Kochi, Shikoku (after Maruyama, 1981). Yokokurayama Formation=Siluro-Devonian sedimentary rocks and tuffs, Shirakidani Formation=Jurassic subduction complex, Ino Formation=Dismembered ophiolite and subduction complex, Takaoka Formation=Jurassic subduction complex, Kochigatani Formation=Triassic sediments, Zohoin Formation=Triassic sediments, Monobegawa Formation=Lower Cretaceous sediments. See Fig. 1 for location.

schist, garnet amphibolite and sheared granite as tectonic lenses (Maruyama and others, 1978; Nakajima and others, 1978; Maruyama, 1981). These anomalous and exotic rocks are surrounded by serpentinite described by Maruyama(1981) as "serpentinite melage" as shown in Figure 3. Within this belt, there are Triassic to Cretaceous shallow marine sediments in elongate and lenticular shape often bounded by faults. The Triassic deposits are shallow marine sandstones and shales yielding such mulluscan fauna as Monotis and Halobia. The Jurassic sediments are composed of shallow marine limestone and calcareous shale (Torinosu Group). The Cretaceous sediments show a cyclic pattern of sedimentation with a upwards fining depositional trend, and are mainly composed of brackish to shallow marine coarse clastic sediments (Tashiro and others, 1980). The overall lithologic and structural characteristics suggest that the southern branch again is the strike-slip mobile zone.

In the northern Kyushu to Chugoku provinces, the Nagato and Maizuru tectonic zones (Fig. 1B) may be strike-slip mobile zones separating the Sangun metamorphic terrane (an older terrane) from the unmetamorphosed terrane of the Tamba belts. The Nagato tectonic zone consists of serpentinite, 370 - 420 m.y. metamorphic rocks, and Triassic to Lower Cretaceous coal bearing clastics with highly deformed and faulted structure(Murakami, 1979; Hirano, 1971; Nishimura, 1979). The Maizuru tectonic zone contains serpentinite, metamorphosed ophiolitic rocks and Triassic to Cretaceous clastic sediments in a highly faulted zone (Nakazawa, 1958, 1973; Ishiwatari, 1978; Igi and others, 1979).

In central Honshu, the Hida metamorphic terrane (Fig. 1) is separated by the Hida border fault zone (Hida marginal belt) from the Tamba (Mino) belt. The Hida border fault zone contains the oldest rock (Ordovician) known in Japan (Igo and others, 1980; Adachi and Igo, 1980) together with Silurian to Devonian sediments, 300 - 350 m.y. metamorphic rocks and abundant serpentinite accompanied by Jurassic to Lower Cretaceous clastic sediments (Chihara and Komatsu, 1981). The Jurassic and Cretaceous sediments show shallow marine to terrestrial environments and in some part reach 10,000 m in lateral accumulation. This fault zone shows characteristics common to the other tectonic zones of Japan and can be interpreted as a strike-slip mobile zone.

The well known Itoigawa-Shizuoka tectonic line (Fossa Magna) offsets the entire structural trend of central Japan but there are still additional similar tectonic zones in the Kanto district. The Tanakura tectonic zone also contributes to separating NE Japan from SW Japan (Otsuki, 1975). This major strike-slip fault zone has been active since Cretaceous time, with 400km of left-lateral displacement (Otsuki and Ehiro, 1979).

The Abukuma massif, separated from the Ashio

belt by the Tanakura tectonic zone, is considered as one of the older terranes of Japan (Kano and others, 1977). Its eastern margin is bounded by the Futaba fault system in which the ultramafics, Siluro-Devonian rocks and Jurassic to Paleogene clastics are exposed. The continuation of this zone reaches to the South Kitakami region where Silurian to Permian sediments with distinctive "exotic" faunal and floral characteristics (Saito and Hashimoto, 1982) are distributed. These deposits are disected by many faults with large strike-slip components (Otsuki and Ehiro, 1978), which in the Hayachine tectonic zones include associated ultramafics and metamorphic rocks. The Triassic to Cretaceous terrestrial to shallow marine sediments have been emplaced and accumulated locally. The South Kitakami region as a whole may be viewed as strike-slip mobile zone in which large exotic rock masses are incorporated.

To the northeast of the South Kitakami region lies the North Kitakami and Iwaizumi belts (Onuki, 1969), possibly Jurassic subduction terranes. They may be separated by strike-slip faults as suggested by trains of Cretaceous and Paleogene clastic basins. The North Kitakami and Iwaizumi belts extend to western Hokkaido. They are separated from the Hidaka belt, similar Jurassic subduction terrane, by tectonic zones of central Hokkaido, including the Kamuikotan belt and Hidaka metamorphic belt.

The Kamuikotan belt consists of ophiolitic rocks of both low and high pressure metamorphism and abundant serpentinites which are cut by many faults showing anastamosing patterns (Asahina and Komatsu, 1979; Katoh and others, 1979; Ishizuka, 1980). Within this belt and west of it, Cretaceous to Paleogene sediments were accumulated in an elongate basin (Ezo, Hakobuchi and Yubari Groups, Matsumoto and Okada, 1971). Between the Kamuikotan and Hidaka belts, there are belts of ophiolites, melanges and metamorphic rocks which include granulite facies (Komatsu and others, 1982). This zone was interpreted as a collision zone between eastern and western Hokkaido, two Mesozoic island arc systems. Thus, the tectonic zones of central Hokkaido as a whole can be considered to constitute a collision suture and associated strike-slip mobile zone formed during Cretaceous or Paleogene (Okada, 1982; Komatsu and others, 1982).

To the east of the Kamuikotan belt lies the Hidaka, Tokoro and Nemuro belts. The Hidaka and Tokoro belts, possible Jurassic subduction complexed (Okada, 1979) may be separated by a strike-slip zone suggested by the presence of localized Cretaceous basins.

Oblique Subduction, Strike-Slip Tectonics and the Evolution of Japan

The fundamental geologic framework of Japan can be interpreted as slices of older terrane and post-Triassic subduction terrane juxtaposed by strike-slip mobile zones. Because the strike-slip mobile zones are associated with subduction terranes, it is possible to consider that the strike-slip motions have been caused by the oblique subduction as has been documented geophysically by Fitch (1972) and geologically by Karig (1979), Page and others (1979), Lewis (1980) and Cole and Lewis (1981).

A generalized relationship of strike-slip patterns and oblique subduction has been presented by Karig (1979). He suggested that a sliced body of subduction complex moves laterally along the trench, causing the erosion and overlap of subduction terranes (Fig. 4A). Adopting this model, we propose a mechanism of emplacement of "exotic" blocks into the strike-slip mobile zones as shown in Figure 4B. Part of the basement rock of the continental margin or island arc can be tectonically sliced and transported laterally along the strike-slip zone as "tectonic lenses". Within such zones, the deep-seated part of the subduction complex such as ophiolites (sometimes metamorphosed to higher grade) are emplaced vertically together with ultramafics (Fig. 4). The main phase of strike-slip motion seems to have took place during early Cretaceous time because although the Jurassic subduction terranes show large scale rearrangement, the late Cretaceous subduction terrane (Shimanto Belt) shows a coherent internal organization. Lateral displacement of the strike-slip mobile zones may be enormous, possibly over 1000 km, because some tectonic blocks are totally unrelated to the geologic terranes of Japan. Such large scale displacement can only be achieved by very oblique subduction or transformed motion between two plates.

To illustrate the relationship between the subduction complex and strike-slip mobile zone in an actual geologic framework, we have examined the late Cretaceous system of SW Japan. The Cretaceous geologic units of SW Japan have been classified into five major geotectonic units by Taira (1979, 1981) and Taira and others (1979, 1981). These are 1) magmatic arc of the inner side, 2) the fore-arc basin (in the sense of Dickinson and Seely, 1979) between the magmatic arc and frontal non-volcanic arc (Izumi Basin) (in Taira and others, 1981, this was called intra-arc basin), 3) non-volcanic frontal arc (uplifted Sambagawa and Chichibu belts), 4) fore-arc shelf basins (Cretaceous sediments along the Kurosegawa tectonic zone) and 5) accretionary fore-arc slope basins (slope basins; in Taira and others, 1981, this was called fore-arc basin) and subduction complex (Shimanto Belt). In this study, zones 2) and 4) can be considered as strike-slip basins.

There is some evidence that the Median Tectonic Line at the time of the Shimanto subduction was undergoing left-lateral slip (Ichikawa, 1980). Furthermore, Otsuki and Ehiro (1978) indicate that many fault zones in Japan show dominantly left-lateral motion during Cretaceous time. The reconstruction of the Kula and Pacific plate indicates that they moved from south to north relative to Japan, possibly oblique to the "Shimanto

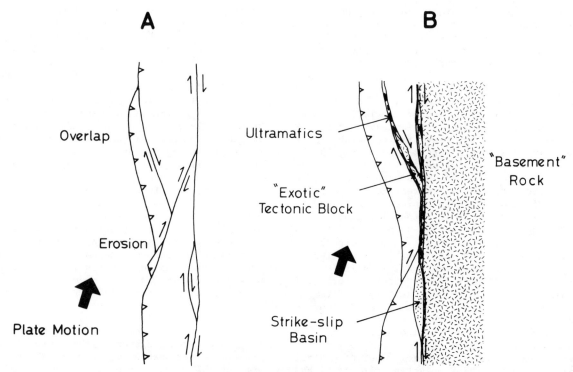

Fig. 4. A: Idealized oblique subduction margin (plan view, after Karig, 1979), B: Model of oblique subduction and strike-slip mobile zone proposed in this paper. Basement rock is part of continental margin or island arc.

trench" (see reconstruction by Uyeda and Miyashiro, 1974; Hilde and others, 1977; Dickinson, 1978).

With these data, we reconstructed a model showing an oblique subduction margin at the time of Late Cretaceous in SW Japan (Fig. 5). In this model two major left-lateral strike-slip fault systems, the Kurosegawa tectonic zone and the Median Tectonic Line, bound the frontal non-volcanic arc in manner similar to the model suggested by Lewis (1980). In the Kurosegawa tectonic zone, a braided pattern of faults caused a series of localized basins where shallow marine clastic sediments were accumulated (Kajisako Formation, Tashiro and others, 1980b). Along this zone, the exotic tectonic slices were emplaced with serpentinite forming the "serpentinite melange". In the Median Tectonic Line mobile zone, a thick pile of clastic sediments was accumulated at a high rate of sedimentation caused by down faulting of the offset block (Izumi Basin). The laterally migrated slice of Jurassic subduction terrane was uplifted along this mobile zone forming a outer non-volcanic arc (Sambagawa and Chichibu belts). Within the inner side of the Izumi Basin, magmatic activity formed the volcanic arc.

Taira (1980) and Taira and others (1981) has proposed the "retroduction model" for explaining the uplift of the subducted Sambagawa belt and "rifting" of the Izumi Basin. The retroduction model considers the reverse process of subduction during tensional periods along convergent margins. In this paper, we emphasize a combination of strike-slip motion together with "rift" motion (transtension, see summary by Reading, 1980) to explain the accumulation of the Izumi basin and uplift of the Sambagawa belt.

The retroduction model was proposed assuming cyclic subduction (Kobayashi and Isezaki, 1976; Niitsuma, 1978). Taira (1979, 1981) has suggested that the Shimanto belt shows a history of episodic development. To explain this episodic development of subduction terranes as well as the cyclic formation of the intra-arc (fore-arc) basins and cyclic sedimentation of the fore-arc shelf basins, the cyclic subduction model has been applied.

Conclusion

During Jurassic time, convergence and subduction along the continental margin of east Asia formed subduction terranes of which the older rocks were a part. The nature of the terranes from which the older rocks came is poorly known; they may have been ancient subduction complexes and/or allochthonous rock mass such as a pieces of Pacifica (Nur and Ben-Avraham, 1978).

Due to apparent very-oblique subduction or

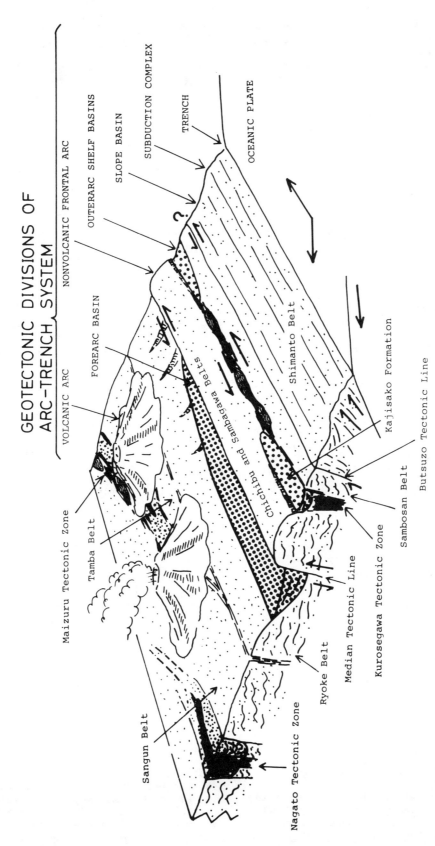

Fig. 5. Reconstruction of Late Cretaceous oblique subduction margin in southwest Japan.

ROLE OF OBLIQUE SUBDUCTION 311

transform motion, strike-slip mobile zone systems were developed within the older terranes and developing subduction terranes. Tectonically sliced geologic bodies migrated along the trenches with accompanying tectonic erosion and overlapping of the subduction terrane. Along the mobile zones, "exotic" tectonic blocks were emplaced due to tectonic erosion of basement complex and uplift of deep-seated ophiolitic rocks. Local sedimentary basins were developed by strike-slip tectonism within the mobile zones. This main phase of strike-slip motion took place during early Cretaceous time.

During late Cretaceous and Paleogene time, continued subduction formed newly developed subduction terranes (Shimanto and Nemuro belts). Strike-slip motion continued and, in some parts, large vertical displacements uncovered metemorphosed parts of Jurassic subduction complexes (Ryoke and Sambagawa belts). In Hokkaido, two island arcs collided during this time. These processes have created the basis geologic framework of Japan.

In this paper, we show that all of Japan has been sliced and reoriented by strike-slip movements which may well be part of a large scale strike-slip system along east Asia (e.g. Allen, 1962, 1965; Fitch, 1972). Because oblique convergence and subduction may be more common than the perpendicular case, the geotectonic framework as presented in Figure 5 may be a fundamental character of convergent margins. Finally we emphasize that a narrow tectonic zone with exotic blocks, serpentinite and sedimentary basins can be viewed as strike-slip mobile zone. Such mobile zones may play important role in reorganization of geologic terranes and lateral and vertical emplacement of small tectonic blocks in orogenic belt.

Acknowledgements. We thank Hisao Nakagawa, Nobuaki Niitsuma, Shigenori Maruyama, Hideo Ishizuka, Yujiro Ogawa and Daniel Karig for discussions. Jean Aubouin also provided helpful suggestions.

References

Adachi, S., and H. Igo, A new Ordovician Leperditiid ostracode from Japan, Proc. Jpn. Acad., 56, Ser. B, 504-507, 1980.

Allen, C. R., Circum-Pacific faulting in the Philippines-Taiwan region, J. Geophys. Res., 67, 4795-4812, 1962.

Allen, C. R., Transcurrent faults in continental areas, Continental Drift Symposium, Phil. Trans. R. Soc., A258, 82-89, 1965.

Asahina, T., and M. Komatsu, The Horokanai ophiolitic complex in the Kamuikotan tectonic belt, Hokkaido, Japan, J. Geol. Soc. Jpn., 85, 317-330, 1979.

Chihara, K., and M. Komatsu, The recent study in the Hida marginal zone, especially the Omi-Renge belt and the Jo-etsu zone---a review (in Japanese), Res. Rep. on the geology and petrology of the Lower Paleozoic and Upper Archeozoic system in and around Japan, edited by H. Kano, Akita Univ., pp. 21-24, 1981.

Cole, J. W., and K. B. Lewis, Evolution of the Taupo-Hikurange subduction system, Tectonophysics, 72, 1-21, 1981.

Crowell, J. C., Probable large lateral displacement on the San Gabriel fault, southern California, Bull. Am. Ass. Petrol. Geol., 36, 2026-2035, 1952.

Crowell, J. C., Sedimentation along the San Andreas Fault, California, in Modern and Ancient Geosynclinal Sedimentation, edited by R.H. Dott, Jr. and R. H. Shaver, Spec. Publ. Soc. Econ. Paleont. Miner. Tulsa, 19, pp. 292-303, 1974a.

Crowell, J. C., Origin of late Cenozoic basins in southern California, in Tectonics and Sedimentation, edited by W. R. Dickinson, Spec. Publ. Soc. Econ. Paleont. Miner. Tulsa, 22, pp. 190-204, 1974b.

Dickinson, W. R., Plate tectonic evolution of north Pacific rim, J. Phys. Earth, 26, Suppl., S1-S19, 1978.

Dickinson, W. R., and D. R. Seely, Structure and stratigraphy of forearc regions, Bull. Am. Ass. Petrol. Geol., 63, 2-31, 1979.

Fitch, T. J., Plate Convergence, transcurrent faults, and internal deformation adjacent to southeast Asia and the western Pacific, J. Geophys. Res., 77, 4432-4460, 1972.

Fujiwara, Y., Changing of the palaeolatitude in the Japanese Islands through the Palaeozoic and Mesozoic, J. Fac. Sci. Hokkaido Univ., Ser. 4, 14, 159-168, 1968.

Hackman, B. D., The Solomon Islands fractured arc, in The Western Pacific, edited by P.J. Coleman, pp. 179-191, Univ. of Western Australia Press, Nedlands, 1973.

Hada, S., T. Suzuki, S. Yoshikura, and N. Tsuchiya, The Kurosegawa tectonic zone in Shikoku and tectonic environment of the outer zone of southwest Japan (in Japanese with English abstract), in The Basement of the Japanese Islands---Professor Hiroshi Kano Commemorative Vol., pp. 341-368, Akita Univ., Akita, 1979.

Hamada, T., The Middle Paleozoic group of Japan and its bearing on her geologic history, J.Fac. Sci. Univ. Tokyo, Ser. 2, 13, 1-79, 1961.

Hara, I., K. Takeda, E. Tsukuda, M. Tokuda, and T. Shiota, Tectonic movement in the Sambagawa belt (in Japanese with English abstract), in The Sambagawa Blet, edited by K. Hide, pp. 307-390, Hiroshima Univ. Press, Hiroshima, 1977.

Hara, I., K. Shoji, Y. Sakurai, S. Yokoyama, and K. Hide, Origin of the Median Tectonic Line and its initial shape (in Japanese with English abstract), Mem. Geol. Soc. Jpn., no. 18, 27-49, 1980.

Harata, T., K. Hisatomi, F. Kumon, K. Nakazawa, M. Tateishi, H. Suzuki, and T. Tokuoka, Shimanto geosyncline and Kuroshio Paleoland, J. Phys. Earth, 26, Suppl., S357-366, 1978.

Hashimoto, M., A review of the petrology of the Sangun metamorphic rocks, Japan (in Japanese with English abstract), Bull. Natn. Sci. Mus., 7, 323-337, 1964.

Hashimoto, M., Two kinds of glaucophanitic terrane in Japan and the environs, Bull. Natn. Sci. Mus., Ser. C, 4, 157-164, 1978.

Hattori, I., and K. Hirooka, Paleomagnetic results from Permian greenstones in central Japan and their geologic significance, Tectonophysics, 57, 211-235, 1979.

Hayashi, S., The Permian conodonts in chert of the Adoyama formation, Ashio mountains, central Japan, Chikyu Kagaku (J. Ass.Geol.Collab.Jpn.), 22, 63-77, 1968.

Hilde, T. W. C., S. Uyeda, and L. Kroenke, Evolution of the western Pacific and its margin, Tectonophysics, 38, 145-165, 1977.

Hirano, H., Biostratigraphic study of the Jurassic Toyora Group, Mem. Fac. Sci. Kyushu Univ., Ser. D, 21, 93-128, 1971.

Hiroi, Y., Geology of the Unazuki district in the Hida metamorphic terrain, central Japan (in Japanese with English abstract), J. Geol. Soc. Jpn., 84, 521-530, 1978.

Hirokawa, O. (Chief Ed.), Geological map of Japan, 1:1,000,000 (2nd Ed.), Geol. Surv. Japan, Kawasaki. 1978.

Horikoshi, E., Plate tectonic aspects of Japanese orogenic belts (in Japanese), Kagaku (Science), 42, 665-673, 1972.

Horikoshi, E., Downward extension of the Kurosegawa tectonic zone based on the distribution of Quaternary volcanoes (in Japanese with English abstract), J. Geol. Soc. Jpn., 85, 427-433, 1979.

Ichikawa, K., Geohistory of the Median Tectonic Line of southwest Japan, Mem. Geol. Soc. Jpn., no. 18, 187-212, 1980.

Ichikawa, K., K. Ishii, C. Nakagawa, K. Suyari, and N. Yamashita, Die Kurosegawa-Zone (Untersuchungen uber Chichibu--Tarrain in Shikoku III) (in Japanese with German abstract), J. Geol. Soc. Jpn., 62, 82-103, 1956.

Ichikawa, K., T. Matsumoto, and M. Iwasaki, Evolution of the Japanese Islands (in Japanese), Kagaku (Science), 42, 181-191, 1972.

Igi, S., K. Nakazawa, and Y. Kuroda, The basement complex in the Maizuru structural belt (in Japanese with English abstract), in The Basement of of the Japanese Islands--Professor Hiroshi Kano Commemorative Vol., pp. 143-152, Akita Univ., Akita, 1979.

Igo, H., Conodont biostratigraphy and restudy of geological structure at the eastern part of the Mino belt, Professor Mosaburo Kanuma Commemorative Vol., pp. 103-113, Tokyo, 1979.

Igo, H., S. Adachi, H. Furutani, and H. Nishiyama, Ordovician fossils first discovered in Japan, Proc. Jpn. Acad., 56, Ser. B, 499-503, 1980.

Ishiga, H., and N. Imoto, Some Permian radiolarians in the Tamba district, southwest Japan, Chikyu Kagaku (J. Ass. Geol. Collab. Jpn.), 34, 333-345, 1980.

Ishiwatari, H., A preliminary report on the Yakuno ophiolite in the Maizuru zone, inner southwest Japan (in Japanese with English abstract), Chikyu Kagaku (J. Ass. Geol. Collab. Jpn.), 32, 301-310, 1978.

Ishizuka, H., Geology of the Horokanai ophiolite in the Kamuikotan tectonic belt, Hokkaido, Japan (in Japanese with English abstract), J. Geol. Soc. Jpn., 86, 119-134, 1980

Isozaki, Y., and T. Matsuda, Age of the Tamba Group along the Hozugawa "anticline", western hill of Kyoto, southwest Japan, J. Geosci., Osaka City Univ., 23, 115-134, 1980.

Kanmera, K., Comparison between ancient and modern geosynclinal sedimentary bodies, I, II (in Japanese, Kagaku (Science), 46, 284-291, 371-378, 1976.

Kanmera, K., M. Hashimoto and T. Matsuda (eds.), The geology of Japan (in Japanese), Iwanami Earth Science Series, 15, pp. 387, Iwanami-shoten, Tokyo, 1980.

Kano, H., Y. Kuroda, K. Uruno, T. Nureki, I. Hara, S. Kanisawa, T. Maruyama, and H. Umemura, The plutonic and metamorphic history of the Abukuma belt from the viewpoint of polymetamorphism (in Japanese with English abstract), in The Sambagawa Belt, edited by K. Hide, pp. 289-296, Hiroshima Univ. Press, Hiroshima, 1977.

Kano, T., Geological study of the Hida metamorphic belt in the eastern part of Toyama Prefecture, central Japan, Pt. 1, 2 (in Japanese with English abstract), J. Geol. Soc. Jpn., 79, 407-421, 81, 535-546, 1973, 1975.

Karakida, Y., T. Oshima, and S. Miyachi, The Kurosegawa zone and the Chichibu terrane in Kyushu (in Japanese with English abstract), in The Sambagawa Belt, edited by K. Hide, pp. 165-177, Hiroshima Univ. Press, Hiroshima, 1977.

Karig, D. E., Material transport within accretionary prisms and the "knocker" problem, J. Geol., 88, 27-39, 1979.

Katoh, T., K. Niida, and T. Watanabe, Serpentinite melange around Mt. Shirikomadake in the Kamuikotan structural belt, Hokkaido (in Japanese with English abstract), J. Geol. Soc. Jpn., 85, 279-285, 1979.

Katto, J., T. Suyari, N. Kashima, I. Hashimoto, S. Hada, S. Mitsui, and I. Akojima, Surface geologic map of Shikoku, Kochi Regional Forestry Office, Kochi, 1977.

Katto, J., and M. Tashiro, A study on the molluscan fauna of the Shimanto terrain, southwest Japan, Part 1: On the bivalve fauna of the Doganaro Formation in Susaki area, Kochi Prefecture (in Japanese with English abstract), Res. Rep. Kochi Univ., 27, Nat. Sci., 143-150, 1978.

Katto, J., and M. Tashiro, A study on the molluscan fauna of the Shimanto terrain, southwest Japan, Part 2: Bivalve fauna from the Murotohanto Group in Kochi Prefecture, Shikoku, Res. Rep. Kochi Univ., 28, Nat. Sci., 1-11, 1979a.

Katto, J., and M. Tashiro, A study on the molluscan fauna of the Shimanto terrain, southwest Japan, Part 3: On the bivalve fauna from the Arioka, Nakamura and Susaki Formations in Shimanto (northern) terrain, Kochi Prefecture (in

Japanese), Res. Rep. Kochi Univ., 28, Nat.Sci., 49-58, 1979b.

Kimura, T., The old 'inner' arc and its deformation in Japan, in The Western Pacific, edited by P. J. Coleman, pp. 255-273, Univ. of Western Australia, Nedlands, 1973.

Kimura, T., Structural development of Japan and plate tectonics (in Japanese with English abstract), Chigaku Zasshi(J. Geography), 86, 54-67, 1977.

Kobayashi, K., and N. Isezaki, Magnetic anomalies in the Sea of Japan and the Shikoku Basin: possible implications, in The geophysics of the Pacific Ocean basin and its margin, edited by G. H. Sutton, M. H. Manghani, and R. Monerly, pp. 235-251, AGU, Washington, D. C., 1976.

Kobayashi, T., Sakawa orogenic cycle and its bearing on the origin of the Japanese Islands, J. Fac. Sci. Imp. Univ. Tokyo, 5, 219-578, 1941.

Kobayashi, T., and T. Hamada, Silurian trilobites of Japan, Palaeont. Soc. Jpn., Spec. Pap., no. 18, 1-155, 1974.

Koike, T., H. Igo, S. Takizawa, and T. Kinoshita, Contribution to the geological history of the Japanese Islands by the conodont biostratigraphy Part 2, J. Geol. Soc. Jpn., 77, 165-168, 1971.

Komatsu, M., S. Miyashita, J. Maeda, Y. Osanai, and T. Toyoshima, Disclosing of a deepest section of continental-type crust up-thrust as the final event of collision of arcs in Hokkaido, north Japan, Proc. Oji Intn. Seminar on Accretion Tectonics, 1982, in press.

Lewis, K. B., Quaternary sedimentation on the Hikurangi oblique-subduction and transform margin, New Zealand, in Sedimentation in Oblique-Slip Mobile Zones, edited by P. F. Ballance and H. G. Reading, Spec. Publ. Int. Ass. Sediment., 4, pp. 171-189, 1980.

Maruyama, S., The dismembered ophiolite belt (in Japanese with English abstract), Chikyu Kagaku (J. Ass. Geol. Collab. Jpn.), 32, 317-320, 1978.

Maruyama, S., Y. Uyeda, and S. Banno, 208-240m.y. old jadeite glaucophane schists in the Kurosegawa tectonic zone near Kochi City, Shikoku, Jpn. Ass. Min. Pet. Econ. Geol., 73, 300-310, 1978.

Maruyama, S., and M. Yamasaki, Paleozoic submarine volcanoes in the high-P/T metamorphosed Chichibu system of eastern Shikoku, Japan, J. Volc. Geotherm. Res., 4, 199-216, 1978.

Maruyama, T., Rb-Sr geochronological studies in the granitic rocks of the southern Abukuma plateau (in Japanese with English abstract), in The Basement of the Japanese Islands----Professor Hiroshi Kano Commemorative Vol., pp. 523-558, Akita Univ, Akita, 1979.

Matsuda, T., and S. Uyeda, On the Pacific-type orogeny and its model-extension of the paired belts concept and possible origin of marginal sea, Tectonophysics, 11, 5-27, 1971.

Matsumoto, T., Fundamental problems in the circum-Pacific orogenesis, Tectonophysics, 4, 595-613, 1967.

Matsumoto, T., Cephalopods form the Shimanto Belt of Kochi Prefecture (Shikoku) (in Japanese with English abstract), in Geology and Paleontology of the Shimanto Belt, edited by A. Taira and M. Tashiro, pp. 283-298, Rinya-kosaikai Press, Kochi, 1980.

Matsumoto, T., and H. Okada, Clastic sediments of the Cretaceous Yezo geosyncline, Mem. Geol. Soc. Jpn., no. 6, 61-74, 1971.

Minato, M., M., Gorai, and M. Hunahashi(eds.), The geologic development of the Japanese Islands, Tsukiji-shokan, Tokyo, 1965.

Miyashiro, A., Metamorphism and related magmatism in plate tectonics, Am. J. Sci., 272, 629-656, 1972.

Miyashiro, A., Paired and unpaired metamorphic belts, Tectonophysics, 17, 241-254, 1973.

Miyata, T., H. Ui, and K. Ichikawa, Paleogene left-lateral wrenching on the Median Tectonic Line in southwest Japan, Mem. Geol. Soc. Jpn., no. 18, 51-68, 1980.

Nakajima, T., S. Maruyama, and K. Matsuoka, Metamorphism of the green rocks of the Ino Formation in central Shikoku, (in Japanese with Enlish abstract), J. Geol. Soc. Jpn., 84, 729-737, 1978.

Nakaseko, K., and A. Nishimura, Upper Triassic radiolaria from southwest Japan, Sci. Rep. Col. Gen. Educ., Osaka Univ., 28, 61-109, 1979.

Nakazawa, K., The Triassic system in the Maizuru Zone, southwest Japan, Mem. Fac. Sci., Kyoto Univ., Geol. Min., 24, 265-313, 1958.

Nakazawa, K., Maizuru structural belt, in The Crust and Upper Mantle of the Japanese areas III, edited by M. Gorai and S. Igi, pp. 92-93, Geol. Surv. Japan, Kawasaki, 1973.

Niitsuma, N., Magnetic stratigraphy of the Japanese Neogene and the development of the island arcs of Japan, J. Phys. Earth, 26, Suppl., S367-S378, 1978.

Nishimura, Y., Metamorphic rocks of 300-400 m.y. ages found in the Inner Zone of southwest Japan (in Japanese with English abstract), in The Basement of the Japanese Islands--Professor Hiroshi Kano Commemorative Vol., pp. 201-216, Akita Univ., Akita, 1979.

Nishimura, Y., T. Inoue, and H. Yamamoto, Sangun Belt, with special reference to the stratigraphy and metamorphism (in Japanese with English abstract), in The Sambagawa Belt, edited by K. Hide, pp. 257-282, Hiroshima Univ.Press, Hiroshima, 1977.

Noda, M., Some inoceramid species (Cretaceous Bivalvia) from the Sukumo-Nakamura area, Shikoku (in Japanese with English abstract), in Geology and Paleontology of the Shimanto Belt, edited by A. Taira and M. Tashiro, pp. 265-282, Rinya-kosaikai Press, Kochi, 1980.

Nozawa, T., Radiometric age map of Japan(metamorphic rocks), Geol. Surv. Japan, Kawasaki, 1977.

Nur, A., and Z. Ben-Avraham, Lost Pacifica continent, Nature, 270, 41-43, 1977.

Nur, A., and Z. Ben-Avraham, Speculations on mountain building and the lost Pacifica conti-

nent, J. Phys. Earth, 26, Suppl., S21-S37,1978.
Okada,H.,Migration of ancient arc-trench systems, in Modern and ancient geosynclinal sedimentation,SEPM, Spec. Publ., no.19, pp.311-320, 1977.
Okada, H., The geology of Hokkaido and plate tectonics(in Japanese), Chikyu (Earth Monthly),11, 869-877, 1979.
Okada, H., Collision orogenesis and sedimentation in Hokkaido, Japan, Proc. Oji Intn. Seminar on Accretion Tectonics, 1982, in press.
Okamura, M., Radiolarian fossils from the Shimanto Belt in Kochi Prefecture, Shikoku (in Japanese with English abstract), in Geology and Paleontology of the Shimanto Belt, edited by A. Taira and M. Tashiro, pp. 153-178, Rinyakosaikai Press, Kochi, 1980.
Onuki, Y., Explanatory text to the geological map of the Iwate Prefecture II (in Japanese), Iwate Pref. Government, 189pp., Morioka, 1956.
Onuki, Y., Geology of the Kitakami massif, northeast Japan (in Japanese with English abstract), Contr. Inst. Geol. Paleont., Tohoku Univ., no. 69, 1-239, 1969.
Ota, M., The Akiyoshi limestone group: a geosynclinal organic reef complex (in Japanese with English abstract), Bull. Akiyoshi-dai Sci. Mus. 5, 1-44, 1968.
Otsuki, K.,and M. Ehiro, Major strike-slip faults and their bearing on spreading in the Japan Sea, J. Phys. Earth,26, Suppl., S537-S555,1978.
Page, B. G. N., J. D. Benett, N. R. Cameron, D. McC. Bridge, D. H. Jeffery, W. Keats, and J. Thaib, A review of the main structural and magmatic features of northern Sumatra, J. Geol. Soc. Lond., 136, 569-578, 1979.
Reading, H. G., Characteristics and recognition of strike-slip fault systems, in Sedimentation in oblique-slip mobile zones, edited by P. F. Ballance and H. G. Reading, Spec. Publ. Int. Ass. Sediment., 4, pp. 7-26, 1980.
Saito, Y., Geology of the younger Paleozoic systems of the southern Kitakami massif, Iwate Prefecture, Japan, Sci. Rep., Tohoku Univ., Ser. 2, 40, 79-139, 1968.
Saito, Y.,and M. Hashimoto,South Kitakami Region: an allochthonous terrane in Japan, J. Geophys. Res., 87, 3691-3696, 1982
Shibata, K., Geochronology of pre-Silurian basement rocks in the Japanese Islands, with speccial reference to age determinations on orthoquartzite clasts (in Japanese with English abstract), in The Basement of the Japanese Islands---Professor Hiroshi Kano Commemorative Vol., pp. 625-639, Akita Univ., Akita, 1979.
Steel, R. J., Devonian basins of western Norwaysedimentary response to tectonism and varying tectonic context, Tectonophysics, 36, 207-224,
Steel, R. J., and T. G. Gloppen, Late Caledonian (Devonian) basin formation, western Norway: signs of strike-slip tectonics during infilling, in Sedimentation in oblique-slip mobile zones, edited by P. F. Ballance and H. G. Reading, Sepc. Publ. int. Ass. Sediment., 4, pp.79-103, 1980.

Sugimoto, M., Stratigraphical study in the outer belt of the Kitakami massif, northeast Japan (in Japanese with English abstract), Contr. Inst. Geol. Paleont., Tohoku Univ., no.74,1-48, 1974.
Suyari, K., On the lithofacies and the correlation of the Izumi Group of the Asan mountain range, Shikoku, Sci. Rep., Tohoku Univ.,Ser. 2, Spec. Vol., 6, 489-495, 1973.
Suyari, K., Y. Kuwano,and K. Ishida, Discovery of the Late Triassic conodonts from the Sambagawa Metamorphic Belt proper in western Shikoku (in Japanese), J. Geol. Soc. Jpn.,86,827-828, 1980.
Suzuki, T., The Kurosegawa tectonic zone and the Chichibu belt in Shikoku (in Japanese with English abstract), in The Sambagawa Belt, edited by K. Hide, pp. 153-164, Hiroshima Univ. Press, Hiroshima, 1977.
Suzuki, T., and S. Hada, Cretaceous tectonic melange of the Shimanto belt in Shikoku, Japan, J. Geol. Soc. Jpn., 85, 467-479, 1979.
Suzuki, T., S. Hada, and S. Yoshikura, A model for the geologic development of the outer zone of southwest Japan (in Japanese), Chikyu (Earth Monthly), 1, 57-62, 1979.
Taira, A., Formation of sediment body in the arc-trench system and cyclic subduction model (in Japanese), Chikyu (Earth Monthly),11, 860-868, 1979.
Taira, A., The Shimanto belt of southwest Japan and arc-trench sedimentary tectonics, Recent Progress in Natural Sci. in Japan, 6, 147-162, 1981.
Taira, A., Y. Saito, and M. Hashimoto, The fundamental processes of geologic evolution of Japan (in Japanese), Kagaku (Science), 51, 508-515, 1981.
Taira, A., J. Katto, and M. Tashiro, The Cretaceous and Cenozoic geologic development of southwest Japan and the tectonism of arc-trench system(in Japanese), Geol. News, 296, 27-40, 1979.
Taira, A., J. Katto, M. Tashiro, and M. Okamura, The geology of the Shimanto belt in Kochi Prefecture, Shikoku (in Japanese with English abstract), in Geology and Paleontology of the Shimanto Belt, edited by A. Taira and M. Tashiro, pp. 319-389, Rinya-kosaikai Press,Kochi, 1980.
Taira, A., K. Nakaseko, J. Katto, M. Tashiro, and Y. Saito, New observations on the Sambosan Group in the western Kochi Prefecture (in Japanese), Geol. News, 302, 22-35, 1979.
Taira, A., H. Okada, J. H. McD. Whitaker, and A. J. Smith, The Shimanto Belt of Japan: Cretaceous-lower Miocene active-margin sedimentation, in Trench-Forearc Geology : Sedimentation and tectonics on modern and ancient active plate margins, edited by J. K. Leggett, pp.5-26,Geol. Soc. London, Spec. Publ., no. 10, 1982.
Taira, A., M. Okamura, J. Katto, M. Tashiro, Y. Saito, K. Kodama, M. Hashimoto, T. Tiba, and T. Aoki, Lithofacies and geologic age relationship within melange zones of northern Shimanto Belt (Cretaceous), Kochi Prefecture, Japan (in

- Japanese with English abstract), in Geology and Paleontology of the Shimanto Belt, edited by A. Taira and M. Tashiro, pp. 179-214, Rinya-kosaikai Press, Kochi, 1980.
- Takai, F., T. Matsumoto, and R. Toriyama (eds.), Geology of Japan, 279 pp., Univ. Tokyo Press, Tokyo, 1963.
- Tanaka, K., Kanoashi Group, an olistostrome, in the Nishihara area, Shimane Prefecture (in Japanese with English abstract), J.Geol. Soc. Jpn., 86, 613-628, 1980.
- Tanaka, K., and T. Nozawa (eds.), Geology and mineral resources of Japan, 430 pp., Geol.Surv. Japan, Kawasaki, 1977.
- Takayanagi, Y., Preliminary notes on the Cretaceous foraminifera from the Uwagumi Formation of the Shimanto Belt, Kochi Prefecture, Shikoku (in Japanese with English abstract), in Geology and Paleontology of the Shimanto Belt, edited by A. Taira and M. Tashiro, pp. 235-239, Rinya-kosaikai Press, Kochi, 1980.
- Tashiro, M., The bivalve fossils from the Shimanto belt of Kochi Prefecture and their biostratigraphic implications (in Japanese with English abstract), in Geology and Paleontology of the Shimanto Belt, edited by A. Taira and M. Tashiro, pp. 249-264, Rinya-kosaikai Press, Kochi, 1980.
- Tashiro, M., T. Kozai, and J. Katto, A biostratigraphical study of the Upper Cretaceous formations at Odochi of Monobe, Shikoku (in Japanese with English abstract), in Geology and Paleontology of the Shimanto Belt, edited by A. Taira and M. Tashiro, pp. 71-82, Rinya-kosaikai Press, Kochi, 1980.
- Teraoka, Y., Cretaceous formations in the Onogawa basin and its vicinity, Kyushu, southwest Japan (in Japanese with English abstract), Rep. Geol. Surv. Japan, no. 237, 1970.
- Teraoka, Y., Cretaceous sedimentary basins in the Ryoke and the Sambagawa belts (in Japanese with English abstract), in The Sambagawa Belt, edited by K.Hide, pp.419-431, Hiroshima Univ.Press, Hiroshima, 1977.
- Umemura, H., Tectonic movement in the Gosaisho-Takanuki district, with special reference to tectonic junction of the Gosaisho and the Takanuki metamorphic rocks (in Japanese with English abstract), In The Basement of the Japanese Islands---Professor Hiroshi Kano Commemorative Vol., pp.491-511, Akita Univ., Akita, 1979.
- Uyeda, S., and A. Miyashiro, Plate tectonics and the Japanese Islands: a synthesis, Geol. Soc. Am. Bull., 85, 1159-1170, 1974.
- Wilcox, R. E., T. P. Harding, and D. R. Seely, Basic wrench tectonics, Bull. Am. Ass. Petrol. Geol., 57, 74-96, 1973
- Yabe, H., The median dislocation line of southwest Japan reconsidered, Proc. Jpn. Acad., 35, 384-387, 1959.
- Yabe, H., Probable position of the outer wing of the Ryoke metamorphics in southwest Japan (in Japanese with English abstract), Chigaku Zasshi (J. Geography), 72, 110-114, 1963.
- Yao, A., T. Matsuda, and Y. Isozaki, Triassic and Jurassic radiolarians from the Inuyama area, central Japan, J. Geosci., Osaka Univ.,23, 135-154, 1980.
- Yoshida, T. (ed.), An outline of geology of Japan (3rd ed.), 61 pp., Geol. Surv. Japan, Kawasaki, 1975.
- Yoshikura, S., and K. Sato, A few evidences on the Kurosegawa tectonic zone near Yokokurayama, Kochi Prefecture (in Japanese), Island-arc basement (Toko-kiban), no. 3, 53-56, 1976.
- Yoshikura, S., and M. Toshida, Kurosegawa tectonic zone in the western Kii Peninsula (in Japanese with English abstract), in The Basement of the Japanese Islands---Professor Hiroshi Kano Commemorative Vol., pp. 319-340, Akita Univ., Akita, 1979.

VERTICAL CRUSTAL MOVEMENTS OF NORTHEAST JAPAN SINCE MIDDLE MIOCENE

Noriko Sugi

Kyoritsu University, Chiyoda-ku, Tokyo, Japan 101

Kiyotaka Chinzei

Geological Institute, University of Tokyo, Tokyo, Japan 113

Seiya Uyeda

Earthquake Research Institute, University of Tokyo, Tokyo, Japan 113

Abstract. The history of the vertical crustal movements of Northeast Japan since Middle Miocene (16 m.y.B.P.) has been examined quantitatively. The movements of the western zone (inner arc) and those of the eastern zone (outer arc) are different. The western zone subsided during the period from 16 to 10 m.y.B.P. and then was uplifted during the subsequent period, the rate of uplift being enhanced during Quaternary time. The eastern zone, east of the Backbone Range, has been uplifted throughout the period. These results have been interpreted in terms of the change in the mode of subduction from the Mariana-type to the Chilean-type. Recent discoveries on the vertical movement in the fore-arc region of the Japan Trench, showing that Neogene subsidence has been followed by Quaternary uplift, have also been incorporated into this history.

Introduction

The trend of the vertical crustal movements during the Late Cenozoic in Japan has been studied through research based on lithostratigraphy and biofacies analyses of the Late Cenozoic deposits (e.g. Kitamura, 1959, 1963; Matsuda et al., 1967). According to these workers, rapid subsidence was predominant during Middle Miocene time in the zone west of the Backbone Range (inner arc) of Northeast Japan, and thereafter these areas were uplifted; in the eastern zone (outer arc) uplift has been continuous since Early Miocene. An acceleration of uplift during Quaternary time has also been stressed by many investigators.

The amount of vertical crustal movement during the Late Cenozoic has also been estimated by several authors. The total vertical displacement since the beginning of Miocene was first summarized by Matsuda et al. (1967) (Fig. 1). The amount of displacement was estimated from the present level of the basal unconformity of the Miocene deposits. Large amounts of subsidence (> 3 km) in central Hokkaido, in coastal zones along the Sea of Japan in northeastern and central Honshu and in the areas facing the Philippine Sea are notable features. A map of the vertical displacements during the Quaternary was published by the Research Group for Quaternary Tectonic Map (1968) (Fig. 2). These displacements were measured mainly by geomorphologic methods, in which the present heights of the denudation surface, formed at the end of Pliocene, were considered to be equal to the displacements during the Quaternary. In contrast to the total post-Miocene movement (Fig. 1), most of Japan has been uplifted during the Quaternary. Kaizuka and Murata (1969) published a map of the vertical displacement during the Neogene, that is during Miocene and Pliocene times (Fig. 3). They obtained the vertical displacement at each point as the difference between those shown in Fig. 1 and Fig. 2.

In this paper, we try to trace the historical changes of the vertical crustal movements in Northeast Japan during the Late Cenozoic with a better time resolution than in previous estimates.

General Features of Northeast Japan

In order to provide a background for the present work, we first review general features of Northeast Japan (Sugimura and Uyeda, 1973; Yoshii, 1979). Figure 4 shows cross-sections of Northeast Japan along about Lat. 40°N, extending from the Sea of Japan to the Japan Trench. Northeast Japan is a typical island arc, with the Japan Trench lying about 200 km oceanward of its east coast. The Pacific plate is considered to

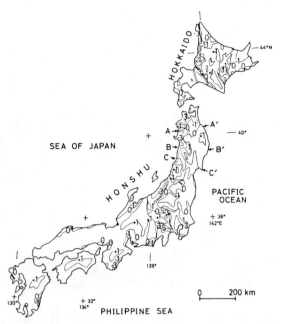

Fig. 1 Vertical crustal displacements of Japan since Early Miocene (Matsuda et al., 1967): contours in kilometers.

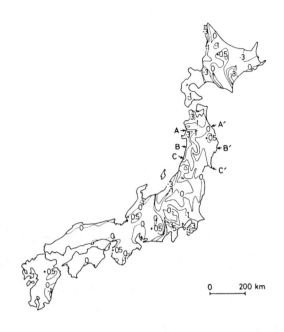

Fig. 3 Vertical crustal displacements of Japan during Neogene time (Kaizuka and Murata, 1969): contours in kilometers.

be subducting under the Asian plate at the Japan Trench. From the direction of the Hawaii-Emperor Seamounts, it is believed that the direction of motion of the Pacific plate changed from NNW to WNW at about 42 m.y.B.P. (Jarrard and Clague, 1977), and that since then the subduction has been nearly perpendicular to the present island arc. Up to this change in plate motion, the subduction along Northeast Japan was probably highly oblique. This view is in harmony with the observation that the present phase of volcanism in Northeast Japan started only after this change (see later). Shallow to deep focus earthquakes forming the landward dipping Wadati-Benioff zone support the existence of the subducting slab (Fig. 4e).

Three chains of highlands: the Dewa Hills (west), the Backbone Range and the Kitakami and Abukuma Mountains (east), extend from north to south along this arc. The Kitakami-Abukuma Mountains are made up of the Mesozoic and Paleozoic sedimentary, granitic and metamorphic rocks. On the other hand, the areas west of the Kitakami-Abukuma Mountains are characterized by gently folded thick accumulation of the Late Cenozoic strata (refer to Fig. 14). The Quaternary volcanic front is situated slightly to the east of the Backbone Range, which divides Northeast Japan into eastern and western zones. These zones differ from each other in their Cenozoic movements as mentioned before, and also in other geophysical features such as the upper mantle seismic wave velocity, gravity anomalies, heat flow and seismic activity. In general, the volcanic front forms a smooth curve, under which lie the earthquakes of 125 to 175 km depth (Sugimura, 1960; Dickinson, 1973). The mechanism of magma production under the arc still remains to be a major unsolved problem. In Northeast Japan, the Miocene volcanic front was located several tens of kilometers east of the Quaternary front (Naka-

Fig. 2 Vertical crustal displacements of Japan during Quaternary time (Research Group for Quaternary Tectonic Map, 1968): contours in kilometers.

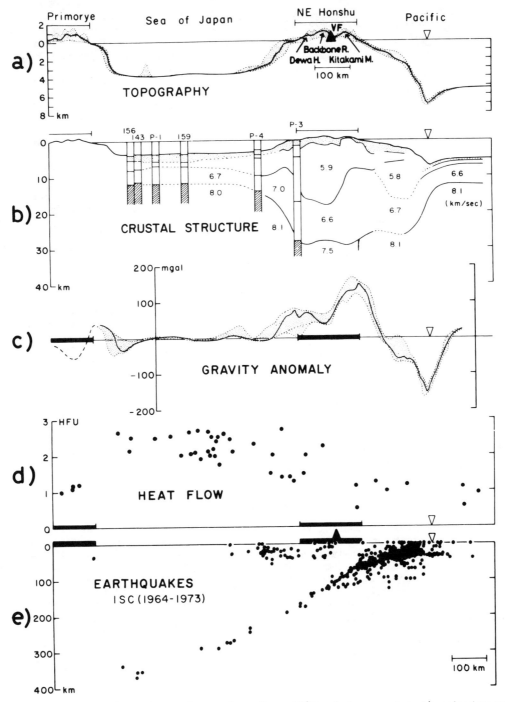

Fig. 4 Cross-sections of Northeast Japan along Lat. 40°N: a) topography, b) velocity structure, c) gravity anomalies, d) heat flow and e) seismic activity (Yoshii, 1979).

mura, 1969; Sugimura and Uyeda, 1973, p. 119; Matsuda, 1978). It seems that the volcanic activity has migrated westward relative to the arc with the average rate of about 0.1 cm/yr. This may represent an increase of the trench-arc gap (Dickinson, 1973) due either to the shallowing of the dip of the Wadati-Benioff zone or to the deepening of magma production zone during the recent past. Alternatively, it may be interpreted as caused by a trenchward motion of the arc (Nakamura, 1969).

There is a region in the uppermost mantle un-

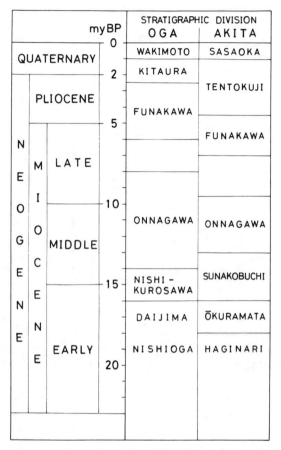

Fig. 5 Stratigraphic divisions of the Late Cenozoic deposits of Oga and Akita, Northeast Japan (Tsuchi et al., 1979).

der the arc, where earthquakes are lacking (Fig. 4e). This aseismic part of the uppermost mantle wedge is considered to be at anomalously high temperatures and, therefore, an upward embayment of the asthenosphere under the arc. The trench-ward front of this aseismic mantle is called the "aseismic front" (Yoshii, 1975). The aseismic front is located more than several tens of kilometers east of the present volcanic front. Kobayashi and Fujii (1981) pointed out an interesting fact that the aseismic front roughly coincides with the Miocene volcanic front: the aseismic uppermost mantle between the present volcanic front and the aseismic front may represent the remnant of high temperatures associated with the Miocene arc volcanism.

Behind the main arc is the Sea of Japan, which is believed to have been formed by sea-floor spreading at some relatively recent geologic time. If the heat flow vs. age relation for the normal oceanic lithosphere holds for the back-arc basin, the age of the Sea of Japan is inferred to be 30-25 m.y. old (Anderson, 1980). However, the exact age still remains to be found. It is one of the difficult problems in plate tectonics to explain why such extensional tectonics exist in the back-arc regions of convergent plate boundaries (e.g. Uyeda, 1977; Uyeda and Kanamori, 1979; Nakamura and Uyeda, 1980).

It is hoped in the present study of investigating the historical changes of the vertical movements in Northeast Japan to find some clue to the above mentioned problems.

Method for Estimation of the Vertical Movements

For the purpose of constructing the history of the vertical movements of Northeast Japan since Middle Miocene, we divided the period from 16 m.y.B.P. to the present into four time-intervals: 16 to 10 m.y.B.P., 10 to 5 m.y.B.P., 5 to 2 m.y.B.P. and 2 to 0 m.y.B.P. First, from the geologic maps, we read the present heights of the strata deposited at around 16, 10, 5 and 2 m.y.B.P., respectively. Assigned ages may have maximum errors of 20 %, and probably 10 to 15 % on average. We generally used the radiometric ages for the stratigraphic divisions of the Late Cenozoic deposits in Northeast Japan by Tsuchi et al. (1979) (Fig. 5). Next we took into account the water depths of deposition of these strata, and estimated the net displacements from their respective deposition times to the present. To estimate the water depth where each stratum was deposited, we used paleontologic data such as benthonic foraminifera and mollusca as well as some lithologic characteristics. Finally we obtained the amount of vertical displacement during each time-interval from the successive differences between them.

We read the present heights of the strata at respective times from the geologic maps along three transect lines across Northeast Japan (Fig. 6). The transect lines were selected so as to represent the geologic structure of Northeast Japan and to pass through the areas where the appropriate information is available. In practice, we marked one value for each rectangular area of 10 km (N-S) by 5 km (E-W) along the transects, by picking out the highest and lowest points of the stratum concerned in each area and taking their mean value. Movements before 16 m.y.B.P. were not treated in the present study because of the insufficient accuracy of the age of strata and the difficulty in reading their heights.

Results

Figure 7a shows the vertical displacements since 16 m.y.B.P. along each transect line. The present heights of the strata deposited at 16 m.y.B.P. are shown by crosses. Double circles are the values inferred from the geologic structure and the heights of nearby strata. All these 16 m.y. old strata are estimated to have been deposited near sea level because they include shallow marine and, in some part, terrestrial facies. Therefore, in this case, the present heights of the strata show directly the net amount of verti-

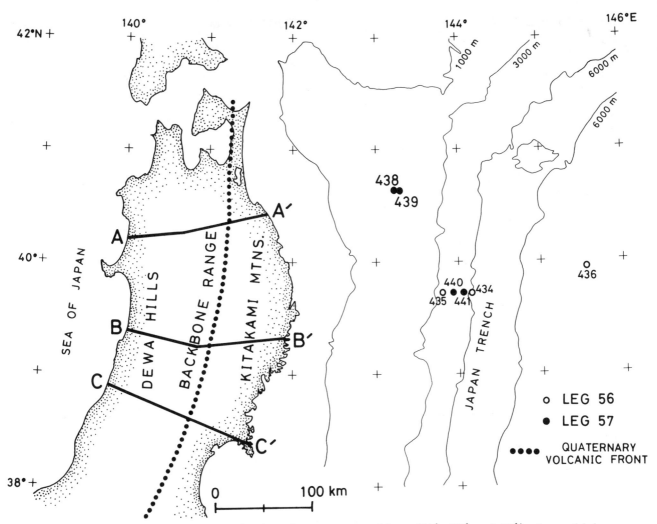

Fig. 6 Map of Northeast Japan showing three transect lines (AA', BB' and CC') along which we obtained crustal movements, and drill site locations for DSDP Legs 56 and 57.

cal displacement since 16 m.y.B.P. The vertical broken lines in Fig. 7 show the position of the Backbone Range.

Figure 7b shows the results obtained for the period since 10 m.y.B.P. In this figure, the estimated depths of deposition of the 10 m.y. old strata are shown by the inclined broken lines. The amounts of true vertical displacements are obtained by the differences between the present heights of the strata and their deposition depths at corresponding positions. The resultant net displacements since 10 m.y.B.P. are shown by open circles. The estimation of the water depths of deposition for these strata presented a difficult problem, because paleontologic data diagnostic of the deposition depth are scarce. For transect line CC', for example, a depth of 2,000 m with a probable error of ± 500 m at the western end was estimated based on rare bathyal benthonic foraminifera (Inoue, 1962). On the other hand, the depth of deposition for the area between the Backbone Range and the east end of the transect line is very well known to be close to sea level based on benthonic mollusca (Chinzei, 1978). We assumed that the deposition depth varied linearly between these two points, because of additional evidence along this transect line which supports the linear variation (Inoue, 1962). For other cases, the depth of deposition has been estimated in the same way.

Similarly, Fig. 7c shows the results for the period since 5 m.y.B.P. The 5 m.y. old strata were estimated to have been deposited at about 1,000 m depth (± 400 m) at the western coastal area based on benthonic foraminifera, and near sea level in the area east of the Backbone Range.

Figure 7d shows the displacements since 2 m.y. B.P., namely during the Quaternary. In this period, the major part of Northeast Japan was above sea level and marine strata are distributed only

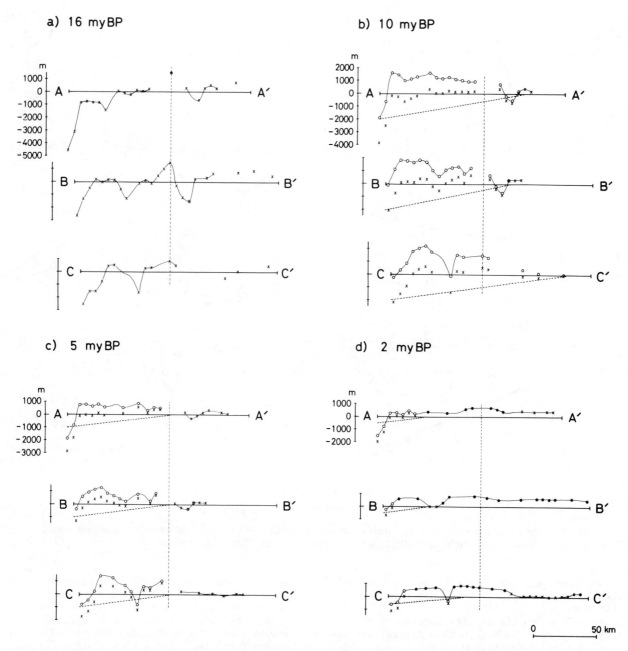

Fig. 7 Vertical crustal displacements along the three transect lines: a) since 16 m.y.B.P., b) since 10 m.y.B.P., c) since 5 m.y.B.P. and d) since 2 m.y.B.P. Crosses: present heights of the strata. Double circles: inferred values of displacements. Inclined broken lines: water depths at the time of deposition. Open circles: displacements since deposition. Solid circles: heights of the denudation surface for the end of Pliocene.

along the coast. Geomorphologic data, namely the heights of the denudation surface at the end of Pliocene, had to be used for the areas above sea level. Solid circles show the results obtained from the geomorphologic data. At the western coast, the deposition depth of the 2 m.y. old strata was assessed to be about 500 m ($\pm$ 300 m) based on benthonic foraminifera and mollusca.

Our second step was the determination of the amount of vertical displacement during each time-interval along the three transect lines. For instance, the displacements during the 16-10 m.y. interval were obtained as the differences between the displacements since 16 m.y.B.P. and those

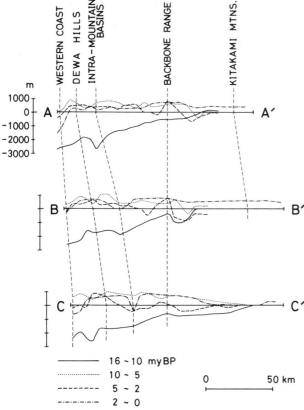

Fig. 8 Vertical displacements during four time-intervals: 16 to 10 m.y.B.P., 10 to 5 m.y.B.P., 5 to 2 m.y.B.P. and 2 to 0 m.y.B.P., along the three transect lines.

since 10 m.y.B.P. The resultant displacements for each time-interval are shown in Fig. 8. These results indicate that very active subsidence took place during the 16-10 m.y. interval, except for the area east of the Backbone Range.

In the subsequent intervals, uplift was dominant but the movements varied from place to place.

The final step, based on the results shown in Fig. 8, was to determine the height changes as a function of time for representative structural and topographic units: the western coast, the Dewa Hills, the intra-mountain basins, the Backbone Range and the Kitakami Mountains. The results are shown in Fig. 9. These curves show the height changes of the 16 m.y. old strata deposited near sea level at the respective places. As for the Kitakami Mountains, the uplift since Early Miocene estimated earlier by geomorphologic methods (Chinzei, 1966) for the northern part of the Mountains is shown in the figure.

In the present exercise, the largest source of error seems to be in the deposition depth. We tried to make a partial assessment of possible errors by allowing a reasonable range for the deposition depth at the western coast for the 10, 5 and 2 m.y. old strata. Figure 10 shows an example of possible ranges thus estimated for transect line BB'. The general trend of the changes appears to be maintained.

Discussion

In the results shown in Fig. 9, the following points may be noticed. The western coast and the intra-mountain basins along the transect lines subsided rapidly during the 16-10 m.y. interval. The maximum amount of subsidence reached nearly 3,000 m at the western coast, giving an average rate in excess of 0.4 mm/yr. Since 10 m.y.B.P., the movements were comparatively small in these structurally low areas (except for the western coast of transect AA', where marked subsidence is recognized during Quaternary time). In more detail, however, it can be seen that a small amount of uplift occurred. The Dewa Hills and the Backbone Range also subsided during the 16-10 m.y. interval, but were uplifted in the subsequent period. Since 2 m.y.B.P. the rate of uplift increased several times. This is consistent with

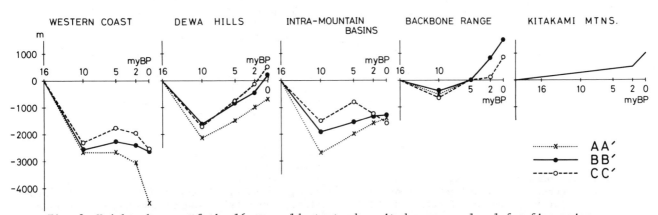

Fig. 9 Height changes of the 16 m.y. old strata deposited near sea level for five major geomorphologic units: the western coast, the Dewa Hills, the intra-mountain basins and the Backbone Range. For the Kitakami Mountains, the uplift estimated by Chinzei (1966) is shown.

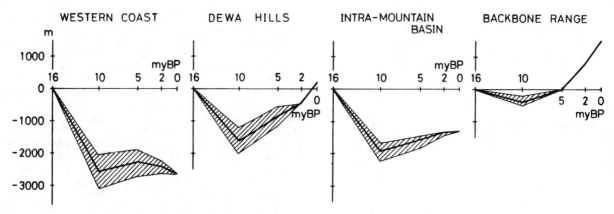

Fig. 10 Probable range of height changes of the 16 m.y. old strata along transect line BB' when a reasonable error is allowed for the deposition depth.

the earlier inference that the major topographic relief of Northeast Japan has been formed during Quaternary time (e.g. Kaizuka and Murata, 1969).

We failed to subdivide the interval between 16 and 10 m.y.B.P. because of a lack of reliable time-marker horizons. However, some paleontologic evidence suggests that the maximum subsidence took place during the Nishikurosawa Stage (Fig. 5), from 16 to 13 m.y.B.P., since the water depths for some parts of the upper horizon of the Nishikurosawa Formation have been estimated as 1,000 to 2,500 m, based on benthonic foraminifera (Kitazato, 1979).

In this study, we have limited ourselves to examining the height changes in narrow bands along three transect lines. However, our results are quite compatible with the displacements previously summarized for the whole of Japan with a lower resolution in time. Comparing Fig. 7a with Fig. 1, we find that the movements from 16 m.y. B.P. to the present are similar to those obtained by Matsuda et al. (1967) for the period since the beginning of Miocene. From this similarlity, taking account of the linear arrangement of highland chains along the Northeast Japan arc, we may conclude that the displacements along our limited bands represent the movements of the major part of Northeast Japan.

During the Deep Sea Drilling Project Legs 56 and 57, the Japan Trench regions east of two of our transect lines (Fig. 6) were drilled (von Huene et al., 1978). As a result of this drilling, remarkable information on the vertical movement of the fore-arc region of Northeast Japan has been obtained (Keller, 1980; von Huene et al., 1978). Analysis of microfossils found in the recovered cores from Sites 438/439 indicates that the water depth at these sites has changed as shown by the curve labeled "Ocean floor" in Fig. 11. When the thickness of the underlying sediments is taken into account, the vertical movement of the basement of the sites can be obtained as the curve "Unconformity" (Fig. 11). These sites were subaerial in Late Oligocene time, and subsided about 3,000 m during the Neogene. The rate of subsidence decreased during the Late Miocene, and finally uplift started at the end of Pliocene. Comparing Fig. 11 with Fig. 9, we notice that the general tendency of verti-

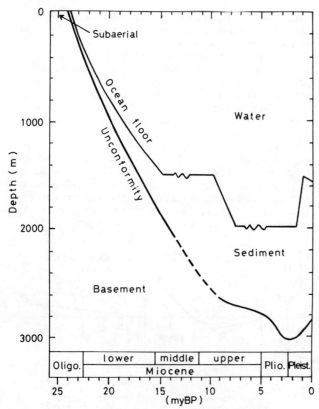

Fig. 11 Diagram of sediment thickness above the unconformity and depth of depositional environments as a function of time for Sites 438/439 since the end of Oligocene, showing subsidence of the deep sea terrace (von Huene et al., 1980).

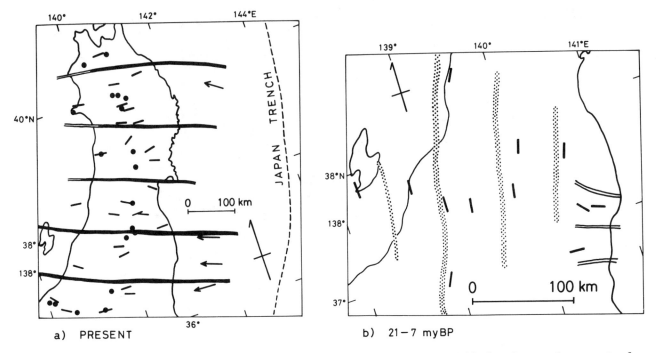

Fig. 12 σ_{Hmax} trajectory map: a) for present-day Northeast Japan and b) for the southern part of Northeast Japan during Miocene time about 21 to 7 m.y.B.P. (Nakamura and Uyeda, 1980). Arrows: slip vectors for low angle thrust events. Dots: active volcanoes. Short bars: dikes. Dark trajectories in a) show σ_{Hmax} which are judged to be σ_1, and dotted trajectories in b) are for σ_{Hmax} which are judged to be σ_2. Parallel lines show that it is not certain whether σ_{Hmax} is σ_1 or σ_2.

cal movement in the fore-arc region is similar to that of Northeast Japan as determined in the present study.

Generally, in a tensional stress field the region will be dominated by subsidence with smaller regional topographic amplitudes. In a compressional stress field, uplift will be dominant with larger topographic amplitudes. Given this relationship and the results shown in Fig. 9, we may estimate the tectonic stress field of Northeast Japan since 16 m.y.B.P. In the eastern zone, i.e. the Kitakami Mountains, the stress has been compressional since Early Miocene; whereas in the western zone, the stress was tensional during the 16-10 m.y. interval. If we take account of the paleontologic evidence, tensional stress may have been dominant during the earlier half of this interval. The compressional tendency became apparent at about 5 m.y.B.P. Therefore, it appears possible that a "paired stress field" (Takeuchi, 1980) existed in Northeast Japan from 16 to 10 m.y.B.P., and the boundary between the different stress fields was located near the volcanic front. During the 10-5 m.y. interval, the vertical movements were insignificant over the whole area. We consider it probable that in this period Northeast Japan was released from intense stress as the stress changed from tensional to compressional. The difference in the stress history between the eastern and western zones may

have some genetic relationship with the obvious changes seen in the geophysical observations near the volcanic front, as shown in Fig. 4. Applying the expected relationship between tectonic stress and vertical movements to the fore-arc region, we can conclude that in the fore-arc region the stress was tensional throughout the Neogene and compressional during the Quaternary.

Previously, the Late Cenozoic regional stress field of Northeast Japan was independently estimated as shown in Fig. 12 (Nakamura and Uyeda, 1980). In this figure, the trajectories of the maximum horizontal stress, σ_{Hmax}, are drawn based on various sources of information such as the directions of dikes, faults, alignment of monogenetic volcanoes and earthquake source mechanisms. According to their results, the present tectonic stress is compressional, but the stress during the 21-7 m.y. period was tensional in the western zone and compressional in the eastern zone. Our stress estimation derived from the vertical crustal movements appears to fit the results of these previous investigations.

Uyeda and Kanamori (1979) classified subduction zones and demonstrated that two modes of subduction exist which represent two basically different processes, namely Chilean and Mariana modes (Fig. 13). According to these authors, the term "mode" represents the strength of mechanical coupling between the subducting slab and the up-

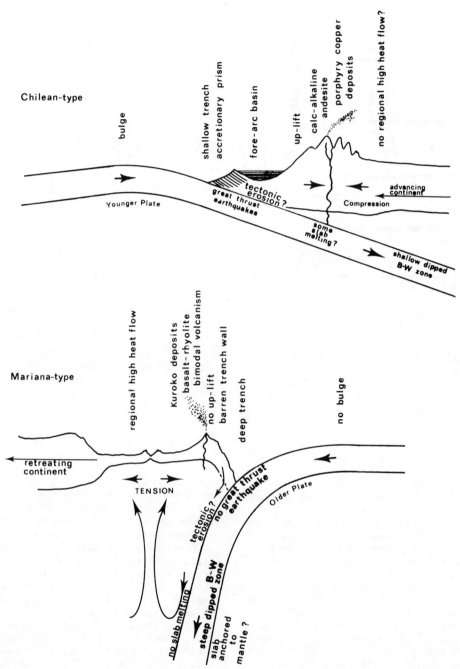

Fig. 13 Two modes of subduction with their possible tectonic implications and causes (Uyeda, 1982).

per, landward plate. In the Chilean-type, the coupling is strong and the oceanic plate subducts at a low angle, receiving large resistance. Truly great earthquakes (Kanamori, 1977a) occur only in this type of subduction. The stress in Chilean-type arcs should be compressional. On the other hand in the Mariana-type, two plates are virtually decoupled and the slab subducts with a high angle without large resistance. In arcs of this type, the stress is tensional and the back-arc region opens by a process similar to sea-floor spreading. Following this argument, we may infer that at the Japan Trench during the 16-10 m.y. interval, Mariana-type subduction was taking place, probably causing the spreading of at least a part of the Sea of Japan. At some time during

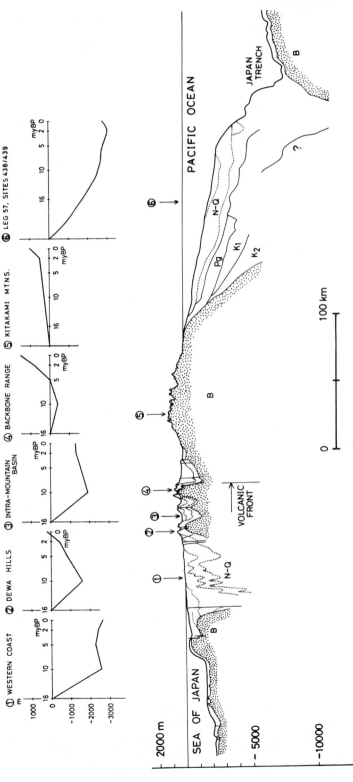

Fig. 14 East-west geologic cross-section of Northeast Japan along Lat. 39°30'N (Ishiwada et al., 1977), and the change of the basement depth determined in this study along transect line BB' and at the fore-arc region (DSDP Sites 438/439). N-Q: Neogene-Quaternary sediments. PG: Paleogene system. K1: Upper Cretaceous system. K2: Lower Cretaceous system. B: basement rocks (pre-Cretaceous and granitic rocks).

VERTICAL CRUSTAL MOVEMENTS   327

the 10-5 m.y. interval, the mode must have changed from the Mariana-type to the Chilean-type. We consider that, at the time of transition, Northeast Japan was released from intense stress as discussed above.

Kanamori (1977b) considered that the two modes of subduction correspond to different stages of an evolutionary process, that is, subduction starts as the Chilean-type, then the coupling between the two plates weakens gradually and subduction changes to the Mariana-type. It has also been suggested that in Mariana-type subduction the slab finally becomes detached and falls into the mantle, and a new cycle of evolution then starts as Chilean-type subduction. Niitsuma (1978), using this evolutionary model of plate subduction, discussed the sedimentary history and development of the island arcs of Japan. Our conclusions on the change in the mode of subduction agree in general with those of Niitsuma. Kobayashi (1982) also discussed the vertical movements of island arcs in terms of similar cyclic models. As an alternative possibility, however, the change in the mode of subduction could have been caused by the change in the direction of motion of the landward plate relative to the trench lines (Uyeda and Kanamori, 1979). Further investigations are needed to differentiate these possibilities.

The volcanic activity of Northeast Japan since Early Miocene was estimated by Sugimura et al. (1963). It was very active during the Early Neogene (13 m.y.B.P. and older), and declined during the Middle and Late Neogene (13 to 2 m.y.B.P.). During Quaternary time it became active again. The western zone of Northeast Japan during the Early Neogene was characterized by so called "Green Tuff volcanism", a submarine and subaerial volcanism dominated by subsidence. This activity appears to be associated with Mariana-type subduction. Konda (1974) reported that this volcanism is characterised by the occurrence of basalt and rhyolite thus showing bimodal modes of chemical composition. The bimodal volcanism is considered as indicative of a tensional stress field (Christiansen and Lipman, 1972). Quaternary volcanic activity in the Japan arc, including the present activity, is of a more typical arc type and probably associated with the present Chilean-type subduction.

Conclusions

We have analyzed the vertical movements of Northeast Japan since 16 m.y.B.P. by measuring the present heights of the strata deposited at 16, 10, 5 and 2 m.y.B.P. along three transect lines and by taking account of possible water depths of deposition. The results, summarized in Fig. 14, indicate rapid subsidence during the 16-10 m.y. interval in the western zone (inner arc). The total subsidence along the western coast has amounted to nearly 3,000 m. After 10 m.y.B.P. the movements switched to uplift. In the eastern zone, east of the Backbone Range, uplift has always been dominant, with an enhancement during the Quaternary. The general tendency of subsidence followed by uplift seems to be similar to the vertical movement in the fore-arc region of the Japan Trench as discovered by the recent DSDP drilling (Fig. 11).

We interpret the above history of the vertical movements as due to a change in the mode of subduction from the Mariana-type to the Chilean-type. The existence, during the Neogene, of a weak compressional stress regime in the eastern zone (Kitakami Mountains) simultaneous with the tensional stress both in the western zone and fore-arc region remains enigmatic. It is interesting to note that a paired stress field similar to that of Middle Miocene Northeast Japan has been reported at other subduction zones, such as Southwest Japan during the Miocene and the present-day Hellenic-Aegean and Alaska-Aleutian regions (Kobayashi, 1979; Angelier, 1978; Nakamura and Uyeda, 1980). Such a paired distribution of regional stresses may be a general characteristic of arc tectonics. This might provide an explanation for the fact that arcs are attached to the trench even when the back-arc region is under a tensional stress and spreading is in progress.

Acknowledgments. We are grateful to Tokihiko Matsuda for useful discussion, and Yasumochi Matoba for providing information on the deposition depths of the strata examined. We are also indebted to Kazuaki Nakamura and Thomas W. C. Hilde for reading the manuscript and giving us helpful comments.

References

Anderson, R. N., 1980 update of heat flow in the east and southeast Asian Seas, in The Tectonic and Geologic Evolution of Southeast Asian Seas and Islands, edited by D. Hayes, Amer. Geophys. Un., 318-326, 1980.

Angelier, J., Tectonic evolution of the Hellenic Arc since the late Miocene, Tectonophysics, 49, 23-36, 1978.

Chinzei, K., Younger Tertiary geology of the Mabechi River valley, northeast Honshu, Japan, J. Fac. Sci., Univ. Tokyo, Sect. II, 16, 161-208, 1966.

Chinzei, K., Neogene molluscan faunas in the Japanese Islands: an ecologic and zoogeographic synthesis, The Veliger, 21, 155-170, 1978.

Christiansen, R. L., and P. W. Lipman, Cenozoic volcanism and plate-tectonic evolution of the Western United States, II. Late Cenozoic, Phil. Trans. R. Soc. Lond., Ser. A, 271, 249-284, 1972.

Dickinson, W. R., Widths of modern arc-trench gaps proportional to past duration of igneous activity in associated magmatic arcs, J. Geophys. Res., 78, 3376-3389, 1973.

Inoue, H., Neogene paleogeography of the surrounding Dewa Hilly Lands (in Japanese with

English abstract), J. J. Assoc. Petrol. Techn., 27, 443-464, 1962.

Ishiwada, Y., Y. Ikebe, K. Ogawa, and T. Onitsuka, A consideration on the scheme of sedimentary basins of Northeast Japan (in Japanese with English abstract), in Geological Papers Dedicated to Prof. K. Huzioka, 1-7, 1977.

Jarrard, R. D., and D. A. Clague, Implications of Pacific island and seamount ages for the origin of volcanic chains, Rev. Geophys. Space Phys., 15, 57-76, 1977.

Kaizuka, S., and A. Murata, The amounts of crustal movements during the Neogene and the Quaternary in Japan, Geogr. Rep., Tokyo Metrop. Univ., No. 4, 1-10, 1969.

Kanamori, H., The energy release in great earthquakes, J. Geophys. Res., 82, 2981-2987, 1977a.

Kanamori, H., Seismic and aseismic slip along subduction zones and their tectonic implications, in Island Arcs, Deep Sea Trenches and Back-Arc Basins, Maurice Ewing Ser. 1, edited by M. Talwani and W. C. Pitman III, Amer. Geophys. Un., 163-174, 1977b.

Keller, G., Benthic foraminifera and paleobathymetry of the Japan Trench area, DSDP Leg 57, in Init. Rep. DSDP, LVI-LVII, pt. 2, U. S. Govern. Print. Off., Wash., D. C., 835-865, 1980.

Kitamura, N., Tertiary orogenesis in northeast Honshu, Japan (in Japanese with English abstract), Contrib. Inst. Geol. Paleont., Tohoku Univ., No. 49, 1-98, 1959.

Kitamura, N., Tertiary tectonic movements of the Green Tuff area (in Japanese), Fossils, No. 5, 123-137, 1963.

Kitazato, H., Marine paleobathymetry and paleotopography of the Hokuroku district during the time of the Kuroko deposition, based on foraminiferal assemblages, Min. Geol., 29, 207-216, 1979.

Kobayashi, K., Cycles of subduction and Cenozoic arc activity in the northwestern Pacific margin, this volume, 1982.

Kobayashi, Y., Early and Middle Miocene dike swarms and regional tectonic stress field in the Southwest Japan (in Japanese with English abstract), Bull. Volcanol. Soc. Jap., 24, 203-212, 1979.

Kobayashi, Y., and N. Fujii, About the "aseismic front" (in Japanese), Abstract, Spring Meeting, Seismo. Soc. Jap., 50, 1981.

Konda, T., Bimodal volcanism in the Northeast Japan arc (in Japanese with English abstract), J. Geol. Soc. Jap., 80, 81-89, 1974.

Matsuda, T., Collision of the Izu-Bonin arc with central Honshu: Cenozoic tectonics of the Fossa Magna, Japan, J. Phys. Earth, 26, Suppl., S409-S421, 1978.

Matsuda, T., K. Nakamura, and A. Sugimura, Late Cenozoic orogeny in Japan, Tectonophysics, 4, 349-366, 1967.

Nakamura, K., Island arc tectonics, a hypothesis (in Japanese), Proceedings, Symposium on Problems Concerning "Green Tuffs", Annual Meeting, Geol. Soc. Jap., 31-38, 1969.

Nakamura, K., and S. Uyeda, Stress gradient in arc-back arc regions and plate subduction, J. Geophys. Res., 85, 6419-6428, 1980.

Niitsuma, N., Magnetic stratigraphy of the Japanese Neogene and the development of the island arcs of Japan, J. Phys. Earth, 26, Suppl., S367-S378, 1978.

Research Group for Quaternary Tectonic Map, Quaternary tectonic map of Japan (in Japanese with English abstract), The Quat. Res., 7, 182-187, 1968.

Sugimura, A., Zonal arrangement of some geophysical and petrological features in Japan and its environs, J. Fac. Sci., Univ. Tokyo, Sect. II, 12, 133-153, 1960.

Sugimura, A., T. Matsuda, K. Chinzei, and K. Nakamura, Quantitative distribution of Late Cenozoic volcanic materials in Japan, Bull. Volcanol., 26, 125-140, 1963.

Sugimura, A., and S. Uyeda, Island Arcs: Japan and Its Environs, Elsevier, Sci. Publ. Co., Amsterdam, 1973.

Takeuchi, A., Tertiary stress field and tectonic development of the southern part of the northeast Honshu Arc, Japan, J. Geosci., Osaka City Univ., 23, 1-64, 1980.

Tsuchi, R. (editor), Correlation and chronology of Neogene sediments in the Pacific side and Japan Sea side of Japan (in Japanese), Rep. Res. Group Biostrat., Japan, 1-156, 1979.

Uyeda, S., Some basic problems in the trench-arc-back arc system, in Island Arcs, Deep Sea Trenches and Back-Arc Basins, Maurice Ewing Ser. 1, edited by M. Talwani and W. C. Pitman III, Amer. Geophys. Un., 1-14, 1977.

Uyeda, S., Subduction zones: an introduction to comparative subductology, Tectonophysics, 81, 133-159, 1982.

Uyeda, S., and H. Kanamori, Back-arc opening and the mode of subduction, J. Geophys. Res., 84, 1049-1061, 1979.

von Huene, R., M. Langseth, N. Nasu, and H. Okada, Summary, Japan Trench transect, in Init. Rep. DSDP, LVI-LVII, pt. 1, U. S. Govern. Print. Off., Wash., D. C., 473-488, 1980.

von Huene, R., N. Nasu, and Scientific Party of DSDP Leg 57, Japan Trench transected, Geotimes, 23, No. 4, 16-20, 1978.

Yoshii, T., Proposal of the "Aseismic front" (in Japanese), Zisin, 28, 365-367, 1975.

Yoshii, T., A detailed cross-section of the deep seismic zone beneath northeastern Honshu, Japan, Tectonophysics, 55, 349-360, 1979.

HIGH MAGNESIAN ANDESITES IN THE SETOUCHI VOLCANIC BELT, SOUTHWEST JAPAN AND THEIR POSSIBLE RELATION TO THE EVOLUTIONARY HISTORY OF THE SHIKOKU INTER-ARC BASIN

Yoshiyuki Tatsumi

Earthquake Research Institute, University of Tokyo, Tokyo, Japan 113

Abstract. Characteristic volcanic rocks called "sanukitoids" along with high magnesian andesites are found in the Setouchi volcanic belt, southwest Japan. In this volcanic belt, extending for about 1,000 km, magmas were produced suddenly at about 13 Ma and the volcanism was extremely short-lived. Detailed petrographic and experimental studies on sanukitoids revealed the follwoing facts: (1) high magnesian andesites cannot be derived from basaltic magmas, (2) an andesite exists which represents the chemical composition of the primary magma, and (3) these andesites are the products of partial melting of uppermost mantle peridotites under both water rich and relatively high temperature conditions. Generation of the above-mentioned characteristic magmas in the Setouchi volcanic belt may have been related to the northward subduction of the newly formed lithosphere of the Shikoku Basin.

1. Introduction

Since first suggested by O'Hara (1965), several experimental petrologists have also predicted the existence of andesitic primary magma generated by the partial melting of hydrous upper mantle. For example, Yoder (1969) pointed out that water plays an important role in the generation of orogenic andesites. It has been shown on both synthetic and natural peridotite-H_2O systems that the partial melting products of peridotite are not basaltic but andesitic under water-saturated conditions (Kushiro, 1972; 1974; Mysen and Boettcher, 1975). However, whether andesites produced by such a mechanism do exist in nature or not has been a matter of much debate.

Magnesium-rich andesites, on the other hand, have been found in some localities. Examples are Papuan enstatite andesites (Dallwitz et al., 1966), boninites in the Bonin Islands (Kikuchi, 1890), and some olivine andesites in southwest Japan (Yamaguchi, 1958). It has been suggested that such magnesian andesites are indeed primary magmas produced by the process of direct partial melting of hydrous upper mantle peridotites (e.g., Kushiro, 1972; Kuroda et al., 1978). However, two problems must be solved to confirm that magnesian andesits are derived from primary andesitic magmas. One is to demonstrate that the magnesian andesites cannot be derived from their associated basaltic magmas. In many localities where magnesian andesites occur, basaltic volcanism is also recognized, and some basalts are more magnesian than the andesites. One may consider, under such circumstances, that magnesian andesites were produced from basaltic magmas by the process of fractional crystallization. The other problem is to prove the very existence of primary andesites in nature. Most magnesian andesites contain considerable amounts of Mg-rich phenocrysts such as olivine and enstatite. The magnesium-rich character may be due to the accumulation of phenocrysts. If we can find a magnesian andesite with only a small phenocryst content, the existence of primary andesites will be established.

For the purpose of solving the above-mentioned problems, detailed studies were made on magnesian andesites in the Setouchi volcanic belt, southwest Japan (Tatsumi and Ishizaka, 1981, 1982). This paper reviews these petrographic studies, and presents the results of melting experiments on magnesian andesites (Tatsumi, 1981, 1982).

2. Setouchi Volcanic Belt

About 200 Quaternary volcanoes exist in Japan along the arc-trench system (Fig. 1). In the Seto Inland Sea area, southwest Japan, there are no Quaternary volcano, but some Miocene volcanic rocks are exposed (Fig. 1 and Fig. 2). These are called the Setouchi volcanic rocks and make up one of the Miocene Volcanic belt in Japan. The following is a description of the Setouchi volcanic belt.

2.1 Sanukite and Sanukitoid

The Setouchi volcanic belt is characterized by the occurrence of unusual volcanic rocks called sanukitoids. Naumann (1885) first pointed out that "ringing stone" are sporadically exposed in southwest Japan and called them "Augitandesit-klingstein". Weinschenk (1890) recognized that

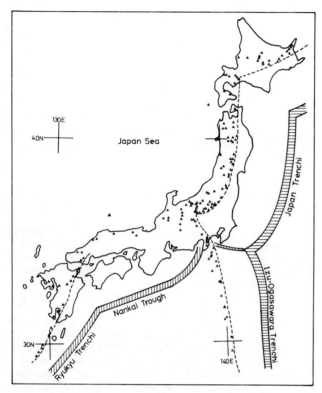

Fig. 1. Quaternary volcanoes (triangles) and volcanic fronts (broken lines) in Japan arc.

these volcanic rocks carry bronzite phenocrysts and coined the term "Sanukit" for them because they typically occur in the Sanuki Province, the old name of the northeastern Shikoku area. He enumerated the following characteristics for sanukite:
(1) Grey to black color and megascopically similar to hornstone.
(2) Especially compact samples show a concoidal fracture and a black luster.
(3) Needle-like bronzite phenocrysts are embedded in a fine-textured groundmass with interstitial glass.
(4) Phenocrystic plagioclase and garnet are rare.

Only a few rocks, however, can be called sanukite on the basis of the Weinschenk's definition. Koto (1916) thus proposed the name "sanukitoid" for all textural variations of consolidated magma to which sanukite belongs. But we cannot easily recognize such magmas. Tatsumi and Ishizaka (1981) redefined the term sanukitoid and used it for plagioclase-free, glassy and relatively aphyric (total phenocrysts <20 volume percent) andesite and basalt in the Setouchi volcanic belt. Sanukitoid usually carries phenocrysts of magnesian olivine, bronzite and augite, and is divided into seven rock types on the basis of its phenocrystic mineral assemblage: (1) augite olivine basalt, (2) augite olivine andesite, (3) bronzite olivine andesite, (4) hornblende olivine andesite, (5) olivine augite bronzite andesite, (6) bronzite andesite, and (7) aphyric andesite. Representative chemical compositions of sanukitoids are listed in Table 1.

Most augite olivine and bronzite olivine andesitic sanukitoids can be chemically classified as "high magnesian andesites" (abbreviated as HMAs hereafter), and have primitive characteristics; that is, they are high in MgO, Cr_2O_3 and NiO contents (Table 1). In comparison with boninite, which is one of the well-known HMAs, HMA sanukitoid has the following characteristics:

(1) More aphyric and poorer in glass than boninite.
(2) Absence of clinoenstatite and pigeonite, and presence of groundmass plagioclase.
(3) Richer in incompatible elements.

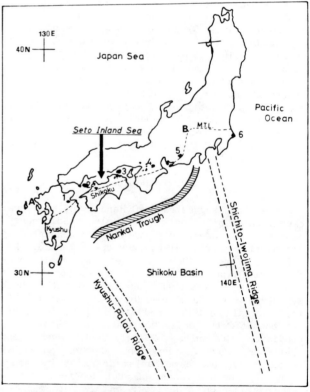

Fig. 2. Distribution of the Setouchi volcanic belt (Tatsumi et al., 1980). 1, northeastern Kyushu; 2, northwestern Shikoku; 3, northeastern Shikoku; 4, Osaka; 5, Shidara; 6, Choshi; M.T.L., Median Tectonic Line. Horikoshi (1972) proposed that the Setouchi volcanic belt was distributed from A (northwestern Kyushu) to B (Kirigamine-Arafune area). Note that Quaternary volcanoes are missing in this region (c.f. Fig. 1).

Table 1 Chemical compositions of representative sanukitoids

	1	2	3	4	5	6	7
SiO_2	49.20	55.73	58.50	55.54	59.10	63.45	61.90
TiO_2	1.01	0.71	0.43	0.62	0.77	0.64	0.97
Al_2O_3	15.31	15.44	13.30	16.41	15.98	16.27	17.54
Fe_2O_3	2.77	2.27	2.50	1.95	2.47	2.46	0.51
FeO	6.28	4.31	3.95	4.54	3.50	1.77	5.41
MnO	0.16	0.13	0.12	0.13	0.09	0.07	0.12
MgO	11.55	7.21	9.47	7.33	5.67	3.26	1.72
CaO	8.81	6.99	6.13	7.35	6.43	4.64	5.42
Na_2O	2.57	2.82	2.61	3.03	3.01	3.57	3.04
K_2O	1.27	2.21	1.28	2.08	1.91	2.37	2.19
H_2O-	0.22	0.42	0.22	0.12	0.35	0.50	0.08
H_2O+	0.76	1.54	1.37	1.09	0.93	1.12	0.76
P_2O_5	0.22	0.14	0.13	0.22	0.18	0.16	0.09
Total	100.13	99.74	100.01	100.41	100.39	100.48	99.75
Ni (ppm)	329	216	184	172	144	93	
Cr (ppm)	483	342	472	296	270	148	6*
$\frac{FeO*}{MgO}$	0.76	0.86	0.65	0.86	1.01	1.15	3.41

1; augite olivine basalt (Tatsumi & Ishizaka, 1982)
2; augite olivine andesite (Tatsumi & Ishizaka, 1982)
3; bronzite olivine andesite (Tatsumi & Ishizaka, 1981)
4; olivine augite bronzite andesite (Tatsumi & Ishizaka, 1982)
6; bronzite andesite (Tatsumi & Ishizaka, 1982)
7; aphyric andesite (Yamazaki & Onuki, 1969)
*; unpublished data of Tatsumi

These features are important when we consider the origin of sanukitoid and boninite (Tatsumi and Ishizaka, 1982).

2.2 Distribution and Age of Setouchi Volcanism

Naumann (1885) found that "Augitandesitklingstein" is distributed along the Seto Inland Sea (Fig. 2). Koto (1916) took notice of the fact that some characteristic volcanic rocks such as pitchsone, biotite andesite and sanukitoid occur together along the Median Tectonic Line (Fig. 2). Morimoto et al. (1957) emphasized that the volcanic rocks along the Seto Inland Sea erupted in Mio-Pliocenen time and called them "the Setouchi Volcanic Series". Much attention has been paid to the fact that eruptions of the Setouchi Volcanic Series took place side by side in five volcanic centers; the Shidara, Osaka, northeastern Shikoku, northwestern Shikoku, and northeasterh Kyushu areas (Fig. 2), which are arranged at intervals of about 100 km distance along the Median Tectonic Line. These authors stated the volcanic activity of the Setouchi Volcanic Series began with effusions of biotite andesite, hornblende andesite and pyroxene andesite and finished with extrusions of aphyric lavas of hypersthene andesite or dacite called sanukite.

Matsumoto (1961) and Matsumoto and Takahashi (1968) place the western end of Setouchi volcanic belt at the northwestern Kyushu area, because sanukitoid and pitchstone occur in that area. Kawachi and Kawachi (1963) found garnet dacite and obsidian in the central Japan region, and included them the Setouchi volcanic belt. Horikoshi (1972) compiled previous data on the occurrence of sanukitoid, pitchstone and garnet dacite, and proposed that the Setouchi volcanic belt extended from western Kyushu to central Japan (A-B in Fig. 2). This interpretation has been, until 1980, widely accepted.

Recently, Tatsumi and his co-workers determined K-Ar and fission track ages for carefully

Table 2 K-Ar and fission track ages of the Setouchi volcanic rocks

Sample	Area	Rock Name	Age	Reference
OT-9.3	NE Kyushu	2px andesite	13.2±0.7	(1)
OS-20	NW Shikoku	andesitic sanukitoid	12.6±0.6	(2)
WS-30	NW Shikoku	bt rhyolite	14.2±0.8	(3)
HJA	NE Shikoku	2px andesite	11.1±0.6	(4)
CHO	NE Shikoku	2px andesite	11.6±0.6	(4)
MDY	NE Shikoku	andesitic sanukitoid	11.2±0.6	(4)
TK-2	NE Shikoku	andesitic sanukitoid	11.6±0.6	(5)
SD-438	NE Shikoku	basaltic sanukitoid	13.6±1.4	(6)
SHD-1	NE Shikoku	bt rhyolite	14.0±1.2*	(3)
YAS-20	NE Shikoku	bt rhyolite	13.9±1.4*	(3)
NJIM	Osaka	andesitic sanukitoid	13.0±0.7	(2)
MIKASA	Osaka	2px andesite	13.1±1.2	(2)
TYO	Choshi	andesitic sanukitoid	11.8±0.6	(7)

* fission trach ages

(1) Tatsumi et al., 1980   (2) Tatsumi et al., 1979
(3) Yamazaki et al., 1981   (4) Tatsumi and Yokoyama, 1978
(5) Tatsumi and Ishizaka, 1978   (6) Tatsumi, unpublished data
(7) Tatsumi and Ishizaka, 1979

selected samples and revised the previous view on the distribution and the age of volcanism of the Setouchi volcanic belt. They found that the volcanic activity in the belt, which was previously considered to have extended to Pliocene, occurred only in Miocene time (Tatsumi and Yokoyama, 1978; Tatsumi and Ishizaka, 1978). Tatsumi and Ishizaka (1979) further indicated that sanukitoids in the Choshi area (Fig. 2) erupted at 12 Ma, and extended the eastern limit of the Setouchi volcanic belt to that area. These findings togethr with some K-Ar dates on sanukitoids in the northeastern Kyushu, northwestern Shikoku and Osaka areas show that the intermediate to basic volcanism, from the northeastern Kyushu to Choshi area, all took place about 12 million years ago (Tatsumi et al., 1980a, b). On the other hand, it has been shown that sanukitoids in the northwestern Kyushu area erupted at times younger than 8 Ma. Furthermore, these rocks have chemical compositions different from sanukitoids in the other areas of the Setouchi volcanic belt (e.g., high La/Sm ratio and K_2O content), suggesting that there is no direct genetic relationship between the sanukitoids from northwestern Kyushu area and other areas. This led us to the conclusion that the northeastern Kyushu area is the western end of the Setouchi volcanic belt (Tatsumi et al., 1981b). Yamazaki et al. (1981) pointed out that the acidic volcanism in the Setouchi volcanic belt occurred at 14 Ma. The age determinations are listed in Table 2. On the other hand, Kaneoka and Suzuki (1970) found that volcanic rocks in the central Japan area, which were formerly considered to belong to the Setouchi volcanic belt by Kawachi and Kawachi (1963), were erupted in Quaternary. Thus, these volcanic rocks do not belong to the Setouchi volcanic belt. It is concluded from these results that short-lived volcanism took place in the Setouchi volcanic belt from the northeastern Kyushu area to the Choshi area at 13 ± 1 Ma.

3. Origin of Sanukitoids

3.1 Non-consanguinity between HMA and Basalt

On Shodo-Shima Island, southwest Japan, which belongs to the northeastern Shikoku area of the Setouchi volcanic belt (Fig. 2), HMAs and olivine tholeiitic basalts closely erupted in both time and space. Thus, it is important to test whether HMAs are differentiation products of basalts. To aid in solving the problem, a detailed petrographical study on these basalts and HMAs was performed, and the consanguinity between them was discussed (Tatsumi and Ishizaka, 1982). These volcanic rocks erupted at 11-12 million years ago (Tatsumi and Yokoyama, 1978).

Representative chemical compositions of the Shodo-Shima Island basalt and HMA are shown in Table 1 (1 and 2). The basalt is more magnesian than the andesite, and the former is plotted on

the more primitive part of the trend in variation diagrams. One might thus concluded that the HMAs were derived from the basalt magma by the process of fractional crystallization. However, four other observations suggest that the HMAs are not the differentiation products of the basalt magmas:
(1) Olivine phenocrysts in the HMAs are more magnesian and have narrower compositional range than those in the basalts (Fig. 3). The magnesian olivine in the HMAs are in equilibrium with the magma on the basis on the Fe-Mg exchange partitioning by Roeder and Emslie (1970), and thus they are not cumulative in origin.
(2) Based on the olivine maximum fractionation model, the primary magma of the HMA was not

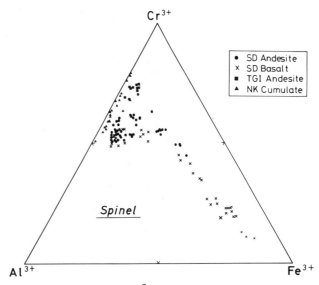

Fig. 4. Cr-Al-Fe^{3+} diagram for spinels in Setouchi volcanic rocks. SD = Shodo-Shima sanukitoids; TGI = Teraga-Ike sanukitoid; NK = Naka-Jima Island (Tatsumi and Ishizaka, 1981).

basaltic but andesitic in composition. The HMA for which the primary magma composition is estimated carries only a few phenocrysts of augite (< 1 modal percent), and the composition of the HMA is plotted on the "olivine control line". Thus, the model gives a realistic composition of the primary magma.
(3) The basalts contain both chromite and titanomagnetite as inclusions in the olivine phenocrysts, while olivine in the HMA contains only chromite (Fig. 4).
(4) The Cr-content of chromite in the HMAs is equal to or higher than that in the basalts (Fig. 4).

3.2 Existence of Primary Andesite

Many HMAs, in general, carry large amounts of Mg-rich phenocrysts (in boninite, the maximum content is 52 modal percent!). Thus, there is the possibility that they are cumulative in origin, and do not represent the chemical composition of a primary andesitic magma in the strict sense. To establish the generation of an andesitic primary magma, we must find a truly primary andesite in nature. A high magnesian Teraga-Ike andesite (TGI) has been shown to reflect the chemical composition of the primary magma (Tatsumi and Ishizaka, 1981).

The TGI andesite, with only 5 modal percent of phenocrysts, has a low FeO*/MgO ratio of 0.65 (3 in Table 1). Furthermore, magnesian olivine phenocrysts (Fo_{91-87}) found in the rock are in equilibrium with the liquid on the basis of the Fe-Mg exchange partitioning, indicating that the

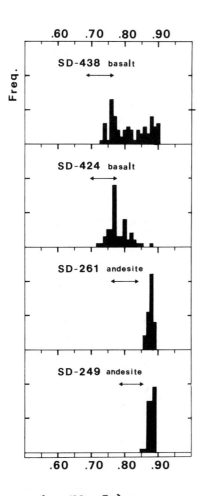

Fig. 3. Frequency distribution diagram for the Mg/(Mg+Fe) ratio of olivine phenocrysts in representative basaltic and andesitic sanukitoids from Shodo-Shima Island, southwest Japan (Tatsumi and Ishizaka, 1982). Olivines in the andesites are more magnesian than those in the basalts.

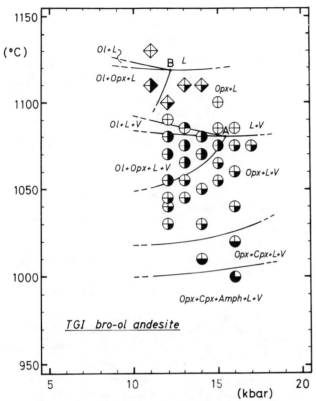

Fig. 5. Melting phase relations of the Teraga-Ike andesite under both water-saturated (circle) and undersaturated (square) conditions (Tatsumi, 1981).

3.3 Melting Experiments on Sanukitoids

As was shown in the previous sections, HMAs in sanukitoids are indeed derived from primary andesitic magmas generated in the upper mantle. There are, however, several unsolved problems concerning the origin of HMA magmas:

(1) It has been believed that HMA magmas are formed under water-saturated conditions. More than 15 wt.% water can be dissolved in an andesitic melt under the P-T conditions in the upper mantle (Hamilton et al., 1972; Sakuyama and Kushiro, 1978; Tatsumi, 1981). It is, however, highly questionable that such a large amount of water exists in the upper mantle. We must examine whether HMA magmas can be in equilibrium with mantle materials under water-deficient conditions.

(2) Two different phenocryst assemblages (olivine + augite and olivine + bronzite) are recognized in HMAs from the Setouchi volcanic belt (abbreviated as cpx-HMAs and opx-HMAs, respectively). The genetic relationship between these two different types of HMAs has to be understood.

(3) In the Setouchi volcanic belt and other areas where HMAs are found, as previously mentioned, basalt and HMA magmas were simultaneously produced. The mechanism for simultaneous production of two magma types must, therefore, be explained.

Melting experiments on HMA and basaltic sanukitoids were conducted in order to solve the above problems (Tatsumi, 1981; 1982).

Experimental results are shown in Figs. 5, 6 and 7; the cpx-HMA SD-261 is in equilibrium with olivine, orthopyroxene and clinopyroxene at 15 kbar and 1030°C under water-saturated conditions, and at 10 kbar and 1070°C in the presence of 7

TGI andesite cannot be a cumulative rock. The high abundance of Cr (472 ppm) and Ni (184 ppm) in the TGI andesite also suggests that it is not a differentiation product.

The chemical composition of the phenocrysts also supports the concept of a primary magma; noncumulative olivine with Fo_{91} has a very high NiO content (0.45 wt.%) and can be a liquidus olivine from the primary magma according to Sato's criterion (1977). The Cr_2O_3 content of orthopyroxene (up to 0.75 wt.%) is nearly equal to that of mantle orthopyroxene. The existence of such primitive minerals suggests that the TGI magma was in equilibrium with the upper mantle peridotite.

The Cr_2O_3 content of spinel can be an indicator of fractionation of chromite from a magma (Tatsumi and Ishizaka, 1982). Because chromite is generally the first liquidus phase crystallizing from a primary magma, a primitive rock may contain Cr-rich spinel. The TGI andesite contains one of the most Cr-rich spinel found in any of the sanukitoids from the Setouchi volcanic belt (Fig. 4). Such high chromium spinel in the TGI andesite supports the concept that the HMA is not a product of differentiation.

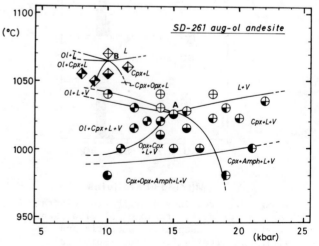

Fig. 6. Melting relations of the HMA SD-261. Symbols are same as in Fig. 5 (Tatsumi, 1982).

wt.% water; the opx-HMA TGI is in equilibrium with olivine and orthopyroxene at 15.5 kbar and 1080°C under water-saturated conditions, and at 11.5 kbar and 1120°C in the presence of 8 wt.% water; on the other hand, the basalt SD-438 is in equilibrium with olivine, orthopyroxene and clinopyroxene at 11 kbar and 1305°C under anhydrous conditions, and at 15 kbar and 1205°C in the presence of 4 wt.% water.

Water content in HMA magmas

The water contents in the above experiments on HMAs (7 and 8 wt.% in the melt) correspond to water-undersaturated conditions based on the solubility limit of water in HMA melts (15 wt.% at 12 kbar; Tatsumi, 1981). Thus, the above experimental results indicate that the HMA magmas, which are saturated with peridotitic minerals under water-saturated conditions, can also coexist with such phases under water-under-saturated conditions; the cpx-HMA SD-261 and the opx-HMA TGI can be generated under a certain P-T-H_2O condition along the line A-B in Fig. 8.

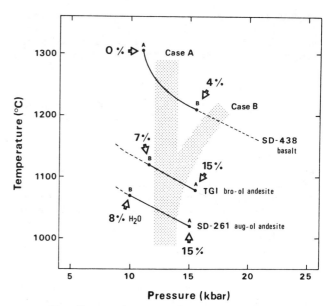

Fig. 8. Conditions of generation of the HMA TGI, the HMA SD-261 and the basalt SD-438. They can be formed under any P-T-H_2O conditions along the solid line AB and its extension (broken line), Case A and B are discussed in the text.

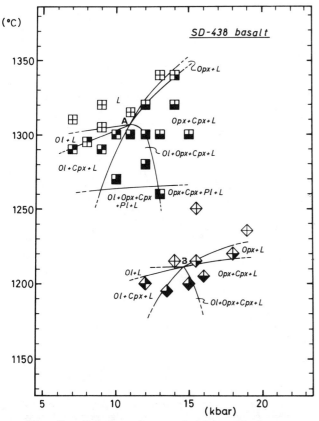

Fig. 7. Melting phase relations of the basaltic sanukitoid SD-438 under anhydrous conditions (squares) and water-undersaturated conditions (diamond) with 4 wt% water (Tatsumi, 1982).

Both the SD-261 and the TGI magmas with lower water contents (< 7 wt.%) may be saturated with peridotitic minerals at lower pressures and higher temperatures than those of Point B in Fig. 8. The depth of Moho beneath southwest Japan is, however, 30-40 km (Yoshii et al., 1974), and the pressure in the upper mantle beneath the area is greater than 10 kbar; therefore, the phase relations at pressures less than 10 kbar are meaningless for discussion of a "mantle" peridotite source for HMA magmas.

Recently, Duncan and Green (1981) suggested that HMA magmas may be produced under anhydrous conditions at 7 to 8 kbar and at about 1360°C. However, the temperature is too high at such shallow levels; furthermore, the absence of plagioclase phenocrysts in HMAs and more silicic sanukitoids with 64 wt.% SiO_2 strongly suggests the existence of considerable amounts of water in HMA magmas based on the experimental results of Yoder and Tilley (1962) and Eggler (1972). The most realistic mechanism for the production of HMA magmas is, therefore, the partial melting of the upper mantle under water-undersaturated conditions.

Origin of two types of HMAs

The melting experiments indicate that the SD-261 cpx-HMA magma is in equilibrium with olivine, orthopyroxene and clinopyroxene at pressures of the upper mantle, suggesting that the cpx-HMA magma is a partial melting product of a mantle peridotite, and the residue after partial melting is lherzolite. On the other hand, the TGI opx-HMA magma coexists with the harzburgitic mineral assemblage (olivine + orthopyroxene). It has been revealed that clinopyroxene is the first phase to disappear among lherzolitic minerals with the progress of partial melting (Mysen and Kushiro, 1977; Jaques and Green, 1980). Therefore, the difference of residual materials between the opx- and the cpx-HMA magmas is considered to depend on that of the degree of partial melting; that is, the opx-HMA magma is produced by higher degree of partial melting than the cpx-HMA magma. This idea is supported by the fact that the temperature of multiple saturation for the opx-HMA magma is about 50°C higher than that for the cpx-HMA magma (Fig. 8).

The chemical compositions of opx- and cpx-HMAs in the Setouchi volcanic belt also suggest different degree of partial melting; the opx-HMAs contain less incompatible elements such as K, Rb, Ti and Zr, and have lower FeO*/MgO ratios than cpx-HMAs, indicating that the degree of partial melting for the former is higher than that for the latter.

Genetic relationship between HMA and basalt magmas

A basaltic liquid can be produced by higher degree of partial melting than HMA magmas. In the simple mantle system, diopside-forsterite-silica-H_2O, the liquid composition of a partial melting product changes from quartz-normative andesite, which corresponds to HMA, to olivine-normative basalt (Kushiro, 1969; Tatsumi, 1982). The basalt magma formed by this process must be coexist with harzburgitic minerals. Melting experiments on the basalt SD-438, however, indicate that the basalt magma is in eqilibrium with lherzolitic minerals. It is thus concluded that the basalt magma is not produced by extensive partial melting of the mantle peridotite which has previouisly produced HMA magmas.

Basaltic magmas can be generated at the same pressure as HMA magmas if the water content in the upper mantle is lower (case A in Fig. 8), and also can be produced at a greater depth than HMA magmas with a similar water content (Case B in Fig. 8). These processes do not contradict the present experimental results. At the present stage, we cannot indicate which process governs the production of the basalt magmas.

4. Setouchi Volcanic Belt and Shikoku Inter-arc Basin

To the south of the Setouchi volcanic belt is the Shikoku Inter-Arc Basin, which is borderd by the Shichito-Iwojima Ridge on the east and the Kyushu-Palau Ridge on the west (Fig. 2). Many studies have been made about the nature and the origin of the Shikoku Basin. Here, some constraints on the tectonic and evolutionary history of the Shikoku Basin can be suggested based on the nature of the Setouchi volcanic belt and its genetic relationship to the subducting portion of the Shikoku Basin lithosphere.

4.1 Previous Work

Murauchi et al. (1968) indicated that the crust of the Shikoku Basin is oceanic in its seismological structure. The existence of linear magnetic anomalies in the area was first pointed out by Tomoda et al. (1968). Correlations of these magnetic anomalies have been made with the reversal time scale, confirming that the basin was formed by a seafloor spreading process (Tomoda et al., 1975; Kobayashi and Nakada, 1978; Shih, 1980). The results of five DSDP drillings support this conclusion and additionally indicate that there was off-ridge volcanism at about 15 Ma (e.g., Klein et al., 1978; Klein and Kobayashi;, 1980). According to Kobayashi and Nakada (1978), rifting behind the Bonin Arc began at approximately 30 Ma in the north and propagated toward the south. At about 25 Ma, the rifting reached the latitude of 25°N. A jump of the spreading axis might have taken place at 22 Ma. The youngest isochron is anomaly 5D, indicating that spreading in the Shikoku Basin ceased at about 17 Ma. However, correlation of the Shikoku Basin magnetic anomalies with the reversal time scale has proved to be difficult, and several tectonic histories, different from that of Kobayashi and Nakada (1978), have been posturated (e.g. Shih, 1980).

4.2 Tectonic Considerations

Related to the tectonic evolution of the Shikoku Basin, the following important aspects about the generation of the Setouchi volcanic belt are noted:
(1) Magmas in the Setouchi volcanic belt were suddenly produced at 13 Ma, based on the results of K-Ar and fission track datings of the volcanic rocks in the belt.
(2) Water played an important role in the generation of the Setouchi volcanic belt. Melting experiments on sanukitoids in the volcanic belt suggest that large amounts of water existed in the upper mantle beneath this area when the magmas were produced. McBirney (1969) proposed a process in which dehydration occurs in the subducting slab and water ascends into the mantle wedge to cause the melting of mantle materials. When magmas were generated in the Setouchi volcanic belt, water may have been supplied by this process. Furthermore, taking the above fact (1) into account, it may be inferred that the subducting slab reached the mantle under the Setouchi volcanic belt at 13 Ma. Kobayashi (1979) indicated that a NNW-SSE horizontal maximum compressional tectonic stress field existed in the southwestern Japan

in middle Miocene time on the basis of analyses of volcanic dyke directions. This supports subduction of the oceanic crust beneath the area at that time. The subducting slab must be lithosphere of the Shikoku Basin.

(3) Sanukitoids were produced at pressures less than 15 kbar (at depths less than 50 km, i.e., in the uppermost mantle), and at temperatures of 1100-1200°C. This temperature seems to be high compared with that in most of the geothermal models for island arcs (Hasebe et al., 1970; Toksöz et al., 1971; Fujii and Kurita, 1978). Such a high temperature can be understood if the subduction of lithosphere began only shortly before that time, so that the uppermost mantle wedge had not been cooled by the slab first reached the mantle beneath the Setouchi volcanic belt at 13 Ma. Furthermore, high temperatures in the uppermost mantle may also have been partially due to the relatively high tempeature of the very young Shikoku Basin lithosphere which was born shortly before the subduction (30-17 Ma; Kobayashi and Nakada, 1978). Also, Klein et al. (1978) and Klein and Kobayashi (1980) concluded that off-ridge volcanism occurred at about 15 Ma in the Shikoku Basin, which would have contributed to making the basin lithosphere hot at that time.

(4) The Setouchi volcanic belt is a narrow zone extending from the northeastern Kyushu area to the Choshi area (Fig. 2). Characteristic sanukitoid magmas were probably generated along the entire length of the volcanic belt. In the area where the Quaternary volcanic belt intersects the presumed Setouchi voclanic belt (central Japan; Fig. 1 and Fig. 2), sanukitoids are missing. Sanukitoids, however, could have been covered by erupted materials of the later volcanic activities, in which case the lithosphere of the Shikoku Basin may have been subducting beneath the entire area.

The above arguments may be summarized as follows: Subduction of newly generated lithosphere of the Shikoku Basin began in Miocene time, the leading edge reaching the mantle beneath the entire Setouchi volcanic belt (i.e., the area from northeastern Kyushu to Choshi in Fig. 2) at 13 Ma. This would suggest that the Shikoku Basin was wider than at present; that is, the Shichito-Iwojima Ridge must have been situated about 100 km east of its present position (Fig. 9). The position of the Shichito-Iwojima Ridge at 5 Ma, as given by the reconstruction by Matsubara and Seno (1980), is close to that suggested by this study. On the other hand, Matsuda (1978) suggested that the Shichito-Iwojima Ridge has been situated at its present position relative to Japan since early Miocene time. This discrepancy is an imporatnt problem to be solved in future studies of this region.

Acknowledgements

The author would like to express his gratitude to Professors Seiya Uyeda and Ikuo Kushiro for their critical comments and careful reading of the manuscript. The author is also benefited greatly from discussions with Professor Kazuo Kobayashi.

References

Dallwitz, W.B., D.H. Green and J.E. Thompson, Clinoenstatite in a volcanic rock from the Cape Vogel area, Papua, J. Petrol., 7, 375-403, 1966.

Duncan, R.A. and D.H. Green, Role of multistage melting in the formation of oceanic crust, Geology, 8, 22-26, 1980.

Eggler, D.H., Water-saturated and undersaturated melting relations in a Paricutin andesite and an estimate of water content in the natural magma, Contrib. Mineral. Petrol., 34, 261-271, 1972.

Fujii, N. and K. Kurita, Seismic activity and pore pressure in island arcs, Abstracts of Papers, International Geodynamics Conference March 1978 Tokyo, 30-31, 1978.

Hamilton, D.L., C.W. Burnham and E.F. Osborn, The

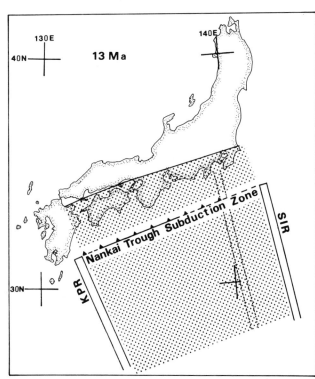

Fig. 9. Possible situation of the Shichito-Iwojma Ridge (SIR) and the Kyushu-Palau Ridge (KPR) at 13 Ma. The shadow area shows the lithosphere of the Shikoku basin. The SIR was situated about 100 km east of its present position (broken line). The leading edge of the subducted oceanic crust reached to beneath the Setouchi volcanic belt.

solubility of water and effect of exygen fugacity and water content on crystallization in mafic magmas, J. Petrol., 5, 21-39, 1964.

Hasebe, K., N. Fujii and S. Uyeda, Thermal process under island arcs, Tectonophys., 10, 335-355, 1970.

Horikoshi, K., On some volcanic rocks of so-called Setouochi (Inland Sea) petrographic province, southwestern Japan, Geosci. Ehime, Special Issue for 60th Birthday Celebration of Dr. Horikoshi, 11-49, 1972.

Jaques, A.L. and D.H. Green, Anhydrous melting of peridotite at 0-15 kb pressure and the genesis of tholeiite basalts, Contrib. Mineral Petrol., 23, 287-310, 1980.

Kaneoka, I. and M. Suzuki, K-Ar and fission tranck ages of some obsidians from Japan, J. Geol. Soc. Japan, 76, 309-313, 1970.

Kawachi, S. and Y. Kawachi, Volcanism of the Kirigamine and Arafune areas after Pliocene -with special reference to the existence of the "Setouchi" factor in the middle part of the "Fossa Magna"-, Chikyu Kagaku, 64, 1-37, 1963.

Kikuchi, Y., On pyroxenic components in certain volcanic rocks from Bonin Island, J. Coll. Sci. Imp. Univ. Japan, 3, 67-89. 1890.

Klein, G. dev., K. Kobayashi, H. Chamley, D.M.Curtis, H.J.B. Dick, D.J. Echols, D.M. Fountain, H. Kinoshita, N.G. Marsh, A. Mizuno, G.V. Nisterenko, H. Okada, J.R. Sloan, D.M. Waples and S.M. Whilte, Off-ridge volcanism and seafloor spreading in teh Shikoku Basin, Nature, 27 746-748. 1978.

Klein, G. and K. Kobayashi, Geoogical summary of the north Philippine Sea, based on Deep Sea Drilling Project Leg 58 results, Initial Rep. DSDP IPOD I, 58, 951-961., 1980.

Kobayashi, K. and M. Nakada, Magnetic anomalies and tectonic evolution of the Shikoku Inter-Arc Basin, J. Phys. Earth, 26, 391-402, 1978.

Kobayashi, Y., Early and middle Miocene dike swarms and regional tectonic stress field in the southwest Japan, Bull. volcanol. Soc. Japan, 24, 203-212, 1979.

Koto, B., On the volcanoes of Japan, J. Geol. Soc. Tokyo, 23, 95-127, 1916.

Kuroda, N., K. Shiraki and H. Urano, Boninite as a possible cald-alkaline primary magma, Bull. Volcanol., 41, 563-575, 1978.

Kushiro, I., The system forsterite-deposide-silica with and without water at high pressure, Am. J. Sci., 267A, 269-294. 1969.

Kushiro, I., Effect of water on the composition of magmas formed at high pressures, J. Petrl., 13, 311-334, 1972.

Kushiro, I., Melting of hydrous upper mantle and possible generation of andesite magma: an approach from synthetic systems, Earth Planet. Sci. Lett., 22, 294-299, 1974.

Matsubara, Y. and T. Seno Paleographic reconstruction of the Philippine Sea at 5 m.y.B.P., Earth Planet. Sci. Lett., 51, 406-414, 1980.

Matsuda, T., Collision of Izu-Bonin arc with Central Honshu; Cenozoic tectonics of the Fossa Magna, Japan, J. Phys. Earth, 26, Suppl. S409-422, 1978.

Matsumoto, Y., Older basalts and sanukitic rocks in the Kishima area, Saga Prefecture, Japan, Rep. Res. Inst. Indus. Sci. Kyushu Univ., 29, 1-25, 1961.

Matsumoto, Y. and K. Takahashi, Welded tuff of the Danjo Island, Dukue City, Nagasaki Prefecture, southwest Japan, J. Geol. Soc Japan, 24, 439-446, 1968.

McBirney, A.R., Compositional variation in Cenozoic calcalkaline suits of Central America, Proceeding fo the Andesite Coference, Bull. 65, Dept. Geol. Mineral Indus., Oregon State, 185-189, 1969.

Morimoto, R., K. Huzita and T. Kasama, Cenozoic volcanism in southwestern Japan with special reference to the history of the Setouchi (Inland Sea) Geologic Province, Bull. Earthq. Res. Inst. Univ. Tokyo, 35, 35-45, 1957.

Murauchi, S. and T. Asanuma, Seafloor spreading in teh Shikoku Basin, south of Japan, Abstracts of Papers, International Geodynamic Conference March 1978 Tokyo, 106-107, 1978.

Murauchi, S. N. Den, S. Asano, H. Hotta, T. Yoshii, T. Asanuma, K. Hagiwara, K. Ichikawa, T. Sato, W.J. Ludwig, J. Ewing, N.T. Edgar and R.E. Houtz, Crustal structure of Philippine Sea, J. Geophys. Res., 73, 3143-3171, 1968.

Mysen, B.O. and A.L. Boettcher, Melting of a hydrous mantle, II, Geochemistry of crystals and liquids formed by anatexsis of mantle peridotite at high pressures and high temperatures as a function of controlled activities of water, hydrogen and carbon dioxide, J. Petrol., 16, 549-593, 1975.

Mysen, B.O. and I. Kushiro, Compositional variations of coexisting phase with degree of melting of peridotite in the upper mantle, Am. Mineral. 62, 845-865, 1977.

Naumann, E., Bau und Entstehung der japanischen inseln, Berlin, 1855.

O'Hara, M.J., Primary magmas and the origin of basalts, Scot. J. Geol., 1, 19-40, 1965.

Roeder, P.L. and R.F. Emslie, Olivine-liquid equilibrium, Contrib. Mineral Petrol., 29, 122-130, 1970.

Sakuyama, M. and I. Kushiro, Vesication of hydrous andesitic melt and transport of alkalies by separated vapor phase, Contrib. Mineral. Petrol. 71, 61-66, 1979.

Sato, H., Nickel content of basaltic magmas; identification of primary magmas and a measure of the degree of olivine fractionation, Lithos, 10, 113-120, 1977.

Shih, T., Magnetic lineation in the Shikoku Basin, Initial Rep. DSDP IPOD I, 783-788, 1980.

Tatsumi, Y., Melting experiments on a high magnesian andesite, Earth Planet. Sci. Lett., 54, 357-365, 1981.

Tatsumi, Y., Origin of high magnesian andesites in the Setouchi volcanic belt, southwest Japan, Part II, melting phase relations at high pressures, Earth Planet. Sci. Lett. in press.

Tatsumi, Y., and K.Ishizaka, K-Ar age of Sanukitoid from Yashima, Kagawa Prefecature, Japan, -Age determination for Setouchi volcanic rocks, No. 2- J. Japan, Assoc. Mineral. Petrol. Econ. Geol., 73, 355-358, 1978.

Tatsumi, Y. and K. Ishizaka, K-Ar age of bronzite andesite from Choshi, Chiba Prefecture, Japan -Age determination for Setouchi volcanic rocks, No. 3, J. Geol. Soc. Japan.

Tatsumi, Y. and K. Ishizaka, Existence of andesitic primary magma: an example from southwest Japan, Earth Planet. Sci. Lett., 53, 124-130, 1981.

Tatsumi, Y. and K. Ishizaka, High magnesian andesite and basalt from the Shodo-Shima Island, southwest Japan, and their bearing on the genesis of calc-alkaline andesites, Lithos, 15, 161-172, 1982.

Tatsumi, Y. and K. Ishizaka, Origin of high magnesian andesites in the Setouchi volcanic belt, southwest Jaspan, Part I, Petrographical and geochemical characters, Earth Planet. Sci. Lett. , in press.

Tatsumi, Y., M. Torii and K. Ishizaka, On the age of the volcanic activity and the distribution of the Setouchi volcanic rocks -Age determination for Setouchi volcanic rocks, No. 5-, Bull. Volcanol. Soc. Japan, 25, 171-179, 1980.

Tatsumi, Y. and T. Yokoyama, K-Ar ages of Tertiary volcanic rocks in the Shodo-Shima Island, Kagawa Prefecture -Age determination for Setouchi volcanic rocks, No. 1-, J. Japan Assoc. Mineral. Petrol. Econ. Geol., 73, 262-266, 1978.

Tatsumi, Y., T. Yokoyama, M. Torii and K. Ishizaka, K-Ar ages of Setouchi volcanic rocks from Osaka and East Yamaguchi areas -Age determination for Setouchi volcanic rocks, No. 4-, J. Japan Assoc. Mineral. Petrol. Econ. Geol., 75, 102-104, 1980.

Toksöz, M.N., J.W. Minear and B.R. Julian, Temperature field and geophysical effect of a downgoing slab, J. Geophys. Res., 76, 1113-1138, 1971.

Tomoda, Y., K. Ozawa and J. Segawa, Measurement of gravity and magnetic force on board a cruising veseel, Bull. Ocean Res. Inst. Univ. Tokyo, 3, 1-70, 1968.

Tomoda, K. Kobayashi, J. Segawa, M. Murauchi, K. Kimura and T. Saki, Linear magnetic anomalies in the Shikoku Basin, northeastern Philippine Sea, J. Geomag. Geoelectr., 28, 47-56, 1975.

Weinschenk, E. , Beitäge zur Petrgraphie Japans, Neues Jahrb., B-Bd, 7, 133-151, 1891.

Yamaguchi, M., Petrography of the Otozan flow on Shodo-Shima Island, Seto-uchi Inland Sea Japan, Mem. Fac. Sci. Kyushu Univ. Ser. D, 3, 217-238, 1958.

Yamazaki, T. and H. Onuki, Differentiation of the calc-alkaline magma at Nijo-san district, Osaka, J. Japan Assoc. Mineral. Petro. Econ. Geol., 62, 249-263, 1969.

Yamazaki, T., M. Torii and k. Ishizaka, Fission track and K-Ar ages of the Setouchi volcanic rocks from the northeastern part of Shikoku, J. Japan Assoc. Mineral. Petrol. Econ. Geol. 76, 276-280, 1981.

Yoder, H.S. and C.E. Tilley, Origin of basalt magma: an experimental study of natural and synthetic rock system, J. Petrol. 3, 342-532, 1962.

Yoshii, T., T. Sasaki, H. Okada, S. Asano, I, Muramatu, H. Hashizume and T. Moriya, The third Kurayoshi explositon and the crustal strucutre in the western part of Japan, J. Phys. Earth, 22, 109-121, 1974.

CROSS SECTIONS OF SOME GEOPHYSICAL DATA AROUND THE JAPANESE ISLANDS

Toshikatsu Yoshii

Earthquake Research Institute, University of Tokyo, Tokyo 113, Japan

Abstract. Cross sections of topography, gravity anomalies, heat flow, earthquakes and focal mechanism are given for 8 regions around the Japanese Islands. The cross sections are constructed by utilizing the data files made from published maps and data listings which have been reported in a previous publication. The Japanese Islands consist of several island arcs that are dissimilar to each other in even their most basic geophysical characteristics, according to the data given here. It is important to define these dissimilarities in order to understand the present tectonics and the tectonic evolution of this island arc complex.

Introduction

Island arcs, the Japanese Islands for example, are interesting regions for geophysical research because they are the tectonically most active zones on Earth and because of their structural complexity. Sugimura and Uyeda [1972] have shown the following simple criteria for defining island arcs; (1) recent volcanic activity, (2) existence of a deep-sea trench, and (3) existence of deep earthquakes. These are regarded as the most essential features which are commonly found in every island arcs. Among other geophysical data of various island arcs, such as earthquake distribution, heat flow and focal mechanisms, we can also see additional common features.

When we investigate those data in detail, however, we recognize that they show remarkable variety from arc to arc. For example, the Japanese Islands consist of several island arcs whose characteristics are not the same even in the most basic aspects such as topography, gravity anomaly, heat flow and earthquakes. Among these arcs, the northeastern Japan arc with a wealth of geophysical data has been often considered to be a 'typical' island arc, but this idea is quite questionable. For example, this arc has a double-planed deep seismic zone [e.g. Hasegawa et al., 1978]. Recent research [e.g. Fujita and Kanamori, 1981],however, has shown that the double-planed seismic zone is a quite special feature and is not always found beneath island arcs.

Careful investigation of dissimilarities between island arcs, as well as their common features, is essential to our understandings of island arc tectonics. In this paper, I show various geophysical profiles of the Japanese Islands based on my earlier compilation of some basic geophysical data [Yoshii, 1979a].

Compilation of Geophysical data

Although many compilations of various geophysical data have been made, they vary greatly in size and form. Some are so voluminous that they are quite difficult to handle. In order to avoide the above difficulties, I made compact files of topography, gravity anomalies, terrestrial heat-flow, earthquakes and focal mechanisms around the Japanese Islands from published maps and data listings along the following basic lines [Yoshii, 1979a].

1. The area is from 25°N to 48°N latitude and from 125°E to 150°E longitude.
2. The data are filed on cards, magnetic tapes and disks so as to be readable by a computer.
3. The data formats were unified for each file and were chosen for convenience in presentation of the data utilizing a plotter.

The outline of the files is as follows.

The file of topography, file TP-1, was constructed by digitizing the land heights and the sea depths in hundreds of maters from published contour maps and charts. The sampling was made at mesh points of one fifth degree for both latitude and longitude. The digitized data, 14614 elevations and depths in all, are arranged in a direction of increasing longitude for each latitude. The computer cards for this file total 812.

The file of gravity anomalies, file GA-1, was made in the same way as the topography file from the published contour maps. Bouguer anomalies on land areas and free-air anomalies on sea areas were digitized in mgal. The size of this file is the same to that of TP-1, but data are lacking for about 30 % of the area due to the lack of gravity observations.

Heat flow data, file HF-1, were compiled from about 40 published papers. Positions and heat-flow values are filed for about 550 stations together with values of thermal gradient and thermal

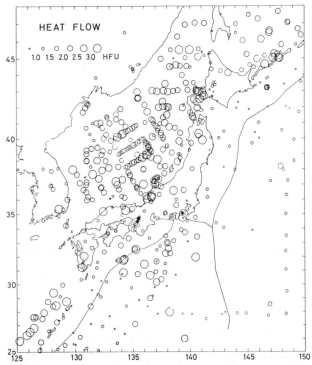

Fig. 1. Mercator map display of, file HF-1, heat flow around the Japanese Islands [Yoshii, 1979a].

conductivity. A few data were read directly from figures because of the absence of tables in the papers referred to. For these data, the thermal gradient and conductivity are, in many cases, missing.

File EQ-1, for earthquake data, was constructed by selecting the events with well-determined depths from the Bulletines of the International Seismological Centre. The period is from 1964 to 1975 and the events thus selected total about 2800. Since the reported depths of events for ocean regions determined from the pP phase are considered to be overestimated due to the low velocity water layer, corrections for water depth were applied [Yoshii, 1979a; Yoshii, 1979b]. This correction is made because the pP phase in ocean regions is usually pwP, which reflects at the sea surface, not the sea bottom.

File ME-1, for focal mechanisms, was made on the same basis as EQ-1. This comes from about 50 published papers, and the number of events and mechanism solutions are 564 and 839, respectively. Two nodal planes and the P- and T-axes for each event were filed together with the origin times and hypocentral parameters.

Detailes descriptions of these files are found in Yoshii [1979a] in which all the files are displayed on maps and the contents of HF-1 and ME-1 are also presented in tables. A stereographic representation of EQ-1 was also shown. In this paper, examples of the map presentation are shown for HF-1 and GA-1 (Figs. 1 and 2). Some of these files, because of the way in which they were constructed, represent a smoothing of much of the actual field recorded data. They allow, however, comparative analysis of different regions without mispresenting the major geophysical characteristics.

Cross-sectional Views of the Japanese Islands

In this paper, computer plotted cross sections for 8 regions (Fig.3) are presented in order to examine the variations in the geophysical features of the various island arcs which make up the Japanese Islands. The southwestern Honshu arc is not treated here, mainly due to sparse earthquakes in file EQ-1. Two or three cross sections are constructed for each of the northeastern Honshu, Izu-Bonin and Ryukyu arcs in order to show remarkable lateral changes within these arcs.

The cross sections are drawn based on all the data in the files within 8 strips as shown in Fig.3. Of these strips, the narrower ones, about 100 km wide, are for topography and gravity anomalies and the wider ones, about 200 km wide, are for the other data. All cross sections are machine plotted and inked over for clarity.

The cross sections of topography, gravity anomalies, heat flow and earthquakes in the 8 regions are shown in Fig.4a through Fig.4h. The

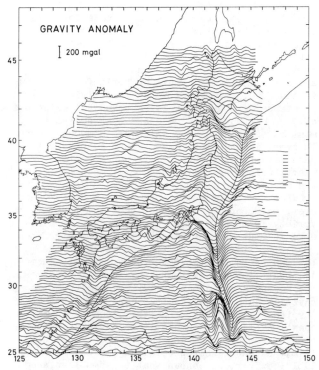

Fig. 2. Mercator map display of, file GA-1, gravity anomalies around the Japanese Islands [Yoshii, 1979a].

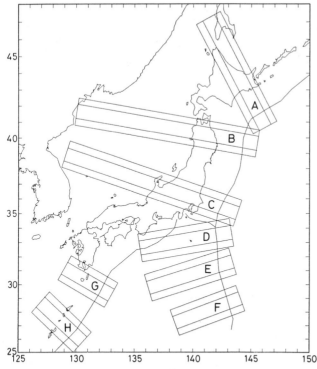

Fig. 3. Limits of the 8 regions for the cross sections presented in Figs. 4a - 4h and 5a -5h. The narrow strips of about 100 km width are for topography and gravity anomalies; the 200 km widths are for the other data.

following is a blief description of these cross sections.

A. The Kurile Arc. This profile crosses the Kurile trench and eastern Hokkaido and reaches to the southwestern part of Sakhalin. The topography is rather simple except at the trench axis due to a seamount. The gravity cross section shows a pattern which is typically observed in most island-arc regions; namely, a large negative trough near the trench axis and a large positive peak near the coastline. The trough, however, is not very sharp and its minimum position is slightly towards the land. Heat flow is low in the Pacific and Sakhalin and high in Hokkaido and the Sea of Japan. The deep seismic zone in this region shows a well-concentrated nature and its dip is about 40°. The shallow swarm found on the left hand side of the cross section is that of rather large back-arc earthquakes occurring in 1971.

B. The Northeastern Honshu Arc. The detailed discussion on the cross sections in this region is found in the previous papers [e.g. Yoshii, 1979b]. We can see a pair of gravity high and low, low heat flow near the trench axis, high heat flow in the Sea of Japan, a clear deep seismic zone and the other features which are considered to be 'typical' of the island arc regions.

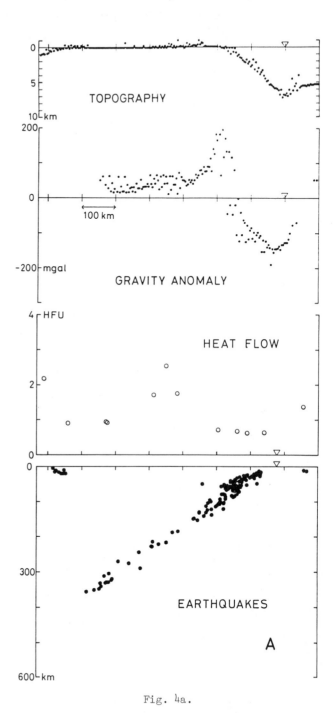

Fig. 4a.

Fig. 4. Cross sections of topography, gravity, heat flow and earthquakes in the 8 regions shown in Fig. 3. Triangles indicate approximate positions of the trench axis.

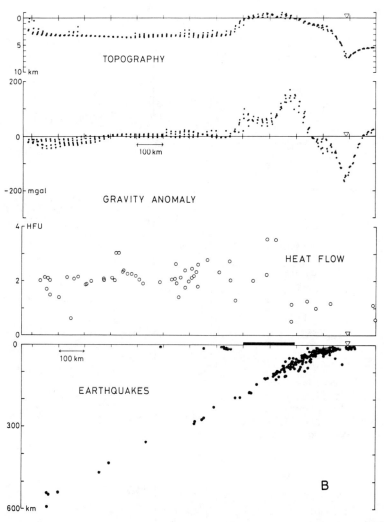

Fig. 4b.

C. *The Northeastern Honshu Arc.* This profile passes through the Japan trench, the central part of Honshu and the southern part of the Sea of Japan. The cross section of topography clearly shows large topographic features in the Sea of Japan, and corresponding variations are found in the gravity profile. Heat flow in the eastern part of the island arc is very low while that in the western region it is rather high, about 2 HFU (10^{-6}cal/s/cm^2). The shallow part of the seismic zone is complicated, apparently by possible northward subduction of the Philippine Sea plate.

D. *The Northern Izu-Bonin Arc.* Water depth at the trench axis exceeds 9000 m and slopes of both side of the axis are steep. The negative gravity anomaly at the trench axis is very large, exceeding -300 mgal. Heat flow on the western side of the trench is generally high. The dip of the seismic zone is rather small in the shallow part but increases gradually with depth and reaches about 45°. In a depth range between about 30 to 70 km, we can see clustered earthquakes which occurred in 1972.

E. *The Central Izu-Bonin Arc.* The cross sections in this region resemble those in region D. The positive gravity anomaly over the arc, however, is not very large and, unlike region D, shows a similar pattern to that of the topography. Heat flow in the Shikoku basin west of the Izu-Bonin arc is highly variable. The deepest portion of the seismic zone bends, suggesting that the subducting plate has encountered a zone it can not penetrate. Shallow seismicity west of the trench is not as great compared to that in region D.

F. *The Southern Izu-Bonin Arc.* This profile crosses the Ogasawara trough which is clearly shown as a flat area in the cross section of topography. The negative gravity anomaly at the trench axis is slightly small than to the north

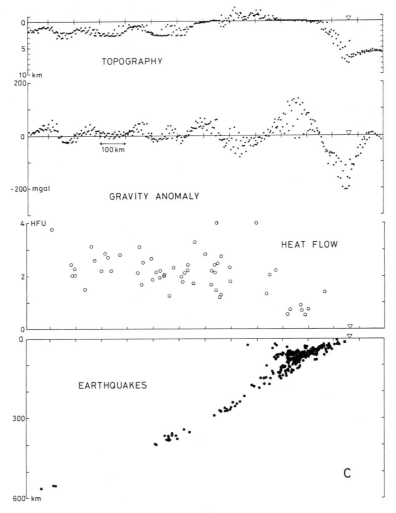

Fig. 4c.

despite the very deep topography exceeding 9000 m, while a large positive anomaly exists over the arc. Gravity values in the Ogasawara trough are negative and show a remarkable contrast to the positive values in the adjasent region west of the trough. Heat flow in the Ogasawara trough is not high although there are few observations. The deeper portion of the seismic zone is nearly vertical and flattens at the bottom like in the profile to the north. Seismicity at shallow to intermediate depths is rather low as region E.

G. The Northern Ryukyu Arc. Water depth at the trench axis is only about 5000 m and the topographic profile is rather gentle. A negative gravity anomaly of about -100 mgal is centered about 100 km westward of the trench axis. The western portion of this profile crosses Okinawa trough where the gravity anomalies are positive. Heat flow is very high around the northern end of Okinawa trough. The seismic zone shows continuous activity from shllow to intermediate depths and reaches at a maximum depth of about 250 km.

H. The Central Ryukyu Arc. Topography in this region is essentially the same as that in the region G except for a greater depression of the Okinawa trough. Two negative gravity anomalies exist between the trench axis and the arc. Heat flow is high in this part of Okinawa trough as it is to the north. The seismic zone in this region is similar to that in region G, except that no events are found deeper than 200 km.

Figures 5a through 5h show cross sections of P- and T-axes of focal mechanism for the eight regions defined in Fig.3. The dip angle is indicated by a bar if the horizontal direction of the P- or T-axis is included within a range between the direction of the profile plus or minus 45°. The reading scheme for the focal mechanisms on the cross sections is given in Fig.6.

In region A some low-angle thrust fault events and a normal fault event are found near the trench. Almost all the events on the deep seismic

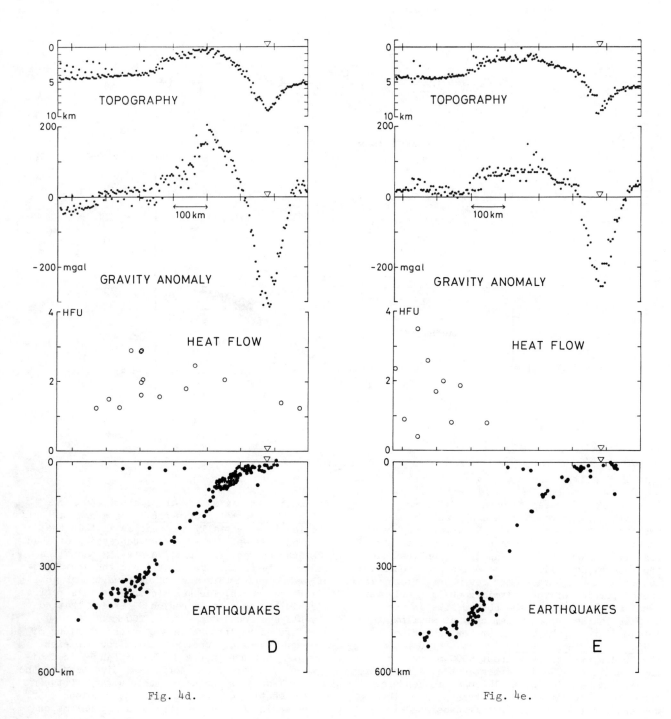

Fig. 4d.   Fig. 4e.

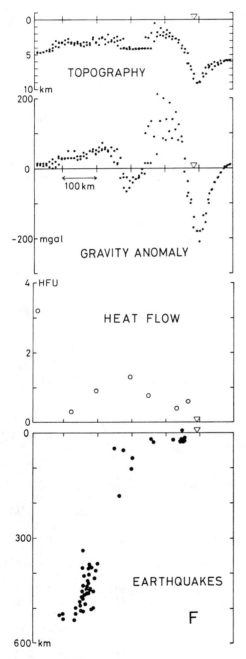

Fig. 4f.

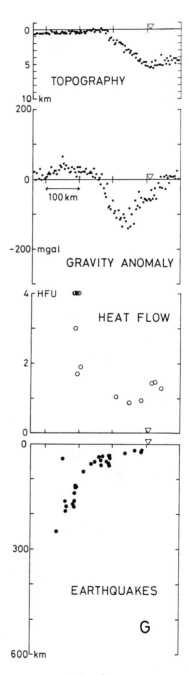

Fig. 4g.

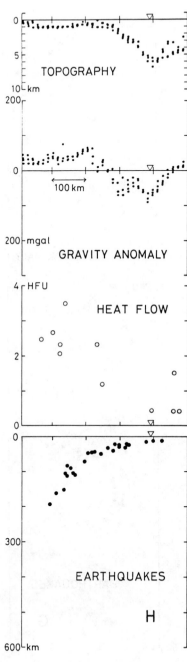

Fig. 4h.

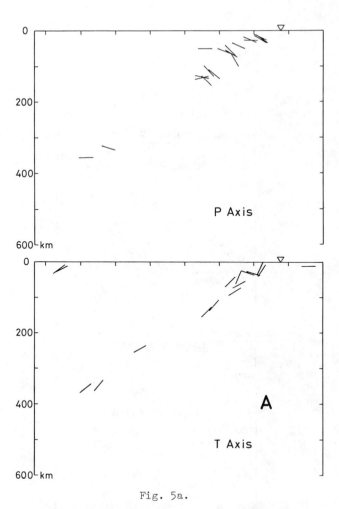

Fig. 5a.

Fig. 5. Cross sections showing dip angles of P- and T-axes of focal mechanism solutions for the 8 regions shown in Fig. 3.

zone are down-dip extension. Shallow events west off Sakhalin, occurring in 1971, are characterized by east-west horizontal compression [Yoshii, 1979a], and only their T axes are plotted.

The deep events in region B are typically of down-dip compression but two events dated in the figure are of down-dip extension which are located on the lower plane of the double-planed seismic zone. The focal mechanism solution of one of them is given in Yoshii [1979a]. At about 200 km west of the trench axis, we can see a sudden change in the mechanism from low-angle thrust to down-dip compression. This position is identical to the 'aseismic front' [Yoshii, 1975; Yoshii, 1979a]. Cross sections of region C are similar to those of region B except for some complexity at about 250 km west of the trench axis, which is presumably an effect of the Philippine Sea plate subduction.

In regions D, E and F of the Izu-Bonin arc, the

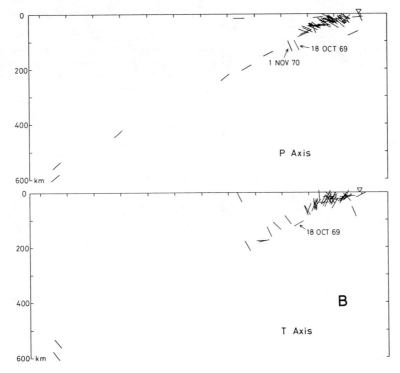

Fig. 5b.

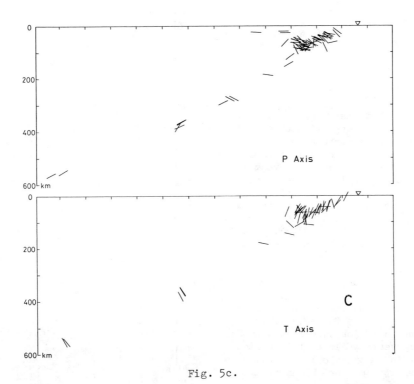

Fig. 5c.

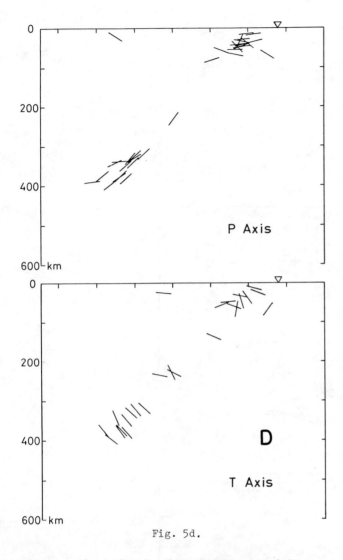

Fig. 5d.

Discussion and Conclusion

As illustrated in Figs. 4 and 5, the geophysical nature of the island arcs which form the Japanese Islands shows large variation. This is true even for the most basic feature of the arcs, such as submarine topography. Variation in the pattern of gravity anomalies is also large. For example, the gravity low at the Izu-Bonin trench is twice as large as that at the Ryukyu trench. In the Ryukyu region, no predominant gravity high has been observed.

The distribution of heat flow is also complicated and it is difficult to construct a single representative profile for the island arcs. This, of cource, may indicate actual variation in thermal state of the various island arcs, but it is also partly due to sparse heat flow data in the areas where observations are difficult to make, such as land regions, deep-sea trenches and shallow ocean areas.

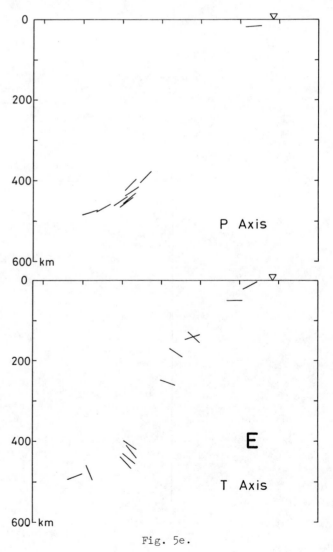

Fig. 5e.

cross sections clearly show that events of down-dip compression make up the deep seismic zone. A gradual change in the P axis corresponds to the bend in the deepest part of the seismic zone in region E. Most of the shallow clustered events found in region D occurred in 1972 as described before. Although the pattern of focal mechanisms for these events is somewhat complicated, the two biggest ones, of m_b 6.5 and 6.7, are clearly of the low-angle thrust type [Yoshii, 1979a]. The other solutions for these clustered events are less accurate because they were determined from data recorded at Japanese stations only. There are very few inter-plate thrust events in regions E and F.

The patterns of mechanisms in regions G and H of the Ryukyu arc are quite confusing; namely, the deep seismic zone in region G is formed by down-dip extensional events while that in region H is formed by down-dip compressional events. This has also been pointed out by Shiono et al. [1980].

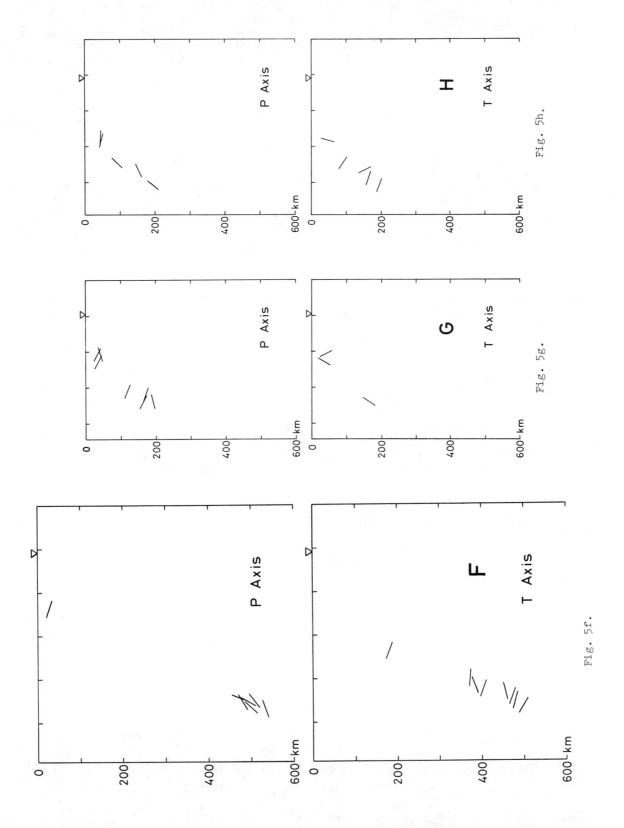

Fig. 5f.

Fig. 5g.

Fig. 5h.

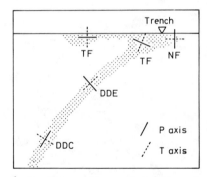

Fig. 6. Reading scheme of focal mechanisms on the cross sections in Fig. 5. NF = normal faulting; TF = thrust faulting; DDE = down-dip extension; and DDC = down-dip compression. The stippled area schematically indicates the seismic zone.

Configuration of the deep seismic zone or the Wadati-Benioff zone also varies from arc to arc. In the Kurile region, the seismic zone appears to bend sharply near the trench axis and its dip angle is nearly constant. In the Izu-Bonin and Ryukyu regions, on the other hand, the seismic zone increases gradually in dip angle with depth. Peculiar upward bending of the deepest part of the seismic zone is found only in the Izu-Bonin arc.

Variation in focal mechanisms is very interesting as well. The dominant mechanism of the main part of the deep seismic zone in the Kurile and northern Ryukyu arcs is down-dip extension while in the other regions it is down-dip compression. The situation in the Ryukyu area is especially intriguing.

The aim of this paper is to show cross sections of the major geophysical characteristics around the Japanese Islands so that this information is available to others who are trying to understand the tectonics of island arcs.

Acknowledgment. The author wishes to thank T.W.C. Hilde and S. Uyeda for critical reviews of the manuscript.

References

Fujita, K., and H. Kanamori, Double seismic zones and stresses of intermediate depth earthquakes, Geophys. J. R. Astron. Soc., 66, 131-156, 1981.

Hasegawa, A., N. Umino, and A. Takagi, Double-planed structure of the deep seismic zone in the northeastern Japan arc, Tectonophysics, 47, 43-58, 1978.

Shiono, K., T. Mikumo, and Y. Ishikawa, Tectonics of the Kyushu-Ryukyu arc as evidenced from seismicity and focal mechanism of shallow to inter-mediate-depth earthquakes, J. Phys. Earth, 28, 17-43, 1980.

Sugimura, A., and S. Uyeda, Island Arcs, Japan and Its Environs, 247 pp., Elsevier Scientific Publishing Company, Amsterdam, 1972.

Yoshii, T., Proposal of the 'aseismic front', Zisin 2, 28, 365-367, 1975 (in Japanese).

Yoshii, T., Compilation of geophysical data around the Japanese Islands (I), Bull. Earthq. Res. Inst., 54, 75-117, 1979a (in Japanese).

Yoshii, T., A detailed cross-section of the deep seismic-zone beneath northeastern Honshu, Japan, Tectonophysics, 55, 349-360, 1979b.

COMPRESSIONAL WAVE VELOCITY ANALYSES FOR SUBOCEANIC BASEMENT REFLECTORS IN THE JAPAN TRENCH AND NANKAI TROUGH BASED ON MULTICHANNEL SEISMIC REFLECTION PROFILES

Yutaka Aoki, Takeshi Ikawa, Yoichi Ohta and Toshiro Tamano

Japan Petroleum Exploration Co., Ltd., Ohtemachi, Tokyo, Japan

Abstract. On multichannel reflection seismic profiles obtained from the subduction zones around Japan, reflection events are often observed below the oceanic sedimentary layers at about 2 seconds two-way time. From the results of detailed velocity analyses, it is concluded that these events on the seismic profiles correspond to the Moho discontinuities. The boundary between the oceanic layer 2 and oceanic layer 3 appears less distinctively than the Moho discontinuity on the reflection profile.

Introduction

The method used to find seismic wave velocities in the crust utilizes, in general, travel time curves of refracted waves. Using this method, crustal structures in the Pacific Ocean side of the Japanese Islands have been investigated in detail by Ludwig et al. (1966), Yoshii et al. (1973), Murauchi et al. (1968) and others. Refraction seismic studies obtain records over a few to several tens of kilometers, over 100 km at times, and the velocity measured is the average wave velocity in that distance. On the other hand in the multichannel reflection seismic method, the seismic wave velocity is obtained from the travel time of the reflection waves using records which have common reflection points (CRP). The velocity obtained here is a root mean squares (RMS) velocity as discussed by Taner and Koehler (1969) and the formula given by Dix (1955) is usually applied to obtain respective interval velocities. This method is effective where the dip of a reflector is negligible or moderate. Since the multichannel reflection seismic method has been developed for oil prospecting, the maximum offset distance between the seismic energy source and detector is approximately 3 to 5 km in most cases and this is remarkably short compared to the offset used in the refraction method. Therefore with the reflection method it is possible to determine the velocity over a short distance or spatially in detail in comparison to the refraction method.

In recent years, the multichannel seismic reflection survey has been used for velocity determination under oceanic ridges, continental margins and subduction zones. Some of these studies are reported by Matsuzawa et al. (1980), Talwani et al. (1977), Herron et al. (1978), Grow et al. (1979), and Nasu et al. (1977, 1978, 1979). We have recently been working over subduction zones around Japan and one of the aspects under investigation is reflectors which are often found below the oceanic basement reflections. These conspicuous reflections characteristically appear on the reflection profile about 2 seconds two way time below the oceanic basement. In this paper, this reflector is provisionally called the suboceanic basement reflector. In order to study its true character detailed velocity analyses using reflection seismic data are made as part of our investigations of the characteristics of the oceanic crust around Japan.

Data Acquisition and Processing

In order to investigate the crustal structure below the oceanic basement depth, records are selected which indicate distinct reflections below basement. The locations of selected seismic profiles are shown in Fig. 1. These seismic records were obtained by M/V KAIYO of JAPEX. Bolt air guns of 1,500 - 2,000 cubic inch capacity with an operating pressure of 2,000 PSI were used as the energy source. Seismic signals were received by an SEC 48 channel streamer cable with a hydrophone group interval of 50 m. Data recording was made digitally using a Texas Instruments DFSV with a sample interval of 4 milli seconds. Shooting interval was 50 meters and all seismic sections treated in this paper are 24 fold records. The seismic records were processed in the following way: Demultiplex, CRP sorting, gain recovery, prestack deconvolution, NMO correction, stack, post stack deconvolution and filtering. Stacking velocity analysis was done after prestack deconvolution. This basic data processing is common to all of the profiles, but paprameters such as band pass filter cutoff frequency and deconvolution window, vary between individual lines depending on their frequency characteristics.

Velocity Analysis

The velocity analysis method used to determine the optimum stacking velocity is similar to that

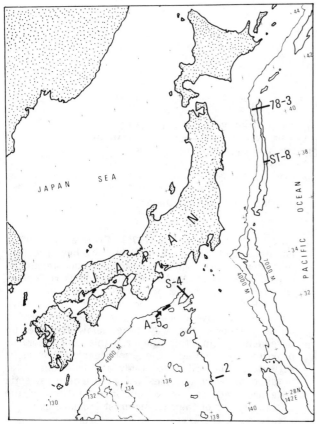

Fig. 1. Location of multichannel reflection seismic profiles over the Japan trench, Nankai trough, and in the eastern Shikoku basin.

described by Taner and Koehler (1969). One CRP gather of 24 traces is used to get velocity function at constant intervals. An example is shown in Fig. 2. The ordinate is two-way travel time and the abscissa is the CRP traces and the NMO corrected stacked trace for each given constant velocity. In this method, the maximum value of the amplitude of the velocity spectrum, which should give correct stacking velocity, appears regardless of the wave nature if the CRP traces possess a coherent hyperbolic event. For instance, even random noise may cause an apparent maximum amplitude value in the velocity spectrum if records of each trace become inphase accidentally. At the same time, even a maximum of small amplitude may indicate a real reflector. Therefore, to minimize errors in velocity estimation caused by these effects and furthermore to correlate the events on a stacked profile where a conspicuous reflection is observed with the velocity spectrum maximum, revelocity analysis stacking eight CRP gathers was conducted using the 24 trace CRP gathers from the previous velocity processing. The first stage of this secondary velocity analysis is to obtain a stacked record using a portion of the actual seismic section, (8 CRP 24 trace gathers) giving a constant stacking velocity. If the velocity used coincides with the true velocity, the NMO correction will be done properly and the amplitude after stacking gets large, similar to the case of ordinary velocity analysis using one 24 trace CRP gather. Then in the second stage, events of large amplitude are picked up from the records of a constant velocity stack (CVSK) section automatically using a computer. One thing to be noted here is that while the velocity analysis using one CRP gather is a peak search in two dimensions of the stacking velocity-time domain, the second method described here is a three demensional maximum search of stacking velocity-time-space (trace) domain. Hence, in this method if an event of large amplitude appeared on one trace due to an accidental coherent noise stack, it will not be picked as the velocity maximum, unless it is coherent spatially. If it is an event which is coherent spatially, it is possible to calculate the dip in the coherent direction and by indicating this dip with the maximum of the velocity spectrum on the same section, discrimination of a diffraction from a reflection becomes very easy. Fig. 3a and Fig. 3b are the velocity analysis results of Line S-4. Fig. 3b is the computer picked diagram of velocity spectrum amplitude maxima and dip shown together with the interpreted velocity curve. Line S-4 was recorded across the Nankai trough where the sediment is very thick, approximately 2 seconds on the time section. Reflections in the sediment are very conspicuous showing gradual increase of velocity value from the sea bottom to the basement at 7.4 seconds.

In Fig. 4, a stackd profile is shown with P-wave interval velocities derived from the veocity analysis illustrated in Fig. 3. The Nankai trough corresponds to the depression between rises on both sides. The right-hand rise on the seismic section is an outer ridge called Zenisu ridge and is running parallel to the Nankai trough (Hydrographic Department M.S.A. Japan, 1976). The left-hand rise is a lower part of inner trench slope and is composed of accreted sediments forming subduction complex (Beck,1972). This is clear from the asymmetrical structures at both ends of the sediments which have filled the trough. The sediment in the trough is thickening toward the left (westward) as are the most recent strata nearest to the sea bottom. From this fact it is inferred that the right side outer ridge on the profile has been continuing to rise up to the present.

Underlyng the sediment is oceanic basement. As is the case for other trenches around the world(e.g. Seely and Vail, (1974); Von Huene, (1979), Ladd and Watkins (1979), Seely (1979)), reflection from the top of oceanic basement is evident even under the inner trench slope. The subducting oceanic basement studied by single channel seismic reflection techniques by Inoue (1978) and Okuda et al. (1976) has been traced landward on the full section of Line 4 up to about 20 km from the trough axis using the multichannel method described here. By observing the sediment (trench fill) filling the trough, it can be seen that ponded sidiment is overlapping probable pelagic sediment at the foot portion of the Zenisu ridge. This is due to repeated tubidity flow in the trench-axial-direction. Consequently, the lower

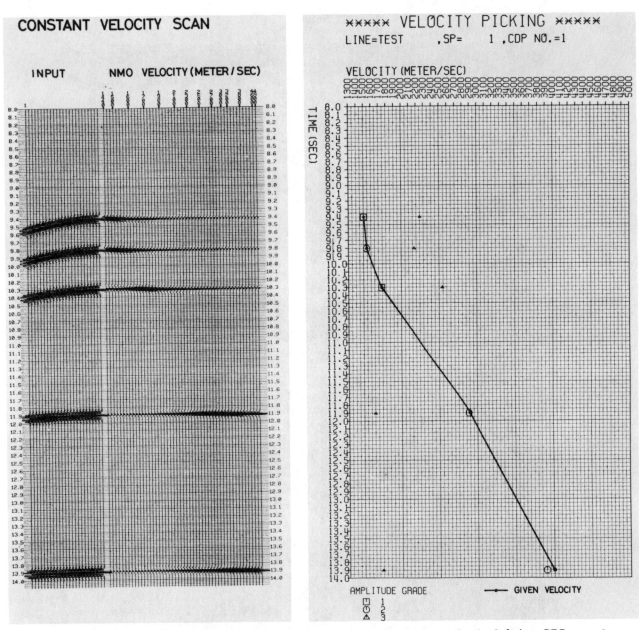

Fig. 2. An example of a single CRP velocity analysis on a synthetic record. On the left is a CRP record for ordinary offset (400-2700m) and stacked traces with different stacking velocities. On the right is the output of the automatic velocity picking program for the stacked traces on the left and the solid line indicates the velocity curve. Note that moveout for the bottom reflector is less than 18 milliseconds at the maximum offset but the error in velocity picking is 2.5 %.

portion of trench fill of the Nankai trough has different origin from the overlying turbidite. The structure of the inner trench slope is very similar to that at DSDP drilling site 298 (Ingle et al., 1975) and is described in an earlier single channel reflection study by Hilde et al. (1969). That is, trench fill is deformed and accreted which is associated with subduction of the Philippine plate. From results achieved by the above velocity analysis, it is ascertained that the sediment layer has a velocity of 1.83-3.34 km/s and the layer beneath has a velocity of 3.34-5.66 km/s.

By application of Dix's formula (1955), interval velocities can be calculated for all individual reflectors. (as in Fig. 4). However the major purpose of this investigation is to obtain reflection seismic velocities below the sediment layer (oceanic layer 1). Accordingly, to minimize errors introduced into

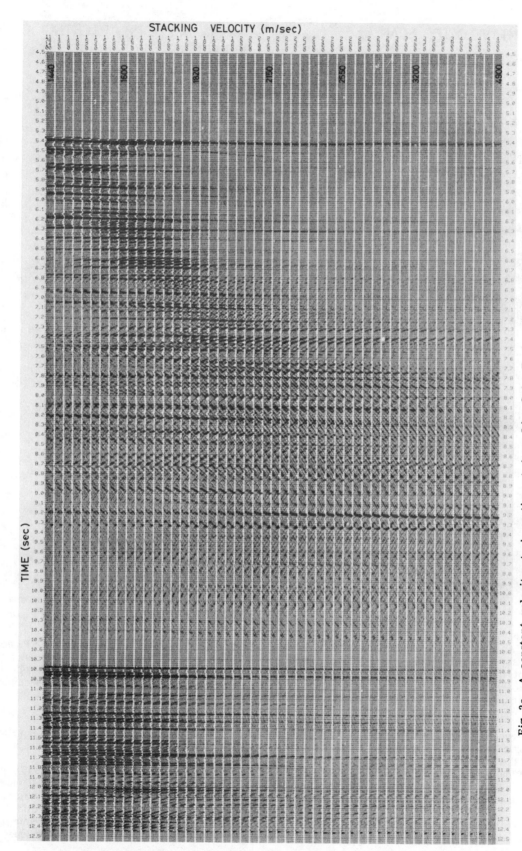

Fig. 3a. A constant velocity stack section at A on Line S-4. Each strip, consists of eight traces, is a portion of a profile with a different stacking velocity assumption. Large amplitude indicates that the correct velocity has been assumed.

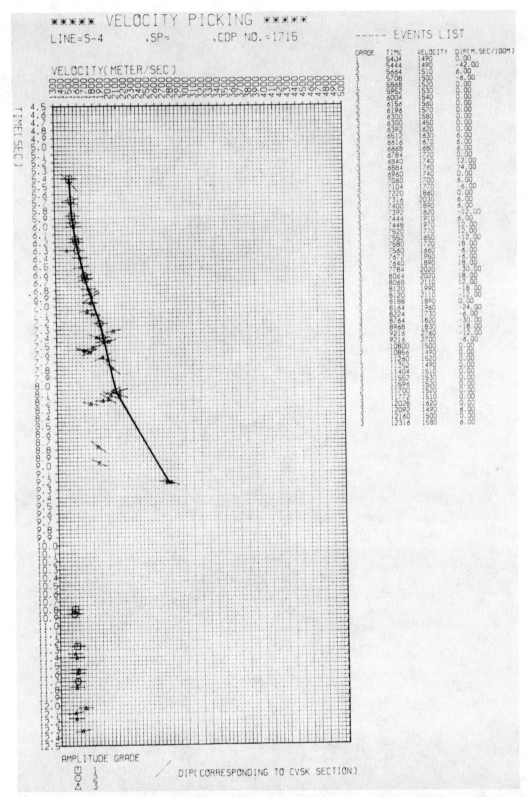

Fig. 3b. Computer picking of velocity amplitude maxima with dip for the CVSK section in Fig. 3a. Spline interpolation is made before the search for velocity amplitude maxima. The solid line indicates the interpreted velocity curve. Dip of the picked event is shown with reverse direction to the CVSK section in this case.

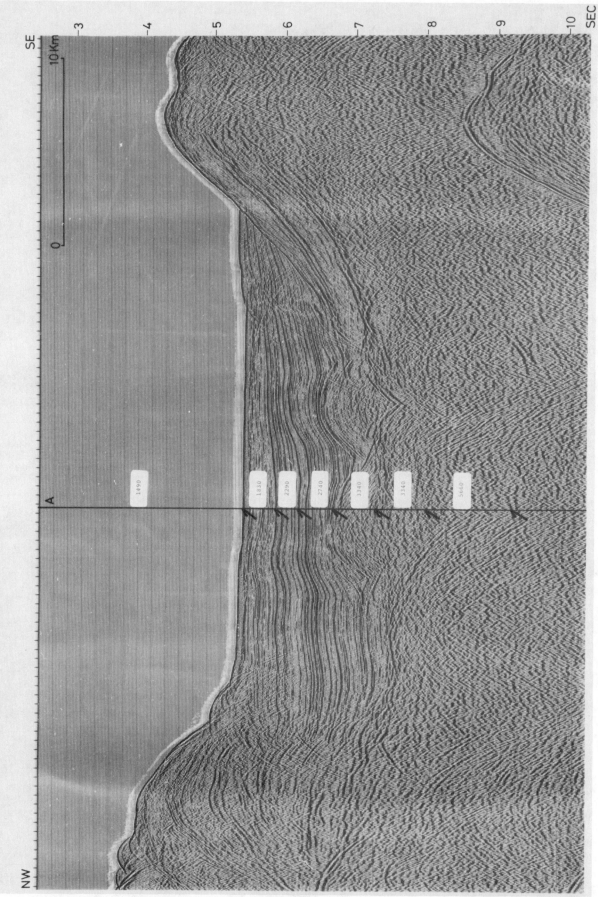

Fig. 4. Multichannel reflection seismic profile of Line S-4 at the Nankai trough with interval velocities at A. The arrow indicates the layer boundary used for velocity determination.

interval velocities (which are defined as the second order quantity from RMS velocities) by picking reflections at small intervals, it was decided to adopt only major boundaries. In particular, boundaries between the sea water, sedimentary layer (layer 1) and the suboceanic basement layer which is defined here as that bounded by layer 1 and the suboceanic basement reflector are selected in the following study. Therefore, it should be noted that P-wave velocities obtained for these intervals are deemed to be average values of selected interval and it does not imply that each interval has a uniform velocity.

A similar analysis was made of A-5, a line parallel to the Nankai trough axis. Results of the velocity analysis is shown in Fig. 5 and the seismic profile and the velocity structure in Fig. 6. Line A-5 also showes thick sediment compared to other trenches in the Northwestern Pacific. However, as shown by the section of Line S-4 (Fig. 4), the lower half is pelagic sediments and the upper half is a turbidite. On Line A-5 the velocity of layer 1 is 1.74-2.0 km/sec and the velocity of the suboceanic basement layer is 4.66-6.04 km/sec. In spite of scattering of the velocities obtained, the time difference between the top of the oceanic basement and the suboceanic basement reflector is nearly 2 seconds and is almost identical with the times on Line S-4.

Results of the velocity analysis on Line 78-3 across the northern Japan trench are shown in Fig. 7 and Fig. 8. This line crosses the Japan trench at a position about 100 km north of the IPOD holes 434, 435, 436, 438, 439, 440 and 441, but about 15 km south of 438 and 439 which were drilled on Leg 56 and Leg 57 of IPOD (Von Huene et al., 1978). The thickness of sediment is about 500 m and uniform on the Pacific side. There is ponded sediment on the trench bottom and beneath it oceanic sediment dips toward the continent. Velocities obtained from Line 78-3 are 1.87-2.19 km/sec for the sedimentary portion on the oceanic side, 2.41 km/sec for the subduction complex and 5.47-6.22 km/sec for the suboceanic basement. One thing to be noted here is that the suboceanic basement layer velocity is nearly 6.1-6.2 km/sec at positions A and C on Fig. 8. This coincides with the velocities of Matsuzawa et al. (1980). However, for position B, the suboceanic basement reflection is observed at about 0.6 second below the oceanic basement but the velocity above the reflector is 3.46 km/sec. The velocity of the lowest layer at B is 5.47 km/sec and is distinctively lower in comparison to positions A and C. Again the time difference from the oceanic basement in the time secion is about 2 seconds.

Fig. 9 and Fig. 10 show results of the Line ST-8 velocity analysis which was obtained by traversing the Japan trench south of Line 78-3 (Fig. 1). The suboceanic basement reflector on this line is evident. The velocity of oceanic layer 1 is 1.78-2.1 km/sec and the velocity in the suboceanic basement layer is 5.93-6.21 km/sec, similar to the results from Line 78-3 and in Matsuzawa et al. (1980), both of which give a velocity of about 6.2 km/sec at two positions. A most important aspect of Line ST-8 is a thrust fault shown in the trench bottom. This fault disrupts even the suboceanic basement reflector.

As stated above, velocities were obtained in the suboceanic basement layers at the Nankai trough and the Japan trench, both subduction zones around Japan. When the crustal structure is not disturbed such as by faults, the velocity is about 6.0-6.2 km/sec. Referring to this result an experiment was performed. The section on the left in Fig. 11 is the seismic section of Line 2, in the eastern Shikoku basin at 29.5°N obtained by conventional exploration data processing. No reflector is observed below the oceanic layer 1. If we assume the existence of a suboceanic basement layer which has a thickness equivalent to 2 seconds of two-way travel time and also assume that the P-wave interval velocity of that layer is 6.2 km/sec, then restacking as described here is shown on the right section of Fig. 11. The reflector marked by an arrow appears about 2 seconds below the oceanic layer 1. For this part of the profile the signal to noise ratio was poor and determination of a velocity was not possible even by CVSK velocity analysis. But results of this experiment can be considered to indicate that this portion of the Philippine plate has almost the same structure as that part of the plate at the northern Japan trench and the oceanic crust of the Shikoku Basin near the Nankai trough.

Discussions and conclusions

Factors governing the accuracy of the reflection seismic velocities obtained in this study are variation of distance between the energy source and detectors, decline of event detectability accompanied by decrease of normal movement, increase of the apparent stacking velocity induced by dip of the reflection event and signal to noise ratio. Variation of distance between the energy source and detector may be brought about by elongation and drift of the streamer cable. Except the so called "elastic section", a streamer cable contains internal steel wire (piano wire) as the stress member and if there is any elongation in the cable it is the effect of the elastic section. Measurement of the offset distance has been performed using water breaks and elongation of the elastic section has been taken into account. Drift of the cable is less than 10° and the errors resulting from this are within a few percent. Furthermore, the effect due to dip, (θ), of a reflection event is of the order of $1/\cos\theta$ and the effect of dip where $\theta < 5$ is less than 2-3 % (Kametani et al., 1975). There remains a question whether the reflection seismic velocity analysis method currently used for oil prospecting could be effective for the derivation of reliable velocities at the depth of suboceanic basement reflectors. However for the synthetic reflection records (Fig. 2) obtained for a structure similar to the profiles in this paper, the velocity error at the deepest reflector (depth about 15 km) was 5 %. Consequently the overall velocity estimation error for suboceanic basement reflectors will be less than 10 % at most, even taking into account the unknown factors such as signal to noise ratio.

Even though the calculated average velocity of layer 1 is an approximation to the real situation,

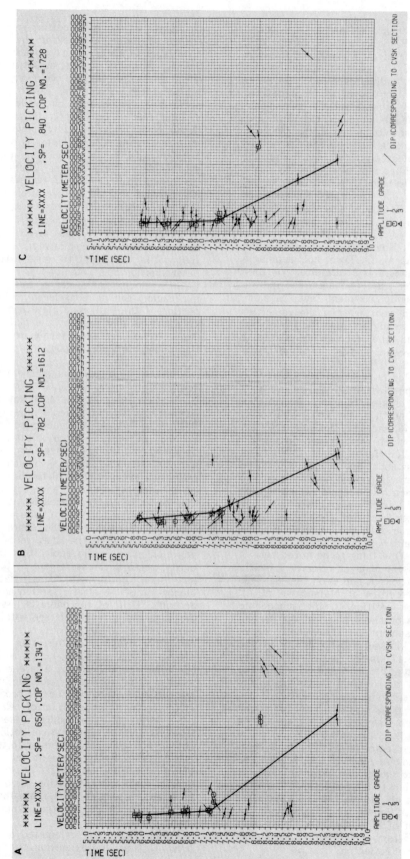

Fig. 5. Computer picked velocity amplitude maxima with a dip display from the CVSK sections at three locations (A, B, and C) on Line A-5.

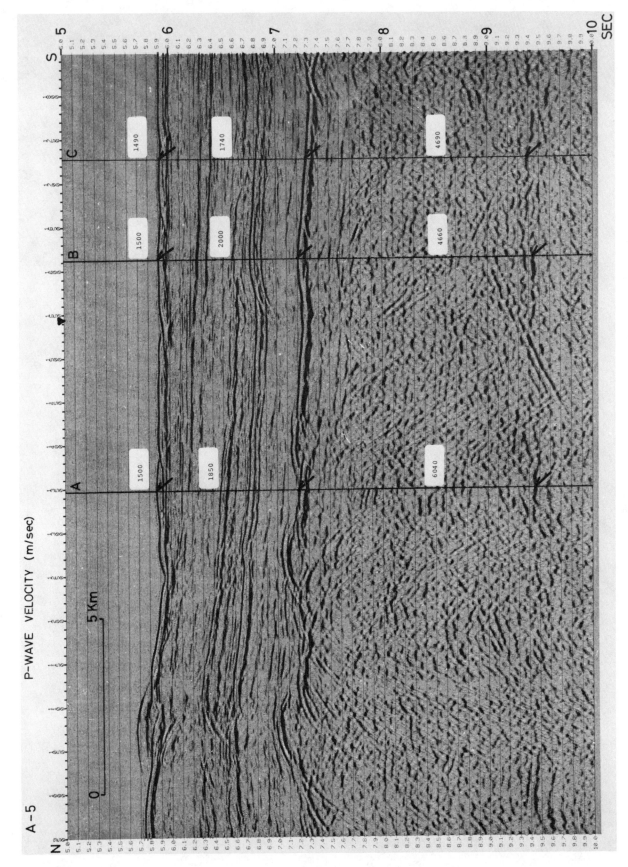

Fig. 6. Seismic profile of Line A-5 parallel to the Nankai trough axis with interval velocities at A, B, and C.

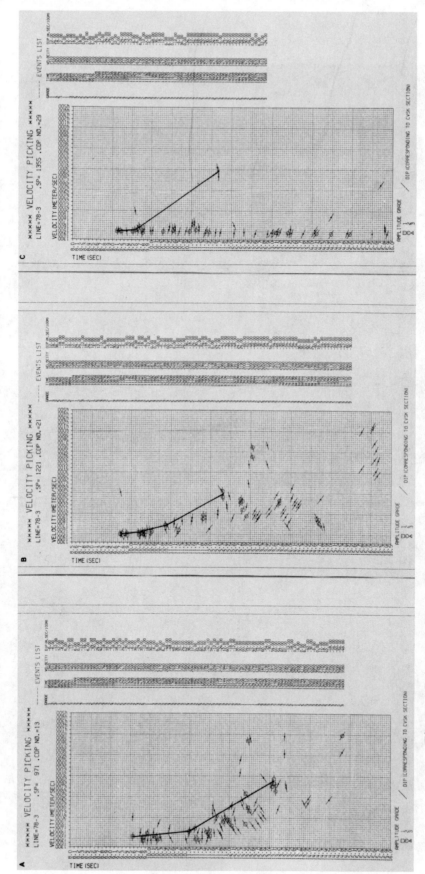

Fig. 7. Computer picked velocity amplitude maxima with a dip display from the CVSK sections at three locations (A, B, and C) on Line 78-3.

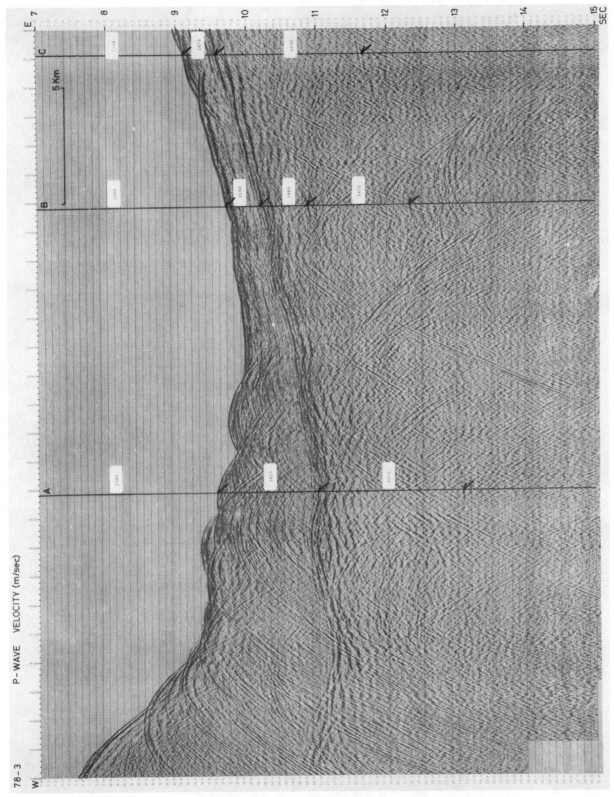

Fig. 8. Seismic profile of Line 78-3 over the Japan trench with interval velocities at A, B, and C.

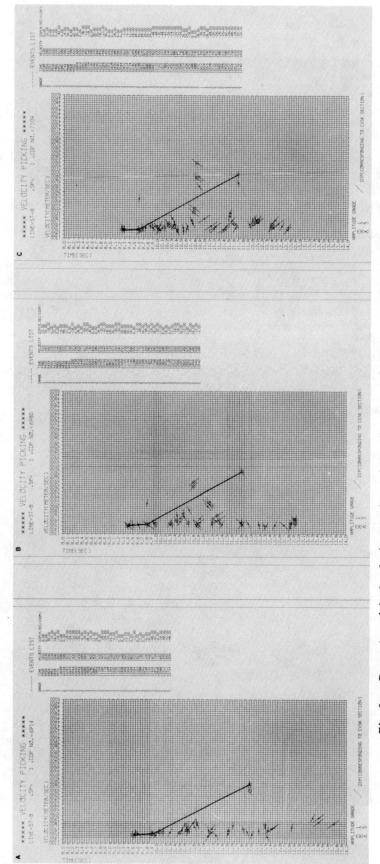

Fig. 9. Computer picked velocity maxima with a dip display from the CVSK sections at three locations (A, B, and C) on Line ST-8.

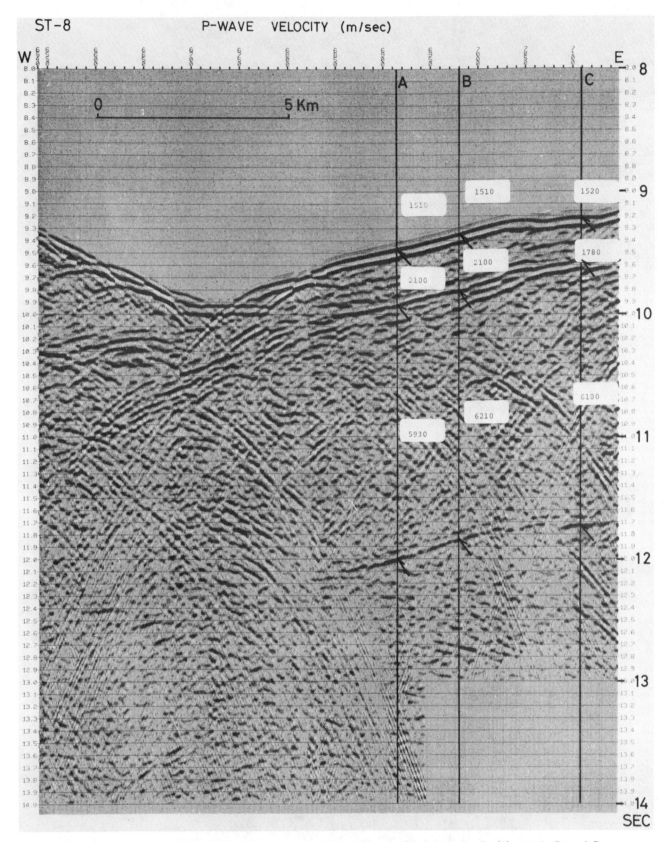

Fig. 10. Seismic profile of Line ST-8 over the Japan trench with interval velocities at A, B, and C.

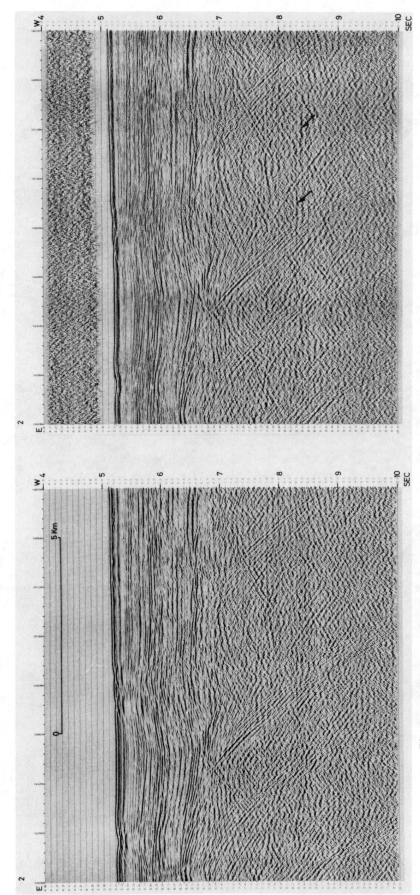

Fig. 11. Seismic profiles with different stacking velocities for an assumed depth of the suboceanic basement layer. On the left, a lower interval velocity of about 3 km/sec has been used while on the right, 6.2 km/sec has been assumed.

since there are numerous reflectors in the layer, the approximation is adequate for investigation of deep crustal structure. This is also endorsed by the fact that all velocity analyses carried out for each CRP show similar values for layer 1 in the same area.

At the northern Japan trench, a value of 5.47-6.22 km/sec is obtained for the suboceanic basement layer, and this velocity corresponds to the velocity of layer 2 or layer 3 of the crustal sructure deduced from refraction studies by Ludwig et al. (1966) and the Research Group for Explosion Seismology (1973, 1977). A comparison with their results shows that the suboceanic basement layer defined in this paper is too thick to be regarded as their oceanic layer 2. As can be seen from the two-way travel time of almost 2 seconds, the thickness of this layer is 5-6 km. Also, the compressional seismic wave velocity of 5.47-6.22 km/sec is too slow for their layer 3 velocity. However, the refraction method detects the highest velocity whereas the reflection method detects the mean velocity, hence a refraction velocity higher than the reflection velocity is not unreasonable. This kind of disparity in velocity can also be seen in the studies of Buffler et al. (1979) and Talwani et al. (1977). When the thickness and velocity of the layer above are considered, it is most reasonable to consider that the suboceanic basement reflector defined in this paper corresponds to the combination of oceanic layers 2 and 3 defined by the refraction investigations in the Japan trench. The boundary between refraction layers 2 and 3 is not apparent in the reflection seismic profile, perhaps due to a thick transition layer. However, there is a possibility that the reflection (Fig. 8) observed at B on Line 78-3 is the boundary between refraction layers 2 and 3, hence by expanding the survey area and further studying reflections at this depth, the character of this boundary on the reflection profile may be clarified.

In the Nankai trough, interval velocities of the suboceanic basement layer scatter widely from 4.66-6.04 km/sec. According to Murauchi et al. (1964, 1968), the thickness of refraction layer 2 near the Nankai trough is about 3.8 km and the velocity is 4.0-5.1 km/sec. Therefore, it seems possible that the suboceanic basement layer may correspond to layer 2 (using the lowest velocity), but as in the Japan trench it is more probable that it corresponds to the combination of refraction layers 2 and 3. As the thickness on the reflection profile is almost constant and equivalent to 2 seconds, the scattering of the velocities is considered to have resulted chiefly from velocity estimation error induced by decline of signal to noise ratio rather than due to scattering of actual velocity.

When the profile of the Philippine sea is compared with Murauchi et al. (1968), the suboceanic basement layer in Fig. 11 is also considered to correspond to the oceanic refraction layers 2 and 3.

Summping up the above, it is concluded that the suboceanic basement reflectors shown here are the crust-mantle boundary, namely the Mohorovicic discontinuity. A distinctive feature of this reflector on the reflection seismic profile is that it appears about 2 seconds below the oceanic layer 1.

Acknowledgements. We are grateful to Japan Petroleum Exploration Co., Ltd. and MITI Japan for their permission to use their data. We are also grateful to Roland Von Huene, Gregory F. Moore, Yasufumi Ishiwada, and Seiya Uyeda for discussions and John E. Shirley for his review of this manuscript. Thanks are also due to the field parties and our colleagues at the data processing center of JAPEX for the data acquisition and processing.

References

Beck, R. H., The oceans, the new frontier in exploration, Austral Petroleum Explor. Assoc. Jour., 12, 5-28, 1972.

Beck, R. H., P. Lehner, P. Diebold, G. Bakker, and H. Doust, New geophysical data on key problems of global tectonics, Proc. 9th World Pet. Congr., 5, 3-32, 1975.

Buffler, R. T., J. S. Watkins, and W. P. Dillon, Geology of the offshore southeast Georgia Embayment, U. S. Atlantic continental margin, based on multichannel seismic reflection profiles, in Geological and Geophysical Investigations of Continental Margins, AAPG Memoir 29, 1979.

Dix, H. C., Seismic velocities form surface measurements, Geophysics, 20, 68-86, 1955.

Grow, J. A., R. E. Mattick, J. S. Schlee, Multichannel seismic depth sections and internal velocities over outer continental shelf and upper continental slope between Cape Hatteras and Cape Cod, In Geological and Geophysical Investigations of Continental Margins, AAPG Memoir 29, 65-83, 1979.

Herron, T. J., W. J. Ludwig, P. L. Stoffa, T. K. Kan, and P. Buhl, Structure of the East Pacific Rise crest from multichannel seismic reflection data, J. Geophys. Res., 83, 798-804, 1978.

Hilde, T. W. C., J. M. Wageman, and W. T. Hammond, The structure of Tosa terrace and Nankai trough off southeastern Japan, Deep-Sea Research, 16, 67-75, 1969.

Hydrographic Department M. S. A. Japan, Bathymetric chart of Iro saki to Muroto saki, Hydro, Drpt., Tokyo, Japan, 1976.

Ingle, J. C., Jr., Karig, D. et al., Site 298, in Karig, D. E., Ingle, J. C., Jr. et al. Initial Reports of the Deep Sea Drilling Project. U. S. Gov. Printing office, Washington, D. C., 31, 317-350, 1975.

Inoue, E. ed., Investigation of the continental margin of soutwest Japan, Cruise Rep., No. 9, Geol. Sur. Japan, 1978.

Kametani, T., A. Umedo, Y. Ishii, and N. Asakura, Problems of seismic techniques in areas of complex geology, Proc. 9th World Pet. Congr., 5, 259-268, 1975.

Ladd, J. W., and J. S. Watkins, Tectonic development of trench-arc complexes on the northern and southern margins of the Venezuela basin, in Geological and Geophysical Investigations of Comtinental Margins, AAPG Memoir 29, 363-371, 1979.

Ludwig, W. J., J. I. Ewing, M. Ewing, S. Murauchi, N. Den, S. Asano, H. Hotta, M. Hayakawa, T. Asanuma, K. Ichikawa, and I. Noguchi, Sediments and structure of the Japan trench, J. Geophys. Res., 71, 2121-2137, 1966.

Matsuzawa, A., T. Tamano, Y. Aoki, and T. Ikawa, Structure of the Japan trench subduction zone, from multichannel seismic-reflection records, Marine Geology, 35, 171-182, 1980.

Murauchi, S., N. Den, S. Asano, H. Hotta, J. Chujo, T. Asanuma, K. Ichikawa, and I. Noguchi, A seismic refraction exploration of Kumano nada, Japan, Proc. Japan Acad., 40, 111-115, 1964.

Murauchi, S., N. Den, S. Asano, H. Hotta, T. Yoshii, T. Asanuma, K. Hagiwara, K. Ichikawa, T. Sato, W. J. Ludwig, J. I. Ewing, N. T. Edgar, and R. E. Houtz, crustal structure of the Philippine sea, J. Geophys. Res., 73, 3143-3171, 1968.

Nasu, N., et al, Multichannel seismic reflection data across the Shikoku basin and the Daito ridges, 1976, IPOD-Japan Basic Data Series, Part 1, No. 1, Ocean Research Institute, University of Tokyo, 1977.

Nasu, N., et al, Multichannel seismic reflection data across the Shikoku basin and the Daito ridges, 1976, IPOD-Japan Basic Data Series, Part 2, No. 2, Ocean Research Institute, University of Tokyo, 1978.

Nasu, N., et al, Multichannel seismic reflection data across the Japan trench, IPOD-Japan Basic Data Series, No. 3, Ocean Research Institute, University of Tokyo, 1979.

Okuda, Y., E. Inoue, T. Ishihara, Y. Kinoshita, K. Tamaki, M. Joshima, and Y. Ishibasi, Marine Geology of the Nankai trough and its northern slopes, Marine Geology, 8, 48-58, (Japanese with English abstract), 1976.

Research Group for Explosion Seismology, Crustal structure of Japan as derived from explosion seismic data, Tectonophysics, 20, 129-135, 1973.

Research Group for Explosion Seismology, Regionality of the upper mantle around northeastern Japan as derived from explosion seismic observations and its seismological implications, Tectonophysics, 37, 117-130, 1977.

Seely, D. R., P. R. Vail, and G. G. Walton, Trench slope model, in The Geology of Continental Margins, New York, Springer-Verlag, 249-260, 1974.

Seely, D. R., The Evolution of Structural High bordering major forearc basins, in Geological and Geophysical Investigations of Continental Margins, AAPG Memoir 29, 1979.

Talwani, M., C. C. Windisch, P. L. Stoffa, P. Buhl, and R. E. Houtz, Multichannel seismic study in the Venezuelan basin and the Curacao ridge, in Island Arcs, Deep-sea Tranches and Back-arc Basins, Am. Geophys. Union Maurice Ewing Series 1, 83-98, 1977.

Taner, M. T., and F. Koehler, Velocity spectra digital computer derivation and applications of velocity functions, Geophysics, 34, 859-881, 1969.

Von Huene, R., N. Nasu et al., On Leg 57 Japan trench transected, Geotimes, 23, No. 4, 16-21, 1978.

Von Huene, R., Structure of the outer convergent margin off Kodiak island, Alaska, from multichannel seismic records, in Geological and Geophysical Investigations of Continental Margins, AAPG Memoir 29, 1979.

Yoshii, T., W. J. Ludwig, N. Den, S. Murauchi, M. Ewing, H. Hotta, P. Buhl, T. Asanuma, and N. Sakajiri, Structure of southwest Japan margin off Shikoku, J. Geophys. Res., 78, 2517-2525, 1973.

DEEP SEISMIC SOUNDING AND EARTHQUAKE PREDICITION AROUND JAPAN

Masami Hayakawa and Susumu Iizuka

Faculty of Marine Science and Technology, Tokai University, Shizuoka, Japan 424

Abstract. In the Tokai area (facing the Pacific Ocean in the central part of Japan), it is predicted for various reasons, that a great earthquake of Magnitude eight class will take place in the near future. The Japanese Government authorities are planning and carrying out various observations based on this prediction including measurements of seismic wave velocity changes with time. Along with these observations, it is important to know the subterranean and sub-seabottom structure to the likely hypocentral depth (30 - 40 km) location of this predicted great earthquake. The three dimensional seismic velocity structure in the area, obtained by the explosion seismic refraction method, is shown. It is very clear that from land to the offshore area southward, the depth of Moho becomes shallower very quickly. The estimated hypocenter of the predicted great earthquake is in the uppermost upper mantle, consequently the material around the proposed focus consists of ultrabasic rocks. On the other hand, this area is in a zone of comparatively high heat flow (2.0 HFU). It is obvious from the thermal gradient in such high heat flow zones that temperatures at anticipated hypocentral depths approach the melting point of wet peridotite, or even that of dry peridotite. Under such circumstances, the partial melting proceeds by the effect of either a pressure decrease or a temperature increase associated with tectonic forces acting on the rocks. The partial melting will yield some changes in volume, consequently some stresses will be added to the tectonic forces and this may trigger the predicted earthquake. Estimations of the changes in seismic wave velocities due to partial melting are also considered through the application of experimental data and theoretical considerations.

Introduction

One of the very important geophysical problems in Japan is earthquake prediction, especially in the Tokai area (facing the Pacific Ocean in the central part of Japan, Fig. 1). It is presumed that a great earthquake of Magnitude eight class will take place here in the near future, based on various considerations including the long period since the great earthquake of 1854 in this area, and the subsidence of Omaezaki (southmost cape of this area) by 40 cm during the last 80 years.

For the purpose of prediction, various observations have been carried out including relevelling, retriangulation, extensometer measurements, observations of great and micro-earthquakes, tide-gauge observations, gravity and magnetic variation measurements and seismic wave velocity change measurements are considered in the latter half of this paper.

Although these measurements are important for the prediction of earthquakes, it is also important to know the subterranean and sub-seabottom structure to the depth of the hypocentral area (30 - 40 km) of the predicted great earthquakes.

According to the plate tectonic theory, in the Tokai (offshore) area, subduction of the Pacific plate acts to drag down the margins of the main Japanese islands (overlying plate) and earthquakes occur along the plane of the contact between the two plates. As explained above, due to the long time lapse, strain energy has probably accumulated in the overlying plate and an earthquake can be expected to occur before long. However, it is not yet clear how much stress the earth's crust can bear before breaking.

To estimate the limit of the strength in the focal region, it is necessary to know the rock composition and properties because the limits of breaking stress are different from one type of rock to another.

On the other hand, it has recently been suggested that a low velocity zone might exist in the deeper crust or in the uppermost mantle in this area, based on the observations at Nagoya of the earthquakes that have occurred near Oshima (Fig. 1). To determine if a low velocity zone exists, it is also necessary to know the subterranean geological structure.

To explain the earthquake occurrences, it is also important to take the thermal effects into consideration. For this purpose it is indispensable to know the thicknesses of granitic and basaltic layers and the depth of Moho. Unless we know these values, it is difficult to discuss the possible effect of partial melting on earthquakes.

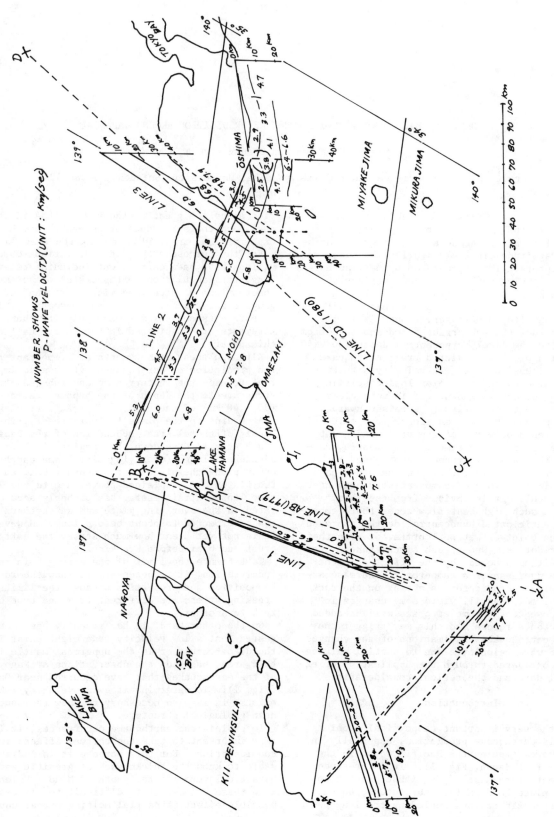

Fig. 1. Three dimensional seismic wave distribution around Tokai area.

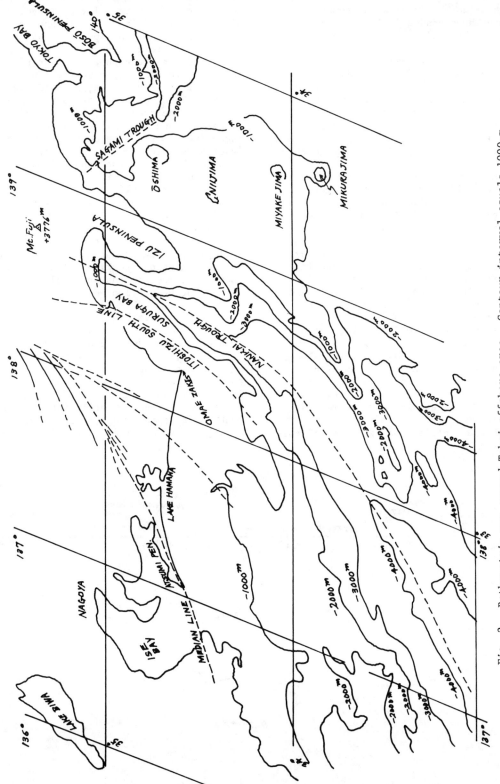

Fig. 2. Bathymetric map around Tokai offshore area. Contour interval equals 1000 m.

Crustal Structure in the Tokai Area

In and adjacent to the Tokai area, several seismic surveys have been conducted in the last several years. The three dimensional seismic wave velocity distribution in this area, obtained by the refraction explosion seismic method, is shown in bird's eye view in Fig. 1. The data sources include reports by the Japan Research Group for Explosion Seismology (JRGES) and others (Aoki et al., 1972; Ikami, 1978; Hotta et al., 1964). Fig. 2 shows the bathymetry in this area based on echo-sounder surveys (mostly by the Hydrographic Department of Japan), in the same bird's eye view.

In Fig. 1, the symbols I_1, I_3, I_2 and T_1 on the JMA line indicate the permanent seabottom seismographs which were installed in 1979 by the Japan Meteorological Agency (JMA). These were emplaced for the purpose of recording seabottom earthquakes, including microearthquakes, which may be expected before the predicted great earthquake, and also for the purpose of forecasting the Tsunami expected to be associated with the earthquake. In Fig. 1, the dotted lines AB and CD are seismic refraction traverses which were made in the summers of 1979 and 1980 using the ocean bottom seismographs (OBS) and conventional seismographs on land. Explosives were suspended at an optimum depth with a buoy and fired electrically. The seismic wave velocities of 6.0 km/sec and 6.6 - 6.8 km/sec and 7.5 - 8.0 km/sec in lines 1, 2 and 3 may correspond to the granitic, basaltic and ultrabasic rock layers respectively.

Fig. 3 shows the profile along the line AB collected in 1979, as the joint work by the members of the Earthquake Research Institute and Geophysical Institute, University of Tokyo, and the Faculty of Marine Science and Technology, Tokai University. This was sponsored by the Scientific and Technological Agency of Japan (Yoshii et al., 1980). Under the assumption that the average seismic wave velocities in the crust and the uppermost mantle are 6.3 km/sec and 8.0 km/sec, the depth of Moho below can be obtained (Fig. 3). By combining the results of line AB and the previous lines 1, 2 and 3, it is clear that from land to the offshore area, the depth of Moho becomes shallower very quickly. For example, the depth of Moho in the line 2 (on land), is about 30 - 35 km, but at 100 - 150 km off the coast in the line AB, only 15 km or less.

The seismic wave velocities of the uppermost mantle in the lines 2 and 3 seem to be lower than normal values. It is not clear at this stage whether this lower velocity is due to the original geological structure or to other reasons such as the beginning of breaking. In addition to the above presumed reasons, possible thermal effects should not be overlooked.

Possible Thermal Effect

When we consider the mechanism of earthquake occurrence, besides the factors of stress, fissure formation and water penetration, there is another important element; the existence of heat. As already explained in the previous paragraph, for the occurrence of great inter-plate earthquakes off Japan with hypocentral depths of 30 - 40 km, the subduction of the Pacific and Philip-

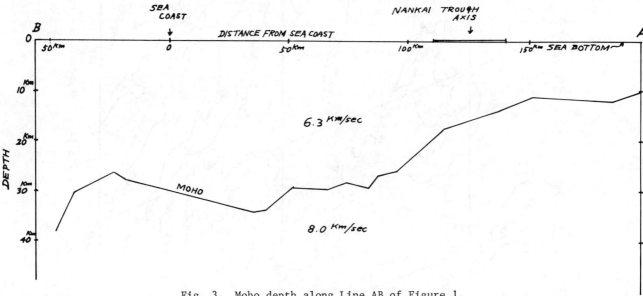

Fig. 3. Moho depth along Line AB of Figure 1.

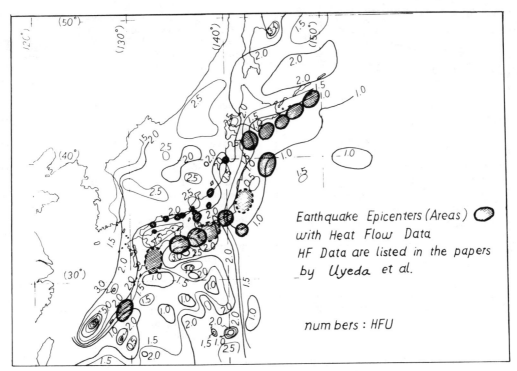

Fig. 4. Earthquake focal areas and heat flow distribution in the Japan region.

pine plates will act to drag down the margins of the main Japanese islands, the leading part of the overlying plate. These great earthquakes are said to occur in the low heat flow zones. However, as seen in Fig. 4, it is clear that the great earthquakes have not always taken place in the low heat flow zones. Some of them have taken place in comparatively high heat flow zones of 2.0 HFU or more, such as the sea just off southwestern Japan, including the present area. It is obvious from the thermal gradient curves that in such high heat flow zones the temperature at the hypocentral depths (30 - 40 km or slightly deeper) approaches the melting point of wet peridotite, or even to that of dry peridotite (Fig. 5 & 6). These thermal gradient curves are based upon the energy balance between the energy discharge by heat flow at the surface and the heat supply from below.

Under such circumstances, the partial melting proceeds by the effect of some pressure decrease or temperature increase possibly associated with tectonic forces acting on the rocks. The partial melting of dry rocks will yield some increase in volume, consequently some stresses will be generated and added to the larger tectonic forces (i.e. due to mantle convection) and this may be the trigger for an earthquake. Detailed estimations of changes in seismic wave velocities due to partial melting were discussed in a previous paper (Hayakawa & Iizuka, 1978). The main results are in the Appendix.

As already shown in Fig. 5, it is quite possible to estimate the subterranean and sub-sea-bottom temperature in the crust and uppermost mantle by using the heat flow data. Fig. 6 is the three dimensional temperature distribution obtained by the above method for a grid whose east side corresponds to the N - S Izu volcanic zone and west side to the N - S direction passing through Lake Hamana. This area includes the epicentral area of the predicted Magnitude eight class earthquake. It's probable hypocenter is indicated by E. As will be seen from this figure, the 1000°C isotherm in the area off Izu peninsula, for example, is raised to a very shallow depth of 30 kilometers. This is reasonable because the Izu peninsula and its southern parts belong to the N - S striking Izu volcanic zone. In the Tokai offshore area where heat flow is nearly 1.7 HFU, the 1000°C isotherm lies at about 50 kilometers depth. Even this depth is considerably shallower than usual. As can be judged from the seismic profiles (Fig. 1 & 3), the presumed hypocenter is in the uppermost mantle, consequently the material around it will be ultrabasic rocks. In the slightly deeper area than this presumed hypocenter (40 - 50 km), the temperature will approach the melting point of wet peridotite, then even to that of dry peridotite (Fig. 5 & 6). Consequently partial melting could proceed if there is some pressure decrease or temperature increase.

We consider here the thermal effect related to plate subduction. In this area, the plate boundary at the bottom of the sea is along the Nankai

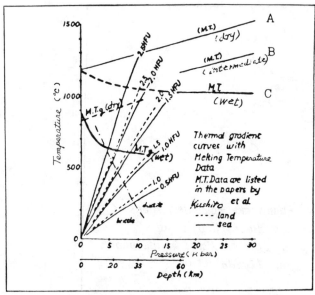

Fig. 5. Thermal gradient curves and melting temperature data.

trough (NE - SW), and subduction is taking place in a NW direction. We can estimate the stress increase accompanied by the volume increase due to partial melting, and the earthquake will take place slightly shallower than the depth of partial melting. Of course, in this case, the direction of brittle fracture will be controlled by the geological structure--the subduction direction. According to the results of calculations, the details of which are not given here (Hayakawa and Iizuka, 1978), the following may be stated. Due to the time lapse, following the process of partial melting, the longitudinal and transverse wave velocities, and the ratio of these velocities first show a decrease. Then they show recoveries, after which an earthquake may take place. For detecting the velocity change with time by the waves transmitting from the predicted hypocenter of 30 - 40 km depth, it is necessary to have the seismic stations at a distance of more than 140 km from the point of the presumed epicenter. Observations are now being continued using both artificial and natural earthquakes. Of course, for the changes in seismic wave velocities and the occurrence of earthquakes the dilatancy effect should also be taken into consideration, but it will be of small importance for earthquakes with focal depths of 30 - 40 km, since there will be limits to the downward penetration of water. In contrast, the deeper the location of hypocenters, the more significant will be the thermal effect. It is also difficult to explain the observable change of seismic wave velocity with time by plate subduction only, even taking finite strain theory into account. By taking the thermal process explained above into consideration, however, it appears possible to produce the velocity change of the observed magnitude.

As an additional remark, it may be pointed out that the hatched areas on the land side and west

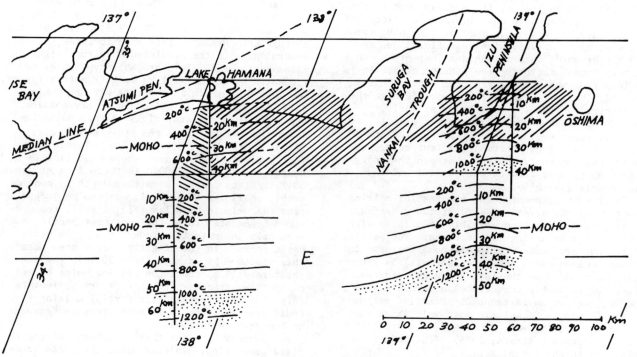

Fig. 6. Sub-seabottom temperature distribution in the Tokai offshore area.

side walls of the grid (Fig. 6) show the distributions of microearthquakes. Their lower (deeper) limits are deeper both westward in the E - W section and northward in the N - S section. This tendency is very interesting because it corresponds to the isotherm distribution.

Appendix

The main points of the results of the previous paper (Hayakawa and Iizuka, 1978) are as follows:
Changes in seismic wave velocities owing to partial melting are calculated using Lindemann's equation based upon Debye's theory of specific heat. Fig. 7 shows the change of seismic wave velocities with temperature, taking the pressure effect into consideration. Next, we consider the changes of seismic wave velocities under partial melting. For example, using Green's experiment (Fig. 3, Green, 1972), the percentage of melt is plotted against temperature at approximately 15 kbar for pyrolite composition under anhydrous and hydrous conditions. It is important that the solidus temperature for each material consisting of pyrolite increases as the melt percentage increases. In order to convert these elements to the response of seismic wave velocities, we apply Lindemann's equation between melting temperature and seismic wave velocities, i.e.

$$T_m = C \times M \times \frac{1}{\left(\frac{1}{V_p^3} + \frac{2}{V_s^3}\right)^{2/3}}$$

where T_m, C and M are melting temperature, constant and atomic weight of the material, and V_p and V_s are longitudinal and transverse wave velocities, respectively. It is clear that when

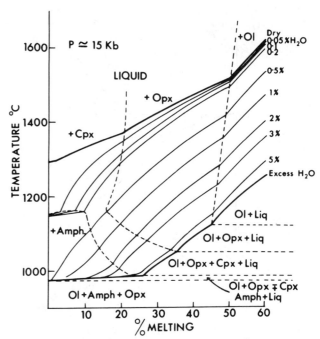

Fig. 8. Temperature increase against progressive partial melting (Green, 1982).

the melting temperature increases, the denominator must decrease. Consequently V_p and V_s must increase. At a glance, it may seem strange that the velocity increases with melting, but after careful consideration, it is found reasonable because the gradual increase of melting temperature corresponds to the solidus temperature increase for the different materials consisting of

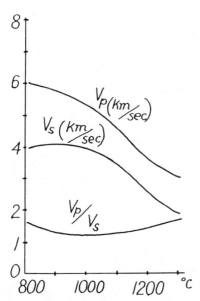

Fig. 7. Change of seismic wave velocities with temperature, taking pressure effect into consideration.

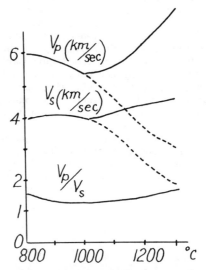

Fig. 9. Changes of seismic wave velocities by progressive partial melting based on the results of Figure 8. The dotted lines are the same as in Figure 7.

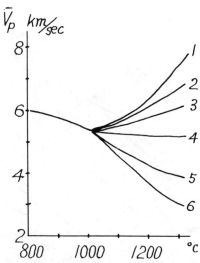

Fig. 10. Modification of Figure 9 with the velocity of melt taken into consideration.

pyrolite. Assuming the ratio of V_p to V_s is maintained, we find that the velocity starts to increase after the start of the partial melting as shown by the solid lines in Fig. 9.

However, strictly speaking, in the above calculation, the effect of the seismic wave velocity of the melt was not taken into consideration. Therefore, we consider here the velocity of the bulk which contains the melt by using a solid-liquid mixture model. Fig. 10 shows the results of this calculation, that is the relationship of longitudinal wave velocity and temperature with progressive partial melting. Curve 1 shows the velocity change in the solid part, curves 2 - 5 show the velocity changes in the partial melting media in the case of $\tilde{V}_p$ = 5, 4, 3, 2 km/sec, respectively, and curve 6 corresponds to the dotted line in Fig. 9. ($\tilde{V}_p$ is the longitudinal wave velocity for the liquid part). This figure shows that the tendency of the solid line in Fig. 9 will not contradict with the case when the velocity of melt is taken into consideration, unless the wave velocity in the liquid part is very low (less than 3 km/sec).

If we suppose here that the area in question was under high temperature but slightly below the melting point in the beginning, and the temperature has gradually increased to start the partial melting, by diapirism or by the pressure decrease due to the dynamical effect, the abscissa in Fig. 9 and 10 may be taken to correspond to the time. On the other hand, as partial melting proceeds, the volume increases unless the rock has excess H_2O. Then, the stress will increase accompanied by the volume increase, and earthquakes will take place at slightly shallower depths than the melting zone.

Acknowledgment. We wish to express our hearty thanks to Professor Thomas W.C. Hilde for giving us the opportunity to present this paper and for reading the manuscript and improving our English.

References

Aoki, H., T. Tada, Y. Sasaki, T. Ooida, I. Muramatsu, H. Shimamura and I. Furuya, Crustal structure in the profile across central Japan as derived from explosion seismic observations, J. Phys. Earth, 20, 197-224, 1972.

Green, D.H., Magmatic activity as the major process in the chemical evolution of the earth's crust and mantle, Tectonophysics, 13, Upper Mantle Sci. Rep. No. 41, The upper mantle, 47-71, 1972.

Hayakawa, M. and S. Iizuka, A mechanism to explain the earthquakes around Japan by the process of partial melting, J. Phys. Earth, 26, Suppl., 571-578, 1978.

Hotta, H., S. Murauchi, T. Usami, E. Shima, Y. Motoya and T. Asanuma, Crustal structure in central Japan along longitudinal line 139°E as derived from explosion seismic observations, Bull. Earthq. Res. Inst., 42, 533-541, 1964.

Ikami, A., Crustal structure in the Shizuoka district, central Japan as derived from explosion seismic observations, J. Phys. Earth, 26, 299-331, 1978.

Kushiro, I., Y. Syono, and S. Akimoto, Melting of a peridotite nodule at high pressures and high water pressures, J. Geophy. Res., 73, 6023-6029, 1968.

Uyeda, S. and K. Horai, Terrestrial Heat Flow in Japan, J. Geophy. Res., 69, 2121-2141, 1964, and many related papers by Uyeda et al.

Yoshi, T., T. Yamada, S. Iizuka, K. Suyehiro, S. Asano, M. Hayakawa, Y. Misawa and Y. Ichinose, Structure of earth's crust in the Tokai offshore area by refraction survey using OBS (ocean bottom seismograph), Preliminary Report of Autumnal meeting of Seismological Society of Japan, 1980 (in Japanese).

GEOTECTONICS OF TAIWAN -- AN OVERVIEW

V.C. Juan, H.J. Lo and C.H. Chen

Institute of Geology, National Taiwan University

Abstract. Taiwan has been tectonically greatly modified by the convergence of the Philippine Sea plate with the Asian continent along the Ryukyu-Taiwan-Luzon Arc System during the Tertiary, after the island-arc formation along the western Pacific region in late Mesozoic time. An east-facing arc system which existed along the eastern edge of Luzon was responsible for the formation of the Chimei-Lutao-Lanhsu Volcanic Arc in the east offshore region of Taiwan during late Eocene to Miocene time, as a result of the first phase of subduction of the Philippine Sea plate. Migration of this west-dipping subduction zone to the collision suture of the Longitudinal Valley in late Miocene time gradually turned the Benioff zone into a nearly vertical position by the Pliocene, while squeezing up a glaucophane schist slab and the oceanic Kuanshan Igneous Complex from the basement of the Coastal Range. The present relative positions of Taiwan and Luzon indicate that the North Luzon Trough is the trace of the southern extension of this late Miocene to Pliocene subduction zone. The last phase of Philippine Sea plate convergence along Taiwan occurred in late Pliocene to Pleistocene time as a north-dipping East Hualien Subduction Zone at latitude 24°N, near the termination of Ryukyu Arc. This was accompanied by accumulation of the Lichi Melange along the southern section of the Longitudinal Valley and may account for the contemporaneous volcanism of the northern offshore area of Taiwan.

Introduction

The geotectonics of Taiwan shows several anomalies in the Ryukyu-Taiwan-Luzon Arc System along the convergent plate boundaries between the Asian and Philippine Sea plates. Taiwan represents the most intriguing segment of this boundary. It shows no recognizable bathymetric trench that is usually associated with island arcs, and is dominated by a nearly vertical subduction zone with arc features concave toward the east. A presently active plate boundary can be clearly defined in the Longitudinal Valley (Fig. 1) from the contact between the continental and oceanic crusts, but this is not the only feature indicating the evolution of the island. Multiple intense crustal deformation events in the vicinity of Taiwan appear to have been closely related to the formation of the island and illustrate the complexities of the convergence and subduction process in this particular area.

Taiwan island in the Ryukyu-Taiwan-Luzon Arc System has been studied by many geologists (Juan and Wang, 1971; Big, 1971; Jahn, 1972; Chai, 1972; Karig, 1973; Yen, 1973). More recently the interpretation of the tectonics of Taiwan has been influenced by new data. Juan (1975) proposed late Miocene west-dipping subduction of Philippine Sea plate beneath the Coastal Range and that the Benioff zone progressively approached a vertical position at the beginning of Pliocene time due to the shifting of direction of movement to WNW of the Philippine Sea plate. Murphy (1973) and Bowin et al., (1978) have suggested that Taiwan has been deformed by a flipped polarity of subduction along Taiwan and Luzon in late Miocene-Pliocene time. Collision of the Luzon arc with Taiwan since Pliocene time has been cited as the cause of transformation of a non-volcanic arc into the present Central Range of Taiwan (Bowin et al., 1978; Liou et al., 1977). However, the Lutao-Babuyan Ridge northeast of Luzon has been interpreted as evidence for an old subduction zone (Karig and Wageman, 1975) and the Palau Ridge with its northern extension, the Gagua Ridge (Fig. 1), was considered as an extinct spreading center (Bowin et al., 1978; Lu and Liaw, 1979). A recent seismicity study in the Ryukyu-Taiwan region indicates present day northward subduction of Philippine Sea plate along an E-W line at latitude 24.0°N, down to a depth of 130 km (Tsai et al., 1977). Wu (1978) in a study of focal mechanism of earthquakes along recently active faults indicated that the Ryukyu arc is terminated at about longitude 123°E by a number of NNW striking right-lateral faults and that the intensity of the very young arc-continental margin collision decreases toward the south, off eastern Taiwan.

The purpose of this paper is to review the various interpretations of the tectonics of Taiwan and attempt to re-organize the geological and geophysical data into a consistent history of tectonic movements in Taiwan.

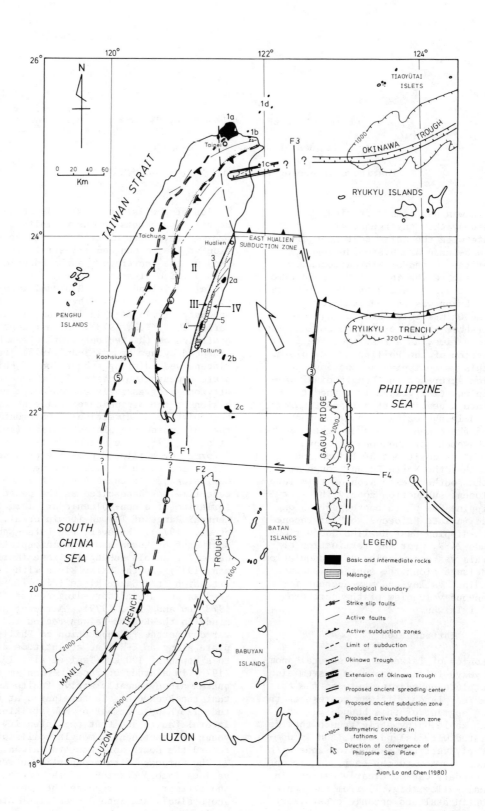

Major Tectonic Provinces of Taiwan

Tectonically, Taiwan can be divided into four provinces: the western foothills, the Central Range, the Longitudinal Valley and the Coastal Range (Fig. 1). They trend nearly N-S coinciding almost exactly with the stratigraphic provinces. The Coastal Range and the Central Range are separated by a striking feature, the Longitudinal Valley (Fig. 1, III), where an active left-lateral strike-slip fault operates along the valley with a N-S displacement of at least 170 km from Hualien to Taitung (Tsai et al., 1977). It is a boundary between the Asian continent and the Philippine Sea plate and shall be discussed in detail in later sections.

The pre-Tertiary metamorphosed basement rocks of schist, crystalline limestone and gneisses exposed on the eastern flank of the Central Range are believed to be of late Paleozoic to Mesozoic age (Yen and Roseblum, 1964). The western flank of the Central Range province is occupied by Paleogene rocks consisting of phyllitic shales, phyllites, slates and semischist of Eocene (53 m.y.) to Oligocene age (Juan et al., 1972) (Fig. 1, II).

The Coastal Range province (Fig. 1, IV), a separate tectonic unit bordering the Philippine Sea, is composed of Neogene marine sedimentary and volcanic rocks (Table 1). In this province, the middle Miocene Tulanshan formation of andesitic lava and pyroclastics is overlain conformably by the Takangkou shale and conglomerates of upper Miocene (Hsu, 1956) and the Chimei formation, a sequence of alternating thin beds of typical turbidites of sandstone and shale (Wang and Chen, 1966).

An allochthonous pile of non-stratified mud and clay with massive exotic blocks, known as Luchi formation and Pliocene in age, is exposed in the southern and southwestern part of the Coastal Range (Ernst, 1977; Ho, 1977). The Pinanshan conglomerate, probably upper Pleistocene, is the youngest formation found in this province.

The Neogene formations in the Coastal Range province have been folded into a series of subparallel anticlines and synclines. Several thrust faults that dip to the east have been recognized. Eruption of andesites and pyroclastics and the exposure of a peridotite-gabbro-dolerite-taiwanite complex in Miocene and Pliocene time in the southern part and glaucophane schist in the northern part, along the strike of the Longitudinal Valley, are the main activities of the province.

In the western foothills province (Fig. 1, I), Miocene rocks of alternating sandstone and shale are widely exposed. They are often coal-bearing in the northern part, but no coal seams are found and shale beds dominate in the southern part of the island. Due to westward migration of the miogeosynclinal axis, the deposition in Neogene time involved a transformation from shore facies at the north to basin facies to the south. Thus a thick mudstone series representing continuous deposition from Miocene up to the Pliocene time is found on southern Taiwan. The province is characterized by tight, parallel and NE-striking flexural-slip folds bounded by thrust faults of the same strike. The imbricated thrust series dipping southeast may indicate that the province is tectonically a superstructure involving only Neogene sediments (Ho, 1976).

The Neogene sediments thin out westward in the Taiwan Strait as evidenced by geophysical subsurface exploration. Small tensional block faulting has been recorded by seismicity (Chiu, 1970). Alkali and tholeiitic basalt flows in the Penghu Island (Chen, 1973; Juan et al., 1979) are thought to have been extruded through the process of early Miocene to Pleistocene rifting in the Taiwan Strait (Juan, 1958).

SYNTHESIS OF THE PACIFIC-TYPE OROGENY

Juan (1958) was the first to recognize Taiwan as a coastal range of continental Asia and indicated that the Tertiary sediments of the island were characterized by miogeosynclinal features in the western continental part and the eugeosynclinal features in the Coastal Range. The eugeosyncline of the Coastal Range was considered as the outer oceanic belt, in contrast to the inner continental belt of the so-called Pacific-type orogeny (Matsuda and Uyeda, 1970), with high P/T type metamorphism and entirely free from granitic intrusions throughout its evolution. Juan and Wang (1971) attempted to correlate the geologic events and tectonic belts of Taiwan with that of the Ryukyu Arc and Luzon Island of the Philippines. The Longitudinal Valley and the Coastal Range of Taiwan were correlated with Cagayen Valley and Sierra Madre of Luzon Island respectively. They recognized the Longitudinal Valley as a facies boundary, reactivated by a transcurrent fault that can be extended to join the mighty Philippine fault as a great dislocation on the border of the western

Fig. 1. A synthesized geotectonic map of Taiwan and its surroundings. Tectonic provinces: I) Western foothills, II) Central Range, III) Longitudinal Valley, IV) Coastal Range. Active fault system: F1) Longitudinal Valley Fault, F2) Philippine Fault, F3) Trace of Central Basin Fault (ancient spreading center), F4) Bashi Fault (transform fault). Proposed tectonic lines: ①) Central Basin Ridge (Central Basin Fault), ②) Gagua Ridge, ③) Extension of Philippine Trench, ④) Extension of Manila Trench (Bowin et al., 1978), ⑤) Extension of Manila Trench (Murphy, 1973). Lithologic units: 1) Pleistocene andesites (1a: Tatun volcano group, 1b: Chilung volcano group, 1c: Kueishantao, 1d: Pengchiahsu, Mienhuahsu and Huapinghsu), 2) Miocene andesites (2a: Chimei, 2b: Lutao, 2c: Lanhsu), 3) glaucophane schist, 4) Kuanshan Igneous Complex, 5) Lichi Melange.

TABLE 1. Stratigraphy of the Coastal Range

Formation	Age	Lithology
Pinanshan conglomerate	Pliocene-Pleistocene	a conglomerate composed largely of pebbles derived from the Central Range
Lichi formation	Pliocene	an allochthonous pile of muddy matrix and massive exotic blocks
Chimei formation	Pliocene	a sequence of alternating thin beds of sandstone and shale
Takangkou formation	late Miocene-Pliocene	a thick clastic series of shale and conglomerate
Tulanshan formation	early Miocene	an agglomerate composed of andesitic lava and pebbles of pyroclastic debris

Pacific. The extension of the Central Basin Fault and the Bashi Fault (Fig. 1), as a transform fault zone, cuts the Ryukyu structures in the north and dislocates the Philippine Fault in the south of Taiwan.

This correlation of the tectonics belts has been received very favorably among Chinese geologists in dealing with the framework of the western Pacific. However, Murphy (1973) postulated that the Manila Trench, a young east-dipping subduction zone was flipped in Pliocene time from a former west-dipping position on the east side of Luzon and Taiwan, where it probably formed a continuous subduction zone with the Ryukyu and Philippine Trenches. He concluded that a Neogene andesitic island arc formed in the eastern offshore of Taiwan on a dormant west-dipping subduction zone. Juan (1975) proposed west-dipping subduction along the Coastal Range under the Asiatic continent assuming westerly movement of the Philippine Sea plate in late Miocene. As the process of convergence at the plate boundary progressed, the Benioff zone progressively approached a vertical position at the beginning of Pliocene time. A collision of Luzon Arc with the continental margin of Asia was visualized by Bowin et al. (1978) during late Miocene and Pliocene. Along Luzon Island itself, a west-dipping subduction zone was formed on the east side during Eocene to early Oligocene. When this activity ceased in late Oligocene, flipping occurred, with east-dipping subduction beginning along the Manila Trench and which continues to be active to the present (Bowin et al., 1978, p. 1647).

A large volume of geological and geophysical data have recently been accumulated which are relevant to the interpretation of the tectonics of Taiwan. Gravity and seismic data (Tsai and Liu, 1977; Lu and Wu, 1974) clearly show a discontinuity in crustal and upper mantle structure beneath the Longitudinal Valley and the Coastal Range. The velocity structure under the main part of Taiwan is of normal continental crust with a mantle velocity of 7.75 km/sec. While the crust under the Coastal Range lacks the 3A layer of Pacific Ocean structure with a velocity of 6.8 km/sec, it shows a shallow 8.0 km/sec velocity typical of upper mantle as found in the oceans (Shor et al., 1970).

The seismicity of Taiwan has been carefully documented in historic and recent years. Instrumentally determined large (M > 6) earthquakes from 1900 to 1976 make it evident that eastern Taiwan is highly active (Wu, 1978, Fig. 6b). The plot of 1962-1974 hypocenters shows that toward the south end of the Philippines, the maximum depth of foci reaches 650 km but becomes shallower going north. Beneath Luzon, the maximum depth of foci is 250 km and further north toward Taiwan, between latitudes of $23°N$ and $24.2°N$, along the whole length of the Coastal Range, only earthquakes shallower than 70 km have occurred. It is clear that the shallow seismicity shows no tendency of forming either an east-dipping or a west-dipping trend of propagation (Wu, 1978, Fig. 7b). Many investigators must have been misled with wrong information by assuming east-dipping subduction along the Coastal Range in Plio-Pleistocene time (such as profiles 3 and 4 in Fig. 6 cited by Murphy, 1973; Chai, 1972; Bowin et al., 1978). It appears also that the shallow seismicity along the manila Trench, according to focal mechanism studies, terminates at about latitude $21°N$, and there exists a gap between that point and Taiwan (Wu, 1978, Fig. 8). The gap is exactly where the Bashi Fault should be located, between the magnetic and free-air gravity anomaly profiles 6-7 and 8-9 in Fig. 4 of Bowin et al.'s paper (1978). Thus the Manila Trench does not extend further north to reach Taiwan.

The correlation of free-air gravity anomalies indicates that the free-air minimum along the West Luzon and North Luzon Troughs terminate at the Bashi Fault (Fig.1, F4) and can be inferred to lie above the location of the plate contact between the Philippine Sea and South China Sea plates (Bowin et al., 1978) or Philippine Sea plate and the Asian continental margin. It is evident that both could be interpreted as the trace of a fossil subduction zone.

Between Luzon and Taiwan there is a submarine volcanic ridge dotted with islands. The Lutao and Lanhsu islands (Chen, 1976) (Fig. 1) situated on the ridge can be extended northward to join the Chimei volcanic group (J.C. Chen, 1975) on the Coastal Range of Taiwan, and southward to the Babuyan Island group just north of Luzon. Karig (1973, Fig. 6b) favored the formation of this volcanic ridge in late Miocene while Bowin et al., (1978, Fig. L) concluded that it should be much younger. However, the andesitic rocks of Chimei have been radiometrically dated at 22.2 to 17 m.y. and are mainly Miocene (Ho, 1969). The Palau Ridge, named by Karig (1973), is a north-trending structure near longitude $123°E$, southeast of Taiwan, that continues unbroken into the eastern flank of the Chimei-Lutao-Lanhsu Volcanic Arc. A prominent ridge segment north of about latitude $21°N$ and north of the Palau Ridge was re-named Gagua Ridge because, though the alignment is good, the structural continuity of the two segments is doubtful. Gagua Ridge is considered to be an extinct spreading center of probably late Eocene to Miocene time (Bowin et al., 1978; Lu and Liaw, 1979) with recent earthquakes of well defined strike-slip solutions (Wu, 1978, Events 14, 20 and 21). Since Gagua Ridge and its extention, Palau Ridge, can be considered as a northern extension of the Philippine Ridge on the east of Luzon Island (Irving, 1950), a straight forward explanation of the formation of Chimei-Lutao-Lanhsu Volcanic Arc is now proposed. It is, apparently, the result of subduction of Philippine Sea plate beneath the continental margin of Asia during an earlier phase of convergence with Taiwan, as was visualized by Karig (1973, Fig. 6a).

Based on paleomagnetic studies of DSDP sediments and the phase shifting of magnetic anomalies, Louden (1977) suggested a $60°$ clockwise rotation of the west Philippine Basin since Miocene time. A new plate boundary likely developed to the west associated with subduction zone migration in response to the northwesterly movement of the Philippine Sea plate. This is recognized as the second episode of subduction, in late Miocene to Pliocene time, as evidenced by the occurrence of the Kuanshan Igneous Complex (Juan et al., 1976, 1978; Liou et al., 1976, 1977; Chou et al., 1978) and glaucophane schist (Juan et al., 1972; Jahn and Liou, 1977) along the Longitudinal Valley. The conspicuous magnetic anomalies extending southward and southeastward from the east coast of Taiwan to the Luzon trough (Hu and Lu, 1979) indicate an important location of the boundary between the Philippine Sea and the East China Sea plates.

Wu (1978) studied the microearthquakes in the vicinity of Taiwan and indicated there is a compressive tectonic stress to the southeast of Taiwan in a direction of $S46°E$ to $S76°E$ with a plunge between $-2°$ and $15°$ to the SE. The events in or around the Longitudinal Valley produced solutions with significant strike-slip components, consistent with an NNE striking left-lateral fault, the only recognized fault on the eastern side of the Longitudinal Valley (Wu, 1978, Fig. 16; Tsai et al., 1977; York, 1976).

Several thousand earthquakes above magnitude 2.0 have been reliably located since 1972. A large number of earthquakes occurred in a planar zone about 50 km thick with a dip of $45°-50°$ northward from an E-W line at about latitude $24°N$ and down to a depth of 130 km (Tsai et al., 1977). This subduction of Philippine Sea plate along a north-dipping Benioff zone was interpreted as the cause of Plio-Pleistocene andesitic lavas at Tatun volcanoes and others in the north offshore of Taiwan (C.H. Chen, 1975; Huh and Chen, 1978; Chang and Chen, 1979; Chen and Lin, 1979). The northwestward advancing Philippine Sea plate at the same time compressed against the east coast of Taiwan producing the Lichi Melange that is distributed along the collision suture of the Longitudinal Valley (Wang, 1976; Hsu, 1976; Ernst 1977; Liou et al., 1977). Tsai (1978) proposed that the Lichi Melange was formed side-by-side with "Kenting Melange" at the southern tip of the island when the Coastal Range was located 150 km south of its present position and before the northward East Hualien subduction started. However, the lithologic composition of the exotic blocks of these two melange units is different and the socalled "Kenting Melange" could possibly have resulted from large-scale submarine gravitational sliding (Ho, 1977). It is impossible to visualize that the Coastal Range could have moved such a long distance in Plio-Pleistocene time and in fact the proposal is indeed contrary to the geologic make-up of Miocene eugeosynclinal nature of the Coastal Range.

A microseismicity study in the Ilan Plain on the northeastern coast of Taiwan has shown that a subsurface vertical fault of no deeper than 20 km extends from the Ilan Plain northeasterly into the Okinawa Trough passing through the island of Kueishantao (Fig. 1). It is apparently the extension of the Median Fault that forms the boundary approximately between the tectonic units of Central Range and the foothills (Tsai et al., 1975). However, Bowin et al. (1978) suggested that this may represent a seismicity zone of the Okinawa Trough extending on shore to Taiwan to form the Ilan Plain. Wang and Hilde (1973) inferred the Okinawa Trough to be an early stage in the development of a marginal sea and traced a magnetic anomaly of Ryukyu Arc to approach very near Taiwan. They interpreted the residual magnetic contours between longitude $122°E$ and $123°E$

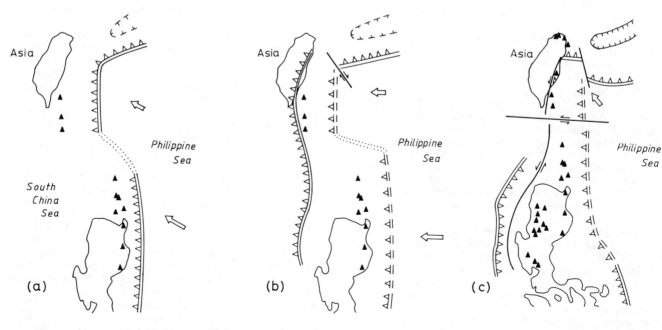

LATE EOCENE – EARLY MIOCENE   LATE MIOCENE – PLIOCENE   LATE PLIOCENE – PLEISTOCENE

Fig. 2 Tectonic history of the Ryukyu-Taiwan-Luzon Arc System. (a) Late Eocene-early Miocene: An east-facing arc system with a subduction zone offshore of East Taiwan forming the Chimei-Lutao-Lanhsu Volcanic Ridge. (b) Late Miocene-Pliocene: Migration of west-dipping subduction zone to the Longitudinal Valley in Taiwan and to the North Luzon Trough. (c) Late Pliocene-Pleistocene: East Hualien Subduction Zone joined with the Ryukyu Arc by the trace of Central Basin Fault. Initiation of east-dipping Manila Trench subduction in South China Sea.

to be the result of a major fault offset. It seems that the presence of a right-lateral fault (Wu, 1978, Fig. 18) would cause difficulties in extending the Okinawa Trough all the way to the Ilan Plain. If this fault terminates at the East Hualien Subduction Zone, it would be a trench-trench-transform fault, and the Okinawa Trough may extend into Ilan Plain.

Tectonic Evolution of Taiwan in the Ryukyu-Taiwan-Luzon Arc System

The Cenozoic tectonics in the western Pacific, especially for the Ryukyu-Taiwan-Luzon Arc System must deal with the interaction of the Asiatic continent and the Philippine Sea plate. The early creation of island-arc systems along the western Pacific region is generally believed to be of late Mesozoic time (Jahn et al., 1976). Taiwan, a segment of the Ryukyu-Taiwan-Luzon Arc System, has been greatly modified structurally by the convergence of the Philippine Sea plate in the offshore area of Taiwan for the first phase of evolution in late Eocene to early Miocene time as evidenced by andesitic volcanism. Subduction of Philippine Sea plate beneath East Taiwan in the second phase in late Miocene to Pliocene time is indicated by the presence of Kuanshan Igneous Complex and the glaucophane schist. An East Hualien northward subduction zone was initiated in late Pliocene to Pleistocene time when the Philippine Sea plate moved in a northwesterly direction, thus completing the present configuration of the whole system (Fig. 2).

An east-facing arc system existed along the eastern edge of Luzon in the early Tertiary, as evidenced by the presence of peridotite-gabbroic sheets and other related rocks in the basement of the Sierra Madre Mountains (Irving, 1950). The northward extension of this system has been recognized in recent magnetic lineation studies along the Gagua Ridge of Eocene (Bowin et al., 1978), representing possibly a segment of the Central Basin Ridge (Lee and Hilde, 1971) before the movement of the Philippine Sea plate turned in a northwesterly direction. Arc-type volcanics, the Chimei-Lutao-Lanhsu andesite ridge (Chen, 1976) formed accompanying the subduction near Taiwan (Fig. 2, a). However, Karig (1973) stated that subduction of this east-facing arc system ended in early Miocene time, possibly based on the cessation of andesitic volcanism.

A second phase of west-dipping subduction, at the collision suture of Longitudinal Valley in Taiwan, which started in late Miocene time (Louden, 1977), could have been accomplished by the simple rotation of direction of plate motion without resorting to polarity reversal. Owing to the uplifting of the whole island of Taiwan during Pliocene (Lee, 1977; Peng et al., 1977) and the difficulties of underthrusting of the sediments of the Coastal Range, the Benioff zone gradually turned to a nearly vertical position; meanwhile facilitating the squeezing up the glaucophane schist slab (Jahn and Liou, 1977) and the Kuanshan Igneous Complex (Chou et al., 1978) along this Valley from the basement. The finding of nannofossils of early to middle Miocene age in the red shale cementing matrix of the plutonic breccias in the Kuanshan Igneous Complex not only fixed the age for the Complex as a part of the basement but also implied that from the biochronological study the Complex is indeed representing a tropical Pacific Ocean crust (Huang et al., 1979). As both the Longitudinal Valley Fault and the Philippine Fault along the Luzon Trough are left-lateral strike-slip faults in present-day configuration, the latter fault may be the trace of the boundary between the Philippine Sea plate and the South China Sea plate, that is, the trace of late Miocene to Pliocene west-dipping subduction, a continuation of the Longitudinal Valley and East Taiwan west-dipping subduction at that time (Fig. 2, b).

In the last phase of Philippine Sea plate convergence, there was a northwesterly inter-plate shortening in the Ryukyu-Taiwan-Luzon Arc System as a result of subduction of the Philippine Sea plate. If we take the down-going slab of the Philippine Sea plate at the East Hualien subduction zone at a present rate of 5-6 cm/year (Tsai et al., 1977) and also consider that it was responsible for supplying material for the eruption of the northern volcanic groups, such as Tatun volcanoes and others, then the inception of subduction would have to be not later than Pliocene time. This necessitates a slower rate of movement along the strike-slip fault in the Longitudinal Valley during Pliocene time to accomodate the subduction depth of 130 km. The Lichi Melange of the Coastal Range which contains mixed Miocene and early Pliocene faunas are materials formed from large-scale deformation of Central Range since the late Pliocene. The required shortening for this deformation has undoubtedly occurred across the Longitudinal Valley by the approach of the Chimei-Lutao-Lanhsu ridge to the east coast of Taiwan (Fig. 2, c). However, neither seismic reflection profiles nor earthquake epicentral plots reveal any subduction beneath the Longitudinal Valley. In the Central Range and the foothill region there is an imbricate thrust system, subparallel to the fold axes which could have been initiated by a west to east compression and is believed to involve only the Neogene sedimentary cover separated from the underlying rocks by a decollement zone (Ho, 1976). Such a structural setting is not compatible with the proposed extension into western Taiwan of the Manila Trench (Murphy, 1973; Bowin et al., 1978), which belongs to an entirely different tectonic regime.

Further constraints on the northward prolongation of the Manila Trench are the occurrence of the Bashi Fault in Pliocene time (Figs. 1 and 2) and the doubtful association of the "Kenting Melange" to subduction of the South China Sea plate. In fact, since Pliocene, the southern part of Taiwan has experienced rapid uplifting, much more than at the north (Lee, 1977; Peng et al., 1977). The striking similarity of uplift history between northern Taiwan and the Ryukyu Islands indicates the consistency of the observation that Taiwan is in the convergence zone between the Asian and Philippine Sea plates.

References

Big, C., A fossil subduction zone in Taiwan, Proc. Geol. Soc. China, 14, 146-154, 1971.

Bowin, C., S.R. Lu, C.H. Lee and H. Schouten, Plate convergence and accretion in Taiwan-Luzon region, Am. Asso. Petrole. Geol. Bull., 62, 1645-1672, 1978.

Chai, B.H.T., Structure and tectonic evolution of Taiwan, Am. J. Sci., 272, 389-422, 1972.

Chang, M.T. and J.C. Chen, Geochemistry of andesites from Kueishantao, Acta Oceano. Taiwan., 9, 39-49, 1979.

Chen, C.H., Petrological and chemical study of volcanic rocks from Tatun volcano group, Proc. Geol. Soc. China, 18, 59-72, 1975.

Chen, J.C., Geochemistry of basalts from Penghu Islands, Proc. Geol. Soc. China, 16, 23-36, 1973.

Chen, J.C., Geochemistry of andesites from the Coastal Range, eastern Taiwan, Proc. Geol. Soc. China, 18, 73-88, 1975.

Chen, J.C. Geochemistry of Lanhsu andesites, Acta Oceano. Taiwan., 6, 77-87, 1976.

Chen, J.C. and P.N. Lin, Geochemistry of andesites from Pengchiahsu, Acta Oceano. Taiwan., 10, 132-144, 1979.

Chiu, H.T., Structural features of the area between Hsinchu and Taoyuan, northern Taiwan, Proc. Geol. Soc. China, 13, 63-75, 1970.

Chou, C.L., H.J. Lo, J.H. Chen and V.C. Juan, Rare earth element and isotopic geochemistry of Kuanshan Igneous Complex, Taiwan, Proc. Geol. Soc. China, 21, 13-24, 1978.

Ernst, W.G., Olistrostromes and included ophiolitic debris from the Coastal Range of eastern Taiwan, Mem. Geol. Soc. China, 2, 97-114, 1977.

Ho, C.S., Geological significance of potassium-argon ages of the Chimei igneous complex in eastern Taiwan, Bull. Geol. Surv. Taiwan, 20, 63-74, 1969.

Ho, C.S. Foothill tectonics of Taiwan, Bull. Geol. Surv. Taiwan, 25, 9-28, 1976.

Ho, C.S., Melange in the Neogene sequence of Taiwan, Mem. Geol. Soc. China, 2, 85-96, 1977.

Hsu, T.L., Geology of the Coastal Range, eastern Taiwan, Bull. Geol. Surv. Taiwan, 8, 39-63, 1956.

Hsu, T.L., Neotectonics of the Longitudinal Valley, eastern Taiwan, Bull. Geol. Surv. Taiwan, 25, 53-62, 1976.

Hu, C.C. and R.S. Lu, Downward continuation of magnetic field and the magnetic anomalies of offshore Taiwan, Acta Oceano. Taiwan., 9, 1-8, 1979.

Huang, T.C., M.P. Chen and W.R. Chi, Calcareous nannofossils from the red shale of the ophiolite-melange complex, eastern Taiwan, Mem. Geol. Soc. China, 3, 131-138, 1979.

Huh, C.A. and J.C. Chen, Geochemistry of dacites from Keelungtao, northern Taiwan, Acta Oceano. Taiwan., 8, 63-79, 1978.

Irving, E.M., Review of Philippine basement geology and its problem, Phili. J. Sci., 79, 267-307, 1950.

Jahn, B.M., Reinterpretation of geologic evolution of the Coastal Range, east Taiwan, Bull. Geol. Soc. Am., 83, 241-248, 1972.

Jahn, B.M. and J.G. Liou, Age and geochemical constraints of glaucophane schist of Taiwan, Mem. Geol. Soc. China, 2, 129-140, 1977.

Juan, V.C., Continental rifting and igneous activities in the Neogene marginal geosynclines of Taiwan, Proc. Geol. Soc. China, 1, 27-36, 1958.

Juan, V.C., Tectonic evolution of Taiwan, Tectonophysics, 26, 197-212, 1975.

Juan, V.C., T.J. Chou and H.J. Lo, K-Ar ages of the metamorphic rocks of Taiwan, Acta Geol. Taiwan., 15, 113-118, 1972.

Juan, V.C., H.J. Lo and C.H. Chen, Crystallization-differentiation of taiwanite, Proc. Geol. Soc. China, 19, 87-97, 1976.

Juan, V.C., H.J. Lo and C.H. Chen, Petrochemistry and origin of taiwanite and dolerite, east Taiwan, in Studies and Essays in Commemoration of Golden Jubilee of Academia Sinica, pp. 71-101, 1978.

Juan, V.C., H.J. Lo and C.H. Chen, Genetic relationship of the Neogene alkali and tholeiitic magmas and the nature of the upper mantle beneath the continental shelf of the western Taiwan, Proc. Geol. Soc. China, 22, 24-38, 1978.

Juan, V.C. and Y. Wang, Taiwan in relation with the tectonic framework of the western Pacific, Acta Oceano. Taiwan, 1, 1-14, 1971.

Karig, D.E., Plate convergence between the Philippines and the Ryukyu Islands, Marine Geol., 14, 153-168, 1973.

Karig, D.E. and J.M. Wageman, Structure and sediment distribution in the northwest corner of the west Philippine basin, in Initial Report of the DSDP, Vol. 31, Washington D.C., U.S. Govt. Printing Office, pp. 615-620, 1975.

Lee, C.S. and T.W.C. Hilde, Magnetic lineations in the western Philippine Sea, Acta Oceano. Taiwan, 1, 69-76, 1971.

Lee, P.J., Rate of the early Pleistocene uplift in Taiwan, Mem. Geol. Soc. China, 2, 71-76, 1977.

Liou, J.G., C.Y. Lan, J. Suppe and W.G. Ernst, The east Taiwan ophiolites: occurrence, petrology, metamorphism and tectonic setting, Min. Res. Serv. Organ. Spe. Rept., 1, 1977.

Liou, J.G., J. Suppe and W.G. Ernst, Conglomerates and pebbly mudstones in the Lichi Melange, eastern Taiwan, Mem. Geol. Soc. China, 2, 115-128, 1977.

Louden, K.B., Paleomagnetism of DSDP sediments, phase shifting of magnetic anomalies and rotation of the west Philippine basin, J. Geophys. Res., 82, 2989-3002, 1977.

Lu, C.P. and F.T. Wu, Two-dimensional interpretation of a gravity profile across Taiwan, Bull. Geol. Surv. Taiwan., 24, 125-132, 1974.

Lu, R.S. and J.Y. Liaw, Origin of the Gagua Ridge, Proc. Geol. Soc. China, 22, 9-23, 1979.

Matsuda, T. and S. Uyeda, On the Pacific-type orogeny and its model-extension of the paired belts concept and possible origin of marginal seas, Tectonophysics, 11, 5-27, 1970.

Murphy, R.W., The Manila Trench-West Taiwan foldbelt: A flipped subduction zone, Geol. Soc. Malaysia Bull., 6, 27-42, 1973.

Peng, T.H., Y.H. Li and F.T. Wu, Tectonic uplift rates of the Taiwan Island since the early Holocene, Mem. Geol. Soc. China, 2, 57-69, 1977.

Shor, G.G., Jr., H.W. Menard and R.W. Raitt, Structure of the Pacific Basin, in The Sea, vol. 4, Pt. II, pp. 3-28, Wiley-Interscience, 1970.

Tsai, Y.B., Plate subduction and the Plio-Pleistocene orogeny in Taiwan, Petrole. Geol. Taiwan, 15, 1-10, 1978.

Tsai, Y.B., C.C. Feng, J.M. Chiu and H.B. Liaw, Correlation between microearthquakes and geologic faults in the Hsintien-Ilan area, Petrole. Geol. Taiwan, 12, 149-167, 1975.

Tsai, Y.B. and H.L. Liu, Spatial correlation between hot springs and microearthquakes in Taiwan, Petrole. Geol. Taiwan, 14, 263-278, 1977.

Tsai, Y.B., T.L. Teng, J.M. Chiu and H.L. Liu, Tectonic implications of the seismicity in the Taiwan region, Mem. Geol. Soc. China, 2, 13-41, 1977.

Wang, C. and T.W.C. Hilde, Geomagnetic interpretation of the geologic structure in the northeast offshore region of Taiwan, Acta Oceano. Taiwan., 3, 141-156, 1973.

Wang, C.S., The Lichi Formation of the Coastal Range and arc-continent collision in eastern Taiwan, Bull. Geol. Surv. Taiwan, 25, 73-86, 1976.

Wang, C.S. and T.T. Chen, Turbidite Formations around the southern plunge of the Eastern Coastal Range near Taitung, Proc. Geol. Soc. China, 9, 46-54, 1966.

Wu, F.T., Recent tectonics of Taiwan, J. Phys. Earth, 26, S265-S299, 1978.

Yen, T.P., Plate Tectonics in the Taiwan region, Proc. Geol. Soc. China, 16, 7-22, 1973.

Yen, T.P. and S. Rosenblum, Potassium-argon ages of micas from the Tananao schist terrain of Taiwan--a preliminary report, Proc. Geol. Soc. China, 7, 80-81, 1964.

York, J.E., Quaternary faulting in eastern Taiwan, Bull. Geol. Surv. Taiwan, 25, 63-72, 1976.

COMPLICATIONS OF CENOZOIC TECTONIC DEVELOPMENT IN EASTERN INDONESIA

John A. Katili and H.M.S. Hartono

Ministry of Mines & Energy, Jakarta, and Geological Research & Development Centre, Bandung, Indonesia

Abstract. The tectonic development of the Indonesian archipelago as the southeastern margin of the Eurasian plate can be followed since Late Paleozoic from a continental nucleus which was located approximately between Sumatera and Kalimantan. The archipelago has developed eastward until it attained the present position as represented by the Banda volcanic arc. Generally the development shows growth towards the fore-arc, although some deviation occurred with minor development backward. The tectonic process is here considered as an active marginal development as it has always been associated with volcano-plutonic and subduction processes. The latest tectonic development of eastern Indonesia with the presently active Banda arc is considered as part of the development of the Sunda arc system.

Eastern Indonesia is the site of a megatriple junction where the eastern part of the Sunda arc trench system, as part of the Greater Eurasian plate, is affected by the combined process of the northward drift of the Australian plate and the westward thrust of the Pacific plate. Thus the complication of the tectonic process exists mainly in eastern Indonesia and occurred throughout the Cenozoic.

During the late Paleozoic and throughout the Mesozoic the development of the Sunda Arc system is considered as regular and always had an arcuate shape of volcanic arc around the continental margin. The tectonic processes affected the evolution of the islands of Sumatera, Kalimantan and part of Java.

During the Tertiary the complication began with the birth of the Sumatera, Java, Lesser Sunda and Banda Arc Tertiary volcanics. Another Tertiary volcanic arc was formed in West Sulawesi. The complication was amplified by the northward drift of the India - Australian plate and westward thrust of the Pacific plate. As a consequence huge transcurrent faults were created, such as the Sorong fault, Palu - Koro fault, Matano fault and the supposed Sumba transform. As a result of this large translational movement, or probably by detachment processes, small continental slivers or part of island arcs, such as Sumba, Sula, Banggai, Bacan, Buru and Buton, were transported to geologically unrelated neighbouring terrains in certain parts of Eastern Indonesia. Several types of collisions may be observed: oceanic plate - island arc, island arc - continental plate and island arc - island arc collisions.

Introduction

It was recognized decades ago that the Indonesian island arc represents a nascent mountain belt. It was also realized that the archipelago possesses a dual character, namely as the place of intersection of the Alp and Circum Pacific orogenic belts and as an intra continental zone between Australia and Eurasia which is known to have been occupied by the ancient Thethys ocean. This ocean, the development of the southeastern margin of Eurasia, and northward drift of the Australian continent together with westward drift of the Pacific plate are responsible for the present shape of the archipelago.

Proponents of the new global tectonic theory find that the Indonesian archipelago can be regarded as a natural laboratory. Many workers have carried out research in this region, supported also by marine geological and geophysical cruises as well as land geological studies. Many reports have been published expressing views on the tectonic development of the archipelago. Nevertheless, many problems remain unsolved.

This presentation expresses the authors' views of the tectonic development of the archipelago, which will, we believe, contribute to the present understanding of the tectonic development of the archipelago.

Figure 1 shows the area discussed in this paper and important faults.

Regular tectonic development during Pre-cenozoic

The following Pre-Cenozoic tectonic have occurred at the following times in western Indonesia : (1) Late Carboniferous - Early Permian,

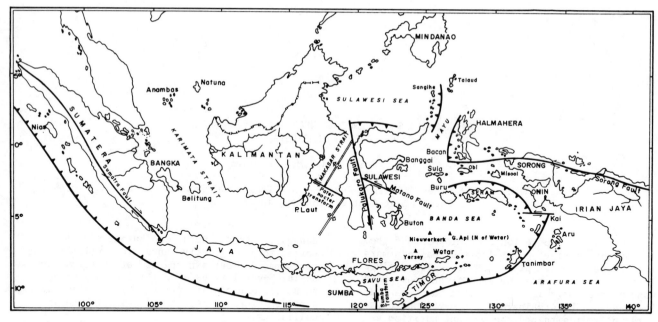

Fig. 1 Area discussed in this paper and important faults

(2) Permian - Early Triassic, (3) Late Triassic-Jurassic, (4) Cretaceous - Early Tertiary. A simplified geometry of Pre-Cenozoic subduction zones in West Indonesia is given in Figure 2.

The shapes of the arcs are quite prominent and it seems that the arcs had acquired their arcuate shape around the Sunda continental margin by Carboniferous time, and maintained it through the Permian and Mesozoic to the present day.

It seems that there is no significant shift of past subduction position from the Carboniferous to the Jurassic. The position was located along the crest of Sumatera, Java Sea, South and West Kalimantan, Anambas and Natuna Island.

The following table shows the correlation of Late Paleozoic-Mesozoic tectonic events with the emplacement of ophiolites and melanges as a result of subduction and with corresponding intrusive and extrusive igneous events.

Development of perpendicular arc trench system during Early Tertiary

During the Tertiary a significant change in the arc trench system took place. The Cretaceous - Early Tertiary arc, called the Sunda - Banda arc, continued to develop normally while a new arc trench system, the Sulawesi arc, came into being perpendicular to it (Fig. 3). The Sunda - Banda Tertiary subduction zone coincides with the Sunda - Banda nonvolcanic outer arc comprising the islands off West Sumatra, submarine ridge south of Java, Timor, Tanimbar, Seram, Buru and Buton, whereas the magmatic arc approximately coincides with the present position of the Sunda-Banda volcanic inner arc (fig.4). The Tertiary Sulawesi magmatic arc extends from West Sulawesi to Mindanao, whereas the Tertiary Sulawesi subduction zone follows the Talaud ridge, submarine Mayu ridge and East Sulawesi. The extent of the subduction zone is confirmed by the occurence of ophiolites or melanges in the islands mentioned, whereas the extent of the magmatic arc is confirmed by dated granites and volcanics (Figs. 4 and 5).

It is postulated that the subducting plates at the Tertiary Sunda - Banda subduction zone had its origin in a spreading center situated in the Indian ocean and that at the Tertiary Sulawesi subduction zone had its origin in the spreading center situated in the Pacific ocean.

The Neogene volcanics in central Kalimantan need some explanation because of their location far in the hinterland of the Sulawesi arc. It is postulated that these volcanics were not generated by the Tertiary Sulawesi arc system, but instead by subduction from the north. Fig.3 includes a subduction zone in north Kalimantan facing towards the northwest and a corresponding volcanic arc in the interior part of Kalimantan.

The Tertiary volcanics north of the Bird's Head and West Halmahera, and accompanying ophiolites occuring in Waigeo and East Halmahera are not related to the island arc skirting the Sunda margin; they belong to the northern New Guinea collision zone.

Complication by northward drift of the Australian continent

From geophysical data it is evident that Australia drifts northward at an approximate rate of 70 mm/yr (70 Km/Ma) in the vicinity of Timor. This movement of Australian plate caused collision of

Tabulation of Tectonic Development of Sunda Arc during
Late Paleozoic - Mesozoic

Time	Subduction	Volcanics/granites
Late Carboniferous - Early Permian	-	Padang Highland, Batang Sangir, Jambi, West Kalimantan and East Malaya volcanics, East Malaya granite belt, Granites from Jambi (298 m.y.).
Permian - Early Triassic	Natuna, Kembayan and Kapuas ophiolites	Volcanics in Sumatra & West Kalimantan, Trengganu & Pahang, granites in East Coast and Main Range of Malaya, Singapore, South Johor, Jambi (251 m.y.). North of Bangka (276 m.y.) and Biliton (278 m.y.).
Late Triassic - Jurassic	Gumai, Garba, Aceh and Natuna ophiolites	Serian, Matan & Ketapang volcanics; granites from Singkep (155 m.y.), Berhala (167 m.y.) and Biliton (180 m.y.).
Cretaceous - Early Tertiary	Ophiolites in island off West Sumatra, Melange in Jampang and Luk Ulo, ophiolites of Meratus and Pulau Laut	Granites from : Kotanopan (89 m.y.), Lassi (112 m.y.), North of Jakarta (100 m.y.), North of Madura (100 m.y.), Natuna (74 m.y.), Anambas (87 m.y.) and West Java (58 m.y., 94 m.y. and 65 m.y.).

Australia with the existing island arc north of it, i.e. the Sunda - Banda Island Arc system.

Indian Ocean plate-Sunda Island Arc convergence

This convergence began in Late Paleozoic time and persisted until the present day as evident from active volcanism and plutonism in the chain of islands comprising the Sunda Inner Arc islands (Sumatra, Java, Bali, Lombok, Sumbawa, Flores and Lomblen). The position of the Tertiary subduction zone more or less coincides with the position of the present volcanic inner island arc, whereas the present subduction zone is positioned slightly towards the fore-arc and coincides with the Java trench and its submarine topographic continuation to the west and east.

Attention should be given to a transition zone within the Sunda - Banda volcanic inner arc which is situated approximately in the eastern part of Flores Island. This transition is a consequence of the different nature of the converging plate, which is oceanic in the west and continental in the east. This difference results in discontinuous physical expression longitudinally along the arc at the transition zone. Nishimura et al. (1979) indicated that the discontinuity involves the offsets of gravity anomaly, distribution pattern of earthquakes, submarine morphology, strike of the active volcanic zone, chemical character of the Quaternary volcanic rocks and other geologic setting.

It is postulated that the northward displacement from the normal East-West trend of the submarine volcanoes Nieuwerkerk, Jersey and Gunung Api North of Wetar is a result of the difference of converging plate accommodated by the sinistral north-south trending Sumba transform fault. It could also be postulated that this transform may join the Palu-Koro fault which has the same trend and sense of displacement.

Although some discontinuous characteristics are present at the transition zone, in general it can be considered that the Sunda arc continues eastward to the Banda arc.

Banda Island Arc - Australian continental crust collision

The present actual collision is between the South Banda island arc system and the Australian continental crust. Prior to continental crust col-

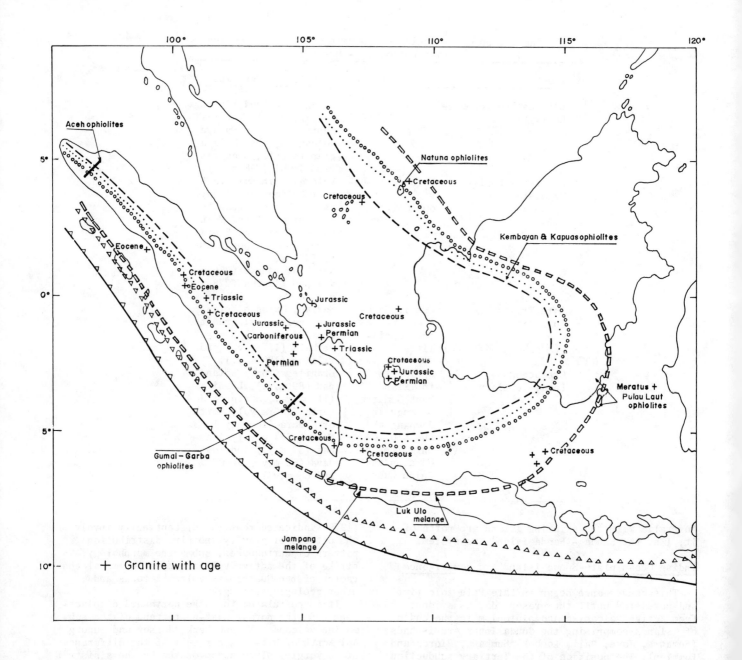

Fig. 2 Lineament of subduction zones

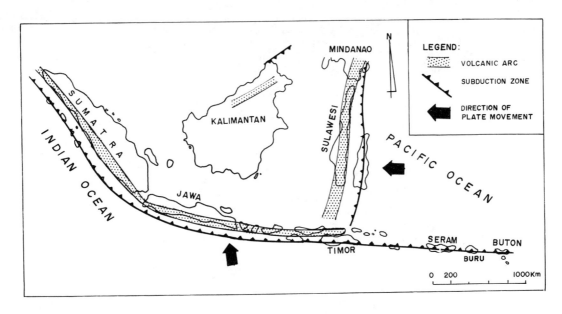

Fig. 3 Evolution of the Indonesian island arcs in early Tertiary time (Katili, 1974)

lision oceanic crust was subducted under the Banda Arc (Fig. 6).

An exhaustive discussion and debate on the geology of the collision complex resulting from the collision of the Banda Island Arc system and the Australian continental crust can be observed in literature. The debates involve the position of subduction zone which has implications for the interpretation of the origin of the stratigraphic units occurring in the Banda nonvolcanic outer arc.

Based on recent data such as seismic reflection, seismic refraction, gravity and the onshore geology of Timor, we place the locus of subduction south of Timor at the trough. Timor is underlain by Australian crust with a structural break separating the Australian shelf sediments from the accretionary wedge and structurally complicated rocks now occurring in the outer Banda arc. Audley Charles et al. (1975) and Chamalaun & Grady (1978) believe that the position of subduction is between the

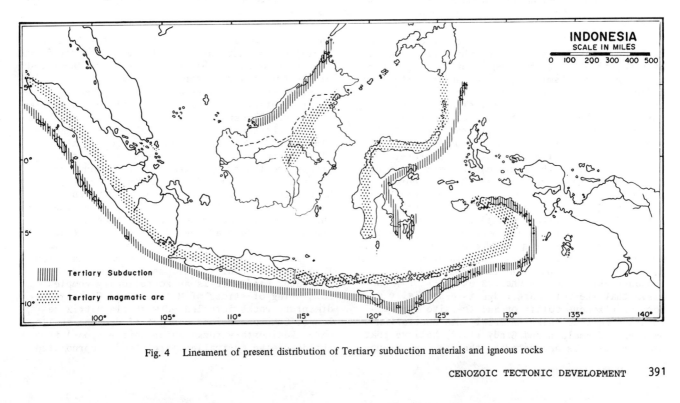

Fig. 4 Lineament of present distribution of Tertiary subduction materials and igneous rocks

CENOZOIC TECTONIC DEVELOPMENT   391

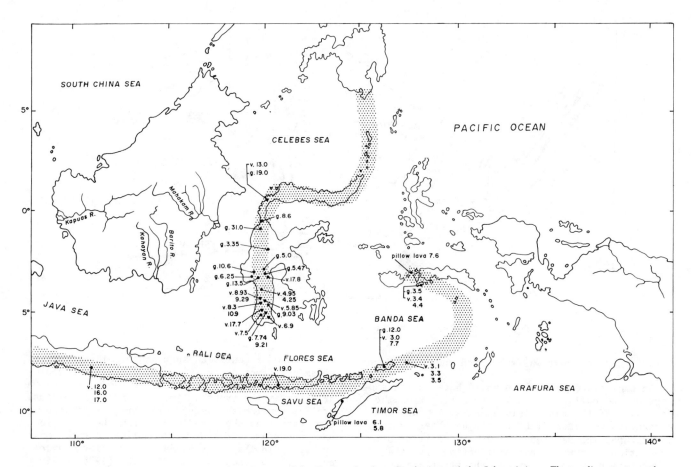

Fig. 5. Radiometric data supporting the existence of the Tertiary Sunda — Banda Arc and the Sulawesi Arc. The two lineaments are the Tertiary Sunda — Banda Magmatic Arc and the Tertiary Sulawesi Magmatic. The extension to North Sulawesi, Sangihe and Mindanao is inferred based on the presence of old volcanics. v= volcanics+absolute age in million years; g=granite+absolute age in million years.

outer and inner Banda Arcs (between Timor and Wetar). Hamilton (1979) believes that the position of subduction is at the trench outside the outer arc. Although seismic refraction and gravity data indicate that continental crust is present beneath the Timor trough, seismic reflection data crossing the trough show that Australian continental shelf sediments dip under the accretionary wedge of the opposing side of the trough. The position of subduction has an implication for the interpretation of the geology of the nonvolcanic outer arc ridge. Position at the trough implies that the outer arc ridge is a subduction melange, an idea held by Hamilton (1979), Katili (1974,1975) and others. Subduction at inter-arc position (between outer and inner arc) implies the presence of Australian material at the outer arc ridge and further interpretation on the origin of the stratigraphic units as autochthonous, par -autochthonous and allochthonous series. Audley Charles et al. (1975) believe that the outer arc ridge is essentially a zone of Pliocene collision and the site of overthrust sheets of Asian material onto Australian basement. Chamalaun and Grady (1978) believe that overthrusting as advocated by the previous group does not occur in Timor. Their structural interpretation is that normal faulting has occurred and the whole stratigraphic units have been deposited at the Australian continental shelf.

Based on the results of seismic refraction, showing outer-arc ridge is underlain by Australian continental crust and of seismic reflection, showing observed down going slab under the accretionary wedge at the trough, Jacobson et al. (1978) support the idea that subduction is at the trough and the trough lineation represents the surface trace of a subduction zone (Fig. 6). Later, in their interpretation of the structure of Seram, Audley Charles et al. (1979) reconsidered the problem of the subduction position and gave two zones, A-zone at the trench and B-zone at inter-arc position (Fig. 7).

According to the imbrication model, (Hamilton, 1979), the tectonic development of the ridge of the Outer Banda Arc is an accretionary complex consisting of slices of Mesozoic and Cenozoic sedimentary rocks interleaved with claymatrix melange. Coherent slices consist of deep to shallow water sedimentary rocks which were stripped from the downgoing Australian margin and incorporated

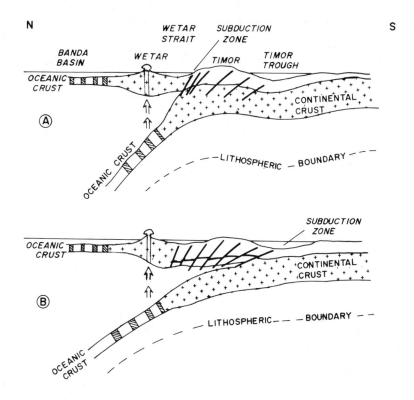

Fig. 6A Schematic crustal cross-section across Timor according to overthrust model (Carter et al., 1976).

Fig. 6B Schematic crustal cross-section across Timor with Timor Trough as surface trace of Subduction (Jacobson et al., 1978).

into the imbricate wedge, as well as pelagic sediments of oceanic origin. The sheared, scaly-clay melange contains blocks and lenses of all sizes up to tens of kilometers long. It is considered as an integral part of the deformed accretionary wedge. According to this hypothesis, it would not be possible to distinguish and map out units of Asiatic and Australian origin. The whole thing appears as a chaotic melange as a result of collision between an island arc and oceanic/continental crust from the South.

The overthrust model (Audley Charles et al., 1975) considers that the outer arc ridge contains three principal groups of formations: autochthonous Australian-derived strata, par-autochthonous materials which originated at distal position from the Australian continent, and allochthonous South-East-Asian derived strata. This model implies that the material termed melange is a surficial olistostrome emplaced during the Late Miocene, and forms a thin veneer over the underlying thrust sheets (Fig. 7). Based on the fossil record, pelagic foraminifera found in the clay matrix, the age of the olistostrome is Late Miocene. This model considers that it is possible to delineate formations within the outer arc ridge and identify the origin, whether it is of Australian (autochthonous), distal Australian (par-autochthonous), or Asiatic (allochthonous) origin.

Chamalaun and Grady (1975) consider that the outer Banda Arc belongs to the Australian continental shelf, based on similarities with formations found in the NW part of the Australian continent; hence, it is all considered as autochthonous. It entered a subduction zone at a trench located in the vicinity of Wetar Strait, later became uplifted by isostatic rebound and is now exposed in the island of Timor.

Recently a new model has been proposed by Johnston (1981) indicating that a detachment of the northern part of Australia had taken place during Late Jurassic. This block drifted away from Australia and attached itself to the Sunda Island Arc system and became part of it during the Middle Cretaceous.

Structural continuity along the collision zone

Bathymetry, topography, seismic reflection profiles crossing the trough, seismic refraction data, gravity Bouguer anomaly show that there is a structural continuity along the arc.

A bathymetric depression or trench may be traced along the arc and can be considered as a continuous feature from the Java trench. Its depth is 1500-3000 meters and it terminates near Buru. A topographic ridge along the arc comprising the non volcanic outer arc with complicated geology can be followed from the Sunda Arc on to the Banda Arc. The inter-arc basins do not maintain the same

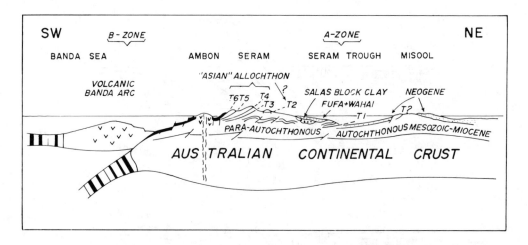

Fig. 7 Schematic cross-section through the volcanic Banda Arc, Ambon, Seram to Misool; showing Australian autochthone goes under Seram (A zone) and Australian continental crust goes under the volcanic Banda Arc (B zone). According to Audley-Charles (1979).

width and depth, the wide Savu Sea, narrow strait between Timor and Wetar, and the wide and deep Weber basin. Recent investigations show that the Savu Sea and Weber basin are underlain by oceanic crust. The volcanic arc follows the islands of Flores, Wetar, Damar, Teon, Nila, Manuk, Banda Api and Ambon. However, the lineament between eastern Flores and Damar is not very smooth and an ambiguous position is represented by the submarine volcanoes Emperor of China, Nieuwerkerk and Gunung Api North of Wetar.

Seismic reflection profiles transecting the trench are numerous. All the profiles at the Timor, Aru and Seram troughs indicate downwarping of the Australian margin sedimentary section towards the trench. This whole complex dips under the accretionary wedge confirming the presence of a subduction zone, hence a prevalance of compressional conditions. Bowin et al. (1980) observed that on the profiles of the Aru trough at the Australian continental side down-to-basin faulting is prominent, and is interpreted to indicate that extensional conditions rather than compressional conditions prevail.

Seismic refraction studies combined with gravity data confirmed continental crust beneath the trench and also under the outer arc ridge. Its boundary coincides approximately with the inner limit of the outer arc ridge and this boundary is also curved, following continuously the curvature of the arc. It ends at the western end of Seram.

The Bouguer gravity anomaly map (Bowin et al., 1980) shows nice continuity along the arc, however it does not follow exactly the lineament of the arc as shown by topography and bathymetry. If one follows the gravity lows by tracing the 0 milligal contour, or more conspicious + 25 milligal contour, it is clear that the gravity low forms a continuous zone. In Timor and Seram it covers partly the land areas and partly offshore areas. Tanimbar is embraced totally in this zone, whereas in the Kai Island the zone is situated west of it coinciding with the eastern flank of the Weber Deep (Fig. 9).

In spite of the smooth continuity represented by various physical features there are some features showing discontinuity along the arc. As already mentioned, Bowin et al. (1980) based on seismic reflection profiles came to the conclusion that compressional conditions vary along the arc, Timor and Seram Troughs show compressional conditions, whereas Aru trough indicates extensional condition. The free air gravity anomaly map also shows irregular patterns of the gravity lows. This could be clearly shown in gravity low areas lower than 100 milligal. The following low gravity zones may be distinguished : Savu Sea, Timor Trough, Aru Trough, Weber Deep, area between Onin Peninsula and East Seram, north of Seram and north of Manipa Strait (between Seram and Buru). A positive free air gravity anomaly extends from the Onin Peninsula to the Kai Islands, separating the gravity lows of the Weber Deep and the area between Onin Peninsula and East Seram (Fig. 8).

Complication by the westward drift of the Pacific plate

In plate tectonics it is recognized that the Pacific plate moves westward, which will have its consequences on the eastern end of the Asiatic continental margin. The velocities of underflow into the subduction zones reach at least 12 mm/yr (120 km/Ma). This westward drift creates conspicuous geological features which will be discussed in this paper.

The Sorong, Palu-Koro and Matano fault zones

The Sorong fault is a result of the northward drift of New-Guinea and Australia and westward thrust of the Pacific plate, resulting in slicing

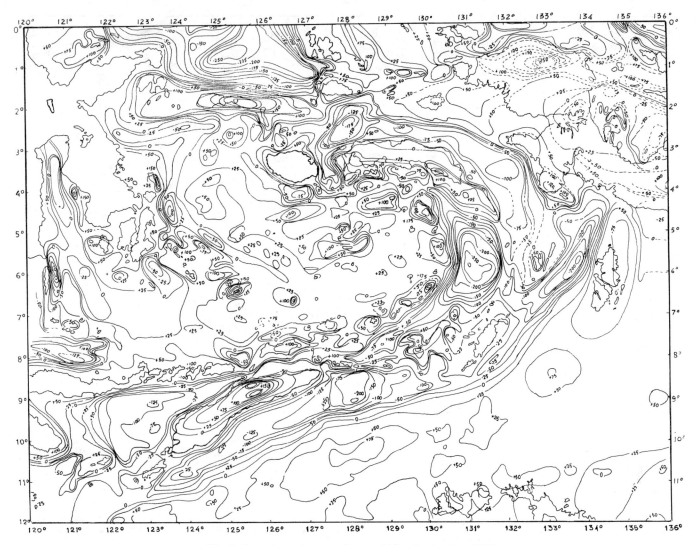

Fig. 8. Free-air gravity anomaly map (After Bowin et al, 1980)

of the northernmost portion of the New Guinea-Australia continental crust. The slices of continental crust are further moved translationally to the west and emplaced as microcontinents in their present position. Such microcontinents are Obi, Bacan, Banggai, Sula and Misool (Fig. 1). This conclusion is based on stratigraphic similarities between the islands and the New Guinea-Australian continent.

The general trend of the Sorong fault is East-West and it has a sinistral displacement. Onshore it stretches for a distance of 1300 kilometers between Sorong and Wewak. The fault extends westward beneath the sea and could be held responsible for the emplacements of the above mentioned microcontinents. Visser & Hermes (1962) have interpreted the Sorong fault to have a displacement of 350 km, based on similarities of the detritus of the Late Miocene Klasafet beds in the Bird's Head with Jurassic rocks occurring in Obi. The dis-

placement should be more if the present position of the microplate of Banggai is accounted for and it may reach a distance of approximately 600 km.

The deep water Mesozoic limestones occurring in parts of Buton, East Sulawesi, Buru, Seram and Misool could be explained as having been deposited in a basin, presumably the Tethys Ocean north of the New Guinea-Australian continent, during the Mesozoic. They became part of the New Guinea-Australian continental crust and later on attached to the existing island arcs such as Sulawesi, Buru and Buton.

Metamorphic and igneous rocks occurring in Banggai and Sula, allochthonous for Southeast Asia are here considered also to be derived from the New Guinea-Australian continent. The Banggai-Sula microcontinent has a greater effect on the K shape of Sulawesi than the other microcontinents.

Katili (1978) published a paper on the past and present geotectonic position of Sulawesi. Accor-

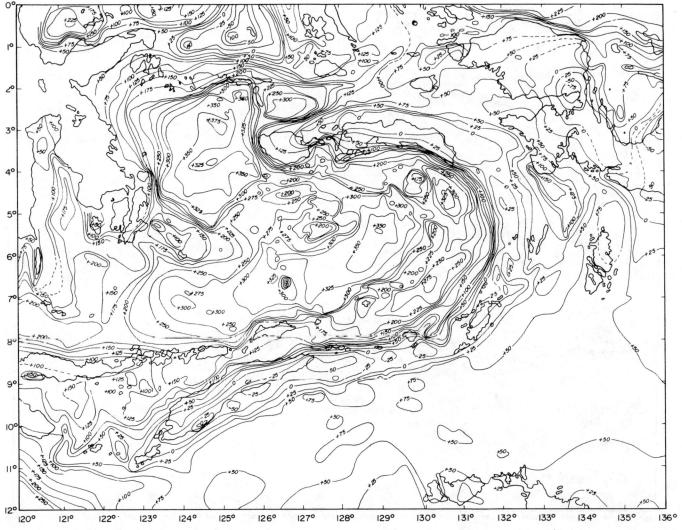

Fig. 9. Bouguer anomaly map (After Bowin et al, 1980).

ding to him at the end of Miocene a double arc (subduction and magmatic arcs) i.e. the Sulawesi arc with a north-south trend developed east of Kalimantan (Fig.10). During the Pliocene or perhaps even earlier this arc was severely deformed. The northward advancing Australian continent coupled with the counter-clockwise rotation of New Guinea and accompanied by the spear-heading westward thrust of the Sula Spur along the Sorong transform fault system severely transformed the east-facing Sulawesi Arc into a K-shaped pattern. This collision caused obduction of the ultrabasic rocks of the eastern and southeastern arms and thrusting of these rocks over the molasse deposit. Continuous westward directed tectonic forces along the Sorong transform fault system and the Matano fault zone in Sulawesi gradually pushed Sulawesi towards the Asiatic continent against Kalimantan and thus closed the southern part of the ancient Sulawesi Sea at the end of the Pliocene. The closure of the ancient Southern Sulawesi sea further pushed the Cretaceous-Early Tertiary Meratus and Pulau Laut subduction materials to an obduction mountain range in the Meratus Range. The rise of this mountain is not accompanied by plutonic activity; it is wholly caused by compressive forces as there is no record of plutonic rocks of this age (van Bemmelen, 1949).

The South Sulawesi sea reopened, starting at the end of Pliocene. The opening is accommodated by spreading centres and transform faulting represented by the Pater Noster and Palu-Koro faults. Some components of the Cretaceous-Early Tertiary Meratus and Pulau Laut subduction materials were separated by the speading and occupy a new position in West Sulawesi. They are the Bantimala melange which is of the same ages as the Meratus melange, the Lariang and Karama sedimentary basins which possess a predominant north-northeast structural pattern, the same feature which characteri-

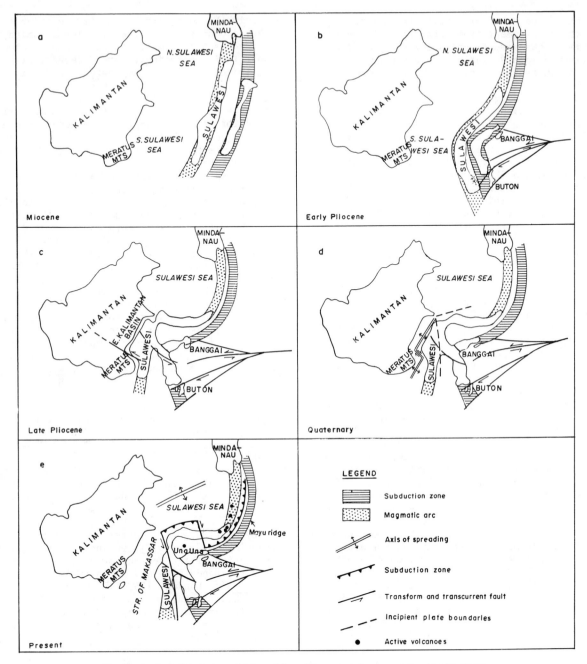

Fig. 10 Geological evolution of Sulawesi from Miocene to the present time (Katili, 1978).

zes the tectonics of eastern Kalimantan.

Eastward spreading South of the Pater Noster transform fault caused subduction and consequently created the Quaternary volcanoes of Lompobatang and Barupu which are now no longer active. The cessation was caused by subsequent spreading of the Sulawesi Sea which moved Sulawesi to the south-southeast along the Palu-Koro transform fault, simultaneously destroying the spreading centres in the Makassar Strait and thus cutting off the magmatic source of the Lompobatang and Barupu volcanoes.

The scenario of closing and opening the Makassar Strait is shown by the consecutive illustrations as given in Figure 10.

Concluding remarks

1. There exists a remarkable difference between the tectonic development of the areas west

and east of the Makassar Strait. In western Indonesia the development from Permian until the present time shows some progress towards the forearc although some deviation could happen with minor development backward. In eastern Indonesia, complication began when in early Tertiary time a north-south trending east facing arc-trench system (the Philippine-Sulawesi arc) developed perpendicular to the existing Sunda-Banda arc. This complication has been amplified by the severe interuption of the regular zonal development when the east-west trending south facing Banda arc was bent westward and the north-south striking Sulawesi arc was pushed westward toward the Asian continent. Responsible for the rolling up of the island arc was the northward drift of the Australian Continent coupled with the westward thrust of the Pacific Plate.

2. The origin of the loop-shaped Banda arc, the age of the Banda Sea and the peculiar K-shape of Sulawesi and Halmahera remain topics of controversy. Two main hypotheses for Banda arc evolution differ regarding the age assumed for the oceanic crust of the Banda Sea and in the inferred past spatial and tectonic relationship between the Banda arc and Sulawesi. The hypothesis of the rolling up of the Banda arc and the westward thrust of the Sulawesi arc (Katili, 1978) has been discussed in this paper. This hypothesis entails that the Banda Sea represents an old trapped oceanic crust. In contrast Hamilton (1977,1979) and Carter et al. (1976) suggest a late Tertiary age for part of the Banda Sea crust. They maintain that the Banda volcanic arc was originally continuous with the Sulawesi volcanic arc to the north, and associated with a west dipping subduction zone. Late Tertiary back-arc spreading west of the Banda arc caused the eastward migration of this arc away from the Sulawesi arc creating the Banda Sea. The curvature of the arc reflects the progressive collision of the arc with the curved continental margin of northwestern Australia and Irian Jaya (West New Guinea). Hamilton's hypothesis was supported by Silver et al. (1979) based on their interpretation of the late Cenozoic tectonic history of Buton, but the present authors believe that this hypothesis fails to explain in a comprehensive way the peculiar shape and the complicated geology of Sulawesi and Halmahera.

The present authors also contended that Buton does not belong to the east arm of Sulawesi, since the Triassic autochthonous series present in the islands of Timor, Seram, Buru and Buton do not occur in the eastern arm of Sulawesi. These Triassic autochthonous series are widely recognized as representing the source of hydrocarbon indications occurring in Timor (mudvolcanoes), Seram (oil) and Buton (asphalt).

3. To test Katili's hypothesis (Katili, 1978) on the past and present geotectonic position of Sulawesi, Sasajima et al. (1980) have been conducting paleomagnetic studies on the north arm of Sulawesi. They discovered that between early and late Miocene a rotation of nearly $40°$ had taken place in the north arm. The interpretation is either a counter clockwise rotation assuming the older rocks are normally magnetized or a clockwise rotation assuming reversed magnetization. Sasajima et al. (1980) suggested a clockwise rotation about an axis near the neck of Sulawesi, between the S and N arms, similar to the suggestion made earlier by Katili (1978) but for slightly earlier in time. E. Silver (personal communication) suggested that the hinge was in the east and that the north arm rotated clockwise along the Palu-Koro fault zone.

4. In spite of a large number of new data our present understanding of the geology of Timor, and other outer arc islands is not very conclusive. This is reflected, as shown in the literature, by differences of opinions advocated by workers on the geology of the Banda Arcs. Seismic reflection profiles crossing the trench show the position of the subduction zone to be at the trench, while seismic refraction studies indicate that the extent of the Australian continental crust continues from the trench under the outer arc ridge and ends at the inner margin of the outer arc ridge. It is therefore likely that under this outer arc ridge an important structural break is present separating the autochthonous from the allochthonous series.

References

Abbott, M.J., and Chamalaun, F.H., Geochronology of some Banda Arc volcanics, Bandung Geol. Res. and Dev. Centre, Spec. Publ. No. 2, 253-271, 1981.

Audley Charles, M.G., Carter, D.J., and Barber, A.J., Stratigraphic basis for tectonic interpretation of the outer Banda Arc, eastern Indonesia, Indonesian Petroleum Assoc., 3^d Ann. Convention, Jakarta 1974, Proc., 25-44, 1975.

Audley Charles, M.G., Carter, D.J., Barber, A.J., Norvick, M.S., and Tjokrosapoetro, S., Reinterpretation of the geology of Seram, implications for the Banda Arcs and northern Australia, J. Geol. Soc. London, 136, 547-568, 1979.

Bowin, C.O., Purdy, G.M., Johnston, C.R., Shor, G., Lawver, L., Hartono, H.M.S., and Jezek, P., Arc-continent collision in the Banda Sea Region, Bull. Am. Assoc. Petrol. Geol., 64, 868-915, 1980.

Carter, D.J., Audley Charles, M.G., and Barber, A.J., Stratigraphical analysis of island arc-continental margin collision in Eastern Indonesia, J. Geol. Soc. London, 132, 179-198, 1976.

Chamalaun, F.H., and Grady, A.E., The tectonic development of Timor - a new model and its implications for petroleum exploration, Jour. Australian Petroleum Explor. Assoc., 18, 1, 102-108, 1978.

Hamilton, W.H., Tectonics of the Indonesian region, U.S. Geol. Surv. Prof. Paper, 1078, 345 pp., 1979.

Hutchison, C.S., Tectonic evolution of Sundaland: A Phanerozoic Synthesis, Proc., Regional conference on the geology of Southeast Asia, Geol. Soc. Malaysia, 6, 61-86 pp., 1972.

Jacobson, R.S., Shor, G.G., Jr., Kieckhefer, R.M., and Purdy, G.M., Seismic refraction and reflection studies in the Timor-Aru trough system and Australian continental shelf, Memoir Amer. Assoc. Petrol. Geol., 29, 209-222 pp., 1978.

Johnston, C.R., A review of Timor tectonics, with implications for the development of the Banda Arc, Bandung Geol. Res. and Dev. Centre, Spec. Publ. No. 2, 199-216, 1981.

Katili, J.A., Geological environment of the Indonesian mineral deposits, Indonesia, Geol. Surv, Econ. Geol. Ser., 7, 1-18 pp., 1974.

Katili, J.A., Volcanism and plate tectonics in the Indonesian island arcs, Tectonophysics, 26, 165-188 pp., 1975.

Katili, J.A., Past and present geotectonic position of Sulawesi, Indonesia, Tectonophysics, 45, 289-322 pp., 1978.

Nishimura, S., Otofuji, Y., Ikeda, T., Abe, E., Yokoyama, T., Kobayashi, Y., Sapri, Sopheluwakan, J., and Hehuwat, F., Physical geology of the Sumba, Sumbawa and Flores Islands, Bandung Geol. Res. and Dev. Centr., Spec. Publ., No. 2, 105-114, 1981.

Sasajima, S., Nishimura, S., Hirooka, K., Otofuji, Y., van Leeuwen, T., Hehuwat, F., Paleomagnetic studies combined with fissiontrack datings on the western arc of Sulawesi, East Indonesia, Tectonophysics, 64, 163-172 pp., 1980.

Silver, Eli A., Joyodiwiryo, Y., and Mc Caffrey, R., Gravity results and emplacement geometry of the Sulawesi ultramafic belt, Indonesia, Bandung Geol. Res. and Dev. Centre, Spec. Publ., No. 2, 313-320, 1981.

van Bemmelen, R.W., The Geology of Indonesia, v. IA, Govt. Printing Office, The Hague, 732 pp., 1949.

Visser, W.A., and Hermes, J.J., Geological results of the exploration for oil in Netherlands New Guinea, Koninkl. Nederlands Geol. Mijnbouw Genoot. Verh., geol. Ser., 20, spec. no., 1-265 pp., 1962.

GEOLOGICAL-GEOPHYSICAL PARADOXES OF THE EASTERN INDONESIA COLLISION ZONE

J. Milsom and M.G. Audley-Charles

Department of Geological Sciences, Queen Mary College, London E1 4NS

A.J. Barber

Department of Geology, Chelsea College, London W6 9LZ

D.J. Carter

Department of Geology, Imperial College, London SW7 2BP

Abstract. Despite a long history of geological study culminating in some very intensive work during the past decade, many aspects of the geology of the Banda arc remain poorly understood and have been subjects of much controversy. A number of outstanding anomalies and paradoxes in the distribution of islands and of bathymetric features, in the patterns of seismicity and volcanicity and in the observed geological relationships remain to be resolved. Two of the most important future lines of research are likely to be marine geophysical studies to determine the history of the Banda sea and paleomagnetic measurements to verify suggested rotations of the various islands.

Introduction

Studies in the Indonesian archipelago have played an important part in the formulation of modern tectonic theory, since it was in Indonesia that many of the fundamental characteristics of island arcs were first identified.

In particular, it was in relation to the Banda and Sunda arcs that belts of positive and negative gravity anomalies were first recorded and described (Vening Meinesz, 1932, 1954) and that van Bemmelen (1949) drew attention to the distinction between, and near parallelism of, volcanic and non-volcanic arcs. The high seismicity of the area has long been known and almost inevitably the Sunda arc was one of those selected by Benioff (1954) for his classic study of dipping earthquake zones.

With such a background, it might be supposed that the geology of the Sunda and Banda arcs would now be well understood and would be in accord with classical plate-tectonic models. However, this is not so. The Sunda arc now tends to be viewed as an active continental margin rather than a "normal" arc, while the Banda arc (Fig. 1) is still an area of geological confusion and heated controversy.

In general terms the anomalies and paradoxes can be seen to arise from the comparatively recent and still continuing collision between southeast Asia and the Australian continental mass, but in detail the results of this process are often unclear and in some cases surprising. A major recent publication by Hamilton (1979) argues forcefully for one form of tectonic interpretation, involving thick belts of chaotic melange produced by "scraping-off" material from the downgoing plate in a subduction zone and accreting it to the opposite side of the zone. In a series of papers Audley-Charles and his colleagues favor instead a more limited role for the chaotic deposits, which are regarded as superficial olistostromes, and see the geology of the outer arc islands as dominated by the emplacement of large and coherent thrust sheets followed, in the Quaternary, by a phase of intensive block faulting (Carter et al, 1976; Audley-Charles et al, 1979). A third group, based at the Flinders University, Adelaide, regard thrusting as of minor importance and normal faulting as paramount (cf. Chamelaun and Grady, 1978). Several other workers and groups of workers have proposed different tectonic schemes and reconstructions, all disagreeing in minor, and sometimes major respects (e.g. Cardwell and Isacks, 1978; Bowin et al, 1980; Papp, 1980). Conclusive arguments are rare, since access to many crucial areas is difficult and some of the basic field relationships are still in doubt.

The authors of the present paper belong to one of the competing groups noted above. However, in writing it we have tried to withdraw a little from the controversy and to point to those aspects

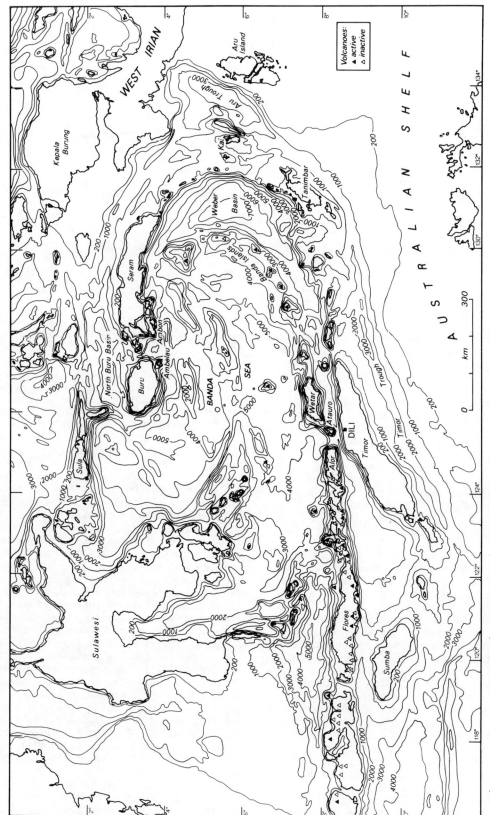

Figure 1. Eastern Indonesia. Bathymetric depths in metres. Bathymetry and distribution of volcanoes after Hamilton (1978)

of the geology which seem to us most interesting and most difficult to interpret, without favoring any particular set of explanations. By this approach, concerned more with questions than with answers, we hope to identify for ourselves and others the most fruitful fields for further study. One factor that very quickly emerged during the writing was the extent to which the various problems are inter-related; many of our difficulties arise from the need to assimilate and incorporate data from an ever widening area of interest.

The Most Arcuate Arc

The term "island-arc" is now well established in the literature, even though one of the simplest and best known examples, Tonga-Kermadec, is scarcely curved at all, while others (e.g. the Sunda arc) can be resolved into a number of straight line segments. The immediate impression given by the Banda arc as a topographic and bathymetric feature is, however, of a very definite arc with a total curvature amounting to nearly $180°$.

The subaerial geology does little to contradict this impression, since the two largest islands, Timor and Seram, are strikingly similar and are arranged 'back to back' as mirror images across the Banda sea. The difficulties arise when attempts are made to formulate a plate tectonic compatible theory within which such a curvature could have arisen.

A fundamental tenet of plate tectonics is that island arcs mark the zones where lithospheric plates are being reabsorbed into the asthenosphere. Such structures can only exist, therefore, where there is or has been some degree of convergence between the lithospheric plates separated by the arc. For an active arc to curve through $180°$ there must clearly be more than two plates involved, i.e. either there is expansion in the back-arc basin forcing different parts of the arc in different directions, or there is more than one plate external to the arc. There may be a nascent spreading ridge in the Banda sea south of Seram (R.S. Jacobson, reported in Barber et al, 1981), but Bowin et al (1980) consider that the generally low to normal heat flow implies that any crustal creation there must have taken place before the Neogene, which seems to eliminate the possibility that the present configuration has been acquired by expansion behind the arc. Many authors have recognized the consequent need for more than one external plate, but the boundaries proposed have been widely different. Cardwell and Isacks (1978) regard a major strike-slip fault in central West Irian as a transform-type plate boundary, while Bowin et al (1980) see in the Aru trough a zone of extension which has contributed to the increasing curvature of the Banda arc. Both groups propose a 'two-slab' model for the subduction zone, with different and quite distinct plates being absorbed beneath Flores and the Banda Islands and beneath Seram. It is also conceivable that subduction does not and never has taken place beneath Seram and that the trough there simply marks the location of a zone of strike-slip faulting which defines the southern margin of the Sula spur. All of these solutions have the unhappy result of removing much of the significance from the very obvious continuity of the arc, since they imply that its tectonic role changes completely from one end to the other. It would indeed be paradoxical if the high degree of curvature, the very feature which first focussed attention on the Banda arc, were ultimately proven to be a random and short-lived arrangement of almost unrelated elements. Hamilton (1979, p. 118) strongly rejects any such an interpretation, and continues 'to infer general continuity of subducted lithosphere around the Banda arc.' Such an inference would seem almost to demand a corresponding continuity of island-arc volcanism.

Continuity of the Volcanic Line

The existence of an essentially continuous line of volcanoes around the entire length of the Banda arc from Flores to Ambelau has been an important element in many theories of the evolution of the area. Virtually all recent treatments which consider the point at all regard the volcanics of the Ambon islands, south of Seram, as evidence of subduction at some time, on account of their acid-intermediate composition. However, Bowin et al (1980) have pointed out that there is no bathymetric continuity between the northernmost of the currently active Banda islands and the easternmost of the islands of the Ambon group (none of which is now the site of more than weak solfataric activity). Furthermore, Cardwell and Isacks (1978) have been able to trace a Benioff zone beneath Seram to depths of the order of only 100 km, which is the usual depth of generation of island-arc andesites, and it is surprising in these circumstances that the associated volcanoes should apparently be nearing the end of their eruptive history, rather than just beginning it. The geology of the islands themselves, particularly Ambon, the largest, is also anomalous when compared to the near-normal island arc relationship between Timor and Alor or Wetar.

The fullest published description of Ambon is still that given by van Bemmelen (1949) but recent investigations by a number of workers have generally confirmed his summary. The island is divided almost in two by the deep indentations of Ambon and Baguala bays (Fig. 2). The larger, north-west section is very similar to the adjacent smaller islands, being built up of late Tertiary and possibly also Quaternary intermediate lavas and tuffs to which the name ambonites has been given. The south-eastern peninsula, however, includes a wide variety of sediments, granites and basic and ultrabasic rocks, all of which have counterparts on mainland Seram, from which they are separated by the volcanic line. The intermixing of outer and inner arc elements extends also to southern Seram, where hot springs similar to

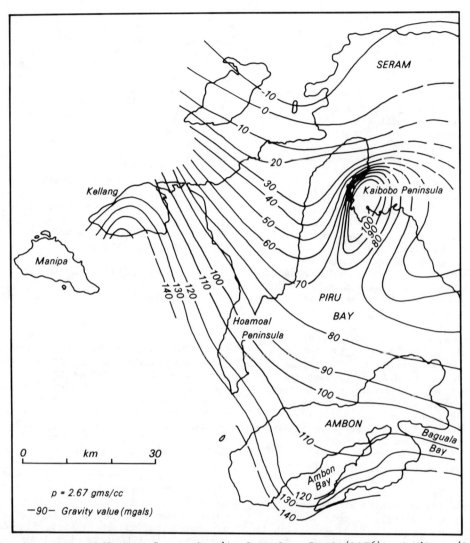

Figure 2. Ambon and Western Seram. Gravity data from Jezek (1976) and Milsom (1977)

those which constitute the last remaining signs of volcanic activity on Ambon are found, and where the volcanic rocks of the Hoamoal peninsula closely resemble ambonites.

The gravity field in the Ambon area departs from the relatively simple pattern found on Timor and its adjacent volcanic islands, and indeed on most of Seram. On Timor, Bouguer anomaly values are lowest along the south coast, facing the Timor trough, and remain low over most of the southern half of the island (Chamelaun et al, 1976). Further north they rise steeply, with maximum values and maximum gradients occurring close to the north coast. The absolute maximum presumably lies somewhere in the Wetar strait, since the Bouguer anomaly on Atauro is similar to those in north-eastern Timor (Milsom and Richardson, 1976). By analogy, northern Seram is expected to be a region of low Bouguer anomaly and low gravity gradients, while in the southern part of the island a rapid increase in levels is expected towards the south coast. Broadly this is the pattern actually observed, except in the vicinity of Ambon (Milsom 1977). In western Ambon the trend of the gravity contours is more nearly north-south, with values increasing rapidly in a westerly direction. This is perhaps explicable in terms of an area of oceanic crust between Seram and Buru, which may have been formed during extensions and fragmentation of the northern Banda arc and the Sula spur, but two additional facts have to be taken into account. First, the highest Bouguer anomalies do not occur between the volcanic group and Seram, but values increase across the whole width of the smaller islands, the highest being found on the extreme southern point of Ambon. Secondly, ultramafic rocks, which are usually associated with the outer arc or the inter-arc gap and with which some of the gravity anomaly features may be correlated, are commonplace on south-eastern Ambon.

Although non-volcanic sections are known to exist in many areas, it is quite hard to envisage a mechanism whereby subduction could produce considerable volcanic activity near Ambon and none at any points more than 150 km to east or west. An alternative explanation is that the ambonites, although intermediate and in an island-arc environment, are only very indirectly subduction related, and instead mark a line of transform faulting associated with crustal extension in Piru Bay and between Seram and Buru. Three thousand kilometers further east, andesitic rocks are being erupted on the Papuan peninsula in an extensional environment in which no present-day or even Neogene subduction seems likely (Jakes and Smith, 1970). The cases are not closely analogous, but make Hamilton's (1979) observation that some of the ambonites contain more than 5% K_2O interesting.

Almost inevitably, consideration of possible origins of the ambonites leads to consideration of the seismic zonation of the Banda arcs.

The Banda Seismic Zone

Geological mapping of the Banda arcs is still not complete, even at a reconnaissance level, and differences in interpretation therefore remain in many cases unresolved. It might be thought that in contrast the physical facts of the locations of earthquake foci, and thus the definition of the Benioff zones, could be established without ambiguity. Even in this field, however, argument continues. In a very clear and concise paper Cardwell and Isacks (1978) have described and contoured the pattern of distribution of foci, have presented fault plane solutions and have related these data to a lead sheet model of the subducted slab.

Their conclusions, that it is not necessary to suppose the dipping slab undergoes significant areal strain, but that it is necessary to assume two separate plates, one subducted beneath Seram and the northern limb of the arc and the other in the east and south, are however vehemently attacked by Hamilton (1979) who considers that the data have been miscontoured to fit the assumptions and that in some cases important trends are concealed because of the positioning and orientation of the planes of projection for the various cross-sections. This latter point emphasises an obvious difficulty that is inevitable when dealing with a sharply curved zone. When, in addition to having a small radius of curvature, an arc bends through 180°, some confusion, particularly as regards deeper events which might be assigned to a slab subducted from almost any point on the circumference, is not surprising. Papp (1980) has attempted to overcome this limitation by modelling the focal distribution in three dimensions, using small spheres and strong wires. His photographs hardly clarify matters, but they do illustrate the problems involved in defining the Benioff zones with the data at present available. His conclusion, which rather contradicts both Hamilton and Cardwell and Isacks, seems to be that there is no subduction at present beneath Seram, but there may be a detached and sinking fragment of lithosphere beneath Buru. His seismic flux maps show a concentration of energy release in western Seram and eastern Buru, which he regards as due to an embryonic subduction zone and which Bowin et al (1980) presumably regard as due to crustal extension.

A further disappointment is that Timor and its surroundings, which are the areas on which the most controversy has been centered and where seismic data might be expected to provide some useful constraints on theory, are almost aseismic to depths of 100 km. This is in itself surprising since although subduction as such might have ceased because of the arrival at the trench of the Australian continental margin, convergence presumably continues. Elsewhere on the earth's surface, continental collision zones in much more advanced stages of evolution then Timor are highly seismic. The apparent lack of activity is perhaps merely an effect of the limited time during which observations have been recorded. Papp (1980), for instance, labelled one zone 'aseismic' in his paper but had to admit that in the twelve years after his compilation was prepared there were twenty-four shocks within it. The same may be true of Timor, on a longer time scale, but it is doubtful even so if any major new insights are to be looked for from seismology. On any realistic time scale seismological studies must be regarded as representing merely 'snapshots' of tectonic processes at one instant of time. Paradoxically, it seems that in what is, on a regional scale, one of the most seismically active parts of the earth's crust, within which subduction is believed to have occurred and be continuing, the nature and even, in some cases, the existence of Benioff zones is a matter of uncertainty (see also discussion in Barber et al, 1981). Past processes in particular can be investigated only on the basis of their more lasting consequences, the observable stratigraphic successions and their facies patterns, structural configurations and geomorphological expressions.

The Weber Basin

Geomorphologically, the most striking feature of the entire Banda arc area is the very deep Weber basin. It is also one of the features which most contributes to the impression of continuous curvature given by the arc. Indeed, Hamilton (1979) argues very strongly that the curvature at the northern end of the basin disproves any hypothesis which involves two distinct plates subducted beneath the northern and eastern parts of the arc respectively, since it demonstrates the essential continuity of the two sections.

The Weber basin illustrates in its most dramatic form one of the major anomalies of the Banda arc in that, in contrast to most other arcs, the outer "trench" is quite shallow whereas the inter-arc basin is quite deep and in some cases very deep. The comparative shallowness of the seas where deep trenches might be expected can be simply explained as the consequence of the collision between the subduction zone and buoyant continental crust which cannot be subducted. The really surprising features are the anomalously great depths and very variable widths of the inter-arc basins. In the case of the Weber basin two alternative explanations have been proposed, in one of which it is considered to be the actual subducting trench (Audley-Charles and Milsom, 1974). The more generally favored alternative is that the Aru trough is the subduction trench and the depth of the Weber basin is a consequence of the intense stresses imposed by the curvature of the arc. (Fitch and Hamilton, 1974; W. Hamilton reported in Barber et al, in press). The most decisive evidence in this respect might be expected to come from seismic reflection profiling across the features involved. A number of such profiles have now been published, notably in a compilation by von der Borch (1979) which draws on data from a number of different sources. The "typical" profile shows a very asymmetric outer trough, with a steep wall on the arc side composed of acoustically opaque material and a less steep outer flank which is effectively the dip slope of the sediments beyond the arc. This pattern is generally regarded as evidence of subduction. Profiles across the inter-arc basins usually show them to be more symmetrical, with evidence of formation by normal faulting. There are some variations in this pattern around the arc. The trough northeast of Seram shows the asymmetry expected of a subduction zone but also a thick sequence of flat-lying reflectors which are hard to reconcile with continuing subduction. The profile crossing both the Weber basin and the Aru trough is even more interesting. Although there is a very faint suggestion that there could be a subduction zone emerging in the bottom of the trough, this is extremely muted and indeterminate, and in another paper Jacobson et al. (1979) note that the seismic evidence for subduction is very weak in this area. The profile given by von der Borch (1979) shows so great a depth contrast between the trough and the basin that is almost seems that if there is a slab dipping west from the bottom of the trough, it must subcrop again in the deepest part of the basin. The basin itself has the sort of asymmetry associated with subducting trenches, but, like the Seram trough, is floored with flat-lying sediments. Bowin et al (1980), who describe the results of a major investigation in this area, came to the conclusion that there is no subduction associated with either Aru trough, which they regard as a site of crustal extension, or with Weber basin. This leaves the young volcanoes of the Banda islands without any satisfactory role, an anomaly which may be linked to their unusual pattern of chemical variation (Barber et al, 1981). The potassium content of the volcanic rocks varies from island to island along this part of the arc, although the volcanoes are at similar distances above the Benioff zone and might be expected to be similarly potassic. In view of all these difficulties, it might seem easier to look to the exposed geology of the islands of the outer arc for clues to the past and present tectonic processes.

The Australian Margin

It is now widely accepted that in many convergent sutures where oceanic lithosphere descends in a subduction zone, parts of the uppermost layers of the crust are stripped from it and accreted as a wedge on the inner wall of the trench. As a consequence, the trench migrates oceanward and the accreted ridge develops between it and the volcanic line and may eventually rise above sea level to form an outer arc of non-volcanic islands. Formation of the outer-arc ridge south of Sumatra in this way has been documented by Moore and Karig (1980), who interpret it as a tectonic melange.

The Banda arc differs from typical island arcs in that its outer rim is bordered by continental rather than oceanic crust. There has been a considerable amount of discussion as to whether, under these circumstances, accretionary wedges (which may or may not be tectonic melanges) will still be formed or whether, because continental crust cannot be subducted very far and so the scope for tectonic scraping is limited, some other process takes place.

Since convergence apparently continues, large-scale imbrication of the continental margin with the emplacement of complexly inter-related but still essentially coherent thrust sheets is an obvious alternative possibility. This corresponds closely to what Bally (1975) and Bally and Snelson (1980) refer to as A-type subduction, in which at the most a few tens of kilometers of continental crust underthrusts a mountain belt. They take as their type area the southern part of the Canadian Rocky Mountains, flanked on the cratonic side by a deep asymmetric basin across which the thrust sheets of the foothills zones have travelled cratonwards. The flanking basin is thus the foredeep of classical Alpine geology, and extension of Bally's ideas from the Rockies to the Banda arc could provide a vital link between classical and plate tectonic concepts. Coherent thrusting is supported by, amongst others, Carter et al (1976) and Norvick (1979), who regard the chaotic deposits of the outer-arc islands as superficial olistostromes and not as accretionary wedges in the normal sense. Fortunately, the disagreement between the two concepts should be resolvable by straightforward high quality geological mapping, aided perhaps by geophysical data from the inter-island areas.

The exactness of the fit of the curved rim of

the Banda arc into the socket formed by the Australian continental shelf and the Kepala Burung peninsula of West Irian is less easily explained. Broadly, this could have arisen in one of three ways. Either both the Banda arc and the Australian continental margin had roughly their present shapes prior to collision (Norvick, 1979), which seems to be asking a great deal of coincidence, or the arc has been 'bent' by its collision with Australia, or the shape of northern Australia and West Irian are partly results of the collision with the Banda arc. The geological similarities around the arc, and particularly the occurence there of Australian margin facies rocks, might be considered to provide some support for the second of these alternatives, while the faulted and apparently rotated structures of westernmost West Irian would seem to be arguments in favor of the third. The most attractive possibility may involve a combination of these two, with movements in West Irian forcing the western part of the island still further west and forcing the Banda arc back in the process. This would seem to require some loss of Banda sea crust, possibly in the subduction zone proposed by Hamilton (1979) east of south-eastern Sulawesi. If to this already complex pattern of events is added extension in the Aru basin as suggested by Bowin et al (1980), the range of possible movements becomes very wide. As a further complication, some workers in the area have recently claimed that the role of thrust faulting from the direction of the Banda sea has been greatly overestimated.

Paleomagnetism, Kaibobo and the Reality of Thrusting

There is at least little dispute that the Banda arc is the site of a collision between an island arc system and a continental margin, and in such an environment most workers are agreed in expecting thrust faulting, whether the emplacement of large scale overthrusts or the finer slicing of tectonic melanges, to play a dominant role. However, workers from the Flinders University in South Australia have claimed that block-faulting and not overthrusting is the dominant tectonic style in Timor.

Carter et al (1976) described a series of allochthonous thrust sheets on Timor resting on autochthonous or para-autochthonous basement. One of these thrusts consists largely of the Maubisse formation, a dominantly calcareous suite which includes some shales and volcanics, which is dated as Permian and is interpreted as resting tectonically on Triassic and Jurassic rocks. There is also a para-autochthonous Permian formation, the Cribas, in the same general area. The type area of the Maubisse was re-mapped by Grady (1976) who found no evidence of thrust faulting and that contacts with other formations seemed to be steep normal faults. This apparent discrepancy is explained by Barber et al (1977) as the result of major Quaternary block faulting which has displaced the Pliocene thrusts.

Grady and Berry (1977) also claimed that the Maubisse and Cribas formations interfinger, although Barber et al (1977) disputed this interpretation, pointing to the two formations different metamorphic and structural states. As a further stage in the Flinders University investigations, Chamelaun (1977) obtained paleomagnetic pole positions for both the Maubisse and Cribas formations. He argued that if the two formations had been widely separated at the time of deposition, these positions should be quite different and not, as he found, the same within the limits of error, and compatible with the positions for mainland Australia at the same period. This result he considers as evidence (although admittedly, because of the small number of sites sampled, not yet very strong evidence) that the Maubisse is not allochthonous either as part of a large scale overthrust or as part of a tectonic melange. Blundell (1979) has shown that the paleomagnetic data allow a separation of as much as $20°$ of latitude between the Maubisse and Cribas at their time of formation.

Evidence that thrusting is unimportant in the outer arc is so contrary to most theories that the conclusions would need to be very fully substantiated. Recent geological mapping in Seram (Audley-Charles et al, 1979) has located important overthrusts, disrupted by normal faults, directed away from the Banda sea. There is at least one other indication of the importance of normal faulting, in the local but very strong gravity high on the Kaibobo peninsula (Fig. 2) which is one of the most geologically diverse parts of Seram. A magnetic complex consisting mainly of very basic igneous rocks has been described, as well as suites of high grade and low grade metamorphics (Audley-Charles et al, 1979). To the north and south of the peninsula there are two small Tertiary basins, but the density contrast between the sediments and crystalline rocks is not of itself sufficient to explain the anomaly. Similar metamorphic and igneous rocks are found on the Hoamoal Peninsula on the western side of Piru bay but it is clear that the gravity anomaly does not extend into that area.

Igneous rocks on Manipa and Kellang islands are also similar to those on Kaibobo, and high gravity fields, insufficiently defined by the readings obtained so far, are associated with these also (Jezek, 1976). At Kaibobo the anomaly can be easily explained by a dense pipe or plug-like body with considerable depth extent but is difficult to reconcile with an environment supposedly dominated by thrusting. Morphologically western Seram seems dominated by horst and graben tectonics, perhaps associated with crustal extension in Piru Bay. Recent movements of this type could have obscured much evidence of earlier thrusts on Timor, but the paleomagnetic evidence awaits confirmation or refutation.

Magnetic Soundings and Volcanic Offsets

It is not only in paleomagnetism, but in some aspects of present day magnetism that Timor seems anomalous. The deep electrical structure of the earth's crust near Timor has been probed by a single geomagnetic depth sounding on the north coast of the island (Chamelaun and White, 1976). The results of such work are normally displayed as vectors pointing towards areas of higher conductivity. In many cases a sea or ocean provides the high conductivity, but the importance of such effects can to some extent be estimated from frequency analysis of the observations.

Prior to the work at Dili, the only relevant results available were for one station on the north coast of Java and two on the north coast of Australia facing the Timor trough. The Java vector points southwards across the island towards the deep water of the Java trench and not towards the immediately adjacent but much shallower Java sea. It is thought that the existence of a high conductivity zone associated with the subduction zone has a major influence on this orientation. With this in mind, it might be supposed that the Australian vectors, both pointing towards Timor, indicate that a similar zone is present, since the water layer in the Timor trough is neither deep enough nor close enough to the observation points to produce the effect on its own. If so, the vector measured on the north coast of Timor would be expected to point south, although the fact that the Banda sea to the north is much deeper than the Java sea might reduce its amplitude or even produce a weak reversal.

The results obtained on Timor were quite unexpected, since the vector not only points north but, filtered on an hourly basis, is approximately twice the strength of the vector on Java. For longer periods, related to deeper structure, the vector weakens and swings west of north. An explanation for this pattern is lacking but it does cast some doubt on the idea that an active subduction zone emerges in the Timor trough. Chamelaun and White (1975) suggest there may be a subduction zone developing close to the north coast of Timor, but this seems to find little support at present. One consistent feature linking Java and Timor is that in each case the vector points towards the volcanic line, which lies north of Timor and south of the north coast of Java. At longer periods the correlation is good, since the westward rotation of the vector means that it points towards the still active section of the arc west of the island. The strong north-directed short-period vector remains a paradox, which may be connected with an anomaly in the present day volcanism of the area.

The long line of active volcanoes extending through Sumatra and Java to the Banda arcs is interrupted for a distance of about 400 km in the area north of Timor. Volcanic islands occur along the line of the arc in this region, but there has been no recent activity. The extinct volcanism appears to have been replaced further to the north, where young submarine or barely emergent volcanoes are found. A curve drawn through the known centers roughly parallels the north coast of Timor, which is convex northwards.

The cessation of volcanism immediately to the north of Timor is commonly regarded as an effect of the collision between the southern part of the Banda arcs and the Australian continental margin. Cessation of subduction implies that no fresh material will be thrust or drawn down to depths at which island arc magmas can be generated, but this does not explain the newer activity to the rear of the arc. An analogy has been drawn with the New Britain arc, north-east of New Guinea, where a rather similar, but smaller scale, offset occurs (Milsom and Richardson, 1976). In the latter case the offset has been explained by Johnson (1976) as a consequence of 'slicing' or large-scale imbrication of the downgoing plate where a part of it which includes continental crust reaches a subduction zone. Although the analogy may be valid, the explanation offered is not wholly satisfactory since the New Britain offset is restored a little further to the west in the north coast area of New Guinea, where continental collision is still further advanced (Jaques and Robinson, 1977).

The offset of the volcanic line is sometimes seen as marking the start of the Banda arc proper. This would place the large island of Sumba outside the scope of this discussion, but it seems impossible to consider the outer arc as a whole without paying some attention to what many would regard as the first island of it.

Sumba and The Start of the Arc

The outer-arc basin which separates the volcanic and non-volcanic islands of the Sunda arc is abruptly terminated by the island of Sumba which straddles the gap between the two ridges. Audley-Charles (1975) interpreted the island as marking the site of an important fracture in the Earth's crust and the commencement of the outer Banda arc. In this latter respect he supports Brouwer (1925), who considered that there was geological and structural continuity between Sumba and Timor, against more recent authors (e.g. Katili, 1971) who deny the existence of such a link. There seems to be no direct evidence from Sumba itself for the fracture, and indeed Audley-Charles (1975) considered that faults pass either side of the island and that it has been affected only to the extent that it has been moved away from the Australian continental margin and rotated. The main fracture is seen as extending along the north-south trending bathymetric contours which define the boundary between the Australian continental mass and the Indian Ocean abyssal plain. Gravitationally, Sumba lies beyond the western end of the Bouguer anomaly low which continues around the entire length of the Banda arc through Seram, Kai, Tanimbar and Timor. This, and the corresponding isostatic anomaly low, reappears again west of Sumba, in the Sunda arc. The Bouguer Anomaly map of Sumba (Untung, 1980) shows no strong trends on the island itself and only low to moderate gradients. The frequency of occurrence of shallow focus earthquakes also decreases sharply to the east of

Sumba and the volcanic line is displaced slightly northwards. The volcanoes immediately to the west of this displacement (which is quite distinct from the major offset in present day activity north of Timor) are atypically potassic in relation to their position relative to the Benioff zone (Foden and Varne, 1980).

Although there is clearly something very unusual about Sumba, a number of objections can be raised against the fracture zone concept. No disturbances of the sedimentary layers around Sumba have been found on seismic reflection profiles (Barber et al, 1981) and the offset of the volcanic line, which should be a reliable guide to deep structure, is a very minor feature amounting, at the most, to about 50 km. Audley-Charles (1978) countered these objections by identifying the fracture as a transform associated with a local spreading episode in the north-east Indian Ocean (Wharton basin) during the Jurassic and Cretaceous and supposing that only minor rejuvenations of strike-slip movement have occurred during the Cenozoic.

The geology of Sumba itself is of little help in resolving the anomaly of its existence. Most writers seem to agree that it is a continental fragment, a conclusion based largely on the acid to intermediate nature of the igneous complex. Its crustal thickness has recently been estimated as 24 km from seismic refraction studies (Barber et al, 1981) which is certainly not oceanic but is thinner than normal continental. Prior to recent mapping by the Indonesian Geological Research and Development Center (Barber et al, 1981) only the Neogene sediments had been systematically studied and these are remarkable for presenting a section about 1000 meters thick which has been only very gently deformed. The Paleogene and older rocks have been subject to rather greater tectonism and now dip at angles of about 60° (Hamilton, 1979). None of the explanations for such an island's presence in the inter-arc gap seems readily acceptable. Audley-Charles (1975,1978) regards it as a rifted and rotated fragment of the Australian continent and is supported in this by recent paleomagnetic results (Otofuji et al, 1980), while Hamilton (1979) sees it as a fragment torn from the continental shelf of the Java sea. Bowin et al (1980) compromise by considering it to have formed part of a 'greater Sula spur' which, while originally part of the Australian continental margin, was rifted away at an early date and so collided with the subduction zone flanking southeast Asia well before the main mass of Australian did so. Hamilton's explanation seems to require that the island somehow passed through the volcanic arc while the other two have to explain why the thrusts and melanges which are seen on the other islands to which they attribute a generally similar geological history are absent on Sumba. That Sumba, unlike Timor, has oceanic crust to the south is probably important in this respect. The problems posed by the lack of deformation where intense tectonism might be expected are repeated at the other end of the Banda arc, on the equally anomalous island of Buru.

Buru and the End of the Arc

The geology of Buru has only recently been systematically studied, although the scattered information available prior to recent mapping by the Indonesian Geological Research and Development Center was compiled by van Bemmelen (1949). The island is reported to consist of a metamorphic basement complex consisting of schist, gneisses and amphibolites, and a sequence of younger rocks, ranging in age from Permian to Paleocene, displayed in sequence by westward tilting. No large thrusts or melanges have been identified on the island, which seems to be structurally remarkably simple, having closer affinities with the Australian-New Guinea continental margin than with the neighboring island of Seram (S. Tjokrosapoetro, reported in Barber et al, 1981). The Banda arc thus provides another major surprise by ending in an island which, while clearly geologically related to the other outer-arc islands, is, by their standards, remarkably undeformed.

The surroundings of Buru provide the final problem of the Banda arc. Hamilton (1979), in noting that the island is virtually unique in the arc in having deep ocean both to north and south, actually understates the case. In fact, with water depths reaching 4000 meters only a short distance from the west coast and with even the comparatively narrow strait which separates it from Seram reaching a depth of more than 3000 meters, the island can be said to be surrounded by deep ocean on all sides. Nonetheless, the gravity maps compiled by Bowin et al (1980) show that the arc-related belt of gravity anomaly reaches Buru and terminates on or near it. The geological similarities to Seram are strong, particularly as regards the basement complex, but paradoxically for an island in such an 'oceanic' setting, basic and ultra-basic rocks are rare or absent.

The major problem presented by Buru is how a strongly developed arc and trench system, whether at present dominated by subduction or by strike-slip motions, can simply and abruptly terminate in an area of deep water which appears to lack any of the allowable plate tectonic boundaries of ridge, trench or transform. Most authors seem agreed in emphasizing the similarities with the islands of the Sula spur to the north, and Hamilton (1979) suggests that Buru may have been "torn from New Guinea by strike-slip faults, carried westward in the Banda sea region when New Guinea lay south of its present latitude and swept up by the Banda arc as the arc migrated eastward in middle Tertiary time". It seems to have survived these experiences with considerable composure. The solution proposed by Bowin et al (1980) is essentially similar, in that they regard Buru as having formed a part of a greater Sula spur, from which it was separated by the opening up of the North Buru basin. They attribute a similar history to Seram, which is seen as simply having undergone more internal structural disruption. If so, the separation of the two islands was an important event, the time and mode of which certainly merits further investigation. For example, we need to know whether it postdates the eruption of volcanics on Ambon and Ambelau. It is thus evident that to understand the Banda arc it is also necessary to understand the Sula spur, and the range of information to be absorbed is still further widened, this time beyond the scope of this paper.

Conclusion

The Banda arc lies within one of the most complex and heterogeneous parts of the Earth's crust and presents investigators with a series of problems and paradoxes. Almost always, attempts to explain the geological relationships within one area lead to consideration of a rather wider zone, which introduces new problems which require data from still further afield.

Thus it eventually seems to be necessary, in order to understand the structure of a few kilometers of, say, Timor or Seram, to understand not only the geology of the whole of the Banda arc but also of the Sula spur to the north and thence, almost certainly, of New Guinea, the Moluccas and Sulawesi. Even in the present state of very partial information, the sheer magnitude of the task of simply ordering and assimilating all the data is formidable. Probably the greatest contributions to our understanding of the area in the future will be made by more comprehensive and detailed studies of the Banda sea, ideally with deep drilling at a number of selected sites, and by the acquisition of paleomagnetic data with very tight geological control from formations of various ages in varying structural positions throughout the arc.

References

Audley-Charles M.G., Sumba fracture, a major discontinuity between eastern and western Indonesia, Tectonophysics 26, 213-228, 1975.

Audley-Charles, M.G., Indonesian and Phillipine Archipelago, in 'Phanerozoic Geology of the World, Vol.2, The Mesozoic', edited by M. Moullade and A.E.M. Nairn, pp 165-207, Elsevier Publishing Co., Amsterdam, 1978.

Audley-Charles, M.G., and J. Milsom, Comment on 'Plate convergence, transcurrent faults and internal deformation adjacent to south east Asia and the western Pacific' by T.J. Fitch, J. Geophys. Res., 79 4980-4981, 1974.

Audley-Charles, M.G., D.J. Carter, A.J. Barber, M.S. Norvick and S. Tjokrosapoetro, Reinterpretation of the geology of Seram: implications for the Banda arcs and northern Australia, J. Geol. Soc. London, 36, 547-568, 1979.

Bally, A.W., A geodynamic scenario for hydrocarbon occurrences, Proc. 9th World Petroleum Congr., Tokyo, 2, 33-44, 1975.

Bally, A.W. and Snelson, J., Realms of subsidence in 'Facts and Principles of World Petroleum Occurrence', edited by A.D. Miall, Can. Soc. Petrol. Geol. Memoir 6, 9-14, 1980.

Barber, A.J., M.G. Audley-Charles and D.J. Carter, Thrust tectonics in Timor, J. Geol. Soc. Australia, 24, 51-62, 1977.

Barber, A.J., H.L. Davies, P.A. Jezek, F. Hehuwat and E.A. Silver, Geology and tectonics of eastern Indonesia: Review of the SEATAR workshop 9-14 July, 1979, in 'The Geology and Tectonics of Eastern Indonesia', G.R.D.C. Indon. Spec. Pub. No. 2, edited by A.J. Barber and S. Wiryosujano, 7-28, 1981.

Benioff, H., Orogenesis and deep crustal structure, additional evidence from seismology, Geol. Soc. Amer. Bull. 65, 385-400, 1954.

Blundell, D.J., discussion of a paper by M. Norvick, 'Tectonic history of the Banda arcs, eastern Indonesia: a review, J. Geol. Soc. London, 136, 527, 1979.

Bowin, C., G.M. Purdy, C. Johnson, G. Shor, L. Lawver, H.M.S. Hartono and P. Jezek, Arc-continent collision in the Banda sea region, Amer. Assoc. Pet. Geol. Bull., 64, 868-915, 1980.

Cardwell, R.K. and B.L. Isacks, Geometry of the subducted lithosphere beneath the Banda sea in eastern Indonesia from seismicity and fault plane solutions, J. Geophys. Res., 83, 2825-2838, 1978.

Carter, D.J., Audley-Charles, M.G. and A.J. Barber, Stratigraphical analysis of island arc continental margin collision in eastern Indonesia, J. Geol. Soc. London, 132, 179-198, 1976.

Chamelaun, F.H., Paleomagnetic reconnaissance result from the Maubisse Formation and its tectonic implications, Tectonophysics, 32, 17-26, 1977.

Chamelaun, F.H. and A. White, Electromagnetic induction at Dili, Portuguese Timor, J. Geophys., 41, 537-540, 1975.

Chamelaun, F.H., K. Lockwood and A. White, The Bouguer gravity field and crustal structure of eastern Timor, Tectonophysics, 30, 241-259, 1976.

Chamelaun, F.H. and A.E. Grady, The tectonic development of Timor: a new model and its implications for petroleum exploration, Aust. Pet. Explor. Assoc. J., 18, 102-108, 1978.

Fitch, T.J. and W. Hamilton, Reply to comments by M.G. Audley-Charles and J.S. Milsom on paper 'Plate convergence, transcurrent faults and internal deformation adjacent to south east Asia and the western Pacific', J. Geophys. Res. 19, 4982-4985, 1974.

Foden, J.D. and R. Varne, Petrology and tectonic setting of Quaternary - Recent volcanic centres of Lombok and Sumbava, Sunda arc, Chemical Geology 30, 201-226, 1980.

Grady, A.E., Reinvestigation of thrusting in Portuguese Timor, J. Geol. Soc. Australia, 22, 223-227, 1976.

Grady, A.E. and R.F. Berry, Some Paleozoic - Mesozoic stratigraphic - structural relationships in east Timor and their significance in the tectonics of Timor, J. Geol. Soc. Australia 24, 203-214, 1977.

Hamilton, W., Tectonic map of the Indonesian Region, U.S. Geol. Surv. Map 1-875-D, Department of the Interior, Reston, Virginia, 1978.

Hamilton, W., Tectonics of the Indonesian Region,

U.S. Geol. Surv. Prof. Paper. 1078, 345 pp, 1979.

Jacobson, R.S., G.G. Shor, R.M. Kieckhefer and G.M. Purdy, Seismic refraction and reflection studies in the Timor-Aru trough system and Australian continental shelf, in 'Geological and Geophysical Investigations of Continental Margins', Amer. Assoc. Pet. Geol. Memoir 29, 209-222, 1979.

Jakes, P. and I.E. Smith, High potassium cal-alkaline rocks from Cape Nelson, eastern Papau, Contr. Mineral. and Petrol., 28, 259-271, 1970.

Jaques, A.L. and G.P. Robinson, Continent/island-arc collision in northern Papua New Guinea, Bur. Miner. Res. Aust. J. Aust. Geol. and Geophys. 2, 289-303, 1977.

Jezek, P., Gravity base stations in Indonesia and in the Southwest Pacific, Tech. Rept. 76-55, Woods Hole Oceanographic Inst. Massachusetts, 75 pp, 1976.

Johnson, R.W., Late Cainozoic volcanism and plate tectonics at the southern margin of the Bismark Sea, Papua New Guinea, in "Volcanism in Australia", Edited by R.W. Johnson, Elsevier Publishing Co., Amsterdam, 406 pp, 1976.

Katili, J., A review of the geotectonic theories and tectonic maps of Indonesia, Earth Sci. Rev. 7, 143-163, 1971.

Milsom, J., Preliminary gravity map of Seram, eastern Indonesia, Geology, 5, 641-643, 1977.

Milsom, J., and A. Richardson, Implications of the occurrence of large gravity gradients in northern Timor, Geologie en Mijnbouw, 55, 175-178, 1976.

Moore, G.F. and D.E. Karig, Structural geology of Nias island, Indonesia: implications for subduction zone tectonics, Amer. J. Sci. 280, 1980.

Norvick, M.S., Tectonic history of the Banda arcs, eastern Indonesia: a review, J. Geol. Soc. London, 136, 519-526, 1979.

Otofuji, Y., S. Sasajima, S. Nichimura, S. Hadiwisastra, T. Yokoyama and F. Hehuwat, Paleomagnetic evidence for the paleoposition of Sumba island, Indonesia, in 'Physical Geology of the Indonesia Island Arcs', edited by S. Nishimura pp. 59-66, Kyoto University, 1980.

Papp, Z., A three-dimensional model of the seismicity in the Banda sea region, Tectonophysics 69, 63-83, 1980.

Untung, M., Regional gravity surveys and base station network in Indonesia, United Nations Economic and Social Commission for Asia and the Pacific CCOP Technical Bulletin No. 13, 1-9, 1980.

Van Bemmelen, R.W., The geology of Indonesia, Government Printing Office, the Hague, 732 pp, 1949.

Vening Meinesz, F.A. Gravity expeditions at sea, Vol. I, Neth. Geol. Comm., Delft, 1932.

Vening Meinesz, F.A., Indonesian archipelago - a geophysical study, Geol. Soc. Amer. Bull. 65, 143-164, 1954.

Von der Borch, C.C., Continent-island-arc collision in the Banda arc, Tectonophysics, 54, 169-193, 1979.

EARTHQUAKE STRESS DIRECTIONS IN THE INDONESIAN ARCHIPELAGO

H.D. Tjia

Department of Geology, The National University of Malaysia
Kuala Lumpur, Malaysia

Abstract. The compressive stresses of shallow earthquakes that occurred between 1929 and 1973 are approximately perpendicular and occasionally approximately parallel to the structural grain of Sumatra, the Lesser Sunda islands, Irian Jaya, Halmahera, and the northernmost Moluccas. The Gulf of Gorontalo is characterized by either tension normal or compression parallel to the structural trends of the adjacent arms of Sulawesi. The Indonesian region east of the Makassar Straits consists of three or four tectonic domains, each of which is characterized by rather uniform earthquake stress directions. The distinct boundaries between the domains are the east-west striking Irian fault zone and the north-northwest trending Palu-Koro fault zone. Fractures, their patterns, and millimeter to centimeter long offsets along fractures that developed by recent earthquakes in the Lawu area, Java, in Lahad Datu, Sabah, and in western Timor were studied in detail in order to determine the generating stress directions. All three earthquakes had shallow foci. Their stress directions were essentially horizontal. However, a few earthquake-developed structures indicate that sometimes vertical compressive stresses may occur as well. The earthquakes of the Lawu area and Timor suggest horizontal compression directions within a narrow 25 degrees range. The earthquake damage at Lahad Datu, however, suggests compression along two mutually perpendicular azimuths. The lateral compression directions were either approximately parallel to the structural grain (western Timor) or perpendicular to it (Lawu area and Lahad Datu).

Introduction

The Indonesian Archipelago sensu lato is that region where three large lithospheric plates meet and interact (Figure 1). Active subduction zones and major transcurrent fault zones subdivide the region into platelets. A review of geotectonic theories that are essentially based on the geosynclinal concept, for instance by R.W. van Bemmelen, G.L. Smit Sibinga, J.H.F. Umbgrove, and J. Westerveld, was published by Katili (1971). In the last decade many authors have applied the plate tectonics concept to tectonic analyses of the region. Foremost among the resident geologists has been J.A. Katili, whose publications on the subject have now been collected into one volume (Katili, 1980). Hamilton (1979) expounded his views on the plate-tectonic framework of the archipelago, and the surrounding areas; the southeast Asian continental region, Australia and the Philippines. Earlier Hamilton (1974, 1975, 1978) published 1 : 5 000 000 scaled maps of sedimentary basins, earthquakes, and tectonics of the region. Other relevant publications are listed in the Katili volume and in Hamilton. Older literature is listed in van Bemmelen (1949).

This paper discusses earthquake stress directions that were interpreted from first-motions of shallow earthquakes that took place in the period 1929 to 1973 and compiled by Denham (1977). Further, the paper presents three detailed structural studies on damage from earthquakes in the Lawu area, Java, in 1979; in Lahad Datu, Sabah, in 1976; and in western Timor, in 1975. In the past, such detailed structural analyses of earthquake damage in Indonesia were not undertaken.

Earthquake Stress Directions (1929-1973)

Figures 2 and 3 show the stress directions of shallow earthquakes (focal depth in the order of 50 km or less) that occurred in Indonesia during 1929 to 1973. The stress directions (compression and/or tension) were compiled by Denham (1977) and only those that plunge at angles of 45 degrees or less were included in the figures. As Ritsema (1956) already discovered, most stress directions of shallow earthquakes in the Archipelago had been mainly horizontal. In Denham's list only a few shallow earthquakes had compression or tension that plunged steeper than 45 degrees. The importance of the lateral stress component is indicated in the figures by the length of the arrows (full line for compression and dashed line for tension). Three categories of stress plunges are considered: those that plunge at 20 degrees or less, those plunging between 21 and 35 degrees, and a third category plunging between 36 and 45 degrees.

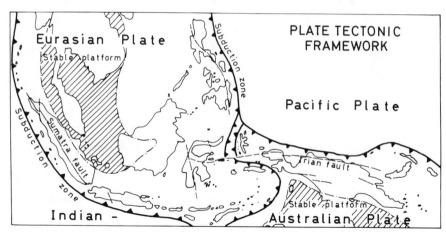

Figure 1. Three major lithospheric plates meet and interact in the Indonesian Archipelago, Southeast Asia. Fragmentation of eastern Indonesia into small platelets has apparently taken place through minor subduction zones in the Molucca Sea and large transcurrent faults. Data from various sources.

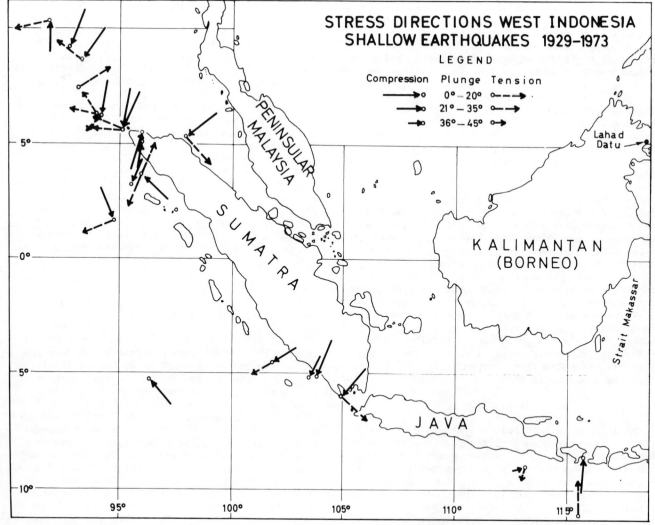

Figure 2. Compressive and tension directions of shallow earthquakes that occurred between 1929 and 1973 in western Indonesia. Data from Denham (1977).

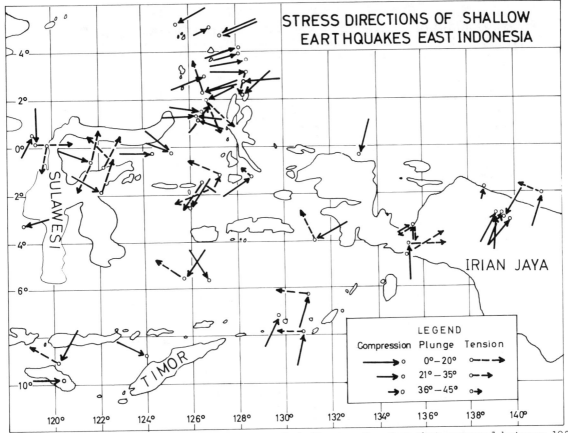

Figure 3. Compressive and tension directions of shallow earthquakes that occurred between 1929 and 1973 in eastern Indonesia. Data from Denham (1977).

Earthquake Stresses, West Indonesia

Sumatra, Java, and the Lesser Sunda islands are the surface expressions of the volcanic Sunda Arc. Figure 2 shows that the earthquake compression directions have strong lateral components. Most of these compressions are slightly oblique and a few are parallel to the strike of Sumatra. The oblique compressions are consistent with right-lateral slip movement along the Sumatra fault zone that runs along the entire western side of the island (see Katili and Hehuwat, 1967; Tjia, 1977). At the east end of the Sunda Arc, compression and tension are normal to it.

In the Sumatra section, some lateral extension directions are oblique and in one case normal to the island arc. In a few localities, no unique solution is possible and the first motion may represent compression in one direction or extension in a direction normal to it. Both types of stresses are shown for these localities.

Earthquake Stresses, East Indonesia

Generally compression directions are perpendicular or almost normal to the structural grain of various islands in the eastern part of the Archipelago (Figure 3). The exception is in the Gulf of Gorontalo, Sulawesi, where the compressive stresses are parallel or subparallel to the regional strikes of the north and east arms of Sulawesi, while tension is generally perpendicular to the regional structures. The following discussion will show that this exceptional situation is only apparent.

In Figure 4 eastern Indonesia is divided into four tectonic domains on the strength of known major geologic discontinuities and in one case based on areally uniform seismic stress directions. The northern tectonic domain (N) lies to the north of the Irian fault zone and to the east of the Palu-Koro fault zone. The Irian fault zone was suggested by Tjia (1973) based on a paper by Visser and Hermes (1962) who recognized the Sorong fault zone in Irian Jaya. Ahmad (1978) confirmed the existence of the left-lateral Matano fault zone in eastern Sulawesi. A modified version of the Irian fault zone also appears on Hamilton's plate tectonic map. The western tectonic domain (W) is the area to the west of the Palu-Koro fault zone, described by Katili (1970) and Tjia and Zakaria (1974). The large southern tectonic domain (S) in-

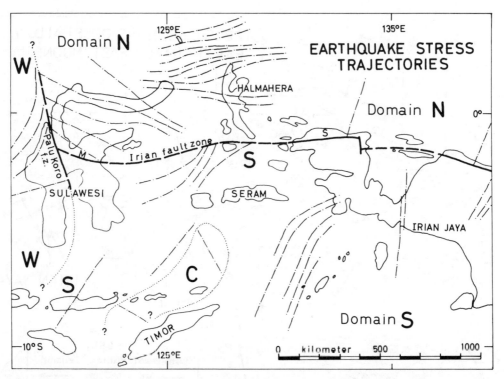

Figure 4. Earthquake stress trajectories constructed based on interpreted first-motion compressive stresses of shallow earthquakes that occurred between 1929 and 1973 and that pitch 45 degrees or less from the horizontal. Compare with Fig. 3. Four tectonic domains are distinguished: (1) Domain N (north) which may consist of subdomains west (North Sulawesi and Halmahera), and east (north of Irian Jaya); (2) Domain W (west); (3) Domain S (south), and the small (4) Domain C (center). The Irian fault zone includes the Matano fault (M) in Sulawesi and the Sorong fault (S) in Irian Jaya.

cludes the region to the south and to the east of the Irian and Palu-Koro fault zones. Within this domain, there seems to be a central tectonic domain (C) that is roughly bordered by Buru, Flores, Sumba and Timor, and that has seismic stress directions different from those of the surrounding domain. The boundaries of the central tectonic domain do not represent any known geological discontinuities. Perhaps the two compressive stress directions of this domain only represent local reorientations of the general northeasterly trending stresses of the surrounding areas. The compressive stress trajectories of Figure 4 are roughly westerly for the western part of domain N but almost southerly along the Irian Jaya section. In domain W the stress trajectories suggest easterly to northerly compression. None of these compression directions are consistent with left-lateral slip along the Palu-Koro fault. On the other hand, the compressive stresses from the east are consistent with left-lateral motion along the fault. In domain S the stresses trend between north and northeast; the latter direction being dominant.

Lawu Earthquake Swarm

Beginning in December 1978 and throughout 1979, an earthquake swarm rocked an elongated zone measuring approximately 30 km by 8 km around Mount Lawu and to its east. Gunung (G.) or Mount Lawu is a dormant volcano with solfataric and fumarolic activity that issues from Candradimuka, a crater on its south flank (Figure 5). The earthquake swarm culminated during May 14 and 15, 1979. Throughout the period of tremors, no changes in volcanic activity took place.

The stronger shocks of the swarm range between 2 and 3.4 on the Richter Scale. Focal depths were between 5 and 15 kilometers. Only one person was killed by falling stones, but damage to man-made structures in a 4-km wide ESE-trending zone was extensive. Van Bemmelen (1949) had suspected this zone to be a major fault zone. In mid-July 1979, together with S. Hamidi of the Indonesian Volcanological Survey, I made detailed field observations of the earthquake damage within the epicentral zone. At a number of places, fractures,

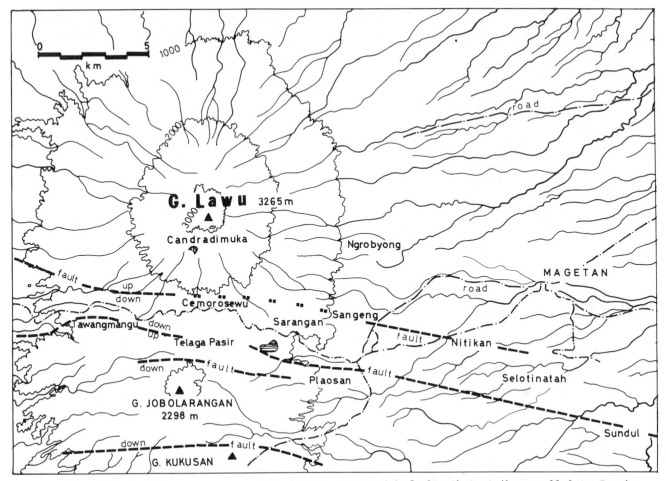

Figure 5. G. Lawu (G. = gunung = mountain) area, Java, with faults that strike parallel to Java's axis. The 1979 earthquake swarm caused damage to villages and towns located along the Sangeng-Nitikan and Tawangmangu-Sundul faults and those situated within the enclosed graben. Candradimuka is a solfataric center at approximately 2400 m elevation.

fracture patterns, displacements along the fractures on the ground, road deck, and on brick-walled houses were measured.

All observed fractures possess irregular surfaces. This was caused by the brittle nature of cement, bricks, soil, and the road deck in addition to the fact that some of the fractures are tensional. Most fractures dip vertical or almost vertical. A few fractures dip 70 to 45 degrees (especially those that are arranged in en echelon fashion), and dip around 25 degrees. Many fractures show definite average directions, but the fractures often consist of segments whose trends may deviate as much as 40 degrees from the general strike. Several fractures are arranged en echelon. This staggered arrangement was used to determine the sense of motion. Continuous and straight fractures may reach legths of 10 meters. Fractures that occur in bundles may attain zonal widths of 65 centimeters. En echelon fractures occur in zones up to 15 cm wide. Along certain fractures, distinct lateral offsets amounting to 6 millimeters or less were recorded; vertical displacements reaching 20 millimeters were also noted. At one locality, bricks at two corners of a house showed rotation around vertical axes. Fissures were opened to widths of 30 millimeters.

The following detailed descriptions of earthquake damage at three localities serve as examples used for the structural analysis.

Pass And Hamlet At Cemorosewu

At the pass of Cemorosewu (Figure 5), fractures extending over a distance of more than a hundred meters cut along the west side of the main road. The fractures strike 315° and are subparellel to the strike of the steep slope that descends from the road into a ravine. Here the rocks are tuff and volcanic conglomerate.

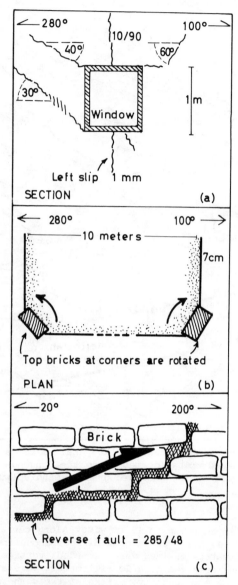

Figure 6. Earthquake damage by the Lawu swarm, 1979.
(a) Window in brick wall of the primary school at Selotinatah village. The en echelon arrangement of fractures suggest normal faulting along a 10/30 (strike 10°, dip 30° towards 100°) fracture in the wall.
(b) Rotations of bricks at the south side of a house near the primary school of Selotinatah. These bricks are approximately 2.25 m above the floor and pivoted about vertical axes in the directions indicated by the arrows.
(c) NNE-striking brick wall at the west side of the same house. A 285/48 reverse fault is suggested by the disalignment of bricks and wider horizontal gashes between the bricks.

The following fractures occur in the walls of a brick house in the vicinity of the pass. Vertical fractures perpendicular to a 300°-striking wall show right (4 mm; middle and top portion) and left (2 mm; lower portion) lateral offsets of the wall. A small reverse fracture (strike 294°, dip 24° to north) in a wall caused the upper portion to jut out 2 mm relative to the lower portion. Another vertical fracture striking 320° has 2 mm right lateral displacement. The various fracture trends and motions are plotted on Figure 7 CMS. A lateral compression in 24° direction explains all offsets. The right and left-lateral motions along the 30°-striking fault are to be expected, for this direction is almost parallel to the direction of horizontal compression, and a slight deviation of the fracture surface may produce the opposite movement.

Trunk Road Near Sangeng Village

A 65-cm wide fracture zone developed in the asphalt road deck of the trunk road from Sarangan

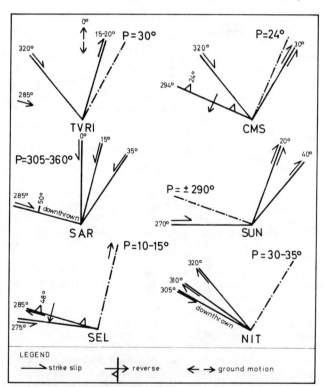

Figure 7. Plans of fracture directions and various types of offsets (see legend) developed by the Lawu earthquake swarm, mainly on May 14-15, 1979. Dash-dot-dash lines represent deduced lateral compression directions. Localities: TVRI = Television relay station at Cemorosewu; CMS = brick house at Cemorosewu; SAR = Sarangan town; SUN = Sundul village; SEL = Selotinatah village; NIT = Nitikan village area. The dip of some fractures is shown by a short dash (normal faulting) or small triangle (reverse faulting).

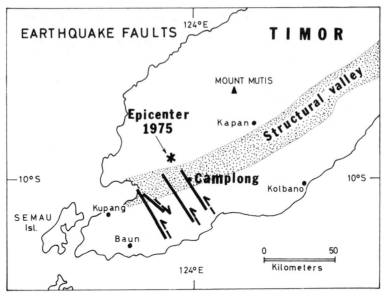

Figure 8. Four major fault zones developed by the 1975-earthquake in Timor (magnitude 6.1). The median structural valley is a graben-like feature but it possesses indications of left-lateral fault slip.

to Magetan near Sangeng village. The fracture zone strikes 305-310°, while individual fractures are vertical, strike 330° and are arranged en echelon, thus indicating right-lateral slip along the strike of the main fracture zone. Approximately 60 meters to the ESE of the road, a landslide in an 8-meter high, vertical valley wall indicates the continuation of the zone. The valley wall exposes semi-consolidated lahar. Farther towards the ESE in the expected continuation of the fault, no surface deformation could be detected. This may be due to the fact that most of the expected continuation of the fault traverses wet rice fields.

Selotinatah Village And Surroundings

The brick-walled primary school of Selotinatah village was damaged by the earthquake swarm. En echelon fractures in the walls indicate that the compression was vertical. Another fracture (strike 275°, dip vertical) shows 10-mm left-lateral offset. A fracture pattern around a window of the school is shown in Figure 6(a). Note normal faulting along the 10/30 fracture (strike 10°, dip 30° to east) to the west of the window and the slight left-lateral movement along a 10/90 fracture.

A house, a score of meters to the east of the school shows interesting damage. Figure 6(b) is the plan and shows that the topmost bricks at two corners on the frontal side were rotated in mirror-image fashion. The rotation caused one of the bricks to jut out 7 cm from the general position of the east wall. The rotations suggest differential motion between the roof that moved southward relative to the bottom part of the house. This differential movement was probably the result of sudden northward ground motion while the roof lagged in this motion.

On the east wall of the same house, a gash-like fracture among a number of bricks possesses wider separations between the horizontal segments compared to separations between vertical sides, as shown in Figure 6(c). Moreover, the non-alignment of the rows of bricks is suggestive of reverse fault motion along this fracture, whose general attitude is 285/48. A right-lateral component of 5 to 8 mm motion is also suggested by the displacement of the bricks on either side of the fault. Near the north end of the west wall of the same house, a zone of en echelon sigmoid gashes (strike 105°, dip 43° towards south) represents normal faulting.

In Figure 7 SEL the lateral and reverse motion directions along the earthquake fractures at Selotinatah are consistent with lateral compression in a (10-15°) to (190-195°) direction.

Lahad Datu Earthquake 1976, East Sabah

The epicentral area of the Lahad Datu earthquake of July 25-26, 1976, was in the western part of Darvel Bay, Sabah (locality indicated in Figure 2). The earthquake consisted of two main shocks; one took place on the night of the 25th, followed by a stronger (magnitude 5.8) shock the next morning. Its focal depth was 33 km. Until September 1976, more than 200 aftershocks were recorded (see Tjia, 1978). The epicenters were located in a 30-km wide zone trending northeast through the Dent Peninsula and

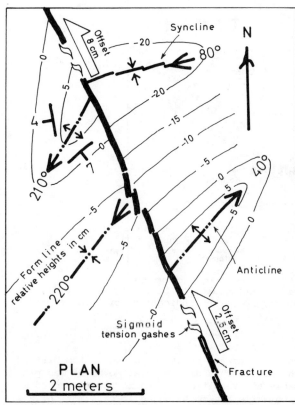

Figure 9. Earthquake deformation on an alluvial terrace surface along the Oesau river, north of Babau village, Timor. En echelon, sigmoid tension gashes, actual offsets, and the warped surface are consistent with sinistral fault motion in a northwest direction. The drag anticlines and synclines are schematically indicated by form lines with heights in centimeters.

part of Darvel Bay. A new mudcone ("mud volcano") erupted during the earthquake. Surface damage consisted of severe fracturing of two five-storey tall buildings that were under construction and of a new business center that consists of three-storey high cement and brick structures supported by concrete pillars. Stilted wooden houses also became slanted. Field investigation showed that damage was sustained only by artificial structures that stood on an olistostrom deposit, known as the Ayer Formation. Taller buildings in the center of Lahad Datu town were not damaged. These buildings stood on the so called Crystalline Basement, a series of metamorphosed Mesozoic rocks (Tjia, 1978).

Fractures, fracture patterns, and millimeter to centimeter long displacements along fractures in the walls and floors of the damaged buildings were recorded in detail. Structural analysis of these fractures showed that their patterns and offsets were the result of north-south compression probably alternating with tension, and also east-west compression that seemed to have alternated with tension. The fact that structures resulting from compression and tension coexist in both north-south and east-west directions, may be explained if ground motion was wavelike; at times traveling north-south, at other times propagating east-west. When the ground surface arched up, tension developed normal to the wave crest. Compression in the same direction took place when the ground surface was downwarped.

Camplong Earthquake of 1975, Western Timor

The earthquake that occurred in Timor on July 30, 1975 had a magnitude of 6.1 and a focal depth between 30 and 50 km. The epicenter was at 123.9° E. and 9.9° S. (Sutardjo, 1975). The main shock occurred at 0917h GMT, and aftershocks were felt up to October of the same year.

The earthquake caused linear fracture zones, many of which contained en echelon arrays of fractures. The fractures displaced parts of buildings and a weir. Some fractures could be individually traced for distances exceeding 150 meters, even two months after the main event. The general trend of the fracture zones was northwest (Figure 8). At Camplong, one fracture zone was mapped striking 10°.

The fracture zones ranged in width between 25 and 60 meters (see Tjokrosapoetro and Tjia, 1978). Each fracture zone consisted of 20 to 100 cm wide bundles of fracture strands that separated wider bands of unfractured soil or rock.

Approximately 700 m north of the main road from Kupang to the interior and that passes Babau village, a 335°-striking fracture zone straddled the Oesau river. The zone was 60 m wide and could be traced easily over a distance of more than 400 meters. The majority of fracture strands, shape of sigmoid gashes, and millimeter to centimeter-sized offsets in the alluvial surface suggested left lateral movement. Adjacent to fracture strands were undulations of the ground surface with amplitudes of less than a half meter and with long axes striking in 40-50° direction, consistent with sinistral motion on the fracture strands (Figure 9).

One kilometer to the east of the above locality, a concrete weir was traversed by a vertical fracture with a strike of 330° and 5 cm of right-lateral displacement. Alluvial sediments at the same locality were also cut by the fault. On the vertical fracture plane were fault striations pitching at 14 degrees or less and other small scale fault markings (see its usage in Tjia, 1972) that also indicated dextral slip.

A small hospital built of brick walls at Lili was destroyed by the earthquake. Fracture patterns and actual offsets in the millimeter range along the fractures may be explained as the result of sinistral motion along a northwest-striking fracture zone that crossed the building (Tjokrosapoetro and Tjia, 1978).

In west Camplong, a fracture zone could be traced over 150 m distance in a 340-350° direction. En echelon fractures in soil and displacements of a few millimeters on brick walls indicate left lateral movement.

The ground surface at Camplong market had zones in which fractures were arranged in en echelon fashion that suggested dextral motion in a 10° direction.

Figure 8 is a summary of the earthquake fracture zones and their lateral motions. These fault movements are consistent with compression acting in the sector of 70-95°. Such compression, however, does not explain dextral motion along 330°-striking faults at the Oesau weir. Perhaps this fault movement was the result of elastic rebound or of local reorientation of the regional stress.

The are structural indications in the raised, Quaternary reef limestone terraces of Timor, that compression in the sector of 70-95° (or alternatively, tension in the sector normal to it) have persisted throughout the Quaternary (Tjokrosapoetro and Tjia, 1978).

Conclusions

Intitial compressions of shallow earthquakes that took place during 1929 to 1973 in the Indonesian Archipelago had mainly horizontal components that were commonly directed slightly oblique to perpendicular to the structural grain of Sumatra and Java, and most parts of eastern Indonesia, except in the Gulf of Gorontalo. There the earthquake compressions were parallel to the structural strikes of the north and east arms of Sulawesi. However, the compressive stress trajectories for that part of Sulawesi are consistent with those characterizing a tectonic domain bordered to the south and the west by the Irian fault zone and the Palu-Koro fault zone, respectively.

Detailed structural analyses of earthquake damage in the Lawu area, Java, showed that the 1979 earthquake swarm occupied an epicentral zone that had already been recognized as a major fault zone striking parallel to Java's axis. The fractures, fracture patterns and offsets at various localities within the epicentral zone were consistent with horizontal compression that acted in the sector of 10-35°, or in other words, normal to slightly oblique to the tectonic grain of Java.

A similar structural study on damage in the Lahad Datu area, Sabah, by the 1976 earthquake showed that the surface deformation resulted from compression alternating with tension in north-south as well as east-west directions. The coexistence of compressional and tensional phenomena along the same azimuth may be explained if ground motion was undulatory. The earthquake probably developed north-south (which is normal to the structural grain of that part of Sabah) and east-west propagating surface undulations.

A third structural analysis of earthquake damage was made on fractures by the 1975 earthquake in Timor. The majority of lateral fault motions along northwest-striking fracture zones were probably the result of lateral compression that acted in the sector of 70-95° (which is parallel to subparallel to Timor's axis) or, alternatively, were developed by tension acting in the sector perpendicular to that of compression. Fractures, fracture patterns, and displacements in raised, Quaternary reef limestone terraces of Timor suggest that the stress field exhibited by the Camplong earthquake has persisted throughout the Quaternary.

References

Ahmad, Waheed, Geology along the Matano Fault zone, East Sulawesi, Indonesia, Proc. Reg. Conf. Geol. Min. Res. S.E. Asia, Jakarta, 1972, 143-150, 1978.

van Bemmelen, R.W., The geology of Indonesia Vol. IA, The Hague, Martinus Nijhoff, 1949.

Denham, D., Summary of earthquake focal mechanism for the western Pacific-Indonesian region, 1929-1973, Nat. Oceanic and Atmos. Admin., Environm. Data Service Report SE-3, 108 pp., 1977.

Hamilton, W., Map of sedimentary basins of the Indonesian region, U.S. Geol. Survey, Misc. Inv. Ser., Map I-875-C, 1 : 5 000 000, 1974.

Hamilton, W., Bathymetric map of the Indonesian region, U.S. Geol. Survey, Misc. Inv. Ser., Map I-875-B, 1 : 5 000 000, 1975.

Hamilton, W., Tectonic map of the Indonesian region, U.S. Geol. Survey, Misc. Inv. Ser., Map I-875-D, 1 : 5 000 000, 1978.

Hamilton, W., Tectonics of the Indonesian region, U.S. Geol. Survey, Prof. Paper 1078, 345 pp., 1979.

Katili, J.A., Additional evidence of transcurrent faulting in Sumatra and Sulawesi, Bandung, Nat. Inst. Geol. and Mining, Bull., vol. 2(1), 29-31, 1970.

Katili, J.A., A review of geotectonic theories and maps of Indonesia. Earth-Science Rev., vol. 7, 143-163, 1971.

Katili, J.A., Modern views on the geotectonics of Indonesia, Jakarta, Dep. Pertambangan R.I., 1980.

Katili, J.A. and F.H. Hehuwat, On the occurrence of large transcurrent faults in Sumatra, Indonesia, Geosc. J. Osaka City Un., vol. 10, art. 1-1, 1-17, 1967.

Ritsema, A.R., The mechanism in the focus of 28 South-East Asian earthquakes, Lembaga Meteorol. dan Geof., Dep. Perhubungan, Publ. no. 50, 76 pp., 1956.

Sutardjo, Laporan bencana alam gempabumi di daerah Kupang, Timor (Nusa Tenggara Timur) tanggal 30 Juli 1975, Pusat Meteorol. dan Geof., Dep. Perhubungan R.I., 1-6, 1975.

Tjia, H.D., Fault movement, reoriented stress field and subsidiary structures, Pacific Geology, 5, 49-70, 1972.

Tjia, H.D., Irian fault zone and Sorong melange, Sains Malaysiana, vol. 2, 13-30, 1973.

Tjia, H.D., Tectonic depressions along the transcurrent Sumatra fault zone, Geologi Indonesia, vol. 4, 13-27, 1977.

Tjia, H.D., The Lahad Datu (Sabah) earthquake of 1976; surface deformation in the epicentral region, Sains Malaysiana, vol. 7, 33-64, 1978.

Tjia, H.D. and Th. Zakaria, Palu-Koro strike-slip fault zone, Sulawesi, Indonesia, Sains Malaysiana, vol. 3, 67-88, 1974.

Tjokrosapoetro, S. and H.D. Tjia, Gejala-gejala tektonik Kwarter di Timor, Geologi Indonesia, vol. 5 (1), 11-26, 1978.

Visser, W.A. and J.J. Hermes, Geological results of the exploration for oil in Netherlands New Guinea, Koninkl. Ned. Geol. Mijnb. Genoot., Verh., Geol. Serie, vol. 20, 265 pp., 1962.

NEW ZEALAND HORIZONTAL KINEMATICS

H.W. Wellman

Geology Department, Victoria University
of Wellington, New Zealand.

Abstract. New Zealand data on horizontal kinematics from active faulting, nodal-plane studies, and retriangulation analysis are shown by maps and set out on tables. The common parameter is the strike of the "b" axis of the horizontal tectonic ellipse.

Interpretation by relative displacement vectors for two large (Indian and Pacific) and five micro-plates (Kaweka, Eastland, Fiordland, Puysegur Bank, and Macquarie) is set out on tables and illustrated by a map. In addition the bending which has produced the open "Z" recurved-arc shape of New Zealand is continuing, the movement being superposed on that of the plates.

Introduction

The very existence of New Zealand as a land area is a result of the convergent part of the dextral-convergent strain for the last 20 million or more years between the Indian and Pacific plates. (See McKay, 1890 for the active faults and uplift of the Kaikoura Mountains, and Cotton, 1922 for a general account of New Zealand geomorphology). Data from active faulting, nodal-plane studies, and retriangulation analysis directly relevant to the surface strain and direction within New Zealand are set out in tables. As far as possible the same parameters are used throughout. Active faulting provides most information and is used as a standard.

Three angles are needed to define the nature of faulting. Strike and dip are universally used and do not require explanation, but there is no general agreement on which one of several possibilities should be used for the third angle. The one used here, and termed the "rake", is readily and accurately determined in the field and has the further advantage that it is independent of the dip. Having a fault, its strike and a fault-displaced line, the "throw" is the vertical displacement, and the "wrench" the horizontal displacement in the line of the fault. The tangent of the rake equals wrench/throw. In the same way if the heave is the horizontal displacement normal to the strike then the tangent of the dip equals throw/heave. The accuracy of the rake (or actually its tangent) relative to the dip has long been emphasized by Lensen (1964), and in the tables below it is listed immediately after the strike and before the dip.

Geometry for homogeneous strain

The geometry of faulting (including nodal-plane solutions) is greatly simplified by assuming movement to be uniform. More explicitly it is assumed that all sets of lines irrespective of direction if straight and parallel remain straight and parallel after deformation. Such deformation is termed homogeneous and it changes a sphere into the well-known strain ellipsoid (Fig. 1).

It is relevant that except for the Alpine Fault, faulting in New Zealand is within fault-zones, and is thus distributed although not necessarily uniformly.

For the strain ellipsoid produced by distributed faulting there are two sets of perpendicular circular sections each with unchanged area, one set representing the fault-plane, and the other set being conjugate and representing the auxilliary-plane. In active faulting the auxilliary-plane is the plane normal to the slickenside direction, and movement is on the fault-plane and not on the auxilliary-plane. In nodal-plane studies the two planes are defined but the fault-plane cannot be distinguished from the auxilliary-plane. Other important planes are the "motion-plane" which is parallel to the slickensides and normal to fault-plane and auxilliary-plane, and the horizontal central section through the strain ellipsoid which is generally an ellipse and is called the "strain ellipse". The strain-ellipse is used on the maps below to define the strain direction.

It should be noted that in the general case neither the fault-plane nor the

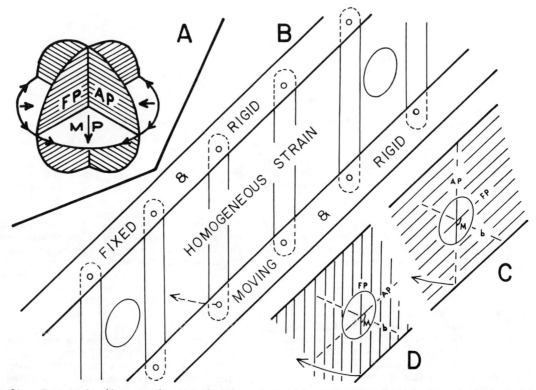

Fig. 1A - Isometric diagram showing the three mutually perpendicular planes of the strain ellipsoid:- FP, fault-plane (circular); AP, auxiliary-plane (circular); MP, motion-plane (elliptical), and the slickensides on the fault-plane and auxiliary-plane. The ellipsoid can have any orientation. As drawn it represents a north-east-striking pure-dextral fault.

Fig. 1B - Simplest possible "working" model for the boundary through New Zealand between the Indian Plate (fixed and rigid) and the Pacific Plate (rigid and moving eastwards). The part between the two rigid bars is supposed to be an elastic sheet homogeneously deformed lying above the rigid connecting bars that control the movement. Note that lengths in the direction of the two rigid bars (parallel to plate boundary) and lengths in the direction connecting bars (at right angles to plate movement) do not change.

Fig. 1C - The area between the two rigid bars of "B" covered by a series of north-east-striking slats. Strain ellipse shows fault plane (FP) striking NE parallel to slats, and auxiliary plane (AP) striking north parallel to the connecting bars. With time the line AP swings around so as to increase the value of the angle "M". The faulting is reverse-dextral and <u>parallel</u> to the plate boundary, as are the Marlborough faults. "b" = strike of minor axis.

Fig. 1D - Like B except that the slats are parallel to the connecting bars and it is the fault (reverse-sinistral) and the slats that rotate and not the NE striking auxiliary-plane. The faulting is <u>oblique</u> to the plate boundary as are the White Creek Fault and the Ostler Fault. "b" = strike of minor axis.

auxilliary plane are vertical, consequently in general although the two planes are at right angles their strikes on a map are not at right angles.

The three components of the fault displacement t, w, & h.
 t = vertical component (throw)
 w = horizontal component parallel to fault-strike (wrench component) either dextral (right-lateral) or sinistral (left-lateral)
 h = horizontal component at right angles to fault-strike (heave) either widening (normal-) or narrowing (reverse-faulting)
 n = fault displacement (net).
 $n^2 = t^2 + w^2 + h^2$

The three faulting angles S, R, & D.
 S = strike of fault-plane (downthrow to right in tables)
 R = Rake = $\tan^{-1} w/t$
 D = Dip = $\tan^{-1} t/h$

Fig. 1A shows the mutually perpendicular relation between the fault-plane, the auxilliary-plane, and the motion-plane.

For convenience of representation, the motion-plane is shown horizontal and the other two planes vertical. In the general case all three planes are inclined and faulting is either partly dextral or partly sinistral. Thus in the general case for nodal-plane studies one solution is partly dextral and the other partly sinistral, but which is the fault-plane and which the auxilliary-plane is unknown. In the tables there are two rows for each nodal-plane solution, the upper row being for the dextral plane, and the lower for the sinistral.

f = fault-plane   d = dextral
a = auxilliary-plane   s = sinistral

The following formulae are used to calculate values given in tables

Nodal-Planes		Active Faults	
Known	Required	Known	Required
S_d		S_f	S_a
S_s	R_d	R_f	R_a
M	R_s		D_a
D_d	b	D_f	$M = S_f - S_a$
D_s			b

($\cos M \tan D_d \tan D_s = 1$)
(check using redundancy)

$\tan R_d = \tan D_d \tan M$ $\tan M = \tan R_f \cot D_f$

$\tan R_s = \tan D_s \tan M$ $\cot D_a = \cos M \tan D_f$

$M = S_d - S_s$ $\tan R_a = \tan D_a \tan M$

$b = 1/2 (S_d + S_s)$ $S_a = S_f + M$

$b = S_f + 1/2 M$

Terms used to describe the nature of faulting

Provided there is an appropriate change in the dip the following terms suffice: normal, normal-dextral, dextral-normal, dextral, dextral-reverse, reverse-dextral, reverse, reverse-sinistral, sinistral-reverse, sinistral, sinistral-normal, normal-sinistral, and back to normal. Unfortunately faults having appreciable throw and near-vertical dip are not uncommon and need a name. "Throw" is used here and can replace either normal or reverse in the above sequence. Thus the essentially dextral Wairarapa Fault at Waiohine River with appreciable throw and near vertical dip is termed a "dextral-throw" fault. Faults with zero or near-zero dip are equally anomalous and are termed "horizontal". For nodal-plane solutions if one plane is horizontal, then the other will be a pure "throw" fault.

The recommendations of Reid et al (1913) are used for distributed faulting, the shift (total displacement) being equal to the drag (distributed displacement) plus the slip (the displacement on the fault-plane). Drag, being distributed movement, is usually difficult to see and its meaning is extended here to include the unrecorded faulting in unfavourable terrain. Bedding-plane movement is like the movement between the leaves of a book, and is termed "bedding faulting". Drag and bedding faulting are of major importance for New Zealand tectonics.

The strain-ellipse as a mapping symbol (Figs.3, 4 & 5).

For active faulting the strain-ellipse shows the direction of strain and the line across the ellipse the strike of the fault. The darkened side of the ellipse defines the downthrown side. For nodal-plane solutions there are two strike lines and a darkened quadrant to indicate downthrow. For retriangulation analysis the ellipse gives the strain direction and the included number the strain rate. The rate is given in units of 10^{-9} radians (n rad) per year.

The advantage of the strain-ellipse method for showing strain is that the ellipse is small enough to be plotted on the site itself, whereas if the strain ellipsoid is used, as is generally done for nodal-plane solutions, then it is too big for the site and has to be "lined" in to its site from the side of the map. The disadvantage of the ellipse method is that the fault-dip is not shown directly. For extreme cases where either the fault-plane or the auxilliary-plane are within say 15° of horizontal the strain direction is undefined and a circle is used.

Retriangulation analysis deals with angular changes only. (This is not true for remeasurement with electronic distance measuring instruments, but retriangulation is effectively all there is in New Zealand at present). However, if there is a dominant fault direction it can be assumed to define one of the two cross sections with unchanged area, and the change in the area of the strain ellipse can then be determined.

Geology

The covering strata of New Zealand (cover for brevity) are less than, and the undermass is more than 100 million years old. At most places the cover has basal coal measures that rest on a peneplain cut across the undermass, and the cover dips less steeply and is far less metamorphosed than the undermass. However, on the east coast from 150km northeast of Christchurch to Eastland, there is

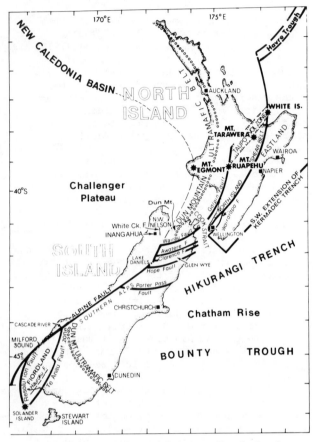

Fig. 2. Sketch map of New Zealand showing the more important localities mentioned in the text.

no peneplain and no basal coal measures. The east coast strata are strongly deformed and contain a high proportion of turbidites.

The undermass on the outside (east and north) of the line of the Dun Mountain ultramafics (Fig. 2) consists almost entirely of greywacke, or of greywacke metamorphosed to schist, and the age is from 280 to 100 million years (Permian to lower Cretaceous). On the inside of the line the undermass is more varied and mostly older and is like the Australian undermass of which it was once a part.

Active Faults (Tables 1 & 2 and Fig. 3).

Of the 68 fault displacements listed 90 per cent are from river terraces, and for 30 per cent throw, wrench, and thus rake were determined independently from multiple reference lines, a reference line being either the back of a terrace tread or a channel cut in a tread. The main feature is that the active faults be crossed by an energetic but stable river that retains its gradient irrespective of the faulting. For multiple features the fault itself is generally a well defined single line that extends evenly for 5km or more. River flow will be fairly uniform throughout the year with a flood discharge in each ten years of some $50m^3/sec$. The terraces will be cut in river-gravel alluvium that built up between 10 and 20 thousand years ago, is less than 10m thick, and generally rests directly on undermass (not cover). Few are in forest, and there are probably a few multiple features still to be discovered in New Zealand.

New Zealand appears to have far more active faults than would be expected from its earthquake frequency. As an explanation it is suggested that many countries are too dry to have effective rivers, that in New Guinea the active faults are hidden by forest, and that in Japan many of the faulted river terraces have been levelled off by farmers to grow rice.

Fault Zones (Tables 1 & 2, Figs. 2, 3 & 6).

As already mentioned, most of the active faults of New Zealand lie within fault-zones. The first from north to south is the Taupo Volcanic Zone, which is about 20km wide and is well-defined from the Bay of Plenty coast south to Mt. Ruapehu. There are numerous discontinuous faults that lie on an arc with a centre at $40°S\ 178.5°E$. They have been closely examined on the ground and in air photographs, and have been drilled for in the search for geothermal steam. They are shown in maps and cross-sections as having throw, being near vertical, and being without appreciable wrench displacement. (Grindley, 1960). According to Lensen (1981) a fault displacing a lava flow at Kakaramea, 35km NNE of Mt. Ruapehu, is normal-dextral in nature. On oblique air photographs Mt. Ruapehu can be seen to lie within a graben formed by two parallel-striking normal faults. Most of the zone is covered by thick volcanic ash and the only dikes known are the basalt dikes at Mt. Tarawera intruded during the 1886 eruption (Cole & Hunt, 1973) with a total width of about 3m. Geological data thus favours normal faulting, probably with some dextral displacement. There must be some kind of southerly extension for the Taupo Volcanic Zone. An estimate made by Wellman (in press) is adopted here, and it is assumed to extend south along the Galpin Fault to join the Wairau Fault. (Fig. 6, southern part of a/K).

The North Island Shear Belt also extends south from the Bay of Plenty coast. It lies to the east of the Taupo Volcanic Zone and is curved in the opposite direction, having a centre of curvature at about $37.6°S\ 171.3°E$. For its whole length the belt is close to the axial ranges of the North Island, and it ends at Cook Strait near Wellington. For most of its length it includes two or three active faults, and there are an equal number

TABLE 1 - Data for the active faults of the North Island at those places where the nature of the faulting is reasonably well known. See Table 3 for faulting during historical earthquakes.

Item No.	Plate Bdy. (Fig.6)	Fault and place	Net slip (m)	Lat. S.	Long. E.	Quality see expln.	Strike (down to right)	Rake (see expln.)	Dip	Strike of "b"	Ref. & (item no.) see below
1	2	3	4	5	6	7	8	9	10	11
1	-	Hauraki Plain, 3km S.E. Waitoa	5.1	37.61	175.65	C	133	D79	R80±10	179	4(2),6
2	-	Hauraki Plain, 6km S.E. Waitoa	4.1	37.63	175.67	C	123	D76	R80±10	169	4(1),6
3	aK	Kakaramea. Omoto Stream, 9km W. Turangi	18.4	39.00	175.71	C	040	D45	N50±20	065	4(3),6
4	-	Inglewood, 4km W. Inglewood Rly.Stn.	14.4	39.16	174.16	C	035	D76	N60±20	076	4(4),6
5	-	Moumahaki F.,5km E. Waverley Rly.Stn.	26.1	39.77	174.68	C	026	D73	N60±10	066	4(5),6
6	-	Wai-inu F., 1km S. Wai-inu Beach	3.0	39.87	174.74	B	038	D040	N70	071	6
7	-	Pull-aparts, Wanganui Coast	Ind.	39.80	174.70	131	054±16	X00	X90±10	54	6
8	KE	Ruatahuna, 2km N.E. Ruatahuna	6.1	38.61	176.98	B	195	D81	R70±10	062	4(6),6
9	KE	Mohaka, 4km W. Mid. Mohaka Bridge	34.1	39.18	176.59	C	199	D88	R80±5	064	4(7),6
10	KE	Branch Mohaka F.,13km W.Mid Mohaka Brdg.	6.1	39.30	176.62	A	200	D81	X90±5	065	4(8),6
11	KE	Mohaka F., track to Big Hill, 43km W. Hastings	7.1	39.59	176.39	A	205	D82	R80±10	071	4(9),6
12	-	Raetihi, F.	6.0	39.43	175.30	B	095±180	S81	R60±10	047	5,6
13	KE	Alfredton, 7km N.N.E. Alfredton	18.1	40.62	175.90	A	040	D85	X90±10	085	4(10)
14	Ep	McKay (bedding F.) 1 Aug.1942 E.Q.	1	40.72	175.83	B	030	X00	R63±10	120	3
15	KE	Wairarapa, 6km N. Alfredton	6.2	40.62	175.86	A	040	D76	X90±10	085	4(11)
16	KE	Wairarapa, Mangatariri River	76.1	40.95	175.52	A	040	D88	X90±10	085	4(13)
17	KE	Wairarapa, Waiohine River, N.Bank	100	41.05	175.40	5	045	D82	X90±10	090	2,4(14)
18	Ep	Wharekauhau F., at coast	20	41.38	175.05	B	025	X00	R45	115	6
19	Ep	Ngapotiki F., at coast	6	41.57	175.38	A	010	X00	R20	100	6
20	KE	Wellington, 27km due W. Alfredton	18.1	40.67	175.54	A	040	D84	X90±10	085	4(12)
21	KE	Wellington, Hutt River, Harcourt Park	10	41.11	175.09	3	060	D85	X90±20	105	1
22	KE	Wellington, Silver Strm., 3km N.coast	48.1	41.33	174.72	A	040	D86	X90±10	085	1,4(15).
23	KE	Oteranga, 3km S.S.W. Makara Beach	7.1	41.24	174.69	A	045	D82	X90±10	090	4(16)
24	KE	Oteranga, 9km S.S.W. Makara Beach	25.1	41.29	174.65	A	041	D84	X90±10	086	4(17).
25	KE	Owhariu. Makara Strm.	5.0	41.26	174.71	A	039	D83	X90±10	084	4(18).
26	aK	Basalt dikes, Mt.Tarawera 1886 (3m)	Ind.	38.24	176.50	A	050	X00	X90±10	140	8.

1. Lensen, 1958.
2. Lensen & Vella, 1971.
3. Neef, 1976.
4. Lensen pers.com. Canberra, Dec. 1979. (18 North Island items). Published 1981.
5. Sissons, 1979.
6. Wellman, Field observations (unpublished).
7. Wellman, From air photographs, July 1980.
8. Cole & Hunt, 1973.

NEW ZEALAND

TABLE 2 – Data for the active faults of the South Island at those places where the nature of the faulting is reasonably well known.

Item No. & Plate Bdy.	Fault and place	Net slip (m)	Lat. S.	Long. E.	Quality (see expln.)	Strike (down to right)	Rake (see expln.)	Dip	Strike of "b"	Ref. & (item no.) see below
1	Column No.	3	4	5	6	7	8	9	10	11
	Giles Creek, 8 parallel bedding faults									
1 Z?	Waimea F., Wai-iti River	15	42.05	171.80	8	017±4	Z00	Z90±20	107	1(7-15)
2 –	Wangamoa F., Wangamoa Valley	32	41.53	172.94	4	213±6	D50±11	R80±10	092	5
3 aK	Wairau F., Waihopai River	6	41.17	173.52	A	055	D90	N60	095	1(24)
4 aK	Wairau F., Branch River	30	41.52	173.74	A	240	D88.7	Z90	105	1(121),5(1)
5 aK	Wairau F., Lake Rotoiti, peninsula	56	41.68	173.19	6	243±1	D87±0.2	Z90±5	108	1(102-8),3
6 aK	S. branch Wairau F., Kiernans Creek	49	41.81	172.83	B	240	D76	R70±10	107	1(95)
7 aK	S. branch Wairau F., Centre Strm to Black Strm	75	41.57	173.54	2	251	D79	Z90	116	1(123)
8 aK		93	41.54	173.65	2	253±2	D90	Z90±20	118	1(125-130)
9 Kp	Awatere F., Blackbirch Strm. "Seddon"	14	41.71	173.88	2	231±1	D85±1	R70±10	097	5(28)
10 Kp	Awatere F., Grey River	35	41.89	173.58	2	242±2	D82±12	R70±10	108	1(184-5),2
11 Kp	Awatere F., Kennet River	5	42.20	173.36	2	60	D84±2	R70±10	106	1(176-7)
12 Kp	Awatere F., Saxton River	20	42.09	173.14	2	247	D88±4	R60±20	112	1(165-6),5(32)
13 Kp	Clarence F., Coverham	14	41.88	173.83	2	249	D88±0.2	R70±10	114	5(31-32)
14 Kp	Kekerengu F., 1km S. Kekerengu River	4	41.98	173.99	A	065±5	D85	Z90±5	110	6
15 Kp	Hope F., 1km E. Glenwye	45	42.59	172.53	A	080	D88	R80±10	125	1(236)
16 Kp	Ashley F., W.Bank Grey River	7	43.18	172.43	?	265	D66	N70±10	125	5(33)
17 Kp	Porters Pass F., 1km E. Lake Lyndon	34	43.30	171.22	2	065	D82±1	R80±10	111	5(34-5)
18 ap	Alpine F., Matakitaki River	9	42.02	172.51	A	226	D76	R60±20	95	5(8)
19 ap	Alpine F., Maruia River. Dip from drill-hole	9	42.35	172.22	A	226	D75	R60	95	1(83),5(9)
20 ap	Alpine F., W.Bank Ahaura River	9	42.56	171.84	2	233	D80±0.3	R70±10	100	5(10-11)
21 ap	Alpine F., 2km S.W. Ahaura River "Haupiri"	33	42.57	171.82	A	239	D78	R70±10	106	5(12)
22 ap	Alpine F., Haupiri River, N.Bank	4	42.60	171.73	A	240	D55	R70±10	112	1(75)
23 ap	Alpine F., Haupiri River, S.Bank	6	42.60	171.73	A	240	D29	R70±10	121	1(77)
24 ap	Alpine F., Inchbonnie	8	42.72	171.48	4	236	D82±0.5	R70±10	102	5(13-16)

Table 2 cont.

Item No. & Plate Bdy.	Fault and place	Net slip (m)	Lat. S.	Long. E.	Quality (see expln.)	Strike (down to right)	Rake (see expln.)	Dip	Strike of "b"	Ref. & (item no.) see below
Column No.	2	3	4	5	6	7	8	9	10	11
25 ap	Alpine F., Taipo River	3	42.76	171.41	B	230	D63	R70±10	100	1(69)
26 ap	Alpine F., Taipo River	3	42.76	171.41	A	235	D80	R70±10	102	5(17)
27 ap	Alpine F., Toaroha River	7	42.90	171.13	2	233	D82±2	R70±10	99	5(18–19)
28 ap	Alpine F., Okuru River	14	43.94	168.99	2	239	D82±2	R70±10	105	5(20–21)
29 aF	Alpine F., Martyr River (moraine)	300	44.13	168.71	B	235	D87	R70±10	101	1(59)
30 aF	Alpine F., Kaipo River (moraine)	–	44.48	167.96	B	230	D90±10	Z90±10	95	4
31 –	Lake Wakatipu, Whites Bay	11	45.11	168.43	3	260	D75±5	N80±10	126	5(47–49)
32 Z	As above (both given as Moonlight F.)	3	45.11	168.43	5	172±2	S52±11	R80±10	123	5(50–54)
33 Z	25km N.W. Mt.Somers "Lake Heron F."	7	43.59	171.12	3	000	S61±12	R80±10	132	5(36–38)
34 Z	Ostler F., at Dry Strm.	4	44.20	170.05	A	005	S72	R80±10	138	5(42)
35 Z	Ostler F., 5km to 7km S. Dry Strm.	24	44.27	170.04	2	002±2	S58±2	R80±10	134	5(43–44)
36 Z	Ostler F., 16km S. Dry Strm. (Lake Ohau Road)	8	44.34	170.01	A	015	S50	R80±10	146	5(45)
37 Z	Ostler F., S.bank Ahuriri River	13	44.50	169.86	A	020	S18	R80±10	141	5(46)
38 Z	Akatore F., Big Creek	8	46.16	170.11	2	185	S62±1	R80±10	137	5(55–56)
39 Fp	Hollyford F., Upukerora River, Mudstone Ridges	27	45.30	167.97	B	015±10	S64±5	R75±5	146	7:57
40 Fp	Hauroko F., 2km S.E. Oblong Hill	180	45.97	167.36	B	030	S80±10	R70±10	165	1(304)
41 –	Little Mt. John F.	18	43.97	170.43	3	250	D82±2	R80±10	115	4(39–41)
42 –	Fowlers F.	12				250	D82	R80±10		8(42)
43 –	Manuherikia Valley "Dunstan F."	17	45.04	169.03	2	062±2	D63±7	R80±10	110	8(57–8)
44 –	Clinton River, 2km N.W. Puhipuhi	90	42.26	173.73	1	000±20	Z00±20	R30±20	090	1(228)

1. Wellman, 1953 (Table 1; 309 items).
2. Lensen, 1964.
3. Lensen, 1968.
4. Wellman & Wilson, 1964.
5. Berryman, 1979 (Table 2; 56 items).
6. Wellman, field and air photo observations.
7. Grindley, 1958.
8. Lensen, per com. Canberra, Dec. 1979 (Table 1; 42 South Island items).

NEW ZEALAND

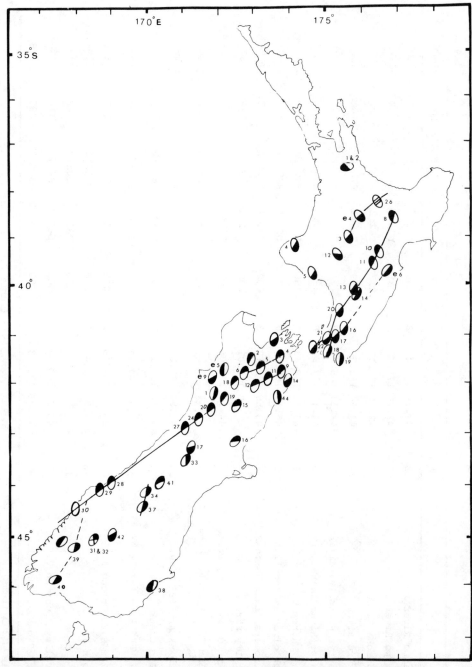

Fig. 3. Sketch map of New Zealand with strain ellipses superposed on active faults to show nature of faulting. Number beside ellipses give item number in tables. The prefix "e" refers to the earthquake faults listed in Table 3. The black quadrant in the ellipses defines the down-throw side of fault-plane and auxiliary-plane.

to the west that were probably active a million or so years ago. When crossing hills and valleys those faults that are well-defined are straight, and fault-dip is thought to be steep. Although the faults are crossed by many streams there are few places where the nature of faulting is shown, but all are consistent and indicate that faulting is essentially dextral throughout. The throw although small relative to the wrench displacement determines the height of the mountain ranges and the direction in which they slope. At the south the faults are downthrown to the east, the Wellington Fault and the easterly downthrow forming Wellington Harbour being the best known example (Wards, 1976: 232).

A series of growing anticlines with north-east striking axes extend along the 120km-wide belt between the Wellington Fault and the line of extension of the Kermadec Trench. The rate the belt is narrowing is inferred from the rate the folds are tilting.

The four main faults of Marlborough were the first active faults to be mapped in New Zealand (McKay, 1890). They strike east-north-east and are extremely well-defined between 173°E and 173.9°E. They are fairly well on line with the north-north-east striking dextral faults of the North Island Shear Zone described above, but two of the faults appear to die out and two branch to the east and a cross fault or faults striking north-west through Cook Strait is likely (Ghani, 1978). To the west (Freund, 1971) the Marlborough faults branch and "die out" within the Southern Alps before reaching the Alpine Fault.

The Alpine Fault is the main fault of the South Island, and is one of the best defined major faults of the world. From a height of 80km in satellite "photographs" it is an outstanding feature (Wards, 1976) but from 10km in airplane photographs it is far less spectacular than the Marlborough Faults. For the 400km at 235° from Lake Daniells (42.57°S 172.33°E) to the Cascade River (44.29°S 168.33°E) the trace of the fault is nowhere more than 5km from the straight line between the two points. To the south-west of Cascade River the strike decreases by 5° to 230° and the fault extends as an equally straight line for a further 100km at least.

From Lake Daniells to Cascade River the Alpine Fault shows dextral displacement and north-westerly downthrow. The downthrow is represented by the steep north-west facing scarp of the Southern Alps and by the north-west facing scarps on displaced river terraces. The dextral displacement is shown by the 1.5km offset of many of the valleys cut across the fault, at three places by dextrally displaced moraine of the Last Glaciation (Wellman, 1953) and by the dextral displacement of the youngest river channels (Berryman, 1979). The rake decreases from about 85° at the two ends to about 45° in the middle opposite the highest part of the Southern Alps. Total throw is shown by the 25km-wide belt of up-ended schist that extends for the whole of the 400km and for an extra 30km to the north from Lake Daniells to Wairau Fault. The total dextral displacement is shown by the 500km offset of the Dun Mountain ultramafic belt (Hunt, 1978).

For the 45km from Cascade River to the coast at Milford Sound there is no indication from either topographic features or differences in regional rock metamorphism that one side has been uplifted more than the other, and for this stretch the fault movement appears to have been pure dextral (Wellman & Wilson, 1964). To the south of Milford Sound there is a 2 km scarp between the Fiordland mountains and the sea bed to the west. It is attributed here to dextral movement having brought oceanic crust into direct contact with continental crust.

The Te Anau Fault Zone is the name given here to a set of faults that extends south from the Alpine Fault at Cascade River to the south coast at the east side of the Fiordland Block. At Cascade River the uplift on the eastern side of the Alpine Fault ends and faults that are mapped as sinistral branch off and strike south (Grindley, 1958). The line of the faults extends through the Te Anau Depression as a fault complex to join the Hauroko Fault near the coast on the eastern side of the Fiordland Block. According to the sense of displacement of "forested spurs" the Hauroko Fault is also sinistral (Wellman, 1953). The throw on the Alpine Fault to the north of Cascade River is thus thought to be transformed into the sinistral displacement of the Te Anau Fault Zone.

The name Resolution Fault is given here to the fault that extends around the south-western end of Fiordland. It is well-exposed on the east side of Five Fingers Peninsula (Wellman, 1954) and causes active faulting (sense unknown) at West Cape (Wellman, 1953).

Historical Earthquakes with surface faulting (Table 3, Fig. 3).

According to Eiby and Reilly (1976) there have been 17 historical earthquakes within New Zealand with magnitude greater than M7. But for 6 only is there faulting data useful for stress analysis, and one of these - Taupo, 1922 - is an earthquake swarm with a maximum magnitude of considerably less than M7. Contemporary accounts of earthquake movement, other than landslides, are entirely from the sea coast or from man-made features such as fences, roads, survey lines, etc. About half the earthquake faults reported are bedding faults that represent distributed movement during folding. The lack of recorded faulting is thus attributed to unfavourable conditions - forest, mountains, swamp, etc. - and to most faulting being distributed and thus difficult to find.

The Wellington earthquake of 1855 with an estimated magnitude of about M8 (Eiby, 1968) is New Zealand's largest historical earthquake. Contemporary accounts (Lyell, 1856) describe faulting on what is now known as the Wairarapa Fault, and uplift of 2.5m at Turakirae Head 20km east of Wellington, decreasing in 10km to zero on the west coast. From displaced river terraces the Wairarapa

TABLE 3 – Data for the New Zealand faults active during historical earthquakes.

Item No. & Plate Bdy.	Year, Name & Magnitude of Earthquake Fault, Place	Length (km)	Net slip (m)	Lat.S Long.E	Strike FP Strike AP	Rake FP Rake AP	Dip FP Dip AP	"b" M	References
1	Column No. 2	2A	3	4	5	6	7	8	9
1 KE	1855 Wellington E.Q. M8±0.1 Wairarapa F. Waiohine R. River channel.	?	12	41.05 175.40	045 135	D86 S90	90±10 N86	090 90	Lyell, 1856; Lensen & Vella, 1971; Wellman, 1972
2 Kp	1888 Glenwye E.Q. M7±0.3 Hope F. Fence. Length 20±10km.	20±10	2	42.60 172.50	080 170	D90±5 S90	X90±5 X90	125 90	McKay, 1888; Cotton, 1922
3 aK	1922 Taupo Swarm. M6.7±0.2 Whangamata F. 15km W. Taupo.	10	2	38.65 175.20	040 220	X00 X00	N45±30 N45±30	040 00	Grange, 1932; Sissons, 1979
4 aK	1922 Taupo Swarm. Kaipo F. 5km W. Taupo. Length 10 km.	10	1	38.70 176.00	220 040	X00 X00	N45±30 N45±30	040 00	as above
5 Z	1929 W. Nelson E.Q. M.7.6 White Creek F. Water race, etc.	10	5	41.77 172.18	182±2 043	S27±1 D38	R60±2 R48	112 139	Fyfe, 1929; Henderson, 1937
6 Ep	1931 Hawkes Bay E.Q. M.7.75 L.Poukawa F. (Bedding F.) Fences.	2	2	39.75 176.75	045 225	X00 X00	R20 R70	135 180	Fyfe & Grange (Field Sheet); Henderson, 1933
7 –	1931 Hawkes Bay E.Q. Cross F. Road & Railway.	2	2	39.73 176.76	090 173	D72 S70	N70±15 N71	131 083	as above
8 Z	1931 Hawkes Bay E.Q. Awanui F. Fences.	3	1	39.70 176.76	000 053	S26 D66	R70±15 R31	116 127	as above
9 –	1968 Inangahua E.Q. M7.1 Rotokohu F. (Bedding F.) Rd. & Rly.	2	2	41.94 171.90	245 163	D61 S75	R75 R62	114 98	Lensen & Otway, 1971; Boyes, 1971 (Appendix 1)
10 –	1968 Inangahua E.Q. Rough Ck. F. (Bedding F.) Rd.	1	5	41.90 171.99	225 045	X00 X00	R20 R70	135 180	Lensen & Otway, 1971; Nathan, 1978 for 20° dip
11 Z	1968 Inangahua E.Q. Inangahua F. Rd. & Rly.	1	0.4	41.87 171.93	190 047	S25 D30	R58 R52	118 143	Lensen & Otway, 1971; Boyes, 1971 (Appendix 1)

Fault is now known to be essentially dextral (rake = D 86°) with uplift of the western side (Lensen & Vella, 1971), but there was no mention of dextral displacement in the contemporary accounts (no fences across the fault?). At Turakirae Head there are five uplifted beach ridges, the youngest being that uplifted in 1855 (Wellman, 1967). There is excellent numerical correlation (Wellman, 1972) between the cumulative beach ridge uplifts and the cumulative throws of the river terraces displaced by the Wairarapa Fault at Waiohine River 60km north-north-east of Wellington, and the displacement of the youngest river terrace is listed in Table 3 as being that of the 1855 earthquake.

The Glenwye earthquake of 1888 produced simple dextral displacement. Fences across the Hope Fault were dextrally displaced by 2m (McKay, 1888), and fault-displaced river terraces show that the 1888 earthquake displacement was merely the last of a long series of dextral displacements. The Hope is one of a set of 4 parallel faults that make up the Marlborough Fault Zone. The faults were discovered in the 1880's, but in spite of the dextral faulting in the 1888 earthquake, it was not until they were viewed in air photographs in 1950 that their dextral nature was discovered.

The 1922 Taupo earthquake swarm caused a maximum of 4m subsidence of the north shore of Lake Taupo relative to lake level, and subsidence to the north along two of the main faults of the Taupo Volcanic Zone. According to retriangulation data the zone widened by 3m during the earthquake swarm, and corresponding narrowing took place on 39km-wide belts on each side of the zone (Sissons, 1979). Retriangulation analysis mostly north of Lake Taupo indicates long-term widening of the zone at 7mm/yr (Sissons, 1979). The 1922 earthquake widening provides important confirmation for the sense of the long-term movement.

For the North-west Nelson earthquake of 1929 with a magnitude of 7.6 a well-established displacement of 5m took place on the White Creek Fault at the Buller River, 60km north of Lake Daniells. On the north side there was a water-race and a road and on the south side a survey line. The water-race is on the up-hill side of the road and survives to show the faulting. The White Creek Fault was not an active fault prior to 1929. It is a major fault and displaces Tertiary strata by several hundred metres but a river terrace on the south side of the river at least 10,000 years old was not displaced prior to the earthquake (Fyfe, 1929).

It will be shown later that within the South Island there is a line of central faults - the Marlborough and the Alpine - that is flanked by two belts that are crossed by many north-south-striking oblique faults. The total movement for the two flanking belts is less than for the central faults, and in addition, movement takes place on many faults. Consequently the interval between successive earthquakes is much greater for any one of the oblique than for any one of the central faults. The White Creek is one of the oblique faults and its earthquake recurrence interval is probably 20,000 years or more compared with the 700 years or so for many of the central faults.

The earthquake caused innumerable landslides over an uninhabited elliptical region 100km long and 40km wide (40.9°S to 41.8°S and 172.1°E to 172.6°E, air photos), but the faulting of White Creek fault extended for 15km only and is at the south-western margin of the landslide ellipse. Total faulting was probably much greater than reported.

The Hawke's Bay (or Napier) earthquake of 1931 with a magnitude of M7.75 is New Zealand's second largest historical earthquake. The main feature was a dome of uplift (38.9°S to 39.9°S and 176.5°E to 177.3°E) 125km long and 30km wide with the major axis striking at 030°, with a gentle north-west and a steep south-east flank. The maximum uplift of 2.7m was 20km north-east of the centre of the dome. A narrow belt of "negative uplift" extended along the steep side of the dome. The surprisingly little faulting reported was entirely within outcrops of late Cenozoic limestone at the southern part of the steep flank of the dome. To the north the steep flank extends along the thick unconsolidated sediments of the Heretaunga Plains and across the sea bed of Hawke Bay. The field sheet of Ongley and Fyfe - the basis for Henderson's (1933) account - is the best map of the faulting. A total of 8 small faults were mapped, all except 2 being bedding thrusts in the moderately-dipping late Cenozoic limestone. One bedding fault and the two cross-cutting faults are listed in Table 3. The 'b' strike of the faults is almost exactly at right angles to the strike of the long axis of the dome of uplift. Unfavourable terrain consisting of young sediments that subsided by compaction is the most likely explanation for the small amount of faulting reported.

The 1932 Wairoa earthquake with magnitude of M7.1 took place just beyond the north-east end of the uplift dome of the Hawke's Bay earthquake. Ongley et al. (1937) reported downthrow on the east side of the Trig. BB fault which strikes at 045° and extends for some 5km. The centre of the fault is 5km ENE of the town of Wairoa and 70km NE across Hawke's Bay from

TABLE 4 — Data for nodal-plane solutions for crustal New Zealand earthquakes, aftershocks or groups of micro-earthquakes.

Item No. & Plate Boundary	For macro-EQs: locality, magnitude & date. For micro-EQs: locality description only.	Ref. item no.	Depth (km)	Number of points: right / wrong	Lat.S. / Long.E.	Strike D / Strike S	Rake D / Rake S	Dip D / Dip S	"b" / M	Author(s) and date of publication
1	2	3	4	5	6	7	8	9	10	11
1 Z	30km N.W. Westport, S.I. EQ M5.9. 1962 May 10.	–	0–12 restrained	29 / 3	41.7 / 171.3	244 / 011	D30 / S64	R69 / R33	127 / 127	Adams & Le Fort, 1963.
2 –	Inangahua Junction 1968 May 23. M7.1	–	0–12 restrained	32 / 3	41.8 / 172.0	045 / 135	000 / 000	R45 / R45	090 / 90	Adams & Lowry, 1971.
3 Z	Inangahua Junction 1968 May 23. M7.1	–	21 restrained	? / ?	41.72 / 172.03	045 / 190	D35 / S30	R46 / R50	117 / 145	Johnson & Molnar, 1972.
4 Z	Centre 20km S.W. Arthur's Pass.	A	2–20 mode=10	67 / 10	43.0±0.2 / 171.5±0.3	072 / 166	D73 / S75	R77 / R75	128 / 94	Scholz, Rynn, Weed, Frohlich, 1973.
5 Z	Strip 3km wide immediately S.E. of Alpine F., for 10km, N.E. of Haast River.	B	6–13	30 / 8	43.87 / 169.16	237 / 031	D38 / S15	R32 / R61	134 / 154	as above.
6 Z	30km wide strip from N. end L. Wakatipu to S.E. of Alpine F.	C	"shallow"	32 / 5	44.4±0.4 / 168.7±0.2	250 / 355	D60 / S65	R66 / R60	122 / 105	as above.
7 –	EQ 25km W. of Big Bay M7.0 1960 May 13.	D		12 / 0	44.3 / 167.7	000 / 180	X00 / X00	R10 / R80	– / –	as above.
8 –	EQ 25km W. of Big Bay M5.6 1964 March 8.	D			44.3 / 167.7	000 / 180	X00 / X00	R10 / R80	– / –	as above.
9 Z	15km wide strip S.E. of Alpine F. from Big Bay to George Sound.	E	"shallow"	34 / 5	44.6±0.2 / 167.8±0.2	015 / 171	D38 / S13	R30 / R62	093 / 156	as above.
10 Z	Strip 15km wide S.E. of coast from George Sound to Doubtful Sound.	F	"shallow"	25 / 5	45.1±0.3 / 167.1±0.2	210 / 005	D37 / S15	R33 / R60	108 / 155	as above.
11 Z	Northern Fiordland less coastal strip above.	G	"shallow"	36 / 4	45.3±0.5 / 167.4±0.4	056 / 140	D90 / S90	X90 / X90	095 / 90	as above.
12 Z	Lakes Ohau, Pukaki & Benmore and the Ben Ohau & Ewe Ranges.		3–11 mode= 7	33 / 2	44.2±0.2 / 170.0±0.3	058 / 330	D79 / S68	R80 / R85	104 / 92	Adams, Robinson, Lowry, 1974.

Table 4 cont.

Item No. & Plate Boundary	For macro-EQs: locality, magnitude & date. For micro-EQs: locality description only.	Ref. item no.	Depth (km)	Number of points: right / wrong	Lat.S. / Long.E.	Strike D / Strike S	Rake D / Rake S	Dip D / Dip S	"b" / M	Author(s) and date of publication
1	2	3	4	5	6	7	8	9	10	11
13 Z	Inangahua Junction 1968 May 23 plus aftershocks for 20 days.	A	L.c.c.	?	41.8±0.2 / 172.0±0.2	058 / 190	D27 / S40	R58 / R44	128 / 141	Robinson, et al., 1975.
14 –	Inangahua Junction 1968 May 23, late aftershocks.	B	0-20 mode=9	?	41.8±0.2 / 172.0±0.2	323 / 207	D78 / S24	N24 / N78	175 / 64	as above.
15 Z	Whole of Central N.W. Nelson micro-EQ study.	–	0-14 mode=11	20 / 0	41.1±0.3 / 172.5±0.2	079 / 182	D42 / S57	R70 / R58	126 / 112	Robinson, Arabasz, 1975.
16 Z	Below Maui Gas Field (Cape Egmont F.) "Opunake" M6.1 1974 Nov. 5. 0.17m net displacement.	–	12 restrained	23 / 3	39.54 / 173.45	010 / 121	D56 / S61	R60 / R55	065 / 111	Robinson, Calhaem, Thomson, 1976.
17 Z	20km x 10km strip along N.E. end of Clarence F.	A	5-15	35 / 2	41.93 / 173.85	081 / 183	D63 / S67	R67 / R63	132 / 102	Arabasz, Robinson, 1976.
18 Kp	Coastal Belt with Cape Campbell near centre, extends across 4 major faults (not Clarence).	B	5-22	36 /	41.7±0.4 / 174.0±0.3	257 / 166	D84 / S82	N80 / N83	121 / 89	as above.
19 Kp	60km x 30km strip along Clarence & Elliot faults.	C	2-13	50 / 6	42.3±0.2 / 172.9±0.3	076 / 170	D68 / S81	R80 / R65	123 / 94	as above.
20 Z	60km x 20km strip along W. end of Awatere Fault.	D	3-14	34 / 3	42.2±0.2 / 172.5±0.3	057 / 175	D79 / S18	R20 / R80	116 / 118	as above.
21 Z	Maruia Village EQ M5.9 1971 Aug 13 aftershocks only.	E	4-14	74 / 26	42.1±0.1 / 172.2±0.1	265 / 175	D75 / S57	N57 / N75	125 / 80	as above.
22 Kp	30km x 30km block mostly between Clarence & Hope Faults nr. Jollie's Pass.	F	4-15	81 / 12	42.5±0.1 / 172.8±0.2	062 / 152	D90 / S90	X85 / X85	107 / 90	as above.
23 Z	Near Carterton & just east of Wairarapa Fault.	M	"Upper Crust"	? / ?	41.0 / 175.5	074 / 170	D59 / S59	R80 / R80	122 / 96	as above from Arabasz & Lowry in prep.
24 Kp	Near Clarence Fault & Hanmer.	–	0-13		42.6 / 172.7	075 / 175	D87 / S73	R80 / R60	125 / 100	Kieckhefer, 1977.

NEW ZEALAND

Table 4 cont.

Item No. & Plate Boundary	For macro-EQs: locality, magnitude & date. For micro-EQs: locality description only.	Ref. item no.	Depth (km)	Number of points: right / wrong	Lat.S. / Long.E.	Strike D / Strike S	Rake D / Rake S	Dip D / Dip S	"b" / M	Author(s) and date of publication
1	2	3	4	5	6	7	8	9	10	11
25	Ostler Fault EQ M4.6 Dec. 17 1978	-	?	9 / 0	44.14 / 170.05	000 / 090	D90 / S90	X90 / X90	135 / 90	Calhaem, 1979
26	Alpine Fault near Haast. (Results indeterminate).	-	6-13	many points inconsistent	43.87 / 169.16	---- Indeterminate ----				Caldwell, Frohlich, 1975.
27 Z	Twizel Network of 10 stations. 51 evenly distributed micro-EQs.	A	0-32 mode=9	175 / 17	44.2±0.5 / 170.1±0.5	075 / 165	D90 / S90	X90 / X90	120 / 90	Haines, Calhaem, Ware, 1979.
28 Z	as above. 6 evenly distributed micro-EQs.	B	0-32 mode=9	27 / 5	44.2±0.2 / 170.0±0.3	230 / 350	D55 / S55	R50 / R50	110 / 120	as above.
29	Ostler Fault. 1978 cluster of 4 micro-EQs.	C	8-14 mode=9	17 / 0	44.00 / 170.05	010 / 100	D96 / S45	X35 / X90	80 / 90	as above.
30	All less than 40km deep micro-EQs located by Dannevirke array.	pA	15,23,38	94 / 22	40.1±0.4 / 176.5±0.4	225 / 109	D54 / S52	N56 / N58	077 / 64	Reyners, M., 1978 Ph.D thesis, V.U.W.
31	25km (E-W) x 5km (N-S) Centre 30km W. Mt. Ruapehu.	iO	5-25	44 / 8	39.3±0.2 / 175.2±0.4	090 / 270	X00 / X00	N45 / N45	090 / 0	as above.
32	15km x 15km with centre 4km S.E. of Waiouru. Alternatively, all shocks nr. Waiouru.	iQ	13-24	71 / 22	39.5±0.1 / 175.8±0.1	120 / 308	D18 / S04	R24 / R66	34 / 172	as above.
33a	25km x 25km centre nr. Mangaweka ("near Taihape").	iR	20-30	32 / 5	39.8±0.2 / 175.8±0.2	014 / 283	D00 / S79	N90 / N85	058 / 89	as above.
33b	As above, alternative solution.	iR	20-30	32 / 5	as above.	045 / 302	D58 / S68	N70 / N60	084 / 77	as above.
34	Mean of 33a and 33b above.	iR	20-30	32 / 5	as above.	030 / 292	D26 / S67	N86 / N72	11 / 82	as above.
35	EQ 1973 Feb. 21 M5.7 nr. Hastings.	iS	12	? / ?	39.71 / 176.79	334 / 243	D63 / S87	N88 / N72	19 / 89	Reyners, M., 1978 from Robinson per com.

436 WELLMAN

Table 4 cont.

Item No. & Plate Boundary	For macro-EQs: locality, magnitude & date. For micro-EQs: locality description only.	Ref. item no.	Depth (km)	Number of points: right / wrong	Lat.S. Long.E.	Strike D / Strike S	Rake D / Rake S	Dip D / Dip S	"b" M	Author(s) and date of publication
1	2	3	4	5	6	7	8	9	10	11
36 Z	Wellington array 120km x 120km	iB	20-30	40 / 5	41.4±0.5 174.9±0.8	236 / 334	D38 / S68	R78 / R57	105 / 105	Robinson, 1978.
37	Ep EQ 1966 March 3 MB5.6	Denham 31 108:9		? / ?	38.56 177.66	030 / 210	X00 / X00	R30 / R60	120 / 180	Johnson & Molnar, 1972.
38	Ep EQ 1966 March 3 (as above). ML6.2	Denham 33 108:11		? / ?	38.77 178.17	106 / 208	X00 / X00	R01 / R88	indeterm. indeterm.	Harris, per com. in Denham, 1977. World Data Centre List.

Napier. Strike-slip displacement is not mentioned and according to a fence shown on two of Ongley's photographs it was small or zero. However, retriangulation indicated the kind of station displacement that would have been expected had there been dextral displacement of about 2m on the Trig. BB Fault. The earthquake thus provides an example of fully distributed dextral movement. In view of the uncertainty of the rake and dip, the Wairoa faulting is not listed in Table 3.

The 1968 Inangahua earthquake of M7.1 is the latest in New Zealand for which there was surface faulting. The centre of the earthquake is 58km north-west of Lake Daniells. There is a nodal-plane solution, and resurveys for vertical and horizontal displacement. The vertical resurveys indicate that a dome about 20km wide was uplifted 1m and the horizontal resurveys indicate that the uplifted part widened. The corresponding narrowing took place on two sets of faults, one set on each side of the dome of uplift. At depth the two sets are thought to correspond to the faulting of the two planes of the nodal-plane solution, the western set being reverse-sinistral and similar in character to the White Creek faulting of the 1929 earthquake, and the eastern set reverse-dextral and conjugate to the western set. Most of the terrain is forested and much is mountainous, and the western set of faults was not observed. The eastern set is represented by bedding faulting on a monocline in Cenozoic strata that is downthrown to the south-east. The three faults for which there is displacement data are listed in Table 3. Two are bedding faults, and all three were discovered because of the displacement of roads or other man-made features.

Appendix 1 of Boyes (1971) is a 1: 100 000 map showing vertical and horizontal earthquake displacements and in red the active faults. Geology, attributed to G.J. Lensen of the Geological Survey, is shown in grey. The Rotokohu Fault is mapped 2km too far south and the Rough Creek Fault is unmapped. The Inangahua Fault is correct. On the western side of the Inangahua Valley steeply dipping Tertiary strata rest unconformably on basement, (Lensen & Otway, 1971 and Nathan, 1978). In the map the unconformable contact is mapped as being a fault contact, and this has had unfortunate consequences. Robinson et al. (1975) considered the wrongly mapped unconformity as being the southern extension of the Glasgow Fault and stated: "During the main event the movement was predominately thrusting with a small left-lateral component on the eastward-dipping Glasgow Fault". No faulting whatever was reported on the unconformity or on the Glasgow Fault itself which is downthrown east not west. The report of left-lateral displacement may have been influenced by the dextral Rotokohu Fault being described as being sinistral by Lensen & Otway (1971: 113) under the heading "Rotokohu Traces". As shown in the list of earthquake faults and in order of decreasing net-slip: the Rough Creek Fault is a pure reverse bedding fault and dips gently south-east; the Rotokohu is a dextral-reverse bedding fault that dips steeply south-east; and the Inangahua is a reverse-sinistral fault that dips moderately steeply east, and not west as shown by Nathan (1978) in his cross-section E-E'.

Nodal-plane solutions (Table 4, Fig. 4).

The solutions in nodal-plane studies are based on the assumption that the P waves from an earthquake have a quadrantal distribution determined by the movement of the fault-plane and auxilliary-plane (Fig.1). If the strain ellipsoid is at the earthquake focus, say at a depth of 10km, then the facing quadrant and its opposite are narrowing and moving outwards, while the quadrants to the right and left are widening and moving inwards. The outward movement produces compressional P waves and the inward dilatational P waves. The two kinds of waves travel out without changing their nature and each can be recognized at the ground surface by its seismogram. From a depth of 10km, the common depth of shallow earthquakes, to the ground surface there is a 30 per cent decrease in P-wave velocity. The decrease causes the waves from the upper half of the strain ellipsoid to bend upwards as they move out from the earthquake focus, and because of the upward bending the movements from the upper half of the ellipsoid reach the ground surface within a circle of 100km radius, whereas the movements from the lower half are spread over the remainder of the earth's surface. The movements from the two halves are symmetrical and either half will provide a solution. Provided a sufficient number of stations exist within the 100km radius circle, the movements from the upper half of the hemisphere will be more easily obtained and more consistent than those from the lower half. For New Zealand 75 per cent of the nodal plane solutions are from micro-earthquake studies, where temporary stations have been used and set out to maximum advantage.

Retriangulation analysis (Table 5, Fig. 5).

Suppose the Pacific/Indian plate boundary to be a uniformly elastic strip 200km wide striking north-east and the Pacific Plate to be moving due west relative to the Indian Plate at say 50mm/yr. Relative plate movement will change the lengths in all except two directions: 1. that parallel to the two edges of the strip, and 2. that normal to the direction of relative plate movement. Thus figures that were circles prior to the plate movement will become ellipses after movement, each ellipse having the two directions of unchanged lengths, and two directions - the "a" and "b" axes with maximum increase and decrease in length. The maximum angular changes are the right angles subtended by the "a" axis (decrease) and "b" axis (increase) (Fig. 1).

Length remeasurements derived from baselines are too inaccurate to be useful for determining strain, and in New Zealand the total length of the lines remeasured by electro-magnetic methods is too short to be useful, but angular changes inferred from retriangulations are just good enough to record the rate of angular change according to the simple assumptions given above.

For the above values the rate of angular change for a right angle facing in the "a" or "b" direction will be 250 n rad/yr (250 10^{-9} rads/yr). The same value is given by retriangulation errors assuming no error now (1980), a 2.5 second of arc error in 1930, and a 5 second of arc error twice as long ago in 1880. Hence given the highest accuracy that can be expected and a set of say <u>nine</u> triangles (27 angles) then the supposed uniform strain could be inferred with fair certainty, because the survey error would be reduced to a third ($1/\sqrt{9}$) of the strain.

If the strike of the edges of the boundary zone is known, from say earthquake epicentres, and the "b" (and "a") directions are known from retriangulations, the strike of the line normal to the relative plate movement can then be found and the rate and direction of relative plate movement inferred.

For New Zealand the retriangulation results show that the rates, and in the North Island the directions ("b" aximuth), are far from being uniform, and this is not surprising as there is no good geological analogue for the uniform elastic strip. The closest is to assume creep on innumerable parallel faults.

Observations show values (in n rads/yr) of 1,000 or more for 5km widths across some of the more active faults, dropping to 300 or less at the margin of the strip defining the plate boundary. (cf. Thatcher, 1979 for California).

Earthquake movements, provided the earthquake areas are spanned by the retriangulations, should be major features of retriangulation analysis provided that the resurveyed stations are sufficiently closely spaced. The results are good for the Taupo Earthquake of 1922, and for the Inangahua Earthquake of 1968 (electro-magnetic distance remeasurements), fair for the Wairoa Earthquake of 1932, and poor for the Hawke's Bay Earthquake of 1931, probably because survey points were too far apart.

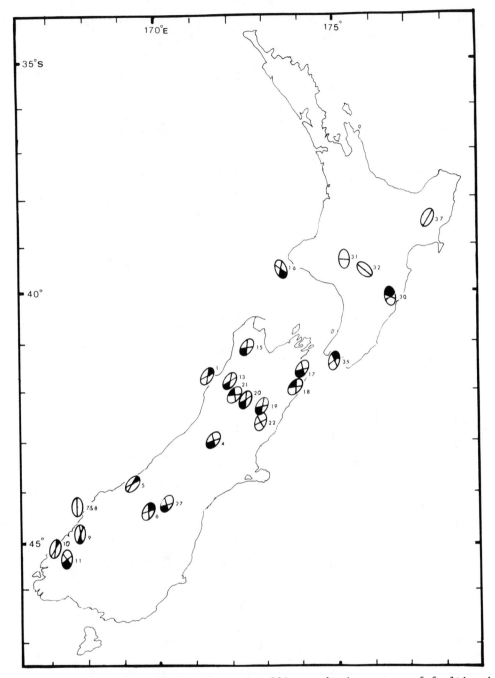

Fig. 4. Sketch map of New Zealand with strain ellipses showing nature of faulting inferred from nodal-plane solutions. Numbers beside ellipses gives item numbers in Table 4. The blackened half in the ellipses define the downthrow side of the fault-plane.

The main results of the analysis of the retriangulation data is to confirm for the South Island the relative plate movement inferred from sea-floor spreading to within about 10 per cent. In the North Island there is confirmation of the widening for the Taupo Volcanic Zone, and confirmation of the dextral displacement on the North Island Shear Zone.

All the available results are listed in Table 5. It will be seen from the table that two consecutive resurveys across the Wellington Fault appear to indicate that strain does not increase gradually but

TABLE 5 - Data for retriangulation analyses within New Zealand. In chronological order of analysis.

Item No. & Plate Bdy.	Fault and place	No obs. (see expln. below)	Av. length sides (km)	Lat. S.	Long. E.	First survey	Last survey	Rate in 10^{-9} radians per year	Strike of "b" axis	Ref. & (item no.) see below
Column No.	2	3	4	5	6	7	8	9	10	11
1 aK	Spanning Wairau F.	5	7	41.6±0.1	173.5±0.1	1880	1952	1110±280	091±8	1:255
2 aK	Spanning Awatere F.	10	6	41.7±0.1	173.9±0.1	1880	1952	830±280	126±13	1:255
3 KE	Spanning Wellington F. 10km from coast	8	10	41.3±0.05	174.7±0.1	1929	1969	480±110	097±8	2 (A)
4 KE	as above	8	10	41.3±0.05	174.7±0.1	1969.6	1971.2	6400±1400	100±30	2 (B)
5 KE	as above	8	10	41.3±0.05	174.7±0.1	1971.2	1971.9	10600±3000	012±17	2 (C)
6 aK	From 4km N.Awatere F. to 12km N. Wairau F.	13	30	41.5±0.2	173.8±0.2	1938	1970	420±110	097±7	3 (B)
7 aK	as above	13	30	41.5±0.2	173.8±0.2	1938	1970	440±100	094±7	4 (4)
8 aK	From Awatere F. to 10km N. of Wairau F.	?	?	41.5±0.2	173.8±0.2	1880	1953	290±140	090±11	4 (3)
9 aK	From 10km to 30km N. of Wairau F.	?	?	41.3±0.1	173.6±0.2	1880	1960	1100±200	145±6	4 (7)
10 aK	N. of Wairau F. (Richmond Range)	40	-	41.4±0.1	173.8±0.2	1880	1953	310±1400	100±2	5 (1)
11 aK	Spanning Wairau F.	23	-	41.5±0.1	173.6±0.2	1880	1953	110±250	129±48	5 (2)
12 aK	From Wairau F. to Awatere F.	59	-	41.6±0.1	173.8±0.3	1880	1953	460±110	107±8	5 (4)
13 Kp	Spanning Awatere F.	25	-	41.7±0.1	174.0±0.2	1880	1953	570±120	113±8	5 (5)
14 Kp	From Awatere F. to Clarence F.	59	-	41.8±0.2	173.9±0.3	1880	1953	670±120	128±6	5 (6)
15 Kp	Spanning Clarence F.	16	-	41.8±0.1	174.1±0.1	1880	1953	690±210	113±9	5 (7)
16 Kp	From Clarence F. to Hope F.	14	-	41.9±0.1	173.1±0.1	1880	1953	610±240	109±12	5 (8)
17 Kp	Spanning Hope F.	13	-	42.3±0.1	173.7±0.1	1880	1953	330±320	118±28	5 (9)
18 ap	Spans Alpine F. Jackson Bay, Haast River	20	20	43.9±0.1	168.8±0.3	1883	1975	510±90	107±5	6
19 Z	"Waikato"	32	-	37.5	175.0	1838	1972	210±90	095±13	7 (1)
20 -	"Taranaki"	62	-	39.3	174.3	1897	1974	100±100	-	7 (2)
21 KE	Spans Wellington F.	51	-	41.2	174.8	1911	1971	330±80	110±7	7 (3)
22 Ep	"Wairarapa"	11	-	41.2±0.2	175.4±0.2	1929	1971	500±90	116±2	7 (4)
23 Ep	"Aorangi Range"	20	-	41.4±0.1	175.4±0.1	1884	1971	290±150	127±11	7 (5)
24 Ep	"Ormondville"	20	-	40.2	175.3	1877	1956	260±250	128±30	7 (6)

Table 5 cont.

Item No. & Plate Bdy.	Fault and place	No obs. (see expln. below)	Av. length sides (km)	Lat. S.	Long. E.	First survey	Last survey	Rate in 10^{-9} radians per year	Strike of "b" axis	Ref. & (item no.) see below
Column No.	2	3	4	5	6	7	8	9	10	11
25 KE	"Kaimanawa"	16	-	39.3	175.7	1898	1965	240±190	057±20	7(7)
26 ap	Coast 3 M. Lagoon to Wanganui River	36	-	43.2	170.2	1879	1973	240±70	119±9	7(8)
27 ap	Spans Alpine F. S.E. of above	22	-	43.3	170.4	1879	1973	670±100	117±4	7(9)
28 Ep	"Ashley"	36	-	43.0	172.5	1877	1973	230±120	145±14	7(10)
29 S	"Teviotdale"	51	-	43.1	172.8	1878	1970	040±130	-	7(11)
30 S	"Kakahu"	44	-	44.0	170.8	1879	1970	030±170	-	7(12)
31 Z	Ostler F. to Lake Tekapo	71	-	44.2	170.0	1880	1970	310±80	104±7	7(13)
32 Z	"Manoiototo"	42	-	45.0	170.0	1870	1970	180±190	-	7(14)
33 Z	Tokomairiru & Lake Waihola	96	-	46.1	170.1	1860	1973	280±150	120±16	7(15)
34 Fp	"Waiau-Takitima"	72	-	45.7	167.7	1878	1972	120±90	040±20	7(16)
35 X	N. side Gisborne Plain	9	-	38.6±0.1	178.1±0.2	1926	1937	-	-	8(1),11
36 X	Gisborne & Waipaoa	12	-	38.6±0.1	177.9±0.2	1937	1931	1170±220	034±5	8(2),11
37 X	as above	15	-	38.6±0.1	177.9±0.2	1878	1937	470±230	129±14	8(3),11
38 X	Hills around Gisborne Plain	67	-	38.7±0.1	177.9±0.2	1878	1937	150±150	108±29	8(4),11
39 X	Gisborne & Waipaoa	28	-	38.5±0.2	178.0±0.2	1878	1971	250±120	028±14	8(5),11
40 X	Mt.Hikurangi E. to coast	32	-	37.8±0.3	178.3±0.3	1886	1975	340±120	004±11	8(6),11
41 X	Waipaoa Valley	30	-	38.4±0.2	177.8±0.2	1875	1974	-	-	8(7),11
42 X	Mt.Hikurangi S.E. to coast	41	-	38.1±0.2	178.1±0.2	1875	1974	230±170	023±20	8(8),11
43 X	41 and 42 above	93	-	38.2±0.3	178.0±0.3	1875	1974	150±100	015±18	8(9),11
44 X	Mt.Hikurangi N.E. to coast	7	-	37.8±02	178.2±0.2	1925	1974	320±70	054±8	8(10),11
45 X	Mt.Hikurangi to 50km S.E. & S.W.	10	-	38.1±0.2	178.1±0.3	1925	1974	340±200	040±11	8(11),11
46 X	Arowhana-Matawai to coast	9	-	38.2±0.1	178.1±0.2	1925	1977	300±90	058±13	8(12),11
47 X	Willow Flat to Wairoa	19	-	39.0±0.1	177.1±0.1	1934	1974	230±220	018±26	8(13),11
48 X	Willow Flat to Raupunga	34	-	32.0±0.1	177.1±0.1	1877	1931	760±400	148±15	8(14),11
49 X	Nuhaka Valley to E. Coast	8	-	38.9±0.1	177.8±0.1	1937	1973	990±400	067±12	8(15),11
50 X	Raupungato Waihua	28	-	39.1±0.1	177.2±0.1	1884	1932	1340±630	030±12	8(16),11
51 X	Elsthorpe to Cape Kidnappers	44	-	39.8±0.2	176.9±0.1	1883	1968	610±200	050±9	8(17),11

Table 5 cont.

Item No. & Plate Bdy.	Fault and place	No obs. (see expln. below)	Av. length sides (km)	Lat. S.	Long. E.	First survey	Last survey	Rate in 10^{-9} radians per year	Strike of "b" axis	Ref. & (item no.) see below
Column No.	2	3	4	5	6	7	8	9	10	11
52 X	as above but discontinuous	12	-	39.8±0.2	176.9±0.1	1937	1968	630±390	027±18	8(18),11
53 -	Hihitahi,Bulls,Woodville,Norsewood	15	-	40.0±0.3	175.8±0.3	1931	1976	290±250	071±25	8(19),11
54 KE	Upper Hutt,Carterton,Woodville, Shannon	33	-	40.7	175.6±0.3	1875	1911	700±180	071±8	8(20),11
55 Ep	Carterton,Pahiatua,Akitio	29	-	40.8±0.2	176.0±0.3	1875	1975	210±120	138±12	8(21),11
56 KE	Hills around Wellington City	18	-	41.2±0.1	174.8±0.1	1870	1912	930±380	076±12	8(22),11
57 KE	as above	17	-	41.2±0.1	174.8±0.1	1912	1931	450±260	110±16	8(23),11
58 KE	as above	13	-	41.2±0.1	174.8±0.1	1931	1972	490±130	114±7	8(24),11
59 KE	as above	12	-	41.2±0.1	174.8±0.1	1912	1972	370±130	116±10	8(25),11
60 KE	as above	8	-	41.2±0.1	174.9±0.1	1947	1973	780±210	115±7	8(26),11
61 KE	Wellington to Mt.Matthews	30	-	41.2±0.2	174.9±0.2	1914	1929	740±190	092±7	8(27),11
62 aK	T.V.Z. 0-80km S. coast	6	20	38.1±0.3	176.5±0.5	1925	1976	200±60	030±8	9(BP2)
63 aK	T.V.Z. 70-100km S. coast	8	6	38.5±0.2	176.2±0.2	1950	1974	150±110	060±21	9(NT16)
64 aK	T.V.Z. 150-180km S. coast	6	7	39.1±0.2	175.3±0.2	1888	1972	250±180	003±15	9(TO3)
65 aK	T.V.Z. 160-200km S. coast	8	15	39.3±0.2	175.3±0.2	1911	1972	180±210	037±40	9(TO2)
66 aK	T.V.Z. 160-190km S. coast	10	7	38.4±0.2	175.7±0.1	1898	1970	270±150	179±12	9(TO9)
67 aK	Taupo Volcanic Zone, average	-	-	-	-	-	-	210±22	026±11	9
68 KE	N.I.S.B. 0-70km S. coast	8	25	38.2±0.3	177.2±0.3	1925	1976	200±50	052±7	9(BP4)
69 KE	N.I.S.B. 170-220km S. coast	9	12	39.6±0.2	175.8±0.2	1898	1972	150±120	090±21	9(TO10)
70 Z	40km S. coast, 40km W. T.V.Z.	8	20	38.0±0.4	176.0±0.5	1929	1976	040±50	-	9(BP1)
71 Z	50km S. coast between T.V.Z. & N.I.S.B.	5	20	38.2±0.3	176.8±0.3	1925	1976	060±50	165±23	9(BP3)
72 Z	150km S.E. coast, 30km W. T.V.Z.	12	7	38.8±0.2	175.3±0.2	1888	1972	080±130	-	9(TO4)
73 Z	Line of T.V.Z., 200km S.E. coast	10	20	39.6±0.2	175.4±0.3	1911	1972	000±50	-	9(TO1)
74 Z	Cape Farewell to Bainham	16	5	40.65±0.15	172.55±0.15	1895	1976	210±110	139±17	10(1)
75 Z	S. end Pikikuruna Range	9	8	41.1±0.1	172.90±0.1	-	-	340±150	100±13	10(2A)
76 -	N. end Moutere Depression	29	5	41.3±0.1	172.90±0.1	-	-	350±110	046±7	10(2B)

pulsates, the rate for 1.5 years after 1969.6 being 13 times as great as the average for the previous 40 years, and the rate for the 0.7 years after 1971.2 being almost twice the previous rate and at right angles.

Notes on tables of data on present-day faulting and stress within New Zealand

Table 1 gives details for those active faults of the North Island for which the strike and rake are reasonably well known. The table is based on that presented at Canberra by Gerald Lensen in December 1979. Column 1 is the item number. Column 2 gives the name of the fault and the locality. Column 3 gives the slip displacement in metres. Where the displaced features are multiple, or where there are several closely spaced parallel faults, the largest displacement is listed. Column 4 gives the latitude south and Column 5 the longitude east. Column 6 gives an estimate of the reliability of the rake. If there are multiple features their number is given and reliability is good. The single-feature-sites are ranked "A", "B" and "C" according to appearance in air photographs. Column 7 gives the strike direction in which the right hand side is downthrown. For a fault downthrown to the west, strike is given as $180°$ and not as $000°$. The same down-throw strike convention is used throughout the tables. For Item 9 the air photo strike is $350°$, Lensen lists $019°$. Column 8 gives the rake and column 9 the dip. Dip for most faults has been estimated from air photographs. Rake is prefixed "D" or "S" for dextral or sinistral; and dip is prefixed "N" or "R" for normal or reverse. An "X" is used where the wrench or dip-slip component is zero or near zero. The same convention is used throughout. Column 10 gives "b", the strike of the minor axis of the strain ellipse. Column 11 refers to the list of references at the foot of the table, the number in brackets being the item number of the published list, irrespective of whether the item was numbered when published.

Table 2 gives details of the active faults of the South Island for which strike and rake are reasonably well known. The list is largely based on that of Wellman, 1953; the order being across the island from north-west to south-east. The lay-out of the table is the same as that of Table 1 except that one standard deviation is given for the rake where there are multiple reference lines.

Table 3 gives details of the faulting during historical earthquakes. For each item there are 9 columns, and for most columns there are two rows for each item.

Table 5 cont.

Item No. & plate Bdy.	Fault and place	No obs. (see expln. below)	Av. length sides (km)	Lat. S.	Long. E.	First survey	Last survey	Rate in 10^{-9} radians per year	Strike of "b" axis	Ref. & (item no.) see below
Column No.	2	3	4	5	6	7	8	9	10	11
77 aK	Near Havelock	17	7	41.3±0.1	173.7±0.1	-	-	150±100	078±19	10(3)
78 Kp	Kaikoura to Rakautara	15	7	42.6±0.2	173.7±0.8	-	-	320±110	137±10	10(9)
79 -	From 10km to 30km N. of Wairau F.	Now disregarded because number of common stations too few.								10=4(7) above

1. Wellman, 1955 A.
2. Otway, 1972, Fig. 6.
3. Otway, 1973, Fig. 4.
4. Bibby, 1975, Table 1.
5. Bibby, 1976, Table 1.
6. Bibby & Walcott, 1977.
7. Walcott, 1978A, Table 2.
8. Walcott, 1978B, Table 2.
9. Sissons, 1979, (Tables 3.1, 3.2, 3.3).
10. Bibby per com. 26th Aug. 1980 (Table 3).

T.V.Z. = Taupo Volcanic Zone. N.I.S.B. = North Island Shear Belt.

No obs. = Number of common angles for 1, 2, 3, 6, 7 & 8 and number of common points for 4, 5, 9 & 10.

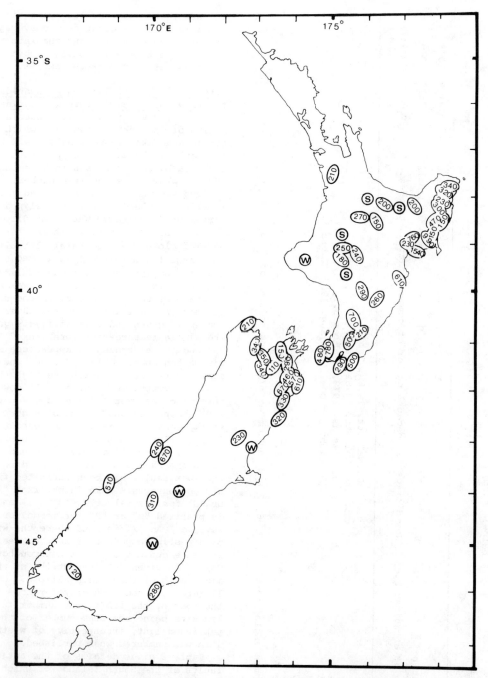

Fig. 5. Sketch map of New Zealand with strain ellipses from retriangulation analysis showing direction and rate of shape-change. Internal numbers give rate of increase in 10^{-9} radians/year of right angles facing in the "b" direction. The external numbers are the item numbers of Table 5.

Column 1 is the number of the fault. Column 2, first row gives year and place of earthquake, and its magnitude and magnitude range; second row gives name of fault, nature of feature(s) displaced by the faulting, and finally the length of the faulting in kilometres. Column 3 gives the net-slip in metres. Column 4 gives the latitude south and the longitude east. Columns 5 to 8 give in the upper row data for the fault plane, and in the lower row the calculated data for the auxilliary-plane: column 5 gives the strike, column 6 the rake, and column 7 the dip. Column 8 upper row gives the strike of the minor axis of the strain ellipse (b), and lower row the angle (M) between the strike of the fault-plane and the strike of the auxilliary plane that is subtended by the "b" direction. Column 9 gives selected references to the earthquake faulting.

Table 4 gives all the nodal-plane solutions for crustal New Zealand earthquakes up to August 1980. The table includes all the relevant solutions from Denham's 1977 list. Solutions are set out in approximate order of publication, and there are two rows for each item. Column 1 is the item number. Column 3 gives the letter references used in the original publication. Reyners made two sets of his own and other solutions depending on whether he thought they belonged to the Indian or the Pacific plate, and they are prefixed "i" or "p" accordingly. Column 4 gives the estimated depth range either for an individual earthquake, for an aftershock sequence, or for a group of micro-earthquakes that have been grouped to provide a solution. "Restrained" indicates that the depth range is one of a limited number used in a computer programme. Column 5 is my estimate of reliability based on a comparison of the solutions. Column 6 gives the latitude south and the longitude east, the range being that of aftershocks or micro-earthquakes used in the solution. Columns 7, 8 and 9 give the strike, rake, and dip; values for the dextral plane being given in the upper row, and values for the sinistral in the lower. Column 10 in the upper row gives "b" the strike of the minor axis of the strain ellipse, and in the lower the angle between the two strike directions subtended by "b". Column 11 gives the references.

Table 5 lists the retriangulation analyses that have been made in New Zealand up to June 1980 in approximate order of analysis. Column 1 gives the number of the analysis, and column 2 the general locality, and the names of the faults spanned by the surveys. Column 3 gives the number of angles used and the approximate line-length in kilometres. Columns 5 and 6 give the latitude and range south and the longitude and range east; the range being the area covered by the retriangulation. Column 7 gives the date of the first triangulation and column 8 that of the second. Column 9 gives the mean annual increase in 10^{-9} radians (n rads) of a right angle facing in the "b" direction, and its standard error. Column 9 gives the mean strike of the "b" direction (the minor axis of the strain ellipse) and its standard error. Column 10 gives the references together with the letters or numbers used to identify the analyses in the original publications.

In order to simplify interpretations of the data, the plate boundary symbols of Fig. 6 have been added to column 1 of Tables 1 to 5.

Finally it is of major importance to estimate the proportion of the plate movement that has been, and will be, released during earthquakes. As already mentioned "drag", defined here as either distributed faulting, or movement on undiscovered faults, is of greater importance in New Zealand than faulting itself. This is not surprising because drag includes all the movements related to folding, and folding is equally if not more important than faulting in determining the changing shape of New Zealand. The other important term is "creep" - gradual movement that takes place without appreciable earthquakes. To contrast with "creep" the term "snap" is used for sudden movement during earthquakes. Thus all long-term tectonic movements belong to one of the four kinds:-

   1A Snap-faulting   1B Creep-faulting
   2A Snap-dragging   2B Creep-dragging

and the proportion of "creep" movement to "snap" movement will determine the amount of plate movement that is released harmlessly. Snap-faulting and snap-dragging have been described above for New Zealand earthquakes and the relative importance of snap-dragging emphasised. The first surveys for creep in New Zealand were made in the 1930's (Mackie, 1971) and at present all the major faults have been monitored for creep for several years. It is clear that surface fault-creep, if it exists, is far less rapid than the long-term faulting shown by the displaced river terraces. (See Earth Deformation Section Reports of N.Z. Geological Survey, 1966 to 1973, etc.) Thus there is every reason to fear that all the plate movement has been and will be released during earthquakes.

Interpretation of data (Fig. 6).

The data on New Zealand horizontal kinematics are interpreted in terms of the continuing bending of an open "Z" shaped recurved arc superposed on the individual movement of seven plates:- major: Indian (A) and Pacific (P); and micro: Eastland (E), Kaweka (K), Fiordland (F), Puysegur Bank (Y) and Macquarie (M).

The "Z" bending movement is taken as being proportional to the total Indian/Pacific movement, and the seven plate movement is calculated by subtracting the "Z" bending rotation from total rotation.

Fig. 6 summarizes the interpretation, the boundaries of the "Z" bending being dash-dot lines, and each of the seven plates being defined with its key letter. Circles enclose vector velocity polygons that show the velocity and bearing of relative plate movement, the circle diameter representing about 50mm/yr. Note that "A" and "P" represent the full movement of the major plates, and "a" and "p" the movement reduced by "Z" bending. A letter pair

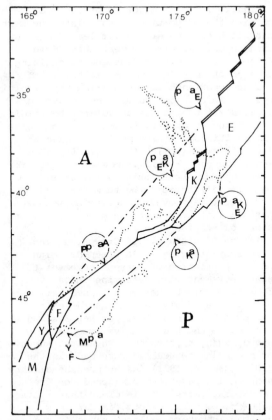

Fig. 6. Map of New Zealand on Mercator's projection with a 1° grid, and two macro- and five micro-plates. A = Indian; P = Pacific; E = Eastland; K = Kaweka; F = Fiordland; Y = Puysegur Bank; M = Macquarie. Of the direct Pacific/Indian plate movement about 20% is by "Z" bending within the 250km-wide belt of continuous minor shallow seismicity. The remaining 80% is shown as "pa". Hence Pa - pa = bending = Z. Vector polygons show relative plate velocities at six critical places. Read "p a" as Indian (a) Plate moving east relative to Indian Plate. Diameter of circles represent a velocity of about 50mm/yr.

using the key letters above is used for plate boundaries, the letter order being from west to east.

Within the New Zealand land area all the plate boundaries are belts, the narrowest being the ap boundary along the straight part of the Alpine Fault (10km), and the widest the southern part of the Ep boundary which extends from the line of the Kermadec Trench to the east side of the KE boundary (100km). The belts contain parallel faults having the same nature, and are considered as being a single fault in vector calculation, and are shown as a line in Fig. 6.

As mentioned above, most of the items in Tables 1 to 5 are identified in column 1a either with the "Z" symbol for bending or by a letter pair for plate boundaries. The other items are marked "-" when the cause of movement is uncertain, with an "S" when the region is thought to be tectonically rigid, and with an "X" when landsliding is thought likely.

Micro-plates are described below in order from north to south, and each clockwise from its most northerly point.

E Eastland micro-plate. Bounded on the east by the Kermadec Trench and its continuations to the KEp triple junction in Cook Strait. On the west by the North Island Shear Belt to the aEK triple junction in the Bay of Plenty, then by an unnamed fault to the south-west end of the Havre Trough, and then north-north-east along the Havre Trough to join the Kermadec Trench at a point well to the north-east of New Zealand.

K Kaweka micro-plate. Bounded by the North Island Shear Zone from the aEK triple junction in the Bay of Plenty to the KEp triple junction in Cook Strait. Bounded to the south-south-east by the Marlborough faults (less the Wairau Fault) to the aKp triple junction at the north-east end of the Alpine Fault. Bounded to the east by the north striking Glenroy Fault, and then to the north to the west and to the north-west by the Wairau Fault, the Galpin Fault, and the Taupo Volcanic Zone to the starting point at the aEK triple junction.

a/p Indian/Pacific plate movement reduced according to "Z" bending. Extends as main part of the Alpine Fault by an almost straight line for 450km from the aKP triple junction near Lake Daniells to the apF triple junction at Cascade River.

F Fiordland micro-plate. Bounded on the east by the Te Anau fault-zone (with a sinistral jog at the lake itself) from the apF triple junction to the FpM triple junction at Solander Island, to the west by Resolution Fault to the aFY triple junction just off the Fiordland coast, and then north-east by the southern part of the Alpine Fault to the starting point at the apF triple junction.

Y Puysegur Bank micro-plate. Bounded to the east from the aFY triple junction by the Resolution Fault to the FMY triple junction just off the south-west corner of Fiordland, then by a curved line that is probably a low angle fault to an unnamed fault almost on the extension of the Alpine Fault to the aYM triple junction, and then on the west by the unnamed fault to the aFY triple junction starting point. The micro-plate is entirely out to sea but was fairly clearly outlined by the aftershocks of the Puysegur Banks earthquake of October 12 1979 with local magnitude of 6.5 (Ware, 1980).

M Macquarie micro-plate. Bounded to the north-east by the southern part of the Resolution Fault to the FpM triple junction, then to the east by the Auckland Slope on the western side of the Campbell Plateau probably to the region of the Indian/Pacific/Antarctic triple junction, then to the west by the Macquarie Ridge to the aYM triple junction, and then to the north by the curved MY boundary to the YFM triple junction starting point. Although shown for simplicity as a single unit, the Macquarie micro-plate is probably divided into two by a central spreading axis.

The four kinds of faulting for calculating vector bearings.

The four cases are set out below and consist of two pairs, one pair being the faults of the upper crust and the other those of the lower crust and upper mantle. The first pair are considered secondary with respect to the second, which represent primary faulting directly related to the forces driving the plates.

	Fault Type	
Secondary faults Fault-set single	"O" Fault strike fixed	Vector bearing normal to auxilliary-plane. (45° dip assumed for fault-plane). Fault strike defines plate boundary.
	"Z" Fault strike rotates	Vector bearing normal to fault-plane. Strike of auxilliary-plane gives strike of plate boundary.
Primary faults Fault-set double	"Ph" Null-axis is horizontal (slab is thinning)	Vector bearing is in direction of lengthening of slab. Bearing subtends larger angle between dextral and sinistral strikes.
	"Pv" Null-axis is vertical (slab is narrowing).	Vector bearing is in direction of lengthening of slab. Bearing subtends clockwise angle from sinistral to dextral strike.

"O" type faults. The "O" is for ordinary, and the faults are the ordinary ones of the plate boundaries, their strike being fixed relative to their shift poles. They cause most of the major, but not the minor and year by year crustal earthquakes. According to standard nodal-plane practice the vector bearing is taken as being normal to the strike of the auxilliary-plane. However, fault-dip is taken in all cases as being 45° and heave thus made to equal throw, surface dips being disregarded except for determining normal or reverse sense of throw. The 45° assumption has long been used by Lensen (1958) and the Alpine Fault was considered to flatten out with depth in order to explain the up-ending of the Alpine schist by Wellman in 1979.

The following rules are used to obtain the vector bearings set out in Table 6 from the basic data in Tables 1 to 5:- Add 90° plus rake to normal-dextral faults; add 270° minus rake to dextral-reverse faults; add 270° plus rake to reverse-sinistral faults; and add 90° minus rake to sinistral-normal faults. For most faults the bearing differences caused by changing from the observed to the 45° dip is a few degrees or less. It is zero for a rake of 00° or 90°, but is 45° for a vertical fault with a 45° rake. The 45° assumption implies that faults flatten out with depth. Consequently, the surface of a series of blocks separated by active west-dipping reverse faults will dip progressively more steeply westward. Commonly drag produces anticlinal axes with steep east limbs. Rotation rate can be determined from the rate of faulting, and more important, the velocity of narrowing can be determined if the rate of block rotation (rate of tilting) and width of fault belt is known. Rate of narrowing for southern end of the Ep boundary is estimated in this way in Table 7, item 14.

"Z" type faults. The "Z" is for the open "Z" bend, and the faults are those of the middle and bending arm of the "Z".

The fault itself rotates and by contrast with the "O" type faults the vector bearing is normal to fault strike and it is the plate boundary that is normal to the strike of the auxilliary-plane. The rate of "Z" bending and the "Z" faults are discussed in detail later.

"Ph" type faults. The "P" is for the primary type of faulting and the "h" for the near-horizontal attitude of the null axis. The faults are those of the upper part of the subducting slab in the northern half of New Zealand, good examples being the faulting at depths of 20km to 40km in the southern part of the North Island (41.0°S 175.5°E) described by Arabasz and Lowry (1980 Fig. 6: B, C, D). The pressure axis is near vertical and the slab is lengthening and thinning. The vector bearing of the displacement of the lower plate relative to the upper, in this case p relative to E, is in the down-dip direction of lengthening.

"Pv" type faults. Similar to the "Ph" type except that the null axis is vertical and the subducting slab is narrowing instead of thinning. The only New Zealand case is below Fiordland (Scholz et al. 1973, Fig. 4: H). As with "Ph" type faulting, the vector bearing of the displacement of the lower relative to

TABLE 6 - Averaged vector bearings for relative plate movement from sense of fault displacement of active faults in Tables 1, 2 and 3 and nodal-plane solutions in Table 4 plus bearings from seven unlisted primary faults.

Item Number.	Plate Boundary.	Nominal Latitude.	Active faulting (A) Nodal Plane (N).	Number of observations.	Strike of fault.	Rake of fault. D = Dextral S = Sinistral.	Dip of fault. N = Normal R = Reverse.	Bearing of vector (0° to 180°).	Comments
1	Z	40°S	A	3	006±3	S26±2	R63±4	096±3	Three EQs NE of Alpine F.
2	Z	44°S	A	7	006±3	S53±6	R80±10	096±3	SE of Alpine Fault.
3	Z	-	A	10	006±3	S45±5	R75±7	096±3	Mean for active faulting.
4	Z	-	N	19	352±5	S53±6	R65±3	082±5	Whole of New Zealand.
5	aK	39°S	A	4	042±3	D11±11	N47±10	143±10	Taupo Volcanic Zone.
6	KE	39°S	A	4	020±2	D83±2	R80±4	027±2	North Island Shear Belt.
7	KE	40°S	A	9	043±2	D83±1	X90±?	050±3	As above.
8	Ep	40°S	A	4	028±14	X00	R37±18	119±14	Wairarapa & H.B. thrusts.
9	Ep	39°S	N	1	030	X00	R30	120	Johnson & Molnar, 19
10	Ep	40°S	N	6	-	-	-	119±5	Primary faulting.
11	ak	41°S	A	5	065±3	D84±3	R86±4	071±5	Wairau Fault.
12	Kp	41°S	A	8	065±3	D85±1	R75±3	070±3	Marlborough Fts. less Wairau.
13	Kp	41°S	N	4	072±4	D82±5	R81±1	080±7	As above.
14	ap	44°S	A	12	054±2	D72±5	R68±1	072±5	Alpine Fault.
15	Fp	45°S	A	2	022±8	S72±8	R75±10	007±6	Te Anau Fault.
16	aF	45°S	A	1	050±2	D90±10	Z90±10	050±2	Alpine Fault.
17	MF	45°S	N	1	-	-	-	31.5	Primary faulting.

the upper plate is down-dip in the direction of lengthening. The relatively small width of the subducting Fiordland plate is a possible explanation for the subduction slab narrowing instead of thinning.

Vector bearings from active faulting and nodal-plane solutions (Table 6).

For active faulting the data are from Tables 1, 2 and 3, and for nodal-plane solutions from Table 4.

Column 1 is the item number and column 2 the plate boundary symbol. Column 3 Nominal latitude - latitude when the site is projected on to the line of the Alpine Fault and Column 4 whether data is from active faulting or nodal-planes. Column 5 gives the number of items averaged in each set. Of the 80 listed items on active faulting 69 have been averaged, given in Table 6, and discussed below and 11 have been disregarded. Of the 38 listed items on nodal-plane solutions 25 have been averaged, etc. and 13 have been disregarded. Columns 6, 7 and 8 give the average value, and the standard error of the mean for the strike, the rake, and the dip. Column 9 gives the calculated vector bearing. The faults are all "ordinary" except for the "Z" bending faults items 1 to 4, and the primary faults items 10 and 17, and vector bearings are calculated accordingly.

Items 1 to 4. "Z" bending. Values are surprisingly consistent and it was found that there is no significant difference from north to south. "Z" bending makes up 19 of the 25 nodal planes used and probably causes most of the minor seismicity in New Zealand, thus explaining the good match between the belt of bending and the belt of minor and year by year seismicity (Eiby & Reilly, 1976).

Item 5. Active faulting indicates that

the Taupo Volcanic Zone is normal-dextral in nature, but there are no nodal-plane solutions to provide a check.

Items 6 and 7. Vector bearings for the North Island Shear Belt are well-defined at two latitudes, but again there are no nodal-plane solutions for comparison.

Items 8 to 10. Vector bearing for the southern part of the Ep boundary is well-defined (119 ± 0.3) and about 5° less than the strike of the growing fold blocks at the south end of the boundary according to Ghani 1978 (Fig. 10).

Items 11 to 13. For the Marlborough faults the nodal-plane bearing is about 10° greater than that from active faulting, the difference being one of sampling site.

Item 15. The Alpine Fault vector bearing from 12 slip observations at the fault itself is not significantly different from the shift-vector-bearing determined from uplift of the Southern Alps and discussed later.

Item 16. Sense of displacement on the extreme southern landward end of the Alpine Fault is shown by a photograph in the Geology of New Zealand (Fig. 10.3). The photograph "looks" south and shows apparent uplift of the seaward side of the fault caused by dextral displacement. There is no uplift of the inland side.

Item 17. At Fiordland the nodal-plane solution for intermediate-depth faulting is remarkably consistent and is taken as representing the south-easterly displacement of Fiordland relative to the Macquarie Plate.

Relative plate velocities from active faulting (Table 7).

Column 1 is the item number, and column 2 the fault, place, etc.

Column 3 gives generally accepted estimates on the relative importance of a particular fault to the full fault zone or plate boundary, together with the letter-pair for the plate boundary.

Column 4 gives the net displacement in metres, the heave being made equal to the throw. The values are generally accepted to within 5 per cent or less.

Column 5 gives the range of ages now commonly accepted for the listed displacements. There is no general agreement, the values being up to 70 per cent different. For the terrace sequences, which represent the most important items, the average rate of dextral displacement is generally taken as being uniform. Dating is indirect and by con-trasting theories.

1. Wellman, 1955. Uniform rate of river downcutting, no oscillations, all terraces post-glacial, the oldest 10 kyr old.

2. Suggate, 1960. Rate of river downcutting decreasing. Oldest terrace 20 kyr old and related to main glacial advance.

3. Lensen, 1968. Rate of river downcutting oscillating, periods of aggradation or still-stand being equated with dated stadials, oldest estimate being 35 kyr.

4. Wellman, 1972. Successive earthquake uplifts at Turakirae Head correlated with successive river terraces at Waiohine River. By extrapolation from 6.3 kyr the oldest terrace is 10 kyr old.

5. Adams, 1979. Western side of Southern Alps. Terraces post-glacial and oldest assumed to match 8 kyr stadial.

6. Vella, P. pers. comm. 1.12.80. Well-dated 20 kyr old volcanic ash layer used for age control. Oldest terrace of Waiohine sequence about 10 kyr old.

Column 6 gives the velocity range according to the range of ages commonly accepted. Column 7 gives the velocity adopted for the fault, and Column 8 that adopted for the plate boundary.

Item 1. Wairarapa Fault, Waiohine River. A 35 kyr age by Lensen, being too large, is disregarded, and as mentioned above the 20 kyr old age by Vella in Lensen and Vella (1971) is now thought by Vella to be about 10 kyr old.

Items 2 and 3. Wairau Fault near Branch River. There is no general agreement for the age of the top of the terrace sequence, but the 720m displacement for the 120m-high terrace, which was disregarded by Suggate in 1965, if taken as last interglacial and 120 kyr old, gives a 6mm/yr velocity, in agreement with a 10 kyr age for the top of the terrace sequence.

Items 4 to 7. The displacements are from Wellman (1953), from Suggate (1978), and from Berryman (1979) and are generally accepted. The ages are as for Item 2.

Item 8. A sum representing all the aK and almost all the Kp plate boundary, ages being as above.

Items 9 and 10. Velocity vectors determine the movement of the Alpine Fault from the Marlborough faults and vice-versa, but Freund in 1971 thought the dextral displacement of the Marlborough fault died out before reaching the Alpine Fault having confused the signs of strike-slip displacement and inadvertently cancelled out the dextral displacement. The cancelling-out was not noticed by Suggate and Lensen in 1973, or by Suggate in 1978 (Wellman, 1981).

Drag makes up about half of the total net displacement on the Alpine Fault. Adams in 1979 showed that the drag-throw, confined to the 5km south-east of the fault, equalled the slip-throw at the fault itself, and, as shown below, drag is slightly more than half of the dextral displacement. Full Alpine uplift includes all the drag-throw, 120m being a generally accepted value for the 100km length south-west from the Hope

TABLE 7 - Vector velocities for relative plate movement from the rate of displacement of the active faults.

Item Number.	Nominal Latitude.	Fault, plate, etc.	Estimated proportion of total plate movement and plate boundary letter pair.	Net displacement in metres.	Generally accepted age range in thousand years (kyrs)	Velocity range in millimetres per year (mm/yr)	Fault velocity adopted (mm/yr).	Relative plate velocity adopted (mm/yr)
1	40°S	Wairarapa Fault, Waiohine R.	0.7 KE	124	20-10	6-12	12	17
2	41°S	Wairau Fault (W), Branch R.	0.8 aK	59	17-10	3-6	6	7
3	41°S	Wairau Fault, high terrace	0.8 aK	720	120	6	6	7
4	41°S	Awatere Fault (A)	0.29 Kp	100	17-10	6-10	10	-
5	41°S	Clarence Fault (C)	0.20 Kp	70	17-10	4-7	7	-
6	41°S	Hope Fault (H)	0.42 Kp	147	17-10	9-15	15	-
7	41°S	Porter Pass Fault (P)	0.09 Kp	30	17-10	2-3	3	-
8	41°S	Sum of A, C, H & P	1.0 Kp	347	17-10	21-35	35	35
9	41°S	Sum of W, A, C, H & P	aK + Kp	406	17-10	24-41	41	42
10	41°S	As above from S. Alps uplift	1.0 ap	350	10-8	35-44	35	35
11	43°S	Alpine F. slip at Okuru R.	(0.4) ap	14	2-1	7-14	14	35
12	43°S	Alpine F. slip. Moraine	(0.4) ap	210	17-15	12-14	14	35
13	43°S	Alpine F. slip. Valley sides	(0.4) ap	1800	120	15	15	35
14	40°S	NW dipping fold flanks, 2° dip. South coast of North Island.	0.6 Ep	1700	100	17	17	28

Fault intersection. Adams (1979: Fig. 3) gives uplift of 130m for each of two rivers. The 120m value is the average from Wellman (1979) of Items 2 to 12 in Table 1 and the mean value from the upper uplift contours in Fig. 2, both being normalized to 10 kyr.

Taking the 120m value and an angle of 20° between the Hope and Alpine faults then a vector triangle gives net displacement as being 120m/sin 20° = 350m, and the dextral shift for the Alpine Fault as 120m/tan 20° = 330m. The calculated 350 value includes the Wairau, Awatere, Clarence, and Hope displacements but not that of the Porter Pass Fault.

Items 11, 12 and 13. Three consistent velocities for 1 kyr to 120 kyr show that the net slip on the Alpine Fault (mostly dextral) is slightly less than, and drag slightly more than half the net shift determined from the Alpine uplift. Dextral drag on the Alpine Fault shows clearly on the 1:1M Geological Map. At the north, the west end of the Hope Fault has been dragged 30km across a width of 15km. At the south, the ultra-mafic belt near Cascade River has been dragged 35km over a width of 5km before being cut off and displaced 500km from the Dun Mountain belt in Nelson. Note that brackets enclosing the proportions for Items 11, 12 and 13 indicate that the 0.4 ratio is a dependent value and not an independent

TABLE 8 - Average vectors for relative plate movement from retriangulation data (Table 5).

Item number.	Plate Boundary.	Nominal Latitude.	Number of observations.	Assumed width of fault-zone in kilometres. (A)	Strike of fault-zone. (B)	Rate of shape change in 10^{-9} rad/yr (n rad). (C)	Strike of "b" axis of horizontal ellipse. (D)	Vector velocity in mm/yr. $(A \times C \times 10^{-3})$.	Vector bearing. $(2D - B \pm 90°)$.	Fault Zone
1	aK	39	5	40	040	210±22	026±25	8±1	102±50	Taupo Volcanic Zone.
2	KE	39	2	50	000	200±50	052±7	10±2	014±14	N.I. Shear Belt.
3	KE	41	10	40	050	740±190	092±7	30±10	054±14	As above.
4	Ep	41	5	70	040	289±52	131±5	20±4	132±10	Wairarapa Thrust Zone.
5	aK	42	7	40	065	311±53	099±6	12±2	043±12	Wairau Fault Zone.
6	Kp	42	6	70	065	532±68	120±4	37±6	085±8	Marlborough Faults.
7	ap	44	3	70	055	473±125	114±4	33±8	083±8	Alpine Fault.
8	Fp	45	1	40	022	120±90	040±20	5±1	148±40	Te Anau Fault Zone.
9	Z	--	10	100	000	171±38	120±11	17±4	090	"Z" Bending.

estimate like the other ratios in the column.

Item 14. In 1978 Ghani used differences in uplift to determine the rate of tilting of last interglacial marine benches along the south-east coast of the North Island. He found the average north-west slope to be $2°$ for a width of 50km and for a normalized age of 100 kyr. The south-east slopes are about twice as narrow and about twice as steep. Lewis in 1971 had already shown that the belt of active folding extends seaward to the line of the Kermadec Trench, giving the total width as 150km. Narrowing velocity for the width sampled by Ghani is 50km x sin $2°$ = 1 700m, and as the rate of folding decreases eastward 0.6 is taken as being the sampling ratio.

Relative plate vectors from retriangulation analyses (Table 8).

Table 8 shows the vectors for relative plate movement calculated from the individual sets of retriangulation analyses listed in Table 5. Column 1 gives the item number and Column 2 the key letters for the plate boundary. Column 3 gives the latitude when the resurveyed area is projected on to the line of the Alpine Fault (nominal latitude), and Column 4 the number of observations for each set. Of the 79 items listed in Table 5, 50 are averaged and summarized in Table 8, and the following 27 are disregarded:-

Items 1 to 5 have rates that are too high for the adopted fault-zone width.

Items 29 and 30 (marked "S") lie on the stable part of the Pacific Plate.

Items 20, 53 and 76 (marked "-") have uncertain values or plate coverage.

Items 36 to 52 (marked "X") are thought to represent surface sliding.

At the San Andreas Fault the strain rate decreases fairly evenly from at least 400 n rad/yr at the fault itself to about 100 n rad/yr 40km out from the fault (Thatcher, 1979).

Items 1 and 2 are from narrow surveys directly above the Wairau and Awatere faults, and their high rates are consistent with their central position.

The items marked "X" are entirely from the east coast of the North Island where active bentonite intrusions and mud volcanoes are common (Ridd, 1970). Most "X" items are consistent with seaward sliding causing extension inland and compression near the coast.

Column 5 gives the width assumed for strain accumulation on the fault zone, and Column 6 the adopted strike. Strain accumulation is unlikely to have a sharp boundary and widths are uncertain to 20 per cent or more.

Column 7 gives the rate of shape-change in n rad/yr and Column 8 the strike of the "b" axis of the horizontal ellipse, the two numbers being averages from Table 5. Column 9 gives the calculated vector velocity in mm/yr and Column 10 the calculated vector bearing. Column 11 gives the name of the fault zone.

The anomalous vectors of Table 8 are discussed under Table 9 which sets out the several averaged vectors for each of the plate boundaries.

"Z" bending: the continuing growth of the Recurved Arc of New Zealand.

Macpherson in 1946 showed that the undermass and cover of New Zealand have the recurved-arc shape of an open "Z" bend. For the centre of New Zealand the upper arm of the "Z" strikes at about $120°$; the middle arm - the arm bent - at about $180°$; and the southern arm at $120°$. Bending is thus about $60°$. In the above estimate it is assumed that the amount of shiftpole rotation equals the amount the arc was originally curved. As mentioned, the belt of minor seismicity coincides with the belt of bending. According to Eiby and Reilly (1976) it strikes at $050°$, has a N-S width of 420km, an E-W width of 500km, and a true width of 320km.

Rate of bending from amount of bending: Over almost the whole of the South Island there was stability with peneplanation on both sides of the Alpine Fault until about 60 Myr ago, then sinistral-normal faulting, widening, and crustal thinning until about 25 Myr ago, and then a major change to the sinistral-reverse faulting, narrowing, and crustal thickening that is taking place today. The belt of bending was widest (420km) 25 Myr ago when the "Z" faults struck at right angles to the belt of bending.

Bending is thus thought to have started 60 Myr ago. It amounts to $60°$ today and the average bending rate is $1° \pm 0.3°$/Myr, the error being an estimate.

Rate of bending from dip of 10 Myr old strata (Upper Miocene): According to the average of 20 nodal-plane values and 10 active fault values the dip of the "Z" faults is now $70° \pm 4°$. The dip of 10 Myr old strata on the blocks between the "Z" faults is estimated as being $20° \pm 10°$, the steeper dips caused by drag on the "Z" faults being disregarded. Thus the dip of the "Z" faults 10 Myr ago was $50° \pm 8°$. The present E-W width of the bending zone is now 500km. The E-W width 10 Myr ago was 500km x $\sin 70°$ / $\sin 50° \pm 8° = 613 \pm 70$km. Accordingly, the width decrease during the last 10 Myr was at 11 ± 7mm/yr, and the rate of bending at $1.4° \pm 0.9°$ Myr.

Rate of bending from uplift rate: Estimated uplift rates for the South Island have been given by Wellman (1979: Fig. 1). The uplift of the Southern Alps and Kaikoura Mountains is attributed to plate boundary movement and disregarded. The remaining uplift within the belt of bending equals 0.7 ± 0.4mm/yr. For a steady-state situation with erosion equalling uplift and no height increase, then with a crustal thickness of 30km and an E-W width of 500km the rate of narrowing is 0.7 ± 0.4 x $500/30 = 12 \pm 6$mm/yr and the rate of "Z" bending $1.6° \pm 0.8°$ / Myr. Heights are probably increasing and the inferred rates should perhaps be doubled to allow for the growth of mountain roots.

Rate of bending from nature of faulting during the five historical earthquakes with significant displacement: There are two "Z" bending earthquakes: 1929 West Nelson (5m x 100km); 1968 Inangahua (1m x 20km); and three others: 1855 Wellington (12m x 240km); 1888 Glenwye (1m x 20m); 1931 Hawke's Bay (2m x 60km). The value in brackets are estimated average net displacement in metres and average length of displacement in kilometres. The "Z" bending earthquakes represent 1/5 of the total and if 40mm/yr is taken as being the total E - W narrowing rate, then "Z" bending makes up 8mm/yr, and the inferred rate of "Z" bending from historical earthquakes $1.1°$ / Myr.

Historical earthquakes provide the most immediate value for the bending rate and the 8mm/yr is adopted in the estimates that follow, in which it is assumed that the rate from "Z" bending is 1/5 of total movement, the shiftpole rotation, which is discussed next, being decreased accordingly from $1.5°$/Myr to $1.2°$/Myr.

Indian/Pacific shiftpole determined from uplift of Southern Alps.

Within New Zealand the straight part of the Alpine Fault is the simplest part of the Indian/Pacific plate boundary and the shiftpole position and rate is calculated from displacement vectors at two points on the straight part of the fault: "K" ($42.90°$S $171.12°$E) the point where the extension of the Hope Fault reaches the Alpine Fault, and "O" ($43.95°$S $168.97°$E) the point where the Okuru River crosses the Alpine Fault.

The adopted uplift value for "K" (11.5mm/yr) is the mean of the maximum value 11.0mm/yr and the "crustal" value 12.0mm/yr. The maximum value is the mean of the values in entries 2 to 12 inclusive in Table 1, Wellman, 1979. The crustal value (12 = 5.5 x 70/32) is from the uplift contours in Fig. 2, Wellman, 1979, using a width of 70km and a crustal thickness of 32km.

The adopted uplift value for "O" (7.0mm/yr) is the mean of the maximum value 6.2 and the crustal value 7.7. The maximum value is from entry 23, Table 1, and the crustal value (7.7 = 3.5 x 70/32) is from the uplift contours in Fig. 2.

The Hope, the largest of the Marlborough faults, is pure dextral in nature, reasonably

TABLE 9 - Summary of total data, being a comparison of normalized vectors from vector polygons with vector data from active faulting (Tables 6 and 7) and with vectors from retriangulation (Table 8). First value in vector pair is velocity in mm/yr, second is the bearing in the 000° to 180° range.

Nominal latitude.	Plate Boundary.	Normalized vectors from displacement polygons.	Vector from active faults. Tables 6 and 7.	Vector bearing from nodal-plane solutions. Table 6.	Vector from retriangulation analysis. Table 8.	Fault Zone
38°S	ap	42/095	--/---	--/---	--/---	ap direct
	aE	20/135	20/135	--/---	--/---	Havre Trough
	Ep	47/120	--/---	--/---	--/---	Kermadec Trench
39°S	ap	40/095	--/---	--/---	--/---	ap direct
	aK	8/143	--/143	--/---	8/102	Taupo Volcanic Zone
	KE	14/015	--/027	--/---	10/014	North Island Shear Belt
	Ep	47/120	--/---	--/120	--/---	Extension of Kermadec Trench
40°S	ap	37/090	--/---	--/---	--/---	ap direct
	aK	8/015	--/015	--/---	--/---	Galpin Fault
	KE	20/050	17/050	--/---	30/054	North Island Shear Belt
	Ep	28/137	28/110	--/119	20/132	Wairarapa Thrusts
41°S	ap	35/089	--/---	--/---	--/---	ap direct
	aK	7/071	7/071	--/---	12/043	Wairau Fault
	Kp	29/095	35/070	--/080	37/085	Marlborough Fts., less Wairau
43°S	ap	31/080	35/072	--/---	33/083	ap direct. Alpine Fault shift
	ap	--/080	14/072	--/---	--/---	Alpine Fault slip
45°S	ap	29/066	--/---	--/---	--/---	ap direct
	aF	75/050	--/050	--/---	--/---	Alpine Fault
	Fp	48/040	--/007	--/---	5/148	Te Anau Fault Zone
	YF	25/010	--/170	--/---	--/---	Resolution Fault
	MF	35/030	--/---	--/032	--/---	Resolution Fault
	YM	13/066	--/---	--/---	--/---	Puysegur Thrust
43° ± 3°S	Z	8/090	8/096	--/082	17/090	Recurved arc. "Z" bending.

straight, and when extended intersects the Alpine Fault at an angle of 20° ± 0.5° (Freund, 1971). Thus the net-shift for the Alpine Fault from the Hope Fault intersection to the Porter Pass intersection 130km to the southeast is 33.6mm/yr (11.5/sin 20°) and the corresponding dextral shift 31.6mm/yr (11.5/tan 20°).

For point "O" the net shift for "K" is multiplied by 13/12 to include the dextral displacement of the Porter Pass Fault and multiplied by 13.5/14.1 to allow for decrease in distance to shiftpole giving a slight increase to 34.6mm/yr. The adopted uplift rate of 7.0mm/yr gives an angle of 11.7° between the direction of net-shift and the strike of the Alpine Fault at point "O".

From the geographical co-ordinates the distance K-O is 1.88° (209km) and the true bearing of the straight Alpine Fault is 055.42° at "K" and 056.90° at "O". Total displacements were then calculated for the two points by adding the "Z" bending vector of 8mm/yr.

At "K" 33.6 at 75.4° plus 8 at 095° = 41.2 at 79.2° giving 169.2° as the bearing to the Indian/Pacific shiftpole. At "O" 34.6 at 68.6° plus 8 at 095° = 41.9 at 73.5 giving 163.5° as the bearing to the shiftpole. Intersection gives a shiftpole at 56.7°S 175.9°E and net-shift and distance to shiftpole gives 1.52°/Myr as the relative rotation rate at the shiftpole.

The shiftpole value adopted - 57°S 176°E rotation 0.3° + 1.2° = 1.5°/Myr the 0.3° being for "Z" bending - lies within the range of values listed by Walcott (1979, Table 1).

Comparison and Summary (Fig. 9).

Table 9 gives normalized vectors for five nominal latitudes, and when available, the values determined from active faulting, from nodal-plane solutions, and from retriangulation. It also gives the values for "Z" bending. Errors are thought to be about 20 per cent for velocities and about 15° for bearings.

Column 1 gives the nominal latitude, and Column 2 the plate boundary symbol. Column 3 gives normalized vectors that "close" and are self-consistent. The "ap" vectors are calculated from the uplift shiftpole 57°S 176°E, 1.2°/Myr, the other vectors are those that agree best with the values listed in the next three columns. Column 4 gives full vectors and vector bearings from active faulting taken from Tables 6 and 7. Column 5 gives vector bearings from nodal-plane solutions taken from Table 6, and Column 6 full vectors from retriangulation analysis **taken** from Table 8.

38°S. No check values known (each full extra vector provides a pair of check values). The 20/135 rate and direction across the Havre Trough is taken from Malahoff (1977) and is not listed in Table 6.

39°S. Three check values listed, both lie within the expected errors.

40°S. Six check values listed, and all, except the 30mm/yr retriangulation rate which is 50 per cent too high, lie within the expected errors.

The "--/015" vector (not listed in Table 6) is based on the strike of the Galpin Fault and its angle with tensional joints.

41°S. Four check values listed, and all, except the "12/043" retriangulation vector, with a rate that is twice too high, and a bearing that is 28° too low, lie within the expected errors.

43°S. Two check values listed, there being good agreement between the active faulting vector from uplift and that from retriangulation.

46°S. Least well-determined and most complex polygon. Based largely on the good nodal-plane solution from the tight cluster of intermediate-depth Fiordland earthquakes (Scholz et al., 1973, Fig. 4:H) which show extension south-west with north-west/south-east narrowing for the FM boundary. Because of lack of data, the vectors for MY, Mp, and ap are taken as being collinear. As mentioned, the Macquarie Plate (M) is probably a plate-pair separated by a slow spreading axis.

43° + 3°S. Values for "Z" bending. The retriangulation value is twice too high.

ACKNOWLEDGMENTS

I would like to thank Drs. Euan Smith and Russell Robinson with help in understanding nodal-plane techniques, and Drs. Hugh Bibby and Dick Walcott for discussion regarding retriangulation. In particular, I am grateful to Gerald Lensen and Hugh Bibby for supplying unpublished data.

References

Adams, J., Vertical drag on the Alpine Fault, New Zealand, in Royal Soc. of N.Z. Bulletin, 18, 47-54, 1979.

Adams, J., Paleoseismicity of the Alpine fault seismic gap, New Zealand, in Geology, 8, 72-76, 1980.

Adams, R. D., and M. A. Lowry, The Inangahua earthquake sequence, 1968, in Royal Soc. of N. Z. Bulletin, 9, 1290135, 1971.

Adams, R. D., and J. H. Le Fort, The Westport Earthquakes, May 1962, in N.Z. Journal of Geology and Geophysics, 6 (4), 487-509, 1963.

Adams, R. D., R. Robinson, and M. A. Lowry, A micro-earthquake survey of the Benmore-Pukaki region; February-March 1973, in Report No. 86, Geophysics Division, Department of Scientific and Industrial Research, Wellington, N.Z., 1974.

Adams, R. D., and D. E. Ware, Subcrustal earthquakes beneath New Zealand; locations determined with a laterally inhomogeneous velocity model, in N.Z. Journal of Geology and Geophysics, 20 (1), 59-83, 1977.

Anderson, E. M., The Dynamics of Faulting, Oliver and Boyd, Edinburgh and London, 1951.

Arabasz, W. J., and M. A. Lowry, Microseismicity in the Tararua-Wairarapa area: depth-varying stresses and shallow seismicity in the southern North Island, N.Z., in N.Z. Journal of Geology and Geophysics, 23, 141-154, 1980.

Arabasz, W. J., and R. Robinson, Microseismicity and Geologic Structure in the northern South Island, New Zealand, in N.A. Journal of Geology and Geophysics, 19 (5), 569-601, 1976.

Berryman, K., Active faulting and derived PHS directions in the South Island, New Zealand, in Royal Soc. of N.Z. Bulletin, 18, 29-34, 1979.

Bibby, H. M., Crustal strain from triangulation in Marlborough, New Zealand, in Tectonophysics, 29, 529-540, 1975.

Bibby, H. M., Crustal strain across the Marlborough Faults, New Zealand, in N.Z. Journal of Geology and Geophysics, 19, 407-425, 1976.

Bibby, H. M., and R. I. Walcott, Earth deformation and the triangulation of New Zealand, in N.Z. Surveyor, 28, 741-751, 1977.

Boyes, W. S., Horizontal and vertical crustal movement in the Inangahua earthquake of 1968, in Royal Soc. of N.Z. Bulletin, 9, 61-72, 1971.

Calhaem, I. M., The Pukaki earthquake of 17th December 1978, in Bulletin of Earthquake Engineering, 12, 7-10, 1979.

Calhaem, I. M., A. J. Haines, and M. A. Lowry, An intermediate-depth earthquake in the central region of the South Island used to determine a local crustal thickness, in N.Z. Journal of Geology and Geophysics, 20, 353-361, 1977.

Clark, R. H., R. R. Dibble, H. E. Fyfe, G. J. Lensen, and R. P. Suggate, Tectonic and Earthquake Risk Zoning, in Transactions of the Royal Soc. of N.Z., (General Section), 1 (10), 113-126, 1965.

Cole, J. W., T. M. Hunt, Basalt dykes in the Tarawera Volcanic Complex, New Zealand, in N.Z. Journal of Geology and Geophysics, 11, 1203-1206, 1973.

Cooper, A. F., D. G. Bishop, Uplift rates and high level marine platforms associated with the Alpine Fault at Okuru River, South Westland, in Royal Soc. of N.Z. Bulletin, 13, 35-43, 1979.

Cotton, C. A., (1888 Earthquake) in Geomorphology of New Zealand, p. 174, N.Z. Board of Science and Art, Wellington, 1922.

Denham, D., Summary of earthquake focal mechanisms for the western Pacific-Indonesian Region, 1929-1973, World Data Center A for Solid Earth Geophysics, Boulder, Colorado 80302, U.S.A., 1977.

Eiby, G. A., An annotated list of New Zealand earthquakes, 1460-1965, in N.Z. Journal of Geology and Geophysics, 11, 630-647, 1968.

Eiby, G. A., and W. I. Reilly, Gravity Magnetism and Seismicity in New Zealand Atlas, Wards, I., Ed., Government Printer, Wellington, 1976.

Evison, F. F., Seismicity of the Alpine Fault, New Zealand, in Royal Soc. of N.Z. Bulletin, 9, 161-165, 1971.

Evison, F. F., R. Robinson, and W. J. Arabasz, Late Aftershocks, Tectonic Stress and Dilatancy, in Nature (London), 246 (5434), 471-473, 1973.

Frank. F. C., Deduction of earth strains from survey data, in Bulletin of the Seismological Soc. of America, 56, 35-42, 1953.

Freund, R., The Hope Fault, a strike-slip fault in New Zealand, in N.Z. Geological Survey Bulletin (n.s.), No. 86, 1971.

Fyfe, H.E., Movement on White Creek Fault, New Zealand, in N.Z. Journal of Science and Technology, 11 (3), 192-197, 1929.

Ghani, M. A., Ph.D Thesis, Victoria University of Wellington, 1974.

Ghani, M. A., Late Cenozoic vertical crustal movement in the southern North Island, New Zealand, in N.Z. Journal of Geology and Geophysics, 21 (1), 117-125, 1978.

Grindley, G. W., The Geology of Eglinton Valley, Southland, in N.Z. Geological Survey Bulletin (n.s.), 58, 1958.

Grindley, G. W., Geological Map of New Zealand 1:250 000, Sheet 8, Taupo, 1st Ed., N.Z. Department of Scientific and Industrial Research, Wellington, 1960.

Grange, L. I., Taupo earthquakes, 1922, in N.Z. Journal of Science and Technology, 14, 139-141, 1932.

Gutenberg, B., C. F. Richter, Seismicity of the earth and associated phenomena, Princeton University Press, Princeton, N.J., 1949.

Haines, A. J., I. M. Calhaem, and D. E. Ware, Crustal seismicity near Lake Pukaki, South Island, New Zealand, between June 1975 and October 1978, in Royal Soc. of N.Z. Bulletin, 18, 87-94, 1979.

Healy, J., J. C. Schofield, and B. N. Thomson, Geological Map of New Zealand 1:250 000, Sheet 5, Rotorua, 1st Ed., N.Z. Department of Scientific and Industrial Research, Wellington, 1964.

Henderson, J., The geological aspects of the Hawke's Bay Earthquakes, in N.Z. Journal of Science and Technology, 15 (1), 38-75, 1933.

Henderson, J., The West Nelson Earthquakes of 1929 (with notes on the geological structure of West Nelson), in N.Z. Journal of Science and Technology, 19 (2), 65-144, 1937.

Hunt, T., Stokes Magnetic Anomaly System, in N.Z. Journal of Geology and Geophysics, 21 (5), 595-606, 1978.

Isacks, B., L. R. Sykes, and J. Oliver, Focal mechanisms of deep and shallow earthquakes in the Tonga-Kemadec region and the tectonics of island arcs, in Bulletin of the Geological Soc. of America, 80, 1443-1470, 1969.

Johnson, T., and P. Molnar, Focal mechanism and plate tectonics of the Southwest Pacific, in Journal of Geophysical Research, 77 (26), 5000-5032, 1972.

Kieckhefer, R. M., Microseismicity in the vicinity of the Clarence Fault, New Zealand, in N.Z. Journal of Geology and Geophysics, 20, 165-177, 1977.

Lensen, G. J., Rationalized fault interpretation, in N.Z. Journal of Geology and Geophysics, 1, 307-317, 1958.

Lensen, G. J., The Wellington Fault from Cook Strait to Manawatu Gorge, in N.Z. Journal of Geology and Geophysics, 1 (1), 178-196, 1958.

Lensen, G. J., The faulted terrace sequence at Grey River, Awatere Valley, South Island, New Zealand, in N.Z. Journal of Geology and Geophysics, 7, 871-876, 1964.

Lensen, G. J., Analysis of progressive fault displacement during downcutting at Branch River terraces. South Island, New Zealand, in Geological Soc. of America Bulletin, 79 (5), 545-555, 1968.

Lensen, G. J., and P. M. Otway, Earthshift and post-earthshift deformation associated with the May 1968 Inangahua earthquake, New Zealand, in Royal Soc. of N.Z. Bulletin, 9, 107-116, 1971.

Lensen, G. J., and P. Vella, The Waiohine River faulted terrace sequence, in Royal Soc. of N.Z. Bulletin, 9, 117-119, 1971.

Lensen, G. J., Tectonic strain and drift, in Tectonophysics, 71, 173-188, 1981.

Lewis, K. B., Growth rate of folds using tilted wave-planed surfaces: coast and continental shelf, Hawke's Bay, New Zealand, in Royal Soc. of N.Z. Bulletin, 9, 225-231, 1971.

Lyell, C., Sur les Effets du Tremblement de Terre du 23 Janvier, 1855, a la Nouvelle Zelande, in Bull. Soc. Geol. Fr., ser. 2, No. 13, 661-667, 1856.

McKay, A., On the Earthquakes of September 1888 in the Amuri and Marlborough Districts of the South Island, in N.Z. Geological Survey Bulletin, 1 (1st Series), 1888.

Mackie, J. B., Geodetic studies of crustal movement in New Zealand, in Royal Society of N.Z. Bulletin, 9, 121-125, 1971.

Macpherson, E. O., An outline of Late Cretaceous and Tertiary diastrophism in New Zealand, in N.Z. DSIR Geological Memoir, 6, 1946.

Malahoff, A., R. H. Feden, and H. F. Fleming, Crustal extension processes in the Havre Trough, South Fiji and New Caledonia basins, Abstract in EOS Transactions, 58, American Geophysical Union, 379, 1977.

Nathan, S., Geological Map of New Zealand 1:63 360, Sheet S31 and pt. S32, Buller-Lyell (1st Ed.), N.Z. Department of Scientific and Industrial Research, Wellington, 1978.

Neef, G., Probable faulting on a minor, reverse, bedding fault adjacent to the Alfredton Fault, north Wairarapa, during the 1 August 1942 earthquake (Note), in N.Z. Journal of Geology and Geophysics, 19, 737-742, 1976.

Ongley, M., Surface Trace of the 1855 Earthquake, in Transactions of the Royal Soc. of N.Z., 73 (2), 84-89, 1937.

Ongley, M., H. E. Walshe, J. Henderson, and R. C. Hayes, The Wairoa Earthquake of 16th September 1932, in N.Z. Journal of Science and Technology, 18 (12), 845-865, 1937.

Otway, P. M., Geodetic monitoring of earth deformation in the Wellington area, in N.Z. Geological Survey Report, 62, N.Z. Department of Scientific and Industrial Research, Wellington, 1972.

Reid, H. F., W. M. Davis, A. C. Lawson, and F. L. Ransome, Report of Committee on Nomenclature of Faults, in Bulletin of Geological Society of America, 24, 1913.

Reyners, M. E., A microearthquake study of the plate boundary, North Island, New Zealand, Ph.D thesis, Victoria University of Wellington, N.Z., November 1978.

Ridd, M.F., Mud volcanoes in New Zealand, in The American Association of Petroleum Geologists Bulletin, 54 (4), 601-616, 1970.

Robinson, R., Seismicity within a zone of plate convergence - the Wellington region, New Zealand, in Geophysical Journal of the Royal Astronomical Soc., 55, 693-702, 1978.

Robinson, R., and W. J. Arabasz, Microearthquakes in the north-west Nelson region, New Zealand, in N.Z. Journal of Geology and Geophysics, 18, 83-91, 1975.

Robinson, R., W. J. Arabasz, and F. F. Evison, Long-term behavior of an aftershock sequence: The Inangahua, New Zealand, earthquake of 1968, in Geophysical Journal of the Royal Astronomical Soc., 41, 37-49, 1975.

Robinson, R., T. M. Calhaem, and A. A. Thomson, The Opunake, New Zealand, earthquake of 5 November 1974, in N.Z. Journal of Geology and Geophysics, 19 (3), 335-345, 1976.

Rynn, J. M. W., and C. H. Scholz, Seismotectonics of the Arthur's Pass region, South Island, New Zealand, in Geological Soc. of America Bulletin, 89, 1373-1388, 1978.

Scholz, C. H., J. M. W. Rynn, R. W. Weed, and C. Frolich, Detailed seismicity of the Alpine Fault zone and Fiordland region, New Zealand, in Geological Society of America Bulletin, 84, 3297-3316, 1973.

Sissons, B. A., The Horizontal Kinematics of the North Island of New Zealand, Ph.D thesis, Victoria University of Wellington, New Zealand, August 1979.

Smith, E. G. C., A micro-earthquake survey of the Rangitikei and Manawatu Basins, in N.Z. Journal of Geology and Geophysics, 22 (4), 473-478, 1979.

Suggate, R. P., The interpretation of progressive fault displacement of flights of terraces, in N.Z. Journal of Geology and Geophysics, 3 (3), 364-372, 1960.

Suggate, R. P., Late Pleistocene geology of the northern part of the South Island, New Zealand, in N.Z. Geological Survey Bulletin, 77, 1965.

Suggate, R. P., and G. J. Lensen, Rate of horizontal fault displacement in New Zealand, in Nature (London), 242, p. 518, 1973.

Suggate, R. P. et al. (Eds.), The Geology of New Zealand, Government Printer, Wellington, 2 vols. 820 p., 1978.

Thatcher, W., Crustal movements and earthquake-related deformation, in reviews of geophysics and spacephysics, in American Geophysical Union, 17 (6), 1403-1411, 1979.

Walcott, R. I., Geodetic strains and large earthquakes in the Axial Tectonic Belt of North Island, New Zealand, in Journal of Geophysical Research, 83 (B9), 4419-4429, 1978A.

Walcott, R. I., Present tectonics and Late Cenozoic evolution of New Zealand, in Geophysical Journal of the Royal Astronomical Soc., 52, 137-164, 1978B.

Walcott, R. I., Plate motion and shear strain rates in the vicinity of the Southern Alps, in Royal Soc. of N.Z. Bulletin, 18, 5-12, 1979.

Walshe, H. E., The Wairoa Earthquake of 16th September 1932. 2. Earth Movements in Hawke's Bay District Disclosed by Triangulation, in N.Z. Journal of Science and Technology, 18 (12), 852-854, 1937.

Ward, R. H., A Note on the Significance of the Recent Subsidence of the Shore of Lake Taupo, in N.Z. Journal of Science and Technology, 5 (5), 280-281, 1922.

Ware, D. E., Ed., N.Z. Seismological report for 1979, in N.Z. Seismological Observatory Bulletin, E-161, 1980.

Wards, I. (Ed.), New Zealand Atlas, Government Printer, Wellington, N.Z., 1976.

Wellman, H. W., The Alpine Fault in detail: River terrace displacement at Maruia River, in N.Z. Journal of Science and Technology, B33 (5), 409-414, 1952.

Wellman, H. W., Data for the study of Recent and Late Pleistocene faulting in the South Island of New Zealand, in N.Z. Journal of Science and Technology, B34 (4), 270-288, 1953.

Wellman, H. W., New Zealand Quaternary Tectonics, in Geologische Rundschau, 43, 238-257, 1955A.

Wellman, H. W., The Geology between Bruce Bay and Haast River, South Westland, in N.Z. Geological Survey Bulletin, (n.s.) 48, (2nd Ed.), 1955B.

Wellman, H.W., Tilted marine beach ridges at Cape Turakirae, N.Z., in Journal of Geosciences, Osaka City University, Japan, 10, 123-129, 1967.

Wellman, H. W., Rate of Horizontal Fault Displacement in New Zealand, in Nature (London), 237 (5353), 275-277, 1972.

Wellman, H. W., The Stokes Magnetic Anomaly, in Geological Magazine, 110, 419-429, 1973.

Wellman, H. W., New Zealand 60 Million Years Ago (abstract) in Bulletin of the Australian Soc. of Exploration Geophysicists, 6, 55-56, 1975.

Wellman, H. W., An uplift map for the South Island of New Zealand, and a model for uplift of the Southern Alps, in Royal Soc. of N.Z. Bulletin, 18, 13-20, 1979.

Wellman, H. W., and R. W. Willett, The Geology of the West Coast from Abut Head to Milford Sound, in Transactions of the Royal Soc. of N.Z., 71 (4), 282-306, 1942.

Wellman, H. W., and A. T. Wilson, Notes on the geology and archaeology of the Martin's Bay district, in N.Z. Journal of Geology and Geophysics, 7 (4), 702-721, 1964.

Wellman, H. W. Marine Pliocene at Resolution Island, Dusky Sound, Fiordland, in N.Z. Journal of Science and Technology, B35 (5), 378-389, 1954.

Wellman, H. W., The necessity for using "cut-outs", in Newsletter, Geological Society of N.Z., No. 54, 49-50, 1981.